Advances in Intelligent Systems and Computing

Volume 337

Series editor

Janusz Kacprzyk, Polish Academy of Sciences, Warsaw, Poland
e-mail: kacprzyk@ibspan.waw.pl

About this Series

The series "Advances in Intelligent Systems and Computing" contains publications on theory, applications, and design methods of Intelligent Systems and Intelligent Computing. Virtually all disciplines such as engineering, natural sciences, computer and information science, ICT, economics, business, e-commerce, environment, healthcare, life science are covered. The list of topics spans all the areas of modern intelligent systems and computing.

The publications within "Advances in Intelligent Systems and Computing" are primarily textbooks and proceedings of important conferences, symposia and congresses. They cover significant recent developments in the field, both of a foundational and applicable character. An important characteristic feature of the series is the short publication time and world-wide distribution. This permits a rapid and broad dissemination of research results.

More information about this series at http://www.springer.com/series/11156

Suresh Chandra Satapathy · A. Govardhan
K. Srujan Raju · J.K. Mandal
Editors

Emerging ICT for Bridging the Future - Proceedings of the 49th Annual Convention of the Computer Society of India (CSI) Volume 1

Springer

Editors
Suresh Chandra Satapathy
Department of Computer Science and Engineering
Anil Neerukonda Institute of Technology and Sciences
Vishakapatnam
India

A. Govardhan
School of Information Technology
Jawaharlal Nehru Technological University Hyderabad
Hyderabad
India

K. Srujan Raju
Department of CSE
CMR Technical Campus
Hyderabad
India

J.K. Mandal
Faculty of Engg., Tech. & Management
Department of Computer Science and Engineering
University of Kalyani
Kalyani
India

ISSN 2194-5357 ISSN 2194-5365 (electronic)
Advances in Intelligent Systems and Computing
ISBN 978-3-319-13727-8 ISBN 978-3-319-13728-5 (eBook)
DOI 10.1007/978-3-319-13728-5

Library of Congress Control Number: 2014956100

Springer Cham Heidelberg New York Dordrecht London

Printed on acid-free paper

Springer International Publishing AG Switzerland is part of Springer Science+Business Media (www.springer.com)

Preface

This AISC volume-I contains 73 papers presented at the 49th Annual Convention of Computer Society of India: Emerging ICT for Bridging the Future: held during 12–14 December 2014 at Hyderabad hosted by CSI Hyderabad Chapter in association with JNTU, Hyderabad and DRDO. It proved to be a great platform for researchers from across the world to report, deliberate and review the latest progresses in the cutting-edge research pertaining to intelligent computing and its applications to various engineering fields. The response to CSI 2014 has been overwhelming. It received a good number of submissions from the different areas relating to intelligent computing and its applications in main tracks and four special sessions and after a rigorous peer-review process with the help of our program committee members and external reviewers finally we accepted 143 submissions with an acceptance ratio of 0.48. We received submissions from seven overseas countries.

Dr Vipin Tyagi, Jaypee University of Engg and Tech, Guna, MP conducted a Special session on "Cyber Security and Digital Forensics ". Dr. B.N. Biswal, BEC, Bhubaneswar and Prof. Vikrant Bhateja, Sri. Ramswaroop Memorial Group of Professional colleges, Lucknow conducted a special session on "Recent Advancements on Computational intelligence". "Ad-hoc Wireless Sensor Networks" special session was organized by Prof. Pritee Parwekar, ANITS, Vishakapatnam and Dr. S.K. Udgata, Univeisity of Hyderabad. A special session on "Advances and Challenges in Humanitarian Computing" was conducted by Prof. Sireesha Rodda, Dept of CSE, GITAM University, Vishakapatnam.

We take this opportunity to thank all Keynote Speakers and Special Session Chairs for their excellent support to make CSI2014 a grand success.

The quality of a referred volume depends mainly on the expertise and dedication of the reviewers. We are indebted to the program committee members and external reviewers who not only produced excellent reviews but also did in short time frames. We would also like to thank CSI Hyderabad Chapter, JNTUH and DRDO having coming forward to support us to organize this mega convention.

We express our heartfelt thanks to Mr G. Satheesh Reddy, Director RCI and Prof Rameshwar Rao, Vice-Chancellor, JNTUH for their continuous support during the course of the convention.

We would also like to thank the authors and participants of this convention, who have considered the convention above all hardships. Finally, we would like to thank all the volunteers who spent tireless efforts in meeting the deadlines and arranging every detail to make sure that the convention runs smoothly. All the efforts are worth and would please us all, if the readers of this proceedings and participants of this convention found the papers and event inspiring and enjoyable.

We place our sincere thanks to the press, print & electronic media for their excellent coverage of this convention.

December 2014

Volume Editors
Dr. Suresh Chandra Satapathy
Dr. A. Govardhan
Dr. K. Srujan Raju
Dr J.K. Mandal

Team CSI 2014

Chief Patrons

Dr. G. Satheesh Reddy, Director RCI
Prof. Rameswara Rao, VC JNTU-H

Advisory Committee

Sri. S. Chandra Sekhar, NASSCOM
Sri. J. Satyanarayana, A.P. Govt
Sri. Haripreeth Singh TS Govt
Sri. Sardar. G.S. Kohli GNIT
Prof. D.V.R. Vithal, Fellow CSI
Maj. Gen. R.K. Bagga, Fellow CSI
Dr. Ashok Agrwal, Fellow CSI
Dr. D.D. Sharma, GNIT Fellow CSI
Sri. H.R Mohan, President CSI
Sri. Sanjay Mahapatra, Secretary CSI
Sri. Ranga Raj Gopal, Treasurer CSI
Prof. S.V. Raghavan
Prof. P. Trimurty

Conference Committee

Sri. Bipin Mehta
Sri. Raju L. Kanchibotla
Sri. Gautam Mahapatra
Dr. A. Govardhan
Dr. Srujan Raju K.

Conveners

Sri. K. Mohan Raidu Main
Prof. C. Sudhakar
Sri. Chandra Sekhar
Dr. Chandra Sekhar Reddy

Organizing Committee

Sri. J.A. ChowdaryTelent, Sprint
Sri. R. Srinivasa Rao, Wipro
Dr. H.S. Saini, GNIT

Programme Committee

Dr. A. Govardhan
Sri. Bipin Chandra Ikavy
Dr. T. Kishen Kumar Reddy
Prof. N.V. Ramana Rao
Dr. Srujan Raju K.
Prof. I.L. Narasimha Rao
Dr. J. SasiKiran
Prof P.S. Avadhani
Sri. Venkatesh Parasuram
Prof. P. Krishna Reddy IIIT

Finance Committee

Sri. GautamMahapatra
Prof. C. Sudhakar
Sir. Raj Pakala

Publication Committee

Dr. A. Govardhan
Dr. S.C. Satapathy
Dr. Subash C. Mishra
Dr. Anirban Paul
Dr. Srujan Raju K.
Dr. Vishnu
Dr. J.K. Mandal

Exhibition Committee

Sri. P.V. Rao
Sri. Rambabukuraganty
Sri. Balaram Varansi
Sri. VenkataRamana Chary G.
Mr. Hiteshawar Vadlamudi
Dr. D.V. Ramana

Transport Committee

Sri. RambabuKuriganti
Sri. Krishna Kumar
Sri K.V. Pantulu
Smt. P. Ramadevi
Sri. Amit Gupta

Hospitality Committee

Sri. Krishna Kumar Tyagarajan
Sri. P.V. Rao

Sponsorship Committee

Sri. Raj Pakala
Sri. Srinivas konda
Sri. PramodJha

Marketing & PR Committee

Sri. KiranCherukuri
Ms. Sheila P.

Registrations Committee

Sri Ramesh Loganathan
Smt. Rama Bhagi
Sri. Rajeev Rajan Kumar
Sri. Krishna Kumar B.
Sri. Sandeep Rawat
Sri. Ram Pendyala
Sri T.N Sanyasi Rao
Sri D.L. Seshagiri Rao
Sri G. Vishnu Murthy
Prof. Ramakrishna Prasad
Sri. Vijay Sekhar K.S.
Dr. SaumyadiptaPyne

Cultural Committee

Smt. Rama Bhagi
Smt. Rama Devi

International Advisory Committee/Technical Committee

P.K. Patra, India
Sateesh Pradhan, India
J.V.R. Murthy, India
T.R. Dash, Kambodia
Sangram Samal, India
K.K. Mohapatra, India
L. Perkin, USA
Sumanth Yenduri, USA
Carlos A. Coello Coello, Mexico
S.S. Pattanaik, India
S.G. Ponnambalam, Malaysia
Chilukuri K. Mohan, USA
M.K. Tiwari, India
A. Damodaram, India
Sachidananda Dehuri, India
P.S. Avadhani, India
G. Pradhan, India
Anupam Shukla, India
Dilip Pratihari, India
Amit Kumar, India
Srinivas Sethi, India
Lalitha Bhaskari, India
V. Suma, India
Pritee Parwekar, India
Pradipta Kumar Das, India
Deviprasad Das, India
J.R. Nayak, India
A.K. Daniel, India
Walid Barhoumi, Tunisia
Brojo Kishore Mishra, India
Meftah Boudjelal, Algeria
Sudipta Roy, India
Ravi Subban, India
Indrajit Pan, India
Prabhakar C.J, India
Prateek Agrawal, India

Contents

Advances and Challenges in Humanitarian Computing

Fuzzy System, Image Processing and Software Engg

Cyber Security, Digital Forensic and Ubiquitous Computing

E-Commerce, ICT Applications, Big Data and Cloud Computing

A Hybrid Approach for Image Edge Detection Using Neural Network and Particle Swarm Optimization

D. Lakshumu Naidu, Ch. Seshadri Rao, and Sureshchandra Satapathy

Department of Computer Science and Engineering, Anil Neerukonda Institute of Technology and Sciences, Vishakhapatnam, Andhra Pradesh, India
{lakshumu,sureshsatapathy}@gmail.com,
seshadri.rao.cse@anits.edu.in

Abstract. An Edge of an image is a sudden change in the intensity of an image. Edge detection is process of finding the edges of an image. Edge detection is one of the image preprocessing techniques which significantly reduces the amount of data and eliminates the useless information by processing the important structural properties in an image. There are many traditional algorithms used to detect the edges of an image. Some of the important algorithms are Sobel, Prewitt, Canny, Roberts etc. A Hybrid approach for Image edge detection using Neural Networks and Particle swarm optimization is a novel algorithm to find the edges of image. The training of neural networks follows back propagation approach with particle swarm optimization as a weight updating function. 16 visual patterns of four bit length are used to train the neural network. The optimized weights generated from neural network training are used in the testing process in order to get the edges of an image.

Keywords: Image edge detcion, Image processing, Artificial Neural Networks, Particle swarm optimization.

1 Introduction

Edges in images are the curves that characterize the boundaries (or borders) of objects. Edges contain important information of objects such as shapes and locations, and are often used to distinguish different objects and/or separate them from the background in a scene. In image processing, edge detection can be employed to filter out less relevant information while preserving the basic structural properties of an image. This image data can be used for further preprocessing steps like extraction of features, image segmentation, interpretation and registration of an image.

A sudden change in intensity of an image is an edge. Edge detection is a process of finding the sudden intensity discontinues in an image. Slow changes refer to small value of derivatives and fast changes refer to large values of derivatives. Dimensional spatial filters or the gradient operator uses this principle to find the edges. This type of filters detects the gradient of image intensity without considering the uniform regions (i.e., the area with contrast intensity) in the image. There are different types of filters [1] have been developed to detect different types of edges in the image. The classical operators like Sobel, prewitt, kirsch detect edges in all directions such as horizontal,

S.C. Satapathy et al. (eds.), *Emerging ICT for Bridging the Future – Volume 1*,
Advances in Intelligent Systems and Computing 337, DOI: 10.1007/978-3-319-13728-5_1

vertical and diagonal etc. Laplacian of Gaussian (LOG) operator finds the correct places of edges but cannot find edges in corners and curves. Though canny and Shen-Castan detects edges even in noise conditions which involves complex computations and more time to execute. Identifying the correct threshold value is the crucial part for all these algorithms. There is no appropriate method to find threshold value.

2 Related Work

Artificial neural networks (ANN) used in many area and also used for edge detection. Neural networks are used a Non linear filter to detect the edges of an image. In [2], Terry and Vu introduced a multi-layer feed forward neural network which is used to detect the edges of a laser radar image of a bridge. Synthetic edge patterns are used to train the networks. The network can detect different types of edges like horizontal, vertical and diagonal and so on. Li and Wang developed a new neural network detector which is applied on 8 bit sub image of an image. After completion of the entire image the results combined and produce the final edges of image. This approach does not provide good training set and increases the overhead.

Hamed Mehrara[4], Mohammad Zahedinejad [5] & [6] developed and produced new training data set with 16 different visual patterns which can be used to detect any type of edges. Initially the gray-scale images are converted into binary form and these binary images are sent to neural networks. Since binary image have only two values 1 and 0, specifies high and low intensities so that we can easily detect the edges. This method suffers from two problems called initialization of weights to the network and selection of threshold for converting image into binary image.

In this research we are training the neural networks using Particle Swarm Optimization instead of Using Back propagation algorithm. As Neural network is good localization and PSO is good for exploration of search space. Neural network produces good results based on selection of initial parameters like connected weights etc. So we can use Particle Swarm Optimization to get the best optimized weights. No need to concentrate on initial weights of ANN.

2.1 Neural Networks

An Artificial Neural Network (ANN) is an interconnected network of neurons which process the data in parallel.ANN is motivated from human brain which is highly complex parallel computer. According to [7] & [8] network has set of inputs and outputs which are used to propagate the calculations at each neuron i.e. output of the one neuron can be input to some other neuron. Each neuron is associated with an input and weight. The weights of the network are represents knowledge and the learning of the network involves updating of these weights. Neural networks are used to detect the patterns of the data.

The performance of any neural network training depends on many parameters like activation function used, type of network architecture, learning algorithm used etc.

2.2 Particle Swarm Optimization

Particle swarm optimization is a population based algorithm which improves the overall performance of a system by interactions of all the particles.PSO was first proposed and developed by Kennedy J. and Eberhart R. C [9]. They explained that each particle in the population maintains its best solution that has been achieved so far by that particle called as pbest. The overall best value of the neighboring particles so far is called gbest.

According to James Kennedy [9] the particle position can be modified by using the following equations.

$$V_i(t+1)=w*V_i(t)+c1*Rand1()*(pbest(i)-X_i(t))+c2*Rand1()*(gbest-X_i(t)) \quad (1)$$

$$X_i(t+1)=X_i(t)+V_i(t+1) \quad (2)$$

Here c1 and c2 are constants and Rand1 () and Rand2 () are automatically generated random numbers within the range from 0 to 1.Eq(1) is used to find the velocity of particle which is used to speed up the process by adding it to current position as in eq(2). Eq(1) contains two parts first part is for personnel enhancement of particle and second part for overall or global enhancement of the population.

3 Proposed System

The proposed System contains two phases. First Phase contains the training of the neural networks using particle swarm optimization. In this phase a 4 bit length 16 visual patterns are used as a training data. The weight updating is done using particle swarm optimization. Second phase contains the testing phase. In this phase we will submit the binary image of window size 2X2 to the trained neural network. It produces the edges of an image.

The novel algorithm contains the following modules to train and produce the edges of an image.

3.1 Network Architecture

There are different types of neural network architectures exist. We are using a multi layer feed forward network of four input neurons, one hidden layer with ‘n’ no of user defined neurons and four output neurons.

3.2 Learning Algorithm

A Supervised learning algorithm is used to train the neural networks. It follows a back propagation approach and which uses bipolar function as activation function. Here we are using Particle swarm optimization for weight updating. For the better performance the size of the swarm is 10. Mean squared error is used as fitness function. The basic working of how we update the weights of neural networks is explained by the following diagram.

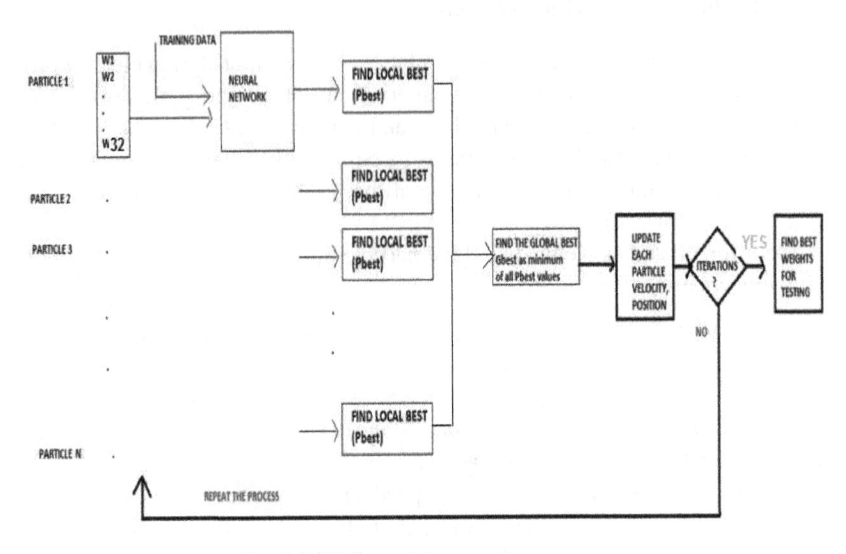

Fig. 1. PSO for weight updating

3.3 Training Data

Training data set is the basic key for any neural network training process. The following 16 visual patterns are used to train the neural networks to get the edges of an image.

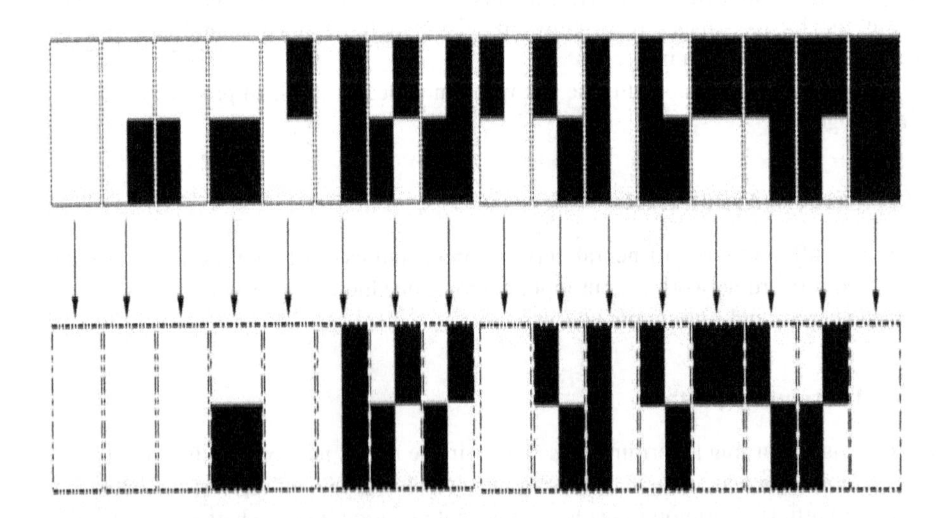

Fig. 2. Visual patterns for training neural networks

The training pattern and its corresponding edges are explained in the following table.

Table 1. visual patterns for training

Training Inputs	Detected As	Training Output
0 0 0 0	None edge	1 1 1 1
0 0 0 1	Corner edge	1 0 0 1
0 0 1 0	Corner edge	0 1 1 0
0 0 1 1	Horizontal edge	0 0 1 1
0 1 0 0	Corner edge	0 1 1 0
0 1 0 1	Parallel edge	0 1 0 1
0 1 1 0	Diagonal edge	0 1 1 0
0 1 1 1	Pseudo noise	1 1 1 1
1 0 0 0	Corner edge	1 0 0 1
1 0 0 1	Diagonal edge	1 0 0 1
1 0 1 0	Parallel edge	1 0 1 0
1 0 1 1	Pseudo noise	1 1 1 1
1 1 0 0	Horizontal edge	1 1 0 0
1 1 0 1	Pseudo noise	1 1 1 1
1 1 1 0	Pseudo noise	1 1 1 1
1 1 1 1	None edge	1 1 1 1

3.4 Testing Phase

The updated weights are used for testing and applied on binary image to get the edges. Testing Phase involves two major components. Converting the given image into binary image and eliminating the noisy edges.

3.4.1 Conversion of Original Image into Binary Image

The original image is first converted into gray scale image and then global threshold is applied on it and a binary image is produced as a result. This binary image is send to neural network with 2X2 window size.

3.4.2 Elimination of Noise

The proposed algorithm eliminates the noisy edges of an image by using White wasshing method. White washing is nothing but replacing the white(1) bits for all noisy edges and none edges. All the nosiy edges and none edges are shown in the above table.

3.5 Architecture of Proposed System

The following diagram explains the working process of proposed system.

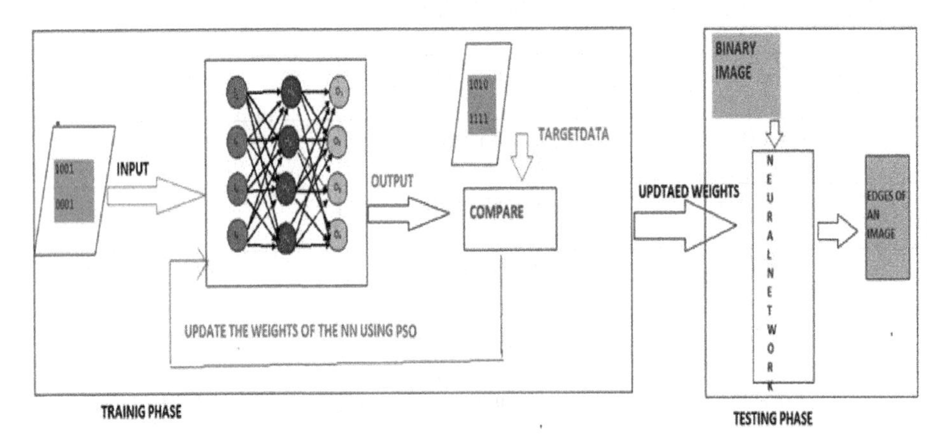

Fig. 3. Proposed System Architecture

3.6 Algorithm for the Proposed System

Input: 16 visual patterns and binary image.

Output: Optimized weights, Edges of an image.

Step 1: Initialize the positions (weight and bias) and velocities of a group of particles randomly.

Step 2: The PSO-NN is trained using the initial particles position.

Step 3: Calculate the mean Square error (MSE) produced from BP neural network can be treated as particles fitness value according to initial weight and bias.

Step 4: The learning error is reduced by changing the particles position, which will update the weight and bias of the network.

(i) The "*pbest*" value (each particle's MSE so far) and

(ii) The "*gbest*" value (lowest learning MSE found in entire learning process so far) are applied to the velocity update equation (Eq. 1) to produce a value for positions adjustment to the best solution or targeted learning error.

Step 5: The new sets of positions (NN weight and bias) are produced by adding the calculated velocity value to the current position value using movement equations (Eq. 1 & Eq. 2). Then, the new sets of positions are used to produce new Learning error in feed-forward NN.
Step 6: This process terminates if the stopping conditions either minimum learning error or maximum number of iteration are met.
Step 7: Save the optimized weights, this can be used for further processing.
Step 8: Find the edges of binary image by using the optimized weights in NN testing process.

4 Experimental Results

Initially we find the edges of an image using Neural network using general back propagation algorithm and after that we use NN-PSO detector to find the edges of an image. Here we take the value for the constants c1=c2=2.0,w=0.729844.After some iterations the the fitness becomes constant and the detector produces optimized set of weights.The output for the basic lena image as shown in the following figure.

Fig. 4. The edges of lena image using BP-NN and PSO-NN edge detectors

The following table shows how the entropy and no edges for lena image for both edge detectors.

Table 2. comparision of BP-NN and PSO-NN edge detectors.

Edge detector	Entropy	No of edges
BP-NN	1.2335	50625
PSO-NN	0.2892	50625

In the above table the entropy is the basic operator to differentiate the efficiency of two algorithms. An entropy is a statistical measure of randomness that can be used to characterize the texture of the input image. Entropy can be calculated by using the following formula

Entropy= $-\sum_{i=1}^{n} pilog(pi)$

Where p_i is the probability of difference between two adjacent pixels is equal to i.

The entropy value using PSO-NN is very less when compared to entropy value of BP-NN (clearly seen from Table 2), indicating that the noise is removed and edges are nicely detected with more information of the pixel as discussed in the proposed edge-detection process.

5 Conclusion

The particle swarm optimization enhances in exploration of the search space and eliminates the weight intialization problem of neural networks. The PSO-NN algorithm produces the optimized set of weights which are used to detect the edges of an image. This algorithm produces same no of edges as BP-NN with less entropy. In this paper global threshold is used for converting the given image into binary image.we can also extend our future work for converting the image into binary using local threshold and can find the edges of color images too. As the neural networks with backpropagation involves more complex operations the optimiztion techniques clearly reduce mathematically computational overhead so we can also extend the work to recent optimization techinique like TLBO.

References

1. Maini, R., Aggarwal, H.: Study and Comparison of Various Image Edge Detection Techniques. International Journal of Image Processing (IJIP) 3(1)
2. Terry, P., Vu, D.: Edge Detection Using Neural Networks. In: Conference Record of the Twenty-seventh Asilomar Conference on Signals, System99999s and Computers, pp. 391–395 (November 1993)
3. Li, W., Wang, C., Wang, Q., Chen, G.: An Edge Detection Method Based on Optimized BP Neural Network. In: Proceedings of the International Symposium on Information Science and Engineering, pp. 40–44 (December 2008)
4. He, Z., Siyal, M.: Edge Detection with BP Neural Networks. In: Proceedings of the International Conference on Signal Processing, pp. 1382–1384 (1998)
5. Mehrara, H., Zahedinejad, M., Pourmohammad, A.: Novel Edge Detection Using BP Neural Network Based on Threshold Binarization. In: Proceedings of the Second International Conference on Computer and Electrical Engineering, pp. 408–412 (December 2009)
6. Mehrara, H., Zahedinejad, M.: Quad-pixel edge detection using neural network. Journal of Advances in Computer Research 2(2), 47–54 (2011)
7. Graupe, D.: Principle of Artificial Neural Networks. World Scientific Publishing Co. Pte. Lte. (2007)
8. Du, K.L., Swamy, M.N.: Neural Network in Soft Computing. Springer (2006)
9. Kennedy, J., Eberhart, R.C.: Particle Swarm Optimization. In: Proceedings of IEEE International Conference on NN, Piscataway, pp. 1942–1948 (1995)

10. Lu, D., Yu, X.-H., Jin, X., Li, B., Chen, Q., Zhu, J.: Neural Network Based Edge Detection for Automated Medical Diagnosis
11. Settles, M., Rylander, B.: Neural Network Learning using Particle Swarm Optimizers. In: Advances in Information Science and Soft Computing, pp. 224–226 (2002)
12. Shi, Y.: Particle Swarm Optimization. Electronic Data Systems, Inc. Kokomo, IN 46902, USA Feature Article, IEEE Neural Networks Society (February 2004)
13. Zhao, L., Hu, H., Wei, D., Wang, S.: Multilayer forward artificial neural network. Yellow River Conservancy Press, Zhengzhou (1999)
14. Marcio, C., Teresa, B.L.: An Analysis Of PSO Hybrid Algorithms For Feed-Forward Neural Networks Training. Center of Informatics, Federal University of Pernambuco, Brazil (2006)
15. Pärt-Enander, E.: The MATLAB handbook. Addison-Wesley, Harlow (1996)
16. Stamatios, C., Dmitry, N., Charles, T., Alexander, C., et al.: A Character Recognition Study Using a Biologically Plausible Neural Network of the Mammalian Visual System, pp. D2.1-D2.10-D12D12.10. Pace University (2011)
17. Gonzalez, R., Woods, R.: Digital image processing, 2nd edn., pp. 567–612. Prentice-Hall Inc. (2002)
18. Argyle, E.: Techniques for edge detection. Proc. IEEE 59, 285–286 (1971)
19. Vincent, O.R., Folorunso, O.: A Descriptive Algorithm for Sobel Image Edge Detection. Clausthal University of Technology, Germany

Studying Gene Ontological Significance of Differentially Expressed Genes in Human Pancreatic Stellate Cell

Bandana Barman[1,*] and Anirban Mukhopadhyay[2]

[1] Department of Electronics and Communication Engineering, Kalyani Government Engineering College, Kalyani, Nadia, West Bengal, India
[2] Department of Computer Science and Engineering, University of Kalyani, Kalyani, Nadia, West Bengal, India
bandanabarman@gmail.com, anirban@klyuniv.ac.in

Abstract. In this paper, we studied and analyzed the significant ontologies by gene ontology in which the differentially expressed genes (DEG) of human pancreatic stellate cell participate. We identified up-regulated and down-regulated differentially expressed genes between dose response and time course gene expression data after retinoic acid treatment of human pancreatic stellate cells. We first perform statistical t-test and calculate false discovery rate (FDR) then compute quantile value of test and found minimum FDR. We set the pvalue cutoff at 0.02 as threshold and get 213 up-regulated (increased in expression) genes and 99 down-regulated (decreased in expression) genes and analyzed the significant GO terms.

Keywords: Microarray data, p-value, false discovery data, q-value, normalization, permutation test, differential gene expression, Gene Ontology (GO).

1 Introduction

If a gene is statistically and biologically significant, then the gene is considered as differentially expressed between two samples. The ratio of expression levels of different samples and variability is a measure of differential expression. The quality control of microarray data improves differentially expressed gene [8] detection and it is also a useful analysis of data. The stratification-based tight clustering algorithm, principal component analysis and information pooling method are used to identify differentially expressed genes in small microarray experiments. To detect DEG for RNA-seq data, the comparison of two Poisson means (rates) are determined [4].

* Corresponding author.

S.C. Satapathy et al. (eds.), *Emerging ICT for Bridging the Future – Volume 1*,
Advances in Intelligent Systems and Computing 337, DOI: 10.1007/978-3-319-13728-5_2

The statistical hypothesis test is a method of statistical inference which uses data from a scientific study. Statistically, a result is significant if it has been predicted as unlikely have occurred by chance alone according to the significance level. The significance level is a pre-determined threshold probability. The hypothesis testings are two types: parametric and non-parametric. The parametric method is based on assuming a particular underlying population distribution. For large sample it is used without that assumption. The nonparametric method can be used without assuming a distribution and often not as "powerful" as parametric methods. Z-test, t-test, One-way ANOVA test are the parametric tests and Wilcoxon rank sum test, Kruskal Wallis H-test[7] are the non-parametric test. In Wilcoxon rank sum test, heuristic idealized discriminator methods are used to identify diffentialy expressed genes [11]. The pre-processing steps for image analysis and normalization is introduced to identify DEG. The DEG are identified based on adjusted p-values. It is a statistical method for identification of diferentially expressed genes in replicated cDNA microarray experiments [5]. To perform testing, data are used after its normalization. Each differentially expressed gene may has univariate testing problem and it is corrected by adjusted pvalue. For finding DEG between two sample conditions, the principal component analysis (PCA) space and classification of genes are done based on their position relative to a direction on PC space representing each condition [10].

Identification of differentially expressed genes (DEG) with understanding wrong or fixed under conditions (cancer, stress etc.) is the main key and message in these genes. It is extensively precessed and modified prior to translation [1]. The DEG may considered as features for a classifier and it is also served as starting point of a model. To identify differentially expressed genes a masking procedure is used. There cross-species data set and gene-set analysis are studied. The ToTS i.e Test of Test Statistics and GESA i.e. Gene Set Enrichment Analysis were also investigated there [3].

The Gene Ontology is three structured, controlled vocabularies (ontologies) which describe gene products in terms of their associated biological processes, cellular components and molecular functions in a species-independent manner. The ontologies develop and maintain ontologies themselves [9]. The gene products annotation entails making associations between ontologies and genes and collaborating databases. We identified differentially expressed genes from microarray gene expression data both dose response and time response gene expression data after retinoic acid treatment of human pancreatic stellate cells. Then we study and analyze the identified both up-regulated and down-regulated significant genes with gene ontologies (GO).

2 Material

The gene expression profile measures the activity of genes at a instance and it creates a global picture of cellular function. The RNA-Seq is the next generation sequencing. We implement our approach in expression profiling by array type microarray data. We collect data set from website, http://www.ncbi.nlm.nih.

gov.in/geodata/GSE14427. The data represents the change in expression of genes after retinoic acid treatment of human pancreatic stellate cells. It is a superseries of the subseries, GSE14426 and GSE14425. In GSE14425 a pancreatic stellate cell line is treated on plastic or matrigel with 1 or 10 micromolar dose of all-trans retinoic acid (ATRA). Then RNA was extracted and hybridized on Illumina Human microarrays. Then the target genes regulated by ATRA and evaluated for dose repsonse. Those RNA expression values are taken and changed due to background culture conditions. In GSE14426 data, same steps are maintained for timepoints of 30 mins, 4 hours, 12 hours, 24 hours and 168 hours.

3 Methods

After collecting data, we first perform normalization. It is done by dividing the mean column intensity and it scales the values in each column of microarray data. Its output is a matrix of normalized microarray data. Several statistical methods are used to identify the differentially expressed genes (DEG). Those methods are Student's t-test, T-test, Linear regression model, Nonparametric test (Wilcoxon, or rank-sums test), SAM etc. We perform statistical t-test to find DEG from the sample microarrays. This is also a statistical hypothesis test. We are describing the mathematical concept of this test as follows:

The size of our used samples are not equal. The number of genes present in each sample is 48701 but in data of dose response the number of expression values for each gene is 4 and in time response data the expressed values for each gene is 6. So the size of sample groups are 48701×4 and 48701×6 respectively. The standard formula for the t-test is as follows:

$$t = \frac{\overline{X_1} - \overline{X_2}}{\sqrt{\frac{var_1}{n_1} + \frac{var_2}{n_2}}} \tag{1}$$

Where, $\overline{X}_1$, $\overline{X}_2$ denote mean of two samples, standard error of the difference between the means is $\sqrt{\frac{var_1}{n_1} + \frac{var_2}{n_2}}$, var_1 and var_2 are the variances of two samples, n_1, n_2 are number of population in samples respectively. We then find false discovery rate (FDR) and estimate the result statistically. To control the expected proportion of incorrectly rejected null hypotheses i.e. false discoveries from a list of findings, FDR procedures are used. It is given by

$$FalseDiscoveryRate = Estimation \frac{no.ofFalseDiscoveries}{no.ofRejectedDiscoveries} \tag{2}$$

The FDR is kept below a threshold value i.e. quantile value (q-value). If $\overline{X}_1$ is larger than , $\overline{X}_2$, then t-value will be positive and negative if it is smaller. In the t-test, the degrees of freedom (DOF) is the sum of population in both sample groups minus 2. A table of significance in t-test is use to test whether the ratio is large enough to say that the difference between the groups is not likely to have been a chance of finding. For testing, the significance risk level (called alpha level) has to set. Mostly, the alpha level is set at .05. The degrees of freedom for the test is also need to determine. The t-test, Analysis of Variance (ANOVA) and a form of regression analysis are mathematically equivalent. The quantile value (q-value) of individual hypothesis test is the minimum FDR at which the test may be called significant.

Our proposed algorithm is as follows:

Input: *Data matrix object (dmo) of human pancreatic stellate cell dose response data and time response data. The number of genes in the sample is 48701.*

Step1: *Normalization is done for data matrix object. It is done to get mean and variance of each population as 0 and 1 rapectively; we named the result as dmoN.*

Step2: *Filtering normalized samples i.e. dmoN*
First, we filter out very low absolute expression values of genes.
Second, we filter out small variance valued genes across the samples.
Then we find total number genes present in the data matrix.
After filtering the data, the total number of genes are 43831.

Step3: *We extract the dose response data samples and time response data samples from dmoN (Set1 and Set2 respectively).*

Step4: *We perform standard statistical t-test to identify significant changes between the expression value measurement of genes in Set1 and Set2.*

Step5: *Then we plot normal quantile plot of t-scores and histograms of t-scores and p-values of the t-test.*

Step6: *We then perform the permutation t-test to compute p-values by permuting columns of gene expression data matrix [6]*

Step7: *We set the p-value cutoff at 0.02; the number of genes with having statistical significance at p-value cutoff are then determined.*

Step8: *The FDR and quantile values (q-values) for each test are then estimated.*

Step9: *The number of genes with having q-values less than cutoff value are determined.*

Step10: *Now false discovery rate adjusted p-values are estimated. It is done by using the Benjamini-Hochberg (BH) procedure [2]*

Step11: *Plot -log10 of p-values Vs biological effect in a volcano plot*

Step12: *Now export Differentially expressed genes (DEG) from volcano plot UI*

Output1: *We got 312 DEG from our sample microarray data matrices. Among the DEGs 213 genes are up-regulated and 99 genes are down-regulated.*

We code the algorithm and the results are shown in the result section.

4 Results

We code the algorithm with Matlab(R2013a) software and then we plot sample quantile plot vs theoretical quantile plot in fig1(a). As histogram plot is the graphical representation of a distribution of data, we plot histogram of t-test (i.e.t-score vs p-value) in fig1(b). This is the estimation of probability distribution of a continuous variable. The diagonal line represents sample quantile values equal to the theoretical quantile in fig1(a). Differentially expressed data points of genes lie farther away from the diagonal line. Data points with t-scores lie within $1 - (\frac{1}{2n})$ and $(\frac{1}{2n})$ display with a circle, where n is the total number of genes.

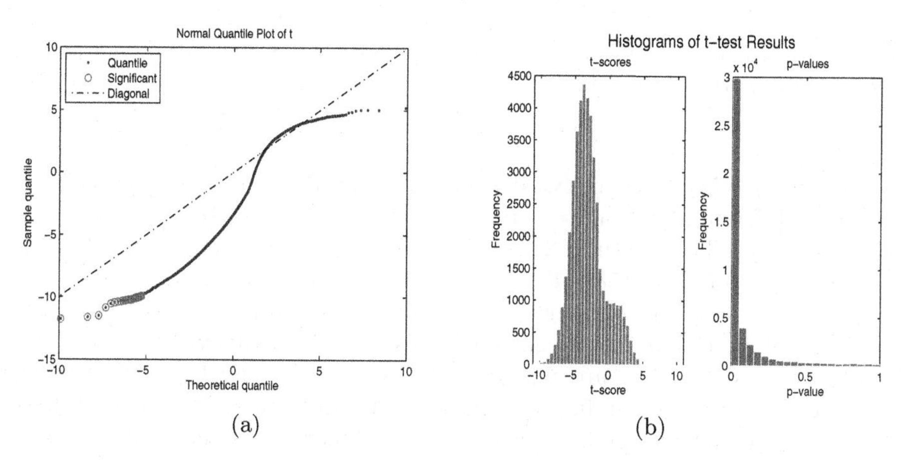

Fig. 1. (a) Plot of sample quantile vs theoretical quantile between dose response microarray data and time response microarray data; (b) Histogram plot of t-test result of samples used in fig1(a)

In fig3(a) estimated false discovery rate (FDR) for results Set1 and Set2 is plotted. Here, lambda, λ is the tunning parameter and it is used to estimate a priori probability of null hypothesis. The value of λ is > 0 and < 1. The polynomial method is used to choose lambda. fig3(b) shows volcano plot of differentially expressed genes (DEG) which we get for pancreatic stellate cell microarray data. Cutoff p-value is displayed as a horizontal line on the plot. The volcano plot shows –log10 (p-value) vs log2 (ratio) scatter plot of genes. Statistically significant genes placed above the p-value line and DEGs are placed outside of the fold changes lines. Both are shown in the plot (fig3(b)).

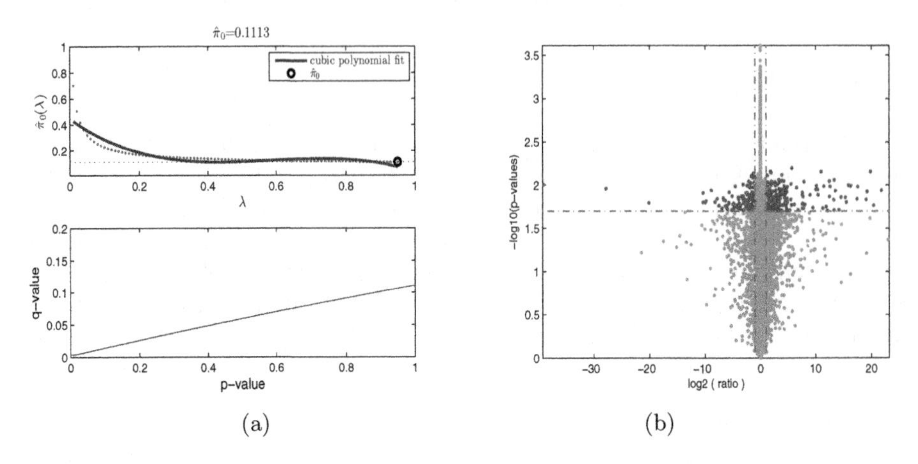

Fig. 2. (a)Estimation with false discovery rate (Set1 and Set2); (b)Volcano plot of differentially expressed genes in Set1 and Set2

Now, we studied and analyzed significant ontologies by gene ontology in which the differentially expressed up-regulated and differentially expressed down-regulated genes. Tne identified 213 UP-regulated gene were analyze with DAVID webtool and found 160, 158, 145 genes are significant in their associated biological processes, cellular components and molecular functions respectively. The analysis is done for 99 down-regulated genes also. After analysis we have seen that 64, 62 and 59 genes are significant in associated biolical process, cellular component and molecular function respectively. In table 1, 2 and 3 we have shown GO-terms, GO-id in biological process, cellular component and molecular function respectively.

5 Conclusion

We identified DEG to show the variations of our data samples in two different conditions. The pancreatic stellate cell data was taken after its retinoic acid treatment. One microarray gene expression was dose reponse and another was time course response. We Identified 312 DEG. Among them 213 gene are found up-regulated and 99 gene are down-regulated. We then analyze all the up and down regulated genes with gene ontology to find their significant GO term and Go id. About 75 percent of differentially expressed genes are found significant. From that GO id we then develop gene interaction network to get a metabolic pathway.

Acknowldgement. A. Mukhopadhyay acknowledges the support received from DST-PURSE scheme at University of Kalyani

References

1. Biology 4361. Differential gene expression. Technical report, Developmental Biology (2008)
2. Benjamini, Y., Hochberg, Y.: Controlling the false discovery rate: a practical and powerful approach to multiple testing. J. Royal Stat. Soc. 57, 289–300 (1995)
3. Chen, Y., Wu, R., Felton, J., et al.: A method to detect differential gene expression in cross-species hybridization experiments at gene and probe level. Biomedical Informatics Insights 3, 1–10 (2010)
4. Chen, Z., Liu, J., Tony Ng, H.K., et al.: Statistical methods on detecting differentially expressed genes for rna-seq data. BMC Systems Biology 5(3), 1–9 (2011)
5. Dudoit, S., Yang, Y.H., et al.: Statistical methods for identifying diferentially expressed genes in replicated cdna microarray experiments. Technical report, Department of Biochemistry, Stanford University, Stanford University School of Medicine, Beckman Center, B400 Stanford, CA (2000)
6. Dudoit, S., Shaffer, J.P., Boldrick, J.C.: Multiple hypothesis testing in microarray experiment. Statistical Science 18, 71–103 (2003)
7. Dudoit, S., Yang, Y.H., Callow, M.J., et al.: Statistical methods for identifying differentially expressed genes in replicates cdna microarray experiments. Statistica Sinica 12 (2002)
8. Kauffmann, A., Huber, W.: Microarray data quality control improves the detection of differentially expressed genes. Genomics 95 (2010)
9. Kumar, A.A., Holm, L., Toronen, P.: Gopargenpy: a high throughput method to generate gene ontology data matrices. BMC Bioinformatics 14, 242 (2013)
10. Ospina, L., Kleine, L.: Identification of differentially expressed genes in microarray data in a principal component space. SpringerPlus 2, 60 (2013)
11. Troyanskaya, O.G., Garber, M.E., et al.: Nonparametric methods for identifying differentially expressed genes in microarray data. Bioinformatics 18(11) (2002)

Software Effort Estimation through a Generalized Regression Neural Network

Parasana Sankara Rao[1] and Reddi Kiran Kumar[2]

[1] Department of CSE,
JNTUK University, Kakinada
priyasankarphd@gmail.com
[2] Department of CS,
Krishna University, Machilipatnam
kirankreddi@gmail.com

Abstract. Management of large software projects includes estimating software development effort as the software industry is unable to provide a proper estimate of effort, time and development cost. Though many estimation models exist for effort prediction, a novel model is required to obtain highly accurate estimations. This paper proposes a Generalized Regression Neural Network to utilize improved software effort estimation for COCOMO dataset. In this paper, the Mean Magnitude Relative Error (MMRE) and Median Magnitude Relative Error (MdMRE) are used as the evaluation criteria. The proposed Generalized Regression Neural Network is compared with various techniques such as M5, Linear regression, SMO Polykernel and RBF kernel.

Keywords: Effort estimation, Constructive Cost Model (COCOMO), Mean Magnitude Relative Error (MMRE) and Median Magnitude Relative Error (MdMRE), Linear Regression, Principal Component Analysis (PCA), Generalized Regression Neural Network.

1 Introduction

Software cost estimation guides/supports software projects planning. Effective control of software development efforts is important in a highly competitive world [1, 2]. Reliable and accurate software development cost predictions challenges perspective accounting for financial and strategic planning [3]. Software cost estimation predicts the probable effort, time, and staffing required for building a software system during a project. Estimates accuracy is low in the starting stages of a project due to limited details available. Traditional/conventional approaches for software projects effort prediction depend upon the mathematical formulae derived from historical data, or experts judgments use lack both effectiveness and robustness.

Expert judgement, algorithmic models and analogy are common approaches for effort estimation. Expert judgement is sought for small projects, and occasionally more than one expert's opinion being pooled for estimation. Algorithmic models like

S.C. Satapathy et al. (eds.), *Emerging ICT for Bridging the Future – Volume 1*,
Advances in Intelligent Systems and Computing 337, DOI: 10.1007/978-3-319-13728-5_3

COCOMO [4], SLIM [5], function points [6] are popular in the literature. Most algorithmic models are derived from:

$$effort = \alpha \ . \ size^{\beta}$$

where α is a productivity coefficient and β refers to the economies of scale coefficient. The size is measured in an estimated code line. Similar completed projects with known effort value are used for predicting effort for a project in analogy estimation.

Algorithmic models predict effort based on accurate estimate of software size with regard to lines of code (LOC) and complexity when the project faces uncertainty. Boehm, the first to consider software engineering economically came up with a cost estimation model, COCOMO-81, based on linear regression techniques, taking the number of lines of code as major input for their models [4]. Algorithmic models are successful within a particular environment, but lack flexibility. They are unable to handle categorical data (specified by a value range values) and lack reasoning capabilities. These limitations led to many studies which explored non-algorithmic methods [7].

Newer cost estimation non-algorithmic computation techniques are soft computing methodologies based on fuzzy logic (FL) [8], artificial neural networks (ANN) [9] and evolutionary computation (EC) [10]. Soft computing techniques were used by researchers for software development effort prediction to offset data imprecision and uncertainty, because of their inherent nature. Such methods are effective when a model has many differing parameters requiring analysis or are too complex to be represented as a normal function.

In this paper, we propose to investigate the Mean Magnitude Relative Error (MMRE) and Median Magnitude Relative Error (MdMRE) using various techniques such as M5, Linear regression, SMO Polykernel and RBF kernel and the proposed Generalized Regression Neural Network. The dataset COCOMO is used for the investigations. The paper is organized as follows: section II deals with related work, section III details the materials and methods used. Section IV gives the results of the experiments and discussion of the same and section V concludes the paper.

2 Related Works

Kaur, et al., [9] proposed Artificial-Neural-Network Based Model (ANN) to improve software estimation for NASA software projects and was evaluated based on MRE and MMRE. Neural Network based effort estimation system performance was compared to the other existing Halstead Model, Walston-Felix Model, Bailey-Basili Model and Doty Model. Results revealed that Neural Network system had lowest MMRE/RMSSE values. Neuro based system proved to have good estimation capability. Neuro based techniques can be used to build a generalized model type for software effort estimation of all types of projects.

Rao et al [11] proposed Functional Link Artificial Neural Network (FLANN) to lower neural net computational complexity to predict software cost and also to ensure suitability for on-line applications. FLANN has no hidden layers, architecture is simple and training does not involve full back propagation. In case of neural network adversity, the neural network works excellently and will first use a COCOMO approach to predict software cost, using FLANN technology with backward propagation. The proposed network processes every neuron crystal clear to ensure that the entire network is completely a "white box". This method provides better accuracy when compared to others as this method involves proper data training of data with back propagation algorithm. Training is simple due to the lack of any hidden layer. With this method using COCOMO as a base model it provides a good estimate for Software Development Approach projects.

Setino et al [12] described rule extraction techniques which derive a set of IF-THEN rules from a trained neural network applied to software effort prediction. Analysis of ISBSG R11 data set showed that rule set performance got through application of REFANN (Rule Extraction from Function Approximating Neural Networks) algorithm on a trained/pruned NN is equal to other techniques. This technique's applicability is compared with linear regression and CART. Experiments showed that good and accurate results are from CART, though the rule volume limits comprehensibility. But, the regression rule extraction algorithm provides end users with an exact rules set (both in terms of the number of rules and the number of rule antecedents) than CART. When comprehensible models alone are considered, extracted rules outperform pruned CART rule set, thereby ensuring that neural network rule extraction was the proper technique for software effort prediction.

Bhatnagar et al [13] employed a neural network (NN) and a multiple regression modelling approach to model/predict software development effort based on a real life dataset. Comparison of results from both approaches is presented. It is concluded that NN can model the complex, non-linear relationship between many effort drivers and software maintenance effort comfortably. The results closely match expert's estimates.

A multi-layer feed forward neural network to model and to estimate software development effort was proposed by Reddy [14]. The network is trained with back propagation learning algorithm through iterative processing a set of training samples. It then compares the network's prediction with actual. A COCOMO dataset trains and tests the network. It was observed that the proposed neural network model improved estimation accuracy. Test results from the trained neural network compared with a COCOMO model recommend that the proposed architecture can be replicated to forecast software development effort accurately. This study aims to enhance COCOMO model estimation accuracy, so that estimated effort is closer to actual effort.

3 Materials and Methods

3.1 Promise Effort Estimation Dataset COCOMO

A total of 63 software project details are included in the COCOMO dataset [15], each being described by 16 cost derivers/effort multipliers. Of 16 attributes, 15 are measured on scale of six categories: very low, low, nominal high, very high, and extra high. The categories are represented by a numeric value. Kilo Delivered Source Instructions (KDSI) is the only numeric attribute. COCOMO dataset usually assesses new techniques accuracy. Fig. 1 shows the effort histogram for COCOMO dataset.

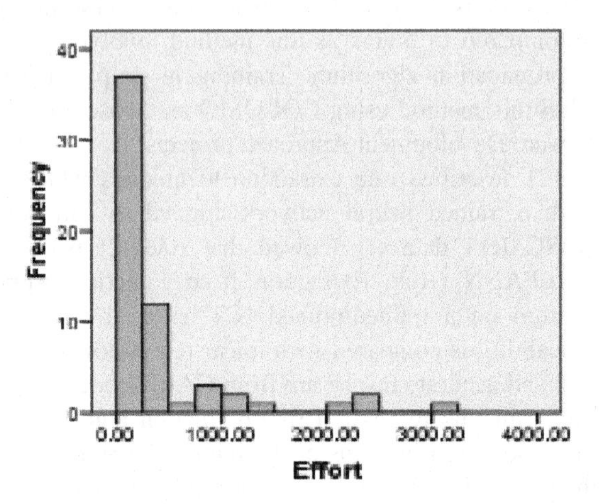

Fig. 1. Effort histogram of COCOMO81

3.2 Regression Analysis with MMRE Formula (Mean Magnitude Relative Error) and MdMRE (Median Magnitude Relative Error)

It is necessary to measure software estimates accuracy for evaluation purposes. In this context, the commonly used evaluation criteria in software engineering [16] are:

Magnitude Relative Error (MRE) computes the absolute percentage of error between actual and predicted effort for each reference project.

$$MRE_i = \frac{|actual_i - estimated_i|}{actual_i} \quad (1)$$

Magnitude Relative Error (MRE) computes absolute error percentage between actual and predicted effort for every reference project. Mean Magnitude Relative Error (MMRE) calculates MREs average over all reference projects. As MMRE is sensitive to specific outlying predictions, median of MREs for the n projects (MdMRE) is adopted when there are many observations as this is less sensitive to the extreme MRE values. Despite wide MMRE use in estimation accuracy, its efficacy in the estimation process has been under discussion. It has often been said that MMRE is unbalanced in many validation circumstances resulting in overestimation [17].

$$MMRE = \frac{1}{n}\sum_{i=1}^{n} MRE_i \tag{2}$$

$$MdMRE = \underset{i}{median}\left(MRE_i\right) \tag{3}$$

3.3 Principal Component Analysis (PCA)

A wide metrics range was proposed for early software project effort estimation. Many authors suggested reduction in the number of sets as standard sets had too many parameters, (large metric sets had high collection costs risking generation of over-fitted models). Principal Component Analysis (PCA) is used for Feature reduction refers to multidimensional space mapping into lower dimension space. Such techniques are usually machine learning pre-process and statistics tasks include prediction and pattern recognition. Feature space contributes to classification, cuts pre-processing costs and reduces classification efforts. This improves overall classifier performance. PCA is a data analysing technique which compresses high dimensional vector data sets into low dimensional ones, being derived from many starting points and optimization criteria. The most important of such criteria is mean-square error minimization in data compression, locating mutual orthogonal directions in data with maximum variance and data de-correlation through orthogonal transformations. PCA aims at data dimensionality reduction while simultaneously retaining much of the original data set variations. This procedure identifies data patterns and highlights similarities and differences [18].

3.4 Linear Regression

Consider the problem of approximating the set of data,

$$D=\left\{\left(x^1,y^1\right),\ldots,\left(x^l,y^l\right)\right\},\quad x\in\Re^n, y\in\Re \tag{4}$$

with a linear function,

$$f(x)=\langle w,x\rangle+b \tag{5}$$

the optimal regression function is given by the minimum of the functional,

$$\Phi(w,\xi)=\frac{1}{2}\|w\|^2+C\sum_i\left(\xi_i^-+\xi_i^+\right) \tag{6}$$

where C is a pre-specified value, and ξ- , ξ+ are slack variables representing upper and lower constraints on the outputs of the system.

3.5 SMO Polykernel and RBF Kernel

SVMs based methods are used for classification [19]. For specific training data $(x_i, y_i),\ i = 1,..,n$, with a feature vector $x_i \in \Re^d$ and $y_i \in \{+1, -1\}$ indicates the class value of x_i when solving the following optimization problem:

$$\min_{w,b,\xi} \quad \frac{1}{2}\|w\|^2 + C\sum_{i=1}^{N}\xi_i \tag{7}$$

subject to

$$y_i\left(w^T\Phi(x_i)+b\right) \geq 1-\xi_i \text{ for i=1...n}$$
$$\xi_i \geq 0 \tag{8}$$

where $\Phi : \Re \rightarrow \text{H}$, H being the high dimensional space $w \in$ H, and b$\in \Re$. $C \geq 0$ is a parameter controlling minimization of margin errors and margins maximization. Φ is selected to ensure that an efficient kernel function K exists.This optimization problem is solved by using Lagrange Multiplier method in practice. Sequential Minimal Optimization (SMO) [21] is a simple algorithm which solves SVM QP problem. Its advantage is its capability to solve Lagrange multipliers sans using numerical QP optimization. The following is the Lagrangian form:

$$\min_{\alpha} \quad \frac{1}{2}\sum_{i,j=1}^{n}\alpha_i\alpha_j y_i y_j K(x_i, x_j) - \sum_{i=1}^{n}\alpha_i \tag{9}$$

subject to

$0 \leq \alpha_i \leq C$ for i=1...n

$$\sum_{i=1}^{n}\alpha_i y_i = 0$$

On solving the optimization problem, w is computed as follows:

$$w = \sum_{i=1}^{n}\alpha_i y_i \Phi(x_i) \tag{10}$$

x_i is a support vector if $\alpha_i \neq 0$. New instance x is computed by the following function:

$$f(x) = \sum_{i=1}^{n_S} \alpha_i y_i K(s_i, x) + b \tag{11}$$

Where si are support vectors and nS is number of vectors.

The polynomial kernel function is given by:

$$K(x_i, x_j) = (\gamma x_i^T x_j + r)^d, \text{ where } \gamma > 0 \tag{12}$$

And the Radial basis function (RBF) kernel:

$$K(x_i, x_j) = \exp\left(-\gamma \left\| x_i - x_j \right\|^2\right), \text{ where } \gamma > 0 \tag{13}$$

3.6 M5 Algorithm

Many tree-building algorithms like C4.5 exist that determine what attributes best classifies remaining data. The tree is then is constructed iteratively. Decision trees advantage is that their immediate version to rules can be interpreted by decision-makers. For numeric prediction in data mining, regression trees or model trees [22] are used. Both build a decision tree structure where each leaf is to ensure local regression for a particular input space. The difference is that while a regression tree generates constant output values for input data subsets (zero-order models), model trees generate linear (first-order) models for every subset.

The M5 algorithm constructs trees where leaves are linked to multivariate linear models and tree nodes are chosen over the attribute which maximizes expected error reduction as a standard deviation of output parameter function. The M5 algorithm builds a decision tree that splits up attribute space in ortho-hedric clusters, with the border paralleling the axis. Model tree advantage is that it can be easily converted into rules; each tree branch has a condition as follows: attribute ≤ value or attribute > value.

3.7 General Regression Neural Network (GRNN)

General Regression Neural Network (GRNN) comes under the probabilistic neural networks category. This network like other probabilistic neural networks needs limited training samples which a back-propagation neural network needs [23]. Data available from operating system measurements is usually not enough for a backpropagation neural network. Hence, use of a probabilistic neural network is specially advantageous as it easily converges to an underlying data function with only limited training samples. Additional knowledge required to get a proper fit is relatively small and is possible without additional user input. Thus GRNN is a useful tool for predictions and comparisons of practical system performance. Fig. 2 shows the architecture of a typical GRNN.

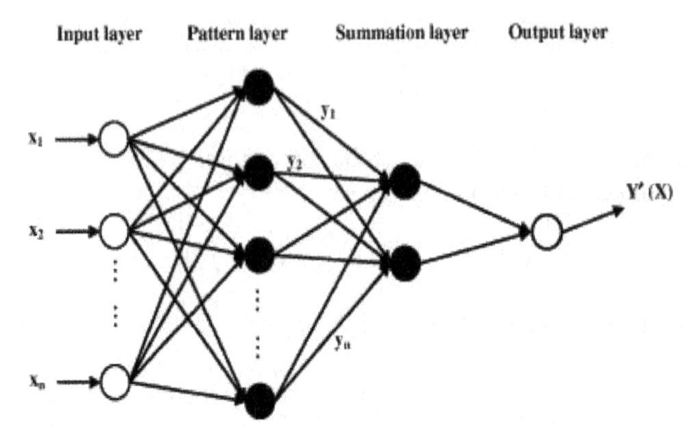

Fig. 2. Architecture of a typical GRNN

The parameters for the proposed GRNN are tabulated in Table 1.

Table 1. Parameters for the Proposed GRNN

Number of input	17
Number of output	1
Number of hidden layer	2
Number of neurons in hidden layer	6
Number of cluster centers	85
Competitive rule	conscience full metric – Euclidean
Activation function	Tanh
Momentum	0.5
Learning rate	0.1

4 Results and Discussion

Attributes of a COCOMO dataset is used along with feature transformation of the attributes using Principal Component Analysis (PCA). Initial experiments using PCA without data transformation, the Mean Magnitude Relative Error (MMRE) and Median Magnitude Relative Error (MdMRE) are evaluated through various techniques like M5, Linear regression, SMO Polykernel and RBF kernel and the proposed Generalized Regression Neural Network. In the next set of experiments, the same is evaluated with feature transformation of the attributes using PCA. Table 2 provides results of average MMRE and MdMRE for various techniques. Fig. 3 and Fig. 4 reveal the same.

Table 2. Average MMRE and MdMRE for various techniques

Technique Used	MMRE	MdMRE
M5	4.674381	185.6412
Linear regression	5.215322	219.7713
SMO polykernel	5.67898	91.95121
SMO RBF kernel	2.675412	84.55606
Proposed GRNN	2.1442	78.4769
M5 + PCA	6.875919	157.6677
Linear regression +PCA	5.21539	219.7753
SMO polykernel + PCA	4.833664	93.8034
SMO RBF kernel + PCA	2.662263	85.94772
Proposed GRNN + PCA	2.0898	79.0105

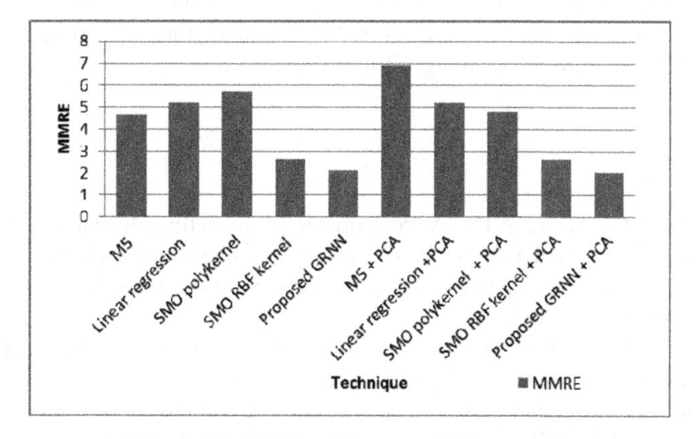

Fig. 3. MMRE for different techniques used

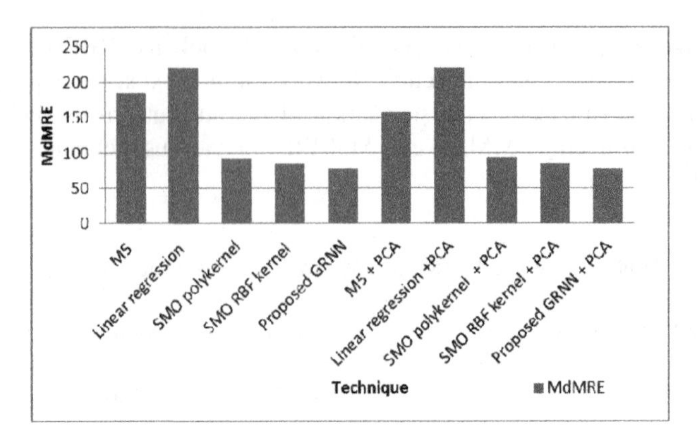

Fig. 4. MdMRE for different techniques used

It is observed that the proposed GRNN improves the MMRE significantly by 19.86% to 62.24% when compared to other algorithms. On using PCA for reduction, it is seen that all the algorithms other than M5 improves in performance. The proposed GRNN with PCA achieves 2.0898 MMRE which is 2.54% better performance when feature reduction is not used.

5 Conclusion

In this paper, a Generalized Regression Neural Network is proposed to utilize improved estimation of software effort for COCOMO dataset. In this paper, the Mean Magnitude Relative Error (MMRE) and Median Magnitude Relative Error (MdMRE) are used as the evaluation criteria.The proposed method was compared with performance of three regression algorithms including Linear regression, M5 and modified Support Vector Machine (SVM) to avoid an quadratic problem. Two kernels were used for the SVM with the first being a polykernel and the second kernel using Radial Basis Function (RBF). Experiments revealed that the proposed method outperformed classical regression algorithms.

References

1. MacDonell, S.G., Gray, A.R.: A comparison of modelling techniques for software development effort prediction. In: Proceedings of the International Conference on Neural Information Processing and Intelligent Information Systems, Dunedin, New Zealand, pp. 869–872. Springer, Berlin (1997)
2. Strike, K., El-Emam, K., Madhavji, N.: Software cost estimation with incomplete data. IEEE Transactions on Software Engineering 27(10) (2001)
3. Hodgkinson, A.C., Garratt, P.W.: A neurofuzzy cost estimator. In: Proceedings of the Third International Conference on Software Engineering and Applications, SAE 1999, pp. 401–406 (1999)

4. Boehm, B.W.: Software Engineering Economics. Prentice-Hall, Englewood Cliffs (1981)
5. Putnam, L.H.: A General Empirical Solution to the Macro Software Sizing and Estimating Problem. IEEE Transactions on Software Engineering 4(4), 345–361 (1978)
6. Albrecht, A.J., Gaffney, J.R.: Software function, source lines of code, and development effort prediction a software science validation. IEEE Trans. on Softi. Eng. 9(6), 639–648 (1983)
7. Ahmed, M.A., Saliu, M.O., AlGhamdi, J.: Adaptive fuzzy logic-based framework for software development effort prediction. Information and Software Technology Journal 47, 31–48 (2005)
8. Kad, S., Chopra, V.: Fuzzy logic based framework for software development effort estimation. Research Cell: An International Journal of Engineering Sciences 1, 330–342 (2011)
9. Kaur, J., Singh, S., Kahlon, K.S., Bassi, P.: Neural Network-A Novel Technique for Software Effort Estimation. International Journal of Computer Theory and Engineering 2(1), 17–19 (2010)
10. Sheta, A., Rine, D., Ayesh, A.: Development of Software Effort and Schedule Estimation Models Using Soft Computing Techniques. In: IEEE Congress on Evolutionary Computation, pp. 1283–1289 (2008)
11. Tirimula Rao, B., Sameet, B., Kiran Swathi, G., Vikram Gupta, K., Ravi Teja, C., Sumana, S.: A Novel Neural Network Approach for Software Cost Estimation Using Functional Link Artificial Neural Network (FLANN). IJCSNS International Journal of Computer Sciencee and Network Security 9(6), 126–132 (1999)
12. Setiono, R., Dejaeger, K., Verbeke, W., Martens, D., Baesens, B.: Software effort prediction using regression rule extraction from neural networks. In: 2010 22nd IEEE International Conference on Tools with Artificial Intelligence (ICTAI), vol. 2, pp. 45–52. IEEE (October 2010)
13. Bhatnagar, R., Bhattacharjee, V., Ghose, M.K.: Software Development Effort Estimation–Neural Network Vs. Regression Modeling Approach. International Journal of Engineering Science and Technology 2(7), 2950–2956 (2010)
14. Reddy, C.S., Raju, K.V.S.V.N.: A concise neural network model for estimating software effort. International Journal of Recent Trends in Engineering 1(1), 188–193 (2009)
15. Boetticher, G., Menzies, T., Ostrand, T.: PROMISE Repository of empirical software engineering data. West Virginia University, Department of Computer Science (2007), http://promisedata.org/repository
16. Mendes, E., Mosley, N., Counsell, S.: A replicated assessment of the use of adaptation rules to improve Web cost estimation. In: International Symposium on Empirical Software Engineering, pp. 100–109 (2003)
17. Shepperd, M.J., Schofield, C.: Estimating Software Project Effort Using Analogies. IEEE Transaction on Software Engineering 23, 736–743 (1997)
18. Ahmad, I., Abdullah, A., Alghamdi, A., Hussain, M.: Optimized intrusion detection mechanism using soft computing techniques. © Springer Science+Business Media, LLC (2011)
19. Vapnik, V., Golowich, S., Smola, A.: Support vector method for function approximation, regression estimation, and signal processing. In: Mozer, M., Jordan, M., Petsche, T. (eds.) Advances in Neural Information Processing Systems 9, pp. 281–287. MIT Press, Cambridge (1997)

20. Chang, C.-C., Lin, C.-J.: LIBSVM: a library for support vector machines (2001)
21. Platt, J.C.: Fast training of support vector machines using sequential minimal optimization. In: Schokopf, B., et al. (eds.) Advances in Kernel Methods: Support Vector Machines. MIT Press (1999)
22. Breiman, L., Friedman, J., Olshen, R., Stone, C.J.: Classification and Regression Trees. Chapman and Hall, New York (1984)
23. Specht, D.F.: A general regression neural network. IEEE Transactions on Neural Networks 2(6), 568–576 (1991)

Alleviating the Effect of Security Vulnerabilities in VANETs through Proximity Sensors

R.V.S. Lalitha[1] and G. JayaSuma[2]

[1] Dept. of C.S.E, Sri Sai Aditya Institute of Science and Technology, Surampalem, India
[2] Dept. of I.T, University College of Engineering, JNTUK, Viziayanagaram, India
{rvslalitha,gjscse}@gmail.com

Abstract. As the rate of road accidents are increasing day by day, an intelligent mechanism is essential to improve road safety. As a solution to this, current researchers focus is to use sensors which is a cost effective and leads to tremendous improvement in Vehicular Ad hoc Networks (VANET) and the need of their existence. Vehicular Ad hoc Network (VANET) is a subset of Mobile Ad hoc Networks (MANETs). VANETs exchange information between vehicles and the Road Side Unit (RSU) for making intelligent decisions spontaneously. Accidents on roads not only lead to the risk of life to victims and also create inconvenience to the public by either traffic jam or traffic diversion. Since VANETs has mobility, GSM communication fails as the messages are overlapped due to coherence. Apart from GSM communication, if the VANETs and typical road junctions are equipped with sensors, provides cost effective solution for reliable communication. In this paper, it is proposed to adopt a proximity sensor approach in VANETs to capture data, transmit it and store it in the local database for future reference if required. The capturing of data is done through Proximity sensors. These sensors will be mainly located at typical junctions and also in secure cars for an immediate response. This work is optimized using Ant-Colony metaheuristic optimization algorithm to trace the shortest path to overcome the inconsistent situations happened during the times of accident occurrences.

Keywords: V2V Communication, Proximity Sensors, Ultrasonic Proximity Sensors, Location Information, Ant-Colony optimization algorithm.

1 Introduction

Mobile Ad hoc Networks (MANETs) consist of mobile nodes with no existing pre-established infrastructure. They connect themselves in a decentralized, self-organizing manner and also establish multi hop routes. If the mobile nodes are vehicles then this type of network is called VANET. VANET is made up of multiple "nodes" connected by "links"[1]. The network is ad hoc because it does not depend on pre existing infrastructure. The connectivity between nodes depends on length-of-link, signal loss and interference. A "path" is a series of links that connects two nodes [4]. Various routing methods use single or multiple paths between any two nodes. Though vehicular networks share common characteristics with conventional Ad hoc sensor

S.C. Satapathy et al. (eds.), *Emerging ICT for Bridging the Future – Volume 1*,
Advances in Intelligent Systems and Computing 337, DOI: 10.1007/978-3-319-13728-5_4

networks, as being self-organized and lack of central control, VANET has unique challenges that impact the design of communication system and its protocol security.

Vehicular Ad hoc Networks (VANET) require cost effective solutions for traffic safety related applications in uplifting passenger safety measures, which results in broadcasting emergency information to the vehicles nearby during mishaps in establishing dynamic routing. A sensor is a device that transforms physical and chemical magnitudes into electrical magnitudes as Temperature, Light, Intensity, Displacement, Distance and Acceleration.

1.1 Vehicle-to-Vehicle (V2V) Communication

V2V is mainly essential during occurrences of accidents. This facilitates in providing information to drivers for taking necessary remedial measures. Normally, this is done with the help of GSM communication to transmit emergency information over the VANET. During the case of accidents, Breakdown car (vCar) causes obstruction on the road. The information as seen by the vCar1 sends to vCar2 via GSM communication as shown in Fig.1. The chief characteristic of VANET users is Location Information which is used to track the position of the vehicle on the road. As already Android mobiles are featured with Geographic Positioning System (GPS), tracking of vehicles is convenient with these mobiles. Also, these mobiles are equipped with sensors, which lead to combine the Location Information with the information sensed by the sensors so as to give a beneficial solution in delivering emergency messages. The data sensed by the sensors is stored in local database for future reference as shown in Fig.2.

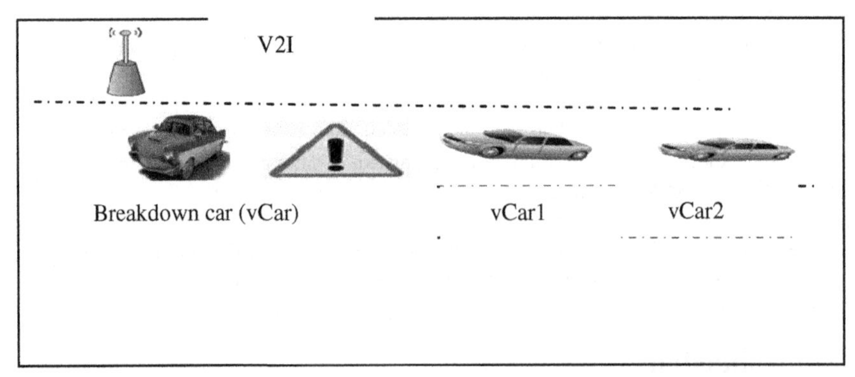

Fig. 1. Conventional messaging through GSM to other vehicles on the lane in case of obstruction ahead

1.2 Ant Colony Optimization (ACO) Algorithm

The goal of Ant Colony Optimization (ACO) algorithm is to appreciate the behavior of ants, in finding shortest path between food and nest without the use of visual information [5]. This concept envisages in making it applied in e-Sensor communication system in building up shortest path dynamically using Location Information of neighborhood of the accident object. The algorithm for tracing the dynamic path is described in Section 3.3.

The e-Sensor Communication is an attempt to achieve the reliable broadcasting mechanism during emergency situations and also provides information about shortest path to cross over using Ant-Colony optimization algorithm.

2 Related Work

Modern cars are oriented towards providing the ultimate accessories to make every journey memorable, safe and comfortable. Mahmoud Abuelela's research is on data dissemination using Smart Opportunistic Data Dissemination Approach [1]. VANETs communication became popular for being its safety alerts in case of accidents, entertainment, data sharing etc. Tamer Nadeem discussed about providing traffic information using GPS to drivers and also about broadcasting of messages [2] over VANET. In 2007, Andreas Festag et al implemented of Vehicle-to-Vehicle communication using C as software and a special type of embedded system as hardware and then tested on VANET[4]. Josiane et al gave detailed analysis and experimentation about Routing and Forwarding in VANETs in 2009. Keun Woo Lim defined Vehicle to Vehicle communication using sensors placed on the road for sensing vehicular information through Replica nodes[3s]. In all the above papers, how data dissemination in VANETs is discussed.

When vehicle met with an accident or broke down in the middle of the journey, the vehicles on the same lane need to wait for a long time, which may result in traffic jam. In normal situations, Vehicle-to-Vehicle (V2V) communication is through OBU (On Board Unit), which is installed in the Car or communicates via wireless link (V2I) to take necessary action. Due to mobility characteristic of the VANET, the communication through GSM fails. As the main concern of VANET is transmitting emergency information in case of accidents occurred, launching proximity sensors in VANETs and at accident prone areas to facilitate automatic messaging gives potential results. In addition to this, the proposed system provides shortest path to crossover the junction using Ant Colony optimization algorithm [5]. In this paper, e-Sensor Communication system is introduced to combine sensor appliances with the VANET so as to obtain quick message communication. This is done in three phases. In Phase I, Broadcasting of messages using GCM services by the user manually is discussed. In Phase II, Identification of the accident object by proximity sensor is discussed. In Phase III, Computing the shortest path to cross over the point of incident using Ant-Colony optimization algorithm and sending that dynamic path information to users are discussed. Finally, in the conclusion how effective communication is done in VANETs by launching this system in Android mobiles is discussed.

3 Methodology

The images captured by Range Sensors are stored as 2D images. The pixel values of the image correspond to the distance between the object from a specified location [2]. Brighter pixel values represent that the object is at shorter distance and the objects that are at longer distances cannot be visualized clearly. Later these pixel values are converted into actual distance in terms of meters for true evaluation. The dynamic range of range sensor is determined by the ratio of largest possible signal to the smallest possible signal. Proximity Sensor is a type of Range Sensor which is already available in Android mobiles and is used for our experimentation. These mobile sensors operate at a distance of 10-15 cm. In this paper, Proximity sensors are used to detect the presence of nearby objects.

The use of sensors [10] in Vehicular Ad-hoc networks plays a predominant role in maintaining information i.e. sent to and from the vehicles by storing information in the local database.

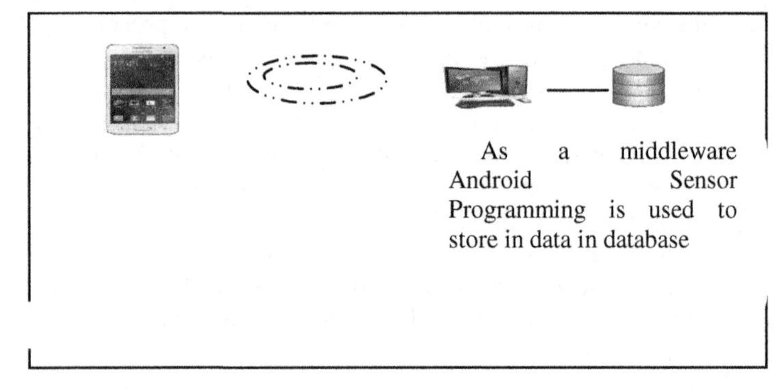

As a middleware Android Sensor Programming is used to store in data in database

Fig. 2. Sensor Communication

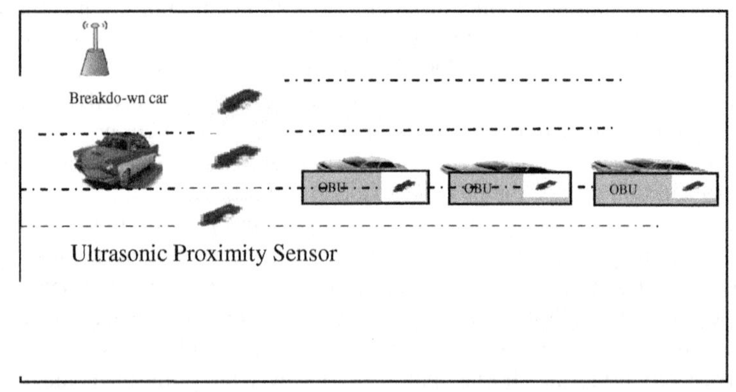

Fig. 3. Warning from the Ultrasonic Proximity Sensor to nodes nearby as there is an obstruction ahead

The proposed approach uses Ultrasonic Proximity Sensors in vehicles and also on the road at danger zones. In this, automatic messages are sent by sensors [6] if an obstruction is detected by it. Similarly, the mobile communication is also used to send messages. This ensures definite attention of the driver that something happened which requires immediate attention. Immediately after sensing the breakdown car, sensors sense this information and broadcast via sensors that are built in sensor equipped cars. This is exemplified in Fig.3.

This paper is mainly focused on three aspects namely i) Transmission of Emergency messages through Web Server using Google Cloud Messaging (GCM)

service, which allows transmission to nearby Android mobiles (range based) manually as shown in Fig.5. ii) Detection of accident objects using Proximity sensors and sending geographic Location Information using GCM service to Web server as shown in Fig.6.iii) In addition to messaging about the incident happened, this e-Sensor Communication system also provide route map to cross over the point of incident, by computing shortest distance using Ant Colony optimization algorithm [5] as shown in Fig.9.This out performs the furnishing clear picture of the information traced.

3.1 Transmission of Emergency Messages through Web Server Using Google Cloud Messaging (GCM) Service, which Allows Transmission to Nearby Android Mobiles (Range Based)

The following procedure is adopted to send information to VANET users about the accident object:

1. Track latitude and longitude positions of the VANET users using GPS services.
2. Update the Location Information of the VANET users after every 20 secs using AsyncTask that runs in background as VANET is dynamic in nature.
3. Send updated coordinates along with the device id which is obtained using GCM service to server for storing in a local database.
4. Use Google Cloud Messaging service to send message to Android-powered devices using Send SMS button manually as shown in Fig.5.and the Location Information of the VANET users is shown in Fig.6. using Google Map APIs.

3.2 Detection of Accident Objects Using Proximity Sensors and Sending Geographic Location Using GCM Service to GCM Server

As the name suggests, e-Sensor Communication uses Proximity Sensors to identify the Accident Object and sends alert message to Web Server for transmitting it over VANET using GCM. As soon as the Accident Object is identified by sensors available in mobile, they send the message automatically to GCM Server without intervention of the user. This reduces latency and delay in transmission.

The steps involved in sensing accident object by the built in proximity sensor available in android mobile:

1. Activation of proximity sensor in the android mobile using SensorManager class via getSystemService (SENSOR_SERVICE).
2. Use proximity sensor via getDefaultSensor (TYPE_PROXIMITY) to identify the accident object.
3. The Above two play key role in identifying the accident object. As soon the object is sensed automatic messaging is to be done to the nearby users as shown in Fig.7.using GCM server. The Location Information of the individual user also can be viewed on the Google Map as shown in Fig.8.
4. GCM server in turn transmits this information to android-powered devices.

3.3 Providing Route Map to Cross over the Point of Incident, by Computing Shortest Distance Using Ant Colony Optimization Algorithm

The establishment of dynamic path is computed using Ant-Colony optimization algorithm. The algorithm for computing shortest path is as follows:

```
Algorithm: To compute Shortest path
Select arbitrary point near by accident object
Do While
    Do Until
        Choose all the possible paths from the current
        point to cross over the accident object by
        choosing arbitrary points randomly
    End do
    Analyze the shortest path among all possible paths
    Update the Location Information based on the
    density of the traffic
End do
```

The shortest path obtained by applying the above algorithm is then displayed on the Google Map by coding. The geographic information about the location of the incident happened is tracked using Location class and sent to the server. The rerouting is computed using Ant-Colony optimization algorithm by finding the neighborhood latitude and longitude positions by comparing the positions to the left and the positions to the right. Thereby minimum is decided and sent to the mobile. Finally, new reestablished path is drawn on the Google Map using Polyline class. This map is sent to the VANET user using GCM Notifications. The route map shown in Fig.9. is computed as follows.

Procedure to compute Geographic positions for the reroute map in the Web server:

- **Obtain Location Information of the current and arbitrary points using GPS:-**

```
1. Compute Geographic distance between two
points using Great Circle formula:
Distance=R*arcos([sin(lat1)*sin(lat2)]+cos(lat1)*cos(l
at2)*cos(lon2=lon1)]
As the built-in proximity sensors of the mobile are
used, it is assumed that current position identified
by the sensor will be the accident object locations.
```

- **To cross over the accident region, apply Ant-Colony optimization algorithm to compute shortest path dynamically. The procedure is as follows:-**

Step 1:

```
Assumptions:
size of the object:0002km
```

```
  lt1,lo1 are latitude and longitude positions.
  s1,s2 are the random positions chosen on the
top/bottom sides.
  x1 :lies to the left of current point
  x2 :lies to the right of current point
  x3: lies to the right of x2
  v1:lies to the top of current point
  v2:lies to the top of x2
  v3:lies to the bottom of current point
  v4:lies to the bottom of x2
Step 2:
  Obtain the left/right positions by
adding/subtracting the accident object size.
  Obtain the top/bottom positions by
adding/subtracting the random numbers computed.
  Current Point c(2,2)          Point v1(2,5)
  Point x1(0,2)                 Point v2(4,5)
  Point x2(4,2)                 Point v3(2,1)
  Point x3(6,2)                 Point v4(4,1)
Dist1=dist(x1,c)+dist(c,v1)+dist(v1,v2)+
      dist(v2,x2)+dist(x2,x3)
Dist2=dist(x1,c)+dist(c,v3)+dist(v3,v4)+
      dist(v4,x2)+dist(x2,x3)
Optimal Path=min(Dist1,Dist2)
```

The paths traced (Red and Green) is represented in the Fig.4.

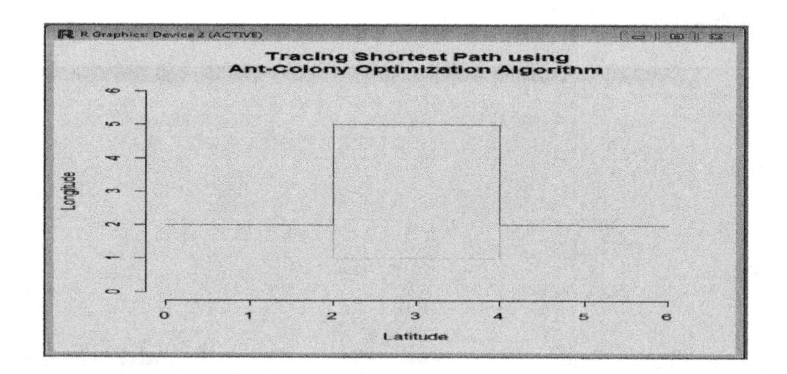

Fig. 4. Graph showing the two possible paths using Ant-Colony algorithm

3.3.1 Displaying Reroute Map on the Google Map on the User's Mobile

The reroute map is drawn on Google Map using Polyline method after computing the values using the method discussed above is shown in Fig.9. The dynamic path is changed dynamically based on the point of accident as the Location Information is updated after every 20Sec.

4 Experimental Results

4.1 Transmission of Emergency Messages through Web Server Using Google Cloud Messaging (GCM) Service by the Mobile User

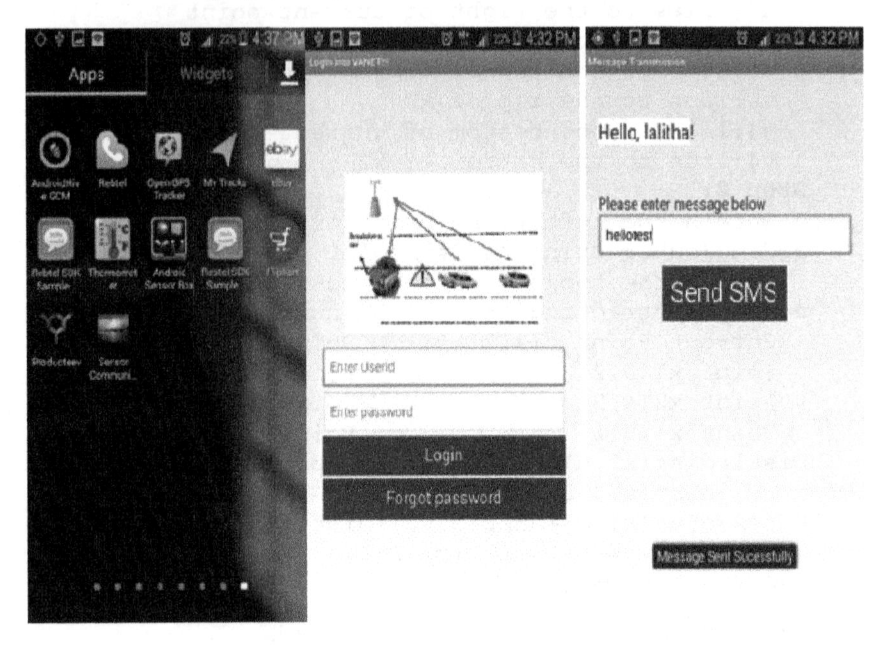

Fig. 5. Message communication by the VANET user to transmit urgent message using GCM by accessing Web server

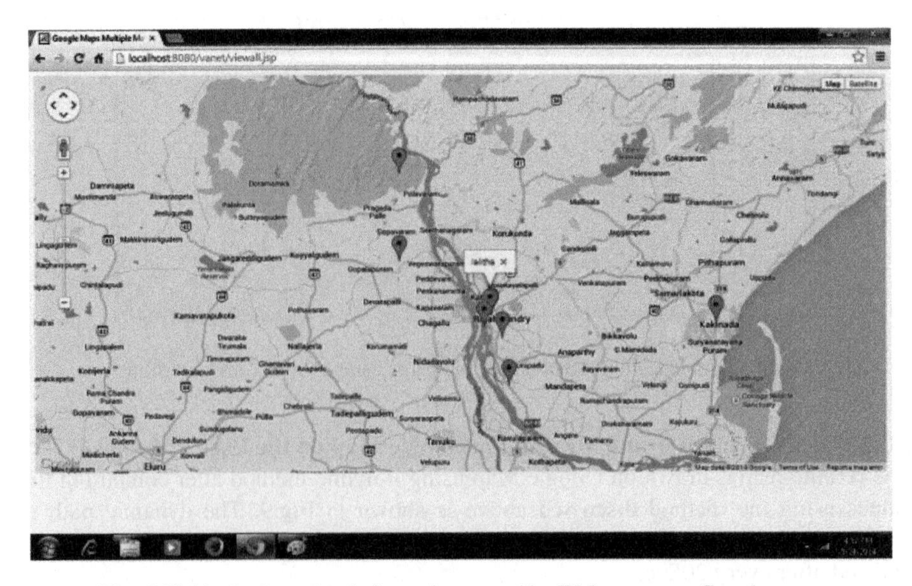

Fig. 6. Vehicular Location Information traced by Web server on Google map

4.2 Automatic Transmission of Messages by Sensors in Android Mobiles through Web Server Using GCM

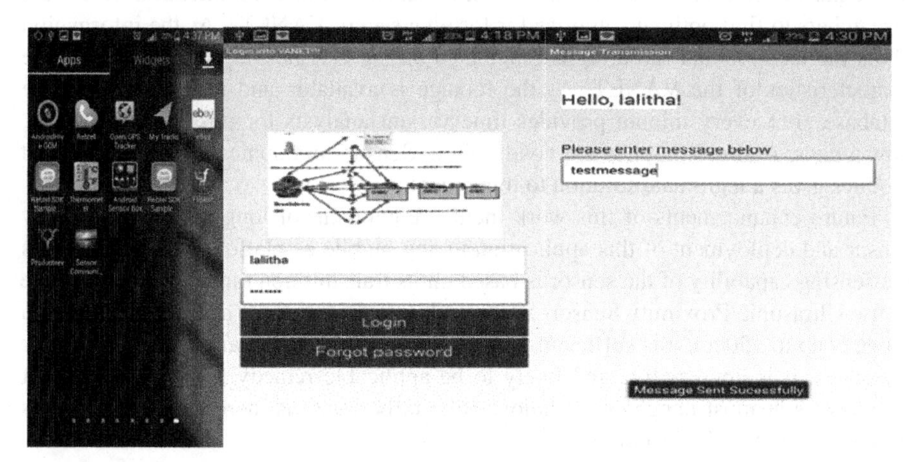

Fig. 7. Message communication by the mobile sensors to transmit urgent message using GCM by accessing Web server

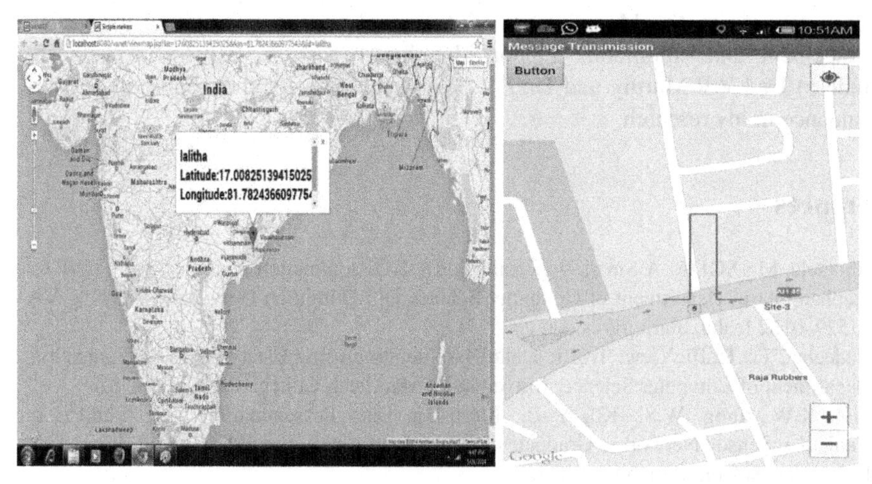

Fig. 8. Vehicle position traced by web server on Google Map

Fig. 9. Reroute map after computing shortest path using Ant-Colony algorithm

4.3 Displaying Reroute Map on the User's Mobile

In addition to messaging about the incident happened, this e-Sensor Communication system also provide route map to cross over the point of incident, by computing shortest distance using Ant Colony optimization algorithm[5]. The reroute is drawn in red color on the Google Map as shown in Fig.9.

5 Conclusions

Vehicular Ad hoc Networks (VANET) have mostly gained the attention of current researchers to find optimal solutions for forming secure VANETs. As the information about the location and rerouting is reticulated on the Google Map this ensembles the characteristics of the VANET. As the footage is available and also updated in the database after every minute provides time variant analysis for post processing. As, now a days, Android mobiles are vividly used, Sensor communication using Android mobiles gives a legitimate solution to implement.

Future enhancements of this work include equipment of long range high quality sensor and deployment of this application to any mobile as platform independent. As the sensing capability of the sensor is based on its transmission range, it is appropriate to use Ultrasonic Proximity Sensors in the real time environment as their transmission range is up to 150cm, it is sufficient to identify the object ahead and to transmit quick messages. It is appreciative and likely to be applicable remedy if the bandwidth of frequencies at most dangerous locations (like hilly areas) is increased which results deriving cost effective solutions.

Acknowledgments. I would like to thank Mr. Mahmoud Abuelela who motivated me to continue this work. I also take this opportunity to express my profound gratitude and deep regards to the Management of Aditya Group of Institutions for giving encouragement and continuous support. Finally, I wish to express my sincere gratitude to Dr.J.V.R.Murthy and Dr.T.V.S.Gireendranath for their constant support and guidance in my research.

References

1. Abuelela, M.: SODA: A Smart Opportunistic Data Dissemination Approach for VANETs. Stephan Olariu Department of Computer Science Old Dominion University, Norfolk, VA 23529–0162 feabu, olariug@cs.odu.edu
2. Nadeema, T.: TrafficView: Traffic Data Dissemination using Car-to-Car Communication. Department of Computer Science, University of Maryland, College Park, MD, USA
3. Lim, K.W., Jung, W.S., Ko, Y.-B.: Multi-hop Data Dissemination with Replicas in Vehicular Sensor Networks. Graduate School of Information and Communication Ajou University, Suwon, Republic of Korea
4. Festag, A., Hessler, A., Baldessari, R., Le, L., Zhang, W., Westhoff, D.: Vehicle-To-Vehicle and Road-Side Sensor Communication for Enhanced Road Safety. NEC Laboratories Europe, Network Research Division Kurfursten-Anlage 36, D-69115 Heidelberg
5. Parpinelli, R., Lopes, H., Freitas, A.: Data Mining with an Ant Colony Optimization Algorithm
6. Zhang, Y., Zhao, Z., Cao, G.: Data Dissemination in Vehicular Ad hoc Networks (VANET)
7. Drawil, N.: Improving the VANET Vehicles' Localization Accuracy Using GPS Receiver in Multipath Environments

8. Ye, F., Luo, H., Cheng, J., Lu, S., Zhang, L.: A TwoTier Data Dissemination Model for Largescale Wireless Sensor Networks. UCLA Computer Science Department, Los Angeles, CA 9000951596, fyefan,hluo,chengje, slu, lixiag @cs.ucla.edu
9. Park, V.D., Corson, S.M.: Temporally-Ordered Routing Algorithm - A Highly Adaptive Distributed Routing Algorithm for Mobile Wireless Networks. In: Proceedings of INFOCOM (1997)
10. Tavli, B.: Broadcast Capacity of Wireless Networks. IEEE Communications Letters 10(2), 68–69 (2006)
11. Wolfson, O., Xu, B., Yin, H.: Reducing Resource Discovery Time by Spatio-Temporal Information in Vehicular Ad-Hoc Networks. In: VANET (2005)
12. Mobile and Portable Radio Group.: Mobile Ad Hoc Networks and Automotive Applications. Virginia Polytechnic Institute and State University
13. Samara, G., Al-Salihy, W.A.H., Sures, R.: Security Analysis of Vehicular Ad Hoc Networks (VANET). National Advanced IPv6 Centre, University Sains Malaysia (2010)
14. Meghanathan, N., Sharma, S.K., Skelton, G.W.: Energy Efficient Data Disseminatio. In: Wireless Sensor Networks Using Mobile Sinks, Usa

A GIS Anchored System for Clustering Discrete Data Points – A Connected Graph Based Approach

Anirban Chakraborty[1] and J.K. Mandal[2]

[1] Department of Computer Science
Barrackpore Rastraguru Surendranath College
Barrackpore, Kolkata-700 120, W.B., India
theanirban@rediffmail.com
[2] Department of Computer Science & Engineering University of Kalyani
Kalyani, Nadia, W.B., India
jkm.cse@gmail.com

Abstract. Clustering is considered as one of the most important unsupervised learning problem which groups a set of data objects, in such way, so that the data objects belongs to the same group (known as cluster) are very similar to each other, compared to the data objects in another group (i.e. clusters). There is a wide variety of real world application area of clustering. In data mining, it identifies groups of related records, serving as the basis for exploring more detailed relationships. In text mining it is heavily used for categorization of texts. In marketing management, it helps to group customers of similar behaviors. The technique of clustering is also heavily being used in GIS. In case of city-planning, it helps to identify the group of vacant lands or houses or other resources, based on their type, value, location etc. To identify dangerous zones based on earth-quake epi-centers, clustering helps a lot. In this paper, a set of data objects are clustered using two connected graph based techniques – MST based clustering and Tree Based clustering. After considering a lot of test cases, at the end of the paper, the second technique is found to be more suitable for clustering than the first one.

Keywords: Clustering, unsupervised learning, data mining, text mining, GIS, connected graph, MST based clustering, Tree Based clustering.

1 Introduction

Clustering is the way of partitioning a set of data points into group of highly similar objects [2]. At the same time, the data objects member of one cluster, are highly dissimilar with the members of other clusters. Clustering is one of the central techniques in spatial data mining and spatial analysis and also heavily used in GIS.

S.C. Satapathy et al. (eds.), *Emerging ICT for Bridging the Future – Volume 1*,
Advances in Intelligent Systems and Computing 337, DOI: 10.1007/978-3-319-13728-5_5

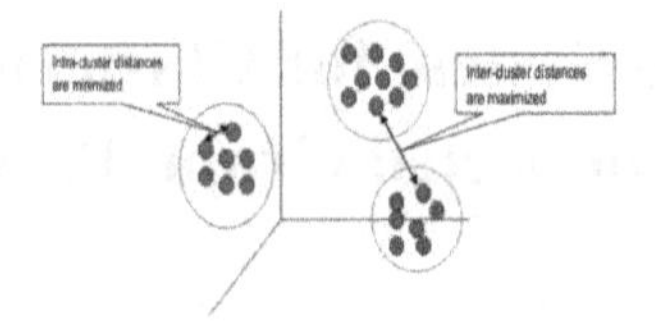

Fig. 1. Clustering concept

The notion of a cluster can be Ambiguous. For example :

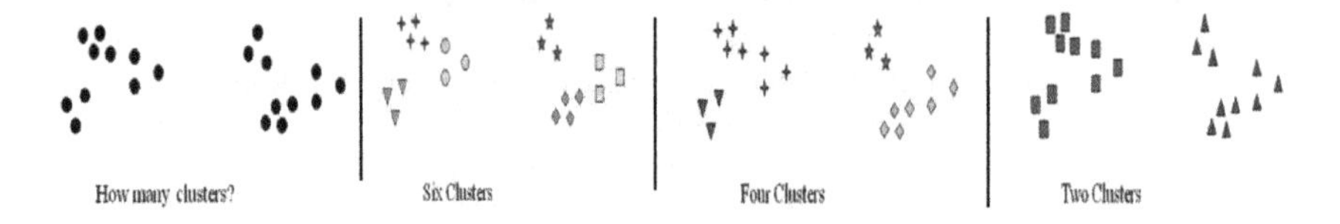

Fig. 2. Ambiguity in the concept of clustering

A good clustering method will produce high quality clusters with

- high intra-class similarity
- low inter-class similarity

For a large number of real world applications, ranging from social networking analysis to crime detection, the technique of clustering is applied and a number of clustering techniques are framed. This paper deals with Minimum Spanning Tree (MST) based method of clustering. But for some irregular shaped clusters this method does not work well and good. That's why another tree based method has been adopted to overcome the problems of MST based clustering.

Section 2 of this paper deals with the proposed technique. The implemented results are given in section 3. Analysis and comparisons are outlined in section 4 and finally conclusions are drawn in section 5.

2 The Technique

MST Based Clustering: Let us consider a complete graph G, in which each node corresponds to a point of the data set X. A Spanning Tree of G is a connected graph that contains all the vertices of the graph & has no loops. The sum of weights of its edges determines the weight of a Spanning Tree.

A Minimum Spanning Tree (MST) of G is a spanning tree with minimum weight (when all weights are different from each other, the MST is unique).

There are two well-known techniques for constructing MST- Prim's Algorithm and Kruskal's Algorithm. Instead of processing the graph by sorted order of edges, as in Kruskal's algorithm, Prim's algorithm works by iterating through the nodes and then finding the shortest edge from the set A to that of set A'(outside set A) where A & A' are two disjoint sets.

The MST based clustering starts with n-data points fed as input with their co-ordinates known. A nxn proximity matrix is created, where cell (i,j) holds the weight (Euclidian distance) between i^{th} and j^{th} point,i.e.

$$\text{Weight} = \sqrt{(x_i - x_j)^2 + (y_i - y_j)^2}$$

where, coordinate of ith point is – (x_i, y_i), coordinate of jth point is - (x_j, y_j).

Now taking the first data point as source and traversing the entire proximity matrix, MST is constructed using Prim's algorithm, as shown below (Fig. 3).

Fig. 3. Original data set Appearance after construction of MST

A Threshold value is now calculated to judge which of the edges of MST will be retained for clustering. Here, threshold value is the average of all the weights of the edges selected in the minimum spanning tree. i.e.

$$\text{Threshold} = 1 / (n - 1) \sum w_{ij}$$

where, n is the total number of vertices and w_{ij} is the weight of the edge connecting the i-th and j-th vertices.

To form the clusters, the weight of each constituent edge of the MST is inspected and all those edges having weight more than threshold are discarded, as shown below (Fig. 4).

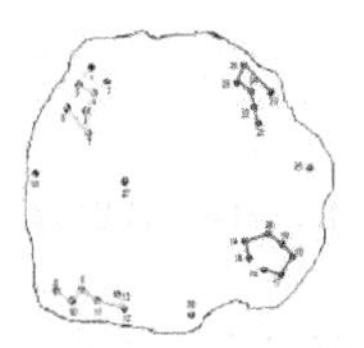

Fig. 4. Discarding illegal edges Appearance of clusters

Here, the edges being discarded are marked by blue cut-lines. Finally, a cluster will contain only those points which are connected by edges. However, if a cluster contains even less than 10% of the total data points (less than 3 points for the above case and pointed as green), then those are not treated as individual clusters, but are regarded as noise points. Thus based on MST based clustering, 4 clusters are obtained here.

As discussed in section 4, MST – based algorithm works well and good for the regular shaped clusters, but it behaves in quite a deteriorated manner in some cases, especially in case of more irregular shaped clusters. In those cases sometimes, more clusters are created than our expectation (expectation by bare eyes), sometimes there exists some points that are neither considered as noise points nor as cluster points. To overcome these demerits, a new tree based clustering technique could be proposed.

Tree Based Clustering: In graph theory, a tree is an undirected graph in which any two vertices are connected by exactly one simple path.

In the tree based clustering technique also, just like the previous one, data points along with their co-ordinates are fed into. Now a proximity matrix is created, where cell (i,j) is storing the value of Euclidian distance between point i and j.

Next, a threshold value is calculated, where the threshold value is the average of all the Euclidean distances in the proximity matrix.

$$\boxed{\text{Threshold} = 1 / (n * n) \sum d_{ij}}$$

here, n is the total number of vertices and d_{ij} is the distance between the i-th and j-th vertices.

The process of cluster formation starts from the first data point fed into, by constructing a tree, with keeping track about the fact that a data point can be the member of any one cluster only. This constructed tree forms the clusters with the basic idea that only those points will lay in the same cluster with point i, which have Euclidian distance less than that of threshold, from i. A cluster is declared as complete when no more points could be added to it and formation of next cluster begins. This method continues until all the data points are visited.

The result of such clustering, with 26 data points, is shown below (Fig. 6)

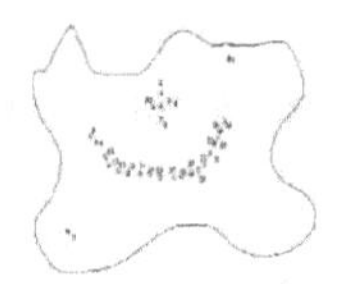

Fig. 5. Digitization of data points

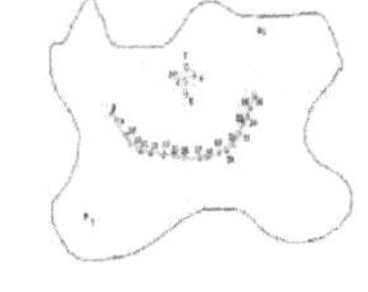

Fig. 6. Appearance of clusters

3 Implementation and Results

This section discusses the implementation results, outputs and its operations. Based on flat-file systems without using any database, the implementation has been done using Net Beans (Java) [4, 5]. As shown in figure 7, new profiles could be created or the user can work with the existing profile. While creating a new profile, the name of the profile has to enter and the map has to choose from any location (Fig. 7). Saving which causes the selected image file to be attached with that profile for future use (Fig. 7).

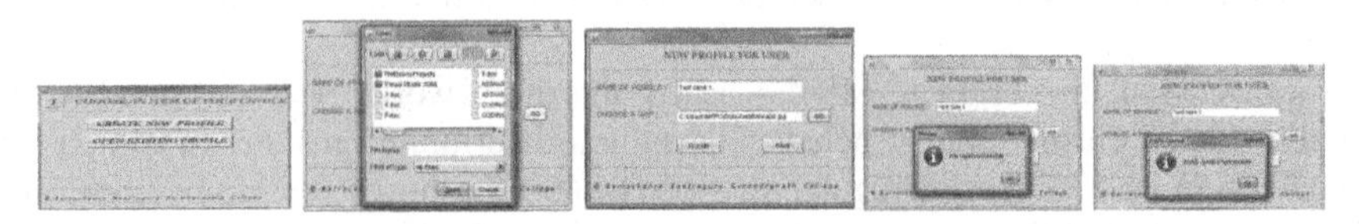

Fig. 7. Steps for new profile creation

While working with the existing profile, the profile name has to select and automatically the associated map will open. As it is the raw raster map, so it has to digitize. The color of the points turns green after digitization [1] (Fig 8). Upon clicking the "Cluster" button, the desired clusters are obtained (Fig 8).

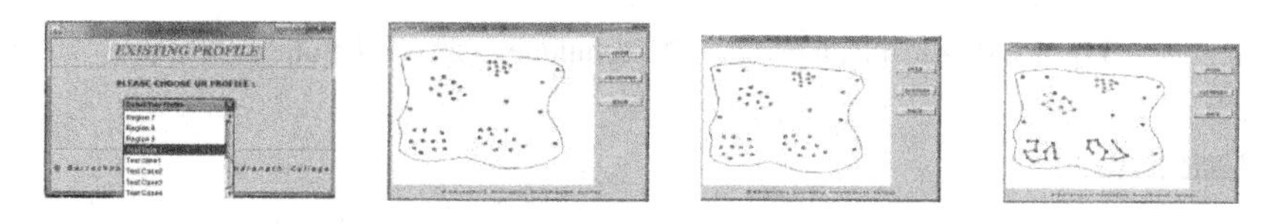

Fig. 8. Obtaining the clusters

4 Analysis and Comparison

The comparison among the two clustering techniques and hence the analysis can be best explained by the following 10 different cluster formations, by these techniques.

<u>**Case study 1:**</u>

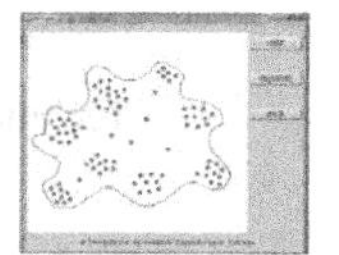
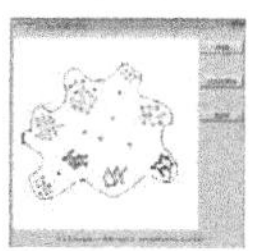

Fig. 9. Original data points 1st Technique 2nd Technique

<u>**Case study 2:**</u>

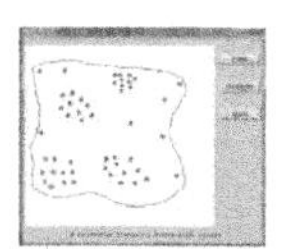
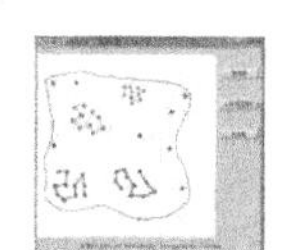
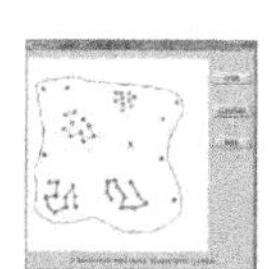

Fig. 10. Original data points 1st Technique 2nd Technique

<u>**Case study 3:**</u>

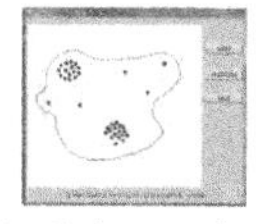
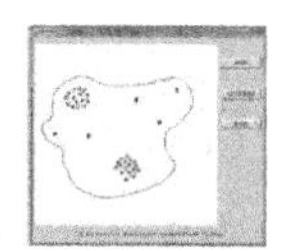
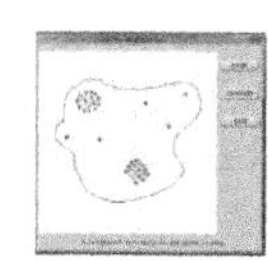

Fig. 11. Original data points 1st Technique 2nd Technique

<u>**Case study: 4**</u>

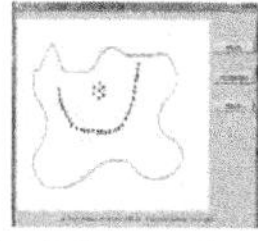
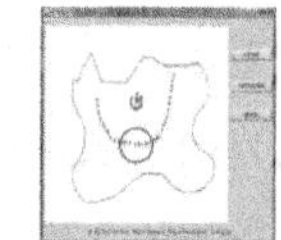
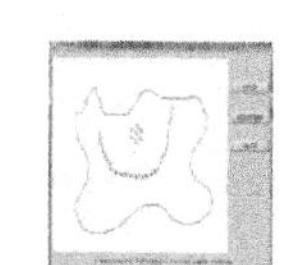

Fig. 12. Original data points 1st Technique 2nd Technique

Case study 5:

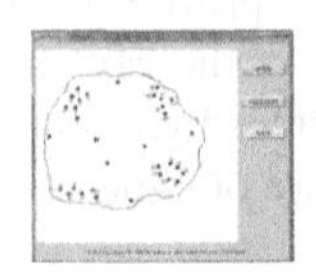
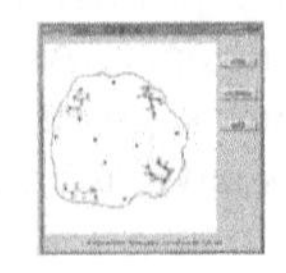
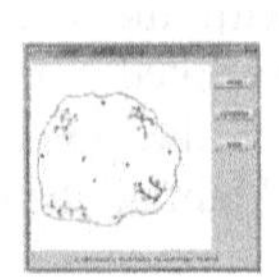

Fig. 13. Original data points 1st Technique 2nd Technique

Case study 6:

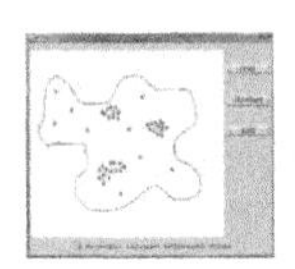

Fig. 14. Original data points 1st Technique 2nd Technique

Case study 7:

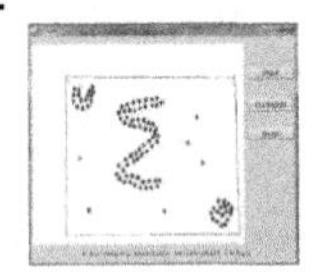
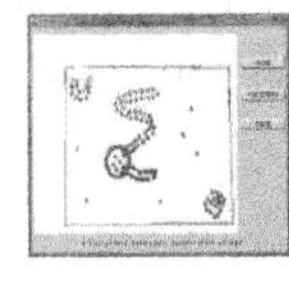
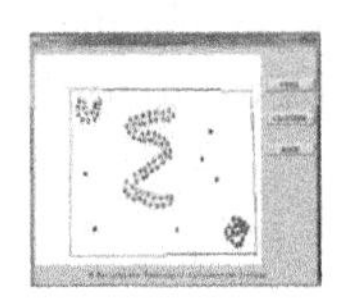

Fig. 15. Original data points 1st Technique 2nd Technique

Case study 8:

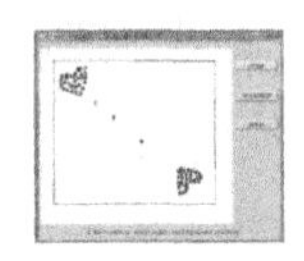
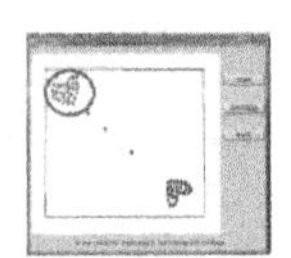
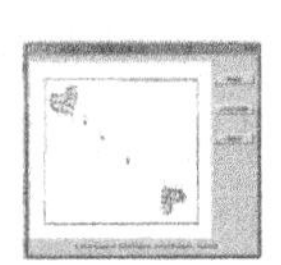

Fig. 16. Original data points 1st Technique 2nd Technique

Case study 9:

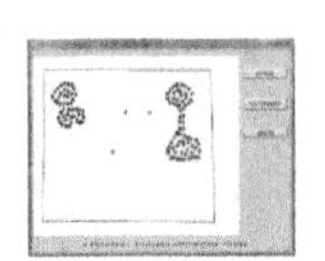

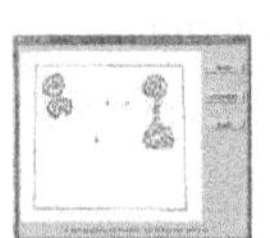

Fig. 17. Original data points 1st Technique 2nd Technique

Case study 10:

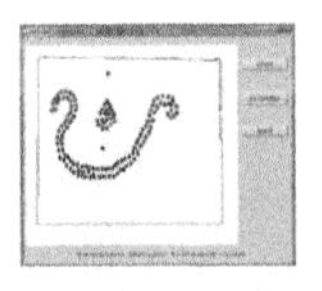
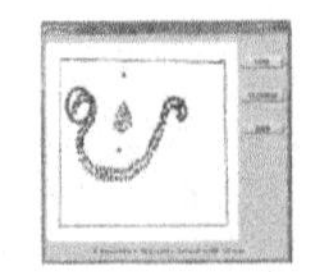

Fig. 18. Original data points 1st Technique 2nd Technique

Among these 10 case studies, for case studies 2, 3 and 5 (Fig. 10, 11 and 13); both the techniques work same.

In the remaining case studies the first technique works quite a deteriorated mannered compared to that of the second one. In these cases (except case 6) the first technique produces more clusters than expected by our bare eyes. These areas are encircled by different colors.

In case 6 (Fig. 14), the first technique includes such a point in one cluster, which is far apart from it. But the second technique solves this problem.

For case studies 7 (Fig. 15) and 10 (Fig. 18), the first technique has not included a number of points in any cluster, treating them as noise. But the problem has solved by the second technique.

Thus it could be inferred that while forming simple regular shaped clusters, although both the technique works same but when it is the question of forming more complex irregular shaped clusters, the second technique is far much better.

The following table (Table I) shows a comparison between MST based clustering and Tree based clustering based on time (executed in a system, with specifications Dual Core Intel processor and 2 GB RAM) taken for the formation of clustering shown in the above mentioned 10 case studies. From this table and from the graph drawn, it could be inferred that the clustering time is less for the 2nd technique.

Table 1. Comparison of the techniques based on clustering time

Case Study Number	Time Taken for MST Based Clustering (in ms)	Time Taken for Tree Based Clustering (in ms)
1	765	740
2	811	593
3	671	610
4	812	609
5	764	674
6	702	648
7	686	652
8	671	650
9	889	858
10	965	717

Based on the above table the following graphical representation is obtained:

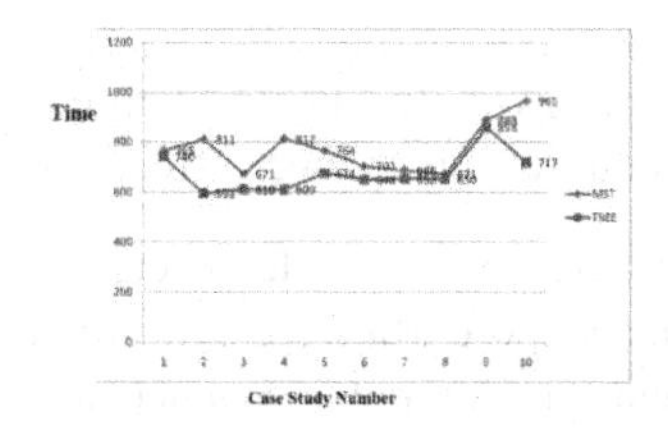

Fig. 19. Comparison graph for clustering time

The above graph shows that, in all the cases, to make clusters, Tree based method needs less time than MST based method, as because the brown line is always below the blue line.

5 Conclusions

There are a number of real world application areas where the above mentioned techniques of clustering are very suitable. Among many some areas are

i) Zachary's karate club network- The first example of application area is taken from one of the classic studies in social network analysis. Over the course of two years in the early 1970s, Wayne Zachary observed social interactions between the members of a karate club at an American university [9]. He constructed networks of ties between members of the club based on their social interactions both within the club and away from it. By chance, a dispute arose during the course of his study between the club's administrator and its principal karate teacher over whether to raise club fees, and as a result the club eventually split in two, forming two smaller clubs, centered on the administrator and the teacher. Here clustering analysis can be implemented to arrange the ties among the members of the clubs based on some clustering parameter.

ii) Plant and animal ecology- The above techniques of clustering could be used to describe and to make spatial and temporal comparisons of communities of organisms in heterogeneous environments. It is also used in plant to generate artificial clusters of organisms (individuals) at the species, genus or higher level that share a number of attributes.

iii) Business and marketing- Cluster analysis is widely used in market research when working with multivariate data from surveys and test panels. Market researchers use cluster analysis to partition the general population of consumers into market segments and to better understand the relationships between different groups of consumers/potential customers, and for use in market segmentation, Product positioning, New product development and Selecting test markets.

iv) Crime Analysis- Cluster analysis can be used to identify areas where there are greater incidences of particular types of crime. By identifying these distinct areas or "hot spots" where a similar crime has happened over a period of time, it is possible to manage law enforcement resources more effectively.

Acknowledgment. The authors express a deep sense of gratitude to the Department of Computer Science, Barrackpore Rastraguru Surendranath College, Kolkata-700 120, India and Department of Computer Science and Engineering, University of Kalyani for providing necessary infrastructural support for the work and to the UGC for financial support under Minor Research Project scheme (sanction number F. PSW-180/11-12 ERO, dated 25.01.2012).

References

1. Chakraborty, A., Mandal, J.K., Chakraborti, A.K.: A File base GIS Anchored Information Retrieval Scheme (FBGISIRS) through Vectorization of Raster Map. International Journal of Advanced Research in Computer Science 2(4), 132–138 (2011) ISSN: 0976-5697
2. http://www.met.edu/Institutes/ICS/NCNHIT/papers/39.pdf (accessed on August 05, 14)
3. Roy, P., Mandal, J.K.: A Delaunay Triangulation Preprocessing Based Fuzzy-Encroachment Graph Clustering for Large Scale GIS Data. In: 2012 International Symposium on Electronic Design Electronic System Design (ISED), Kolkata, India, December 19-22, pp. 300–305 (2012)
4. http://www.zetcode.com/tutorials/javaswingtutorial (accessed on May 17, 2013)
5. http://www.tutorialspoint.com/java/index.htm (accessed on May 14, 2013)

Comparative Analysis of Tree Parity Machine and Double Hidden Layer Perceptron Based Session Key Exchange in Wireless Communication

Arindam Sarkar and J. K. Mandal

Department of Computer Science & Engineering,
University of Kalyani, Kalyani-741235,
Nadia, West Bengal, India
{arindam.vb,jkm.cse}@gmail.com

Abstract. In this paper, a detail analysis of Tree Parity Machine (TPM) and Double Hidden Layer Perceptron (DHLP) based session key exchange tecchnique has been presented in terms of synchronization time, space complexity, variability of learning rules, gantt chart, total number of threads and security. TPM uses single hidden layer in their architecture and participated in mutual learning for producing the tuned weights as a session key. DHLP uses two hidden layers instead of single. Addition of this extra layer enhances the security of the key exchange protocol. Comparisons of results of both techniques has been presented along with detail analysis.

Keywords: Tree Parity Machine (TPM), Double Hidden Layer Perceptron (DHLP), session key, wireless communication.

1 Introduction

These days a range of techniques are available to exchange session key. Each technique has its own advantages and disadvantages. In key exchange the main security intimidation is Man-In-The-Middle (MITM) attack at the time of exchange the secret session key over public channel. Diffie-Hellman key exchange algorithm suffers from this MITM attack. Most of the key generation algorithms in Public-Key cryptography suffer from MITM attack [1]. Where intruders can reside middle of sender and receiver and tries to capture all the information transmitting from both parties. Another noticeable problem is that most of the key generation algorithms need large amount of memory space for storing the key but now-a-days most of the handheld wireless devices have a criterion of memory constraints. In proposed DHLPSKG, problem MITM attack of Diffie-Hellman Key exchange [1] has been set on. In TPM and DHLP based session key generation procedure, both sender and receiver use identical architecture. Both of these DHLP [2] and TPM's [3,4,5,6,7] start with random weights and identical input vector.

The organization of this paper is as follows. Section 2 of the paper deals with the TPM synchronization algorithm. DHLP based protocol for generation of session key

S.C. Satapathy et al. (eds.), *Emerging ICT for Bridging the Future – Volume 1*,
Advances in Intelligent Systems and Computing 337, DOI: 10.1007/978-3-319-13728-5_6

has been discussed in section 3. Section 4 deals experimental results and discussions. Section 5 provides conclusions and future scope and that of references at end.

2 Tree Parity Machine (TPM)

In recent times it has been discovered that Artificial Neural Networks can synchronize. These mathematical models have been first developed to study and simulate the activities of biological neurons. But it was soon discovered that complex problems in computer science can be solved by using Artificial Neural Networks. This is especially true if there is little information about the problem available. Neural synchronization can be used to construct a cryptographic key-exchange protocol. Here the partners benefit from mutual interaction, so that a passive attacker is usually unable to learn the generated key in time. If the synaptic depth (L) is increased, the complexity of a successful attack grows exponentially, but there is only a polynomial increase of the effort needed to generate a key. TPM based synchronization steps are as follows [3, 4, 5, 6, 7].

TPM Synchronization Algorithm

Step 1. *Initialization of random weight values of TPM.*

$$\text{Where, } w_{ij} \in \{-L, -L+1, \ldots, +L\} \tag{1}$$

Step 2. *Repeat step 2 to 5 until the full synchronization is achieved, using Hebbian-learning rules.*

$$w_{i,j}^{+} = g\left(w_{i,j} + x_{i,j}\tau\Theta(\sigma_i\tau)\Theta(\tau^A\tau^B)\right) \tag{2}$$

Step 3. *Generate random input vector X. Inputs are generated by a third party or one of the communicating parties.*

Step 4. *Compute the values of the hidden neurons using (eq. 3)*

$$h_i = \frac{1}{\sqrt{N}} w_i x_i = \frac{1}{\sqrt{N}} \sum_{j=1}^{N} w_{i,j} x_{i,j} \tag{3}$$

Step 5. *Compute the value of the output neuron using (eq. 4)*

$$\tau = \prod_{i=1}^{K} \sigma_i \tag{4}$$

Compare the output values of both TPMs by exchanging the system outputs.

if Output (A) ≠ Output (B), Go to step 3

else if Output (A) = Output (B) then one of the suitable learning rule is applied only the hidden units are trained which have an output bit identical to the common output.

Update the weights only if the final output values of the TPM's are equivalent. When synchronization is finally occurred, the synaptic weights are same for both the system.

3 Double Hidden Layer Perceptron (DHLP)

In DHLP [2] based key exchange method each session after accepting an identical input vector along with random weights both DHLP's, compute their outputs, and communicate to each other. If both outputs are same then both DHLP's starts synchronization steps by updating their weights according to an appropriate learning rule. After the end of the full synchronization procedure weight vectors of both DHLP's become identical. These indistinguishable weight vector forms the session key for a particular session. So, as a substitute of transferring the session key through public channel DHLP based synchronization process is carried out and outcomes of this used as a secret session key for that entire session. That actually helps to get rid of famous Man-In-The-Middle attack.

- DHLP offers two hidden layers instead of one single hidden layer in TPM [3, 4, 5, 6, 7]
- Instead of increasing number of hidden neurons in a single hidden layer DHLP introduces an additional layer (second hidden layer) which actually increased the architectural complexity of the network that in turn helps to make the attacker's life difficult to guessing the internal representation of DHLP.
- Secondly DHLP uses Hopfield Network for generation of random input vector.
- DHLP offers weight vectors of discrete values for faster synchronization.
- DHLP uses frames for connection establishment and synchronization procedure.
- DHLP introduces authentication steps along with synchronization.

DHLP Synchronization Algorithm

Input: - Random weights and Hopfield Network based PRNG generated input vector
Output: - Secret session key through synchronization.
Method:-

Step 1. Initialization of synaptic links between input layer and first hidden layer and between first hidden layer and second hidden layer using random weights values. Where, $w_{ij} \in \{-L, -L+1, \ldots, +L\}$

Repeat step 2 to step 12 until the full synchronization is achieved,

Step 2. The input vector (x) is created by the sender using 128 bit seed of Hopfield Network based PRNG.

Step 3. Computes the values of hidden neurons by the weighted sum over the current input values.

Step 4. Compute the value of the final output neuron by computing multiplication of all values produced by K2 no. hidden neurons at the second hidden layer:

$$\tau = \prod_{p=1}^{K2} \sigma_p^2$$

Step 5. Sender utilizes its 128 first weights as key for encryption of T variable (formerly stored in its memory) $Encrypt_{\text{sender\_weight}}(T)$.

Step 6. Sender constructs a SYN frame and transmitted to the receiver for handshaking purpose in connection establishment phase. SYN usually comprises of command code, ID, Secret Seed, Sender output (τ^{Sender}), $Encrypt_{\text{sender\_weight}}(T)$ and CRC (Cyclic Redundancy Checker).

Step 7. Receiver performs Integrity test after receiving the SYN frame and then Receiver utilize its 128 first weights as key for decryption of $Encrypt_{\text{weight}}(T)$ that was received from the sender.

Step 8. If $(Decrypt_{\text{receiver\_weight}}\,(Encrypt_{\text{sender\_weight}}(T)) = T$ then networks are synchronized. Go to step 12.

Step 9. If $(Decrypt_{\text{receiver\_weight}}\,(Encrypt_{\text{sender\_weight}}(T)) \neq T$ then receiver use the secret seed (SS) received from sender to produce the receiver inputs (x) identical to sender input (x) and calculates the output τ^{Receiver} using step 3 and step 4.

Step 10. If $(\tau^{\text{Receiver}} = \tau^{\text{Sender}})$ then performs the following steps

Step 10.1 Receiver should update their weights where $\sigma_k^{Sender/Receiver} = \tau^{Sender/Receiver}$ using any of the following learning rules:

Anti-Hebbian:

$$W_k^{A/B} = W_k^{A/B} - \tau^{A/B} x_k \Theta(\sigma_k \tau^{A/B})(\tau^A \tau^B)$$

Hebbian:

$$W_k^{A/B} = W_k^{A/B} + \tau^{A/B} x_k \Theta(\sigma_k \tau^{A/B})(\tau^A \tau^B)$$

Random walk:

$$W_k^{A/B} = W_k^{A/B} + x_k \Theta(\sigma_k \tau^{A/B})(\tau^A \tau^B)$$

Step 10.2 At the end of receivers weights update, the receiver sends ACK_SYN to instruct the sender for updating the weights using step 10.1.

Step 10.3 Sender then transmits encrypted message to receiver.

Step 10.4 Receivers then checks

If $(Decrypt_{\text{receiver\_updated\_weight}}(Encrypt_{\text{sender\_updated\_weight}}(T)) = T$ then networks are synchronized. Go to step 12.

Step 10.5 if $(Decrypt_{\text{receiver\_updated\_weight}}(Encrypt_{\text{sender\_updated\_weight}}(T) \neq T$ then networks are still not synchronized. Go to step 10.1.

Step 11. If $(\tau^{\text{Receiver}} \neq \tau^{\text{Sender}})$ then the receiver sends the message NAK_SYN to notify the sender. Go to step2.

Step 12. Finally, the receiver sends the frame FIN_SYN to inform the sender regarding the index vector of the weight to form the final session key.

4 Result and Analysis

In this section result and analysis of the TPM and DHLP based key exchange approach has been presented.

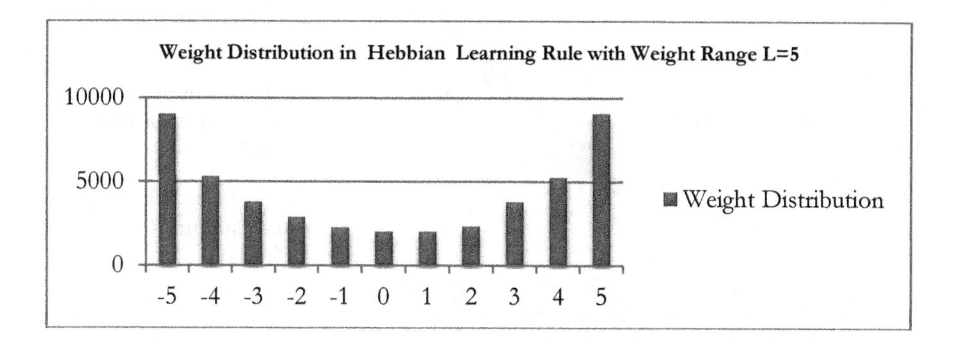

Fig. 1. TPM based weight distribution

Figure 1 shows the weight distribution of TPM based method where are not uniformly distributed.

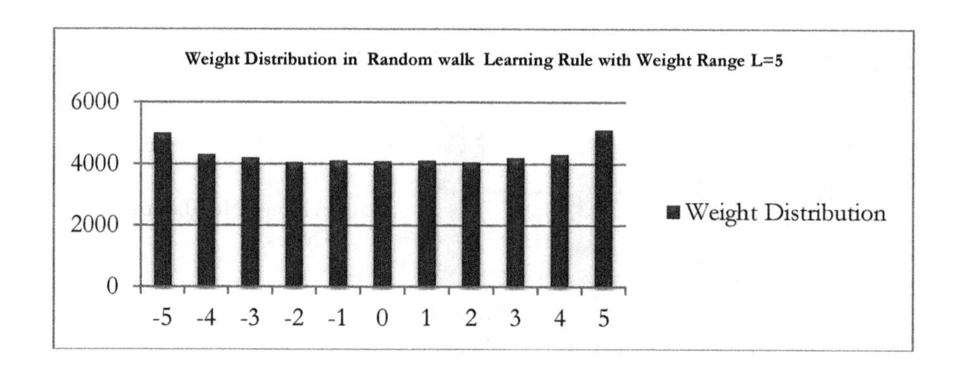

Fig. 2. DHLP based weight distribution

Figure 2 shows the weight distribution of DHLP based method where weight are uniformly distributed.

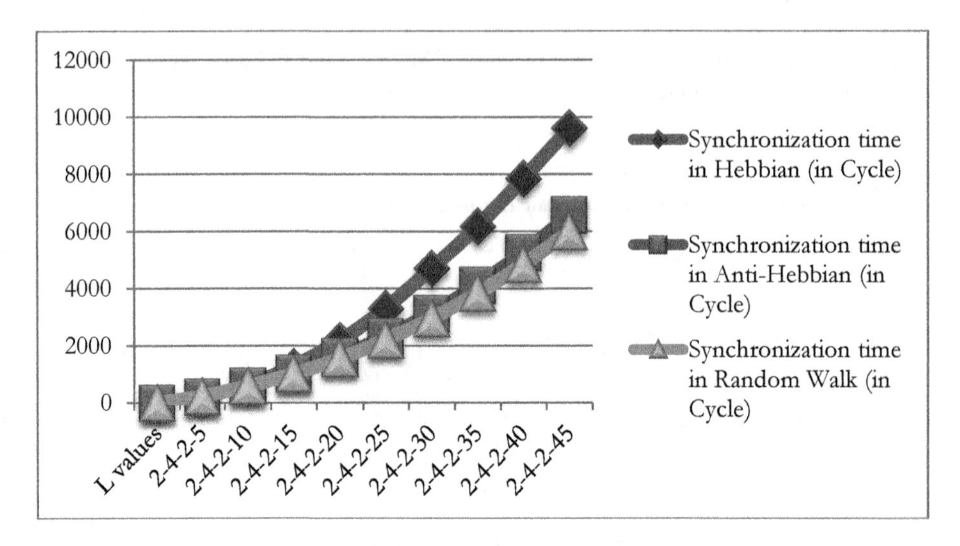

Fig. 3. DHLP based synchronization vs. learning rules

Figure 3 shows the Random-Walks takes minimum synchronization time than other two.

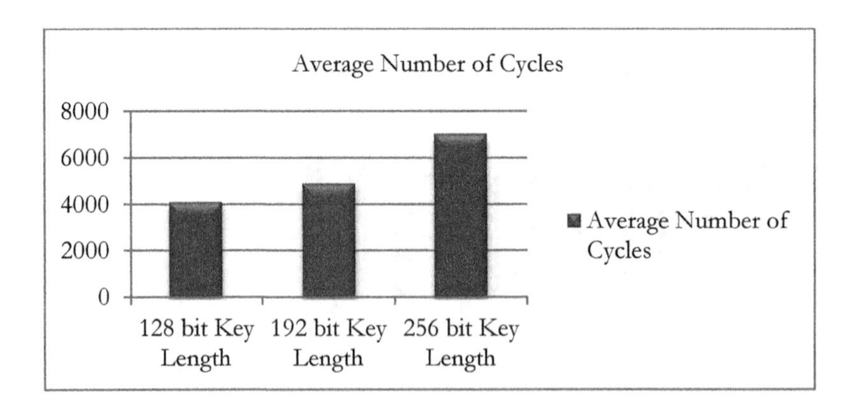

Fig. 4. Key length Vs. Average number of iterations

Figure 4 shows the increase of average number of cycles required for DHLP synchronization for different key length.

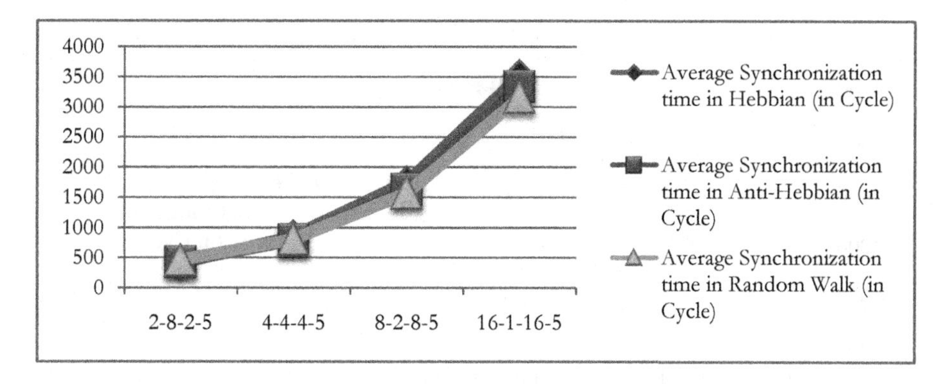

Fig. 5. Generation of 256 bit Session Key using fixed weight Range ($L = 5$) and different number of Neurons in Input and Hidden Layer

Figure 5 shows the generation of 256 bits key using fixed weight range and different number of neurons.

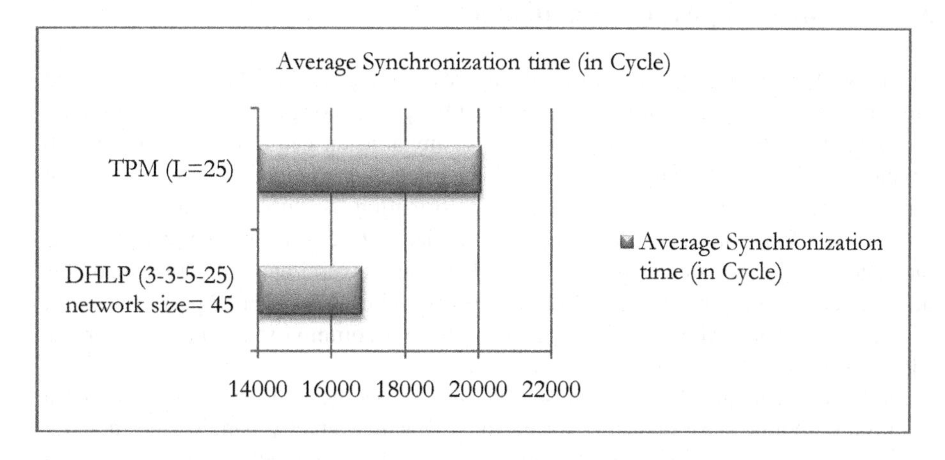

Fig. 6. Comparisons of 192 bit key length vs. average synchronization time (in cycle) for group synchronization (Group size= 4)

Figure 6 shows the Comparisons of 192 bit key length vs. average synchronization time (in cycle) for group synchronization (Group size= 4)

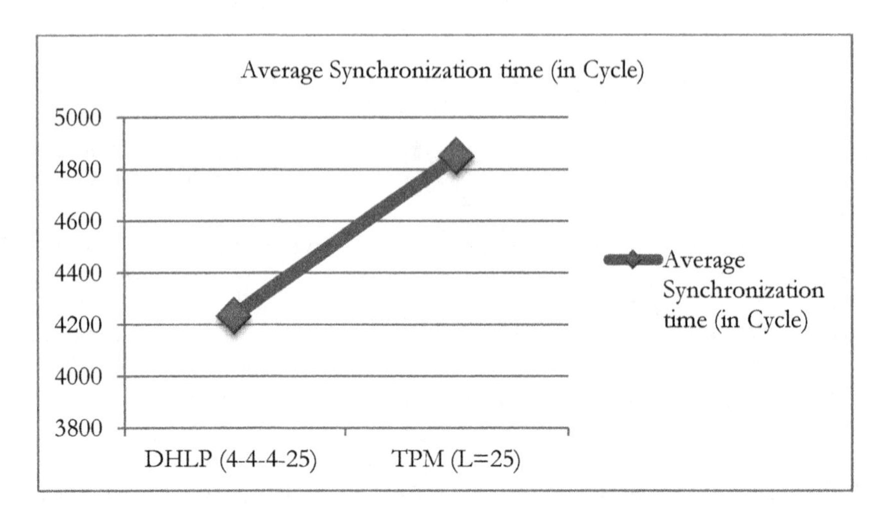

Fig. 7. Comparisons of 256 bit key length vs. average synchronization time (in cycle)

Figure 7 shows the Comparisons of 256 bit key length vs. average synchronization time (in cycle) .

5 Future Scope and Conclusion

Till now all session key generation techniques already devised [1] in concentrated only in session key generation mechanism but the process of generating keys does not guarantee the information security. Therefore, any attacker can also synchronize with an authorized device, because the protocol is a public knowledge. Thus, to ensure that only entities authorized have access to information is necessary authentication service. The function of the authentication service is to ensure the recipient that the message is from the source that it claims. There are several authentication methods, differentiated mainly by the use of secret-keys or public-keys. DHLP performs secret keys authentication where both entities must have a common secret code. Where as TPM has no such facility in terms of authentication.

Both the technique does not cause any storage overhead. Both of them needs a minimum amount of storage for storing the key which greatly handles the memory constraints criteria of wireless communication. The implementation on practical scenario is well proven with positive outcomes. In future group key exchange mechanism DHLP based protocol can be applied

Acknowledgment. The author expressed deep sense of gratitude to the Department of Science & Technology (DST) , Govt. of India, for financial assistance through INSPIRE Fellowship leading for a PhD work under which this work has been carried out, at the department of Computer Science & Engineering, University of Kalyani.

References

1. Kahate, A.: Cryptography and Network Security, Eighth reprint. Tata McGraw-Hill Publishing Company Limited (2006)
2. Arindam, S., Mandal, J.K.: Group Session Key exchange Multilayer Perceptron based Simulated Annealing guided Automata and Comparison based Metamorphosed Encryption in Wireless Communication (GSMLPSA). International Journal of Wireless & Mobile Networks (IJWMN) 5(4), 203–222 (2013), doi:10.5121/ijwmn.2013.5415, ISSN 0975 - 3834 (Online), 0975 - 4679 (Print); Mislovaty, R., Perchenok, Y., Kanter, I., Kinzel, W.: Secure key-exchange protocol with an absence of injective functions. Phys. Rev. E 66, 066102 (2002)
3. Ruttor, A., Kinzel, W., Naeh, R., Kanter, I.: Genetic attack on neural cryptography. Phys. Rev. E 73(3), 036121 (2006)
4. Engel, A., Van den Broeck, C.: Statistical Mechanics of Learning. Cambridge University Press, Cambridge (2001)
5. Godhavari, T., Alainelu, N.R., Soundararajan, R.: Cryptography Using Neural Network. In: IEEE Indicon 2005 Conference, India, December 11-13 (2005)
6. Kinzel, W., Kanter, L.: Interacting neural networks and cryptography. In: Kramer, B. (ed.) Advances in Solid State Physics, vol. 42, p. 383 arXiv- cond-mat/0203011. Springer, Berlin (2002)
7. Kinzel, W., Kanter, L.: Neural cryptography. In: Proceedings of the 9th International Conference on Neural Information Processing (ICONIP 2002) (2002)

Comparative Analysis of Compression Techniques and Feature Extraction for Implementing Medical Image Privacy Using Searchable Encryption

J. Hyma[1], P.V.G.D. Prasad Reddy[2], and A. Damodaram[3]

[1] Department of CSE, Gitam University,Visakhapatnam, India
Jhyma.gitam@gmail.com
[2] Department of CS&SE, Andhra University, Visakhapatnam, India
Prasadreddy.vizag@gmail.com
[3] Department of CSE, JNTUH, Hyderabad
Indiadamodarama@rediffmail.com

Abstract. The secure preservation of biomedical image data is a primary concern in today's technology enabled medical world. The new advancements in the technology are insisting us to outsource our digital data to a third party server and bring as and when needed. In this regard, efficient storage and transmission of this large medical data set becomes an important concern. In this paper we studied different compression techniques as a significant step of data preparation for implementing searchable encryption of medical data privacy preservation. We also shown texture based feature extraction for enabling privacy preserving query search. The simulation results obtained using different modalities of CT and MRI images with the performance comparison of wavelet and contourlet transform in peak signal to noise ratio for different compression ratios.

Keywords: Image privacy, lossless compression, feature extraction, searchable encryption.

1 Introduction

The most advanced and sophisticated medical equipment producing huge amounts of medical images in digital form. Today's health care diagnostic centers need privacy preserving transmission to outsource this medical image data through a public network. In addition to that it also needs to ensure privacy preserving search on this data when it is handover to some third party. Searchable encryption is one of the available techniques to ensure privacy preservation under third party maintenance. In this regard as an initial step of data preparation and when the network storage space and bandwidth are limited, the image data has to be compressed. As we are functioning on the medical image data, the loss of information during compression should be completely ignored. Many compression algorithms have been proposed for medical images earlier based on JPEG and DCT. However, the successful higher compression ratio has been obtained using wavelet transformation.

S.C. Satapathy et al. (eds.), *Emerging ICT for Bridging the Future – Volume 1,*
Advances in Intelligent Systems and Computing 337, DOI: 10.1007/978-3-319-13728-5_7

The need for effective medical image compression quickly becomes apparent. Particularly, medical image dataset size is exploding because of the evolution of image acquisition technology together with changes in usage of medical image dataset. From the comparison point of view, the aim lies in finding coding solutions dedicated to the outsource and transmission of images and associated information that will be complaint with the memory and computation capacities of the final workstations.

In recent years, a significant effort have been made to design medical image compression algorithm in which the main objective is to achieve better quality of decompressed images even at very low bit rates [2][3][4]. The work proposed in [5] highlights a wavelet based set partitioning in hierarchical trees (SPIHT) coder for progressive transmission of DICOM images. Enhancement of image features and its preservation after reconstruction is experimentally proved in [6]. In order to reduce transmission time and storage cost, an efficient data-compression scheme to reduce digital data without significant degradation of medical image quality is needed and it is studied with two region-based compression methods in [7]. A new image compression algorithm is proposed, based on independent embedded block coding with optimized truncation of the embedded bit-streams (EBCOT) and is studied in [8]. This algorithm exhibits state-of-the-art compression performance while producing a bit-stream with a rich set of features, including resolution and SNR scalability together with a “random access” property. Objective and subjective quality evaluation on various medical volumetric datasets is studied in [9] [10] and they show that the proposed algorithms provide competitive lossy and lossless compression results when compared with the state-of-the-art. Due to the great use of digital information, image compression becomes imperative in medical field. In this field the representation of the medical information needs to be efficient. The objective of image coding is to reduce the bit rate for efficient image transmission and outsourcing purpose while maintaining an acceptable image quality for different purposes [11][12][13][14][15].

The design of a new image compression method requires many dedicated services. Medical images generally go with personal metadata that have to remain confidential. In this research, we propose a privacy-preserving medical image data compression method in transformation domain and comparative study of two compression techniques of wavelet and contourlet transforms in peak signal to noise ratio for different compression ratios.

2 Proposed Method

In this section, the proposed method is described in detail. The basic service model in the proposed system includes following: At first data owner acquires raw medical image data from various imaging application contexts. Now this data has to be prepared well for secure outsourcing to Third Party Server (TSP). The data owner may choose this option of outsourcing to reduce the memory to local storage and maintenance overhead. Here the need of compression significantly shown to increase the speed of transmission and encryption for securely transmitting this compressed data. Before we encrypt the data we would like to extract the features of the images. So here we obtain compressed image data set and its corresponding feature set which are encrypted next for ensuring security. Finally data owner outsource the encrypted medical images including with their feature sets to the third party server for stack away and further processing. Figure 1 demonstrates the flow of proposed algorithm.

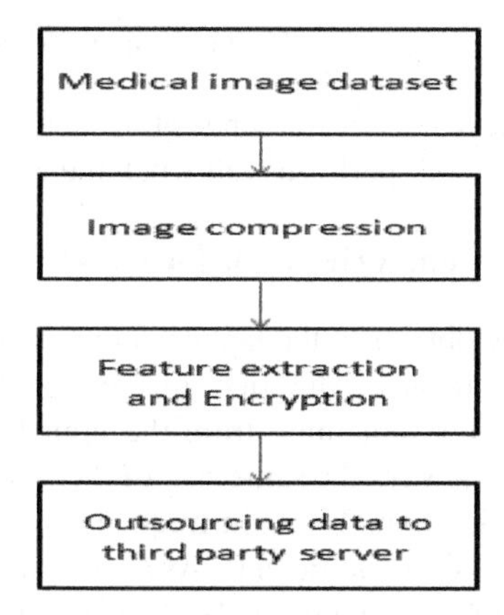

Fig. 1. Schematic diagram of proposed algorithm

2.1 Medical Image Compression

In this section, we review existing compression techniques that can serve as candidate solutions for lossless medical image compression. Two major types of compression techniques will be discussed here. One is wavelet based image compression. Other is based on Wavelet based contourlet transformation. The comparison of these compression techniques in terms of compression performance, peak signal to noise ratio and computational efficiency will be discussed in future sections.

Wavelet Based Image Compression

The two dimensional discrete wavelet transform is a separable transform that is optimal at isolating discontinues at vertical and horizontal edges. In two dimensional wavelet transforms a medical image passes through low and high pass filter banks followed by decimation process, along x and y axis separately. Finally the image has been divided into four parts namely approximations (LL), Detail horizontal (LH), Detail vertical (HL), and detail diagonal (HH). This procedure can be continued and called as image pyramidal decomposition.

Wavelet compression method is a form of data compression well suited for medical image compression. Wavelet compression can be either lossy or lossless. The integer wavelet transform, however provides compression in lossless format. As a result we can reconstruct the reference image perfectly from a compressed set of decomposed parts of IWT. The following figure shows an example of lossless compression using IWT.

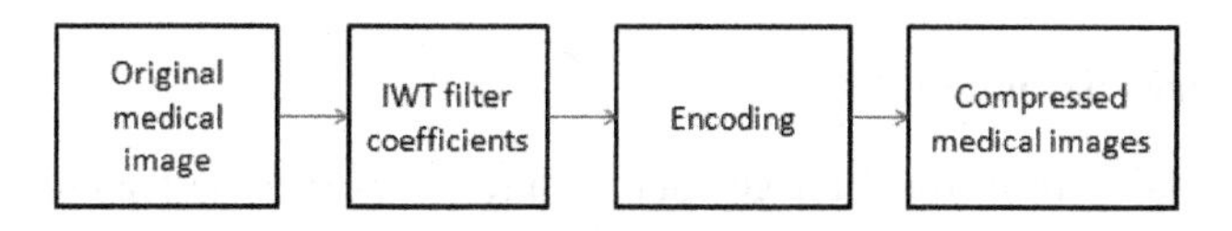

Fig. 2. DWT Image compression

$$X(a,b) = \frac{1}{\sqrt{a}} \int_{-\infty}^{\infty} \psi(t - b|a)\, x(t)dt$$

Although the dwt is effective tool in representing images containing smooth areas separated with edges, but it can't perform well when the edges are smooth.

a. The Wavelet Based Contourlet Transform (WBCT):

In this method, redundancy obtains at the laplacian pyramidal decomposition process. As a result we get two images. The first image resulting from low frequency coefficients, and the other one obtaining from the high frequency coefficients. The detail versions obtained has always the same size of the immediately anterior, because we don't have resolution direction. The high frequency coefficients decomposition is computed with the detail version coefficients. As directional filter bank is not suitable to handle the approximation information (LL), hence it is important to combine the directional filter bank with multi scale decomposition where the approximations of the image are removed before applying the directional filter bank. This is the main idea behind this transform called wavelet based contourlet transform which is called as non redundant transform. For WBCT it is important to know that we can obtain the perfect reconstruction of a medical image for the best cases. The steps compute to WBCT is a follows:

I. Compute the wavelet decomposition (haar) of an image.
II. Design the directional filter bank, for this research we select the haar filter bank.
III. Perform the directional decomposition using the medical images into detail versions. The process is made using the normal 5/3 tap filter which decomposes an image with a maximum three directions or eight sub bands at the finer wavelet sub bands.
IV. Repeat the step III with the next wavelet level of detail versions. In this research, we only do the directional decomposition to the high frequency levels.

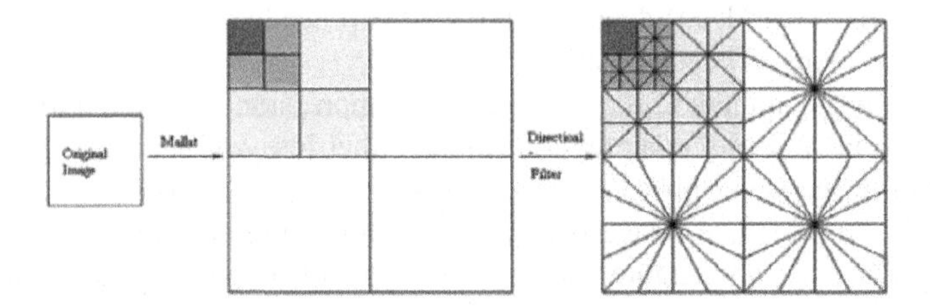

Fig. 3. Proposed WBCT method

2.2 Feature Extraction

After compression of medical image data we have to extract the feature set. Medical images feature extraction process derives local image description, i.e. a feature set values are obtained for each pixel. The image features can be extracted based on their color, shape, texture and others. Here we are proposing a texture based feature extraction and shown how much significant it is for further processing. In this study we have extracted entropy of image set and is given below.

$$S = -\sum_{x}\sum_{y} I(x,y)\log I(x,y)$$

In the next step we encrypt medical image dataset including with feature set samples.

2.3 Medical Data Encryption

Where C = cipher text, P = plaintext, K = secret key, K_0 = leftmost 64 bits of secret key, K_1 = rightmost 64 bits of secret key, $\oplus$ = bitwise exclusive or (XOR), $\boxplus$ = addition mod 264.

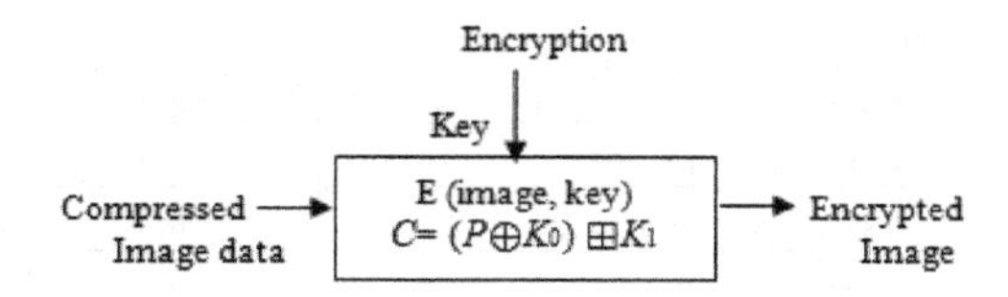

Fig. 4. Block diagram of encryption process

3 Simulation Results

In this research, we have implemented a novel approach for image compression supported by compression using wavelet based contourlet transform in MATLAB. We have considered wavelet based contourlet transform, namely directional filter bank with 5/3 tap filter for image compression. We have conducted experiments using proposed WBCT algorithm for three test images. The input image size is 256*256 and its corresponding compressed image and encrypted images are shown below.

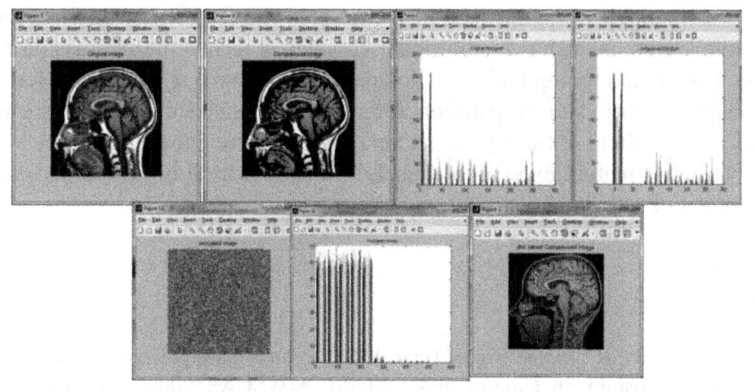

Fig. 5. (a)Original Image (b)Compressed image using WBCT algorithm (c)Original image histogram (d)Compressed image histogram (e)Encrypted image (f)Encrypted image histogram (g)Compressed image using dwt

We have calculated output parameters like PSNR and compression ratio which determines efficiency of proposed system. Compression ratio is defined as,

$$Compression\ Ratio = \frac{Uncompressed\ image\ filesize}{Compressed\ image\ filesize}$$

PSNR function is defined as,

$$PSNR = 10\ log\left(\frac{max^2}{MSE}\right) \quad \text{Where, } MSE = \frac{1}{MN}\sum_{1=0}^{m-1}\sum_{j=0}^{n-1}\left\|I(i,j) - I_q(i,j)\right\|^2$$

The performance comparison of wavelet and WBCT for three test images is given below table.

Table 1. Comparison of performance in compression ratio and PSNR

Image name	WBCT based compression			Wavelet based compression		
	Compression ratio	PSNR	Simulation time	Compression ratio	PSNR	Simulation time
IMAGE1(CT)	24.0189	12.1195	0.96	22.0213	16.2880	0.89
IMAGE2(CT)	10.8523	25.2126	0.96	10.6171	31.5717	0.89
IMAGE3(CT)	23.6238	20.1083	0.96	22.8108	25.9903	0.89
IMAGE4(MRI)	22.3417	20.2288	0.96	19.3003	22.1477	0.89
IMAGE5(MRI)	14.4903	26.9520	0.96	10.4427	33.0017	0.89
IMAGE6(MRI)	26.2351	18.6392	0.96	20.3179	26.4751	0.89
IMAGE3(X-ray)	28.9258	13.1624	0.96	24.3428	17.9873	0.89

4 Conclusion

Now a day's transferring of medical images over public networks have been extended rapidly. Providing security and preserving privacy is most crucial part in present days. In this research we studied two compression algorithms that suits for lossless compression of the medical images and also extracted the features of the image data set. Simulation results have shown significant changes in terms of compression ratio, transmission speed and computation efficiency of the two algorithms. Here we conclude the process of data preparation for secure transmission through a public network and to store them at third party server, and as future enhancement we would like to show how this feature extraction is useful for the retrieval of the similar images under the concept of searchable encryption.

References

[1] Olyaei, A., Genov, R.: Mixed-Signal Haar Wavelet compression Image Architecture. In: Midwest Symposium on Circuits and Systems (MWSCAS 2005), Cincinnati, Ohio (2005)

[2] Dhouib, D., Nait-Ali, A., Olivier, C., Naceur, M.S.: Comparison of wavelet based coders applied to 3D Brain Tumor MRI Images. IEEE (2009)

[3] Miaou, S.-G., Chen, S.-T.: Automatic Quality Control for Wavelet-Based Compression of Volumetric Medical Images Using Distortion-Constrained Adaptative Vector Quantization. IEEE Trans. on Medical Images Using Distortion-Constrained Adaptative Vector Quantization 23(11) (November 2004)
[4] Qi, X., Tyler, J.M.: A progressive transmission capable diagnostically lossless compression scheme for 3D medical image sets. Information Sciences 175, 217–243 (2005)
[5] Ramakrishnan, B., Sriraam, N.: Internet transmission of DICOM images with effective low bandwidth utilization. Digital Signal Processing 16, 825–831 (2006)
[6] Namuduri, K.R., Ramaswamy, V.N.: Feature preserving image compression. Patter Recognition Letters 24, 2767–2776 (2003)
[7] Penedo, M., Pearlman, W.A., Tahoces, P.G., Souto, M., Vidal, J.J.: Region-Based Wavelet Coding Methods for Digital Mammography. IEEE Trans. on Medical Imaging 22(10), 1288–1296 (2003)
[8] Taubman, D.: High performance scalable image compression with EBCOT. IEEE Trans. on Image Processing 9(7), 1158–1170 (2000)
[9] Schelkens, P., Munteanu, A., Barbarien, J., Mihneai, G., Giro- Nieto, X., Cornellis, J.: Wavelet coding of volumetric medical datasets. IEEE Trans. on Medical Imaging 22(3), 441–458 (2003)
[10] Kim, Y.S., Pearlman, W.A.: Stripe-based SPIHT compression of volumetric medical images for low memory usage and uniform reconstruction quality. In: Proc. Icassp, vol. 4, pp. 2031–2034 (June 2000)
[11] Xiong, Z., Wu, X., Cheng, S., Hua, J.: Lossy-to-Lossless Compression of Medical volumetric Data Using Three-dimensional Integer Wavelet Transforms. IEEE Trans. on Medical Imaging 22(3), 459–470 (2003)
[12] Munteanu, A.: Wavelet Image Coding and Multiscale Edge Detection Algorithms and Applications. Ph.D dissertation, Vrije Univ. Brussel, Brussels, Belgium (2003)
[13] Strintzis, M.G.: A review of compression methods for medical images in PACS. International Journal of Medical Informatics 52(1), 159–165 (1998)
[14] Kofidis, E., Kolokotronis, N., Vassilarakou, A., Theodoridis, S., Cavouras, D.: Wavelet-based medical image compression. Future Generation Computer Systems 15(2), 223–243 (1999)
[15] Barlaud, M.: Compression et codage des images et des videos, Hermes (2002)

Offline Detection of P300 in BCI Speller Systems

Mandeep Kaur[1], A.K. Soni[2], and M. Qasim Rafiq[3]

[1] School of Computer Science & Engg., Lingaya's University, India
mandeephanzra@gmail.com
[2] School of Computer Science & Engg., Sharda University, Greater Noida, U.P., India
ak.soni@sharda.ac.in
[3]Dept. Computer Sc. & Engg. AMU, Aligarh
mqrafiq@hotmail.com

Abstract. The paper presents a framework for offline analysis of P300 speller system using seeded k-means based ensemble SVM. Due to the use of small-datasets for the training of classifier, the performance deteriorates. The Proposed framework emphases on semi-supervised clustering approach for training the SVM classifier with large amount of data. The normalized mutual information (NMI) has used for cluster validation that gives maximum 88 clusters on 10 fold cross-validation dataset with NMI approx equals to 1. The proposed framework has applied to the EEG data acquired from two subjects and provided by the Wadsworth center for brain-computer interface (BCI) competition III. The experimental results show the increase in SNR value and obtain better accuracy results than linear, polynomial or rbf kernel SVM.

Keywords: P300 signal, Wavelet Transform, semi-supervised clustering, ensemble, support vector machines (SVMs).

1 Introduction

The P300 is a positive wave with a latency of about 300- 350 ms, was first reported by Sutton and colleagues in 1965 and P300 based Brain Computer Interface system was first introduced by Farwell and Donchin in 1988 [1] [2]. This paradigm also called P300 Speller System that latter improved by Donchin et.al. in 2000, to be used for disabled subjects [3]. In this paradigm, the subject presented a matrix of symbols where each row and column intensifies in random order with predefined interval. The P300 is an event related potential, elicited when the subject pays attention to only that row and column intensifications that contain the desired symbol. With P300 occurrence, the target row-column assumed and the desired symbol inferred [4]. For the offline analysis, BCI competition data sets have been widely used to measure the performance of BCI systems. This is the main reason for testing the novel seeded k-means based ensemble algorithm for predicting the occurrence of P300 in the ongoing EEG on the BCI competition data, so that the results can be directly compared to other available algorithms and their comparative differences explored. Within BCI, the P300 signal plays a significant role of detecting human's intention for controlling

S.C. Satapathy et al. (eds.), *Emerging ICT for Bridging the Future – Volume 1*,
Advances in Intelligent Systems and Computing 337, DOI: 10.1007/978-3-319-13728-5_8

external devices [5]. The major concern is to classify the P300 signal from the ongoing EEG (the background noise) in P300 based BCI systems. As EEG is the process of recording brain activities from cerebral but it also records some electrical activities arising from sites other than the brain. Therefore, during acquisition, the P300 (ERPs) is contaminated with a lot of noise, also known as artifacts in EEG. The unwanted electrical activities arising from some different source rather than from cerebral origin are termed as artifacts. The noise can be electrode noise called extra-physiologic artifacts or may be generated from subject body called physiologic artifacts. The various types of extra-physiologic artifacts are like Inherent Noise, Chemical Noise, Ambient Noise, Thermal or Johnson noise (voltage to 60 Hz), Shot noise, Flicker noise [6]. The various types of physiologic artifacts are like activities of the eye (electrooculogram, EOG), muscles (electromyogram, EMG), Cardiac (electrocardiogram, ECG), Sweating, swallowing, breathing, etc. Due to this, the amplitude of the P300 (ERP) signal becomes much lower than the amplitude of the noise; causes P300 signal presets a very low signal-to-noise ratio (SNR). Therefore, to detect P300 signal from ongoing EEG, the signal-to-noise ratio (SNR) must be enhanced. Coherent Averaging is the most common method to achieve this goal. The artifacts can remove using various techniques like filtering, independent component analysis, principal component analysis, wavelet transform etc. The approximate coefficients of Daubechies4 (db4) have considered as a feature vector for further analysis. As the P300 detection is the primary task of this research, a number of features are required for training the classifier. The P300 based BCI systems have reported the requirement of bigger search space as the number of features increases, the training of the classifier is very time consuming; number of channels increases the cost of the system. To overcome these limitations, this research emphasis on implementing semi-supervised cluster classification using seeded k-means and ensemble SVM classifier for P300 based BCI Speller paradigm [7]. The paper is organized as follows: Section 2 discusses the material and methods used in the work. Then detailed methodology is given in Section 3. The Result analysis has discussed in Section 4. And Section 5 gives the conclusion.

2 Material and Methods

A. BCI PARADIGM

The subject was presented a screen with 36 symbols arranged in a 6 by6 character matrix. The rows and columns of the matrix were intensified randomly at a rate of 5.7Hz. The user's task was to focus attention on letter of the word, prescribed by the investigator (i.e., one character at a time). Twelve intensifications (six row and six columns) were made out of which only two (one row and one column) intensifications contained the desired character. Each character epoch was followed by a 2.5 s period, and during this time the matrix was blank [8].

B. DATA ACQUISITION

The EEG dataset used is acquired from Wadsworth BCI competition 2004 Data set IIb (P300 speller paradigm). The data acquired using BCI 2000 system. The dataset is still available on the competition webpage [9], recorded from two subjects A and B in 5 sessions each. The data set was made up of 85 training characters and 100 testing characters, each composed of 15 trials. The data set was recorded from 64 electrodes that are first band-pass filtered from 0.1-60 Hz and then sampled at 240 Hz.

C. DATA PRE-PROCESSING

Only 10 channels are used for P300 detection in ongoing EEG. The data were latter low-pass filtered at 30 Hz and high-pass filtered at 0.1 Hz using an 8th order butterworth filter [10].

D. COHERENCE AVERAGING

The data set was further pre processed to detect the P300 signal via increasing the signal-to-noise ratio using coherent averaging. Coherent Averaging is the most common method to average the potential registered synchronized with the application of the stimulus. The following equation (2.1) represents coherence averaging:

$$y_K[n] = \frac{1}{K}\sum_{k=1}^{K} x_k[n] = s[n] + \frac{1}{K}\sum_{k=1}^{K} r_k[n] \quad (2.1)$$

Where xk [n] is the k-nth epoch of potential registered, s [n] is the signal P300 with length $1 \leq n \leq N$, rk [n] is the k-nth signal with noise variance σ2 and K is the total number of epochs. For each time instant n, consider noise as an estimator of the average value of a sample of K data. This averaging results in a new random variable, having same average value and a variance σ equals to σ2/K. This also results in signal-to-noise ratio (SNR) improvement, but also contaminated with white noise and other artifacts [10].

E. INDEPENDENT COMPONENT ANALYSIS (ICA)

Due to the contamination of artifacts, the EEG signal is described as x (t) = a (t) + s (t) where x (t) is the EEG signal, a (t) is an artifact and s (t) is the signal of interest. Using vector-matrix notation ICA is defined by equation (2.2) [11, 12], x = As. ICA algorithm applied to the filtered EEG data to remove the artifacts from EEG signal and separate independent sources for each trial.

$$xi\ (t) = ai1 * s1\ (t) + ai2 * s2\ (t) + ai3 * s3\ (t) + ai4 * s4(t) \quad (2.2)$$

ICA algorithm assume that the sources are non-Gaussian or non-white (i.e., the activity at each time point does not correlate with the activity at any other time point), because if the sources are white Gaussian, then there are an infinite number of unmixing matrices that will make the unmixed data independent and the problem is under-constrained [13, 14]. ICA Algorithm to remove artifacts from EEG

1. ICA starts with EEG signals recorded from 10 channels X (t) = {x1(t) . . . x10(t)}
2. As these signals are mixed with (n≤10) unknown independent components (sources) S (t) = {s1(t), s2(t) . . . , sN(t)} containing artifacts say A.
3. Centering the input observation signal x
4. Whitening the centered signal x
5. Initializing weight matrix w, Consider A as unknown mixing matrix defining weights at which each source is present in the EEG signals recorded at the scalp.

6. The number of independent components S (t) is determined by the number of EEG channels recorded and considered.
7. Components were selected using a-priori knowledge that the P300 evoked potential reaches a peak around 300 ms after the stimuli. Therefore, only those ICs with a peak in amplitude between 250 and 400 ms were retained.
8. Remove artifact components and re-compute data X = SW-1 where X=cleaned data, S=components, with artifact components set to zero [14] [15] [16].

ICA applied to de-noising the EEG signal in order to enhance the signal to noise ratio of the P300, and separating the evoked potential from some artifacts, like signals obtained from eye-blinking, breathing, or head motion, cardiac etc. Evoked Related Potential and electroencephalogram analysis are one of the practical applications of ICA [14-19].

F. PRINCIPLE COMPONENT ANALYISIS (PCA)

With ICA, it is possible to reduce the dimension of the data by discarding the eigenvalues of the covariance matrix, obtained during whitening, and using principal component analysis also. PCA assumes that the most of the information is contained in the directions along which the variations are the largest. PCA based on Eigen analysis technique.

PCA Algorithm to remove artifacts from EEG:

1. Acquire the EEG signal after ICA
2. Obtain mean of the input signal
3. Subtract the mean from each of the input data dimensions
4. Calculate the covariance matrix and thus represent EEG as covariance matrix of the vectors.
5. Determine Eigen values and Eigen vectors of the above data on covariance matrix. Consider Eigen values and Eigen vectors of a square symmetric matrix with sums of squares and cross products. The eigenvector related with the largest Eigenvalue has the same direction as the first principal component. The eigenvector associated with the second largest Eigenvalue determines the direction of the second principal component. Note: The sum of the Eigenvalues equals the trace of the square matrix and the maximum number of eigenvectors equals the number of rows (or columns) of this matrix.
6. For each Eigenvector low pass filtering is done using a filter of cutoff frequency 50 Hz and the energy of filtered data is calculated.
7. The energy, calculated corresponding to Eigenvalue, is compared with a threshold. If this energy is greater than that of threshold, then marked these for re-construction, else discard the vector. (Compute the number of eigenvalues that greater than zero (select any threshold)).
8. Reconstruct only on those Eigenvectors, which are marked for reconstruction.

PCA able to remove ocular artifacts, but not completely clean EEG as few artifacts like ECG, EMG have almost same amplitudes [20] [21].

G. WAVELET TRANSFORM: DAUBECHIES (db4)

It has been discussed that the EEG signals are non-stationary in nature and do not allow the accurate retrieval of the signal frequency information. Therefore, to extract frequency information, signal transformations like Fourier Transform (FT), Short-Time-Fourier Transform (STFT), Wavelet Transform (WT); Continuous Wavelet Transform (CWT) and Discrete Wavelet Transform (DWT) are required. It has discussed in [22] that Wavelet transform (WT) is the best signal analysis tool. The 4th order Daubechies is applied to each extracted PCA component to decompose the signal to the required band for the approximated coefficients. Daubechies (Db4) wavelet used as the mother wavelet and the approximated coefficients used as the features for the next step and concatenated into feature vectors [23].

H. SEMI-SUPERVISED CLUSTERING

Semi-supervised clustering, involves some labeled class data or pair wise constraints along with the unlabeled data to obtain better clustering. The advantage of semi-supervised clustering is that the data categories (clusters) can generate from initial labeled data as well as extend and modify the existing ones to reflect other regularities in the data. The distance-based Semi-supervised clustering is implemented using Seeded-K means (S-KMeans) and Constrained-K means (C-Kmeans). Both use labeled data to form initial clusters and constrain subsequent cluster assignment. S-KMeans uses the pre-defined labeling of the seed data may be changed during algorithm run and suitable with noisy seeds [24-26]. Apply Seeded KMeans semi-supervised clustering on the obtained features. Consider random number of features from feature set to label the cluster. Assume that *k=2* is known. On getting seed information, k-means will label the rest of the unlabeled features. [24] developed a generalization of k-means clustering for the problems where class labels are known. Let xi and xi′ be observations from a data set with p features, and x_{ij} represents the value of the jth feature for observation i. Suppose further that there exists subsets $S_1, S_2, \ldots, S_K$ of the xi's such that $x_i \varepsilon S_k$ implies that observation i is known to belong to cluster k. (Here K denotes the number of clusters, which is also assumed to be known in this case.) Let |Sk| denote the number of xi's in Sk. Also let $S = U^K{}_{k=1}Sk$. The algorithm follows:

1. For each feature j and cluster k, calculate the initial cluster means as follows:

$$\bar{x}_{kj} = \frac{1}{|S_k|}\sum_{x_i \in S_k} x_{ij}$$

2. For each feature j and cluster k, calculate $|x_{kj}|$, the mean of feature j in cluster k.
3. Repeat steps 2 and 3 until the algorithm converges [24-26].

I. ENSEMBLE SUPPORT VECTOR MACHINE (SVM)

Support vector machines (SVM) algorithm is originally designed by Vladimir Vapnik et.al. in 1979 for binary or two-class classification [25]. To perform binary classification, SVM finds a hyperplane that that divides the data sets into two classes with the largest margin. However, the data of real problems are also non-separable that requires mapping into high dimensional feature space, where the training is separable [26]. This leads to non-linear SVM that is dividing the non-separable data linearly in a higher dimensional space. This can achieve using a kernel function. The proposed framework uses the polynomial and Gaussian kernel functions [27]. However, it has reported that using these kernels, the number of support vectors increases [28]. A way to decrease these numbers of support vectors for the training of support vector machine (SVM), k-means clustering algorithm has used [29].

3 Methodology

Data recorded from 64 channels; instead of all the datasets from only 10 channels have selected for the proposed work. The Figure 1 depicts methodology and Figure 2 shows a complete framework for the proposed system. For the training of bagging based ensemble SVM classifier, the wavelet features have extracted from pre-processing data and then projected to Seeded K-Means. For testing, the pre-processed unknown data have fed to the classifier for classification.

1. Load the training dataset
2. Pre-processing: it includes low-pass and high-pass Butterworth filtering: 0.1 Hz – 30 Hz using an 8th order Butterworth filter, Coherence averaging, Independent Component Analysis, principal component analysis.
3. Apply wavelet filtering to extract the features to be trained using db4 wavelet.
4. Apply Seeded K-Means semi-supervised clustering on the obtained features.
5. For the training of Bagging based SVM Ensemble, the clusters are fed as input. The Linear, Polynomial, and RBF types of SVM kernels have used.
6. The classification is performed using unknown feature vectors.

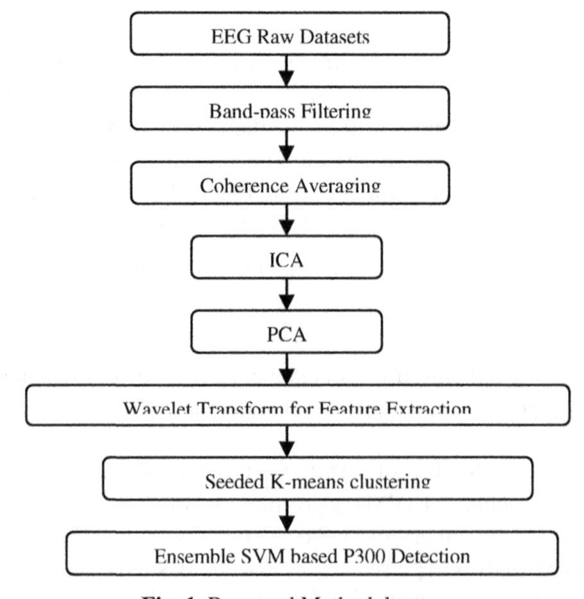

Fig. 1. Proposed Methodology

The system uses two-class non-linear Support vector machine with seeded k-means clustering, that requires three parameters: kernel parameter, γ, the penalty factor C and the number of clusters, Cn to detect whether a specified signal contains P300 or not. The framework uses bagging technique in which each individual SVM has trained using randomly chosen training samples via a bootstrap method. The output of individual trained SVMs has averaged via an appropriate aggregating strategy known as majority voting. The result shows that the ensemble SVM using bagging (majority voting) method procures better accuracy than the independent type of SVMs.

The proposed seeded k-means based ensemble SVM illustrated as:

```
Inputs:
D1: Training set
D2: Testing Set
y: Labels
P: SVM Parameters
k: kernel parameters, C: penalty factor, Cn: number of clusters
For each SVM
{  Initial Phase
    Select parameters k, C and Cn for each SVM
 Bootstrapping Phase and Training Phase
   For each SVM parameter, DO
        {  Run Seeded k-means algorithm and all cluster centers are
      Regarded as input to the classifier
                  {S = Train (D1, yD1, P);}
        }
          Save the best parameters Pbest
                      For I = 1 to 3 (linear, polynomial and rbs SVM)
                            {   S(i) = Train (D1, yT, Pbest);  }
Testing Phase
        For testing set D2
        {   ŷ(i)D2 = Test (S(i), D2, yD2);          }

Decision Phase
        {  Majority vote= ŷ(i)D2 → ŷD2          }
}
Return Accuracy of D2
```

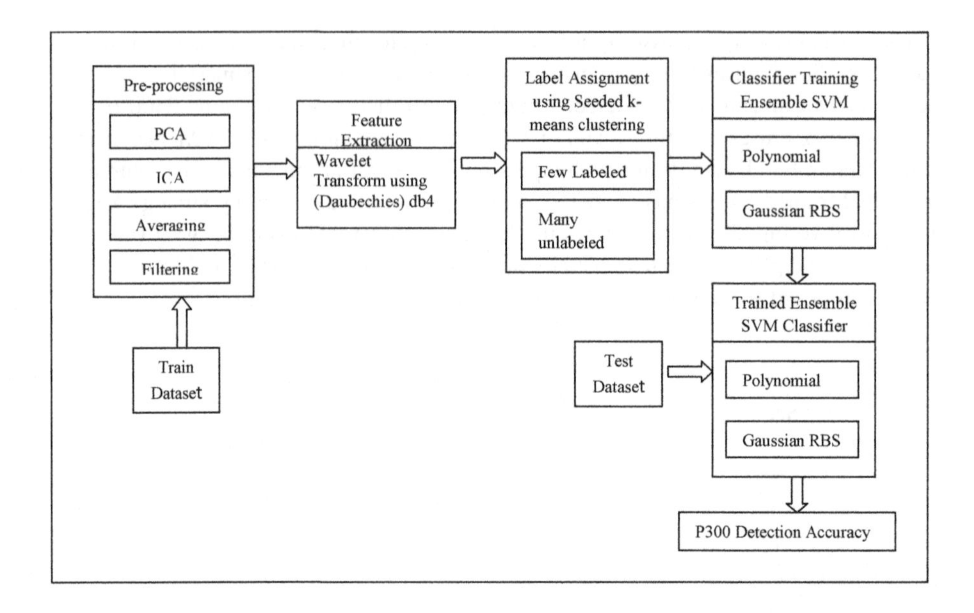

Fig. 2. Illustration of Seeded k-means based Ensemble SVM Framework

4 Result Analysis

The performance of the proposed framework for P300 speller explained in various following ways:

1. **SNR Comparison**: The analysis of P300 waveform in the ongoing EEG performed using Signal-to-noise Ratio (SNR) until pre-processing phase has discussed in [7]. The algorithm applied to BCI Competition 2004 dataset II and obtained the average SNR of 4.7501 dB for Subject A and 4.7339 dB Subject B. In addition, we reported the average SNR of 4.5338 dB with respect to average trial (20 trials).

2. **Cluster Validation**: The Normalized mutual information (NMI) used as the cluster validation for proposed framework. The NMI estimates the cluster quality based on the given true labels of the data sets. It is an external validation metric that have a range of NMI values 0 to 1. The NMI value 1 means best clustering quality. The NMI is a type of information entropy measure that is defined as

$$NMI(\mathcal{C},\mathcal{B}) = \frac{I(\mathcal{C};\mathcal{B})}{\sqrt{H(\mathcal{C})H(\mathcal{B})}} = \frac{\sum_{i=1}^{k}\sum_{j=1}^{k} n_{ij}\log\frac{n \cdot n_{ij}}{n_i \cdot n'_j}}{\sqrt{\sum_{i=1}^{k} n_i \log\frac{n_i}{n}\sum_{j=1}^{k} n'_j \log\frac{n'_j}{n}}}$$

Where H means the entropy, and I mean the mutual information, ni means data points in the ith cluster, n′j denotes the data point in the jth cluster and nij denotes the data point included in ith and jth cluster. The proposed framework used maximum

10-fold cross-validation for each data set with a number of clusters varies from 20 to 88 shown in Table 1. It is not essential that when the number of clusters increases, the value of NMI increase, as with other external clustering validation measures like purity and entropy [34, 35].

Table 1. Cluster Evaluation uses NMI for Subject A and Subject B

Subject A			Subject B		
Seed Fraction (cross validation)	**NMI**	**Number of Clusters**	**Seed Fraction (cross validation)**	**NMI**	**Number of Clusters**
0.2	0.45	67	0.2	0.33	70
0.4	0.56	74	0.4	0.49	81
0.6	0.60	82	0.6	0.57	68
0.8	0.74	80	0.8	0.66	79
1.0	0.8	88	1.0	0.79	82

3. **Accuracy**: The accuracy is determined in terms of whether the P300 exists in a cluster or not i.e. the detection of P300 in a Cluster using ensemble SVM. Suppose C is the clustering result, i.e. clusters and B define the predefined class (cluster) then the accuracy is defined as one to one association between the clusters and the classes:

$$Acc(\mathcal{C}, \mathcal{B}) = \frac{\max\left(\sum_{i=1}^{k} n_{i,Map(i)}\right)}{n}$$

Where n refers to the number of data points in the dataset, i refer to the cluster index, Map (i) is the class index corresponding to cluster index i and ni,Map(i) is the number of data points not only belonging to cluster i, but to class Map(i) [34, 35]. The Table 2 shows the accuracy of different types of SVM with optimal values of C and γ, in 10-fold cross-validation training data set. It concluded that ensemble SVM (using the majority voting method) outperforms better than linear, polynomial or rbf SVM.

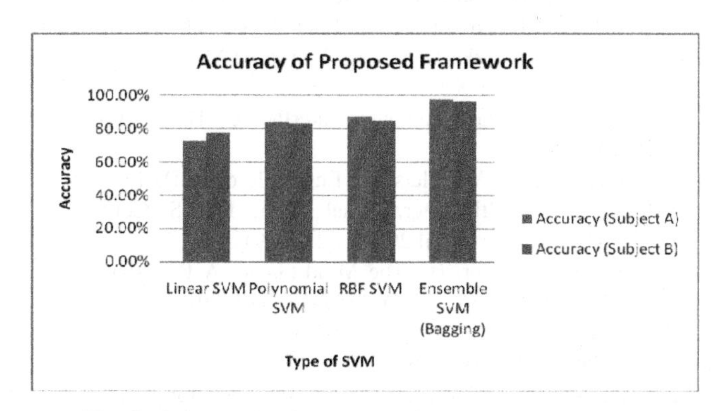

Fig. 3. Accuracy Performance of Subject A and Subject B

The development phases, merits and demerits of various existing P300 based Brain Computer Interface (BCI) system using supervised approach have been discussed in [36]. It appraises limitations of wavelet based BCI systems and compares the various wavelet methods for P300 signals. Some of the limitations motivated us to propose this novel method for discovering knowledge embedded in P300 signals using semi-supervised clustering [36].

5 Conclusion

The aim of this work is to present a novel way of discovering the knowledge embedded in EEG signals. This knowledge is then used to classify an unknown signal into a discovered signal class that the said system will interpret into a device control signal. The work suggests a new framework of extracting the features using wavelet transform and discovering the knowledge (classes) from brain activity patterns using semi-supervised clustering and then classifying these using Classifier Ensemble in EEG based BCI system. The Seeded K-means based semi-supervised cluster classification using Ensemble SVM classifier has used as the basis of the framework. The ensemble SVM has trained from initial pre-defined cluster known as seed and then extends for large datasets to detect the presence of P300 signal. The system used two-class linear Support vector machine for non-linear high dimensional data. The complete illustration of the proposed framework depicted in Section 3. The result shows that the ensemble SVM using bagging (majority voting) method procures better accuracy than the independent type of SVMs. The future work focuses on applying this approach for online P300 data analysis for more number of subjects.

References

[1] Polich, J.: Neuropsychology of P300. In: Luck, S.J., Kappenman, E.S. (eds.) Handbook of Event-related Potential Components. Oxford University Press (2010)

[2] Vahabi, Z., Amirfattahi, R., Mirzaei, A.R.: Enhancing P300 Wave of BCI Systems Via Negentropy in Adaptive Wavelet Denoising. Journal of Medical Signals & Sensors 1(3), 165–176 (2011)

[3] Kaper, M.: P300 based brain computer interfacing. Thesis, pp. 1–149. Bielefeld University (2006)

[4] Salvaris, M., Sepulveda, F.: Wavelets and Ensemble of FLDs for P300 Classification. In: Proceedings of the 4th International IEEE EMBS Conference on Neural Engineering, Antalya, Turkey, April 29-May 2 (2009)

[5] Finke, A., Lenhardt, A., Ritter, H.: The Mind Game: A P300-based Brain Computer Interface Game. Neural Networks 22(9), 1329–1333 (2009)

[6] https://www.bhsu.edu/Portals/91/InstrumentalAnalysis/StudyHelp/LectureNotes/Chapter5.pdf

[7] Kaur, M., et al.: Semi-supervised Clustering Approach for P300 based BCI Speller Systems. In: Accepted for Publication in Fifth International Conference on Recent Trends in Information, Telecommunication and Computing, ITC 2014, will be Held During, Chandigarh, March 21-22 (2014)

[8] El Dabbagh, H., Fakhr, W.: Multiple classification algorithms for the BCI P300 speller diagram using an ensemble of SVMs. In: IEEE Conference and Exhibition (GCC), Dubai, United Arab Emirates, February 19-22 (2011)
[9] http://bbci.de/competition/iii/#datasets
[10] Espinosa, R.: Increased Signal to Noise Ratio in P300 Potentials by the Method of Coherent Self-Averaging in BCI Systems. World Academy of Science, Engineering and Technology, International Journal of Medical, Pharmaceutical Science and Engineering 7(11), 106–110 (2013)
[11] Espinosa, R.: Increased Signal to Noise Ratio in P300 Potentials by the Method of Coherent Self-Averaging in BCI Systems. World Academy of Science, Engineering and Technology, International Journal of Medical, Pharmaceutical Science and Engineering 7(11), 106–110 (2013)
[12] Gerardo, R.-C., et al.: On signal P-300 detection for BCI applications based on Wavelet analysis and ICA Preprocessing. In: Electronics, Robotics and Automotive Mechanics Conference, September 28 -October 1, pp. 360–365 (2010) ISBN: 978-1-4244-8149-1
[13] Presentation by Ata Kaban, Independent Component Analysis & Blind Source Separation. The University of Birmingham
[14] Cristea, P.: Independent Component Analysis (ICA) for Genetic Signals. In: SPIE's International Symposium, San Jose Convention Center, San Jose, California, USA, January 20-26, pp. 20–26 (2001)
[15] Hyvärinen, A., Oja, E.: Independent Component Analysis: A Tutorial. In: Neural Networks, vol. 13, pp. 411–430. Helsinki University of Technology, Laboratory of Computer and Information Science, Espoo (2000)
[16] Srinivasulu, A., Sreenath Reddy, M.: Artifacts Removing From EEG Signals by ICA Algorithms. IOSR Journal of Electrical and Electronics Engineering (IOSRJEEE) 2(4), 11–16 (2012) ISSN: 2278-1676
[17] Schimpf, P.H., Liu, H.: Localizing Sources of the P300 using ICA, SSLOFO, and Latency Mapping. Journal of Biomechanics, Biomedical and Biophysical Engineering 2(1), 1–11 (2008)
[18] Naik, G.R., Kumar, D.K.: An Overview of Independent Component Analysis and Its Applications. Informatica 35, 63–81 (2011)
[19] Lemm, S., et al.: Enhancing the Signal-to-Noise Ratio of ICA-Based Extracted ERPs. IEEE Transactions on Biomedical Engineering 53(4), 601–607 (2006)
[20] Raj, A., Deo, A., Bandil, M.K.: Canvassing Various Techniques for Removal of Biological Artifact in EEG. International Journal of Recent Technology and Engineering (IJRTE) 1(3), 149–154 (2012) ISSN: 2277-3878
[21] Aviyente, S., et al.: Analysis of Event Related Potentials Using Pca and Matching Pursuit on the Time-Frequency Plane. In: IEEE Conference Proceesing on Med. Biol. Soc. (2006)
[22] Kaur, M., Ahmed, P., Qasim Rafiq, M.: Analysis of Extracting Distinct Functional Components of P300 using Wavelet Transform. In: Proceedings of 4th International Conference on Mathematical Models for Engineering Science (MMES 2013) and the 2nd International Conference on Computers, Digital Communications and Computing (ICDCC 2013), Brasov, Romania, June 1-3, pp. 57–62 (2013) ISBN: 978-1-61804-194-4
[23] Khatwani, P., Tiwari, A.: A survey on different noise removal techniques of EEG signals. International Journal of Advanced Research in Computer and Communication Engineering 2(2), 1091–1095 (2013)
[24] Bair, E.: Semi-supervised clustering methods. 1270 © 2013 Wiley Periodicals, Inc. (2013), doi:10.1002/wics (Article first published online: July 23, 2013)

[25] Boswell, D.: Introduction to Support Vector Machines, http://www.work.caltech.edu/boswell/IntroToSVM.pdf (August 6, 2002)
[26] Moore Andrew, W.: Professor School of Computer Science. Lecture Notes on Support Vector Machines. Carnegie Mellon University (November 23, 2001), http://www.cs.cmu.edu/~cga/ai-course/svm.pdf
[27] Matthias, K.: BCI Competition 2003—Data Set IIb: Support Vector Machines for the P300 Speller Paradigm. IEEE Transactions on Biomedical Engineering 51(6), 1073–1076 (2004)
[28] Daniel, M.: Classification and Clustering using SVM. 2nd Ph.D Report, Thesis Title: Data Mining for Unstructured Data, University of Sibiu (2005)
[29] Xia, X.-L., Lyu, M.R., Lok, T.-M., Huang, G.-B.: Methods of Decreasing the Number of Support Vectors via k -Mean Clustering. In: Huang, D.-S., Zhang, X.-P., Huang, G.-B. (eds.) ICIC 2005. LNCS, vol. 3644, pp. 717–726. Springer, Heidelberg (2005)
[30] Sugato, B.: Semi-supervised Clustering: Learning with Limited User Feedback. PhD Proposal, The University of Texas at Austin (November 2003) (Also Appears as Technical Report, UT-AI-TR-03-307)
[31] Sugato, B.: Semi-supervised Clustering: Probabilistic Models, Algorithms and Experiments. Thesis, The University of Texas at Austin (August 2005)
[32] Sugato, B., et al.: Semi-supervised Clustering by Seeding. In: Proceedings of the 19th International Conference on Machine Learning (ICML 2002), Sydney, Australia, pp. 19–26 (July 2002)
[33] Sugato, B., et al.: A Probabilistic Framework for SemiSupervised Clustering. In: Proceedings of the Tenth ACM SIGKDD International Conference on Knowledge Discovery and Data Mining (KDD 2004), Seattle, WA, pp. 59–68 (August 2004)
[34] Wang, J.: An Effective Semi-Supervised Clustering Framework Integrating Pairwise Constraints and Attribute Preferences. Computing and Informatics 31, 597–612 (2012)
[35] Tang, W., et al.: Enhancing Semi-Supervised Clustering: A Feature Projection Perspective. In: Proceedings of the 13th ACM SIGKDD International Conference on Knowledge Discovery and Data Mining, pp. 707–716. ACM (2007)
[36] Mandeep Kaur, P., Ahmed, A.K., Soni, M.: Qasim Rafiq, Wavelet Transform use for P300 Signal Clustering by Self-Organizing Map. International Journal of Scientific & Engineering Research 4(11) (November 2013)

Nanorobotics Control Systems Design – A New Paradigm for Healthcare System

Sankar Karan[1,2], Bhadrani Banerjee[3,4], Anvita Tripathi[3,4], and Dwijesh Dutta Majumder[1,3,5]

[1] Institute of Radiophysics and Electronics, Calcutta University, Kolkata-6, India
[2] Computational Intelligence and Nanotechnology Research Society, Kolkota
[3] Institute of Cybernetics Systems and Information Technology, Kolkata-108, India
[4] Sir J.C. Bose School of Engineering, SKFGI, WB, India
[5] Electronics and Communication Science Unit, Indian Statistical Institute, Kolkata-108, India
{sankar.karan,mamanskfgi2013,
anvianvita93,drdduttamajumder}@gmail.com

Abstract. Nanorobotics is currently emerging as an attractive area of scientific research bridging biological and computational science along with mechanical science at the molecular level. We present a new approach to control the machines at the nano-meter or molecular scale (10^{-9} meter) in the perspective of the theory of cybernetics, the science of control, communication and computation with integration to complex man–machines systems. The problem under study concentrates its main focus on nano-robot control systems design including systems casualty, state notation and automata. We also describe the theory of nano-scale thermodynamically driven self assembly for collaborative information processing in a Bio-nanorobotics systems. A fuzzy shaped based approach is described in context of recognizing a single malignant cell along with its stage, as a target for medical treatment. The synthesis and imaging of magnetic nanoparticles, that can be functionally bind with the medicine and reached the effected regions for targeted drug delivery, such as in cancer treatment is also presented.

Keywords: Nanotechnology, Nano-robotics; Control Systems; Cybernetics; Molecular Self Assembly; Molecular Motors; Automata; Targeted Drug Delivery; System Casualty; Entropy.

1 Introduction

Nowadays medical science is more and more improving with marriage of nanotechnology [1], its automation of atomic manipulation [2] and robotics technology with biological knowledge. Using nano-robot as a manipulation tool at the molecular level requires us to understand at least the essence of the mechanics [3] of materials which has a long and outstanding history in physics and engineering. In order to achieve cost effective molecular manufacturing new methodologies and theories are necessary to explore the nano world through the interaction of traditional sciences, when scientist faced a set of problems concerned with communication,

S.C. Satapathy et al. (eds.), *Emerging ICT for Bridging the Future – Volume 1*,
Advances in Intelligent Systems and Computing 337, DOI: 10.1007/978-3-319-13728-5_9

computation and control in molecular machines systems[4-5], which could be a building blocks of nano devices.

A nano-robot is essentially a controllable electromechanical machine at the nanometer or molecular scale, having the capability of actuation, control, communication, and power, interfacing across special scale and between the organic, inorganic and biotic realms. We illustrate that the components, that consists of a nano-robotic system is made of biological components, such as proteins and DNAs, that perfected by nature through millions of years of evolution and hence these are very accurate and effective.

1.1 Background and Perspective

Nanorobotics is a multidisciplinary field, where physicists, chemists, biologists, computer scientists and engineers have to work towards to a common goal. Bio-nano robots are designed by harnessing properties of biological materials mainly peptides and DNAs, that will explore new horizons in nano-computing, nano-fabrication and nano-electronics, diagnostics and therapeutics [6-7]. While tracing the origins automatic activities of human, animal bodies and their systems of control in the cellular level (molecular operating system) it leads to the machines, particularly for mechanization of thought process and its mathematical counterpart, i.e. logic for nano-computing [8] and the flow of information, it is possible to deal with integrated man–machines systems through a general systems theory [4] also in the fields of nanorobotics. The characteristic abilities that are desirable for Nanorobots to function are swarm Intelligence [9], cooperative behavior, Self assembly and replication, nano Information processing and programmability and nano to macro world interface architecture for its control and maintenance.

2 Self Assembly - Modular Organization of a Bio-Nanorobots

With the individual bio-nano robot in full function, they will now need to collaborate with one another to further develop systems in different stage. We propose a lattice gas model of thermodynamically driven molecular self assembly [10] for such nano-bio systems, which includes the thermodynamic potential for force balance approach. We have calculated total net potential of the entire self-assembly process with a given building unit as

$$\varphi_{total}(x,y,z) = \varphi_0 + \lambda_p [\varphi_{a,p}(x,y,z) + \varphi_{r,p}(x,y,z)] + \lambda_s [\varphi_{a,s}(x,y,z) + \varphi_{r,s}(x,y,z)] + \lambda_t [\varphi_{a,t}(x,y,z) + \varphi_{r,t}(x,y,z)] + \cdots\cdots + \varphi_{ext}(x,y,z) \quad \text{With} \quad \lambda_p + \lambda_s + \lambda_t + \ldots\ldots = 1 \tag{1}$$

Where x,y,z are the position coordinate of the particles , $\varphi 0$ is the initial potential energy of all nanoparticles take part in self assembly process. φa , φr are the attractive and repulsive potentials for the self assembly of primary, secondary and tertiary respectively as represented by p: primary, s: secondary, t: tertiary respectively. λ's are the fractional contribution to the net potential of each self assembly step. φext is the potential contribution of the external force when they are applied. Where φ0 term related to the internal energy of the system that can be calculated from the thermodynamic potential at constant volume and constant pressure.

2.1 Thermodynamic Potential at Constant Volume [φ_{tv0}]

Helmholtz function or total free energy F is

$$F = U - TS \tag{2}$$

For a small change

$$dF = dU - Tds - SdT = pdV - SdT \tag{3}$$

$$dF = - pdV \tag{4}$$

$$\int_A^B dF = \int_A^B - p\,dV = -W \tag{5}$$

This equation can be defined to describe the change of the thermodynamic Potential F, U is the internal energy, S is entropy of the system, i.e. the thermodynamic probability of that state, W is the total work done by the nano-composite system. A and B are initial and finite state of the system. It means that decrease of the function F in a reversible isothermal process is equal to the amount of work obtained in that process.

2.2 Thermodynamic Potential at Constant Pressure [φ_{tp0}]

Gibb's function G is defined as

$$G = U - TS + pV \tag{6}$$

For small change

$$dG = dU - Tds - SdT + pdV + Vdp = Vdp - SdT \tag{7}$$

for isothermal change at constant pressure, dG=0 and G=Constant, i.e. Where S is the entropy of the system, measures the degree of order for a system and it is larger for more disorder system.

3 Nanorobotics Devices: A Protein Based Machines

ATP synthase represents the molecular motor of life on the planet earth. Synthesis of ATP is carried out by an enzyme [11]. Mitochondrial ATP Synthase complex consists of two large structural components called F0 which encodes the catalytic activity, and F1 which functions as the proton channel crossing the inner mitochondrial membrane [12]. The figure also illustrates the subunits inside the two motor components. F1 constitutes of α3β3γδε subunits. F0 has three different protein molecules, namely, subunit a, b and c. The γ-subunit of F1 is attached to the c subunit of F0 and is hence rotated along with it. The α3β3 subunits are fixed to the b-subunit of F0 and hence do not move. Further the b-subunit is held inside the membrane by a subunit of F0 (shown in the above figure below by Walker). The reaction catalyzed by ATP synthase is fully reversible[13], so ATP hydrolysis generates a proton gradient by a reversal of this flux.

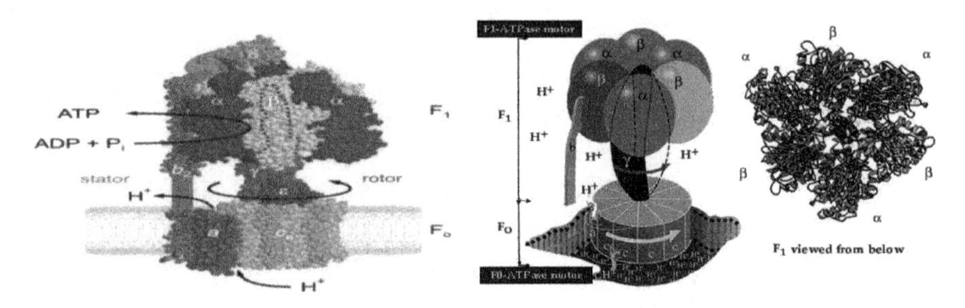

Fig. 1. Basic structure of the ATP Synthase

4 Design Philosophy -A Systems and Cybernetics Approach

We present the design of control systems for the machines at the nano-meter or molecular scale (10-9 meter) in the perspective of the theory of cybernetics, the science of control, communication and computation with integration to complex man–machines systems [4]. Rapid advance in collection of data in the field of engineering (nanorobotics systems) and biology (human body) created problems on a higher level of control and gave rise to the synergy of cybernetics in nanotechnology. Let us consider a Cartesian set X = X1 * X2 * ... * Xn and construct a propositional function L(x) with the individual variable x ranging over the set X. If the proposition L (xi) is true whenever xi ε Xs, then L(x) defines an attribute of the system: B = {L1...., LP}. Two important examples of attributes in case of nano robotic systems are reproducibility and controllability of the systems. The fundamental problems importance in case of nano-robotic systems is the systems casualty and initial state. The output of the system at any time depends only on the input up to that time but not the input beyond that time and the initial state is asymptotically irrelevant, like final state systems, they tends to forget the initial state , which is desirable as otherwise the system tends to become unstable. The concept of state space has played an important role in case of nano robotic system design. The concept of state-space is a great unifying tool in the theory of general dynamical systems and in the theory of cybernetic systems, automata for example, as it is explained below.

Definition 1: A finite state system or an Automaton A is a triple of finite sets Σ, U, and Y, and a pair of mappings .

$$F: \Sigma \times U \times T \rightarrow \Sigma \quad \text{and} \quad g: \Sigma \times U \times T \rightarrow Y \tag{8}$$

Where T is the set of integers, T = {..., —, 0, 1,}. The elements of Σ are states of A; the elements of U are the inputs of A, and the elements of Y are the outputs of A. The time t is assumed to range over T, with the state, input and output of A at time t denoted by St , Ut , and Yt , respectively. The mappings f and g relate the state at time t + 1 and the output at time t to the state and the input at time t

$$St=f(St,Ut,t); \quad Yt=g(St,Ut,t) \text{ and}$$

$$St+1=f(f(f(St0,Ut0,t0),Ut0+1,t0+1)\ldots Ut,t) \tag{9}$$

Following Kalman's definition of dynamical systems it can be shown that Eq. stated above has the group property of a dynamical system. But the expression for the output in the case of a finite-state system is slightly more general as it includes inputs to the system.

Definition 2: An oriented abstract object A is a family {R[t0,t1]}, t0, tl ε (-a, a), of sets of ordered pairs of time functions, with a generic pair denoted by (U, Y), in which the first component, U § U[t0,t1], is called an input segment or, simply an input, and the second component, Y § Y[t0,t1], is called an output segment or simply, an output, where the symbol , §, stands for "is defined to be." A pair of time functions (U, Y), is an input-output pair belonging to A, written as (U, Y) ε A, if (U, Y) ε R[t0,t1] for some t0, t1 in (-a,a). Thus, an oriented abstract object A can be identified with the totality of input-output pairs which belong to A. The members of the family {r[t0,t1]} are required to satisfy the consistency condition, so that every section of an input–output pair of A be in itself an input–output pair of A. The input segment and output segment of space are denoted by R[U] and R[Y]. This implies that R[t0,t1], which is the set of all pairs (U[t0, t1], Y[t0, t1],) belonging to A, is a subset of the product space R[U] x R[Y]. We now introduce the notion of state. Let a be a parameter ranging over a space Σ; usually a is an n-vector and Σ is Rn, the space of ordered n-tuples of real numbers, a parameterizes A if there exists a function Ā on the product space Σ x R[u] such that for all (U, Y) belonging to A and all t0 and t, we can find an a in Σ such that

$$Y = \bar{A}(a; u) \tag{10}$$

For each a in Σ and each u in R[u], the pair (U,Ā(a,u)) is an input–output pair belonging to A. a can now be called the state of A if Ā has the response separation property.

The above cybernetic approach offers a mathematical methodology on the building of theories concerning the historical-social process as a whole, which actually is a synthesis of the so-called hermeneutic–dialectical and the logical–analytical approaches in the theory of science for design of nanorobotics systems.

5 Fuzzy Shape Based Analysis for Cancer Cell and Its Stage

Pap smear image data of Cervix cancer cell as shown in Fig 2 (a, b, d & e) were classified using fuzzy C-Means clustering [14] algorithm (FCM) to separate the nucleus and Cytoplasm area inside the cell. Let X = { X_1 , X_2 , ... , X_n } be a set of samples to be clustered into c classes. Here we consider color as a feature for classification in RGB (red, green, blue) color space. The criterion function used for the clustering process is

$$J(V) = \sum_{k=1}^{n} \sum_{x_k \in C_1} \left| x_k - v_i \right|^2 \tag{11}$$

Where vi, is the sample mean or the center of samples of cluster i, and V = { v1 , ... , vc}. To improve the similarity of the samples in each cluster, we can minimize this criterion function so that all samples are more compactly distributed around their cluster centers. Membership values (μ's) are assigned as per FCM algorithm. For cluster validity, we consider three types of measures: partition coefficient, partition entropy and compactness and separation validity function. We proposed a generalized method of shape analysis [15] and shape based similarity measures, shape distance and shape metric to measure the nucleus and cytoplasm shape. The shape of an object can be defined as a subset X in R2 if (a) X is closed and bounded , (b) Interior of X is non-empty and connected and (c) Closure property holds on interior of X . This representation of shape remains invariant with respect to translation, rotation and scaling. Moreover another object Y in R2 is of same shape to object X ∈ R2 if it preserves translation, rotation and scaling invariance. In term of set these three transformations can be represented as Translation: Y = {(x + a), (y + b): x, y∈ X} , Rotation : Y = {P1(α).P2(β)X} where P1 & P2 are rotation around x and y axes . Scaling: Y = {(kx, ky): x, y ∈ X} . Distance d1 between shape X and Y in F is defined as follows: d1(X,Y)=m2[(X-Y)∪(Y-X)] where m2 is Lebesgue measure in R2 and d1 satisfies following rules: (i) d1(X,Y) ≥ 0 , (ii) d1(X,Y) = 0 if and only if X = Y (iii) d1(X,Y) = d1(Y,X) and (iv) d1(X,Y) + d1(Y,Z) ≥ d1(X,Z).

We defined Carcinoma Ratio (Cr) is the max[(Nh/Ch) ,(Nw/Cw)] where Cd :Cytoplasm Maximum width ,Ch: Cytoplasm Maximum Height , Nw: Nucleus maximum Width, Nh: Nucleus Maximum Height , Cr: Carcinoma Ratio all measurement have been taken in pixel unit.

6 Grading of Tumor Cell Using Image Segmentation

Our design and methodology deals with utilization of nano-robot inside the human body to detect and kill the cancer cell. The size and shape of the nucleus and cytoplasm helps to determine its stage.

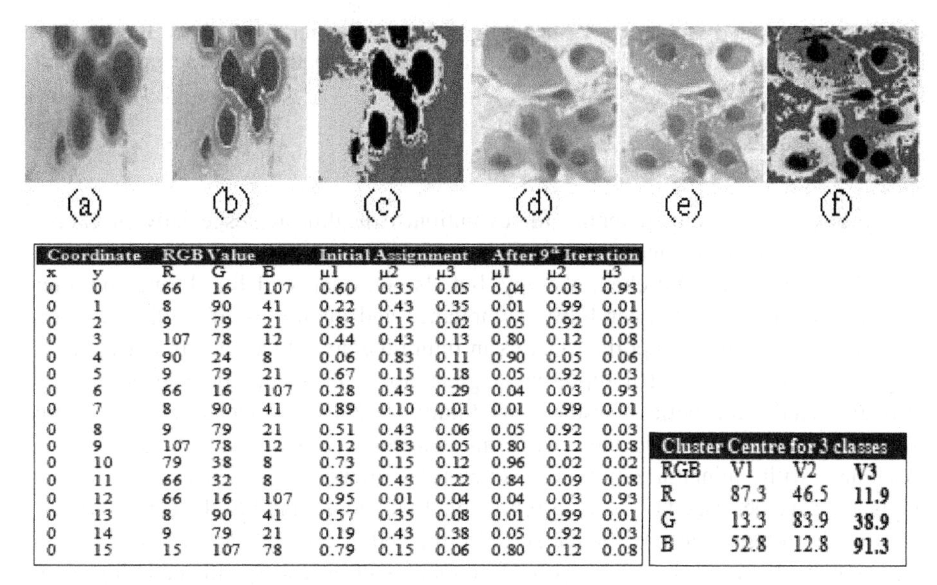

Coordinate		RGB Value			Initial Assignment			After 9th Iteration		
x	y	R	G	B	µ1	µ2	µ3	µ1	µ2	µ3
0	0	66	16	107	0.60	0.35	0.05	0.04	0.03	0.93
0	1	8	90	41	0.22	0.43	0.35	0.01	0.99	0.01
0	2	9	79	21	0.83	0.15	0.02	0.05	0.92	0.03
0	3	107	78	12	0.44	0.43	0.13	0.80	0.12	0.08
0	4	90	24	8	0.06	0.83	0.11	0.90	0.05	0.06
0	5	9	79	21	0.67	0.15	0.18	0.05	0.92	0.03
0	6	66	16	107	0.28	0.43	0.29	0.04	0.03	0.93
0	7	8	90	41	0.89	0.10	0.01	0.01	0.99	0.01
0	8	9	79	21	0.51	0.43	0.06	0.05	0.92	0.03
0	9	107	78	12	0.12	0.83	0.05	0.80	0.12	0.08
0	10	79	38	8	0.73	0.15	0.12	0.96	0.02	0.02
0	11	66	32	8	0.35	0.43	0.22	0.84	0.09	0.08
0	12	66	16	107	0.95	0.01	0.04	0.04	0.03	0.93
0	13	8	90	41	0.57	0.35	0.08	0.01	0.99	0.01
0	14	9	79	21	0.19	0.43	0.38	0.05	0.92	0.03
0	15	15	107	78	0.79	0.15	0.06	0.80	0.12	0.08

Cluster Centre for 3 classes			
RGB	V1	V2	V3
R	87.3	46.5	11.9
G	13.3	83.9	38.9
B	52.8	12.8	91.3

Fig. 2. FCM results for Grading of tumor cell (Cervix Cancer) , Image Segmentation FCM for cervical cancer diagnosis and staging , (a) & (d) HSIL and Atrophic Pap image, (b) & (e) Carcinoma tracing (c)&(f) Segmented Pap images

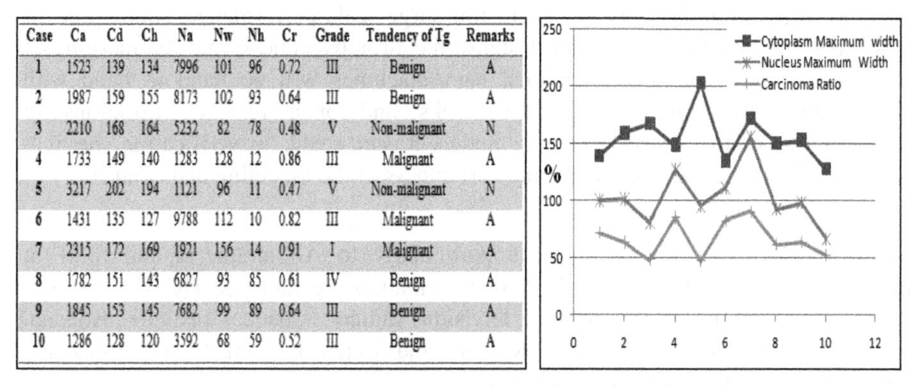

Case	Ca	Cd	Ch	Na	Nw	Nh	Cr	Grade	Tendency of Tg	Remarks
1	1523	139	134	7996	101	96	0.72	III	Benign	A
2	1987	159	155	8173	102	93	0.64	III	Benign	A
3	2210	168	164	5232	82	78	0.48	V	Non-malignant	N
4	1733	149	140	1283	128	12	0.86	III	Malignant	A
5	3217	202	194	1121	96	11	0.47	V	Non-malignant	N
6	1431	135	127	9788	112	10	0.82	III	Malignant	A
7	2315	172	169	1921	156	14	0.91	I	Malignant	A
8	1782	151	143	6827	93	85	0.61	IV	Benign	A
9	1845	153	145	7682	99	89	0.64	III	Benign	A
10	1286	128	120	3592	68	59	0.52	III	Benign	A

Fig. 3. Grading of Tumor Cell (graphically), Ca : Cytoplasm area , Na: Nucleus Area , Tg: Tumor growth ,N: Normal A: Abnormal Carcinoma Ratio (C_r)

Nano-robot use intensified magnetic property of the cancer cell to locate their exact position inside the human body from the force of attraction, the sensor inside the nano-robot isolate it from the normal cell. In addition to magnetic force of cancer cell, temperature of both the cell is considered for monitoring in vivo inter cellular energy transactions. This method of drug targeting provide a solution to eliminate the problems in Chemotherapy like treatment , where cancer treating drugs are delivered to both the healthy and diseased cells during treatment. The already collected data of common cancer cell is used to predict the cancer stage. A survey was done among a sizeable number of cervical cancer patients and their pathological test reports were collected along with the microscopic photographs of the Pap test. The microscopic images are analyzed with fuzzy c-mean algorithm and generalized shape theory. The results as shown in fig.2 and Fig.3 are validated with the pathological test reports and doctor's opinion.

7 Conclusion

Nanorobotics has strong potential to revolutionize healthcare, especially in cancer treatment. Many practitioners believe that cancer treatment should be based not only on prognostic factors and chemotherapy, but also on quality of life during and after treatment, knowing that, tolerability, compliance and quality of life will, therefore, could become the most important factors in future therapy of cancer . In our opinion, magnetic nanoparticles [16] represent a great hope for successful cancer therapy through nano-bio-robotics system in future. Nano robotics is a fruitful interdisciplinary research in the area of man-machines communication, i.e. learning machines with miniaturization of computer communication network. As an application we describe how a nano-robot able to diagnose and kill the cancer cell by controlled drug transfer to particular site. Our method of cervical cancer diagnosis and therapeutics [17] is a robust one and is very successful in assisting the pathologists in the screening process of cervical cancer. The results, we have obtained, are validated with the clinical findings and it proves to be satisfactory with some minor enhancement has to be made. Future work may include addition of a statistical and operational research envelope for evaluating large-scale performance, simulations of new environments and nano-robot designs. We strongly believe that the merging of the new technologies into operational nano-machines will go hand in hand with progressing simulation fidelity. Bottlenecks and open research questions in Nanorobotics are to construct dynamic models at nano-scale, for verification, the right tools or methods to check accuracy and correctness of the modeling and control.

Acknowledgement. We are thankful particularly to All engineers and Staff of Institute of Cybernetics Systems And Information Technology (ICSIT), all our colleagues at the Biological Science Division, Indian Statistical Institute, Kolkata, ECSU, Indian Statistical Institute, Kolkata and Department of Radiophysics and Electronics, Calcutta University, Kolkata.

References

[1] Drexler Eric, K.: Nanosystems: Molecular Machinary, Manufacturing and Computation. John Wiley Publication (1992)

[2] Dutta Majumder, D., Karan, S., Goswami, A.: Synthesis and Characterization of Gold Nanoparticle –a fuzzy Mathematical Approach. In: Kuznetsov, S.O., Mandal, D.P., Kundu, M.K., Pal, S.K. (eds.) PReMI 2011. LNCS, vol. 6744, pp. 324–332. Springer, Heidelberg (2011)

[3] Ikai, A.: The world of Nano-Biomechanics. Elsevier (2007)

[4] Dutta Majumder, D.: Cybernetics and General Systems – a unitary Science. Kybernets 8, 7–15 (1979)

[5] Dutta Majumder, D.: Cybernetics a Science of Engineering and Biology. Cybernetiea XVII, 3 (1975)
[6] Cavalcanti, A., Freitas, R.A.: Nanorobotics control design: a collective behavior approach for medicine. IEEE Transactions on Nano Bioscience 4(2), 133–140 (2005)
[7] Varadan, V.K.: Nanotechnology based point-of-care diagnostics and therapeutics for neurological and cardiovascular disorders. IEEE Sensors, 2334–2337 (2010)
[8] Bourianoff, G.: The future of nanocomputing. IEEE Computer, 44–53 (2003)
[9] Samal, N.R., Konar, A., Das, S., Abraham, A.: A closed loop stability analysis and parameter selection of the particle swarm optimization dynamics for faster convergence. In: Proc. Congr. Evolu. Compu., Singapore, pp. 1769–1776 (2007)
[10] Reif, J.H., LaBean, T.H., Seeman, N.C.: Programmable assembly at the molecular scale: self-assembly of DNA lattices. In: Proc. 2001 ICRA IEEE Int. Conf. on Robotics and Automation, vol. 1, pp. 966–971 (2001)
[11] (1997), `http://nobelprize.org/nobel_prizes/chemistry/laureates/1997/press.html`
[12] Boyer, P.D.: The ATP synthase—a splendid molecular machine. Annu. Rev. Biochem. 66, 717–749 (1997)
[13] Hiroyasu, I., Akira, T., Kengo, A., Hiroyuki, N., Ryohei, Y., Asasuke, Y., Kazuhiko, K.: Mechanically driven ATP synthesis by F1-ATPase. Nature 427, 465–468 (2004)
[14] Bezdek, J.C., Ehrlich, R., Full, W.: FCM: The fuzzy c-means clustering algorithm. Computers & Geosciences 10(2-3), 191–203 (1984)
[15] Dutta Majumder, D.: A Study on a Mathematical Theory of Shapes in Relation to Pattern Recognition and Computer Vision. India Journal of Theoretical Physics 43(4), 19–30 (1995)
[16] Dutta Majumder, D., Karan, S., Goswami, A.: Characterization of Gold and Silver Nanoparticles using its Color Image Segmentation and Feature Extraction Using Fuzzy C-Means Clustering and Generalized Shape Theory. In: Proc. IEEE International Conference on Communications and Signal Processing 2011, pp. 70–74 (2011)
[17] Duttamajumder, D., Bhattacharaya, M.: Cybernetics approach to medical technology: Application to cancer screening and other diagnostic. Kybernets 29(7/8), 817–895 (2000)

A Comparative Study on Subspace Methods for Face Recognition under Varying Facial Expressions

G.P. Hegde[1] and M. Seetha[2]

[1] Dept. Computer Science & Engineering, SDMIT, Ujire, Mangalore, Karnataka, India
[2] Dept. Computer Science & Engineering, GNITS, Hyderabad, India
gphegde123@gmail.com, smaddala2000@yahoo.com

Abstract. Face recognition is one of the widely used research topic in biometric fields and it is rigorously studied. Recognizing faces under varying facial expressions is still a very challenging task because adjoining of real time expression in a person face causes a wide range of difficulties in recognition systems. Moreover facial expression is a way of nonverbal communication. Facial expression will reveal the sensation or passion of a person and also it can be used to reveal someone's mental views and psychosomatic aspects. Subspace analysis are the most vital techniques which are used to find the basis vectors that optimally cluster the projected data according to their class labels. Subspace is a subset of a larger space, which contains the properties of the larger space. The key contribution of this article is, we have developed and analyzed the 2 state of the art subspace approaches for recognizing faces under varying facial expressions using a common set of train and test images. This evaluation gives us the exact face recognition rates of the 2 systems under varying facial expressions. This exhaustive analysis would be a great asset for researchers working world-wide on face recognition under varying facial expressions. The train and test images are considered from standard public face databases ATT, and JAFFE.

1 Introduction

In the context of face recognition, the objective of subspace analysis is to find the basis vectors that optimally cluster the projected data according to their class labels. In simple words, a subspace is a subset of a larger space, which contains the properties of the larger space. Generally, the face images data are high dimensional in nature. This leads to the problem of the so called curse-of-dimensionality and several other related troubles. Especially in the context of similarity and matching based recognition, it is computationally intensive [1-20]. Minimum average correlation energy (MACE) and Independent Component Analysis (ICA) are 2 latest state of the art subspace approaches considered for face recognition under varying facial expressions. In this article considered ICA and MACE approaches for recognizing faces under varying facial expressions. We have used a common set of train and test

S.C. Satapathy et al. (eds.), *Emerging ICT for Bridging the Future – Volume 1*,
Advances in Intelligent Systems and Computing 337, DOI: 10.1007/978-3-319-13728-5_10

images from ATT (formerly 'The ORL Database of Faces'), and JAFFE (Japanese Female Facial Expression), This evaluation gives us the exact face recognition rates of the 2 systems under different facial expressions. The rest of the article is presented as follows: Section 2 Minimum Average Correlation Energy. Section 3 presents Independent Component Analysis. Results are discussed in section 4. Section 5 draws the Conclusion. Section 6 presents Future Work.

2 Minimum Average Correlation Energy (MACE)

Suppose that we have N training images from the true class with each image having d pixels in it. We perform two dimensional (2-D) FFTs on these images and convert the 2-D FFT arrays into one dimensional (1-D) column vectors by lexicographic ordering. These vectors are the column vectors of the $d \times N$ matrix X. Column vector c with N elements contains the pre specified correlation peak values of the training images and the $d \times d$ diagonal matrix D contains along its diagonal the average power spectrum of the training images (i.e., average of the magnitude squares of the columns of X). Note that the synthesized hMACE is a column vector with d elements and the 2-D correlation filter is obtained by reordering the column vector back to a 2-D array. The + symbol represents the complex conjugate transpose [21-25].

Typically pattern matching using correlation filters is based on the value of the largest correlation peak. However, this value is sensitive to illumination changes in the input image. This can be a serious issue as we can expect our face recognition system to operate in variable illumination conditions frequently due to its mobile nature, i.e., it is essential that a user can authenticate themselves to gain access to their data at any time and location regardless of lighting conditions. Therefore, to reduce this dependency on illumination levels, we employ the peak-to-side lobe ratio (PSR) as shown below:

$$PSR = \frac{peak - mean}{\sigma}$$

We use 2 masks centered on the peak. The annular region between the two masks determines the side lobe region. By using more than one correlation peak to compute a match score enhances the confidence of the system as it uses multiple correlation points to calculate this single match score (PSR). The size of sidelobe region and central mask strongly depend on spectrum size. A typical correlation output for an authentic face image is characterized by a sharp correlation peak resulting in a large PSR. On the other hand, a typical response to an impostor is characterized by a low value for correlation output. In feature space all linear features can be expressed as follows: $y_i = v + x_i$ where $v= [v_1, v_2, ..., v_M]$ is a $d \times M$ feature extraction matrix, x_i is 2-dimensional image considered as a $d \times 1$ column vector(d is the number of pixels), y_i is the feature vector of length M. M depends on what feature to use. Feature Correlation Filter and MACE Filter share the same formulation. In this way it is possible to incorporate the feature representations of faces into correlation filters shown in Fig. 1. Given a database of known faces, each image of this data set is filtered using a Gabor filter of filter bank.

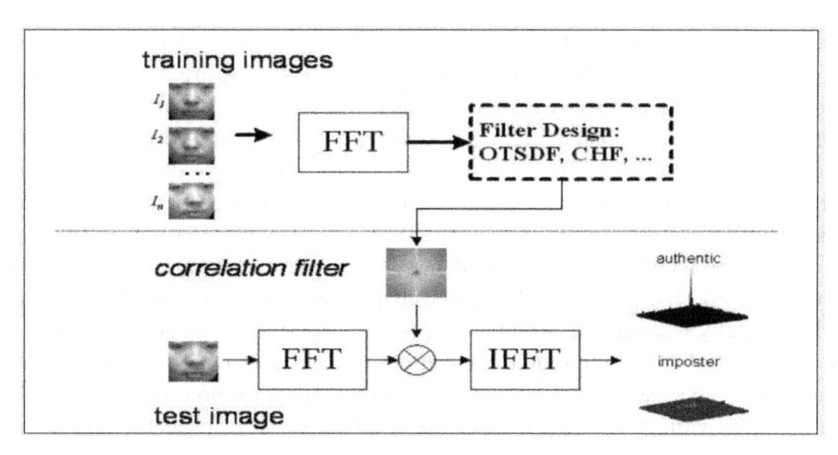

Fig. 1. Training images and correlation filter used for MACE based Face Recognition

Each filter has its own scale and orientation. At the end of filtering process for each facial image we have original image and 40 Gabor filtered images (5 scales and 8 orientations), so in total 41 bi-dimensional matrices. In order to improve computational efficiency spatial 2D convolution is performed in frequency domain. For each class of database then we compute 41 MACE filters. If the total number of class is P we compute in total 41P MACE filters shown in Fig. 2.

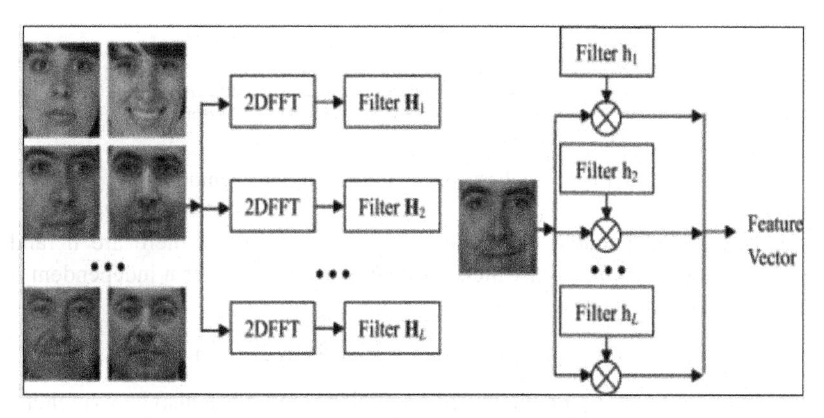

Fig. 2. MACE and Gabor filter bank for Face Recognition

3 Independent Component Analyses

ICA is proposed by Jutten and Herault firstly to solve the cocktail lounge problem; its basic intension is to separate the mixed signals into some independent components by optimization algorithm according to statistical independence principle, and take the separated components as the approximate estimation of source signals [26-30]. The function of ICA is to extract corresponding independent signals from the source mixed signals. One of the applications of ICA is feature extraction. ICA is a method in which statistical characteristics in second order or higher order are considered. Basis vectors decomposed from face images obtained by ICA are more localized in distribution space than those by PCA. Localized characteristics are favorable for face recognition, because human faces are non-rigid bodies, and because localized characteristics are not easily influenced by face expression changes, location, position, or partial occlusion. ICA is a recently developed statistical technique that can be viewed as an extension of standard PCA and does not consider LDA. Using ICA, one tries to model the underlying data so that in the linear expansion of the data vectors the coefficients are as independent as possible. ICA bases of the expansion must be mutually independent while the PCA bases are merely uncorrelated. ICA has been widely used for blind source separation and blind convolution. Blind source separation tries to separate a few independent but unknown source signals from their linear mixtures without knowing the mixture coefficients. Fig. 3 sets the fundamental ground for ICA face recognition.

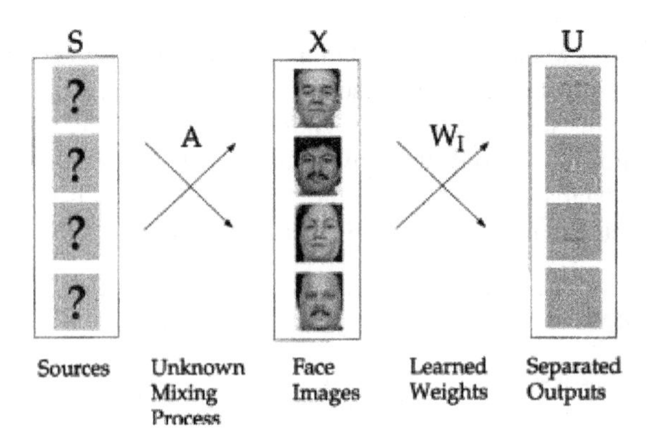

Fig. 3. Image synthesis model for ICA face recognition

The model of independent component analysis is: suppose there are n random variables as x= (x_1, x_2, x_3, ..., x_n) which is obtained from the other n independent non-Gaussian variables s=(s_1, s_2, s_3, …, s_n)

$$x_i = a_{i1}s_1 + a_{i2}s_2 + \ldots, + a_{in}s_n, \; i= 1, 2, \ldots, n$$

where i, j = 1, 2, …, n, a_{ij} are the mixed coefficients and it can also be written as:

$$x = As$$

In order to estimate one of the independent component, considering linear combination with xi as y = bTx where b is the unsure vector. Thus the relation y = bTAs can be got. Assuming q=bTA there is

$$y = bTx = qTs$$

In face recognition, every face image as a column, there are n training set and n observation set. Vector X can be thus obtained, then using the ICA we can get m basic images as:

$$U = WX$$

where W is the mixed matrix. For a given test image y which is being bleached and centered, and then projected on the m basic images, the face feature vector is given as below:

$$z = Uy$$

For optimizing independent components, ICA Component Subspace Optimization or Sequential Forward Floating Selection (SFFS) is used.

4 Results and Discussion

We have considered 2 standard public face databases ATT, and JAFFE to train and test MACE and ICA. The details of train and test images used are tabulated in Table 1.

Table 1. Train and test sample details used to evaluate MACE and ICA face recognition subspace approaches

	ATT Database	JAFFE Database
Description	Major facial expressions (open / closed eyes, smiling / not smiling) and facial details (glasses / no glasses)	7 facial expressions (6 basic facial expressions + 1 neutral)
Total no. of subjects	40	10
Total no. of classes	40	10
Trained images/class	2	2
Total no. of trained images	40 x 2= 80	10 x 2 = 20
Test images/class	10	20
Total no. of test images	40 x 10 = 400	10 x 20 = 200
Dimension	92x112	256x256

Face recognition rate (FRR) is calculated. For example consider ATT database, it contains 40 subjects, each subject is assigned a class number. Thus for ATT database we have 40 classes. For 40 classes we have 40 folder placed in Train_data folder (s_{01}, s_{02}, s_{03}, …, s_{40}). Each class is given a unique class ID, s_{01} is given class ID 1, s_{02} is given class ID 2, s_{03} is given class ID 3 and so on. Now there are total 40 class IDs. Training is performed by taking 2 images per class, 40 x 2 = 80 images are used for training.

Once the system is completely trained then images are picked from Test_data folder, inside Test_folder we will again have 40 folders (s_{01}, s_{02}, s_{03}, …, s_{40}), each of these folders will have 10 images. Totally 40 x 10 = 400 images are used for testing. Now when recognition is performed by picking images from Test_data folder, for example if an image is picked from s01 folder, ideally it needs to match to s01 folder of Train_data and it needs to have class ID 1. If it matches to any other images, for example if s01 image is matched to s02, then the class ID would be 2 and this would be considered as a mismatch. Thus FRR is calculated as:

FRR = (Total number of face images recognized accurately / Total number of face images available in the database) x 100

Similar approach is used for JAFEE face databases. Sample input image and recognized image using MACE for JAFEE database, which has major facial expression variations is shown in Fig. 4.

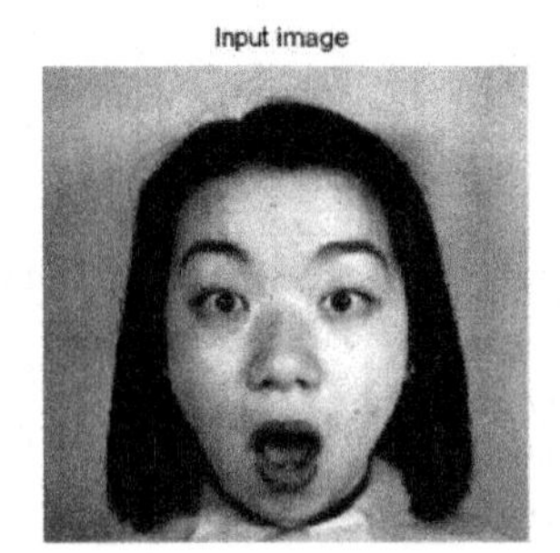

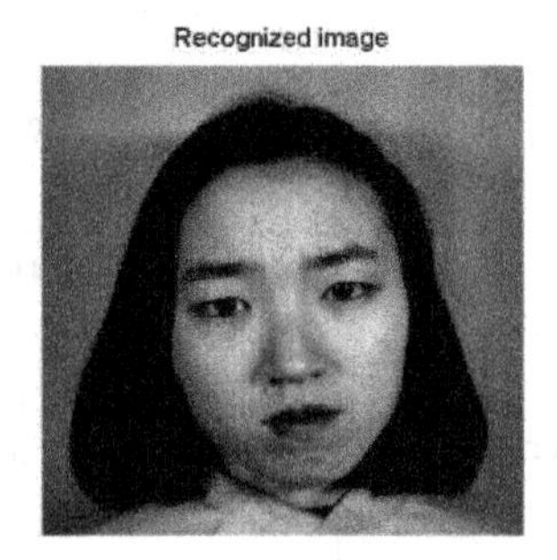

Fig. 4. Recognition using MACE were input image is form JAFFE database

The results are tabulated in Table-2.

Table 2. FRR calculated for MACE and ICA face recognition approaches

Techniques	ATT Database	JAFFE Database
MACE - FRR	380/400 = 95%	193/200 = 96.50%
ICA - FRR	351/400 = 87.75%	142/200 = 71%

5 Conclusion

In this article an attempt is made to develop and analyze the 2 state of the art subspace approaches for Face Recognition under varying Facial Expression. The approaches considered are: Minimum average correlation energy (MACE) and Independent Component Analysis (ICA). Both the systems were trained and tested using standard public face databases ATT, and JAFFE. This evaluation gives us the exact face recognition rates of both systems under varying facial expressions. The face recognition rate of MACE on ATT database was 95% which was better than ICA. Also face recognition rate of MACE on JAFE database, which contains major facial expression variations, was 95.50% which was better than ICA. Thus we conclude MACE is the best subspace technique to recognize faces under varying facial expression variations.

6 Future Work

The 2 state of the art subspace face recognition techniques developed and analyzed for varying facial expression could be used to evaluate the face recognition rate under varying pose and illuminations using standard public face databases: Pointing Head Pose Image Database, YALEB Database.

References

1. Bettadapura, V.: Face Expression Recognition and Analysis: The State of the Art. IEEE Trans. on Pattern Analysis and Machine Intelligence 22
2. Zhao, W., Chellppa, R., Phillips, P.J., Rosenfeld, A.: Face recognition: A literature survey. Assoc. Comput. Mach. Comput. Surv. 35(4), 399–458 (2003)
3. Ahonen, T., Hadid, A., Pietikainen, M.: Face description with local binary patterns: Application to face recognition. IEEE Trans. Pattern Anal. Mach. Intell. 28(12), 2037–2041 (2006)
4. Leonardis, A., Bischof, H.: Robust recognition using eigen images. Comput. Vis. Image Understand. 78(1), 99–118 (2000)
5. Naseem, I., Togneri, R., Bennamoun, M.: Linear regression for face recognition. IEEE Trans. Pattern Anal. Mach. Intell. 32(11), 2106–2112 (2010)
6. Wright, J., Yang, A.Y., Ganesh, A., Sastry, S.S., Ma, Y.: Robust face recognition via sparse representation. IEEE Trans. Pattern Anal. Mach. Intell. 31(2), 210–227 (2009)
7. Wright, J., Ma, Y.: Dense error correction via l1 minimization. IEEE Trans. Inf. Theory 56(7), 3540–3560 (2010)
8. Yang, M., Zhang, L.: Gabor feature based sparse representation for face recognition with Gabor occlusion dictionary. In: Daniilidis, K., Maragos, P., Paragios, N. (eds.) ECCV 2010, Part VI. LNCS, vol. 6316, pp. 448–461. Springer, Heidelberg (2010)
9. Kumar, N., Berg, A.C., Belhumeur, P.N., Nayar, S.K.: Attribute and simile classifiers for face verification. In: Proc. IEEE Int. Conf. Comput. Vis., pp. 365–372 (October 2009)
10. Wolf, L., Hassner, T., Taigman, Y.: Effective face recognition by combining multiple descriptors and learned background statistic. IEEE Trans. Pattern Anal. Mach. Intell. 33(10), 1978–1990 (2011)
11. He, R., Zheng, W.-S., Hu, B.-G., Kong, X.-W.: Two-Stage Nonnegative Sparse Representation for Large-Scale Face Recognition. IEEE Trans. on Neural Networks and Learning Systems 24(1) (January 2013)
12. Duan, C.-H., Chiang, C.-K., Lai, S.-H.: Face Verification With Local Sparse Representation. IEEE Signal Processing Letters 20(2) (February 2013)
13. Yang, J., Chu, D., Zhang, L., Xu, Y., Yang, J.: Sparse Representation Classifier Steered Discriminative Projection With Applications to Face Recognition. IEEE Trans. on Neural Networks and Learning Systems 24(7) (July 2013)

14. Yang, A.Y., Zihan, Z., Balasubramanian, A.G., Sastry, S.S.: Fast l1 -Minimization Algorithms for Robust Face Recognition 22(8) (August 2013)
15. Chen, Y.-C., Patel, V.M., Shekhar, S., Chellappa, R., Phillips, P.J.: Video-based face recognition via joint sparse representation. In: 10th IEEE International Conference and Workshops on Automatic Face and Gesture Recognition (FG) (April 2013)
16. Fernandes, S.L., Bala, G.J.: Time Complexity for Face Recognition under varying Pose, Illumination and Facial Expressions based on Sparse Representation. International Journal of Computer Application 51(12), 7–10 (2012)
17. Fernandes, S.L., Bala, G.J.: A comparative study on ICA and LPP based Face Recognition under varying illuminations and facial expressions. In: International Conference on Signal Processing Image Processing & Pattern Recognition (ICSIPR) (February 2013)
18. Lee, T.S.: Image representation using 2-d Gabor wavelets. IEEE Trans. on Pattern Analysis Pattern Analysis and Machine Intelligence 18(10) (October 1996)
19. Manjunath, B.S., Chellappa, R., von der Malsburg, C.: A Feature Based Approach to Face Recognition. In: Proc. of International Conf. on Computer Vision (1992)
20. Lades, M., Vorbruggen, J., Buhmann, J., Lange, J., von der Malsburg, C., Wurtz, R.: Distortion invariant object recognition in the dynamic link architecture. IEEE Trans. Comput. 42(3), 300–311 (1993)
21. Wiskott, L., Fellous, J.M., Krüger, N., von der Malsburg, C.: Face Recognition by Elastic Graph Matching. In: Intelligent Biometric Techniques in fingerprint and Face Recognition, ch. 11, pp. 355–396 (1999)
22. Pan, Z., Adams, R., Bolouri, H.: Image redundancy reduction for neural network classification using discrete cosine transforms. In: Proc. IEEE-INNS-ENNS Int. Joint Conf. Neural Networks, Como, Italy, vol. 3, pp. 149–154 (2000)
23. Hafed, Z.M., Levine, M.D.: Face recognition using the discrete cosine transform. Int. J. Comput. Vis. 43(3), 167–188 (2001)
24. Pedrycz, W.: Conditional fuzzy clustering in the design of radial basis function neural networks. IEEE Trans. Neural Netw. 9(4), 601–612 (1998)
25. Brunelli, R., Poggio, T.: Face recognition: Features versus templates. IEEE Trans. Pattern Anal. Mach. Intell. 15(10), 1042–1053 (1993)
26. Er, M.J., Wu, S., Lu, J., Toh, H.L.: Face recognition with radial basis function (RBF) neural networks. IEEE Trans. Neural Netw. 13(3), 697–710 (2002)
27. Yang, F., Paindavoine, M.: Implementation of an RBF neural network on embedded systems: Real-time face tracking and identity verification. IEEE Trans. Neural Netw. 14(5), 1162–1175 (2003)
28. Zhang, B., Shan, S., Chen, X., Gao, W.: Histogram of Gabor phase patterns (HGPP): A novel object representation approach for face recognition. IEEE Trans. Image Process. 16(1), 57–68 (2007)
29. Albiol, A., Monzo, D., Martin, A., Sastre, J., Albiol, A.: Face recognition using HOG-EBGM. Pattern Recogn. Lett. 29(10), 1537–1543 (2008)
30. Wolf, L., Hassner, T., Taigman, Y.: Effective unconstrained facerecognition by combining multiple descriptors and learned background statistics. IEEE Trans. Pattern Anal. Mach. Intell. 33(10), 1978–1990 (2011)

Recognizing Faces When Images Are Corrupted by Varying Degree of Noises and Blurring Effects

Steven Lawrence Fernandes[1] and Josemin G. Bala[2]

[1] Dept. Electronics & Communication Engineering, Karunya University,
Tamil Nadu, India
[2] Dept. & Communication Engineering, Karunya University,
Tamil Nadu, India
steva_fernandes@yahoo.com, josemin@karunya.edu

Abstract. Most images are corrupted by various noises and blurring effects. Recognizing human faces in the presence of noises and blurring effects is a challenging task. Appearance based techniques are usually preferred to recognize faces under different degree of noises. Two state of the art techniques considered in our paper are Locality Preserving Projections (LPP) and Hybrid Spatial Feature Interdependence Matrix (HSFIM) based face descriptors. To investigate the performance of LPP and HSFIM we simulate the real world scenario by adding noises: Gaussian noise, Salt and pepper noise and also adding blurring effects: Motion blur and Gaussian blur on six standard public face databases: IITK, ATT, JAFEE, CALTECH, GRIMANCE, and SHEFFIELD.

1 Introduction

Face has its own advantages over other biometrics for people identification and verification-related applications, since it is natural, nonintrusive, contactless etc. Unfortunately, most human faces captured un-controlled environment are corrupted by noises and blurring effects and hence offer low distinctiveness [1-10]. The past decades have witnessed tremendous efforts to recognize images using appearance based techniques which are corrupted by varying degree of noises and blurring effects. Two state of the art techniques considered in our paper are Locality Preserving Projections (LPP) and Hybrid Spatial Feature Interdependence Matrix (HSFIM) based face descriptors. LPP is a linear manifold learning approach. LPP is used to generate an unsupervised neighborhood graph on training data, and then finds an optimal locality preserving projection matrix under certain criterion. LPP aims to disclose the low dimensional local manifold structures embedded in high dimensional feature space [11-15]. Spatial Feature Interdependence Matrix (SFIM) based face descriptor is partially inspired by the success of the statistical object representation models, they are also different. To compute the hybrid SFIM of a face image, we decompose the target image into a number of local regions we then divide every face image into a set of spatially non-overlapped rectangular facial regions having the

S.C. Satapathy et al. (eds.), *Emerging ICT for Bridging the Future – Volume 1*,
Advances in Intelligent Systems and Computing 337, DOI: 10.1007/978-3-319-13728-5_11

same size [16-21]. In this paper we have investigated the performance of LPP and HSFIM we simulate the real world scenario by adding noises: Gaussian noise, Salt and pepper noise and also adding blurring effects: Motion blur and Gaussian blur on six standard public face databases: IITK, ATT, JAFEE, CALTECH, GRIMANCE, and SHEFFIELD. The remaining of the paper is organized as follows: Section 2 provides a brief overview of ICA, Section 3 presents HSFIM, Section 4 presents results and discussions and finally Section 5 draws the conclusion.

2 Locality Preserving Projections

Locality Preserving Projections (LPP) method is to preserve the local structure of image space by explicitly considering the manifold structure, which is transformed into solve a generalized eigenvalue problem using a long. It can be viewed as a linear approximation of Laplacian eigen maps [22-25]. LPP is an unsupervised manifold dimensionality reduction approach, which aims to find the optimal projection matrix W by minimizing the following criterion function:

$$J_1(W) = \sum_{i,j} \left\| y_i - y_j \right\|^2 S_{ij}.$$

Substituting $y_i = WTx_i$ into above objection function. Direct computation yields

$$J_1(W) = 2trace(W^T XLX^T W),$$

Where L=D-S is called Laplacian matrix, D is a diagonal matrix defined by

$$D = diag\{D_{11}, D_{22}, \cdots, D_{NN}\}$$

where $D_{ii} = \sum_i S_{ij}$. Matrix D provides a natural measure on the data points. The bigger the value D_{ii} (corresponding to y_i) more important is y_i. Thereby, the projection matrix W should maximize the following constraint objective function simultaneously:

$$J_2(W) = \sum_{i=1}^{N} y_i^T D y_i = trace(Y^T DY)$$

$$= trace(W^T XDX^T W).$$

Solving problems min $J_1(W)$ and max $J_2(W)$ simultaneously is equivalent to minimizing the following criterion function:

$$J(W) = \frac{trace(W^T XLX^T W)}{trace(W^T XDX^T W)},$$

The optimal locality preserving projection matrix

$$W_{LPP} = \arg \min_{W \in R^{d \times l}} J(W)$$

can be obtained by solving the following generalized eigen-value problem:

$$XLX^T W = (XDX^T) W \Lambda.$$

where is Λ a diagonal eigenvalue matrix. While in most cases, the number of training data is smaller than the dimension of feature vector, i.e. d << N. When this 3S problem occurs, the matrix XDX^T is not full rank.

3 Hybrid Spatial Feature Independence Matrix Based Face Descriptors

A wide range of features are available for describing the content of an image region of interest, e.g., intensity, color, edge, texture and shape. Traditional approaches focus on using self-similarities or correlations across videos or images to capture the dynamic motion patterns in the temporal space, or jointly using shape and appearance information extracted from the images to achieve enhanced discrimination in object classification [26-30]. Considering that the proposed SFIM based face descriptor is a square symmetric matrix whose diagonal entries are zeros, the upper triangular entries excluding the diagonal are used as the components of the feature vector. Let T be a training dataset consisting of P normalized face images collected from K subjects, let p_k be the number of the training images for the kth subject, I_t be a testing face image, the hybrid SFIM based recognition algorithms are as below.

1. Divide each face image into $n_h \times n_w$ spatially non-overlapped rectangular regions.
2. Compute HI, HLBP and HOG features for all rectangular regions of each face image.
3. Compute three respective SFIMs for each face image.
4. Compute hybrid SFIM for each face image.
5. Rearrange the upper triangular entries (excluding the diagonal) of each hybrid SFIM into a vector x_p, where $x_p \in \Re^{(N^2-N)/2}$ and $N = n_h \times n_w$, we obtain a training vector set $x_1, x_2, \ldots, x_P$.
6. Repeat steps 1 - 5 on testing face image I_t, we obtain a testing vector.

4 Results and Discussion

We have considered 6 standard databases to test LPP and HSFIM: IITK, ATT, JAFEE, CALTECH, GRIMANCE, and SHEFFIELD. The details of the number of train and test samples are mentioned in TABLE 1 and TABLE 2. The noise value will indicate the amount of noise added to the face image. The blur value indicates the amount of blurriness' added to the face image.

Table 1. Details of various noises and blurring effects added to face images for IITK, ATT and JAFEE databases

DATABASE	IITK	ATT	JAFEE
Gaussian noise values	50,55,60,65,70	50,55,60,65,70	50,55,60,65,70
Salt & Pepper noise values	50,60,70,80,90	50,60,70,80,90	50,60,70,80,90
Motion blur values	10,20,30,40,50	10,20,30,40,50	10,20,30,40,50
Gaussian blur values	5,10,15,20,25	5,10,15,20,25	5,10,15,20,25
Train and Test sample details			
Total no. of classes	61	40	10
Trained images/class	5	5	5
Total no. of trained images without adding noise and motion blur effects	61 x 5 = 305	40 x 5 = 200	10 x 5 = 50
Test images/class	5	5	5
Total no. of test images after adding noise a motion blur effects	61 x 5 = 305	40 x 5 = 200	10 x 5 = 50
Dimension	640x480	92x112	256x256

Table 2. Details of various noises and blurring effects added to face images for CALTECH, GRIMANCE and SHEFFIELD databases

DATABASE	CALTECH	GRIMANCE	SHEFFIELD
Gaussian noise values	50,55,60,65,70	50,55,60,65,70	50,55,60,65,70
Salt & Pepper noise values	50,60,70,80,90	50,60,70,80,90	50,60,70,80,90
Motion blur values	10,20,30,40,50	10,20,30,40,50	10,20,30,40,50
Gaussian blur values	5,10,15,20,25	5,10,15,20,25	5,10,15,20,25
Train and Test sample details			
Total no. of classes	27	18	20
Trained images/class	5	5	5
Total no. of trained images without adding noise and motion blur effects	27 x 5 = 95	18 x 5 = 90	20 x 5 = 100
Test images/class	5	5	5
Total no. of test images after adding noise a motion blur effects	27 x 5 = 135	18 x 5 = 90	20 x 5 = 100
Dimension	896x592	180x200	200x200

Sample input image and recognized image using LPP for IITK database, in the presence of Salt & Pepper noise is shown in Fig. 1.

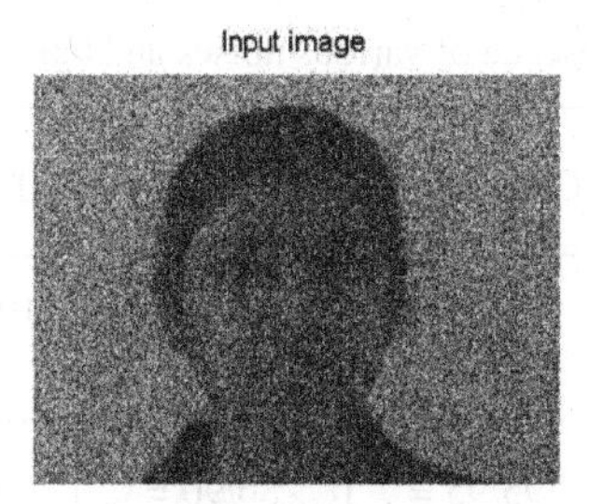

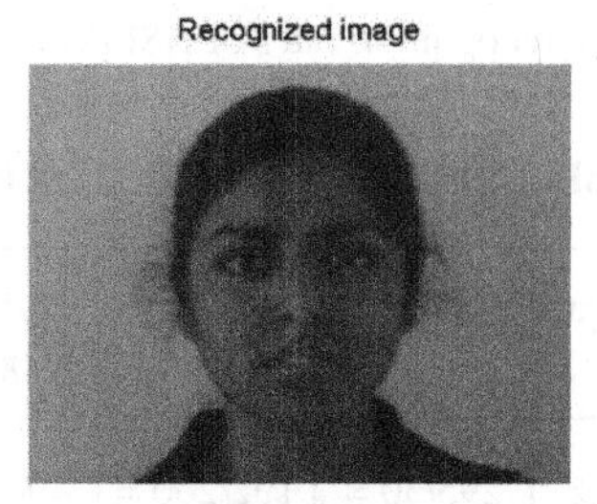

Fig. 1. Recognition using LPP were input image is from IITK database having Salt & Pepper noise

Sample input image and recognized image using HSFIM for ATT database, in the presence of Motion blur is shown in Fig. 2.

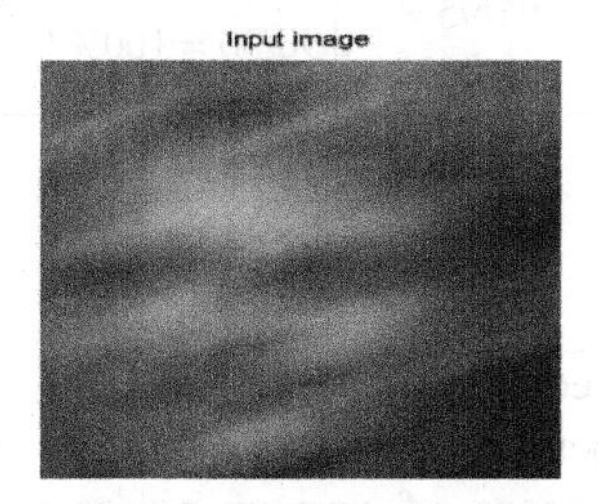

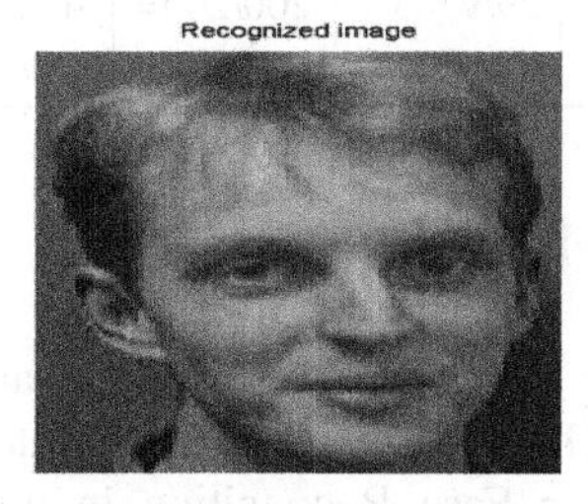

Fig. 2. Recognition using HSFIM were input image is from ATT database having Motion blur

FRR = (Total number of face images recognized accurately / Total number of face images available in the database) x 100. FRR obtained for LPP and HSFIM in the presence of Gaussian noise, Salt & pepper noise, Motion blur, and Gaussian blur on databases IITK, ATT, JAFEE, CALTECH, GRIMANCE, and SHEFFIELD are tabulated in TABLE 3 and TABLE 4 respectively.

Table 3. Face recognition rates of LPP in the presence of various noises and blurring effects

DATABASE	IITK	ATT	JAFEE	CALTECH	GRIMACE	SHEFFIELD
FRR-Gaussian noise	240/305 = 78.6%	160/200 = 80%	39/50 = 78%	75/95 = 79%	80/90 = 88.8%	80/100 = 80%
FRR-Salt & pepper noise	265/305 = 86.8%	174/200 = 87%	34/50 = 68%	83/95 = 87.3%	82/90 = 91.1%	78/100 = 78%
FRR- Motion blur	272/305 = 89.6%	178/200 = 89%	40/50 = 80%	80/95 = 84.2%	79/90 = 87.7%	67/100 = 67%
FRR- Gaussian blur	266/305 = 87.2%	172/200 = 86%	42/50 = 84%	90/95 = 94.7%	75/90 = 83.3%	82/100 = 82%

Table 4. Face recognition rates of HSFIM in the presence of various noises and blurring effects

DATABASE	IITK	ATT	JAFEE	CALTECH	GRIMACE	SHEFFIELD
FRR-Gaussian noise	300/305= 98.3%	185/200= 92.5%	50/50 = 100%	95/95 = 100%	90/90 = 100%	100/100 = 100%
FRR-Salt & pepper noise	295/30 = 96.7%	192/200 = 96%	49/50 = 97.5%	95/95 = 100%	86/90 = 95.55%	98/100 = 98%
FRR- Motion blur	292/305= 95.6%	195/200 = 97.5%	50/50 = 100%	88/95 = 92.6%	90/90 = 100%	97/100 = 97%
FRR-Gaussian blur	296/305 = 97%	200/200= 100%	49/50 = 98%	95/95 = 100%	90/90 = 100%	100/100 = 100%

5 Conclusion

In this paper we have developed and analyzed are Locality Preserving Projections (LPP) and Hybrid Spatial Feature Interdependence Matrix (HSFIM) based face descriptors for Face Recognition in the presence of various noises and blurring effects. LPP is used to generate an unsupervised neighborhood graph on training data, and then finds an optimal locality preserving projection matrix under certain criterion. HSFIM of a face image is composed by firstly decomposing the target image into a number of local regions. For easy implementation, we then divide every face image into a set of spatially non-overlapped rectangular facial regions having the same size. From our analysis we have found that HSFIM provides good face recognition rate in the presence of Gaussian noise, Salt & Pepper noise, Motion blur and Gaussian blur when compared to LPP. Our analysis was carried out on six standard public face databases: IITK, ATT, JAFEE, CALTECH, GRIMANCE, and SHEFFIELD.

References

1. Beveridge, J.R., Bolme, D., Draper, B.A., Teixeira, M.: The csu face identification evaluation system: Its purpose, features, and structure. Mach. Vis. Appl. 16(2), 128–138 (2005)
2. Bowyer, K.W., Chang, K., Flynn, P.J.: A survey of approaches and challenges in 3D and multi-modal 2D + 3D face recognition. Comput. Vis. Image Understand. 101(1), 1–15 (2006)
3. Liu, C.: Capitalize on dimensionality increasing techniques for improving face recognition grand challenge performance. IEEE Trans. Pattern Anal. Mach. Intell. 28(5), 725–737 (2006)
4. Liu, C.: The Bayes decision rule induced similarity measures. IEEE Trans. Pattern Anal. Mach. Intell. 29(6), 1086–1090 (2007)

5. Sanderson, C.: Automatic Person Verification Using Speech and Face Information. Ph.D. thesis, Griffith University, Queensland, Australia (2002)
6. Gyaourova, A., Bebis, G., Pavlidis, I.: Fusion of infrared and visible images for face recognition. In: Pajdla, T., Matas, J(G.) (eds.) ECCV 2004. LNCS, vol. 3024, pp. 456–468. Springer, Heidelberg (2004)
7. Heo, J., Kong, S., Abidi, B., Abidi, M.: Fusion of visual and thermal signatures with eyeglass removal for robust face recognition. In: Proceedings of IEEE Workshop on Object Tracking and Classification Beyond the Visible Spectrum in Conjunction with CVPR, pp. 94–99 (2004)
8. OToole, A.J., Abdi, H., Jiang, F., Phillips, P.J.: Fusing face recognition algorithms and humans. IEEE Trans. Syst., Man, Cybern. 37(5), 1149–1155 (2007)
9. OToole, A.J., Phillips, P.J., Jiang, F., Ayyad, J., Penard, N., Abdi, H.: Face recognition algorithms surpass humans matching faces across changes in illumination. IEEE Trans. Pattern Anal. Mach. Intell. 29(9), 1642–1646 (2007)
10. Zhang, D.: Automated Biometrics-Technologies and Systems. Kluwer Academic Publishers, Dordrecht (2000)
11. Jain, A., Bolle, R., Pankanti, S.: Biometrics: Personal Identification in Networked Society. Kluwer Academic Publishers, Dordrecht (1999)
12. Ross, A., Nandakumar, K., Jain, A.K.: Handbook of Multibiometrics. Springer, New York (2006)
13. Singh, R., Vatsa, M., Noore, A.: Hierarchical fusion of multi-spectral face images for improved recognition performance. Inf. Fusion (2007), doi:10.1016/j.inffus.2006.06.002 (in press)
14. Singh, S., Gyaourova, A., Bebis, G., Pavlidis, I.: Infrared and visible image fusion for face recognition. In: Proceedings of SPIE Defense and Security Symposium, vol. 5404, pp. 585–596 (2004)
15. Jain, A., Ross, A., Prabhakar, S.: An introduction to biometric recognition. IEEE Trans. Circuits Syst. Video Technol. 14(1), 4–20 (2004)
16. Chen, H.-K., Lee, Y.-C., Chen, C.-H.: Gabor feature based classification using Enhance Two-direction Variation of 2DPCA discriminant analysis for face verification. In: IEEE International Symposium on Next-Generation Electronics (ISNE) (February 2013)
17. Rahulamathavan, Y., Phan, R.C.-W., Chambers, J.A., Parish, D.J.: Facial Expression Recognition in the Encrypted Domain Based on Local Fisher Discriminant Analysis. IEEE Transactions on Affective Computing 4(1) (March 2013)
18. Huang, S.-M., Yang, J.-F.: Linear Discriminant Regression Classification for Face Recognition. IEEE Signal Processing Letters 10(1) (January 2013)
19. Fernandes, S.L., Bala, G.J.: Analyzing recognition rate of LDA and LPP based algorithms for Face Recognition. International Journal of Computer Engineering & Technology 3(2), 115–125 (2012)
20. Fernandes, S.L., Bala, G.J.: A comparative study on ICA and LPP based Face Recognition under varying illuminations and facial expressions. In: International Conference on Signal Processing Image Processing & Pattern Recognition (ICSIPR) (February 2013)
21. Ross, A., Nandakumar, K., Jain, A.: Handbook of Multibiometrics. Springer, New York (2006)
22. Jain, A.K., Ross, A., Prabhakar, S.: An introduction to biometric recognition. IEEE Trans. on Circuits and System for Video Technology Special Issue on Image- and Video-based Biometrics 14, 4–20 (2004)
23. Daugman, J.: How iris recognition works. IEEE Trans. Circuits Syst. Video Technol. 14(1), 21–30 (2004)

24. Ross, A., Jain, A.K.: Multimodal biometrics: an overview. In: Proceedings of 12th European Signal Processing Conference, pp. 1121–1224 (2004)
25. Bowyer, K.W., Hollingsworth, K.P., Flynn, P.J.: Image understanding for iris biometrics: A survey. Computer. Vision Image Understand. 110(2), 281–307 (2008)
26. Zhao, W., Chellappa, R.: Beyond one still image: Face recognition from multiple still images or a video sequence. In: Face Processing: Advanced Modeling and Methods, ch. 17, pp. 547–567. Elsevier, Amsterdam (2006)
27. Verma, T., Sahu, R.K.: PCA-LDA based face recognition system & results comparison by various classification techniques. In: IEEE International Conference on Green High Performance Computing (ICGHPC) (March 2013)
28. Maodong, S., Jiangtao, C., Ping, L.: Independent component analysis for face recognition based on two dimension symmetrical image matrix. In: 24th Chinese Control and Decision Conference (CCDC) (May 2012)
29. Yi, L., Nanjing, X.T.: An Anti-Photo Spoof Method in Face Recognition Based on the Analysis of Fourier Spectra with Sparse Logistic Regression. In: Chinese Conference on Pattern Recognition (2009)
30. Hwang, W., Park, G., Lee, J., Kee, S.-C.: Multiple Face Model of Hybrid Fourier Feature for Large Face Image Set. In: IEEE Computer Society Conference on Computer Vision and Pattern Recognition (2006)

Functional Analysis of Mental Stress Based on Physiological Data of GSR Sensor

Ramesh Sahoo and Srinivas Sethi

Departments of CSEA, IGIT, Sarang, India
ramesh0986@gmail.com, srinivas_sethi@igitsarang.ac.in

Abstract. Stress plays a vital role in everyday life. It is mental state and is accompanied by physiological changes. So monitoring of these significant changes are important, which can help to identify the matter of anxiety at an early stage before serious. Various methods have been adopted to detect the stress with various sensors. GSR sensor is one of them to detect the stress at a particular time in different position with moods. In this paper three different positions like lying, sitting and standing have been considered with three moods. Normal, tension, and physical exercise have been considered for three different moods of human life. It has been observed that, the result of GSR value in term of physiological data are constantly varies in respect to surface area contact with body and maximum GSR values observed during tension moods.

Keywords: Mental stress, GSR Sensor, Physiological Data.

1 Introduction

Stress is a physiological response to the emotional, mental, or physical challenges that become a serious problem affecting people of different life situations, and age groups in day-to-day life. It spoils one's devotion, performance, behaviour, responsibility, thinking ability, etc. Stress can pay to illness directly, or indirectly, through its physiological properties [1]. These could ultimately cause fatigue of cardiovascular, hormonal, neural and muscular system, etc., due to lacking recovery and repair. It needs to inspire people to adjust their life style and behavior for which they achieve a better stress balance far before enlarged level of stress results in serious health problems.

Stress plays a character in mental illness such as widespread anxiety sickness and depression and three different kind of stress have been observed like: Chronic stress, Acute Stress (Mental Stress) and episodic stress. Chronic stress is caused by long term stress factor, which is difficult to manage as it cannot be measured in a steady and can be very dangerous on a long run. Under healthy circumstances, the body returns to its usual state after dealing with acute stress. Episodic acute stress that observes more regularly and this type of stress is typically seen in people who make

S.C. Satapathy et al. (eds.), *Emerging ICT for Bridging the Future – Volume 1*,
Advances in Intelligent Systems and Computing 337, DOI: 10.1007/978-3-319-13728-5_12

self-inflicted, impracticable or irrational demands. It is not like a chronic stress. Physiological fluctuations have been associated with stress. By sensing these fluctuations, it can be expected to build a system that can routinely identify stress by distinguishing patterns in these sensor signals those sensor signals capture physiological responses.

In the case of psychological fluctuations, stress inclines to make the glands more active as physical body and mind are interactive [2]. Physical disease may leads mental illnesses and mental anxiety may have adverse impact on health. Physical or mental illnesses may develop, subjected to stress.

Nonstop monitoring of a person's stress levels is vital for understanding and managing personal stress. A number of physiological indicators are widely used for stress assessment, including galvanic skin response(GSR)[4]. Fortunately, wireless devices are existing to monitor these physiological indicators. By using these devices, persons can carefully track variations in their vital signs in order to maintain better health, in day-to-day life.

In this paper an activity-aware system has been developed with GSR[4], sensor to distinguish between physical activity and mental stress. It also often treated as electro dermal activity, which is a measure of the conductivity of the skin and it cannot be controlled by the user. It has been conducted a user study across three different physical activities like, laying, sitting, standing, with normal, tension and exercise.

In the next section, it has been discussed the related work in section 2. Next, stress and its background with GSR have been discussed in section 3. Section 4 describes the methodology and results and discussion have been discussed in section 5. Finally the conclusion has been discussed in the section 6.

2 Related Work

Physical activity has a constructive impact on people's better health and it may also decrease the existence of chronic diseases and it may recognition with wearable sensors which can provide response to the user about lifestyle concerning physical activity. Electro-dermal response of bio-medical system is the transform in electrical properties of skin as per variation in psychological and physiological conditions. Depending upon the activities, stress/emotion can be evoked. Due to the change in stress/emotion, GSR value will varies. The robustness and reliability of the system may increase by using multi physiological parameter [2].

In [5], the authors describe the classification of everyday activities like walking, running, and cycling and study to find out, the recognize activities about the sensors usefulness and the kind of signal processing with classification. The authors collected data library of sensed data and test through a set of wearable sensors while executing numerous activities during the measurement session. Custom decision tree, spontaneously created decision tree, and artificial neural network are three classifiers are shown and different accuracies ranges have been measured. This would increase the users' consciousness of their daily activity level with active lifestyle. It has been also enhanced with the help of sports, and thus, promote a more active lifestyle [6].

Automatic acknowledgment of physical motion can be used to show the user the spreading of daily activities of a person. Further an effective decision tree classifier and a personalization algorithm have been implemented on wireless motion bands and PDA and the online system can detect the most common daily activities with different accuracies after classifier personalization. [7].

The stress level can be refereed based on the investigation of GSR and speech signals. The authors in [9], investigate the classification techniques that can be used to automatically govern periods of acute stress relying on information confined in GSR and/or speech of a person. In [10], the authors implemented the single regression, activity recognition and activity-specific Energy Expenditure models on data collected from different subjects, while performing a set of Physical Activities, grouped into six clusters in the form of lying, sedentary, dynamic, walking, biking and running.

Stress detection may be performed based on fuzzy logic and two physiological signals used in it, they are: Galvanic Skin Response (GSR) and Heart Rate(HR)[11]. This method generates an individual stress template, gathering the behaviour of individuals under different conditions with different degrees of stress. The authors projected the method for real-time applications.

3 Stress and GSR

Skin Conductance (SC)[13] is known as GSR, a method of determining the electrical conductance of the skin and it varies with its wetness level. GSR is used as a sign of physiological/psychological arousal and it measures the electrical conductance between two points, and is basically a kind of ohmmeter. The electrical skin resistance decreases at high level of sweating takes place and increases at dryer skin. Emotions such as stress, excitement, shock, etc., can produce fluctuation in the skin conductivity.

Although several possible signals have been considered to detect stress, it has been consider GSR signal in this paper, which is a good relative indicator of stress. GSR has long been considered a measure of physiological and mental stress [12]. Higher GSR may be noticed for higher levels of stress. Determination of the specific level of GSR for specific emotion is difficult. All the emotions like anger, fear, startle response, sexual feelings, orienting response, etc., may produce alike GSR responses.

4 Methodology

In this section, it has been described, the components of the system used in the experiment and the segmentation of the experimental data with data collection.

4.1 Data Acquisition System

There are three main components have been used for data acquisition as follow.

Arduino Uno Board [14]
The Arduino Uno is a microcontroller mother board and it has 14 digital input/output pins. It can simply connect to a computer with wired USB cable. It may also power with an AC-to-DC adapter or battery to run the board. It has various facilities for communicating with a computer and another Arduino/ other microcontrollers as well. Since Arduino Uno is based on the ATmega328, it offers UART TTL (5V) serial communication, which is exist on digital pins 0 and 1 where 0 stands for RX and 1 stands for TX.

The Arduino software contains a serial monitor that permits text data to be sent to and from the Arduino board. The RX and TX LEDs on the Arduino Uno board will glow when data is being sent/received via the USB-to-serial chip and USB connection to the computer system. The Arduino software can be embedded on the Arduino Uno micro controller. By using this software it can write Arduino sketch and upload it to the Arduino Uno board and it can observe the result using serial monitor.

(a): Arduino Uno

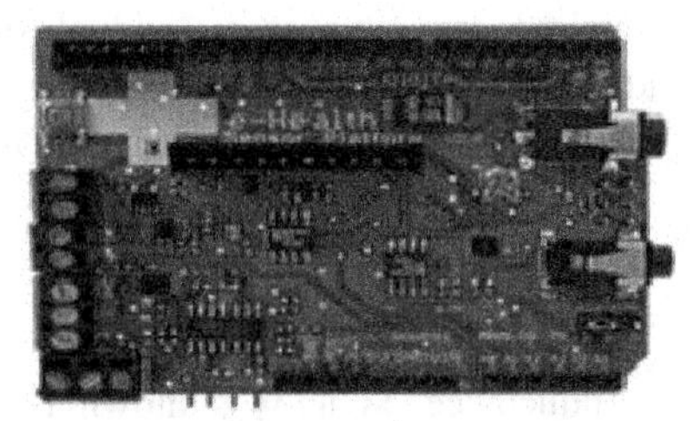

(b): e-Health Sensor platform

Fig. 1.

E-Health Sensor Platform [13]
The e-Health Sensor platform (Shield V2.0) allows Arduino Uno to perform biometric and medical uses for body monitoring through different sensors. The information can be used to observe the state of a human body in real time. Biometric information can be sent wire/ wirelessly using any of the connectivity like: USB, Bluetooth, Wi-Fi, 3G, GPRS, 802.15.4 and ZigBee as per the application. A camera can be attached to the 3G component to send photos and videos of the human body to a computer system or server. It can also be used in cloud environment.

GSR SENSOR [13]
In skin conductance response (SCR) technique, conductivity of skin is measured at index and middle finger of the palm. The working principle of GSR is to measure electrical skin resistance depends on sweat produced by the body. The GSR sensor measures the psycho-galvanic reflex of the body. GSR sensor is one of the components of polygraph devices which can be used in scientific research of emotional or physiological arousal.

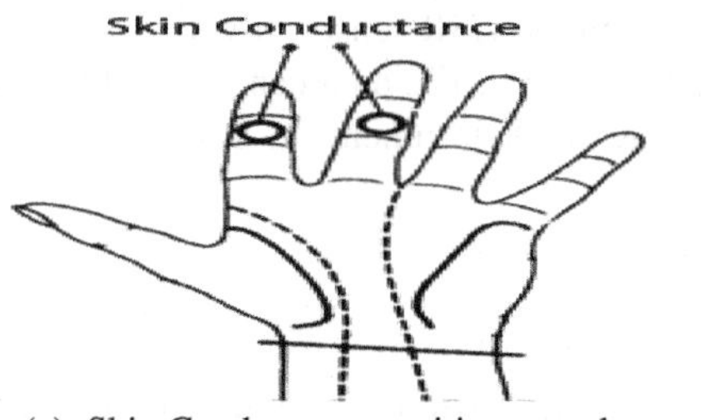

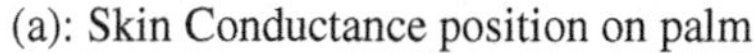
(a): Skin Conductance position on palm

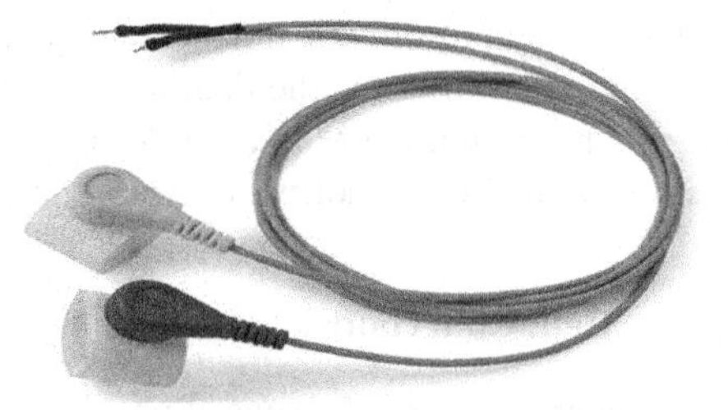
(b): GSR Sensor

Fig. 2.

4.2 Experimental Setup

For experimental setup, it has been used three hardware component such as Arduino uno, E health Sensor Platform and GSR Sensor and two Software component they are Matlab 2013a and serial port monitor. The e-Health sensor platform board is connected with Arduino Uno board and the Arduino Uno connected to a computer system with the help of USB cable. The board is powered up by USB cable. It receives physiological sensor data from Arduino Uno board through serial port. The GSR sensor has two metallic electrodes contacts and it works like an ohmmeter measuring the resistance of the materials. The fingers are tightened with the metallic electrodes with the help of the Velcro as shown in the Fig.-4.

(a): GSR Sensor with e-Health Sensor platform

(b): e-Health Sensor platform with Arduino Uno

Fig. 3.

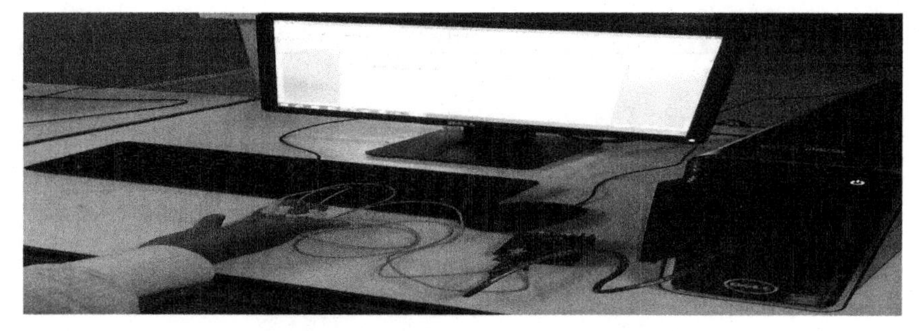

Fig. 4. Complete experimental setup

During the physiological data collection through the GSR sensor, it will send to the serial monitor and save the data in the text file. The text file data can be used to plot the graph through the Matlab or Arduino with KST environment. In this paper it has been used the Matlab software to plot the graph.

4.3 Data Collection

It has been collected the sensed physiological data for different activities in different positions for a period of 10 minutes. The physiological data are segmented in different datasets for normal, tension, and exercise in different positions like lying, sitting and standing with 115200 baud rate.

5 Experimental Results and Discussion

The changes in physical conditions under some stress are measured by considering skin resistance in different situations. It has been considered different physical conditions like lying, sitting and in different moods. Physiological data are acquired and analyzed separately for different background. Data corresponding to GSR have been shown from fig.-5 to 10. The duration of physiological signal acquisition of each experiment is 10 minutes. The stress of a body is not always same and it changes according to positions and the situations. There is a small deviation under different test and changing state of stress values due to external reasons. It does not claim on exact stress, specialty in contrast to other mental states.

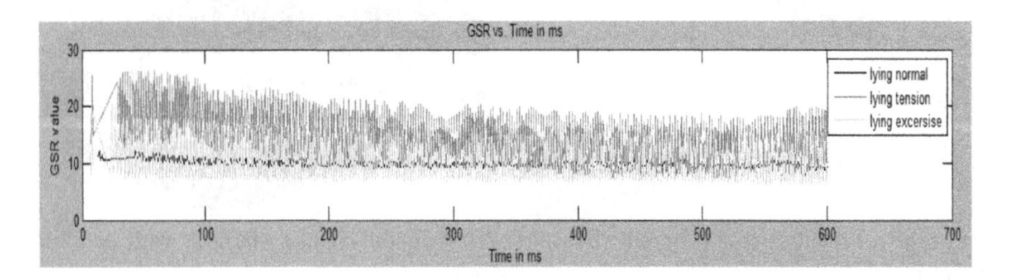

Fig. 5. GSR value vs. time in ms for lying position in different moods

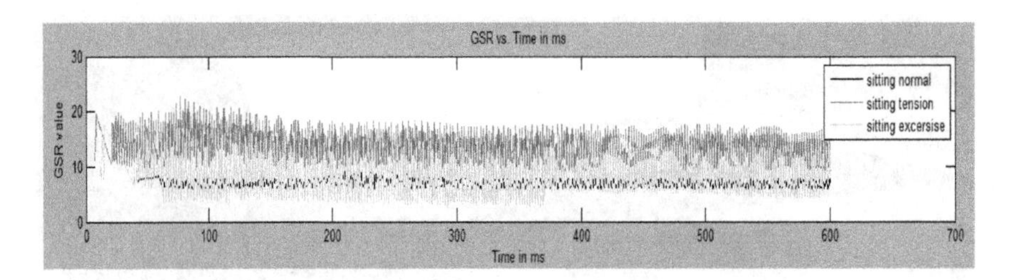

Fig. 6. GSR value vs. time in ms for sitting position in different moods

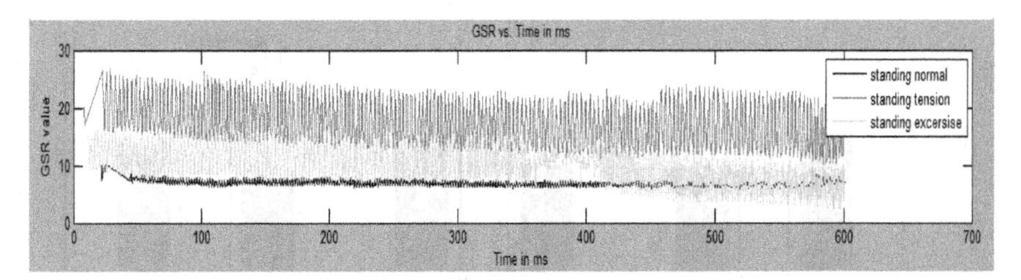

Fig. 7. GSR value vs. time in ms for standing position in different moods

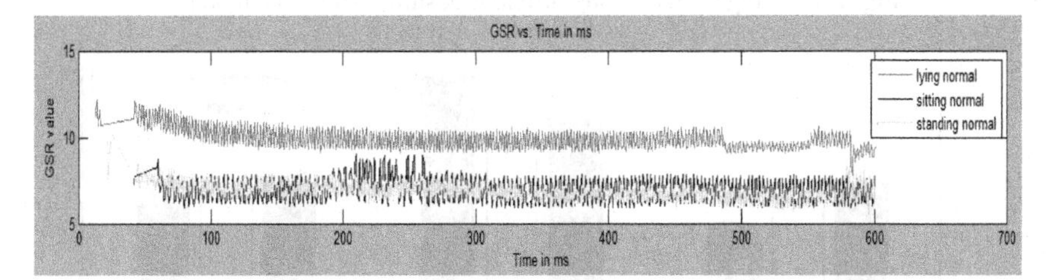

Fig. 8. GSR value vs. time in ms for different positions in normal mood

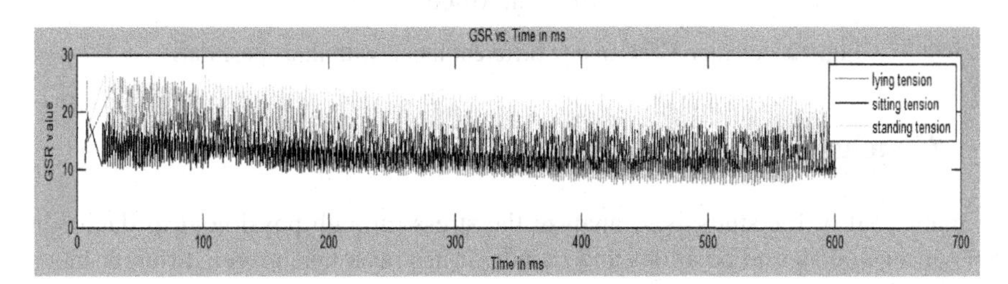

Fig. 9. GSR value vs. time in ms for different positions in tension mood

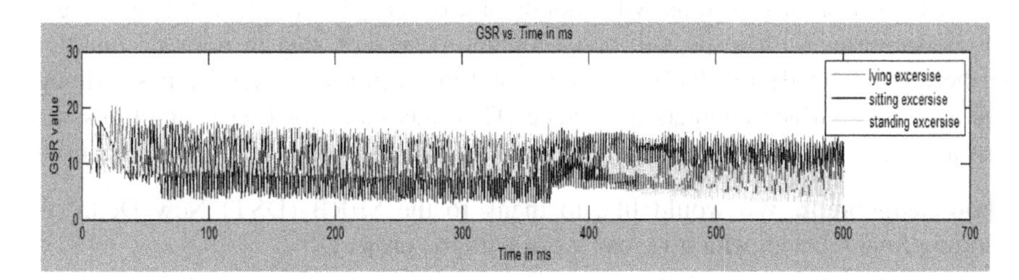

Fig. 10. GSR value vs. time in ms for different positions in exercise mood

From the Fig.11 and 12 it has been observed that in normal condition the average value of GSR is varies as per contact area of the body with surface area. But it is not true in other conditions like tension and exercise. In all cases GSR values are more in tension conditions of a person.

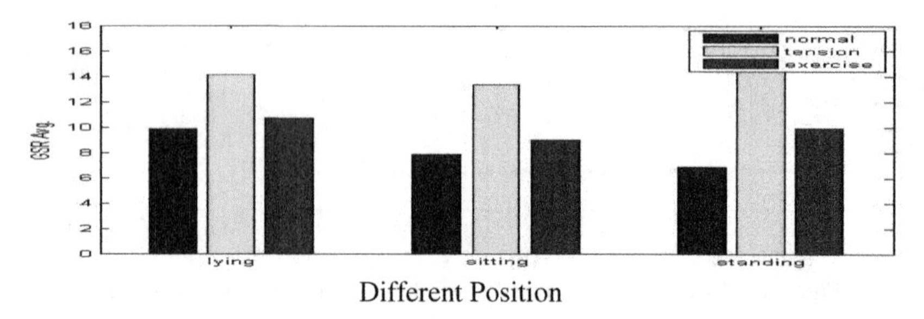

Fig. 11. Average GSR among different position with three moods

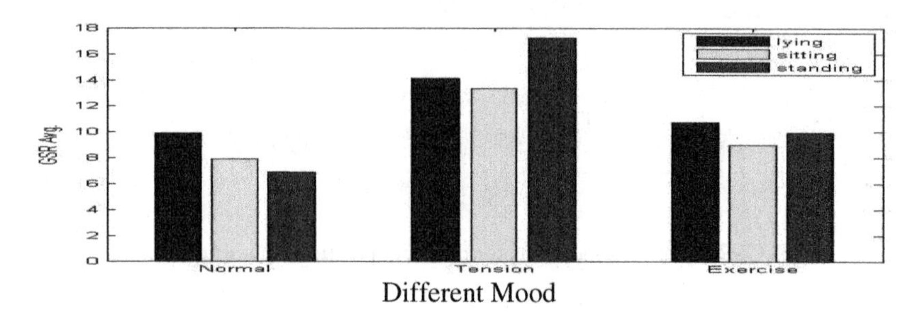

Fig. 12. Average GSR among different mood with three positions

6 Conclusion

The main goal of the study is to analyze the stress through physiological data using GSR sensor in different positions and moods. It has been taken three different moods like normal, tension and exercise. Three different positions like lying, sitting and standing have been consider for experiment. It has been calculated the average value of GSR for different positions with moods and observed that, the average value of GSR is decreases with respect to body contact to surface area in normal condition. But there is no steady results have been found in tension and exercise cases though there are same surface contacts to surface. This may cause of body conditions with stress at those times. This may analyze to state of mind/ stress in real life applications.

Acknowledgement. We would like to thank to the SERB (DST) New Delhi, for providing fund through which we can work with the project.

References

1. Glanz, K., Schwartz, M.: Stress, coping, and health behavior Health behavior and health education: Theory, research, and practice, pp. 211–236 (2008)
2. Vijaya, P.A., Shivakumar, G.: Galvanic Skin Response: A Physiological Sensor System for Affective Computing. International Journal of Machine Learning and Computing 3(1) (February 2013)

3. Juha, P., et al.: Relationship of Psychological and Physiological Variables in Long-term Self-monitored Data during Work Ability Rehabilitation Program. IEEE Transactions on Information Technology in Biomedicine 13(2) (March 2009)
4. Westerink, J.H.D.M., van den Broek, E.L., Schut, M.H., van Herk, J., Tuinenbreijer, K.: Computing Emotion Awareness Through Galvanic Skin Response and Facial Electromyography. In: Probing Experience Philips Research, vol. 8, pp. 149–162 (2008)
5. Juha, P., Miikka, E., Panu, K., Jani, M., Johannes, P., Ilkka, K.: Activity Classification Using Realistic Data From Wearable Sensors. IEEE Transactions on Information Technology in Biomedicine 10(1), 119–128 (2006)
6. Miikka, E., Juha, P., Panu, K., Jani, I.K.: Detection of Daily Activities and Sports With Wearable Sensors in Controlled and Uncontrolled Conditions. IEEE Transactions on Information Technology in Biomedicine 12(1), 20–26 (2008)
7. Juha, P., Luc, C., Miikka, E.: Personalization Algorithm for Real-Time Activity Recognition Using PDA, Wireless Motion Bands, and Binary Decision Tree. IEEE Transactions on Information Technology in Biomedicine 14(5), 1211–1215 (2010)
8. Juha, P., Miikka, E., Mark, G.: Automatic Feature Selection and Classification of Physical and Mental Load using Data from Wearable Sensors. In: Proceedings of the 10th IEEE International ITAB Conference Corfu, Greece, November 2-5 (2010)
9. Kurniawan, H., Maslov, A.V., Pechenizkiy, M.: Stress detection from speech and Galvanic Skin Response signals. In: IEEE 26th International Symposium on Computer-Based Medical Systems (CBMS), pp. 209–214 (2013)
10. Altini, M., Penders, J., Vullers, R., Amft, O.: Combining Wearable Accelerometer and Physiological Data for Activity and Energy Expenditure Estimation. In: Proceedings of the 4th Conference on Wireless Health, USA (2013)
11. de Santos Sierra, A., Avila, C.S., Bailador del Pozo, G., Guerra Casanova, J.: Stress detection by means of stress physiological template. In: Nature and Biologically Inspired Computing (NaBIC), pp. 131–136 (2011)
12. Fenz, W.D., Epstein, S.: Gradients of a Physiological Arousal of Experienced and Novice Parachutists as a Function of an Approaching Jump. Psychomatic Medicine 29, 33–51 (1967)
13. `http://www.cooking-hacks.com/documentation/tutorials/ehealth-biometric-sensor-platform-arduino-raspberry-pi-medical`
14. `http://arduino.cc/en/Main/arduinoBoardUno`

Low Power Affordable and Efficient Face Detection in the Presence of Various Noises and Blurring Effects on a Single-Board Computer

Steven Lawrence Fernandes and Josemin G. Bala

Dept. Electronics & Communication Engineering, Karunya University, Tamil Nadu, India
steva_fernandes@yahoo.com, josemin@karunya.edu

Abstract. Till today face detection is a burning topic for the researchers. In the areas like digital media, intelligent user interface, intelligent visual surveillance and interactive games. Various noises and blurring effects face images captured in real time. Single board computer for efficient face detection system is introduced in this paper which works well in the presence of Gaussian Noise, Salt & Pepper Noise, Motion Blur and Gaussian Blur. Raspberry Pi based single-board computer is used for the experiments, because it consumes less power and is available at an affordable price. The developed system is tested by introducing varying degree of noises and blurring effects on standard public face databases: GRIMACE, JAFEE, INDIAN FACE, CALTECH, FACE 95, FEI – 1, FEI – 2. In the absence of noise and blurring effects also the system is tested using standard public face databases: GRIMACE, JAFEE, INDIAN FACE, CALTECH, FACE 95, FEI – 1, FEI – 2, HEAD POSE IMAGE, SUBJECT, and FGNET. The key advantage of the proposed system is excellent face detection rates in the presence of noises, blurring effects and also in the presence of varying facial expressions and across age progressions. Python scripts are developed for the system, resulted are shared on request.

1 Introduction

Pixel color is taken directly in conventional face detection methods as information cues. For illumination and cue changes, these collected data are very sensitive [1-2]. These problems are addressed by many researches. They introduced transform features. The transform features convert the pixel color (or intensity) with the help of nonlinear transform function. These are classified into two groups. Among them, the first one is intensity based transform feature and the second one is gradient based transform feature. First one convert pixel color (or intensity) into and encoded value. It compares pixel value and neighboring pixels. Papageorgiou and Poggio [3] found out Haar like features. In this, two rectangular regions are taken and their average intensities are calculated. The Haar like feature encodes these differences between average intensities. Irrespective of pixel color (or the intensity) the Haar like feature is applied to extraction of texture. To detect faces, Harr like features are utilized by

S.C. Satapathy et al. (eds.), *Emerging ICT for Bridging the Future – Volume 1,*
Advances in Intelligent Systems and Computing 337, DOI: 10.1007/978-3-319-13728-5_13

Viola and Jones[4], [5]. Integral images [5] is been utilized by them to calculate Haar like feature. The integral images are also used to compute efficient scheme to construct a vigorous classifier. This is achieved by cascading several impuissant classifiers which utilizes AdaBoost training. Binary Haar like feature is proposed by Yen et al.[6] which keep only directional relationship for Haar feature computation. But for only robust classification Haar feature was too impotent. Based on general definition texture in local neighborhood Ojala et al [7] proposed local binary pattern(LBP). But in LBP, since there are many different patterns of local intensity, the sensitivity variation makes the training process more tedious in case of AdaBoost. Local gradient Patterns (LGP) are used to overcome this problem. It produces constant patterns, even though local intensity variations are present along edges. Similar to LBP is Cnesus Transform (CT), introduced by Zabin and Woodfill [8 -14]

In this paper, we made an effort to apply hybrid feature that combines local transform features: LBP, LGP, and BHOG by means of the AdaBoost feature selection method on single-board computer – Raspberry Pi. The main advantage of Raspberry Pi based single-board computer is that it consumes less power and it is available at an affordable price. With a total cost of less than $200, the system can be easily built; the system consumes only 3 watts of power for 17 pages/sec. In the proposed system we have used local transform feature among several local transform features specifically, LBP, LGP, and BHOG having the lowest classification error is sequentially selected until we obtain the required classification performance. The developed system is tested with different facial expressions, varying poses, across age progressions and in presence of various noises: Gaussian Noise, Salt & Pepper Noise and various blurring effects: Motion blur, Gaussian blur. Standard public database used are: GRIMACE, JAFEE, INDIAN FACE, CALTECH, FACE 95, FEI – 1, FEI – 2, HEAD POSE IMAGE, SUBJECT, FGNET.

2 Single Board Computer-RASPBERRY PI

In educational institutes, to give hands on experience, Raspberry Pi was invented by Raspberry foundation [15-30]. It is a single board computer with two models. One model consist of 256MB RAM, single USB port. It doesn't contain any network connection. The second model has 512 MB RAM and two USB port. The second model also has Ethernet port. We have used second model. It contains Broadcom BCM 2835 system on a chip. This chip includes one ARM1176JZ-F700 MHz low processor. The GPU on the board is Broadcom BCM2835 system on a chip which includes an ARM 1176JZ-176JZ-F 700 MHz low processor. The board contains 3.5mm HDMI audio output. It supports SC, MMC, SDIO card slots. Blue ray quality playback can be obtained by GPU using H.264 at 40MBits/s.

3 Local Binary Patterns (LBP)

Original LBP operators are used to label pixels along with decimal numbers. The labelled pixels are called LBPs or as LBP codes. They include other pixels around the selected pixel, encode them and form as local structure. Around the selected pixel, 8 neighbors are selected in a 3x3 matrix form. The center pixel value is deducted with the value that of neighbor pixel values. The results are verified. The pixel negative value is encoded as zero. Others are encoded as one. Now, all the encoded values are concatenated in clockwise fashion. The starting point is left top most neighbor. The result so obtained (binary number) will be taken as label of the center pixel. Like this the derived binary numbers of the region is taken as LBPs or LBP codes.

4 Local Gradient Patterns (LGP)

The eight neighbors gradient values are used by LGP operator. The average gradient value of the neighbors is calculated and are substituted to the given pixel. It is used as threshold value for LBP encoding. The neighbor pixels gradient value is checked with the current pixels gradient value. If the neighbors gradient value is greater than the current then it is assigned a value 1 or else 0 is assigned. The LBP code will be concatenation of binary 1s and 0s. This will form as binary code. This method can be utilized for different sized neighbor hoods. As an example consider a circle with radius r. Let it take p sampling points.

5 Binary Histogram of Oriented Gradients (BHOG)

Following are the steps to generate BHOG features. Take the square of the gradient magnitude. Square it. Identify the block, take the orientation of all the pixels in the block. Square it. In the same now build the orientation histogram HOG(b). Encode the orientation histogram in to 8 bit vector. Here each bit is determined by thresholding. Now if this thresholding is greater than the given threshold, assign 1 bit; else assign 0 bit. Compared to HOG, there are several advantages of BHOG. In BHOG, to calculate gradient magnitude, we need not have to calculate square root. The reason for this is, it just compares the threshold and value of histogram bin. Next is, an essential step in original HOG, normalization of orientation histogram, is not present in BHOG. Instead, it just needs relative comparison of given threshold value and value of histogram bin. By AdaBoost training, BHOG feature can be obtained. This is because the BHOG feature can be represented as scalar value of one dimensional.

6 Hybridized Local Transform Features (HLTF) for Face Detection on Single Board Computer – Raspberry Pi Model B

The HTLF algorithm is as shown below:
Step 1: Start
Step 2: Prepare the positive and negative training image
Step 3: For this positive and negative training image, calculate weight values.
Step 4: For LBP, LGP and BHOG local transform features, obtain positive, negative training feature images.

Step 5: For all feature images, calculate classification errors.
Step 6: The transform feature which has minimum classification error need to be selected as best local transform feature.
Step 7: Large weight values are given to the training images that are incorrectly classified by selected feature. Small weight values are given to the training images that are correctly classified as selected feature in subsequent iterations. This update is done for all training images.
Step 8: Share the weight values among LBP, LGP and BHOG features on Respberry PI Model B in order to prevent reselection of previously selected feature by other feature types.
Step 9: Stop.

7 Results and Discussions

We have considered 7 standard databases to test HLTF on a single board computer – Raspberry Pi. We have introduced various noises: Gaussian noise, Salt & Pepper noise to all the 6 standard databases: GRIMACE, JAFFE, INDIAN FACE, CALTECH, FACE 95, FEI – 1, FEI – 2. This process is followed by introducing blurring effect: Motion blur and Gaussian blur to same 7 standard databases. Varying degree of noises and blurring effects were added to face images using paint.net tool. The values are tabulated in TABLE 1 for GRIMACE, JAFFE, and INDIAN FACE databases and in TABLE 2 for CALTECH, FACE 95, FEI – 1, and FEI – 2 databases. The noise value will indicate the amount of noise added to the face image. Blur value indicate the amount of blurriness added to the face image.

Table 1. Details of various noises and blurring effects added to face images for GRIMACE, JAFEE and Indian face databases

DATABASE	GRIMACE	JAFFE	INDIAN FACE
Gaussian noise values	50,55,60,65,70	50,55,60,65,70	50,55,60,65,70
Salt & Pepper noise values	50,60,70,80,90	50,60,70,80,90	50,60,70,80,90
Motion blur values	10,20,30,40,50	10,20,30,40,50	10,20,30,40,50
Gaussian blur values	5,10,15,20,25	5,10,15,20,25	5,10,15,20,25

Table 2. Details of various noises and blurring effects added to face images for CALTECH, FACE 95, FEI – 1, FEI - 2 databases

DATABASE	CALTECH	FACE 95	FEI - 1	FEI - 2
Gaussian noise values	50,55,60,65,70	50,55,60,65,70	50,55,60,65,70	50,55,60,65,70
Salt & Pepper noise values	50,60,70,80,90	50,60,70,80,90	50,60,70,80,90	50,60,70,80,90
Motion blur values	10,20,30,40,50	10,20,30,40,50	10,20,30,40,50	10,20,30,40,50
Gaussian blur values	5,10,15,20,25	5,10,15,20,25	5,10,15,20,25	5,10,15,20,25

FRR = (Total number of face images recognized accurately / Total number of face images available in the database) x 100. Face detection rate obtained using Hybridized Local Transform Features (HLTF) algorithm on Single Board Computer – Raspberry PI Model B in the presence Gaussian Noise, Salt and Pepper Noise, Motion Blur, Gaussian Blur, Facial Expression variations, Pose variations and Age variations are tabulated from Table 3 to TABLE 9.

Table 3. Face detection rate in the presence of Gaussian noise

DATABASE	Total no. of images in the database	Total no. of face images detected appropriately	Total no. of images detected as face but not faces	Total no. of faces not detected	Face detection rate
GRIMACE	18*4=72	59	00	13	81.94
JAFFE	10*4=40	40	00	00	100
INDIAN FACE	30*5=150	111	00	39	74
CALTECH	19*4=76	74	00	02	97.36
FACE 95	72*5=360	352	00	08	97.77
FEI - 1	100*5=500	403	00	97	80.6
FEI - 2	100*5=500	421	00	79	84.2

Table 4. Face detection rate in the presence of motion blur

DATABASE	Total no. of images in the database	Total no. of face images detected appropriately	Total no. of images detected as face but not faces	Total no. of faces not detected	Face detection rate
GRIMACE	18*4=72	61	00	11	84.72
JAFFE	10*4=40	40	00	00	100
INDIAN FACE	30*5=150	110	00	40	73.33
CALTECH	19*4=76	74	00	02	97.36
FACE 95	72*5=360	357	00	03	99.16
FEI - 1	100*5=500	440	00	60	88
FEI - 2	100*5=500	459	00	41	91.80

Table 5. Face detection rate in the presence of Gaussian blur

DATABASE	Total no. of images in the database	Total no. of face images detected appropriately	Total no. of images detected as face but not faces	Total no. of faces not detected	Face detection rate
GRIMACE	18*4=72	59	00	13	81.94
JAFFE	10*4=40	40	00	00	100
INDIAN FACE	30*5=150	102	00	48	68
CALTECH	19*4=76	74	00	02	97.36
FACE - 95	72*5=360	340	00	20	94.44
FEI - 1	100*5=500	341	00	159	68.20
FEI - 2	100*5=500	365	00	135	73

Table 6. Face detection rate in the presence of facial expression variations

DATABASE	Total no. of images in the database	Total no. of face images detected appropriately	Total no. of images detected as face but not faces	Total no. of faces not detected	Face detection rate
FACE 94	19*20+113*20+20*20= 3040	3020	01	19	99.34
FACE 95	72*20=1440	1415	00	25	98.26
FACE 96	147*20=2940	2895	00	45	98.46
JAFFE	10*20=200	200	00	00	100
GEORGIA TECH	50*15=750	718	05	27	95.73
GRIMACE	18*20=360	352	07	01	97.77
INDIAN FACE	22*11+ 39* 6= 476	356	00	120	74.78
CALTECH	19*18=342	342	00	00	100
BRAZILLIAN FEI	100*2+100*2=400	394	00	06	98.50

Table 7. Face detection rate in the presence of pose variations

DATABASE	Total no. of images in the database	Total no. of face images detected appropriately	Total no. of images detected as face but not faces	Total no. of faces not detected	Face detecti on rate
HEAD POSE IMAGE	13*186=2418	761	00	1657	31.47
SUBJECT	10*74=740	352	10	378	47.56

Table 8. Face detection rate in the presence of age variations

DATABASE	Total no. of images in the database	Total no. of face images detected appropriately	Total no. of images detected as face but not faces	Total no. of faces not detected	Face detection rate
FGNET	1002	921	00	81	91.91

8 Conclusion

This paper introduces a single board computer - Raspberry Pi for efficient face detection system in the presence of Gaussian Noise, Salt & Pepper Noise, Motion Blur and Gaussian Blur. We apply hybrid feature that combines local transform features: LBP, LGP, BHOG by means of the AdaBoost feature selection method on single-board computer – Raspberry Pi because it consumes less power and it is available at an affordable price. The developed system is tested by introducing varying degree of noises and blurring effects on standard public face databases: GRIMACE, JAFEE, INDIAN FACE, CALTECH, FACE 95, FEI – 1, FEI – 2. The developed system is also tested in the absence of noise and blurring effects using standard public face databases: GRIMACE, JAFEE, INDIAN FACE, CALTECH, FACE 95, FEI – 1, FEI – 2, HEAD POSE IMAGE, SUBJECT, and FGNET. Thus, the key advantage of the proposed system is excellent face detection rates in the presence of noises, blurring effects and also in the presence of varying facial expressions and across age progressions

References

1. Togashi, Haruo and Kanagawa, Face recognition apparatus, face recognition method, gabor filter application apparatus, and computer program. U. S. Patent 8077932 (December 13, 2012)
2. Li, S., Gong, D., Liu, H.: Face identification method based on multiscale weber local descriptor and kernel group sparse representation. C. N. Patent 102722699A (October 10, 2010)

3. Papageorgiou, C., Poggio, T.: A Trainable System for Object Detection. Int'l J. Computer Vision 38(1), 15–33 (2000)
4. Viola, P., Jones, M.: Robust Real-Time Face Detection. Int'l J. Computer Vision 57(2), 137–154 (2004)
5. Viola, P., Jones, M., Snow, D.: Detecting Pedestrians Using Patterns of Motion and Appearance. Int'l J. Computer Vision 63(2), 153–161 (2005)
6. Yan, X.C.S., Shan, S., Gao, W.: Locally Assembled Binary (LAB) Feature with Feature-Centric Cascade for Fast and Accurate Face Detection. In: Proc. IEEE Conf. Computer Vision and Pattern Recognition, pp. 1–7 (2008)
7. Ojala, T., Pietikainen, M., Harwood, D.: A Comparative Study of Texture Measures with Classification Based on Feature Distributions. Pattern Recognition 29(1), 51–59 (1996)
8. Zabih, R., Woodfill, J.: Non-Parametric Local Transforms for Computing Visual Correspondence. In: Eklundh, J.-O. (ed.) ECCV 1994. LNCS, vol. 801, pp. 151–158. Springer, Heidelberg (1994)
9. Zhang, L., Chu, R., Xiang, S., Liao, S., Li, S.Z.: Face Detection Based on Multi-Block LBP Representation. In: Lee, S.-W., Li, S.Z. (eds.) ICB 2007. LNCS, vol. 4642, pp. 11–18. Springer, Heidelberg (2007)
10. Lowe, D.: Distinctive Image Features from Scale Invariant Keypoints. Int'l J. Computer Vision 60(2), 91–110 (2004)
11. Ke, Y., Sukthankar, R.: PCA-Sift: A More Distinctive Representation for Local Image Descriptors. In: Proc. IEEE Conf. Computer Vision and Pattern Recognition, pp. 511–517 (2004)
12. Bay, H., Ess, A., Tuytelaars, T., Gool, L.: Surf: Speeded Up Robust Features. Computer Vision and Image Understanding 110(3), 346–359 (2008)
13. Dalal, N., Triggs, B.: Histograms of Oriented Gradients for Human Detection. In: Proc. IEEE Conf. Computer Vision and Pattern Recognition, pp. 886–893 (2005)
14. Zhu, Q., Avidan, S., Yeh, M., Cheng, K.: Fast Human Detection Using a Cascade of Histograms of Oriented Gradients. In: Proc. IEEE Conf. Computer Vision and Pattern Recognition, pp. 1491–1498 (2006)
15. T. R. P. Foundation, About us, `http://www.raspberrypi.org/about` (Online; accessed June 2013)
16. Arcia-Moret, A., Pietrosemoli, E., Zennaro, M.: WhispPi: White space monitoring with Raspberry Pi. In: Global Information Infrastructure Symposium, pp. 28–31 (2013)
17. Ali, M., Vlaskamp, J., Eddin, N., Falconer, B., Oram, C.: Technical development and socioeconomic implications of the Raspberry Pi as a learning tool in developing countries. In: 5th Computer Science and Electronic Engineering Conference (2013)
18. Nagy, T., Gingl, Z.: Low-cost photoplethysmograph solutions using the Raspberry Pi. In: IEEE 14th International Symposium on Computational Intelligence and Informatics (2013)
19. Milenkovic, A.M., Markovic, I.M., Jankovic, D.S., Rajkovic, P.J.: Using of Raspberry Pi for data acquisition from biochemical analyzers. In: 11th International Conference on Telecommunication in Modern Satellite, Cable and Broadcasting Services (2013)
20. Edwards, C.: Not-so-humble raspberry pi gets big ideas. Engineering & Technology 8(3), 30–33 (2013)
21. Mitchell, G.: The Raspberry Pi single-board computer will revolutionise computer science teaching (For & Against). Engineering & Technology 7(3), 26 (2012)
22. Batur, A.U., Hayes, M.H.: Adaptive active appearance models. IEEE Trans. Image Process. 14(11), 1707–1721 (2005)
23. Edwards, C.: ICT lessons get the raspberry. Engineering & Technology 7(4), 76–78 (2012)

24. Tso, F.P., White, D.R., Jouet, S., Singer, J., Pezaros, D.P.: The Glasgow Raspberry Pi Cloud: A Scale Model for Cloud Computing Infrastructures. In: IEEE 33rd International Conference on Distributed Computing Systems Workshops, pp. 108–112 (2013)
25. Sundaram, G.S., Patibandala, B., Gaddam, S., Alla, V.K.: Bluetooth communication using a touchscreen interface with the Raspberry Pi. In: Proceedings of IEEE Southeastcon, pp. 1–4 (2013)
26. Ionescu, V.M., Smaranda, F., Diaconu, A.-V.: Control system for video advertising based on Raspberry Pi. In: RoEduNet International Conference 12th Edition Networking in Education and Research, pp. 1–4 (2013)
27. Abrahamsson, P., Helmer, S., Phaphoom, N.: Affordable and Energy-Efficient Cloud Computing Clusters: The Bolzano Raspberry Pi Cloud Cluster Experiment. In: IEEE 5th International Conference on Cloud Computing Technology and Science, pp. 170–175 (2013)
28. Jain, S., Vaibhav, A., Goyal, L.: Raspberry Pi based interactive home automation system through E-mail. In: International Conference on Optimization, Reliabilty, and Information Technology, pp. 277–280 (2014)
29. Banerjee, S., Sethia, D., Mittal, T.: Secure sensor node with Raspberry Pi. In: International Conference on Multimedia, Signal Processing and Communication Technologies, pp. 26–30 (2013)
30. Shams, E., Thiessen, J.D., Bishop, D.: A PET detector interface board and slow control system based on the Raspberry Pi, pp. 1–3 (2013)

The Computational Analysis of Protein – Ligand Docking with Diverse Genetic Algorithm Parameters

S.V.G. Reddy[1,*], K. Thammi Reddy[1], and V. Valli Kumari[2]

[1] Department of CSE, GIT, GITAM University, Visakhapatnam
venkat157.reddy@gmail.com, thammireddy@yahoo.com
[2] Department of CS & SE, College of Engineering, Andhra University, Visakhapatnam
vallikumari@gmail.com

Abstract. The binding energy is the significant factor which elucidates the efficiency of docking between protein-ligand and protein-protein. To perform the Docking process, Genetic algorithm with its standard parameters is been used very often which will furnish the docking conformations, binding energies, interactions etc. In this proposed work, the parameters of genetic algorithm are variably changed for the docking process and we have observed the enhancement of binding energy, number of interactions etc. which plays a substantial role in the drug design.

Keywords: Genetic Algorithm, parameters, protein-ligand, Docking, binding energy, drug design.

1 Introduction

To perform the Docking process, the drug target neuraminidase (3B7E) and its ligand Zanamivir is taken from the protein data bank [1] (In fact any drug target and a ligand may be picked for docking purpose) . The discovery studio [2] is used to separate the drug target and the ligand and to generate the respective .pdb files.

2 Methods – Docking Process

The Auto dock tools (MGL tools) is the software which is used to perform the docking [3][4][5] between the drug target and the ligand. The docking process will be done in the following steps. The preprocessing is done for the ligand and the drug target preparation.

1) Ligand preparation 2) Drug target preparation 3) Grid Preparation 4) Docking preparation 5) Docking Execution

* Corresponding author.

S.C. Satapathy et al. (eds.), *Emerging ICT for Bridging the Future – Volume 1*,
Advances in Intelligent Systems and Computing 337, DOI: 10.1007/978-3-319-13728-5_14

Ligand Preparation (Preprocessing): The ligand Zanamivir is chosen for which the torsions are applied after tracing the root and sets the aromatic carbons .

Drug Target Preparation (Preprocessing): The drug target neuraminidase (3B7E) is chosen for which the necessary bonds are added of different types. Next, the polar, non polar, histidine hydrogens are added. The kollman and gasteiger charges are added, checks the totals on residues and sets the charge field . Lastly, the atoms are assigned with radii, edit their type and AD4 type is assigned.

Grid Preparation: Here both the ligand Zanamivir and the drug target neuraminidase (3B7E) is chosen and the grid box is generated to perform the docking.

Docking Preparation: Here both the ligand Zanamivir and the drug target neuraminidase (3B7E) is chosen and the genetic algorithm [6][7] is used to perform the docking with some standard parameters [8][9] which are furnished below.

The following are the standard parameters taken and their default values for the genetic algorithm

- Number of GA runs – 10
- Population size – 150
- Maximum number of evaluations
- Medium – 25,00,000
- Maximum number of generations – 27000
- Maximum number of top individuals that automatically survive – 1
- Rate of gene mutation – 0.02
- Rate of crossover – 0.8
- GA crossover mode – Twopt
- Mean of Cauchy distribution for gene mutation – 0.0
- Variance of Cauchy distribution for gene mutation – 1.0
- Number of generations for picking worst individual – 10

The docking process can be executed with the same values of standard parameters as above which results to the docking conformations. But the experimenting is done by changing the values of genetic algorithm standard parameters and perform the docking process.

The parameters changed are 1)Number of GA runs, 2)Population size, 3)Maximum number of evaluations, 8)GA crossover mode. The options for 3)Maximum number of evaluations are medium 25,00,000 , Short – 2,50,000 Long – 250,00,000. And the options for 8)GA crossover mode are twopt, Arithmetic, uniform.

The docking is done by changing the values of genetic algorithm standard parameters randomly which lead to the enhancement of the binding energy and hydrogen interactions.

Docking Execution: First the Autogrid process is run to build the Grid using autogrid4.exe and then the Autodock process is run to perform the docking using autodock4.exe.

3 Results and Discussion

The ligand Zanamivir and the drug target neuraminidase (3B7E) is considered for docking and the complete docking process is applied five times. Few values of standard parameters of genetic algorithm are changed and considered as SET1, few other values of standard parameters are changed and considered as SET2 and likewise SET3, SET4, SET5. The details of the genetic algorithm parameters used for SET1, SET2, SET3, SET4, SET5 are mentioned below.

SET1 is standard parameters with number of GA runs to 15, population size to 200, maximum number of evaluations set to short i.e. 2,50,000 , GA crossover mode as Twopt

SET2 is standard parameters with maximum number of evaluations set to medium i.e. 25,00,000 and GA crossover mode as twopt

SET3 is standard parameters with maximum number of evaluations set to medium i.e. 25,00,000 and GA crossover mode as uniform

SET4 is standard parameters with maximum number of evaluations set to long i.e. 250,00,000 and GA crossover mode as Twopt

SET5 is standard parameters with number of GA runs to 15, population size to 200, maximum number of evaluations set to medium i.e. 25,00,000 , GA crossover mode as twopt

The docking process is performed for the first time using SET1, and the docking process is applied for the second time using SET2, third time using SET3, fourth time using SET4, fifth time using SET5. After the docking process, for each and every SET, we got 10 best conformations (docking poses) and for each and every conformation the corresponding binding energies and hydrogen bond interactions are recorded. The binding energies of all the conformations of all SETs are furnished in Table 1.

Table 1. Binding Energies of the best docking conformations using various SETs

SET	1	2	3	4	5	6	7	8	9	10
SET1	-8.53	-8.37	-7.55	-7.51	-7.08	-6.62	-6.18	-5.82	-5.78	-5.73
SET2	-9.36	-9.34	-8.54	-7.86	-7.57	-7.33	-7.29	-7.17	-7.02	-6.37
SET3	-9.4	-9.28	-9.07	-8.74	-8.46	-8.14	-7.56	-7.47	-7.33	-7.15
SET4	-9.52	-9.5	-9.48	-8.99	-8.87	-7.6	-7.58	-7.54	-7.47	-6.73
SET5	-9.59	-9.33	-9.15	-9.11	-8.9	-8.62	-8.56	-8.55	-7.71	-7.47

When we Analyze the results, the docking with the SET5 lead to high value of binding energy [10] of -9.59 (the negative high value is superior E.g. -14 is superior when compared to -10) and more number of hydrogen bond interactions (11) for the first conformation, and while docking with the SET1 lead to low value of binding energy of -8.53 and less number of hydrogen bond interactions (9) for the first conformation. The binding energies for the first conformation of SET1, SET2, SET3, SET4, SET5 are -8.53, -9.36, -9.4, -9.52, -9.59 and the graph is generated as in Fig 1.

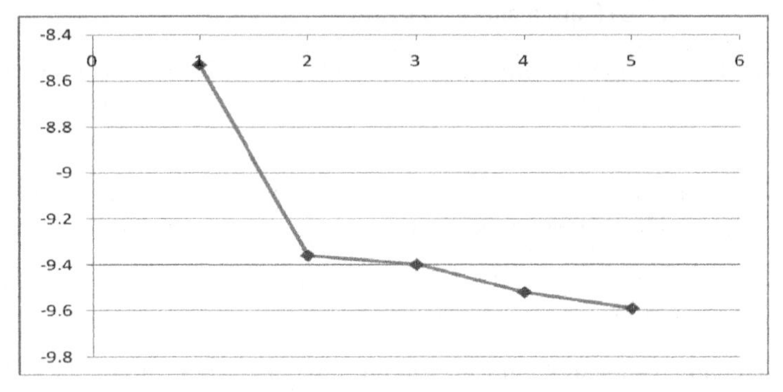

Fig. 1. Binding Energy Graph for all the SETs

The docking with SET5 of first conformation furnishes us the docking attributes and the Hydrogen bonds interactions (11) between the protein (3B7E) – ligand (Zanamivir) which are showed in Table 2 and Table 3 respectively.

Table 2. Docking attributes of 1st conformation (SET 5)

Docking Attribute	value
Binding energy	-9.59
Ligand _efficiency	-0.42
Inhib_constant	94.2
Inhib_constant_units	nM
Intermol_energy	-12.27
Vdw_hb_desolv_energy	-8.32
Electrostatic_energy	-3.95
Total_internal	-3.36
Torsional_energy	2.68
Unbound_energy	-3.36
Filename	P5.dlg
clRMS	0.0
refRMS	0.92
rseed1	None
rseed2	None

Table 3. Hydrogen bond interactions (11) between ligand and protein of 1st conformation (SET 5)

Sno	Hydrogen bond interaction
1	I5 : A : ZMR1001 : H21 - P5 : A : TRP178 : O
2	P5 : A : ARG118 : HH12 - I5 : A : ZMR1001 : O1A
3	P5 : A : ARG152 : HH11 - I5 : A : ZMR1001 : O10
4	I5 : A : ZMR1001 : H12 - P5 : A : GLU276 : OE1
5	I5 : A : ZMR1001 : H20 - P5 : A : ASP151 : O
6	I5 : A : ZMR1001 : H15 - P5 : A : GLU277 : OE2
7	I5 : A : ZMR1001 : H16 - P5 : A : ASP151 : OD1
8	P5 : A : ARG371 : HH12 - I5 : A : ZMR1001 : O1A
9	P5 : A : ARG371 : HH22 - I5 : A : ZMR1001 : O1B
10	P5 : A : ARG292 : HH22 - I5 : A : ZMR1001 : O1B
11	P5 : A : ARG292 : HH12 - I5 : A : ZMR1001 : O1B

4 Protein-Ligand Interactions

The Hydrogen bonds interactions between the ligand(Zanamivir) and protein (3B7E) is showed in Fig 2, 3. For example, the first hydrogen bond interaction is ZMR1001: H21 of ligand and TRP178 : O of protein(3B7E) which can be seen with Red circle in Fig 2,3. Likewise the other hydrogen bond interactions also can be observed.

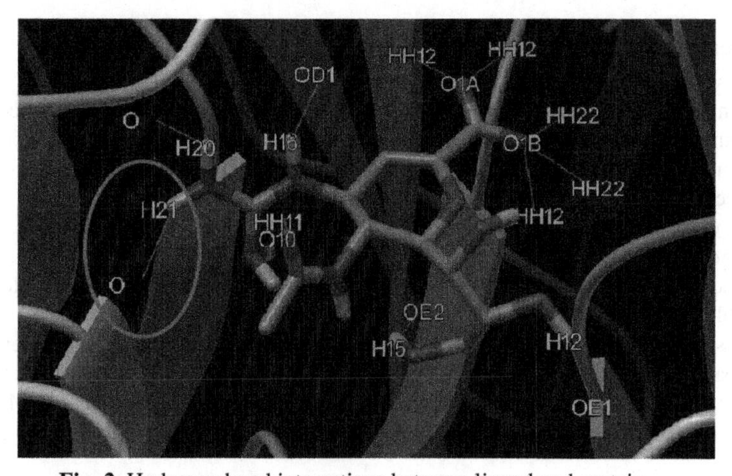

Fig. 2. Hydrogen bond interactions between ligand and protein

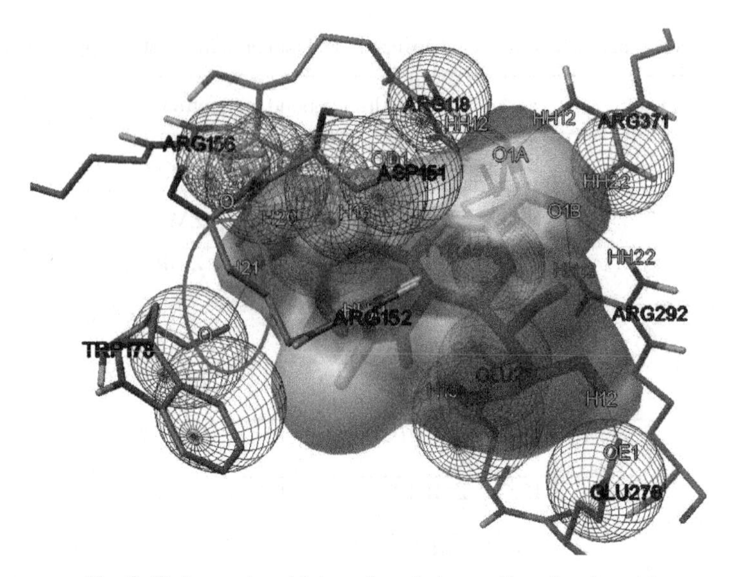

Fig. 3. Hydrogen bond interactions between ligand and protein

5 Conclusion

The docking process is performed using genetic algorithm with standard parameters by the Autodock tools (MGL tools). In this proposed work, the Docking experiment is carried out by changing the values of genetic algorithm standard parameters and achieved good results of binding energy and hydrogen bond interactions between the ligand (Zanamivir) and protein (3B7E). We may use few Optimization techniques for the parameters to have the optimum binding energy and hydrogen bond interactions for a particular protein and ligand in the docking process which is very significant in the drug design.

References

1. Bernstein, F.C., Koetzle, T.F., Williams, G.J., Meyer Jr., E.E., Brice, M.D., Rodgers, J.R., Kennard, O., Shimanouchi, T., Tasumi, M.: The Protein Data Bank: A Computer-based Archival File For Macromolecular Structures. J. of. Mol. Biol. 112, 535 (1977)
2. Accelrys Software Inc. Discovery studio Visualizer 3.0 (2011)
3. http://www.scribd.com/doc/61267608/UsingAutoDock4forVirtualScreening-v4
4. Goodsell, D.S., Morris, G.M., Olson, A.J.: Automated docking of flexible ligands. Applications of AutoDock. J. Mol. Recognit. 9, 1–5 (1996)
5. Friesner, R.A., Banks, J.L., Murphy, R.B., Halgren, T.A., Klicic, J.J., Mainz, D.T., Repasky, M.P., Knoll, E.H., Shaw, D.E., Shelley, M., Perry, J.K., Francis, P., Shenkin, P.S.: Glide: A New Approach for Rapid, Accurate Docking and Scoring. 1. Method and Assessment of Docking Accuracy. J. Med. Chem. 47, 1739–1749 (2004)

6. Morris, G.M., Goodsell, D.S., Halliday, R.S., Huey, R., Hart, W.E., et al.: Automated docking using a Lamarckian genetic algorithm and an empirical binding free energy function. J. Comput. Chem. 19, 1639–1662 (1998)
7. Minaei-Bidgoli, B., Punch, W.: Using genetic Algorithms for data mining optimization in an educational web-based system. In: Cantú-Paz, E., Foster, J.A., Deb, K., Davis, L., Roy, R., O'Reilly, U.-M., Beyer, H.-G., Kendall, G., Wilson, S.W., Harman, M., Wegener, J., Dasgupta, D., Potter, M.A., Schultz, A., Dowsland, K.A., Jonoska, N., Miller, J., Standish, R.K. (eds.) GECCO 2003. LNCS, vol. 2724, pp. 2252–2263. Springer, Heidelberg (2003)
8. DeJong, K.A., Spears, W.M.: An Analysis of the Interacting Roles of Population Size and Crossover in Genetic Algorithms. In: Schwefel, H.-P., Männer, R. (eds.) PPSN 1990. LNCS, vol. 496, pp. 38–47. Springer, Heidelberg (1991)
9. Sarmady, S.: An Investigation on Genetic Algorithm Parameters, School of Computer Science. Universiti Sains Malaysia (2007)
10. Rao, M.S., Olson, A.J.: Modelling of factor Xa-inhibitor complexes: a computational flexible docking approach. Proteins 34, 173–183 (1999)

An Auto Exposure Algorithm Using Mean Value Based on Secant Method

V. Krishna Sameera, B. Ravi Kiran, K.V.S.V.N. Raju, and B. Chakradhar Rao

Anil Neerukonda Institute of Technology and Sciences
Visakhapatnam, Andhra Pradesh, India
{krishnasameerav,brk2007}@gmail.com

Abstract. Auto exposure is the fundamental feature or property for the scenes to be exposed properly. This paper proposes an automatic exposure algorithm for well exposure. This algorithm uses one of the numerical methods, secant method. The correct exposure values are determined using center weighted average metering technique in which the center of the scene is mainly considered. The exposure values are based on the shutter speed and the gain. At a particular range of the mean the scene is said to be properly exposed. The implementation of this algorithm is done using point grey research programmable camera.

Keywords: Auto exposure, center weighted average technique, secant.

1 Introduction

Auto exposure algorithms are widely used now-a-days in all digital cameras. Three categories of exposure are under exposure, proper exposure, and over exposure. As the proper exposure is the major concern in photography these automatic exposure algorithms play major role. There will be different lighting conditions where the image has to be captured. According to the situation the image must be exposed properly. The factors that can make the scene properly exposed are shutter speed and ISO gain. When the image has to be captured in a darker region the factors, the shutter speed has to be increased and the gain also has to be increased to make the scene well exposed. In the same way when he image has to be captured in a brighter region the factors, the shutter speed has to be decreased and the gain also has to be decreased to make the scene well exposed. The adjustment of these factors is performed in several ways as in [1].

There are different methods that are implemented to obtain auto exposure. One of them is numerical method like false position, bisection method are implemented. This paper is implemented using secant method and mean is calculated using center weighted average metering technique as explained in the next section.

2 Related Work

According to Tetsuya Kuno and Narihiro Matoba [4] for a scene a sampling area has taken and given as input to the camera processor, which calculates an integrated value

S.C. Satapathy et al. (eds.), *Emerging ICT for Bridging the Future – Volume 1*,
Advances in Intelligent Systems and Computing 337, DOI: 10.1007/978-3-319-13728-5_15

for the imaging signal. This integrated value is divided by the ISO and shutter speed to obtain the brightness of the scene. The optimum brightness is obtained by varying these shutter speed and ISO and these values are maintained in the lookup table. The drawback of this method is the storage and maintenance of this look up tables.

The auto exposure algorithm is implemented using bisection method as explained below. The disadvantage is that, this method takes more number of iterations.

2.1 Bisection Method

The Bisection method [2][3] is the basic method which is initiated by identifying the initial roots and calculating the bisect value or the mid value of them and by repeating the process of bisecting with new approximation and the initial root basing on the function value of the first approximation. Similarly in auto exposure the initial exposure value is identified and verifies the type of exposure basing the on the mean luminance value of the scene. The new approximation is obtained by bisecting the previous approximation [2] with the initial root as shown in figure 1. This process is repeated until proper exposure is obtained.

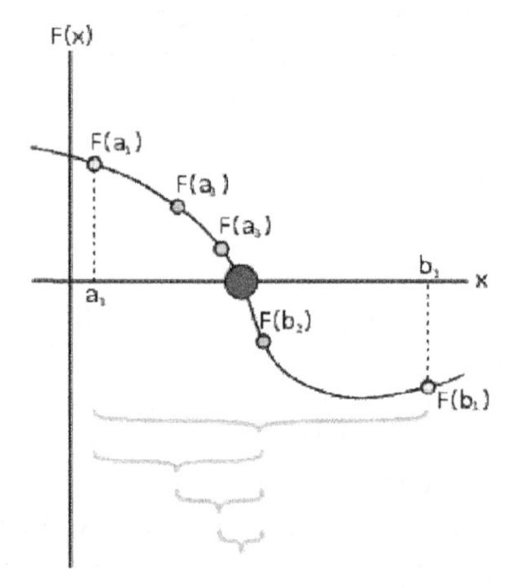

Fig. 1. Diagrammatical representation of Bisection method [8]

The advantage of the bisection method is its reliability, but the disadvantage is it takes number iterations to bisect to the optimum value, as the interval to bisect will decrease so it takes more amount of time to get optimum exposure value. To overcome this disadvantage the secant method is used, as it finds the optimum value faster than the Bisection method. The false position method [5] has the problem of slow divergence of the root. This problem is also solved by the secant method [7] as the initial values are considered so it will reliably leads to the optimum exposure value. Thus the secant method is fast and reliable to find the exposure values.

Metering is a technique how the camera determines the properties of exposure such as shutter speed, aperture based on the light [9]. There are different metering modes. These modes allow control what part of a scene the meter will take its reading from a useful capability when it comes to getting good exposure in different lighting situations. Center weighted metering mode [9] is a type of averaging meter where the focus is made for the center of an image. This mode is adopted in our algorithm to calculate mean for a better exposure control.

3 Secant Method

The secant method [6] [7] is a numerical analysis method, which is used to find the root for a non-linear equation (curve). The root of the equation is a point where the curve cuts the x-axis when y equals 0. This method is used in the proposed algorithm.

In the secant method, the next approximation value is the intersection point of the X-axis and the straight line formed by the initial values. By iteratively performing the same process we will finally end up with the root values, as shown in the Figure2.

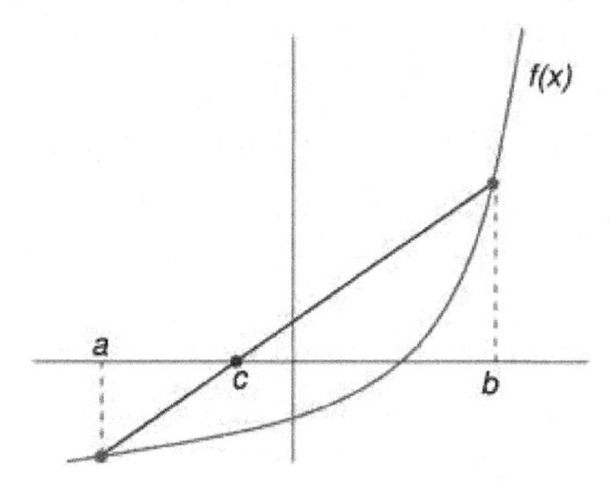

Fig. 2. Diagrammatical representation of Secant method [6][7]

The secant method chooses the point c to be the point where the straight line between $\mathbf{A}(a_i, f(a_i))$ and $B(b_i, f(b_i))$ cuts the x axis.

The equation of the line is

$$Y = f(b_i) + ((x - b_i) * \frac{f(b_i)-f(a_i)}{(b_i-a_i)}) \quad (1)$$

So on solving (1) by substituting y=0 and x=$p_i$we get the root p_i and the formula of secant method [7] which is as in (2) .

$$p_i = a_i - f(b_i) * \frac{b_i-a_i}{(f(b_i)-f(a_i))} \quad (2)$$

Now the root finding formula (2) can be used for the auto exposure algorithm to find the shutter speed and the gain (ISO) as in formula (3) and (4) respectively. In the secant method as in [7] the initial roots has to contain the opposite signs, so in our method we first consider the initial exposures in such a way that one exposure value as the under exposure and the other as the over exposure of the scene, so that the proper exposure will converge between them.

The formula (2) is adapted to the shutter speed and gain as in formula (3) and (4) by making some modifications to it. The basic secant method formula is slightly modified as in the equations (3) and (4), without disturbing the basic methodology of the secant and its performance.

The next approximation of shutter speed $\boldsymbol{Shutter_{x+1}}$ is obtained by using the formula (3) where $\boldsymbol{shutter_x}$ is the approximation of the present iteration and $\boldsymbol{mean_k}$ is the mean value of the scene for $\boldsymbol{Shutter_x}$.The $\boldsymbol{shutter_{x-1}}$ is the previous iteration value of shutter speed, where $\boldsymbol{mean_{x-1}}$ is the mean of the scene for $\boldsymbol{shutter_{x-1}}$ .For the first iteration the initial exposure values are considered for $\boldsymbol{shutter_x}$ and $\boldsymbol{shutter_{x-1}}$.The values of the $\boldsymbol{shutter_x}$ and $\boldsymbol{shutter_{x-1}}$ are further modified in every iteration as mentioned in the algorithm discussed below.

$$\mathbf{Shutter}_{x+1} = \frac{(shutter_x \times mean_{x-1}) + (shutter_{x-1} \times mean_x)}{(mean_{x-1} + mean_x)} \quad (3)$$

The next approximation of gain $\boldsymbol{Gain_{x+1}}$ is obtained by using the formula (4) where $\boldsymbol{Gain_x}$ is the approximation of the present iteration, $\boldsymbol{mean_x}$ is the mean value of the scene for $\boldsymbol{Gain_x}$.The $\boldsymbol{Gain_{x-1}}$ is the previous iteration value of gain, where $\boldsymbol{mean_{x-1}}$ is the mean of the scene for $\boldsymbol{Gain_{x-1}}$ .For the first iteration the initial exposure values are considered for $\boldsymbol{Gain_x}$ and $\boldsymbol{Gain_{x-1}}$.

Similarly as in shutter speed the values of $\boldsymbol{Gain_x}$ and $\boldsymbol{Gain_{x-1}}$ are also modified.

$$Gain_{x+1} = \frac{(Gain_x \times mean_{x-1}) + (Gain_{x-1} \times mean_x)}{(mean_{x-1} + mean_x)} \quad (4)$$

Algorithm

The following is the Algorithm for auto exposure using mean value:

Input: The scene that is to be captured with proper exposure
Output: The scene with proper exposure values.

Steps
1. Start
2. Capture the image with lowest shutter speed and gain ($shutter_{x-1}$,$gain_{x-1}$)
3. Calculate the mean ($mean_{x-1}$) for the shutter speed and gain ($shutter_{x-1}$,$gain_{x-1}$)
4. Capture the image with highest shutter speed and gain ($shutter_x$, $gain_x$)
5. Calculate the mean ($mean_x$) for shutter speed and gain ($shutter_x$, $gain_x$)
6. Calculate the next approximation ($shutter_{x+1}$,$gain_{x+1}$) using formulae (1) and (2)
7. Capture the scene with approximation ($shutter_{x+1}$,$gain_{x+1}$)
8. Calculate mean ($mean_{x+1}$)
9. Check whether the scene is under exposed or properly exposed or over exposed by checking the mean is in between 0 to 100 or 100 to 150 or 150 to 255 respectively.
   a. If the scene is under exposed then change the values of ($shutter_{x-1}$, $gain_{x-1}$) with the present approximation values ($shutter_{x+1}$, $gain_{x+1}$) and go to step 6.
   b. If the scene is over exposed then change the values of ($shutter_x$, $gain_x$) with present approximation value ($shutter_{x+1}$, $gain_{x+1}$) and go to step 6.
   c. If the scene is properly exposed the go to step 10.
10. The shutter speed and gain values for the proper exposure are displayed.
11. Stop.

The classification of the lighting condition in the algorithm is performed as ,if the scene contains the mean value in the limit [100-150] is considered as the proper exposure and the scenes with mean value below 100 is considered as under exposure. Similarly the scenes with above 140 mean value is considered as the over exposure.

4 Experimental Results

In this section we are comparing the secant based auto exposure algorithm with the bisection based auto exposure algorithm.

The iteration wise results of bisection method are in the Table 1 and the results of the secant method are in Table 2.The figure 3 represents the properly exposed image with mean value 117.4 captured with shutter speed 271.9 ms and gain 3.1db which values are obtained by using the auto exposure algorithm using bisection method.

The figure 3 represents the proper exposure obtained by using the auto exposure algorithm implemented by using the bisection method with mean value 117.4.

Fig. 3. Proper exposure of an image using Bisection method

The iteration wise results of bisection method are in the Table 1 as shown below.

Table 1. Iteration wise results of the algorithm based on bisection method

Iterations	Shutter speed (in ms)	Gain (in db)	Type of Exposure (center weighted mean luminance)
Iteration1	2006.5 ms	8.6 db	Over (255.0)
Iteration2	989.8 ms	5.3 db	Over (253.8)
Iteration3	511.9 ms	3.1 db	Over (180.5)
Iteration4	249.9 ms	1.1 db	Under (86.0)
Iteration5	261.9 ms	2.1 db	Under (94.3)
Iteration6	271.9 ms	3.1 db	Proper (117.4)

The Bisection method took 6 iterations to find the proper exposure for the leather sample shown in figure 3, whereas the secant method took only 3 iterations to find the proper exposure of the same image under the same lighting conditions as shown in figure 4.

The following are the results of the auto exposure algorithm based on the secant method.

The iteration wise results of secant method are in the Table 2. The figure 4 represents the properly exposed image with mean value 128.9 captured with shutter speed 360.9 ms and gain 2.5db which values are obtained by using the auto exposure algorithm using secant method.

The figure 4 represents the proper exposure obtained by using the auto exposure algorithm implemented by using the secant method with mean value 128.9.

Fig. 4. Proper exposure of an image using Secant method

The results of the secant method are in Table 2 as shown below.

Table 2. Iteration wise results of the algorithm based on Secant method

Iterations	Shutter speed (in ms)	Gain (in db)	Type of Exposure (mean luminance)
Iteration1	90.2 ms	0.4db	Under (27.9)
Iteration2	589.9 ms	1.7db	Under (86.3)
Iteration3	360.9 ms	2.5db	Proper (128.9)

The usage of the secant method for the auto exposure algorithm than the bisection method reduces 3-5 iterations to converge to the proper exposure value. The secant method is quick to find the proper exposure than the bisection method.

5 Conclusion

An auto exposure algorithm is implemented using secant method based on the center weighted average metering technique. This gives more accurate and well exposed results when compared to other methods. As secant method takes less number of iterations to converge so this method is adopted. Also the center weighted average metering mode gives weight to the main object which is at the center. Therefore the computational effort is reduced. The future work is being extended by adding the concept of entropy which can give best results.

References

1. Muehlebach, M.: Camera Auto Exposure Control for VSLAM Applications. Autumn Term (2010)
2. Valli Kumari, V., Ravi Kiran, B., Raju, K.V.S.V.N., Shajahan Basha, S.A.: Optimal exposure sets for high dynamic range scenes. In: Proc. SPIE 8285, International Conference on Graphic and Image Processing (ICGIP 2011), p. 828511 (September 30, 2011)
3. Vallikumari, V., RaviKiran, B., Raju, K.V.S.V.N.: HDR scene detection and capturing strategy. In: 2011 Annual IEEE India Conference (INDICON), December 16-18, pp. 1–4 (2011)
4. Kuno, T., Sugiura, H., Matoba, N.: A new automatic exposure system for digital still cameras. IEEE Transactions on Consumer Electronics 44(1), 192–199 (1998), doi:10.1109/30.663747
5. Kiran, B.R., Vardhan, G.V.N.A.H.: A Fast Auto Exposure Algorithm for Industrial Applications Based on False-Position Method. In: Satapathy, S.C., Udgata, S.K., Biswal, B.N. (eds.) FICTA 2013. AISC, vol. 247, pp. 509–515. Springer, Heidelberg (2014)
6. Cho, M., Lee, S., Nam, B.D.: Fast auto-exposure algorithm based on numerical analysis. In: International Society for Optics and Photonics Electronic Imaging 1999 (1999)
7. Weisstein, E.W.: Method of Secant. From Math World–A Wolfram Web Resource, http://mathworld.wolfram.com/MethodofSecant.html
8. http://www.math.ust.hk/~mamu/courses/231/Slides/ch02_1.pdf (link referred on May 9th, 2014)
9. Understanding metering and metering modes, http://photographylife.com/understanding-metering-modes

Image Authentication in Frequency Domain through G-Let Dihedral Group (IAFD-D3)

Madhumita Sengupta*, J.K. Mandal, and Sakil Ahamed Khan

Dept. of Computer Science and Engineering, University of Kalyani,
Kalyani Nadia, West Bengal, India
{madhumita.sngpt,jkm.cse,sakil.saky}@gmail.com

Abstract. In this paper a G-Let based authentication technique has been proposed to authenticate digital documents by embedding a secret image/message inside the cover image in frequency domain. Cover image is transformed into G-Let domain to generate 2n number of G-Lets out of which selected G-Let(s) are embedded with secret message for the purpose of authentication or copyright protection. The special feature of IAFD-D3 is the use of Dihedral Group with 'n' equals to three. Thus total six G-Lets are generated out of which only single G-Let is used as a locker to lock the secret. Experimental results are computed and compared with the existing authentication techniques like GASMT, Li's Method, STMDF, Region-Based based on Mean Square Error (MSE), Peak Signal to Noise Ratio (PSNR), Image Fidelity (IF), Universal Quality Image (UQI) and Structural Similarity Index Measurement (SSIM) shows better performance in IAFD-D3, in terms of low computational complexity and better image fidelity.

Keywords: Authentication, Dihedral Group, Steganography, Frequency Domain, G-Lets, PSNR.

1 Introduction

Transmission of delicate information is a major concern in today's world due to open access of network via internet. Vital data on transmission through internet increases the risk of illegal and unauthorized access of data. To overcome this risk and to create a secure environment for transferring important information several methods are available. Massive areas of such methods are covered by cryptography and steganography.

Cryptography cluttered up the vital information in a manner that no intruder can be able to understand secret message. It converts the vital information as plain text of communication to cipher text before transmission through unsecured medium. On the other hand steganography hides the vital information as a secret message in an innocent cover to avoid being seen. Steganography embed secret message inside a piece of unsuspicious information. Secret can be hidden inside any sorts of digital cover such as text file, image, audio or video.

* Corresponding author.

S.C. Satapathy et al. (eds.), *Emerging ICT for Bridging the Future – Volume 1*,
Advances in Intelligent Systems and Computing 337, DOI: 10.1007/978-3-319-13728-5_16

Hiding of information through any existing steganographic algorithm can be classified in two domains, Spatial and Frequency. In case of spatial domain based embedding algorithm [1] computation complexity become very low with low robustness and lesser security. Whereas for frequency domain steganographic algorithm computation complexity increase with the increase of robustness and security.

Existing steganographic algorithm on frequency domain are based on discrete cosine transform [2], discrete Fourier transform[3], discrete wavelet transform [4], Z-transform [5] and many newly developed transform techniques. Before embedding through frequency domain based algorithm the cover medium converts from spatial domain to frequency domain, then the frequency components are used as a pocket for the secret bits. On reverse operation for authentication, the medium again generates frequency components, then using same hash function the secret bits are extracted to generate secret message.

One of the newly developed transformation technique based on the linear transformation of group theory recently used in steganographic algorithm for the purpose of authentication is G-Let [6, 11]. In this paper G-Let based authentication technique has been proposed.

Various parametric tests are performed and results obtained are compared with recent techniques such as Li's Method [7] SCDFT [8], STMDF [9], Region-Based [10] and GASMT [11] to obtain a comparative statement between existing and proposed technique.

Section 2 of this paper explains Dihedral group and G-Let transformation technique. Section 3 deals with proposed scheme, encoding, adjustment and authentication process. Results and discussion are given in section 4 with references cited at end.

2 G-Lets over Dihedral Group

Dihedral group is a group of symmetries of a regular polygon including both rotations and reflections. A regular polygon having n sides provides n-rotation and n-reflection symmetries. The associated rotation and reflection make up Dihedral group D_n.

For this research 'n' is taken as three. Thus D3 may have three rotation and three reflection symmetries. So there are in total six symmetries for D3 available and these are called G-Lets. These G-Lets are calculated using the rotation and reflection symmetries.

The total sum of exterior angle of a regular polygon is 360°. For n sides regular polygon average angle about origin can be calculated by $2\pi k/n$. Where k is 0, 1... n-1. G-Lets can be calculated from matrices of eq 1 and eq. 2. Where R_k is rotation matrix and S_k is refection matrix [11]. The general matrixes of rotation and reflection for G-Lets are given in equation 1 and equation 2

$$R_k = \begin{pmatrix} \cos\left(\frac{2\pi k}{n}\right) & -\sin\left(\frac{2\pi k}{n}\right) \\ \sin\left(\frac{2\pi k}{n}\right) & \cos\left(\frac{2\pi k}{n}\right) \end{pmatrix} \tag{1}$$

$$S_k = \begin{pmatrix} \cos\left(\frac{2\pi k}{n}\right) & \sin\left(\frac{2\pi k}{n}\right) \\ \sin\left(\frac{2\pi k}{n}\right) & -\cos\left(\frac{2\pi k}{n}\right) \end{pmatrix} \tag{2}$$

For G-Lets with n equals to three angles are (2π*0)/3, (2π*1)/3 and (2π*2)/3 i.e. 0, 2π/3 and 4π/3. So for D_3, G-Let matrices are shown in equation 3 and 4 for rotation and reflection respectively.

$$R_0 = \begin{pmatrix} 1 & 0 \\ 0 & 1 \end{pmatrix} \quad R_1 = \begin{pmatrix} -\frac{1}{2} & -\frac{\sqrt{3}}{2} \\ \frac{\sqrt{3}}{2} & -\frac{1}{2} \end{pmatrix} \quad R_2 = \begin{pmatrix} -\frac{1}{2} & \frac{\sqrt{3}}{2} \\ -\frac{\sqrt{3}}{2} & -\frac{1}{2} \end{pmatrix} \tag{3}$$

$$S_0 = \begin{pmatrix} 1 & 0 \\ 0 & -1 \end{pmatrix} \quad S_1 = \begin{pmatrix} -\frac{1}{2} & \frac{\sqrt{3}}{2} \\ \frac{\sqrt{3}}{2} & \frac{1}{2} \end{pmatrix} \quad S_2 = \begin{pmatrix} -\frac{1}{2} & -\frac{\sqrt{3}}{2} \\ -\frac{\sqrt{3}}{2} & \frac{1}{2} \end{pmatrix} \tag{4}$$

3 Proposed Scheme

The proposed IAFD-D3 technique is a G-Lets based authentication technique in frequency domain where only reflection based G-Lets are computed and used for embedding secret information followed by adjustment to generate secure stego images. At receiver side on clam for authenticity or to prove originality the image again generates reflections G-Lets to extract secret bits through the same hash function.

IAFD-D3 technique is divided into two major tasks, one at sender and another at receiver end. At sender side embedding process is again subdivided into three sub tasks. The generation of forward G-Let transformation on reflection over different angle as described in section 3.1, followed by embedding process elaborated in section 3.2 and adjustment in section 3.3. The receiver side operation named as authentication process is described in section 3.4.

3.1 G-Lets Transformation

G-Let transformation [6, 11] technique comes with two set of operation one forward transformation, where image divides into sub bands called G-Lets based on number of rotations and reflections, where the number depends on the value of 'n' selected. Another set of operation is the inverse G-Let transformation, where the summations of G-Lets for rotation or reflection symmetry except at rotation/reflection at zero degree angle with proper negation operation can able to regenerates the image.

Forward G-Lets Transformation

Six different G-Lets are mathematically computed through D3 G-Let transformation with the value of 'n' as three. Those G-Lets coefficients are grouped based on rotation and reflection transformation used for calculation using equation 1 and 2. After computation through rotational transformation R_0, R_1, and R_2 of equation 3 are labeled as G-let 1, G-let 2 and G-let 3 and by using reflection transformation S_0, S_1

and S_2 of equation 4 are labeled as G-let 4, G-let 5 and G-let 6 respectively. The calculation is done recursively on a 2 x 2 non overlapping windows in a row major order of the cover image. Illustration of forward G-Let computation through example of single window is given in figure 1.

126	243
213	97

(a) 2x2 window of cover image

126	243
213	97

(b) G-Let 1

147.438000	-230.61600
-22.498001	-232.957993

(c) G-Let 2

-273.43800	-12.384000
-190.501999	135.957993

(d) G-Let 3

126	-243
213	-97

(e) G-Let 4

147.438000	230.61600
-22.498001	232.957993

(f) G-Let 5

-273.43800	12.384000
-190.501999	-135.957993

(g) G-Let 6

Fig. 1. 2x2 window of original cover image with D3 based six forward G-lets

The pictorial representation of forward G-let transformation over Lena image of dimension 512 x 512 is shown in figure 2. Where G-Let 1, G-Let 2 and G-Let 3 are calculated through rotational transformation and G-Let 4, G-Let 5 and G-Let 6 are calculated through reflection transformation.

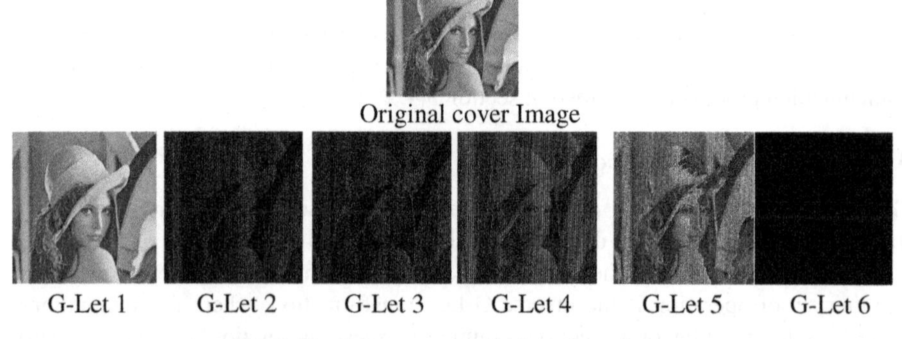

Fig. 2. Six G-Lets constructed from Lena Image

Inverse G-Lets Transformation

Inverse transformation is the reverse procedure to reconstruct the original image. This can be done by simple addition of all the G-Lets of either rotation or reflection without including the angle 0° and 360°. In IAFD-D3 cell by cell addition of reflection G-Lets such as G-let 5 and G-let 6 has been done to reconstruct the image/stego-image with negation of first column as shown through computation in figure 3.

147.438000	230.61600
-22.498001	232.957993

+

-273.438000	12.384000
-190.501999	-135.957993

-126	243
-213	97

Negation of first column

=

126	243
213	97

Original Image

Fig. 3. Generation of original 2 x 2 image matrix through inverse D3 G-Lets

3.2 Embedding Technique

The embedding operation takes G-Let 5 and secret image. The gray scale secret image is converted into array of bits. Cover G-Let 5 is taken to fetch 2nd and 4th components of every 2 x 2 window and secret bits are embedded from the array of bits produced earlier. The operation on G-Let 5 for embedding secret is shown in figure 4.

147.438000	230.61600
-22.498001	232.957993

G-Let 5

147.438000	233.61600
-22.498001	231.957993

G-Let 5 after embedding

Fig. 4. Embedding process on G-Let 5

3.3 Adjustment Technique

The adjustment is required to get the best error free stego image. In proposed IAFD-D3 adjustment on G-Let 6 is made based on the gap generated in G-Let 5 during embedding the secret. If the component in G-Let 5 deviates due to embedding, the calculated deviation needs to be implied on G-Let 6 at same position to keep alive the symmetry. 2nd component of G-Let 5 increases by three after embedding, thus in the same position of G-Let 6 it needs to adjust by increasing the value by three. Whereas the 4th component of G-Let 5 is decreased by one due to embedding thus G-Let 6 needs to adjust accordingly. After adjustment the G-Lets 6 is shown in fig 5.

147.438004	230.615997
-22.498001	232.967993

G-Let 5

147.438004	233.615997
-22.498001	231.967993

G-Let 5 after embedding

-273.437988	12.384001
190.501999	-135.957993

G-Let 6

-273.437988	15.384001
190.501999	-136.957993

G-Let 6 after embedding

Fig. 5. Adjustment on G-Let 6 as per deviation in G-Let5

3.4 Authentication Technique

On receipt of stego-image at the receiver, secret image can be decoded through mathematical calculation. On the receiver side D3 G-Let transformation scheme is applied to create six G-Lets. Then the predefined hash function is used to fetch the bits from 2nd and 4th components of 2 x 2 windows in a row major order from G-Let 5 and to generate an array of bits. From this array of bits the secret image can be finding out by coupling every eight bit to form a byte. The byte values are the required pixel value for secret image.

4 Results and Discussion

Ten pgm images have been taken as cover [13] with two different secret images for payload 0.5bpB and 1.5bpB respectively and IAFD-D3 is applied to compute the results. All cover images are of 512 x 512 in dimension and secret images are 128 x 128 in dimension for payload 0.5bpB and 221 x 221 in dimension for payload 1.5 bpB. Average MSE for payload 0.5 bpB is 1.0 and for payload 1.5 bpB is 21.7. Average PSNR for payload 0.5bpB is 48.1 dB and for payload 1.5 bpB is 34.76 dB. The statistical calculation of image fidelity on average shows 0.99995 and 0.99891 for payload 0.5 bpB and 1.5 bpB respectively. UQI for payload 0.5 bpB and 1.5 bpB are 0.9995373 and 0.9901017 respectively. SSIM for payload 0.5 bpB and 1.5 bpB are 0.9995536 and 0.9904409 respectively. All cover images are shown in figure 6. The detail computation is given in table 1 and 2.

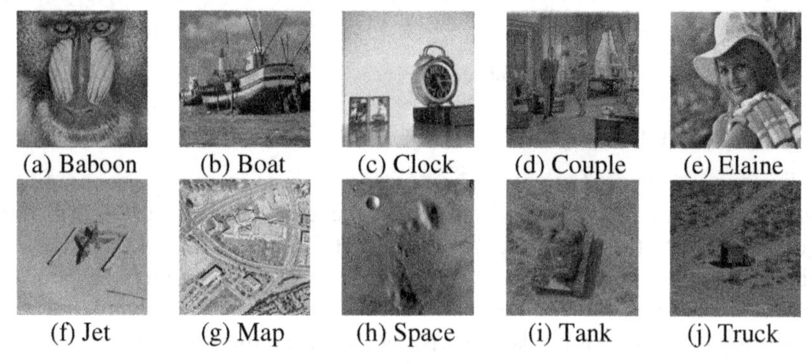

(a) Baboon (b) Boat (c) Clock (d) Couple (e) Elaine
(f) Jet (g) Map (h) Space (i) Tank (j) Truck

Fig. 6. Cover images of dimension 512 x 512 in .pgm format

Table 1. Computation of five parameters for IAFD-D3 technique for payload of 0.5 bpB

Cover Image	MSE	PSNR	IF	UQI	SSIM
Baboon	0.999268	48.133986	0.999946	0.999734	0.999739
Boat	1.004349	48.111958	0.999947	0.999771	0.999774
Clock	0.996078	48.147868	0.999974	0.999847	0.999849
Couple	0.997925	48.139825	0.999939	0.999615	0.999623
Elaine	1.001129	48.125903	0.999952	0.999766	0.999769
Jet	1.010834	48.084006	0.999968	0.998981	0.999039
Map	0.998779	48.136108	0.999971	0.999681	0.999687
Space	1.005508	48.106946	0.999941	0.999327	0.999353
Tank	0.994965	48.152727	0.999945	0.999326	0.999352
Truck	0.999588	48.132593	0.999918	0.999325	0.999351
Average	***1.0008423***	***48.127192***	***0.9999501***	***0.9995373***	***0.9995536***

Table 2. Computation of five parameters for IAFD-D3 technique for payload of 1.5 bpB

Cover Image	MSE	PSNR	IF	UQI	SSIM
Baboon	21.949936	34.716471	0.998819	0.994249	0.994337
Boat	21.992630	34.708032	0.998839	0.995048	0.995113
Clock	19.707016	35.184595	0.999478	0.997021	0.997047
Couple	21.998413	34.706890	0.998648	0.991648	0.991833
Elaine	21.785126	34.749203	0.998945	0.994968	0.995036
Jet	22.148956	34.677271	0.999289	0.978431	0.979621
Map	21.939941	34.718449	0.999356	0.993116	0.993242
Space	21.992538	34.708050	0.998706	0.985602	0.986142
Tank	21.926697	34.721072	0.998795	0.985426	0.985982
Truck	21.987839	34.708978	0.998191	0.985508	0.986056
Average	***21.7429092***	***34.7599011***	***0.9989066***	***0.9901017***	***0.9904409***

A comparative study has been made among Li's Method [7], STMDF [9], Region-Based [10], WTSIC [7] and GASMT [11] with proposed IAFD-D3 technique on the basis of payload verses PSNR in dB. For maximum of 0.5 bits per byte of payload the maximum PSNR is 48.13. That can be achieved through proposed IADF-D3 with robustness in algorithm. The detail data are shown in table 3.

Table 3. Comparison of proposed IAFD-D3 with existing techniques

Technique	*Capacity (bytes)*	*Size of cover image*	*bpB (Bits per bytes)*	*PSNR in dB*
Li's Method [7]	1089	257 * 257	0.13	28.68
STMDF[9]	16900	512 * 512	0.51	38.18
Region-Based [10]	16384	512 * 512	0.5	40.79
WTSIC [7]	16384	512 * 512	0.5	42.25
GASMT [11]	16384	512 * 512	0.5	43.95
IAFD-D3	16384	512 * 512	0.5	48.13

Acknowledgment. The author expressed deep sense of gratitude to the Department of CSE, University of Kalyani where the computational resources are used for the work and DST PURSE, Govt. of India.

References

1. Wu, H.-C., Wu, N.-I., Tsai, C.-S., Hwang, M.-S.: Image steganographic scheme based on pixel-value differencing and LSB replacement methods. In: IEE Proceedings of the Vision, Image and Signal Processing 2005, vol. 152(5), pp. 611–615 (2005)
2. Walia, E., Jain, P., Navdeep: An Analysis of LSB & DCT based Steganography. Global Journal of Computer Science and Technology 10(1) (Ver 1.0), 4–8 (2010)

3. Ghoshal, N., Mandal, J.K.: A Novel Technique for Image Authentication in Frequency Domain Using Discrete Fourier Transformation Technique (IAFDDFTT). Malaysian Journal of Computer Science 21(1), 24–32 (2008) ISSN 0127-9084, Faculty of Computer Science & Information Technology, University of Malaya, Kuala Lumpur, Malyasia (2008)
4. Mandal, J.K., Sengupta, M.: Authentication /Secret Message Transformation Through Wavelet Transform based Subband Image Coding (WTSIC). In: IEEE International Symposium on Electronic System Design, Bhubaneswar, India, December 20-22, pp. 225–229 (2010), doi:10.1109/ISED.2010.50, ISBN 978-0-7695-4294-2, Print ISBN: 978-1-4244-8979-4
5. Sengupta, M., Mandal, J.K.: Image coding through Z-Transform with low Energy and Bandwidth (IZEB). In: The Third International Conference on Computer Science and Information Technology (CCSIT 2013), Bangalore, India, February 18-20 (2013)
6. Rajathilagam, B., Rangarajan, M., Soman, K.P.: G-Lets: A New Signal Processing Algorithm. International Journal of Computer Applications (0975 – 8887) 37(6) (January 2012), doi:10.5120/4609-6591
7. Yuancheng, L., Wang, X.: A watermarking method combined with Radon transform and 2D-wavelet transform. In: IEEE Proceedings of the 7th World Congress on Intelligent Control and Automation, Chongqing, China, June 25-27 (2008)
8. Tsui, T.T., Zhang, X.-P., Androutsos, D.: Color Image Watermarking Using Multidimensional Fourier Transformation. IEEE Trans. on Info. Forensics and Security 3(1), 16–28 (2008)
9. Mandal, J.K., Sengupta, M.: Steganographic Technique Based on Minimum Deviation of Fidelity (STMDF). In: IEEE Second International Conference on Emerging Applications of Information Technology (EAIT 2011), February 19-20, pp. 298–301 (2011), doi:10.1109/EAIT.2011.24, print ISBN: 978-1-4244-9683-9
10. Nikolaidis, A., Pitas, I.: Region-Based Image Watermarking. IEEE Transactions on Image Processing 10(11), 1721–1740 (2001)
11. Sengupta, M., Mandal, J.K.: G-Let based Authentication/Secret Message Transmission (GASMT). In: IEEE 4th International Symposium on Electronic System Design, ISED 2013, NTU Singapore, December 10-12, pp. 186–190 (2013), doi:10.1109/ISED.2013.44, ISBN - 978-0-7695-5143-2
12. Kutter, M., Petitcolas, F.A.P.: A fair benchmark for image watermarking systems. In: Security and Watermarking of Multimedia Contents Electronic Imaging 1999, vol. 3657. The International Society for Optical Engineering, Sans Jose (1999)
13. Weber, A.G.: The USC-SIPI Image Database: Version 5, Original release. Signal and Image Processing Institute, University of Southern California, Department of Electrical Engineering (October 1997), http://sipi.usc.edu/database/ (last accessed on January 25, 2010)

Information Tracking from Research Papers Using Classification Techniques

J.S.V. Sai Hari Priyanka, J. Sharmila Rani, K.S. Deepthi, and T. Kranthi

Department of Computer Science & Engineering, Anil Neerukonda Institute of Technology & Sciences, Sangivalasa, Bheemunipatnam [M], Visakhapatnam
{priyapatnaik.hari,kranti.it16}@gmail.com,
{sharmila.cse,selvanideepthi.cse}@anits.edu.in

Abstract. The research area identification is a challenging issue to solve as there lakhs of millions of research papers available, it is required to classify the papers based on primary and secondary areas. The existing text mining techniques classifies research documents in the static manner. So there is a need to develop a framework that can classify the research documents in dynamic manner. This paper mainly describes a framework that can classify area of research documents arrived to the repository. The proposed frame work consists of two phases where first is to construct word list for each area of the paper. In second phase it continuously updates the word list associated to the new stream of research documents. The experimental results compared with existing techniques are reported and that satisfies the minimum requirement.

Keywords: Research area identification, Text mining, Classification.

1 Introduction

Now a day there is tremendous growth in paper publishing in various streams of research. There are many research disciplines and again we have a lot of many sub areas under a particular discipline. When a person sees a paper, he might not get an idea to which research discipline the paper belongs to by just seeing the title. He has to give a thorough reading to the paper and it consumes a lot of time. It would be better if we have some tool which can specify not only the main research area but also the sub area as the researcher may be interested in a specific area under the main domain. This will saves time. In this paper, a method was proposed for classifying various papers present in the repository into their respective disciplines and also identifying their sub areas. In the first phase i.e. system training phase lot many papers of various disciplines are collected and stored in the repository. Each paper is taken, pre-processing techniques are applied and the paper is classified into a primary research area by taking the frequency count of index terms into account. In the second phase the sub area is identified by using text clustering algorithm.

Text mining techniques [1] used for classification applied on static repository of research documents. Now a day's number of documents in repository increases with

S.C. Satapathy et al. (eds.), *Emerging ICT for Bridging the Future – Volume 1*,
Advances in Intelligent Systems and Computing 337, DOI: 10.1007/978-3-319-13728-5_17

time i.e. new stream of documents of various areas added to the repository and it require identifying and adding feature words in the list. This list is useful at the time of classifying documents of different areas. So there is a need to classify documents using the recent knowledge of the repository after adding new stream of documents.

There are some challenges when processing text applications. First, the proposed technique must extract feature words before the next text of documents arrives to the repository. Second, It should process and extract relevant feature words and then add to word list (if not there in list) as fast as possible. Third, while processing text of documents memory is the constraint, so it should use appropriate processing model among Landmark, Damped and sliding window model.

In this paper we describe a framework that can classify text of research documents arrived to the repository. The proposed frame work consists of two phases where first it construct word list for each area of the paper. In second phase it continuously updates the word list associated to the new stream of research documents.

Remaining sections of the paper describes as follows section 2 describes the related works based on text mining, classification using area tracking. Section 3 describes the problem statement related to stream of documents dynamically added to repository. The proposed architecture and frame work explained in section 4. Experimental results have reported in section 5 and finally conclude the paper in section 6.

2 Related Work

The section mainly divided into three sub sections described as follows

2.1 Background Works Related to Text Mining and Classification

Chen et al. [2] proposed a fuzzy-logic-based model as a tool for document selection. Henriksen et al. [3] tool for ranking and scoring research and development projects basing on required criteria. Ghasemzadeh et al. [4] proposed a framework called project portfolio selection, this is an important decision making aspect for almost every organization with stated objectives and without exceeding or violating the available resources and constraints. Machacha et al's [5] decision making and finding of an optimal alternative from the possible alternatives is a difficult task when the information available is vague, an approach was proposed by Machacha could be applied to address the problem. Butler et al. [6] proposed a theory for research document ranking and selection basing on number of attributes used. Loch et al. [7] proposed for selecting portfolios of a program within a budget constraint. Murad Habib et al. [8] developed an analytic network process model. Greiner et al. [9] proposed a frame work to support project selection.

Choi et al [13] explained text-mining approach for document extraction and filtering. Sun et al. [14] used a hybrid knowledge based and model document identification for assigning respective experts. Hettich explained [15] a text-mining approach to combine documents and assign area to each document. Many text-mining works have been carried out to classify documents [16][17], they are developed with a focus on English text.

2.2 Background Work Related to Text Processing

Stream of documents consists of an ordered sequence of text documents. Each set of documents are usually called as a "transaction". The issue of document processing model here is to find a way to extract transactions for association rule mining from the overall stream of documents. The stream model finds a set of research documents selected for processing .In [10][11][12], the authors use stream processing models in their algorithms to get the frequent item sets of data streams.

3 Problem Definition

A Text document is an ordered sequence of words that arrives in timely order. Document streams are different from documents stored in static repository. They are uninterrupted, limitless come in high speed. Let R be repository of d different areas and m be the maximum number of feature words. Let S^t_n be stream of n documents arrives at time stamp t and F_{nxm} be feature word matrix associated with recently added documents. Let L_{nxm} be feature word list for d areas and each area having maximum m feature words.

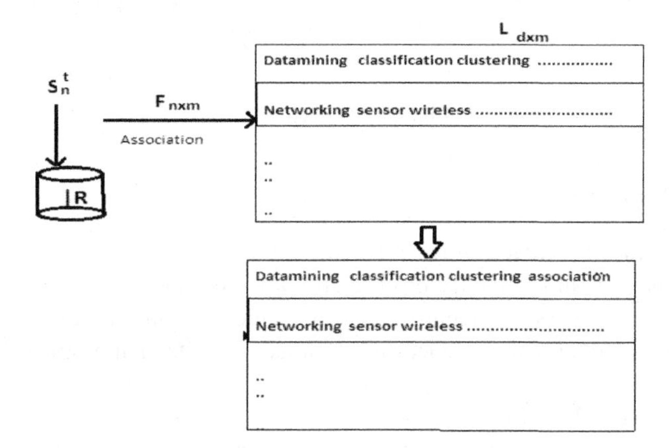

Fig. 1. Extraction of keywords

The document that to be classify (doc_t) feed as input and apply the following steps.

Step 1: Read the text from doc_t
Step 2: Preprocess the text.
Step 3: Remove the stop words from document.
Step 4: Construct list of unique words and their counts.
Step 5: Apply Bayesian classification that most appropriate area match with feature word list.

4 Proposed Work

4.1 Architecture

The above figure represents architecture in which the stream documents arrived, then each every document applied to sequence of activities preprocessing, feature word extraction and word list creation. The documents are moved to repository one by one and storage management is based on any processing models like Land mark, Damped model and sliding window model. The feature words of the particular area updated only if newly generated words are not there in list. User can upload document in parallel to know area of unlabeled document.

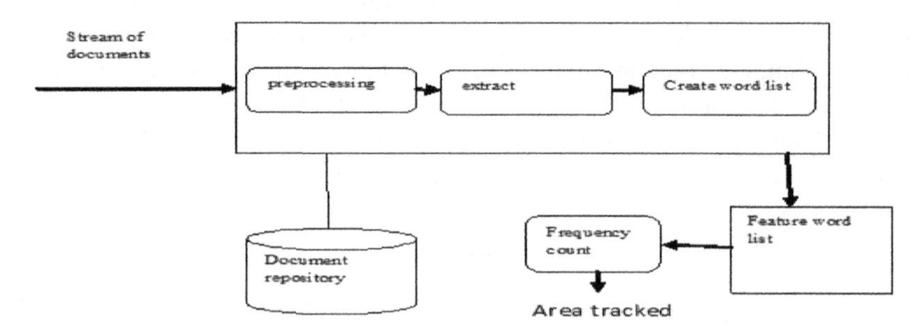

Fig. 2. Architecture

4.2 Text Mining Stages

4.2.1 Information Extraction from Document

First step for the system to classify unstructured text is to use information extraction. It is a software process of identifying key sentences and relationships within document. It uses predefined sequences in document, is called pattern matching.

4.2.2 Tracking of Key Topic

Tracking of key topic is a system functioning by keep set of user profiles and, based on the documents the user views, to forecasts other documents of interest to the user. Keywords are a set of significant words in an article that gives high-level description of its contents to readers.

4.2.3 Document Classification

Categorization involves identifying the main themes of a document by placing the document into a pre-defined set of topics. The next section explains algorithm for classifying documents.

4.3 Naive Bayes Algorithm for Text Classification of Documents

Naive Bayes classifier is a popular method for classification of text which helps to identify whether a document belongs to one area or other by finding the word frequencies as the features. It depends upon the precise nature of the probability model; it can be trained very efficiently by a supervised learning.

Step 1: L : Set of feature words of order dxm ,doc_t : input document to check
Step 2: Let there be‘d’ Classes: C1, C2, C3…Cd
Step 3: Naïve Bayes classifier predicts X belongs to Class Ci if
    P (Ci/X) > P(Cj/X) for 1<= j <= m , j <> i
    P(Ci/X) = P(X/Ci) P(Ci) / P(X)
    Maximize P(X/Ci) P(Ci) as P(X) is constant

The probability of keywords occurrence will be calculated and the keywords with highest frequency count belonging to a particular area will be the primary area.

5 Results and Observation

Experiments conducted static and text based environments. The aim is to comparative study of the efficiencies. Experiments were conducted on 2.7 GHz Intel Dual core with 2GB of primary memory. The results obtained were from naturally randomly generated text documents and implementation of the using java 7 on net bean Integrated Development environment. The following are the parameters and their ranges for the implementation.

Table 1. Experimental parameters with ranges

Parameter	Range
Number of documents in repository	200K-1000K
Number of Areas	5
Maximum number of feature word per area	100-200

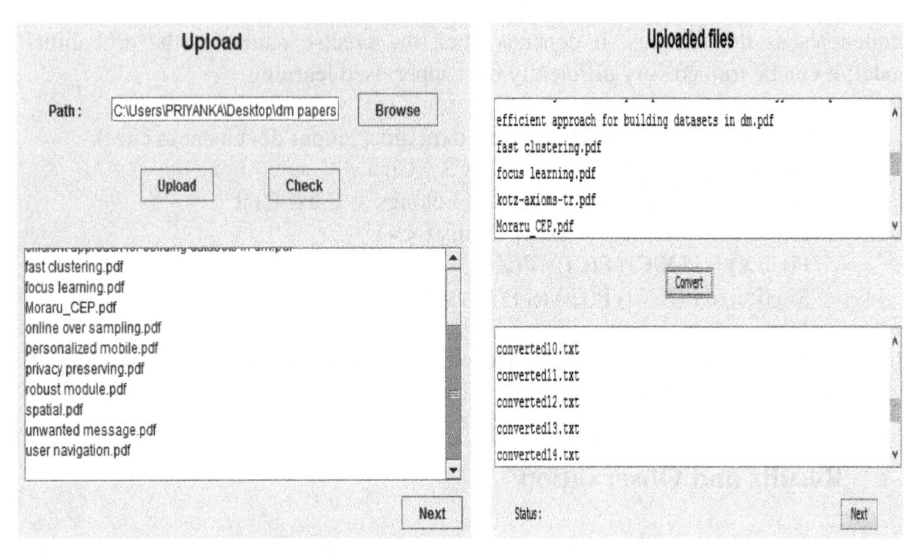

Fig. 4. Uploading documents

Fig. 5. Text normalization

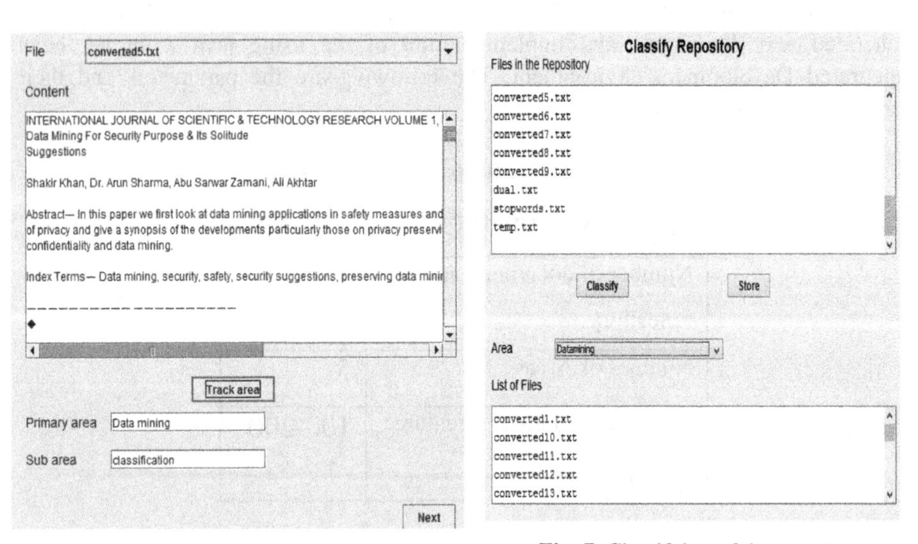

Fig. 6. Tracking of area

Fig. 7. Classifying of documents

Fig.4 depicts uploading of research documents. Fig.5 depicts the text normalization process followed by preprocessing technique will be performed on the particular paper so that the stop words free text was obtained. And the resultant file is matched with the keywords present in the dataset consisting of keywords belonging to different areas. Fig.6 depicts the probability of keywords occurrence will be calculated and the keywords with highest frequency count belonging to a particular area will be the primary area. And for finding the sub area different dataset has been maintained for each area and again we have to calculate the frequency count, the highest keywords count of respective area will be the sub area of a particular paper. Fig.7 depicts classifying of documents.

Table 2. Areas and their Probabilities

Area	Probability of keywords occurance
Datamining	0.57236
Network security	0.21929
Networks	0.13157
Image Processing	0.05921

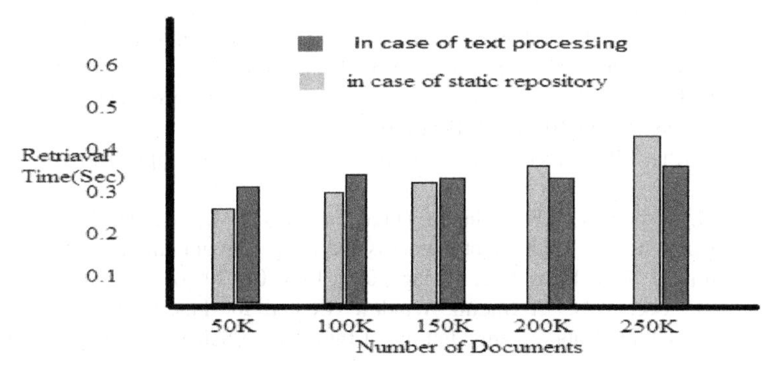

Fig. 3. Performance analysis

The above graph depicts the comparative study between static and dynamic approaches. X axis represent number of documents processed and Y axis represent retrieval time in seconds. It is observed that incremental update processing is better when number of documents increases.

6 Conclusion

This paper has presented a frame work for classifying research documents, along with a dynamically updated repository. A feature word list is constructed to categorize the concept words in different research areas and to form relationships among them. It facilitates text-mining and classification techniques to classify the research documents based on their domain. The experimental results in environment showed that proposed frame work improved the accuracy and speed. In this we described a framework that can classify stream of research documents arrived to the repository. The proposed frame work consists of two phases where first it constructed word list for each area of the paper. In second phase it continuously updated the word list associated to the new stream of research documents.

References

1. Ma, J., Xu, W., Sun, Y.-H., Turban, E., Wang, S., Liu, O.: An Ontology-Based Text-Mining Method to Cluster Proposals for Research Project Selection. IEEE Transactions on Systems, Man, and Cybernetics—Part A: Systems And Humans 42(3) (May 2012)
2. Chen, K., Gorla, N.: Information system project selection using fuzzy logic. IEEE Trans. Syst., Man, Cybern. A, Syst., Humans 28(6), 849–855 (1998)
3. Henriksen, A.D., Traynor, A.J.: A practical R&D project-selection scoring tool. IEEE Trans. Eng. Manag. 46(2), 158–170 (1999)
4. Ghasemzadeh, F., Archer, N.P.: Project portfolio selection through decision support. Decis. Support Syst. 29(1), 73–88 (2000)
5. Machacha, L.L., Bhattacharya, P.: A fuzzy-logic-based approach to project selection. IEEE Trans. Eng. Manag. 47(1), 65–73 (2000)
6. Butler, J., Morrice, D.J., Mullarkey, P.W.: A multiple attribute utility theory approach to ranking and selection. Manage. Sci. 47(6), 800–816 (2001)
7. Loch, C.H., Kavadias, S.: Dynamic portfolio selection of NPD programs using marginal returns. Manage. Sci. 48(10), 1227–1241 (2002)
8. Habib, M., Khan, R., Piracha, J.L.: Analytic network process applied to R&D project selection
9. Greiner, M.A., Fowler, J.W., Shunk, D.L., Carlyle, W.M., Mcnett, R.T.: A hybrid approach using the analytic hierarchy process and integer programming to screen weapon systems projects. IEEE Trans. Eng. Manag. 50(2), 192–203 (2003)
10. Chang, J.H., Lee, W.S.: A Sliding Window Method for Finding Recently Frequent Itemsets over Online Data Streams. Journal of Information Science and Engineering (July 2004)
11. Chi, Y., Wang, H., Yu, P.S., Richard R.: Moment: Maintaining Closed Frequent Itemsets over a Stream Sliding Window. In: IEEE Int'l Conf. on Data Mining (November 2004)
12. Lin, C.-H., Chiu, D.-Y., Wu, Y.-H., Chen, A.L.P.: Mining Frequent Itemsets from Data Streams with a Time-Sensitive Sliding Window. In: SIAM Int'l Conf. on Data Mining (April 2005)
13. Choi, C., Park, Y.: R&D proposal screening system based on textmining approach. Int. J. Technol. Intell. Plan. 2(1), 61–72 (2006)
14. Sun, Y.H., Ma, J., Fan, Z.P., Wang, J.: A hybrid knowledge and model approach for reviewer assignment. Expert Syst. Appl. 34(2), 817–824 (2008)
15. Hettich, S., Pazzani, M.: Mining for proposal reviewers: Lessons learned at the National Science Foundation. In: Proc. 12th Int. Conf. Knowl. Discov. Data Mining, pp. 862–871 (2006)
16. Wei, C.P., Chang, Y.H.: Discovering event evolution patterns from document sequences. IEEE Trans. Syst., Man, Cybern. A, Syst., Humans 37(2), 273–283 (2007)
17. Cheng, T.H., Wei, C.P.: A clustering-based approach for integrating document-category hierarchies. IEEE Trans. Syst., Man, Cybern. A, Syst., Humans 38(2), 410–424 (2008)

Sentiment Analysis on Twitter Streaming Data

Santhi Chinthala, Ramesh Mande, Suneetha Manne, and Sindhura Vemuri

Dept. of IT, V R Siddhartha Engineering College, Vijayawada
{shanthi.ch20,ramesh.welcome,suneethamanne74}@gmail.com,
vemurisindhura@yahoo.com

Abstract. Twitter, an online social networking service is devised so as to treasure trove what is circumstance at any juncture in time, everywhere in the globe and it can provoke the data streams at rapid momentum. In the twitter network all the messages generate a data momentum and handle eminently vigorous behaviours of the actors in the twitter network. Twitter serves an enormous collection of APIs and actors can utilize them without registering. In twitter data information streams are mannered and categorizing issues are concentrated and these streams are evaluated for discovering analysis of sentiment and extracting the opinion. The automatic collection of corpus and linguistic analysis of the collected corpus for sentiment analysis is shown. A sentiment classifier that is able to determine decisive, pesimisive and non-decisive sentiments for a document is performed using the collected corpus. Using various learning algorithms like Naive Bayesian Algorithm, Max Entropy Algorithm, Baseline Algorithm and Support Vector Machine, a research on twitter data streams is performed.

Keywords: Twitter, Naive Bayesian, Max Entropy, Baseline, Support Vector Machine.

1 Introduction

Twitter is an online micro blogging and a social networking service that facilitate users to deliver and interpret 140 length posts known as "tweets" [1]. The twitter service swiftly attained worldwide reputation, as of 2011 with over 300 million users, provoking over 300 millions of tweets and superintendence about 1.6 billions of queries a day. It has been elucidated as "the SMS of the internet". Twitter can be accessed through a website interface, SMS, or mobile device application. Before making a purchase end users have a chance to use sentiment analysis to these research services or products. As well, this research will be useful to the marketers to know the opinion of the public on their company products and user's satisfaction also can be analyzed. Critical feedback can be taken by the organizations for their newly released products [2].

S.C. Satapathy et al. (eds.), *Emerging ICT for Bridging the Future – Volume 1*,
Advances in Intelligent Systems and Computing 337, DOI: 10.1007/978-3-319-13728-5_18

1.1 Definition of Sentiment Analysis

Sentiment Analysis deals with the concept of classifying the decisive and pesimisive emoticons also analyzation of text. Whether there is any absence then the sentiment becomes neutral. Various categories of subjectivity analysis that includes sentiment analysis is as shown in Fig 1.

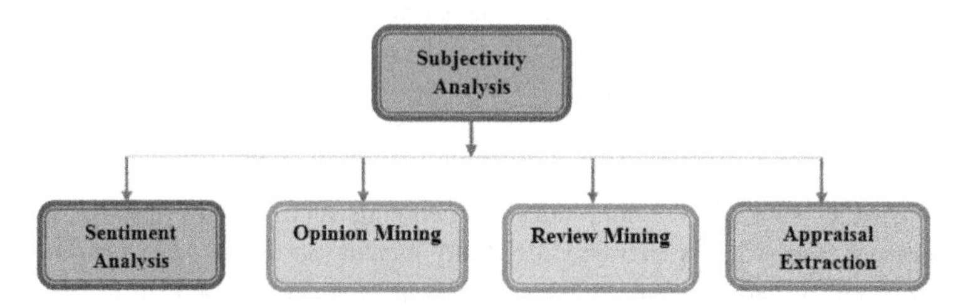

Fig. 1. Categories of Subjectivity Analysis

2 Related Work

Python has an easy way to pick up and learn. Instructions for downloading and installing Python for all platform users can be found at http://www.python.org/download. To unquestionably reckon Python to path at the Windows Command, Windows users are highly recommended to install Active Python.

The project is designed depends on thoughts disclosed in, the classification of tweets is done based on unigram features and the classification is done by classifiers by providing trained data acquired based on distant supervision. The distant supervision is used as labels for decisive or pesimisive and this sentiment labels are adequate for minimizing interdependencies in Machine Learning methods [3] [4]. The major task of sentiment analysis is as shown in Fig 2.

In the Existing techniques non-decisive tweets are not taken into account that points to erroneous distribution. This paper solves this complication by separating non-decisive tweets in the training dataset and uses a feature vector to train the Machine Learning classifier and classifies the generated tweets as decisive/pesimisive/non-decisive.

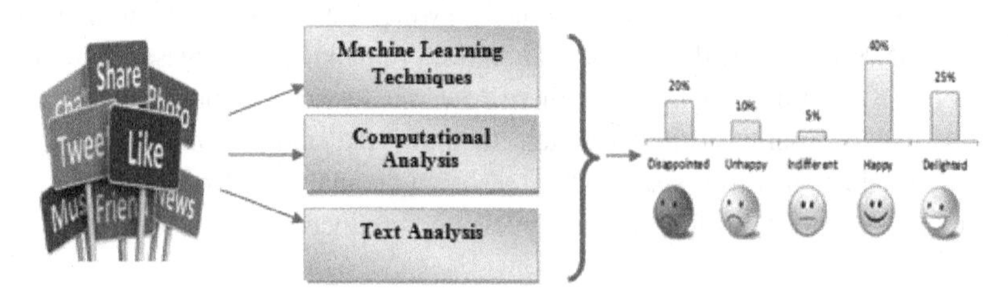

Fig. 2. Sentiment Analysis

3 Approach

Feature extractors and different Machine Learning Classifiers are used here. The unigrams and feature extractors are unigrams, with weighted decisive and pesimisive keywords. A framework is designed that separates feature extractors and classifiers as a pair of components.

3.1 Query Term

Normalizing the key terms should be performed. If user wants to perform sentiment analysis about a product, is classified by the count of positive/negative query terms.

3.2 Emoticons

Emotions are used by the training process as unwanted labels and it plays significant role in categorization. Emoticons are pruned from the training data. If emoticons are excluded, appropriate result cannot be calculated in the case of Support Vector Machine classifiers and MaxEnt, and less impact on Naive Bayes. Classifier is restricted to analyze all other features in the tweet, if all the emoticons are pruned.

3.3 Feature Reduction

There are many unique properties for the twitter language model. To reduce the feature space the following properties are taken:

Usernames: Direct messages are given by using Twitter Usernames. Before the username (@SanthiCh20) @ symbol is used.

Usages of Links: Links are often included by the users in their tweets. For all the URLs a compatibility class is used.

Stop Words: More no. of filler and stop words are present in this. They are like "the", "is", and "a". These are discarded because they do not express the sentiment for a tweet. The complete list of stop words can be found at [4].

Repeated Letters: Tweets contain normal language. A nonempty result set may be generated in twitter, when user take a key word "lucky" with the no. of c's (For e.g., lucccky, luccccccky, lucccccccccccky). If a row contains any letter occurred more than two times then it is swapped with two letters. From the above word, It is modified to "lucky".

3.4 Feature Vector

Unigrams: The extreme of pre-processing, end up with features which are unigrams and each of the features have same weights [5].

Weighted Unigrams: Instead of weighing each of the unigrams equally, bias by weighing the decisive and pesimisive keywords more than the other features present in the feature vector [5].

Feature reductions effect is shown in the Table 1. These reductions shrink 8.96 percentile of original rate which set down the feature.

Table 1. Feature Reduction Effect

Steps for Reduction of Feature	Features (#)	Original Percentile (%)
USER NAME/URL/ Letters which are Repeated	102358	37.10
Stop words ("is", "the", "a")	24311	8.96
Final	24311	8.96

4 Proposed Work

The architecture of proposed work is shown in Fig 3 that mainly focuses on Collection of Twitter Data, Preprocessing, Feature Vector Generation and classification using Machine Learning Techniques.

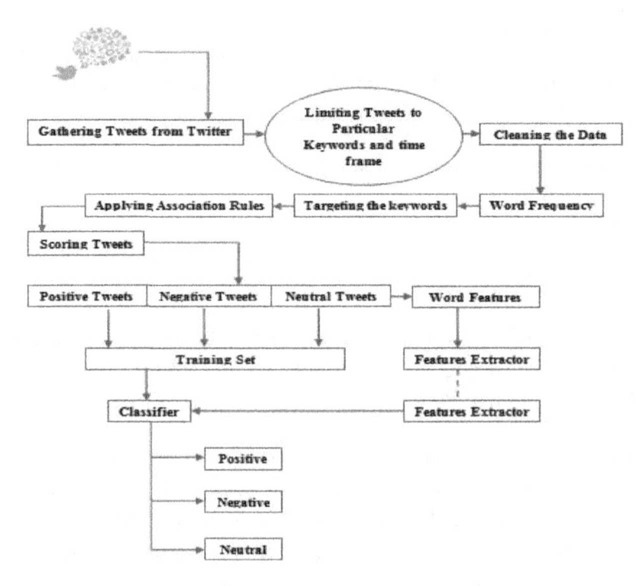

Fig. 3. Proposed Model

4.1 Data Collection

Twitter equips appliance to a mass data through its Application Programming Interface (API). The streaming mechanism grabs the input information Tweets and operates any anatomizing, percolating, or aggregation mandatorily anterior to accumulating the outcome to a data store. The HTTP handling mechanism queries the data store for outcome in reply to user inquiry. JSON(**JavaScript Object Notation**)[14] is recommended due to compactness. Python scripts were written to interact with streaming API and related data can be collected.

4.2 Data Transformation

Twitter data can be stored using SQLite. Python script returns requested data in a JSON file format. JSON is not much compatible file format to process directly on textual data. So the python script can be written in such a way that JSON data is stored into database and a flat file was then generated for the purpose of pre-processing.

4.3 Preprocessing

For further processing flat files that are generated can be used. Scrutinizing information that has not been cautiously quarantined may yield ambiguous outcomes. Accordingly the delegation and endowment of data is prime and foremost heretofore operating dissection. Following are some of the preprocessing steps that are followed [14]. . Fig 4 represents the real time streaming of twitter data.

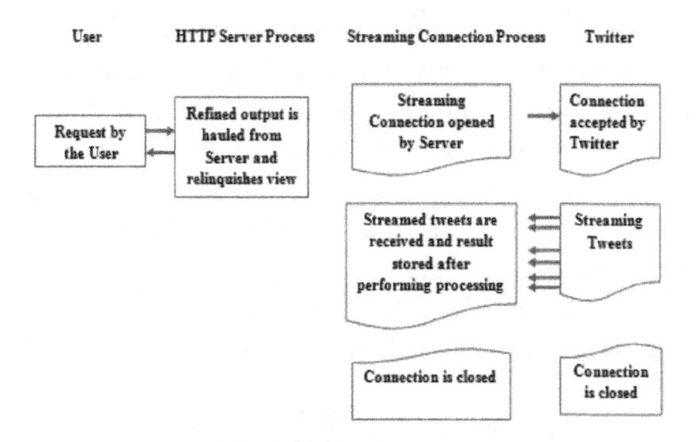

Fig. 4. Real Time Stream of Tweet Data

Tokenization: Tokenization is the process dividing entire text into meaningful words referred as tokens. A token may be a symbol, phrase or simply a word.

Filtering: Filtering tokens process is done in three phases. In first phase the stop words are filtered. Removing such stop words makes text processing easy. Stop words based on different criteria was evaded. Secondly, the tokens based on length to avoid some freaky terms can be neglected.

Stemming: Stemming is the process of replacing each word or a token by its corresponding stem where a stem is a form to which affixes can be attached.

4.4 Machine Learning Methods

The dataset that is collected from twitter is applied for preprocessing and feature extraction/selection is to be done so as to identify the required features later feature reduction mechanism is used to optimize the performance of the machine learning classifier.

Two main methodologies that can be distinguished in Sentiment Analysis are machine learning approaches and the symbolic approaches. Symbolic approaches are referred as approaches that use manually crafted rules and lexicons.

1. **BaseLine:** Baseline uses the decisive and pesimisive keyword list for every individual tweet, and calculate the no. of decisive and pesimisive keywords in order that occur [6]. Thus the classifier get backs the maximum count for the sentiment. Whenever there is a draw, non-decisive polarity is returned.
2. **Naive Bayes:** In data classification Naïve Bayes machine learning technique is one of the best method. In this I'm using the Multinomial Naïve Bayes Technique. Here, d is a tweet and c* is class and that is assigned to d, where [9]

$$c^* = argmac_c P_{NB}(c|d)$$

$$P_{NB}(c|d) := \frac{(P(c)) \sum_{i=1}^{m} P(f|c)^{n_i(d)}}{P(d)} \tag{1}$$

From the above equation, 'f' is a 'feature', count of feature (fi) is denoted with ni(d) and is present in d which represents a tweet. Here, m denotes no. of features. Parameters P(c) and P(f|c) are obtained through maximum likelihood estimates, and add-1 smoothing is utilized for unseen features. To train and classify using Naive Bayes Machine Learning technique the Python NLTK library was used [8].
3. **Maximum Entropy:** The idea behind Maximum Entropy models is that one should prefer the most uniform models that satisfy a given constraint [10]. MaxEnt models are feature-based models. MaxEnt makes no independence assumptions for its features, unlike Naive Bayes. The model is represented by the following [10]:

$$P_{ME}(c|d,\lambda) = \frac{\exp\left[\sum_i \lambda_i f_i(c,d)\right]}{\sum_c \exp\left[\sum_i \lambda_i f_i(c,d)\right]} \tag{2}$$

4. **Support Vector Machine:** SVM is one of the significant techniques for classification [11]. Use libsvm library with a linear kernel [12]. Two sets of vectors with size m are taken as input data. In the vector each entry represents the availability of the feature. In the unigram feature extractor, each feature is a single word found in a tweet. If the feature is present, the value is 1, but if the feature is absent, then the value is 0. Feature presence is used, as opposed to a count, so that there is no necessity to scale the input data, which speeds up overall processing [13].

5 Results and Observations

The accuracy of classifiers based on unigrams and weighted unigrams features is represented graphically as in Fig 5.

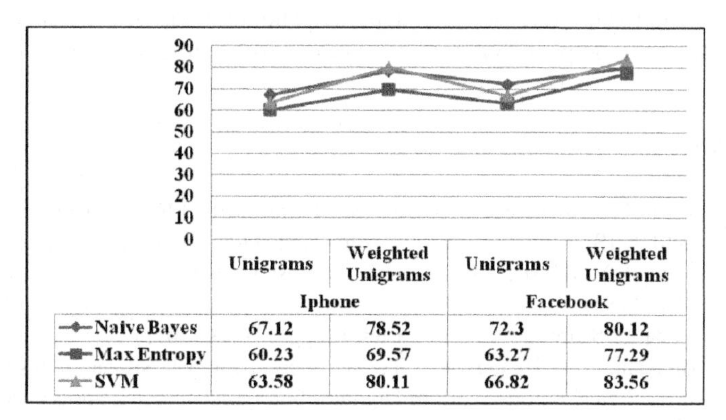

Fig. 5. Accuracy of Classifier

Positive and negative keyword weights are considered than other keywords in classification of sentiment for a tweet and accuracy of classifier is as shown in Fig 5. Support Vector Machine produced the best performance with 83.56% & 80.11% accuracy for the keywords Face book, iPhone and Naive Bayes accuracy exceeded Max Entropy accuracy by a considerable margin i.e. 80.12% and 78.52% for the keywords Face book and iPhone respectively. Fig 6 represents twitter sentiment analysis on keyword Face book using Naive Bayes Classifier.

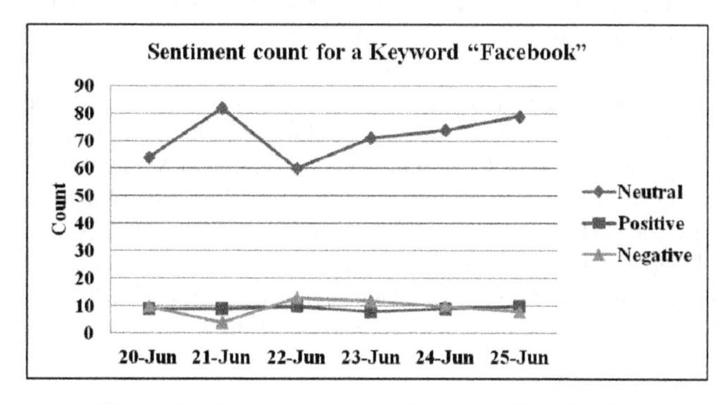

Fig. 6. Sentiment count for a Keyword "Face book"

6 Conclusion and Future Work

A novel feature vector of weighted unigrams is used, along with algorithms such as Naive Bayes, Maximum Entropy and Support Vector Machines to obtain competitive accuracy in sentiment classification of tweet. For classifying tweets in sentiment Machine learning techniques perform well.

Semantics. The comprehensive sentiment of a tweet is classified by the algorithms. Semantic role labeler can be used which indicates which noun is associated with the verb and accordingly the classification occurs.

Internationalization. Here the focus is only on English tweets but Twitter has a large amount of international audience. This approach should be used to classify sentiment with a language specific positive/negative keyword list in other languages.

References

1. Co-founder of Twitter receives key to St. Louis with 140 character proclamation. ksdk.com. KSDK (September 19, 2009). After high school in St. Louis and some time at the University of Missouri-Rolla, Jack headed east to New York University (retrieved September 29, 2009)
2. Gomes, L.: Twitter Search Is Now 3x Faster. Blogger, York (2011)
3. Go, A., Bhayani, R., Huang, L.: Twitter Sentiment Classification using Distant Supervision. Technical report, Stanford Digital Library Technologies Project (2009)
4. Read, J.: Using emoticons to reduce dependency in machine learning techniques for sentiment classification. In: Proceedings of ACL 2005, 43nd Meeting of the Association for Computational Linguistics. Association for Computational Linguistics (2005)
5. Zhu, X., Goldberg, A.B., Rabbat, M., Nowak, R.: Learning Bigrams from Unigrams (2008)
6. Zaman, A.N.K., Matsakis, P., Brown, C.: Evaluation of Stop Word Lists in Text Retrieval Using Latent Semantic (2011)
7. Shih, M., Sanchez, D.T.: Perspectives and Research on the Positive and Negative Implications of Having Multiple Racial (2005)
8. Bird, S.: NLTK-Lite: Efficient Scripting for Natural Language Processing
9. Gamallo, P., Garcia, M.: A Naive-Bayes Strategy for Sentiment Analysis on English Tweets. In: 8th International Workshop on Semantic Evaluation (SemEval 2014), Dublin, Ireland, pp. 171–175 (2014)
10. Nigam, K., Lafferty, J., Mccallum, A.: Using maximum entropy for text classification. In: IJCAI 1999 Workshop on Machine Learning for Information Filtering, pp. 61–67 (1999)
11. Cristianini, N., Shawe-Taylor, J.: An Introduction to Support Vector Machines and Other Kernel-based Learning Methods. Cambridge University Press (March 2000)
12. Chang, C.-C., Lin, C.-J.: LIBSVM: A Library for Support Vector (2013)
13. Computer, D.O., Wei Hsu, C., Chung Chang, C., Jen Lin, C.: A practical guide to support vector classification chihwei hsu, chih-chung chang, and chih-jen lin. Technical report
14. Peng, D., Cao, L., Xu, W.: Using JSON for Data Exchanging in Web Service Applications (2011)

A Secure and Optimal Data Clustering Technique over Distributed Networks

M. Yogita Bala and S. Jayaprada

Department of Computer Science and Engineering, Anil Neerukonda Institute of Technology and Sciences, Sangivalasa, Bheemunipatnam[M], Visakhapatnam
{yogitaaruna,jayaprada.suri}@gmail.com

Abstract. Clustering is an automatic learning technique aimed at grouping a set of objects into subsets or clusters. The goal is to create clusters that are coherent internally, but substantially different from each other. Privacy is an important factor while datasets or data integrates from different data holders for mining over a distributed networks. Secured and optimal data clustering in distributed networks has played an important role in many fields like Information Retrieval, Data mining, Knowledge and Data engineering or community based clustering. Secured mining of data is required in open network. In this paper we are proposing an efficient privacy preserving and optimal data clustering technique over distributed networks.

Keywords: Distributed networks, clustering, privacy preserving data mining.

1 Introduction

In a web of distributed network, communication between nodes is established for transmitting data freely which has led to a steep increase in the department of research so that it could be made safe and secure way of mining. The most important thing being the Privacy Preserving Techniques when the data mining is done and heavy research continues on using Classification, Association Rule Mining or Clustering.

Clustering can be defined as a process of grouping up similar type of objects based on distance (for numerical data) or similarity (for categorical data) between data objects. In distributed environment nodes or data holders, it maintains individual data sets and every node are connected with each other by an edge[3].

The recent and existing Clustering Algorithms are designed for Central Execution. In this the clustering is performed on a dedicated node, and is not suitable for deployment over large scale distributed networks hence specialized algorithms for distributed and Peer-to-Peer clustering have been developed[6],[7],[8],[9]. The limitation to this approach is that it is either limited to a small number of nodes, or focus on low dimensional data only.

An approach for privacy preservation process, like Randomization and Perturbation, are available and can be maintained in two separate ways. The first way is known as Cryptographic; in this approach real data sets can be converted to unrealized datasets by encoding the real datasets. The other way is termed as Imputation Methods; some fake values are imputed between the real dataset and extracted while mining with a set of rules[1][2].

S.C. Satapathy et al. (eds.), *Emerging ICT for Bridging the Future – Volume 1*,
Advances in Intelligent Systems and Computing 337, DOI: 10.1007/978-3-319-13728-5_19

S.Jha, L. Kruger and P. McDaniel proposed the privacy preserving Distributed Clustering Algorithm on which the Distributed Environment is represented by. Here data can be clustered by grouping the similar type of objects and secure transmission through protocols[4][5]. Geometric techniques are being used for privacy preserving Clustering Technique by using Perturbation Method of string transformation[10].

2 Related Work

In social network, nodes can be represented as vertices and those vertices V ($v_1, v_2 \ldots v_n$) connected through set of edges E in a undirected graph G (V,E)..

In Distributed Networks data can range from Numerical to Categorical. Numerical data can be compared with respect to quasi identifier difference whereas Categorical data can be compared with similarity between the data objects. Distributed clustering does group similar type of objects based on minimum distance between the nodes.

Text Clustering methods are divided into three types namely, Partitioning Clustering, Hierarchal Clustering and Fuzzy Clustering. Partitioning algorithm randomly selects *k* objects and defines them as *k* clusters. Then it calculates cluster centroids and makes clusters as per the centroids. It calculates the similarities between the text and the centroids. It repeats this process until the criteria specified by the user is matched.

In this paper we are proposing an architecture in which every holder clusters the documents itself after pre-processing and then computes the local and global frequencies of the documents for calculation of file relevance score or document weights. Our approach makes sure that if a new dataset is placed at data holder, it requests the other data holders to forward the relevant features from other data holders instead of entire datasets to cluster the documents.

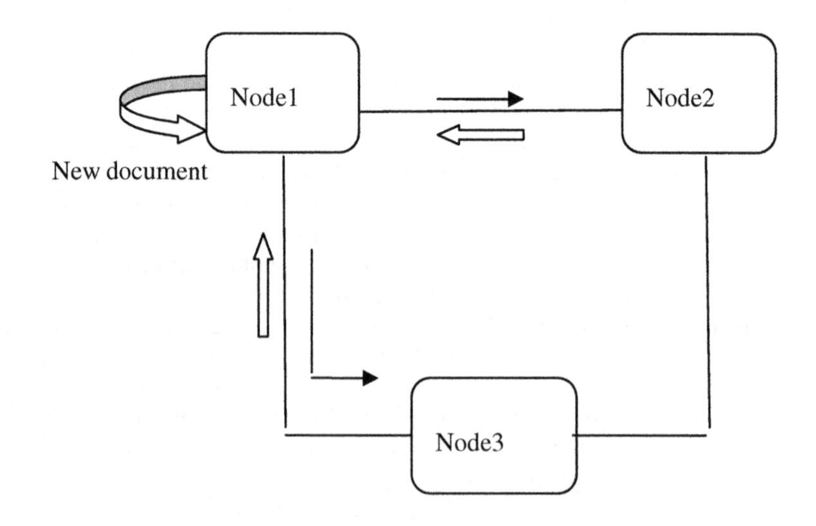

3 Proposed System

In this paper we are proposing an efficient and secure data mining technique with optimized k-means algorithm and cryptographic approach, for cluster the similar type of information, initially data points need to be share the information which is at the individual data holders. In this paper we are emphasizing on mining approach not on cryptographic technique. For secure transmission of data various cryptographic algorithms.

To cluster the similar types of information, initially data points need to be shared the information that is at the individual data holders.

Individual peers at data holders initially pre-process raw data by eliminating the unnecessary features from datasets. After the pre-processing of datasets, computation of the file relevance scores of the datasets or pre-processed features set in terms of term frequency and inverse document frequencies computes the file relevance matrix to reduce the time complexity while clustering datasets. We are using a most widely used similarity measurement i.e Cosine Similarity,

$Cos(d_m,d_n)= (d_m * d_n)/Math.sqrt(d_m * d_n)$
Where
dm is centroid (Document weight)
dn is document weight or file relevance score

In our approach we are enhancing K Means algorithm with recentroid computation instead of single random selection at every iteration, the following algorithm shows the optimized k-means algorithm as follows

Algorithm:

1. Select K points as initial centroids for initial iteration until Termination condition is met (user specified maximum no of iterations)
2. Get relevance(dm,dn)
   Where $d_m$is the document M file relevance score from relevance matrix
   $d_n$is the document N file relevance score from relevance matrix
3. Assign each point to its closest centroid to form K clusters
4. Re-compute the centroid with intra cluster data points (i.e. average of any k data points in the individual cluster).
   Ex: $(P_{11}+P_{12}+....P_{1k}) / K$
   All points from the same cluster
5. Compute new centorid for merged cluster

In the traditional approach of k-means algorithm, it randomly selects a new centroid. In our approach we are enhancing by prior construction of relevance matrix and by considering the average k random selection of document relevance scores for new centroid calculations.

For the transmission of data in between the nodes or data holders we use cryptographic technique.,3DES this will provide secured way of data transmission in distributed network.

Triple DES (3-DES, TDES)[12] is based on the DES algorithm. 3-DES has the benefit of reliability and a longer key length that removes many of the attacks that can be used to reduce the amount of time it takes to break DES. Triple DES[11] runs three times slower than DES, but is very much secure if used correctly. The procedure for decrypting something is the same as the procedure for encryption, except it is executed in reverse. The encryption of the whole data would cost more and is expensive (in term of performance). If we just encrypt the key itself and prepend it to the data, then it will not using much performance because this time the data itself is small (before we encrypt the whole data, but this time we only encrypt the key).

4 Results

The following table shows a simple way to retrieve similarity between documents at individual data holders by computing cosine similarity prior clustering.

Document	d1	d2	d3	d4	d5
d1	1.000	0.180	0.213	0.251	0.974
d2	0.180	0.999	0.521	0.160	0.443
d3	0.213	0.521	1.000	0.510	0.177
d4	0.251	0.482	0.510	0.999	0.205
d5	0.974	0.160	0.177	0.205	1.000

Above table D(d1,d2,d3….dn) represents set of documents at data holder and their respective cosine similarities, reducing the time complexity while computing the similarity between the centroids and documents while clustering each time.

Consider the above scenario, the below table shows the similarity between the new document and nodes or data holder.

Node	**Similarity with a new document at each node**
1	0.82053
2	0.88312
3	0.85916

Here Node2 has the maximum similarity. So we have to send the new document to node2 and to decide to which cluster it is to be placed. For that we have to calculate the similarity measure between the new document and cluster that present in the node2.

Below table shows the similarity between cluster in node2 and a new document.

Node2	Similarty with a new document at each cluster in a particular node
Cluster 1	0.898166
Cluster 2	0.886139
Cluster 3	0.905588

From the above table new document and cluster3 similarity is more. So the new document belongs to the cluster 3 of Node 2.

5 Conclusion

We are concluding our current research work with efficient privacy preserving data clustering over distributed networks. Quality of the clustering mechanism is enhanced with pre-processing relevance matrix. Centroid computation in k-means algorithm and cryptographic technique solves the secure transmission of data between data holders and saves the privacy preserving cost by forwarding the relevant features of the dataset instead of raw datasets. Security is also enhanced by establishing an efficient key exchange protocol and cryptographic techniques while transmission of data between data holders.

References

1. Giannotti, F., Lakshman, L.V.S., Monreale, A., Pedreschi, D., Wang, H.(W.): Privacy - Preserving Mining of Association Rules From Outsourced Transaction Databases. IEEE Systems Journal 7(3) (September 2013)
2. Fong, P.K., Weber-Jahnke, J.H.: Privacy Preserving Decision Tree Learning Using Unrealized Data Sets. IEEE Transaction on Knowlegde And Data Engineering 24(2) (Feburary (2012)
3. Tassa, T., Cohen, D.J.: Anonymization of Centralized and Distributed Social Networks by Sequential Clustering. IEEE Transactions on Knowlegde and Data Engineering 25(2) (Februrary (2013)
4. Privacy Preserving Clustering, `http://siis.cse.psu.edu/pubs/esorics05.pdf`
5. Clifton, C., Kantarcioglu, M., Lin, X., Zhu, M.Y.: Tools for Privacy Preserving Distributed Data Mining. Acm Sigkdd Exploration Newsletters 4(2) (December 2002)
6. Datta, S., Giannella, C.R., Kargupta, H.: Approximation Distributed K-Means Clustering over a peer-to-peer network. IEEE TKDE 21(10), 1372–1388 (2009)
7. Eisenhardt, M., Muller, W., Henrich, A.: Classifying document by distributed P2P clustering. In: INFORMATIK (2003)

8. Hammouda, K.M., Kamel, M.S.: Hierarchically distributed peer-to-peer document clustering and cluster summarization. IEEE Transaction Knowledge Data Engineering 21(5), 681–698 (2009)
9. Hsiao, H.C., King, C.T.: Similarity discovery in structured P2P overlays. In: ICPP (2003)
10. Oliveria, S.R.M., Zaiane, O.R.: Privacy Preservering Clustering by Data Transformation. JIDM 1(1), 37–52 (2010)
11. Triple Data Encryption Standard, http://en.wikipedia.org/wiki/Triple_DES
12. Triple DES algorithm, http://www.vocal.com/cryptography/tdes/

Enabling the Network with Disabled-User-Friendliness

T.K.S. Lakshmi Priya and S. Ananthalakshmi

Faculty of Engineering, Avinashilingam University, Coimbatore-641 108
{tkslp.dr,ananthyy}@gmail.com

Abstract. Large-scale facilities are to be established to help the under-privileged population to make their way up the education ladder and to reach out for high level employment opportunities. Tapping the vast potentials of the Internet and tuning it for this special community can show quick and effective results. In this line, we introduce the concept of 'Disabled-aware Network Infrastructure' that can ease the access to Internet resources, specifically for the disabled users. We present a model based on Deep packet inspection techniques and content adaptation algorithms for applying at the intelligent Networking elements. Using a sample scenario we describe how this model, upon implementation, can deliver a more disabled-friendly Internet content to a differently-abled end-user.

Keywords: Application-aware networks, e-learning, Open Education Resources, accessibility for disabled.

1 Introduction

In India, the disabled population is about 21 million as per 2001 census which has been reported by Central Statistics Office [1]. More than 50% are below the age of 30years. Among the age of 6-10 only about 50-60% attend school. With support from initiatives such as the Sarva Siksha Abhiyan and Inclusive Education model, school level education system has brought in a significant increase in enrollment of persons with limited disabilities.

But the state of higher education for the disabled is significantly less. We desire that India, with its rich heritage and huge Intellectual manpower, must take steps along this direction. With this as motivation, we have proposed a framework for facilitating higher education to the differently-abled persons, based on Application-aware Networking. The major role in this work is in making appropriate provisions in the hardware/software of the network elements, hence the name Disabled-user-friendliness in the network.

The rest of the paper is organized as follows: Section 2 presents the background for this work and derives the motivation based on which the idea was conceived. In Section 3 the work related to Education for the disabled and the benefits reaped by online education as an Internet-based application, from networking processing techniques are highlighted. Section 4 the proposed idea of application-level support in the network is presented with a framework and substantiated with specific cases and pointers for implementation. The paper concludes in Section 5 with open ideas future scope.

S.C. Satapathy et al. (eds.), *Emerging ICT for Bridging the Future – Volume 1*,
Advances in Intelligent Systems and Computing 337, DOI: 10.1007/978-3-319-13728-5_20

2 Background and Motivation

In this Section we present the Higher education scenario for the disabled persons in India and compare this scenario with the advancements in the conventional higher education system. Analyzing these two scenarios we bring out the motivation for this paper.

2.1 Higher Education for Disabled Persons

The 2004 statistics indicate that only two per cent of the disabled in India are enrolled in higher education as indicted in Figure 1. This is true in spite of several governmental initiatives to educate people with special needs. A significant and early initiative has been made by The Ministry of Social Justice and Empowerment via *The Persons with Disabilities Act* in 1995 which mentions that access to education at all levels should be provided to persons with disabilities [2].

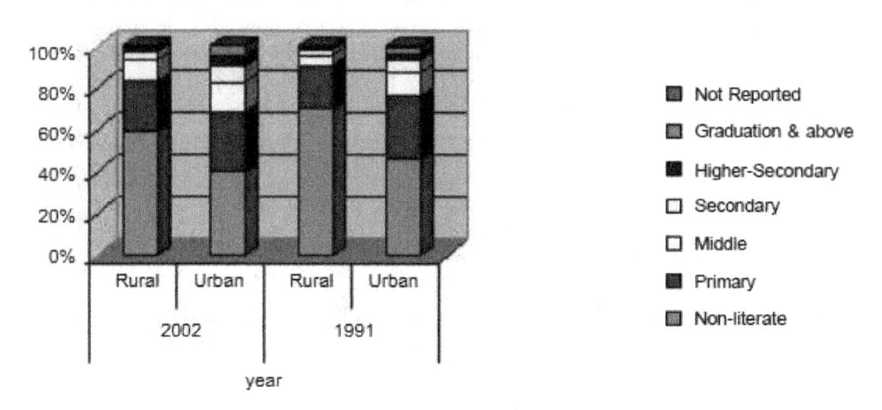

Fig. 1. Disabled Persons Education Status
Source: Report by National University of Educational Planning and Administration

Special education activities at Universities and Technical Institutions are being supported by the University Grants Commission (UGC) and All India Council for Technical Education (AICTE) from the X plan onwards [3,4,5]. Such schemes are meant to provide accessibility to disabled persons, to create a friendly environment with appropriate facilities and equipment, to spread awareness about the capabilities and limitations of such persons, and to assist them in the teaching-learning process.

In spite of such initiatives made at a national level, it is believed that the primary cause for low enrollment is the lack of societal intervention. However, we strongly believe that 'technology' can play a better role. This statement is substantiated by the very fact that we are already witnessing the benefits that the differently-abled are reaping out of the 'Mobile Technology' – talking GPS, touch screens, alarms and reminders, to name a few. One would also agree that a disabled person operates his mobile phone almost as easily as a normal person. It is this skill that we wish to kindle and help them embrace higher education.

2.2 Education and Computer Networks

Turning to the conventional education system, of late one finds the proliferation of the e-learning concept - Online courses, Open Educational Resources, Collaborative learning portals, and Massive Open Online Courses (MOOCs). One thread of thought that traces the reason for their popularity is that they are Internet-based applications and they are available on all kinds of computational devices including mobile phones. Probing deeper, one arrives at Application-aware processing [6]. The related terms are Deep Packet Inspection (DPI) [7], Deep Packet Processing, and Content Inspection (CI).

Being *Application-aware* means, network nodes are capable of operating on the application-level contents of the network packets, unlike the conventional header-only processing. This has been a major break-through in computer networking, because Application-aware processing has facilitated the movement of computation and storage, from end-nodes to network nodes, (aka offloading). This implies that end-nodes can now be lighter on compute power and storage, and thus elevating all hand-held devices to become eligible end-nodes in a networked environment. In other words, access to Internet applications is not restricted to high end computers.

2.3 Motivation

In this section we have presented two inter-disciplinary linkages, which lead to our idea for this paper. One is the connection between differently-abled persons and technology adaptation, and the other is the link between the e-learning phenomenon and innovations in network processing.

The duo: differently-abled community is conversant with technology, and technology drives education; lends its way to provisioning higher education for disabled persons through technology. However, the prevailing facilities for reaching the common learner at his door-step do not directly apply to the differently-abled persons, but requires complex adaptations. To our knowledge, not much work has been done along this line on a large scale for the Indian scenario.

In this paper we introduce the disabled-aware network infrastructure, which means that each network element does its operations that are tuned for the e-learning application for end-users who are differently-abled persons. This is done by implementing techniques such as Content Inspection and Deep Packet Inspection, thus making the node application-aware. This means that the nodes can perform specific operations upon identifying that the packets belong to e-learning application and the end-user is a disabled user.

3 Related Work

In this section, we present the related work related to two aspects: (i) applying Information and Communication technology (ICT) to provide higher education to persons with special needs and (ii) significance of network processing operations in reaching out end-users at a wide-area network level for specific application areas.

3.1 ICT-Based Access to Education for Disabled Persons

Access to education for the disabled has been tremendously eased with the development of PC-based software, Internet services, Websites, smart apps for handhelds, and special-purpose gadgets. A few are listed in Table 1.

The Global Initiative for Inclusive Information and Communication Technologies (G3ict) [8] is a non-profit organization, that aims at facilitating and supporting the implementation of the dispositions to the disabled community on the accessibility of Information Communication Technologies (ICTs) and assistive technologies.

Table 1. Sample ICT-based Assistive aids

Name	Purpose	Example
Optical Recognition Software	Extract text from image, scan, PDFs, and Photographs	ABBYY Fine Reader
Screen Reader	Reads out text on the display screen	Non-Visual Desktop Access (NVDA)
Reading Environment (Talking Book Players & Recorders)	Read books, play music, keep memos, or record lectures	Plex Talk Player
Tactile Image Enhancer	Producing relief and hence tactile images	M/S. Repro-Tronics, USA
Inclusive Publishing standards	Books for reading experience with eyes, ears and fingers	Digital Accessible Information System (DAISY) standard
Screen Magnification Program	converts your computer into a powerful magnifier	Virtual Magnifying Glass
Braille Emulation Software	Training in Braille production	BRL: Braille through Remote Learning
Closed Captioning System	Subtitling audio/video content for interpretation	enCaption

3.2 Network Processing Strategies to Enhance E-learning

The ability of the network elements, to perform application-level processing has been feasible after the introduction of high-speed and intelligent processors in the manufacture of network elements [9]. Since then the application-aware nodes have been able to perform operations such as efficient computer network traffic management and secure network communication [6, 10] for specific applications. In this Section, we present some of the networking processing techniques such as Content Distribution Networks (CDNs), DPI, in-network processing and offloading into the network, which have facilitated online education, and MOOCs.

CDNs are massive globally deployed servers which hold files (i.e., content) that can be accessed over the Internet. The mechanism provided by CDNs enables heavy files like multimedia files, to be shared thus paving way for media-rich online content access for education purposes [11]. Balica et. al. describes a DPI method for Machine to Machine Communication using specialized hardware which can be applied for e-learning [12].

Our experiences in this area involve the use of Network Processors as building blocks for application-aware networks. We have proposed architecture framework and evaluated the architecture and the suitability of Network Processors for this framework, with sample applications [13, 14, 15]. The concept of applying application-awareness to E-learning has been published in our earlier work [16]. The work in this paper is focused on facilitating the differently-abled persons with a more friendly access to the e-learning experience.

4 Disabled-Friendly Network Infrastructure – The Concept

In this Section we introduce our concept of facilitating the network infrastructure with awareness to applications, specifically e-learning applications that are accessed by differently-abled persons. We begin with how content flows in an e-learning environment and identify potential in-network operations along the flow for the e-learning environment and for adaption required for disabled persons. Then, we present a model for such operations and based on this model for a sample scenario, we provide the algorithm for a particular operation.

4.1 The E-learning Environment

During any online interaction in an e-learning environment, information flows between the E-learning server and the differently-abled learner as shown in Figure 2.

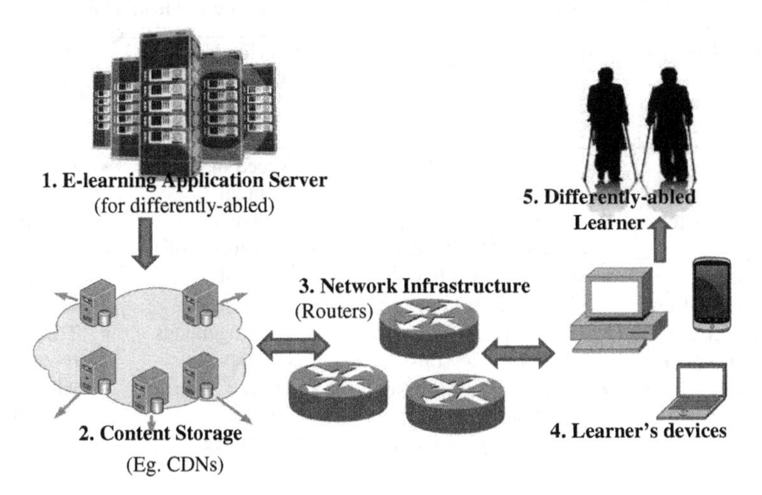

Fig. 2. Components of an E-Learning Environment

E-learning Application Server houses the web-based application that serves content for the disabled persons. Server versions of the assistive aids mentioned in Table 1 may be deployed here.

Content storage consists of large Data Centers that hold volumes of lessons, tutorials, assignments and other e-learning resources which may or may not be disabled friendly.

Network Infrastructure consists of networking elements such as the Network Switches and Routers that operate at the core and at the edge of the network. Typical services provided by these elements include load balancing, traffic management, network security and traffic redirection, for the network packets flowing between users and servers. Presently these device have become *smart* with intelligence built into their operations.

End-users' devices are the computers and smart gadgets possessed by the learners for accessing online material. For example, Personal Computers, laptops, smart phones, netbooks and iPads. Client versions of the assistive aides listed in Table 1 may be deployed at these nodes.

Differently-abled learner is the end-user who wishes to learn through online mode and is keen on using electronic gadgets.

From a network perspective, operations that involve the exchange of packets between the user and the server in the e-learning environment are tabulated in Table 2. Presently, the role of network intelligence in adapting the content for applications for disabled persons has not been explored. The assistive aids (as in Table 1) and the networking operations (as in Table 2) of the e-learning application, if supported from within the network, can provide benefits in terms of accessibility on multiple devices, minimizing end-to-end delay, and ease of use.

Table 2. Networking Operations in an E-learning Environment

Security Checks	Automated response to learners	Service-Level Agreements	Tracking Learner's Activities
• User authentication • User-website permission • Website Authentication • User-website specific checks	• Responding to Simple Queries • User-Device Specific Content Provisioning • Learner-level based Lesson Selection • User's profile-based "see also" lessons • User's location-based "sample scenarios" • User's profile-based file type	• Speed of Content Delivery • Resolution of Content • Support for Heterogeneous end devices • Device-specific quality guarantees	• Frequency of lesson-wise utilization • Time spent at specific lessons • Tracking references while at a specific lesson

4.2 Model of a Disabled-Aware in-Network Operation

Providing in-network support for operations that involve networks can be done by employing DPI / CI techniques at Intelligent networking elements. These network elements are then capable of identifying the application that generated the incoming packets (i.e., e-learning application, in this case), and that the end-user is a differently-abled person. Intelligent algorithms deployed at these smart networking nodes are used to adapt the content appropriately.

Such intermediaries can be programmed to perform certain operations that optimize the resources at the end-user by either offloading or moving certain tasks from end node to the network or by providing additional adaptations from within the network. These nodes are referred to as *disabled-aware network nodes* in this paper and the operation that they perform are the *disabled-aware operations.* The components along the end-to-end flow of traffic in an e-learning application is illustrated in Figure 3.

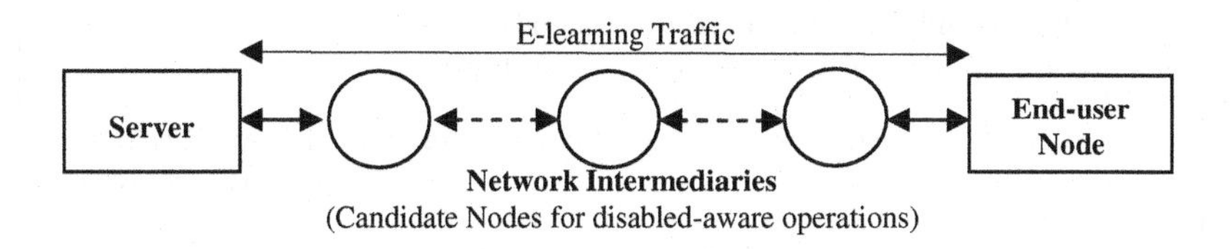

Fig. 3. Disabled-aware Nodes in an E-learning Environment

Each disabled-aware operation is modeled as a three-stage operation: (1) *Ingress*: Regular packet receive operation: during which the network node receives packets for transfer. (2) *Inspect*: (Deep Packet Inspection – DPI): Contents of the packets are inspected to check if this is a candidate packet for the given operation and (3) *App-aware*: the application-level task to be done at the network node with a specific purpose.

It is in the third stage that the operations indicated in Table 1 and/or Table 2 are incorporated. For instance, Security Checks and Automated Response to learners can be performed at the storage nodes, Service-Level Agreements can be made part of the network intermediaries while the Tracking Activities can be deployed at the user edge. Thus e-learning-aware operations can vary from value-addition to Service-Level Agreements.

The proposed model paves way for appropriately applying any of the prevalent network optimization or performance enhancement techniques at each of the stages individually. For instance, employing multiple packet receiver units at the *ingress* can help handle high traffic rates, hardware logics at *inspect* stage can hasten the DPI operations, and offloading content adaptation *app-aware operations* using network co-processors can handle complex operations.

4.3 Sample Scenario with Pseudo Code

Let us consider *Closed Captioning System* indicated in Table 1, as a sample task running at an intelligent router in our proposed environment. This means that whenever a disabled learner sends a query to the E-learning Application Server (Item No. 1 in Figure 2) for a video lesson, the intelligent router can insert sub-titles in the appropriate language. Thus, even if the e-learning server does not provide adaptation for disabled users, the provision can be offered in-the-network. Our previous works on in-network multimedia transcoding [14, 15] are a proof-of-concept for this scenario.

In Figure 4, the algorithm for the sample operation, *inNetworkCaptioning* (), is given. This algorithm is designed in accordance to the model described above.

```
inNetworkCaptioning()
// this module which is deployed at a Router, can respond to queries on the
// video lessons, contained in this server.
{
 //Step 1: INGRESS - Normal Packet Receive Operation
 Receive network packet
 //Step 2: INSPECT - Deep Packet Inspection
 Check if sender is a known learner
 Then
    Check if lesson of the known learner is available
                                         on this server
    Then
       Check if the query on the lesson is a video
                                             lesson
       Then
          // Step 3: APP-AWARE - E-learning -aware Operation
          Identify appropriate response segment
          Detect video content in frame
          Insert suitable caption
          Terminate received network packet here
          Create Egress packet
           Send this response to learner
       Endif
    Endif
 Endif
}
```

Fig. 4. Algorithm for inNetworkCaptioning()

5 Conclusion and Future Work

In this paper we have identified the need for providing Internet-based educational assists for the persons with disability. With this as the motivation, we have indicated how the concept of application awareness can be used to tune certain network operations so as to provide adaptations even for applications that are not already disabled friendly. For this idea, prospective operations have been identified that can be made in-network and disabled-aware, and for a specific scenario pseudo code has been provided.

The concept presented in this paper is to be implemented and evaluated for real-time performances. It is proposed to use NS2 simulator for test results by configuring a node as an Application-aware router, running the above algorithm on this router and generating suitable synthetic packets from the end-node.

References

1. Disability in India – A Statistical Profile, Central Statistics Office Ministry of Statistics & Programme Implementation, Government of India (2011), http://mospi.nic.in (accessed August 24, 2014)
2. The Persons with Disability Act, Ministry of Social Justice and Empowerment (1995), http://socialjustice.nic.in/pwdact1995.php (accessed August 24, 2014)
3. Guidelines for Development Grant to Colleges, XI Plan Period, UGC (2007), http://www.ugc.ac.in/oldpdf/xiplanpdf/amendmendedguidlineforcolegdev290409.pdf (accessed August 24, 2014)
4. Inclusive and Qualitative Expansion of Higher Education, 12TH Five-Year Plan, UGC (2012), http://www.ugc.ac.in/ugcpdf/740315_12FYP.pdf (accessed August 24, 2014)
5. Draft Report of Working Group on Higher Education, XI Five Year Plan, Government of India, Planning Commission (2007), http://www.aicte-india.org/downloads/higher_education_XIplan.pdf (accessed August 24, 2014)
6. Case Study, F5 and Microsoft - Creating an Application- Aware Network with Microsoft Application Center and iControl (2005), http://support-vz.f5.com/pdf/case-studies/microsoft-icontrol-cs.pdf (accessed August 24, 2014)
7. Amir, E.: The Case for Deep Packet Inspection. In: IT Business Edge (2007), http://www.itbusinessedge.com (accessed January 2008)
8. Global Initiative for Inclusive Information and Communication Technologies. G3ict, http://www.g3ict.org (accessed August 24, 2014)
9. Comer, D.E.: Network Systems Design Using Network Processors. Prentice Hall (2003)
10. Cisco Press Release, Cisco Unveils Application-Oriented Networking (2005), http://newsroom.cisco.com/dlls/2005/prod_062105.html (accessed August 2014)
11. McKeown, M.: Enabling Media Rich Curriculum with Content Delivery Networking (2005), http://www.terena.org/activities/schools/workshop-2/Cisco-CDN-mmckeown.pdf (accessed August 24, 2014)
12. Balica, A., Costache, C., Sandu, F., Robu, D.: Deep Packet Inspection for M2M Flow Discrimination - Integration on an ATCA Platform. Review of the Air Force Academy (2014), http://www.afahc.ro (accessed August 24, 2014)
13. LakshmiPriya, T.K.S., Parthasarathi, R.: Architecture for an Active Network Infrastructure Grid - the iSEGrid. In: Hutchison, D., Denazis, S., Lefevre, L., Minden, G.J. (eds.) IWAN 2005. LNCS, vol. 4388, pp. 38–52. Springer, Heidelberg (2009), doi:10.1007/978-3-642-00972-3.
14. LakshmiPriya, T.K.S., Hari, P.V., Kannan, D., et al.: Evaluating the Network Processor Architecture for Application-Awareness. In: Proc. of Second International Conf. On Communication System Software and Middleware, COMSWARE (2007), doi:10.1109/COMSWA.2007.382437, E-ISBN: 1-4244-0614-5
15. Lakshmi Priya, T.K.S., Ranjani, P.: Coordinated Support for Application-Aware Networks. In: Special Issue on New Technologies, Mobility and Security, Ubiquitous Computing and Communication Journal, Volume: NTMS - Special Issue (2008) ISSN Online 1992-8424, ISSN Print 1994-4608
16. Ananthalakshmi, S., Lakshmi Priya, T.K.S.: Extending E-Learning Awareness in the Virtual Education Ecosystem. In: International Convention on Virtual Education: Issues, Challenges and Prospects, Consortium for Educational Communication, CEC, New Delhi, pp. 110–116 (2012)

References

Hierarchical Clustering for Sentence Extraction Using Cosine Similarity Measure

D. Kavyasrujana and B. Chakradhar Rao

Department of Computer Science and Engineering, Anil Neerukonda Institute of Technology and Sciences, Sangivalasa, Bheemunipatnam[M], Visakhapatnam
{kavyasrujana,botchachakradhar16}@gmail.com

Abstract. Clustering is an unsupervised learning technique, grouping a set of objects into subsets or clusters. It forms the clusters that are similar with the data points internally, but dissimilar with the data points that are present in other clusters from each other. Extraction of data efficiently and effectively from the datasets or data holders need enhanced mechanism. Extraction of relevant sentences based on user query plays a big role in data mining and web mining etc. In this paper we propose an efficient and effective way to extract sentences by taking query as input and forming hierarchical clustering with cosine similarity measure. A Threshold value is taken initially, and clusters are divided depending on it. Further clustering is done based on the previous Threshold value.

Keywords: Hierarchical clustering, Information retrieval, Sentence similarity, Web mining, natural language processing.

1 Introduction

With the fast development of technology a huge amount of data are available over Internet and increasing enormously. It is very tough task for the users to go through the entire information and find the relevant data from the sources [1]. Sentence clustering plays an important role to handle the information overload. Sentence clustering is the process of taking the textual document and extracting the relevant information to the users in chronological order. This helps the users to make out the ideal information with in short period.

Clustering can be defined as a process of grouping up similar type of objects based on distance (for numerical data) or similarity (for categorical data) between data objects. Sentence clustering is the process of organizing the extracted data automatically for fast retrieval and quick understanding.

Clustering at sentence level is chosen because it handles many challenges when compared to clustering large segments [2,3]. The existing Clustering techniques are performed on document level than sentence level. The limitation of the document clustering is documents are too big to cluster hence an intermediate level is often useful.

S.C. Satapathy et al. (eds.), *Emerging ICT for Bridging the Future – Volume 1*,
Advances in Intelligent Systems and Computing 337, DOI: 10.1007/978-3-319-13728-5_21

In this paper the clustering is performed on the sentences where the word frequency and similarity between the sentences plays key role. The database maintains a bag of words called keywords. Sentences are extracted from the group of sentences which consists of the keywords[4]. Later the extracted sentences are grouped using the similarity measure called cosine similarity. A threshold value is taken manually and clusters are formed depending on the value.

The paper is structured as section 2 discusses the related work of sentence clustering and extraction. Section 3 shows the overview of proposed work. Section 4 presents the conclusion.

2 Related Work

Sentence clustering is the best technique for effective extraction of information from the databases. There are many clustering techniques for organizing the retrieved data. Some applied agglomerative clustering where the most similar clusters are merged together. This is process is continued until the desired number of clusters are obtained.

Sentence Clustering methods are divided into three types namely, Partitioning Clustering, Hierarchal Clustering and Fuzzy Clustering[5]. Partitioning algorithm[6] randomly selects *k* objects and defines them as *k* clusters. Then it calculates cluster centroids and makes clusters as per the centroids. It calculates the distance between the text and the centroids. It repeats this process until the criteria specified by the user is matched.

The vector space model is another technique used in clustering . It takes the semantic data of the sentences defines that the sentences with words semantically related are similar. The similarity measure[7] called word net. It checks the word to word similarity.

In our work the database maintains of bag of words called keywords. The database consists of set of sentences after pre-processing[8] . It extracts the sentences that match with the keywords[9] and then computes the similarity between the sentences using the similarity measure called cosine similarity. Later it computes the frequency of words in the sentences. A threshold value is chosen to create the cluster. As a starting point the threshold value can be empirically chosen to get the optimal clusters. Advanced techniques like evolutionary approaches can be envisioned to optimally select a threshold value.

3 Proposed System

In this paper a novel method of sentence extraction is proposed. In our proposal hierarchical clustering of sentences are formed after extraction of relevant sentences with the help of user given query which is just a keyword. A simple SQL type primitive is used to extract the sentences from the database of documents. The extracted sentences although may match with the user query but can differ with the relevant meaning. The next goal is to group the sentences having similar meaning or matching. This can be obtained if the sentences are grouped or clustered. Cosine similarity match is performed to check for the similarity among sentences. All sentences are grouped to two clusters initially based on a threshold value chosen by user with the experience. Later the concept of hierarchical clustering is used based on different threshold values down the line for arriving at different groups. The process is repeated until we arrive at the critical threshold value that is set by user to complete the clustering.

The proposed work consists of the following components:

1. Preprocessing
2. Extracting
3. Similarity measure
4. Threshold value
5. Sentence clustering

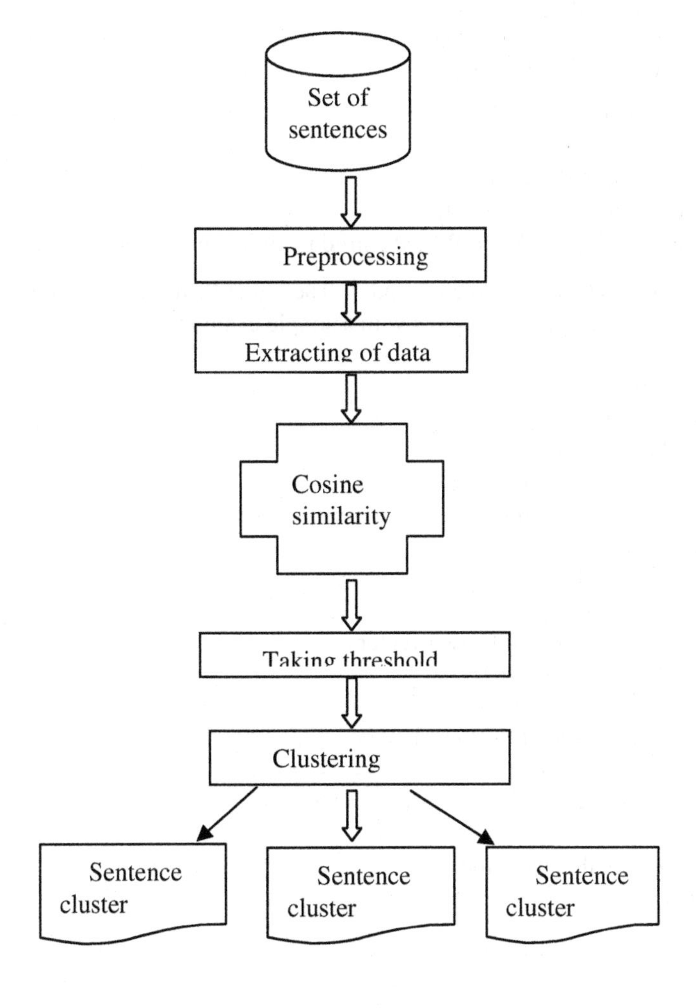

Preprocessing:

From the set of sentences we remove the stop words. The sentences are split into words. The insignificant words like "a","and","are", "the" are removed. T hese stop words do not provide any meaning of the sentences. The sentences are in the form of words.

Extracting:

After preprocessing the sentences the non- stop words are compared with the keywords and the sentences that consist of the keywords are extracted.

Similarity Measure:

The similarity measure used in our work is cosine similarity. It is defined as the measure of similarity between the two vectors.

Cosine Similarity is defined using a formula:

Similarity= Cos(□)=A.B/‖A‖.‖B‖

1. Take the dot product of vectors A and B
2. Calculate the magnitude of vector A
3. Calculate the magnitude of vector
4. Multiple the magnitude of A and B
5. Divide the dot product of A and B by the product of magnitude of A and B.

The result of this is always between 0 and 1. The value 0t means 0% similar and 1 means 100% similar. Let us discuss this with a simple example.

EX: Richard has blue eyes.
Richard has big blue eyes.
The union therefore is:
['Richard', 'has', 'big', 'blue', 'eyes']

2. Create frequency of occurrence vectors for each sentence:

Sentence 1: [1, 1, 0, 1, 1]
Sentence 2: [1, 1, 1, 1, 1]

3. Calculate the dot product of the two vectors:

1*1 + 1*1 + 0*1 + 1*1 + 1*1 = 4

4. Calculate the magnitude of the two vectors:

Sentence 1: √(1^2 + 1^2 + 0^2 + 1^2 + 1^2) = √4 = 2
Sentence 2: √(1^2 + 1^2 + 1^2 + 1^2 + 1^2) = √5

5. Find the product of the magnitudes:

2 * √5 = 4.47213

6. Divide the dot product (step 3) by the product of the magnitudes (step 4):
4 / 4.47213 = 0.8944.

The above sentences are 89% similar.
Thus we calculate the frequency of two sentences.

Threshold Value:

After finding the similarity we take a threshold value. Clustering of the sentences is done depending on the threshold value. The choosing of threshold value is a challenging task and may lead to further studied. However, in our proposal; it can be randomly chosen and with experience of working on the proposal it can be fined tuned with experience. If the similarities of the sentences are less than the threshold value then sentences are chosen to one cluster and if it is greater than threshold value they are put in another cluster.

Clustering:

In this phase sentences are grouped into clusters. The process is continued until all the cluster consists of sentences are of unique theme. The resulting clustering is the hierarchical clustering where the numbers of clusters are not defined earlier.

4 Results

A document is given as an input, for which the sentences are to be clustered based on the given threshold value. First a particular document is uploaded and preprocessing techniques are applied to the document so that, the stop words free document is obtained which in turn divided into sentences and further the similarity of sentences are calculated. The sentences having similarity values greater or equal than threshold value will be clustered into one group and the sentences with less similarity values than the threshold value are clustered into other group. Later the cluster which is having more number of sentences with variable similarity values are further clustered depending on the new threshold value. The cluster with very few sentences is not further clustered. The resultant clusters are called hierarchical clusters.

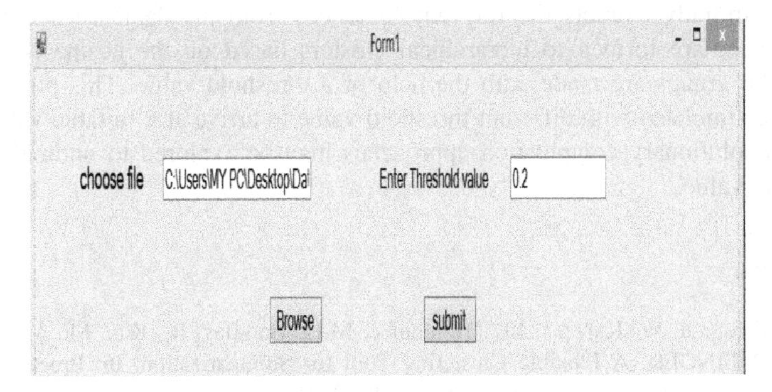

Fig. 1. Uploading a document

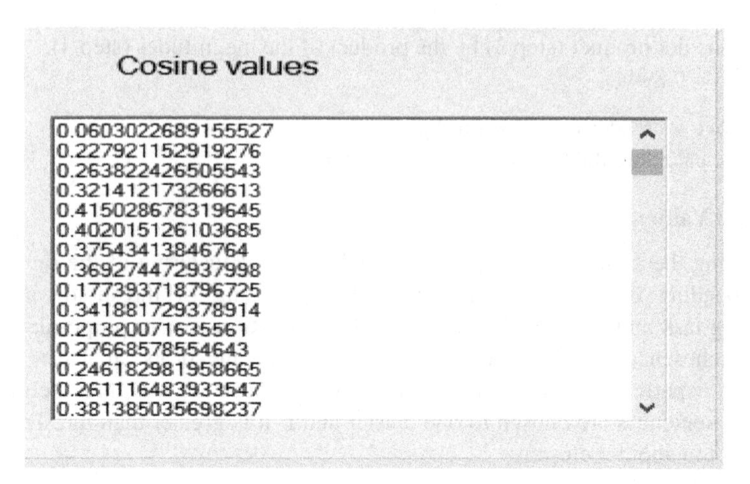

Fig. 2. Cosine similarity of sentences

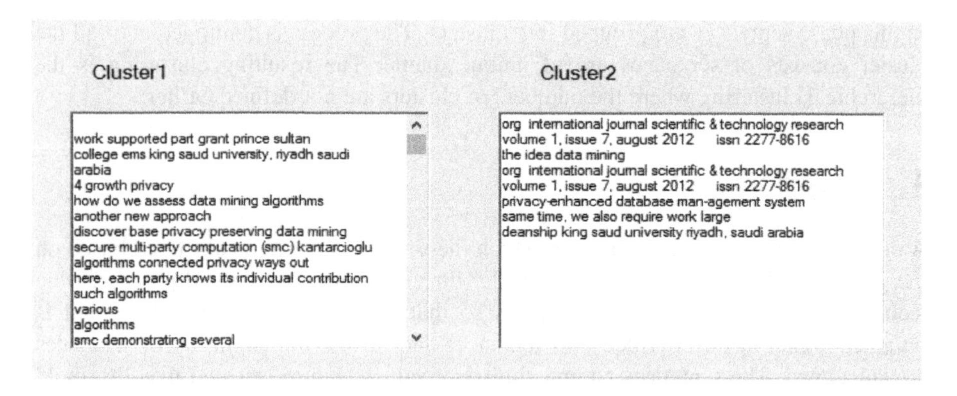

Fig. 3. Clustering of sentences

5 Conclusion

The paper has presented a novel way of extraction of sentences with clustering. The users query initially extracts the relevant sentences from the database and further those sentences are formed to hierarchical clusters based on the cosine similarity measure. The groups are made with the help of a threshold value. This proposal is further to be simulated with different threshold value to arrive at a suitable value and also some evolutionary computation approaches may be explored to optimally find the threshold value.

References

[1] Hatzivassiloglou, V., Klavans, J.L., Holcombe, M.L., Barzilay, R., Kan, M., McKeown, K.R.: SIMFINDER: A Flexible Clustering Tool for Summarization. In: Proc. NAACL Workshop Automatic Summarization, pp. 41–49 (2001)

[2] Zha, H.: Generic Summarization and Keyphrase Extraction Using Mutual Reinforcement Principle and Sentence Clustering. In: Proc. 25th Ann. Int'l ACM SIGIR Conf. Research and Development in Information Retrieval, pp. 113–120 (2002)
[3] Radev, D.R., Jing, H., Stys, M., Tam, D.: Centroid-Based Summarization of Multiple Documents. Information Processing and Management: An Int'l J. 40, 919–938 (2004)
[4] Aliguyev, R.M.: A New Sentence Similarity Measure and Sentence Based Extractive Technique for Automatic Text Summarization. Expert Systems with Applications 36, 7764–7772 (2009)
[5] Skabar, A., Abdalgader, K.: Clustering Sentence-level Text Using a Novel Fuzzy Relational Clustering Algorithm. IEEE Transactions on Knowledge and Data Engineering 25(1) (2013)
[6] Hanyurwimfura, D., Bo, L., Njagi, D., Dukuzumuremyi, J.P.: A Centroid and Relationship based Clustering for Organizing Research Papers. International Journal of Multimedia and Ubiquitous Engineering 9(3), 219–234 (2014)
[7] Wang, D., Li, T., Zhu, S., Ding, C.: Multi-Document Summarization via Sentence-Level Semantic Analysis and Symmetric Matrix Factorization. In: Proc. 31st Ann. Int'l ACM SIGIR Conf. Research and Development in Information Retrieval, pp. 307–314 (2008)
[8] Nasa, D.: Text Mining Techniques- A Survey. International Journal of Advanced Research in Computer Science and Software Engineering 2(4) (April 2012) ISSN: 2277 128X
[9] Gupta, V., Lehal, G.S.: A Survey of Text Mining Techniques and Applications. Journal of Emerging Technologies in Web Intelligence 1(1) (August. 2009)

A Novel Encryption Using One Dimensional Chaotic Maps

Saranya Gokavarapu* and S. Vani Kumari

Department of Computer Science & Engineering,
GMR Institute of Technology, Rajam, Andhra Pradesh, India
saranya.gokavarapu19@gmail.com, vanikumari.s@gmrit.org

Abstract. This paper introduces a simple and efficient chaotic system using one-dimensional (1D) chaotic map. For high security purpose this paper proposing a new encryption algorithm and it consists of uniform density function. This algorithm has more security to different type of images such as gray scale images, color images. The proposed algorithm transform images into different noise like encrypted images with excellent confusion and diffusion properties. Using a security keys this algorithm provides different encrypted image. In this paper we explain about scrambling system using hyper chaotic system it provides RGB color image which splits the three components and then by using hyper chaotic sequence generation we generate the scramble the original image. Using a same set of security keys, novel image encryption algorithm is able to generate a completely different encrypted image each time when it is applied to the same original image, and it has high sensitivity and good ability of resisting statistic attack.

Keywords: Chaotic system, Image encryption, pixel scrambling.

1 Introduction

Image encryption schemes have been increasingly studied to meet the demand for real time secure image transmission over the internet and through wireless networks. How to protect the secret information is an important issue in commercial or military application. When people send their important data to the others over a long distant, to prevent the leakage of original data became a tough and critical problem. To deal with this problem, many algorithms were proposed.

Many papers have been published on chaotic encryption algorithms. None of the papers adequately discusses the problem of security (or) estimate the computational effort required to break the system. Many existing image encryption algorithms have been proposed based on different technologies such as SCAN [2, 3] wave transmission [18] and chaos [7-12]. Among these all Novel image encryption algorithm become one of excellent and eminent encryption method. Chaotic system is defined as behaviour of dynamic system that are highly sensitive to initial conditions.

* Corresponding author.

S.C. Satapathy et al. (eds.), *Emerging ICT for Bridging the Future – Volume 1*,
Advances in Intelligent Systems and Computing 337, DOI: 10.1007/978-3-319-13728-5_22

This algorithm uses one-dimension (1D) chaotic system. On other hand we have a simple and easy structure to implement [13-17]. But they have problems like non-uniform, low security and data distribution.

To overcome the above problems this paper introduces a novel image encryption algorithm using 1D chaotic maps where this consists of high security and have uniform density function, also gives a good performance and complexity. Using a same set of security keys algorithm is able to generate a completely different encrypted image each time when it is applied to original image. Brilliant characteristics of the proposed colour image encryption approach are enough security and good performance.

2 Background

Of all the chaotic maps, the 1D chaotic maps have many advantages and applications because of their nature of simple structures. The following section discusses about three 1D chaotic maps they are Logistic, Tent, and Sine maps and these will be used for our new chaotic system.

3 New Chaotic System

In this section we will explain a new chaotic system to overcome the problems of above section and has better performance compared to above maps, three types will be discussed here

3.1 System Structure

In this a new chaotic system will be explained shown in Fig 1. It is a nonlinear combination of two different 1D chaotic maps which can be shown as seed maps. This can be derived by following equation

$$X_{n+1} = A_{FG} = \left(F(a, X_n) + G(b, X_n)\right) mod\ 1 \tag{1}$$

This can be defined as F & G. Where F (a, X_n) and G(b, X_n) are two 1D chaotic maps(seed maps) with parameters a and b, mod is defined as modulo operation, and n is the iteration number.

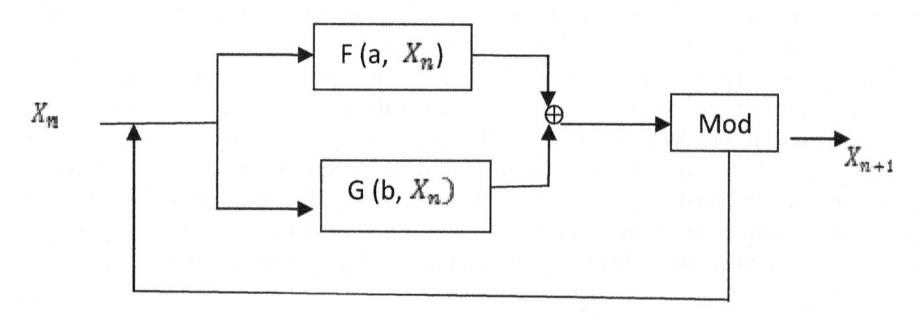

Fig. 1. Chaotic system

This mod operator ensures output data within range [0, 1] By using different type of seed maps this chaotic system will generate many chaotic sequences. Compared to all seed maps, this system has more complex chaotic properties and correct chaotic range.

3.2 Examples of the New Chaotic System

In this new chaotic system efficient and excellent performance will be shown one of its seed maps is out of the chaotic range, this new system can still have efficient chaotic behaviors., by explaining three examples.

3.2.1 The Logistic-Tent System

This example uses the combination of both Logistic and Tent maps as seed maps. So, that this system is called as Logistic-Tent system (LTS). this is defined by the following equation:

$$X_{n+1} = A_{LT}(r, X_n) = \left(L(r, X_n) + T((4-r), X_n)\right) mod\ 1 = \begin{cases} \left(rX_n(1-X_n) + \frac{(4-r)X_n}{2}\right) mod\ 1 & X_i < 0.5 \\ \left(rX_n(1-X_n) + \frac{(4-r)(1-X_n)}{2}\right) mod\ 1 & X_i \geq 0.5 \end{cases} \quad (2)$$

Where parameter r ϵ (0, 4] , Its chaotic range is within (0, 4], which is more greater than these of the Logistic maps or Tent maps. Its output sequences have uniform distribution within [0, 1]. So, the LTS has good chaotic performance than the Logistic and Tent maps.

3.2.2 Discussion

New chaotic system has minimum three advantages and good behavior, simplified structure compared with its corresponding seed maps.

At first, the distribution of its density function is more uniform than its corresponding seed maps where all the three seed maps have the limited data ranges within [0, 1] and their output sequences of the new chaotic system have the data range between 0 and 1. So, this property is good for the different applications where the data will be more secure in different information security purposes. Secondly, the new chaotic system has a great chaotic range. Even if one seed map is out of the chaotic range, the new chaotic systems still have good chaotic behaviors. Lastly, the new chaotic system has good chaotic behaviors which was determined in above three examples are all greater than their corresponding seed maps, shows better chaotic performance.

4 Novel Image Encryption Algorithm

In this algorithm encryption section contains eight steps as follows: RGB components splitting, Hyper chaotic sequence generation, Scramble of original image using hyper chaotic system, Random pixel insertion, Row separation, 1D substitution, Row combination and Image rotation. In this algorithm takes original image and apply above eight steps. The RGB components split the original image into R, G, B three components. Then hyper chaotic sequence disorders the location of R, G, B as three component pixels. These three components R, G, B can be scramble by hyper chaotic system.By their row positions combine 1D matrix into 2D matrix data and then rotate it into 90 degree counter clock wise. This process is continued four times then encrypted image is obtained. To decrypt image again apply R,G,B components splitting, hyper chaotic sequence generation and scramble of original image using hyper chaotic system. This process is known as decryption of image. After this process the final obtain the original image.

4.1 RGB Components Splitting

This method can not only change the place of the image pixel but also change the pixel values. We split the color image into R,G,B three components.

4.2 Hyper Chaotic Sequence Generation

In this section we use 2D hyper chaotic sequence to disorder the place of R,G,B three component pixels. These three different chaotic sequences are determined by a 1D Logistic map. Hyper chaos produces the hyper chaos sequence time of growth, which will likely influence confidential requirement of real time. We use the general hyper chaotic mapping derived by following equation

$$\begin{cases} X_{n+1} = aX_n + bY_n^2 \\ Y_{n+1} = cX_n + dY_n \end{cases} \tag{3}$$

Where a=1.55, b= -1.3, c= -1.1 & d=0.1.

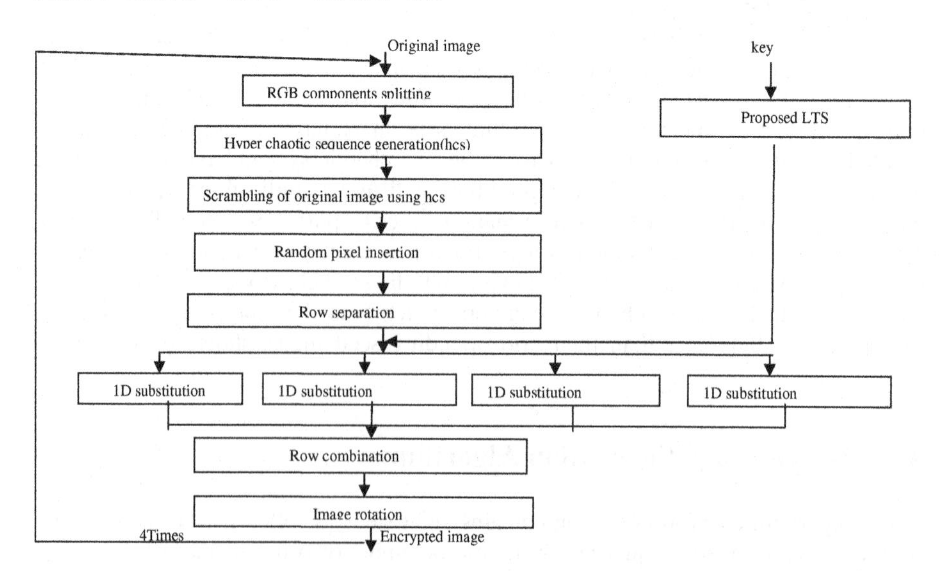

Fig. 2. Novel image encryption algorithm

4.3 Scrambling of Original Image Using Hyper Chaotic Sequence

After splitting the RGB components the pixel values of the three components R,G,B can be scrambled through the three logistic chaotic sequences and then we obtain the encrypted image and it is effective.

4.4 Random Pixel Insertion

The main aim of the random pixel insertion is inserting one pixel random value in the starting of each row in the original image, derived by following equation:

$$I(i,j)=\begin{cases} Rand(i) & if\ j=1 \\ O(i,j-1) & otherwise \end{cases} \tag{4}$$

Here O is the original/input image with the size of M×N, I (i, j) is the processed image with the size of M×(N+1),1 $\le$ i $\le$ M, 1 $\le$ j $\le$ (N+1); where Rand(i) is a random function where we insert the values randomly at starting that produces random numbers.

4.5 Row Separation

In this the image is to transfer as I(i, j) row by row into 1D matrices, derived by following equation

$$R_i(j) = I(i,j) \tag{5}$$

R_i is defined as the ith 1D row matrix with length of (N+1).

4.6 1D Substitution

This process is designed to change data values in each 1D matrix R_i . It is derived by the following

$$B_i(j) = \begin{cases} R_i(j) & if\ j=1 \\ B_i(j-1) \oplus R_i(j) \oplus ([S_k(i,j) \times 10^{10}] mod\ 256) & otherwise \end{cases} \tag{6}$$

Where ⊕ defined as bit-level XOR operation, [.] is the floor function, and $S_k(i,j)$ is the random sequence for the kth (k=1,2,3,4) in encryption round which is determined by the seed map LTS (A_{LT}) derived by the following equation:

$$S_k(i,j) = \begin{cases} S_1(0,0) & for\ i=0, j=0, k=1 \\ S_2(M,0) & for\ i=0, j=0, k=3 \\ S_{k-1}(N,0) & for\ i=0, j=0, k=2,4 \\ A_{LT}(r_0, (S_k(i-1,0)) & for\ i>1, j=0 \\ A_{LT}(r_k, S_k(i,j-1)) & for\ i>1, j>0 \end{cases} \tag{7}$$

Where r_k *and* $S_k(0,0)$ are the parameters and taken as initial value in the kth encryption round, respectively; $S_1(0,0)$, r_0 and $r_k$are defined by users.

4.7 Row Combination

In this it is defined as that after changing the data values in every row matrix, the row combination is the inverse process of the row separation and the random pixel insertion. It combines all the 1D matrices back into a 2D image matrix and then removes the first pixel in each row. This is derived by following equation:

$$C\,(i,j)=B_i(j+1) \quad (8)$$

C is defined as 2D image matrix with size of M×N and $j \leq N$.

4.8 Image Rotation

In this the image will rotate the 2D image matrix 90 degrees conter clockwise, derived by the following equation:

$$E\,(i,j)=C(j,N-i+1) \quad (9)$$

We can also use transpose function for rotating the image.. In this novel image encryption algorithm security keys are derived of six parts: the LTS parameter (r_0) and starting value ($S_1(0,0)$), the LTS parameters in every encryption round (r_1, r_2, r_3, r_4) in Eq.(12).

In image decryption process the authorized users must have correct security keys and follow the inverse method of image described in Fig. Novel image encryption algorithm. The inverse 1D substitution is derived by the following equation

$$R_i(j) = B_i(j-1) \oplus B_i(j) \oplus ([S_k(i,j) \times 10^{10}]mod\ 256 \quad (10)$$

4.9 Discussion

The novel image encryption algorithm designs five above explained processes. It has at least five advantages.

1.Generate a scrambling process of three components RGB color image using hyper chaotic sequence.2.Generate a random and cannot determined encrypted image every time when it is applied to the same original image with the same set of security keys. A new encrypted image is totally different from any previous one.3.Encrypt images with a high speed and security. This is because its 1D substitution process can be implemented parallel, which may require a high computation cost.4.Encrypt images with efficient confusion and diffusion properties, efforting to a high level of security and 5.Withstand the chosen-plaintext, data loss & noise attacks.

5 Simulation Results

The novel image encryption algorithm can provide a high level of security and good chaotic property to different type of images such as gray scale images, color images, and binary images. All encrypted images are noise like one. These can prevent the actual data from leakage.. In this algorithm we use the RGB values for scrambling by using hyper chaotic sequence method.

6 Security Analysis

When developing a new encryption algorithm, its security issues to be consider. In security keys parameters and initial values are set to 14 decimals.

6.1 Security Key Analysis

In encryption algorithm contains security key space a sufficient amount of endure the brute force attacks it has high sensitivity to any changes its security keys.

6.1.1 Securitykeyspace

The security keys of the novel image encryption algorithm are determined of five parameters (r_0, r_1, r_2, r_3, r_4) and the initial value (S1 (0, 0)). Where r_0, r_1, r_2, r_3, r_4 are in range of [0, 4] and S1(0,0)ε[0, 1].. It is completely large to resist the brute force attack.

6.1.2 Key Sensitivity

In the key sensitivity, initial key set (denoted as K1) is $r_{0=}$ 3:997, $r_{1=}$ 3:99, r_2 = 3:96, r_3 = 3:77, r_4 = 3:999, S1 (0; 0)=0:6. We use K_1 to encrypt the original image to obtain the encrypted image (denoted as E1). And then small change is applied to r_0 = 3.997 while keeping others unchanged. In the decryption process, a small change is applied only to r_1 =3.99000000000001.

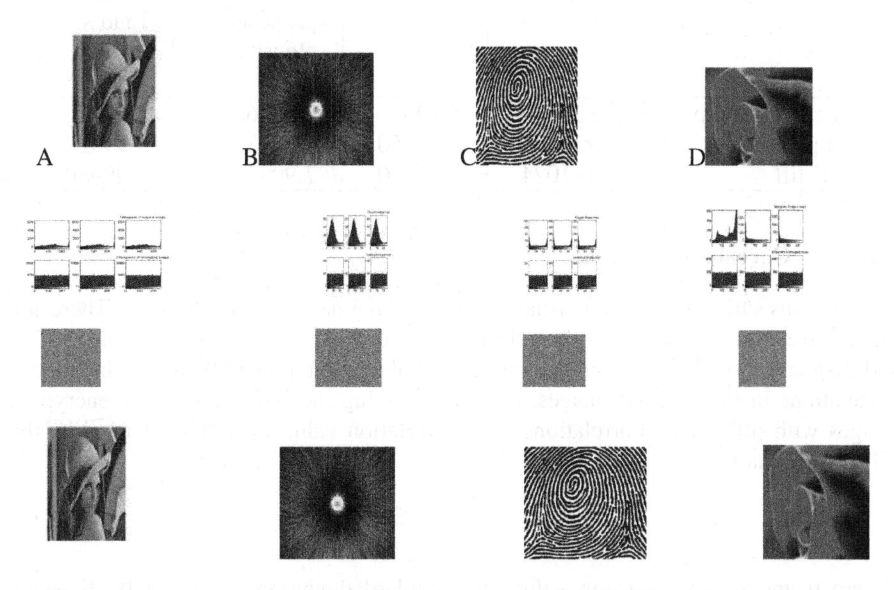

Fig. 3. The proposed algorithm encrypts different type of images. The first, fourth, fifth rows are the original, encrypted, and reconstructed images; the second and third are histograms of original images. (a) Grayscale image (b)binary image (c) fingerprint image (d) color image.

6.2 Statistical Analysis

The main theme of this novel image encryption algorithm is to transform the visually significant images into noise-like encrypted images. There are many statistical techniques for evaluating the noise-like encrypted images, including the information entropy and correlation analysis.

6.2.1 Information Entropy

The Information entropy (IFE) is calculated to evaluate the uncertainty in a random variable derived by the following equation:

$$H_L=\sum_{l=0}^{F-1} P(L=l)\log_2 \frac{1}{P(L=l)} \quad (11)$$

Where F is the gray level and P(L =l) is the percentage of pixels of which the value is equal to l. The IFE can be used for evaluating the randomness of an image. For a gray scale image with a data range of [0, 255], its maximum IFE is 8. Table shows the IFE scores of images before and after applying the novel image encryption algorithm. From these results, the IFE scores of all encrypted image with different sizes are close to8.

Table 1. Information entropy analysis

File name	Image size	Original image H_0	Encrypted image by proposed algorithm H_E^1	Encrypted image by Liao's algorithm [11] H^2_E
Fingerprint.bmp	256×256	5.19630	7.99760	7.99720
Fruit.png	512×512	7.45150	7.99930	7.99920
5.2.03.tiff	1024×1024	6.83030	7.99980	7.99980

6.2.2 Correlation Analysis

The obvious characteristic of visually meaningful images is redundancy. There are more correlations between pixels and their neighbouring pixels at horizontal, vertical and diagonal directions. The image encryption algorithm aims at breaking these pixel correlations in the original images, and transforming them into noise-like encrypted images with little or no correlations. The correlation values can be calculated by the following equation:

$$C\,xy = \frac{E[(X-\mu_x)(Y-\mu_y)]}{\sigma_x\sigma_y} \quad (12)$$

Where μ and σ are the mean value and standard deviation, respectively, E is the expectation value. Hence, a good encrypted image should be unrecognized and have the correlation values close to zero. Table2 compares correlations of the original image with its encrypted versions generated by different encryption algorithms. The original image is the image shown in Fig. 3(a).

6.3 Chosen-Plain Text Attack

The novel image encryption algorithm with the excellent diffusion property is able to resist the chosen-plaintext attack. When more existing image encryption algorithms use the same security keys to encrypt an original image, their encrypted image are duplicate. This security weakness provides the opportunity for attackers to break the encryption algorithms using the chosen-plain text attack. To address this problem, our proposed novel image encryption algorithm designs a random pixel insertion process. It allows our algorithm to generate a totally different encrypted image each time when our algorithm is applied to the same original image with the same set of security keys.

6.4 Data Loss and Noise Attacks

In real applications, images will inevitably experience the data loss and noise during transmission. An novel image encryption algorithm should resist the data loss and noise attacks.

6.5 Speed Analysis

To analyze the computation cost of the novel image encryption algorithm, we compare the encryption speed of the proposed algorithm to two state- of-art algorithms, the Liao's algorithm and Wu's algorithm Our experiments have been conducted under MATLAB7.1.10(R2012a) in a computer with the Windows7operatingsystem,Intel(R)Core(TM)i7-2600 CPU @3.40GHzand4GBRAM. Table shows the encryption time of three algorithms for different image sizes. As can be seen, the novel image encryption algorithm performs faster than the two existing ones. Furthermore, the encryption speed of our algorithm can be further improved by performing the 1D substitution processes in parallel. This says that the novel image encryption algorithm is suitable for real applications.

Table 2. Comparison of encryption time of different algorithms

File Name	Image Size	Proposed algorithm(s)	Liao's algorithm	Wu's algorithm
Fingerprint.bmp	256×256	0.1780	0.5690	7.6410
Fruit.png	512×512	0.6630	2.2510	34.7680
5.2.0.3.tiff	1024×1024	3.1420	8.9860	151.7090

7 Conclusion

This paper consists a chaotic system by combining two existing chaotic maps, chaotic system will produce a large a number of chaotic maps. This system has efficient and good chaotic behaviours, large chaotic range, high security and uniform distributed density function. Three examples are explained to show the efficient performance of the chaotic system.

To find applications of the chaotic system in multimedia security, we have introduced a novel image encryption algorithm. It has efficient confusion and diffusion properties and can also resist the chosen plaintext attack. In novel image encryption algorithm images are random and unpredictable, even using the same set of security keys and the original image. This also uses the scrambling method using hyper chaotic sequence. This can also withstand the data loss and noise attacks.

References

[1] Zhou, Y., Bao, L., Chen, C.L.P.: A new 1D chaotic system for image encryption
[2] AbdEl-Latif, A.A., Li, L., Wang, N., Han, Q., Niu, X.: A new approach to chaotic image encryption based on quantum chaotic system, exploiting color spaces. Signal Process. 93(11), 2986–3000 (2013)
[3] Chen, R.-J., Horng, S.-J.: NovelSCAN-CA-based image security system using SCANand2 Dvon Neumann cellular automata. Signal Process.: Image Commun. 25(6), 413–426 (2010)
[4] Maniccam, S.S., Bourbakis, N.G.: Image and video encryption using SCAN patterns. Pattern Recognit. 37(4), 725–737 (2004)
[5] Kanso, A., Ghebleh, M.: A novel image encryption algorithm based on a 3D chaotic map. Commun. Non Linear Sci. Numer. Simul. 17(7), 2943–2959 (2012)
[6] Xu, S.-J., Chen, X.-B., Zhang, R., Yang, Y.-X., Guo, Y.-C.: An improved chaotic cryptosystem based on circular bitshift and XOR operations. Phys. Lett. A 376(10-11), 1003–1010 (2012)
[7] Behnia, S., Akhshani, A., Mahmodi, H., Akhavan, A.: Anovelalgorithm for image encryption based on mixture of chaotic maps. Chaos Solitons Fractals 35(2), 408–419 (2008)
[8] Chen, C.K., Lin, C.L., Chiang, C.T., Lin, S.L.: Personalized information encryption using ECG signals with chaotic functions. Inf. Sci. 193(0), 125–140 (2012)
[9] Seyedzadeh, S.M., Mirzakuchaki, S.: A fast color image encryption Algorithm based on coupled two-dimensional piece wise chaotic map. Signal Process. 92, 1202–1215 (2012)
[10] Tong, X., Cui, M.: Image encryption scheme based on 3Dbakerwith dynamical Compound chaotic sequence cipher generator. Signal Process. 89, 480–491 (2009)
[11] Zhang, Y., Xiao, D., Shu, Y., Li, J.: A novel image encryption scheme based on a linea rhyperbolic chaotic system of partial differential equations. Signal Process. Image Commun. 28(3), 292–300 (2013)
[12] Cheng, C.-J., Cheng, C.-B.: An asymmetric image cryptosystem based on the adaptive synchronization of an uncertain unified chaotic system and acellular neural network. Commun. Nonlinear Sci. Numer. Simul. 18(10), 2825–2837 (2013)
[13] Bao, L., Zhou, Y., Chen, C.L.P., Liu, H.: A new chaotic system for image encryption. In: 2012 International Conference on System Science and Engineering (ICSSE), pp. 69–73 (2012)
[14] Ye, G.: Image scrambling encryption algorithm of pixel bit based on chaos map. Pattern Recognit. Lett. 31(5), 347–354 (2010)
[15] Wang, X., Teng, L., Qin, X.: A novel color image encryption algorithm based onchaos. Signal Process. 92, 1101–1108 (2012)
[16] Zhu, Z.-L., Zhang, W., Wong, K.-W., Yu, H.: A chaos-based symmetric image encryption scheme using a bit-level permutation. Inf. Sci. 181(6), 1171–1186 (2011)

[17] Bhatnagar, G., Jonathan Wu, Q.M.: Selective image encryption based on pixels of interestand singular value decomposition. Digit. Signal Process. 22(4), 648–663 (2012)
[18] Patidar, V., Pareek, N., Sud, K.: A new substitution–diffusion based image cipher using chaotic standard and Logistic maps. Commun. Nonlinear Sci. Numer. Simul. 14(7), 3056–3075 (2009)
[19] Liao, X., Lai, S., Zhou, Q.: A novel image encryption algorithm based on self adaptive wave transmission. Signal Process. 90, 2714–2722 (2010)
[20] Vani Kumari, S., Neelima, G.: An Efficient Image Cryptographic Technique by Applying Chaotic Logistic Map and Arnold Cat Map. International Journal of Advanced Research in Computer Science and Software Engineering

An Empirical Analysis of Agent Oriented Methodologies by Exploiting the Lifecycle Phases of Each Methodology

E. Ajith Jubilson[1,*], P.M. Joe Prathap[1], V. Vimal Khanna[1], P. Dhanavanthini[2], W. Vinil Dani[3], and A. Gunasekaran[4]

[1] RMD Engineering College, Chennai, India
{ajithjubilsonb.tech,drjoeprathap,vimalkhanna93}@gmail.com
[2] Sri ManakulaVinayagar Engineering College, Puducherry, India
danavanthini@gmail.com
[3] St Xavier's Catholic College of Engineering, Nagercoil, India
vinil@sxcce.edu.in
[4] Tata Consultancy Services, Chennai, India
a.gunasekaran@icloud.com

Abstract. Agent oriented methodologies illustrates the potential steps and offers a promising solution to analyze, design and build complex software systems and this makes these methodologies a significant entity to improvise the current practices that exists in today's software engineering process. Agent paradigm exhibits exigent humanoid properties. Our demand is to produce a wide range of enterprise and mission-critical applications that has to be autonomous, extensible, flexible, robust, reliable and capable of being remotely monitored and controlled. At this juncture the agent oriented methodologies need to be analyzed and chosen based on our application's need. The previous analyzes were purely conceptualized and attribute based and the phases in each methodology were not exposed in the analysis. This paper presents an empirical approach to compare each phase on course of selecting the appropriate agent oriented methodology that satisfies each scenario of our applications. An agent based system for online shopping is developed for analysis. Since software engineering methodologies are a quantifiable approach, we justify that AOSE methodologies must also be quantifiable.

Keywords: Agent oriented methodologies. AOSE.Empirical analysis. Lifecycle phases.Software engineering.

1 Introduction

The rise in interest of automated systems has given focus to agent oriented systems. The earlier attempts to develop automated software systems have ended up in complexities and bottlenecks [1]. Also they don't adapt to the environment in real world scenario [2]. Because of these reasons such a system may not be acceptable as a standard one. Nowadays there are lots of advancements in the technology and the

* Corresponding author.

S.C. Satapathy et al. (eds.), *Emerging ICT for Bridging the Future – Volume 1*,
Advances in Intelligent Systems and Computing 337, DOI: 10.1007/978-3-319-13728-5_23

updates each application face is innumerable. So we need an approach that is complete and learns unsupervised [3].

Agent oriented systemsare able to perform tasks and they provide autonomous services to assist a user without being instructed [4]. Agents collaborate in terms of different competence and functionalities while supporting services to an application. They are responsible for making smart opinions to execute such activities and to provide valid and useful results. Agents can perform solo work, but most importantly, they can comply with other agents. Agents are software components in the system that are viewed as many workers of different designations doing their respective task in order to achieve the overall outcome. Examples of agents include Facebook friend search which gives the apt person who has a kin with us, mobile phone application search like application stores, agents controlling smart home management, system diagnostics, transportation logistics and network management[5][6].

Another description as Agent-Based Software Engineering (ABSE) is a refinement of some aspects of AOSE based upon practical experience in agent building [7]. The agent oriented methodologies are classified according to the development patterns and then the empirical survey is done for one methodology belonging to every pattern [8]. The classification of the agent-oriented methodologies and their description are given below.

2 Literature Survey

Fig.1.illustrates the classification of agent oriented methodologies based on development patterns. The detailed descriptions are given below.

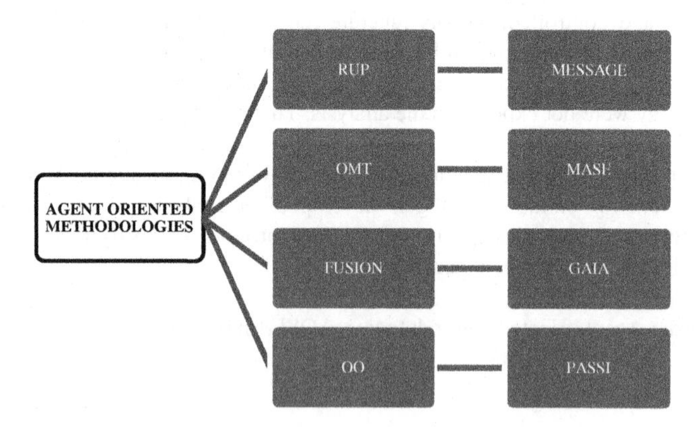

Fig. 1. Agent oriented methodologies classification

2.1 Rational Unified Process (RUP)

The RUP development approach gives a formatted way for companies to foresee creation of software programs. Since it provides a specific plan for each step of the development process, it enhances reusability and reduces unexpected development costs [9]. The phases of RUP include Inception, Elaboration, Construction and Transition [10].

2.1.1 MESSAGE

MESSAGE follows the iterative and incremental approach of RUP that involves progressive refinement of the requirements, plan, design, and implementation. It focuses on the elaboration phase where the core system architecture and high risk factors need to be identified.

2.2 Object Modeling Technique (OMT)

OMT is a modeling and development technique that concentrates on forecasting the product and communicating regarding the product with the stakeholders [11] [12].

2.2.1 MASE

The Multi Agent Systems Engineering is an object modeling technique that relies on visual representation of the system. The roles and goals play a vital role in this system. Roles are abstractions of the system whereas goals are the representations that portray the system [13].

2.3 Fusion

Fusion is a software development approach that enhances reusability. The concepts addressed in this model include dynamic design, development and testing. The components in this model are loosely coupled in order to ensure communication problem between the components. The phases in this approach include project preparation, project blueprint, realization, testing and go live and support. The new phase included here is go live and support, which ensures continuous tracking and monitoring of the system.

2.3.1 GAIA

GAIA (Generic Architecture for Information Availability) was the first complete methodology proposed to guide the process of developing a multi agent system from analysis to design. The models included in GAIA are role model, interaction model, environment model and organization model. It is depicted in Fig. 2.

This approach concentrates on production of components that can be adapted according to the environment. The organization model mainly concentrates on those agents whose presence was not identified in the analysis phase derives directly from the adoption of a given organizational structure. The service model aims to identify the services associated with each agent class or equivalently, with each of the roles to be played by the agent classes.

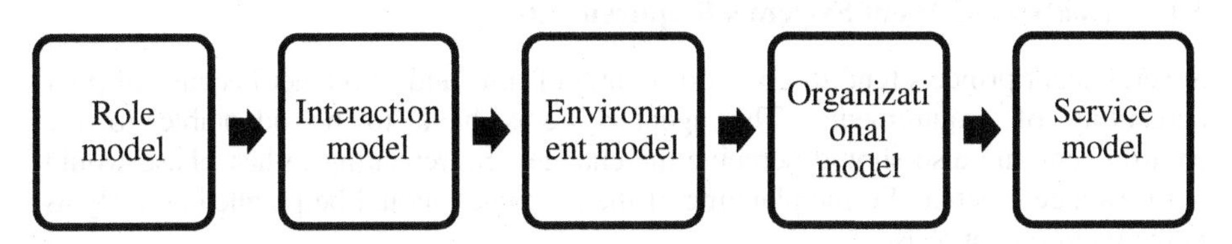

Fig. 2. Models of GAIA

2.4 Object Oriented Approaches

Object oriented approaches are developed by incremental adding of each phase of the lifecycle activity like analysis, design, implementation and testing. These are done iteratively with recursive cycles. The special feature of object oriented design is interaction between several objects. The objects are initialized in different classes. The objects interact by calling the methods of the other object; this is done by sending a message to the other object. This has to be instantiated. But adapting this to agent oriented approaches; message passing should happen without being instantiated.

2.4.1 PASSI

A Process for Agent Societies Specification and Implementation (PASSI) is iterative, reusable and the most widely accepted object-oriented methods. There are two types of iterations in it. The first is characterized by new requirements and involves all the PASSI models. The second iteration involves only the modifications to the Agent Implementation Model. It is defined by a double level of iteration. The PASSI process is composed of five process components that follow the OO approach as shown in Fig. 3.

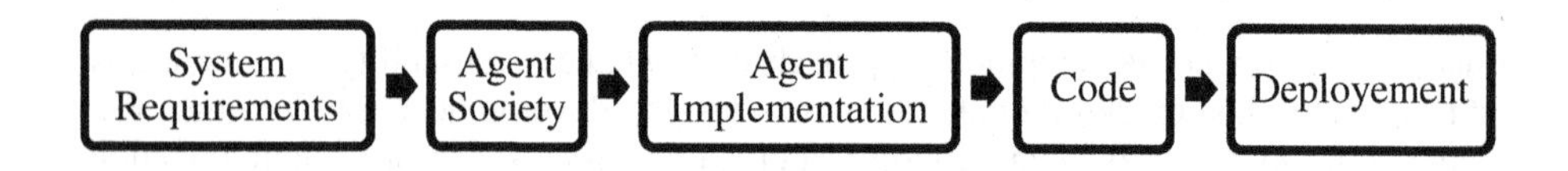

Fig. 3. PASSI Lifecycle

3 Empirical Analysis

To analyze a process quantitatively leads to better planning, controlling and optimizing of it. Quality constraints such as accuracy, security and performance are often crucial to the success of a software system. Manual analysis or theoretical analysis is always not good because due to the natural human tendency of hiding erroneous outputs. By the omission of these erroneous outputs the percentile of choosing the apt model to develop an agent oriented system becomes a failure. The empirical analysis solves this problem of choosing the apt agent oriented model for our system. Also we exploit the various lifecycle phases of the waterfall approach to make our comparison very vivid.

3.1 Analysis of Agent System's Requirements

Several agent projects tend to go out of bound of time and of budget because of poor perceiving of requirements. The agent systems have to be adaptable to the environment and also should perceive the changes. So repenting in last phase would not be a nice practice. So the planning of the outcomes should be planned as early as in the requirement phase.

3.1.1 Agent Volatility Metric

It is the standard deviation of the requirements defined in the SRS to the actual implementation. This metric takes function points to consideration since it is the best potential metric to calculate the agent functionality aspect of the system.

$$\sigma_f = \sqrt{E[X_f - \mu_f]^2}\,. \qquad (1)$$

Where

$\sigma_f$is the standard deviation of the function points.

E denotes the average or expected value of X_f.

X_f is the actual number of function points in the Agent System.

$\mu_f$is the average number of function points defined in the requirements.

3.1.2 Agent Achievability Metric

It is the percentage of the number of agents and their functionalities defined in the requirements to the actual implementation of the agent system.

$$P_{af}= D_{af}/ I_{af}*100. \qquad (2)$$

Where

P_{af} is the percentage of defined agents and the implemented agents

D_{af} is the number of agents and functionalities defined in the requirements

I_{af} is the no of agents and their functionalities implemented and verified

3.2 Analysis of Agent System's Design

With the emergence of agile development methodologies like extreme programming (XP) the requirements and design phase has changed several folds. The time and the effort for these phases at the initial stage have been sheered to a good extent to allow changes to be feasible at the later stages. So these metrics could play a handy role in shaping the structure of the design.

3.2.1 Module Independence Metric

Module is a part of software product that has a specific task. It must be self-standing and independent to other parts. As we have seen the buzz words of agents like autonomous and self-contained the module independence has to play an important role in the agent oriented system design. The module independence is derived from the *module pool programming* employed in SAP. A sequence of screens of a particular module having their inputs, outputs and processing logic just behind them are said to be module independent. This metric range from 0 to 1. 0 is the worst case and 1 is the best case.

$$MI= [MI_1+MI_2+\ldots+MI_N]/N. \qquad (3)$$

Where $MI_N=OC_N/IC_N$

$MI_N$is the module independence of module N.

OC_N is the output component of module N.

$IC_N$is the input component of module N.

The module independence of the complete project is calculated by the average module independence of each module.

3.2.2 The Response for Class Metric

The response set of a class is a metric that can be credibly executed in echo to a message received by an object of that class. The communication between the different objects in the class also corresponds to this metric. The design of a project would be triumphant only if the number of remote communication stands feeble. The communication between objects must be viable betwixt the class. And the final measure should not count scads of responses. The response for class also includes recursive calling within the class. So lower the response for class higher the throughput of the system.ie., lower the response for class higher the caliber of the project

$$RFC = M + R \text{ (First-step measure).} \tag{4}$$
$$RFC' = M + R' \text{ (Full measure).}$$

Where

RFC is the response for class.
M is the number of methods in the class.
R is the number of remote methods directly called by methods of the class.
R' is the number of remote methods called, recursively through the entire tree.

A comparative study is made for an online railway ticket booking system in which the ticket booking module is developed using several languages like C, C++ and JAVA. Here, several inter process communication and reliance on superficial data is induced to check the capability of the languages. The inputs had different lines of codes and many classes. The obtained outputs are tabulated in Table 1.

3.3 Agent System's Implementation Analysis

The disentanglement of the Business Process Management (BPM) strategies of the development paradigm has shifted from conventional agent software development to the business oriented agent development. The implementation is entirely different from requirement and design which are done using predefined business process models.

3.3.1 Agent Functionality Efficiency

We emphasize on the fact that the no of agents used in the agent oriented application is in direct proportion to the functionality addressed by the system. So if more number of agents has been deployed the usability and efficiency of the system is going to rise in a high.

$$AFE \ \alpha \ \sum N(a) \ . \tag{5}$$
$$\text{Usability} \ \alpha \ N(a).$$

Whrere

AFE is Agent Functionality Efficiency.
N (a) is the number of agents used.

Table 1. Comparison of C, C++ and JAVA using Module Independence & response for class metric

Language	*C*	*C++*	*JAVA*
Module Independance	*0.43*	*0.62*	*0.94*
Response For Class	*80.34*	*60.17*	*34.83*
Quality	*Low*	*Medium*	*High*
Bugs	*High*	*Medium*	*Low*

3.4 Analysis of Agent System's Testing

Testing is a vital activity of the software development process. Competence of the testing techniques determines the value of the software. Based on the study made, testing phase in all the agent development methodologies has been given minimum importance. Unless we define measures, it is difficult for us to test a particular system. Thus it becomes mandatory for defining metrics for testing agents system.

3.4.1 Defect Elimination Efficiency (DEE)

It is the calculation of the number of defects that are handled by the developer before the release of the product. It is calculated as a percentage of defects resolved to total number of defects found.

$$DEE\ (\%) = \left(\frac{\sum Defects\ Resolved}{\sum \text{Defects at the moment of measurement}}\right)*100. \tag{6}$$

3.4.2 Defects Per Size (DPS)

The Defects PerSize analysis helps us to find out the flaws according to our development size. The size of the system can be the classical Lines Of Code (LOC) or function points. The Defect Per Agent(DPA) can also be calculated using the same calculation.

$$\text{DPS (\%) = (Defects detected / system size)*100.} \tag{7}$$
$$\text{DPA (\%) = (Defects detected / no of agents used)*100.}$$

3.4.3 Number of Test cases Per Unit Size (TPUS)

Test cases are the smallest component and the basic requirement of a testing scenario. The complexity and the efficiency of the testing depend on the test cases written. Higher the test cases more complex the testing would result in. Lesser the test cases the lower the testing efficiency. So there must be a good balance between the test cases and the complexity of the system which would result in the successful testing of the system.

$$\text{TPUS = Number of test cases / (KLOC (or) FP).} \tag{8}$$

Where

TCPUS is the Number of Tests Cases per Unit Size.

KLOC represents Kilo Lines of Code & FP represents Function Point.

3.5 Analysis of Agent System's Deployment and Maintenance

The evolution of heterogeneous environments and the communicational complexity has raised the difficulty of deployment and maintenance. As the motive of the agent oriented development states that the agent system should be adaptable to the external environment. So our deployment of the system should satisfy all the constraints to achieve this.

4 Experimental Analysis

A software product for Online shopping system which involves agents has been developed using JADE (Java Agent Development Framework). The configuration of the product of agents which are developed using Jade can be changed even in run time.

The proposed analysis has to be made at different stages of abstraction levels of the system such as initialization, system development, integration and acceptance. An agent system for online shopping is developed with the various methodologies like Message, MASE, GAIA and PASSI.

Table 2. Analysis of the methodologies using the requirement metrics

Requirement Metrics	*Message*	*MASE*	*GAIA*	*PASSI*
Agent Volatility Metric	*4.846*	*4.037*	*4.327*	*4.248*
Agent Achievability Metric	*83.762%*	*87.795%*	*86.225%*	*86.127%*

Table 2 tabulates the analysis of the requirement metrics for agent oriented methodologies. Table 3 tabulates the analysis of the design metrics for agent oriented methodologies. With the inquiry of the requirement and design phases of the agent oriented methodologies we find a little deviation of parameters from one methodology to the other. The agent oriented methodologies are well designed in order to satisfy these phases. But on close analysis we find that the Message methodology is lagging in few design metrics such as Response for Class and Module Independence. MASE is best suited for these requirement and design phases of an agent oriented system.

Table 3. Analysis of the methodologies using the design metrics

Deisgn Metrics	*Message*	*MASE*	*GAIA*	*PASSI*
Module Independence	*0.826*	*0.978*	*0.927*	*0.948*
Response For Class	*89.148*	*94.329*	*93.662*	*84.163*

Now the latter phases of the agent oriented system need to be taken into consideration. Since the agent tool does not take into consideration of the functional aspect of the system it cannot be used to analyze the implementation and testing so an activity diagram is drawn which displays the workflows of the system which help us to analyze the working aspects of the system. Also with the help of the workflows the systems can be tested for their performance.

Table 4. Analysis of the methodologies using the implementation metrics

Implementation Metrics (Number of agents used =5)	**Message**	**MASE**	**GAIA**	**PASSI**
Functional efficiency	*72.53%*	*77.97%*	*78.55%*	*96.32%*
Usability	*79.83%*	*79.24%*	*80.32%*	*98.44%*

Table 5. Analysis of the methodologies using the testing metrics

Testing Metrics	***Message***	***MASE***	***GAIA***	***PASSI***
Defect Elimination Efficiency (DEE)	*40.73%*	*46.91%*	*44.17%*	*48.64%*
Defects per Size (DPS)	*66.32%*	*53.47%*	*50.25%*	*46.48%*

With the analysis of the Implementation phase of the agent oriented methodologies, we clearly observe that PASSI methodology stands out because it belongs to the Object Oriented Modeling. The principles of Object Oriented Programming aid the development scenario of the PASSI methodology. At the end of the testing phase we study that the Agent Oriented methodologies clearly lag in the testing of the developed systems. So some new modules need to be added with the testing scenario of these methodologies. Thus for the phases analyzed here are quantifiable with proper metrics. Since these phases are involved with the developer, a quantitative analysis is very feasible. But the final deployment phase heavily relies with the customer end, which involves several external factors. So there is a chance for major deviation from the developer perspective. If we need to analyze the methodologies on consideration with deployment, a case study is a vital one and it will serve our purpose of quantitative approach.

5 Conclusion

Agent Oriented Methodologies is a promising new approach to software engineering that uses the notion of agents as the primary method for analyzing, designing and implementing software systems. The effectiveness of Agent Oriented Methodologies resides in its ability to translate the distinctive features of agents into useful properties of (complex) software systems and to provide an intuitive metaphor that operates at a higher level of abstraction compared to the object oriented model. So in this paper we display an empirical approach to compare each phase of every agent oriented methodology on course detecting the efficiency of it so that they could be utilized to develop an efficient agent application that is cost effective and are of industrial standards.

References

1. Brazier, B., Dunin-Keplicz, N., Jennings, R., Treur, J.: Formal specification of multi-agent systems: a real-world case. In: Proceedings of First Int. Conf. on Mult-Agent Systems (ICMAS 1995), San Francisco, CA, pp. 25–32 (2005)
2. Abdelaziz, T., Elammari, M.: A Framework for evaluation of Agent-OrientedMethodologies. In: International Conference Oninnovations in Information Technology, pp. 491–495 (2007)
3. Bokhari, M.U., Siddiquiq, S.T.: Metrics for Requirements Engineering and Automated Requirements Tools. In: Fifth National Conference Computing for Nation Development (2011)
4. Henderson-Sellers, B., Giorgini, P.: Agent-Oriented Methodologies. Idea Group Publishing (2005)
5. Joe Prathap, P.M., Ajith Jubilson, E.: Facebook Impact on Engineering Students in India – A Case Study. Jokull Journal 63, 54–65 (2005)
6. Alonso, F., Fuertes, J.L., Martinez, L., Soza, H.: Measuring the Social Ability of Software Agents. In: Proceedings of the Sixth International Conference on Software Engineering Research, Management and Applications, Prague, Czech Republic, pp. 321–345 (2008)
7. Bordini, R.H., Moreira, Á.F.: Proving BDI properties of agent-oriented programming languages In the asymmetry thesis principles in AgentSpeak (L). Annals of Mathematics and Artificial Intelligence Special Issue on Computational Logic in Multi-agent Systems 42(1-3), 197–226 (2004)
8. Bergandi, F., Gleizes, M.P.: Methodologies and Software Engineering for Agent Systems. Kluwer Academic Publishers (2004)
9. Belguidoum, M.: Analysis of deployment dependencies in software components. In: ACM Symposium on Applied Computing, pp. 735–736 (2006)
10. Dumke, R., Koeppe, R., Wille, C.: Software Agent Measurement and Self-Measuring Agent-Based Systems. In: preprint No 11. Fakultatfur Informatik, Otto-von-Guericke-Universitat, Magdeburg (2000)
11. Srivastava, P.R., Karthik Anand, V., Rastogi, M., Yadav, V., Raghurama, G.: Extension of Object-Oriented Software Testing Techniques to Agent Oriented Software Testing. Journal of Object Technology 7, 155–163 (2008)
12. Fuxman, A., Liu, L., Pistore, M., Roveri, M., Mylopoulos, J.: Specifying and analyzing early requirements in Tropos: some experimental results. In: Proceedings of the 11th IEEE International Requirements Engineering Conference, Monterey Bay, CA. ACM, New York (2003)
13. Rumbaugh, J.R., Blaha, M.R., Lorensen, W., Eddy, F., Premerlani, W.: Object-Oriented Modeling and Design. Prentice-Hall, Inc. (1991)

An Expeditious Algorithm for Random Valued Impulse Noise Removal in Fingerprint Images Using Basis Splines

Mohit Saxena

APEX Group of Institutions,
Jaipur-303905, India
mohit.saxena234@gmail.com

Abstract. In image forensics, the accuracy of biometric based identification and authentication system depends upon the quality of fingerprint images. The quality of the fingerprint images gets compromised by the introduction of various kinds of noise into them. The fingerprint images may get corrupted by random valued impulse noise mainly during the capturing or transmission process. To obtain noise free fingerprint images, they are processed using noise removal methods and filters. In this paper, a two stage novel and efficient algorithm for suppression of random valued impulse noise using basis splines interpolation is proposed. The algorithm removes noise from the image in the first stage and in second stage, it regularizes the edge deformed during noise removal process.

Keywords: Fingerprint images, noise suppression, basis splines, edge preservation.

1 Introduction

Fingerprint based recognition is most commonly used biometric identification and authentication system. The idea is based on fact that each individual have unique fingerprint as well as its characteristics remain unchanged with time and therefore can play an important role in security, automatic identification and authentication systems [1]. A biometric system is basically a pattern recognition system that works on extracting biometric data such as fingerprints, iris, facial features etc. from a person, making feature set from the extracted data and comparing this feature set with template set stored in database [2]-[5].

Fingerprint images may get corrupted by noise which degrades the quality of the image by replacing original pixel values with new ones and hence result in the loss of contained information. Impulse noise corrupts images mainly during acquisition process or transmission process. It can be broadly classified into two categories i.e. salt and pepper noise because the noisy pixel candidate can take value either 0 (darkest) or 255 (brightest) and random valued impulse noise in which noisy pixel candidate can take value varying in the range of 0-255[6], making it hard to detect because in local neighborhood difference between noisy and noise free pixels are not significant and also, its existence cannot be determined by image histogram [7]-[8].

S.C. Satapathy et al. (eds.), *Emerging ICT for Bridging the Future – Volume 1*,
Advances in Intelligent Systems and Computing 337, DOI: 10.1007/978-3-319-13728-5_24

So, to maintain the efficiency and accuracy of biometric systems by extracting correct details from the images, it is necessary to remove contained noise from the fingerprint images. In this paper, fingerprint image corrupted with random valued impulse noise is considered and is processed using proposed algorithm of noise removal.

The paper is organized as: in section 2; the novel proposed algorithm for suppression of Random Value Impulse Noise (RVIN) in fingerprint images is discussed. In section 3, performance measures like PSNR, MSE and SSIM are discussed. Section 4 comprises of simulation results and discussion. Section 5 concludes the work.

2 Proposed Basis Spline Interpolation Algorithm

The Cubic Basis splines (B-splines) [9] are used for interpolation in the proposed algorithm. B-Spline are used by researchers like Unser in signal and image processing [10].B-spline are generally used for generating closed curves in sections by specifying four out of six control points cyclically for each section

The Basis Functions from [9] include:

$$B_0(u) = (1 - u)^3 / 6 . \quad (1)$$

$$B_1(u) = (3u^3 - 6u^2 + 4) /6 . \quad (2)$$

$$B_2(u) = (-3u^3 + 3u^2 + 3u + 1) /6 . \quad (3)$$

$$B_3(u) = (u^3) /6 . \quad (4)$$

Basis Spline is used for interpolation because they have compact support passes through control point and have local propagation property. Also, the Basis spline is cubic, the polynomial is of order 4, and the degree is 3.It is also known that for a Basis spline polynomial of order k, the continuity condition for interpolation is C^{k-2}. Therefore, for the above cubic Basis spline, we get C^2 continuity which provides the required smoothness and the interpolation is more precise.

Stage 1:

The Basis splines in equations (1)-(4) are used for the interpolation and removal of random value impulse noise from the fingerprint image. For each and every individual pixel of the image, initially the noise free pixels are extracted from the image by checking the pixel value against the threshold value. The noisy pixel candidates have been identified and there indices are placed in a matrix which will be used later for edge restoration. Four pixel values are required for the noisy pixel interpolation and for every pixel the mask size used is of 3x3.

Let original fingerprint image be X(i,j) and Y(i,j) be the fingerprint image corrupted by random valued impulse noise. We initialize with mask $W(x,y) \in Y(i,j)$, for every pixel of corrupted fingerprint image Y(i,j).

The window of size 3x3 for every element of X is applied. The absolute difference of all the elements of a window with center element is calculated.

These are now taken as control points represented by C(1), C(2), C(3), C(4) for Basis spline interpolation. Also, taking difference between adjacent elements and marking the maximum difference value as d_i. So, in case of interior pixel the threshold value of $t_i > d_i$ is used and in the case of edge pixels, threshold value $t_e >= 30$ is used.

```
A. Algorithm :Detection and Elimination of Random
   Valued Impulse Noise

   Input: Original fingerprint image X (i,j)
   ------------------------------------------------
   Image with noise Y (i,i), for all Y(i,j)

   The window Wxy is of size=3x3, where Wxy∈ R.

   Step 1: Calculating absolute difference of each
   element from center element and sorting difference
   in ascending order.
   Step 2: Writing the corresponding elements of
   elements in step 1.
   Step 3: Calculating difference with neighbouring
   element and making groups of three elements.
   Step 4: Adding elements of group and selecting
   minimum sum value.
   Step 5: Corresponding elements are taken as control
   points (C(1), C(2), C(3), C(4)) for interpolation.
   Step 6: Calculating difference between adjacent
   elements of Step 5 and mark the maximum difference
   value as di (threshold for interior pixel).

    Now, for ti> di and te>30
    for four pixels in Z, and u=0.10, Calculating
   {

   b0 ←        (1 - u)^3/6;
   b1 ←        (3u^3 - 6u^2 + 4)/6 ;
   b2 ←        (-3u^3 +3u^2 + 3u + 1)/6;
   b3 ←        (u^3)/6;

   nv=b0*C(1)+b1*C(2)+b2*C(3)+b3*C(4)

   Y(i,j) ←       nv (Replacing old value)
   }
   Replace Onf (i,j)   ←      Y(i,j)
   Output: (Onf) Noise free restored image
```

Stage 2:

Here, the noise free but edge deformed image of stage 1 is taken as input. The noisy pixels detected in stage 1 are only taken into consideration. These noisy pixels are again interpolated using cubic basis spline and final noise free and edge regularized image is obtained.

B. Edge restoration Algorithm

```
∀ O_nf (i, j) and u=0.10
for i = 1 : L4 (Noisy detected pixels)

        C(1)  = upper element of O_nf (i ,j)
        C(2)  = below element of O_nf (i ,j)
        C(3)  = element left of O_nf (i ,j)
        C(4)  = element right of O_nf (i, j)

b0   ←     (1 - u)^3/6;
b1   ←     (3u^3 - 6u^2 + 4) /6;
b2   ←     (-3u^3 +3u^2 + 3u + 1) /6;
b3   ←     (u^3) /6;

nv=b0*C(1)+b1*C(2)+b2*C(3)+b3*C(4)
Op(i,j)   ←     nv (Replacing noisy pixel)
end
Output: Op (Noise free and edge regularized image)
```

3 Performance Measures

In this section, the performance such as PSNR, MSE and SSIM, used for performance evaluation of the proposed algorithm are discussed.

A) Peak-to-Signal Noise Ratio (PSNR) and Mean Square Error (MSE)

It is one of the most commonly used parameter for evaluation and comparison of results of different algorithms. The proposed algorithm has been tested for noise levels starting from 20% and to a maximum of 60%. All the results have been analyzed using MSE and PSNR given below.

$$\mathrm{MSE} = \frac{1}{mn} \sum_{x=0}^{m=1} \sum_{y=0}^{n=1} |A(x,y) - R(x,y)| . \tag{5}$$

Where, A represents the original image and R is the restored image of resolution m*n.

$$PSNR = 10\log_{10}\left(\frac{max^2}{MSE}\right). \tag{6}$$

Where, max represents the maximum possible value of element (pixel) i.e. 255 in case of gray scale images.

B) Structural Similarity Index Measure (SSIM)

PSNR alone is not enough to evaluate the performance of an algorithm. SSIM [11] provides a new and very prominent measure for restored image quality measurement developed by Wang Z. Bovik et.al.Structural similarity index measure [11] calculates image quality by taking original noise free image as a reference and takes into account the luminance, contrast and the structural information of the image. It is found to be an excellent measure in comparison to PSNR and MSE. SSIM is defined as:

$$SSIM(x,y) = \frac{(2\mu_x\mu_y + C_1)(2\sigma_{xy} + C_2)}{(\mu_x^2+\mu_y^2+C_1)(\sigma_x^2+\sigma_y^2+C_2)}. \tag{7}$$

and mean SSIM index (MSSIM) is given by :

$$MSSIM(x,y) = \frac{1}{M}\sum_{m=1}^{M} SSIM(x_m, y_m). \tag{8}$$

where μ_x is the mean of image x and σ_x is the standard deviation of image x, similarly μ_y is the mean of image y and σ_y is the standard deviation of image y, C_1,C_2 are the constants σ_{xy} is the co-variance of x and y, given by :

$$\mu_x = \frac{1}{N}\sum_{i=1}^{N} x_i. \tag{9}$$

$$\sigma_x = \left[\frac{1}{N-1}\sum_{i=1}^{N}(x_i - \mu_x)^2\right]^{1/2}. \tag{10}$$

$$\sigma_{xy} = \left(\frac{1}{N-1}\sum_{i=1}^{N}(x_i - \mu_x)\right). \tag{11}$$

(a) (b) (c)

Fig. 1. (a) Peppers.tiff corrupted with 40% Random Valued Impulse Noise (b)Peppers image restored for 40% noise level using proposed algorithm (c) SSIM Index map of peppers image with 40% RVIN restoration

4 Experimental Results

A fingerprint grayscale sample image of size 512x512 is used for simulation. The simulation is done on MATLAB 7.14 using image processing toolbox and results obtained are mentioned below. The table contains the results obtained by applying proposed algorithm for PSNR (dB), MSE and SSIM at different noise level density ranging from 20 % to 60 %.

Table 1. Results obtained for different RVIN % levels and parameters

RVIN %	Performance Parameters		
	PSNR (dB)	*MSE*	*MSSIM*
20	25.83	169.75	0.9560
30	24.03	256.60	0.9306
40	22.36	377.61	0.8996
50	20.42	589.09	0.8429
60	18.40	938.77	0.7501

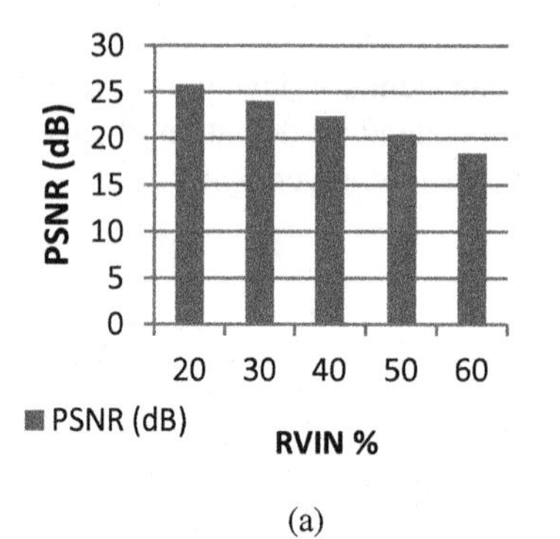

(a)

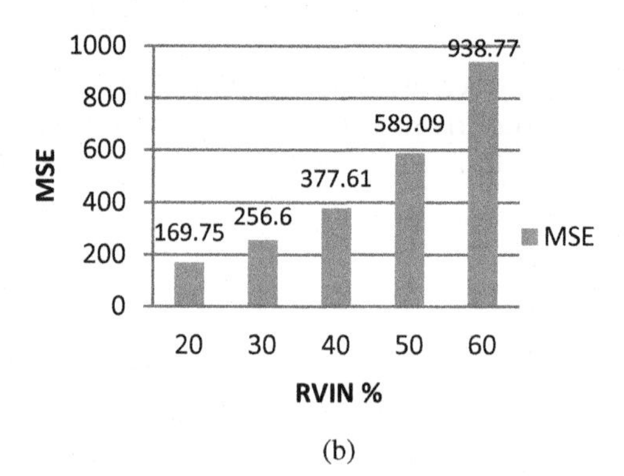

(b)

Fig 2. (a)PSNR at different noise levels (b) MSE at different noise levels

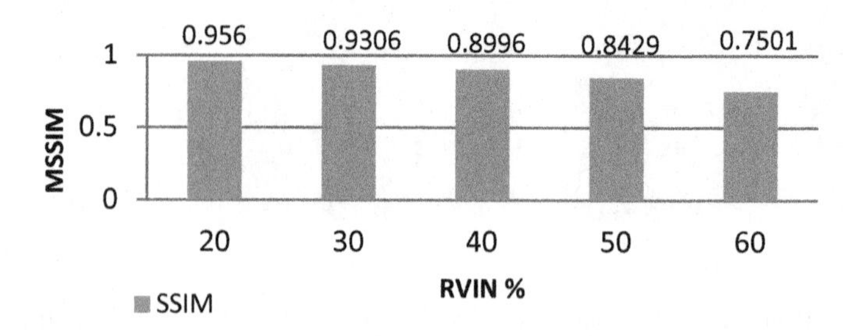

Fig 3. MSSIM of restored image for different RVIN %

Figure 6-8 shows the graphical representation of the quantitative results obtained by applying proposed algorithm to the noise corrupted image. The pictorial outputs for varying noise levels are shown below:

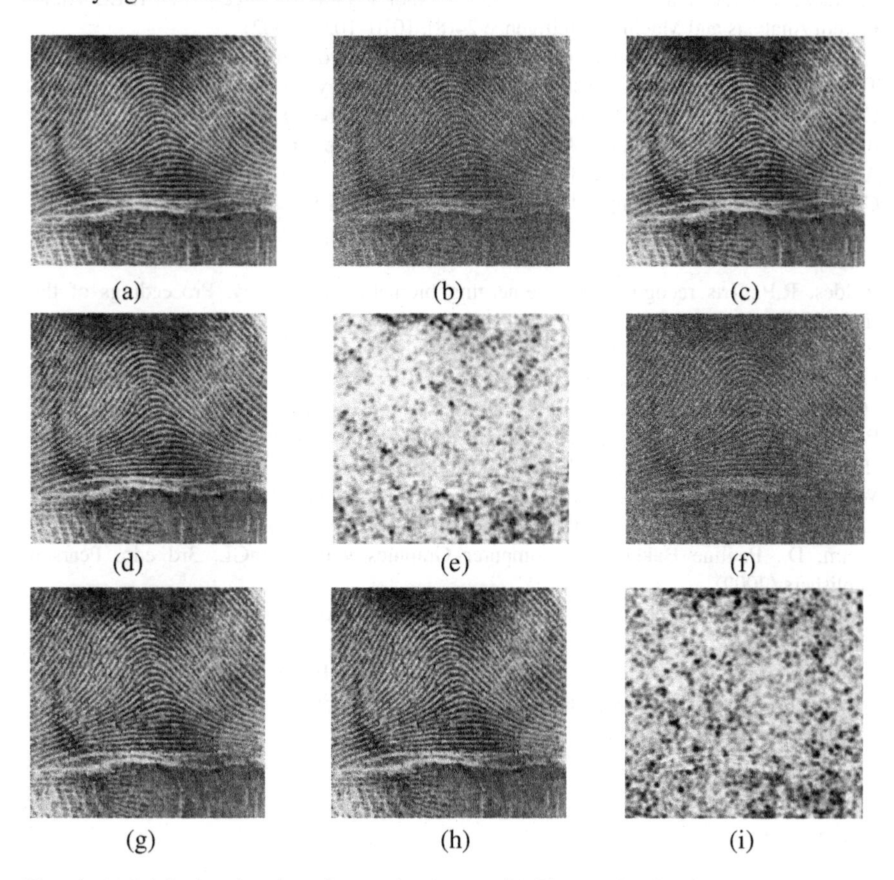

(a) (b) (c)

(d) (e) (f)

(g) (h) (i)

Fig. 4. (a)Original noise free fingerprint image (b) Fingerprint image corrupted with 50% RVIN (c) Noise removed edge deformed image at 50% noise (d) Edge restored image at 50% noise (e) MSSIM of restored image at 50% noise. (f) Fingerprint image corrupted with 60% RVIN (g) Noise removed edge deformed image at 60% noise (h) Edge restored image at 60% noise (i) MSSIM of restored image at 60% noise.

5 Conclusion

In this paper, a novel and efficient algorithm for suppression of random valued impulse noise in fingerprint images using cubic basis spline interpolation is proposed. The algorithm also efficiently removes noise from the image along with edge detail restoration. The quantitative performance of the proposed algorithm is evaluated using measures like PSNR, MSE, and MSSIM and is shown in table I. The figures 2-4 comprises of graphs and simulation result images to provide a clear look of the performance of the algorithm. The algorithm is fast having time complexity of around 50 seconds using MATLAB 7.14 and Core 2 Duo 2.2 GHz processor system.

References

1. Pankanti, S., Prabhakar, S., Jain, A.K.: On the individuality of fingerprints. IEEE Trans. Pattern Analysis and Machine Intelligence 24(8), 1010–1025 (2002)
2. Jain, A.K., Ross, A., Prabhakar, S.: An introduction to biometric recognition. IEEE Transactions on Circuits and Systems for Video Technology 14(1), 4–20 (2004)
3. Clancy, T.C., Kiyavash, N., Lin, D.J.: Secure smart cardbased fingerprint authentication. In: Proceedings of the 2003 ACM SIGMM Workshop on Biometrics Methods and Application, WBMA 2003 (2003)
4. Goh, A., Ngo, D.C.L.: Computation of cryptographic keys from face biometrics. In: Lioy, A., Mazzocchi, D. (eds.) CMS 2003. LNCS, vol. 2828, pp. 1–13. Springer, Heidelberg (2003)
5. Wildes, R.P.: Iris recognition: an emerging biometric technology. Proceedings of the IEEE 85(9), 1348–1363 (1997)
6. Bodduna, K., Siddavatam, R.: A novel algorithm for detection and removal of random valued impulse noise using cardinal splines. In: Proc. INDICON 2012, Kochi, India, pp. 1003–1008 (December 2012)
7. Petrovic, N., Crnojevic, V.: Universal impulse noise filter based on genetic programming. IEEE Trans. Image Process. 17(7), 1109–1120 (2008)
8. Wu, J., Tang, C.: PDE-Based Random-Valued Impulse Noise Removal Based on New Class of Controlling Functions. IEEE Trans. Image Process 20(9), 2428–2438 (2011)
9. Hearn, D., Pauline Baker, M.: Computer Graphics with OpenGL, 3rd edn. Pearson Publishers (2009)
10. Unser, M.: Splines: A Perfect Fit for Signal andImage Processing. IEEE Signal Processing Magazine 16(6), 24–38 (1999)
11. Wang, Z., Bovik, A.C., Sheikh, H.R., Simoncelli, E.P.: Image quality assessment from error visibility to structural similarity. IEEE Trans. Image Process. 13(4), 600–612 (2004)

TERA: A Test Effort Reduction Approach by Using Fault Prediction Models

Inayathulla Mohammed and Silpa Chalichalamala

Sree Vidyanikethan Engineering College
Tirupati, AP, India
{inayathulla512,silpa.c8}@gmail.com

Abstract. It is known fact that testing consumes more than fifty percent of development effort in software development life cycle. Hence it may be advantageous for any organization if the testing effort is reduced. Various fault prediction models have been proposed but how these prediction models reduce the test effort after prediction is rarely explored. An approach is proposed which uses the prediction models in order to estimate the reduction of test effort. Initially by using the prediction models the number of faults are predicted and based on these faults appropriate test effort is allocated to each module. The basic strategy used to allocate the test effort is to let the test effort be proportional to the predicted number of faults in a module. The test effort can be reduced only if the suitable test strategy is used and also the fault prediction must be accurate.

Keywords: Software Testing, Test Effort, Fault Prediction, Prediction Models.

1 Introduction

Now a days software systems have become complex hence testing must be performed efficiently. Whenever project scheduling is tight and resources are less the amount of test effort to be employed is a serious concern Fault prediction models are used in order to select the software modules which are fault prone so as to focus more on only fault prone modules. The basic criteria to consider a module to be fault prone is that a module in current development is considered to be fault prone, if a similar module was fault prone in earlier development. The similarity of modules is identified based on the product and process metrics. In order to implement fault prediction, one need to assess the accuracy of prediction because if the prediction accuracy is less it may lead to an increase in test effort. Till now various prediction models exist but how test effort is allocated after prediction is not addressed. So using six test effort allocation strategies[1] we allocate test effort after prediction. Fig. 1, shows the fault prediction methodology[8].To evaluate our work an open source dataset[7] is used. The two important factors which affect the results of prediction are size of dataset and predictor variables used in prediction[6]. Generally change metrics are efficient in prediction but the dataset used has only static measures so prediction was done with only static measures. The main objective of TERA is to analyse the amount of test

S.C. Satapathy et al. (eds.), *Emerging ICT for Bridging the Future – Volume 1*,
Advances in Intelligent Systems and Computing 337, DOI: 10.1007/978-3-319-13728-5_25

effort that can be reduced when fault prediction is used. Considering this objective the following questions are formulated which are required to be answered at the end:

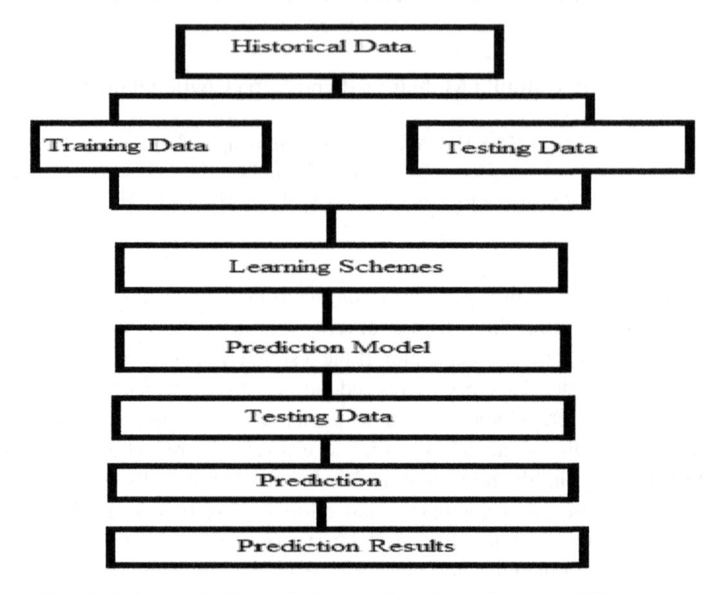

Fig. 1. Software Fault Prediction Methodology(Courtesy:[8])

(Q1) Which is the best strategy to allocate test effort after prediction from the available six strategies?

The basic strategy allocates the test effort proportional to predicted number of faults in a module[1]. But this strategy does not consider the size of a module. In this approach we implemented six strategies[1] among which few strategies consider the size of the modules also. The strategy which discovers 100% faults with less effort among the six strategies is considered the best. The total test effort is assumed based upon the condition that 10 test cases will be executed for every 500 lines of code.

(Q2) How prediction accuracy is calculated for a prediction model?

If the prediction is not accurate test effort may increase. So prediction must be evaluated and P_{opt} chart is used to evaluate the prediction.

(Q3) How much amount of test effort is reduced when this approach is implemented?

This is the primary objective of this approach. To answer this question, an assumption is made that there are still some faults remaining after testing because most software systems are not fault free even after the product is released. Therefore, a parameter called the remaining fault rate is used and a fault discovery model is implemented which guarantees that this model will discover same number of faults as actual testing.

2 Related Work

In [2][5] different prediction models were used. Their prediction results allowed testers to focus their efforts on those files which had high fault rates, which enabled them to identify faults more quickly and with less effort. In[3] ROC(receiver operating characteristic area)curve was used to evaluate the prediction. In[4] the prediction performance was evaluated in terms of F1 which is a combination of precision and Recall values. models were used for prediction but how much of test effort was reduced was not addressed. The test effort was allocated using seven different strategies[1]. The work was implemented on one dataset.

3 TERA

The first step is to build prediction models. Then by using prediction results test effort is allocated with test effort allocation strategies. Then fault discovery model is implemented to find the discoverable faults by each strategy.

3.1 Prediction Models

Until now, various types of fault-prone module predictors have been used. In our approach number of faults in a module are predicted which is a numeric value. So in order to predict a numeric value we use the regression models. Random Forest[9] can predict a number and is one of the popular machine learning technique. Linear regression model and Classification and Regression Trees (CART) are also employed in this approach. All the three prediction models support regression analysis. To build prediction models R-3.0.2(statistical tool) is used.

3.2 Effort Allocation Strategies

Once the faults in a module are predicted the next step is to allocate the test effort. Following strategies[1] are used to allocate test effort.

3.2.1 Equal Test Effort to All Modules

This is the basic strategy. The total test effort is equally divided among all modules.

3.2.2 Test Effort α Module Size

The allocated test effort t_i for the module m_i is as follows

$$t_i = t_{total}.S_i/S_{total} \quad (1)$$

where t_{total} is total test effort, S_i is the size of the i_{th} module, and S_{total} is the total size of all modules.

3.2.3 Test Effort α # of Predicted Faults

Test effort t_i is as follows

$$t_i = t_{total}.F_i/F_{total} \quad (2)$$

where F_i is the number of predicted faults in module i and F_{total} is the sum of predicted faults in all modules.

3.2.4 Test Effort α Predicted Fault Density

Allocated test effort t_i is as follows

$$t_i=t_{total}(F_i/S_i)/\sum_{i=1}^{n}(F_i/S_i) \tag{3}$$

3.2.5 Test Effort α # of Predicted Faults × Module Size

Allocated test effort is

$$t_i=t_{total}.F_i.S_i/\sum_{i=1}^{n}(F_i.S_i) \tag{4}$$

3.2.6 Test Effort α # of Predicted Faults × log(Module Size)

Allocated test effort is

$$t_i=t_{total}.F_i.\log(S_i)/\sum_{i=1}^{n}(F_i.\log(S_i)) \tag{5}$$

3.3 Fault Discovery Model

A fault discovery model[1] is used in order to calculate the number of discoverable faults. This model assures that the number of faults detected by this model will be same as number of faults detected by actual testing also. The size of software modules in testing dataset is also considered. The following equation below is implemented which gives the number of discoverable faults for the test effort allocated by each strategy.

$$H_i(t_i)=a[1-\exp(b_it_i)]\ ,\ b_i=b0/S_i \tag{6}$$

*H(t):*Discoverable faults.
t: Testing time (effort).
b: Probability of detecting each fault per unit time.
a: The number of initial faults before testing
The value of a is estimated as per the equation below

$$a_i=H_i+(R.S_i/1000) \tag{7}$$

H: Actual faults
R: 0.3,0.5,1.0 (Remaining fault rate)
$b0=-S_{total}/t_{total}.\log(1-H_{total}/a_{total})$

3.4 Evaluation Criteria of Fault Prediction

Various evaluation measures such as recall, precision, F-value[4], Alberg diagram and ROC curve exist to evaluate the prediction. The P_{opt} chart is used to evaluate the prediction in this approach. The prediction model which has highest P_{opt} value is considered to be the best model. P_{opt} is defined as 10-Δ_{opt}, where Δ_{opt} is the area between the optimal model and the prediction model. If Δ_{opt}=0 it means that there is no difference between the actual and predicted faults. As much as the value of Δ_{opt} is less that much is accurate the prediction is. In the Fig. 2 a sample graph is shown where x-axis is considered as module number and y-axis represents faults.

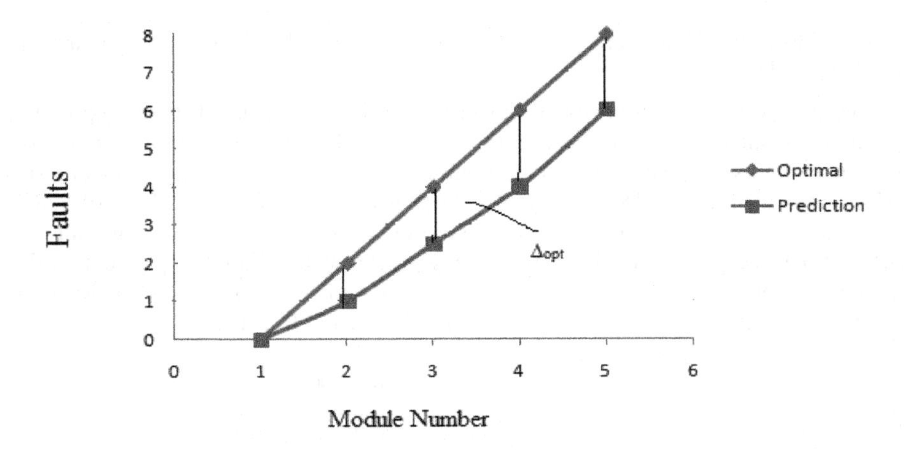

Fig. 2. Example of P_{opt} chart (Courtesy:[6])

4 Results

Fault prediction is meaningless when plenty of resources are available. The dataset[7] used is one of the NASA Metrics Data Program defect data sets. It is a Promise Software Engineering Repository data set made publicly available in order to encourage predictive models of Software Engineering. The dataset had 145 modules information out of which 140 modules are used for training dataset and remaining five are used for testing dataset. The dataset contained only static code measures. For Random Forest all variables were used as predictor variables where as for linear regression and CART only few variables were selected.

(Q1) Which is the best strategy to allocate test effort after prediction from the available six strategies?

After conducting prediction and allocating test effort with three different remaining fault rate values strategy A5 showed best results.

(Q2) How much amount of test effort is reduced when this approach is implemented?

The following table below shows the P_{opt} value of three prediction models.

Table 1. Popt values of prediction

Prediction Model	P_{opt} Value
Random Forest	8.8
Linear Regression	8
CART	6.6

Q3) How much amount of test effort is reduced when this approach is implemented?

The test effort required for discovering faults is calculated based upon the prediction results of Random Forest. Table 2 shows the test required to discover 100 percent of faults by each strategy based on the prediction results of each model. The A5 strategy for Random forest requires 87% of test effort to discover 100% of faults. Therefore 13% of test effort is saved. The test effort is same for Remaining fault rate 0.5 and 0.3.The graphs shown in Fig. 3, Fig. 4 are obtained after implementing the complete approach with three different values of remaining fault rates for Random Forest. From the results it is clear that around 15-20% of effort is reduced.

Table 2. Test Effort Required to Discover 100 Percent Faults

Prediction Model	P_{opt}	Testing Effort(%)					
		A1	A2	A3	A4	A5	A6
Random Forest	8.8	100	87	95	104	87	95
CART	6.6	100	86	101	101	93	103
Linear Regression	8	100	85	98	95	98	101

(a) Remaining Fault Rate = 1

Prediction Model	P_{opt}	Testing Effort(%)					
		A1	A2	A3	A4	A5	A6
Random Forest	8.8	100	85	91	98	81	96
CART	6.6	100	86	102	102	93	103
Linear Regression	8	100	85	98	95	98	101

(b) Remaining Fault Rate = 0.5

Prediction Model	P_{opt}	Testing Effort(%)					
		A1	A2	A3	A4	A5	A6
Random Forest	8.8	100	85	91	91	81	96
CART	6.6	100	86	102	102	93	106
Linear Regression	8	100	85	98	95	98	101

(c) Remaining Fault Rate = 0.3

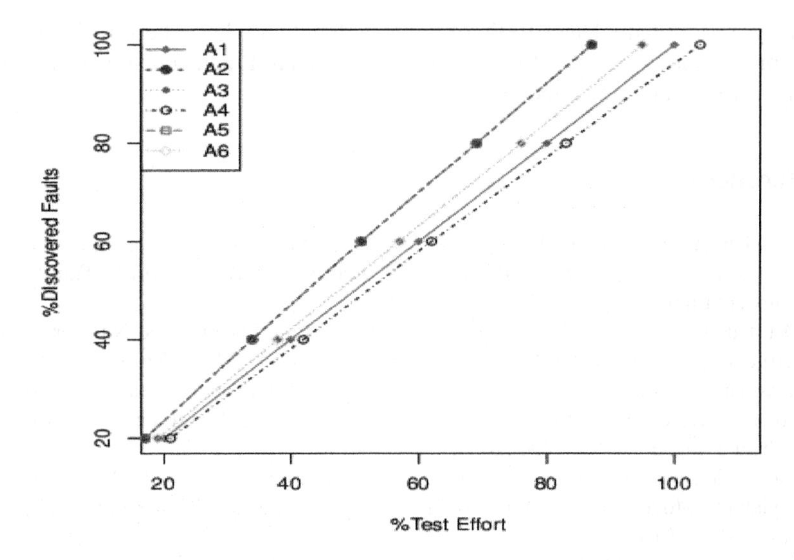

Fig. 3. Comparison with R = 1

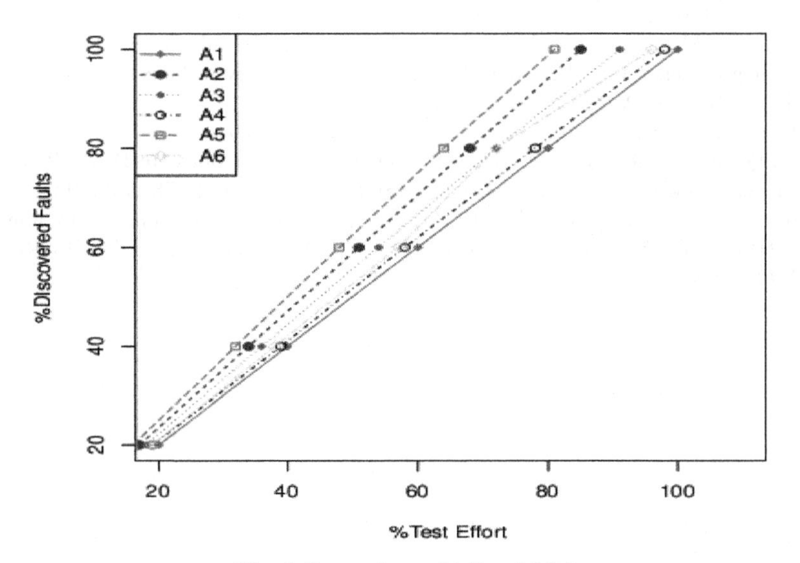

Fig. 4. Comparison with R = 0.5,0.3

5 Conclusion

The results suggest that strategy A5detects more number of defects with less test effort among all strategies. When prediction is evaluated Random Forest showed the best performance. The results also prove that test effort is reduced only when prediction is accurate and suitable effort allocation strategy is used. Also the size of training dataset must be sufficiently large for the prediction to be accurate. The strategy A5 is able to detect 100% of faults with only 80-85% of effort. Thus by using fault prediction 15-20% of effort can be reduced. Fault Prediction allows the testers to focus more on fault prone modules.

References

1. Monden, A., Hayashi, T., Shinoda, S., Shirai, K., Yoshida, J., Barker, M.: Assessing the Cost Effectiveness of Fault Prediction in Acceptance Testing. IEEE Transactions on Software Engineering 39(10) (October 2013)
2. Ostrand, T.J., Weyuker, E.J., Bell, R.M.: Predicting the Location and Number of faults in Large Software Systems. IEEE Trans. Software Eng. 31(4), 340–355 (2005)
3. Lessmann, S., Baesens, B., Mues, C., Pietsch, S.: Benchmarking Classification Models for Software Defect Prediction: A Proposed Framework and Novel Findings. IEEE Trans. Software Eng. 34(4), 485–496 (2008)
4. Kamei, Y., Monden, A., Matsumoto, K.: Empirical Evaluation of SVM-Based Software Reliability Model. In: Proc. Fifth ACM/IEEE Int'l Symp. Empirical Software Eng., vol. 2, pp. 39–41 (2006)
5. Knab, P., Pinzger, M., Bernstein, A.: Predicitng Defect Densities in Source Code Files with Decision Tree Learners. In: Proc. Third Working Conf. Mining Software Repositories, pp. 119–125 (2006)
6. Kamei, Y., Matsumoto, S., Monden, A., Matsumoto, K., Adams, B., Hassan, A.E.: Revisiting Common Bug Prediction Findings Using Effort Aware Models. In: Proc. IEEE Int'l Conf. Software Maintenance, pp. 1–10 (2010)
7. http://promise.site.uottawa.ca/SERepository
8. Song, Q., Jia, Z., Shepperd, M., Ying, S., Liu, J.: A General Software Defect-Proneness Prediction Framework. IEEE Transactions on Software Engineering 37(3) (May/June 2011)
9. Guo, L., Ma, Y., Cukic, B., Singh, H.: Robust Prediction of Fault-Proneness by Random Forests. In: 15th International Symposium Software Reliability Engineering, ISSRE, pp. 417–428 (2004)

A Multibiometric Fingerprint Recognition System Based on the Fusion of Minutiae and Ridges

Madhavi Gudavalli[1], D. Srinivasa Kumar[2], and S. Viswanadha Raju[3]

[1] JNTU Hyderabad & Department
of Information Technology
JNTUK UCE Vizianagaram
Vizianagaram-535003, Andhra Pradesh, India
madhavi.researchinfo@gmail.com
[2] Department of Computer Science and Engineering
Hosur Institute of Technology and Science
Errandapalli Village, Hosur Taluk, Krishnagiri District-635115
srinivaskumar_d@yahoo.com
[3] Department of Computer Science and Engineering
JNTUH CEJ, JNTUniversity Hyderabad
Hyderabad- 500085, Telangana, India
svraju.jntu@gmail.com

Abstract. Fingerprints are widely used since more than 100 years for personal identification due to its feasibility, permanence, distinctiveness, reliability, accuracy and acceptability. This paper proposes a multibiometric fingerprint recognition system based on the fusion of minutiae and ridges as these systems render more efficiency, convenience and security than any other means of identification. The increasing use of these systems will reduce identity theft and fraud and protect privacy. The fingerprint minutiae and ridge features are ridge bifurcations and ridge endings respectively that are combined to enhance overall accuracy of the system. The existence of multiple sources adequately boosts the dimensionality of the feature space and diminishes the overlap between the feature spaces of different individuals. These features are fused at feature level as it provides better recognition performance because the feature set comprises abundant source information than matching level or decision level.

Keywords: Fingerprint, Feature level fusion, Hough Transform, Multimodal Biometrics, Minutiae, Ridges.

1 Introduction

Biometrics are automated methods of recognizing an individual based on their physiological (e.g., fingerprints, face, retina, iris) or behavioral characteristics (e.g., gait, signature). It is the ultimate form of electronic verification of physical attribute of a person to make a positive identification. Each biometric has its strengths and weaknesses and the choice typically depends on the application. No single biometric

S.C. Satapathy et al. (eds.), *Emerging ICT for Bridging the Future – Volume 1*,
Advances in Intelligent Systems and Computing 337, DOI: 10.1007/978-3-319-13728-5_26

is expected to effectively meet the requirements of all the applications. A number of biometric characteristics such as acceptability, universality, permanence, performance, measurability, uniqueness and circumvention are being used in various applications. Fingerprint recognition has a very good balance of all the properties. People have always used the brain's innate ability to recognize a familiar face and it has long been known that a person's fingerprints can be used for identification. Fingerprint uniqueness was discovered years ago, nonetheless the awareness was not until the late sixteenth century when the modem scientific fingerprint technique was first initiated. The English plant morphologist, Nehemiah Grew, published the first scientific paper report in 1684, that explained the ridge, furrow, and pore structure in fingerprints, as a result of his work, a large number of researchers followed suit and invested huge amounts of effort on fingerprint studies[2]. Among all the other kind of popular personal identification methods, fingerprint identification is the most mature and reliable technique.

Identification based on multimodal biometrics that fuses several biometric sources represents an emerging trend [9]. Multibiometric systems are expected to be more accurate that can efficiently enhance the recognition ability of the system as they can utilize, or have capability of utilizing, more than one physiological or behavioural characteristic for enrolment, verification, or identification. Multibiometric system design depends on various factors such as sources of information, acquisition and processing architecture, level of information fusion and fusion methodology. Fusion in biometrics [1], and fingerprint systems in particular, is an instance of information fusion. Ross and Jain [8] have presented an overview of multimodal biometrics with various levels of fusion, namely, sensor level, feature level, matching score level and decision level. The amount of information content generally decreases towards the later modules of a biometric system. For instance, raw data obtained by the sensor module holds the highest information content, while the decision module produces only an accept/reject output. The main advantage of fusion in the context of biometrics is an improvement in the overall matching accuracy. A strong theoretical base as well as numerous empirical studies has been documented that support the advantages of fusion in fingerprint systems.

In the literature, extensive research is carried out for fusion at matching score, decision and rank levels [3] [4]. The feature level fusion is understudied problem inspite of plenty research papers relevant to multimodal biometrics. A method for feature level fusion of hand and face biometrics was proposed by Ross and Govindarajan in the paper [5]. Gyaourova et al. [6] proposed feature level fusion of IR-based face recognition with visible based face recognition. A multibiometric system based on the fusion of face features with gait features at feature level was proposed by Ziou and Bhanu in the paper [7]. Although fingerprint is extensively used and recognized trait, no methods for fusion of various features of fingerprint at feature level have been schemed in the literature. Thus, this paper proposes a multibiometric recognition system that fuses fingerprint minutiae and ridges at feature level. This system yields high matching accuracy better than either of minutiae or ridge based identification systems. The experimental results on FVC2004 database adduce the recognition gain attained with feature level fusion.

2 Proposed System

With the increased use of biometric identification systems in many real time applications, the challenges for large-scale biometric identification are significant both in terms of improving accuracy and response time. If the same biometric measure is used as a pointer to multiple identity records for the same individual across different systems, the possibility of linking these records raises privacy concerns. Moreover, the performance of the identification algorithms needs to be significantly improved to successfully handle millions of persons in the biometrics database matching thousands of transactions per day. Many unimodal biometrics systems that exists in the literature often suffer from limitations such as the inability to tolerate deformed data due to noise deformed data from the sensor device, distorted signal by environmental noise, and variability of an individual's physical appearance and pattern over time. Within this context, multimodal biometrics systems are able to solve some of these limitations by integrating information from multiple biometrics sources. These systems offer significant improvement in the recognition accuracy and reliability as it combines features acquired from various sources at different levels of fusion. The feature level fusion is understudied problem which has began to receive more and more attention from researchers as it is able to provide better recognition performances for the reason that the feature set contains much richer information on the source data.

As fingerprint is the most mature, practical, reliable and widely used biometric technique in personal identification for several centuries, this paper proposes a multi representation fingerprint system that combines minutiae and ridges at feature level retrieved from a single finger. The multi representation systems combine the output of multiple feature extraction algorithms, or that of multiple matchers applied on the same feature set of a single trait. The minutiae and ridges are ridge bifurcations and ridge endings respectively. The feature-level fusion combines multiple feature sets from the same finger irrespective of whether the multiple impressions are acquired using the same or different sensors.

2.1 Methodology

The methodology of the proposed system is as follows:
Step: 1 Binarize the gray scale fingerprint image of size 248*338 from the database FVC2004.

a) Original image b) Binarized image c) Thinned image

Fig. 1. Results of applying binarization and thinning directly to the original image

Step: 2 Ridge thickness of binarized image is reduced to 1-pixel width by thinning of the image.

Step: 3 Minutiae extraction (ridge bifurcations) is done on thinned image by considering the region of interest. The pixels corresponding to minutiae are characterized by Crossing Number (CN). The crossing number CN(P) for a pixel *P* is half the sum of the differences between pairs of adjacent pixels in the 8-neighborhood of *P* and is given by the equation

$$CN(\text{P}) = \frac{1}{2} \sum_{i=1...8} |Val(P_{i \bmod 8}) - Val(P_{i-1})|$$

where P_0, P_1, ...P_7 are the pixels belonging to an ordered sequence of pixels defining the 8-neighborhood of *P* and *Val(P)* ∈{0,1} is the pixel value. The minutiae are the pixels which have CN(P) =3 that corresponds to ridge bifurcations.

Step: 4 Hough Transformation is used to extract ridge information. The ridge pixels are characterized by CN(P)=1 that corresponds to ridge endings.

0	1	0
0	1	0
1	0	1

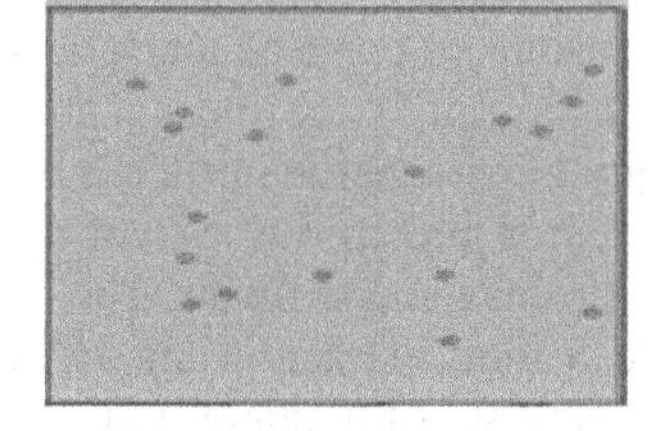

Fig. 2. a) CN(P)=3 Bifurcation Minutiae b) Minutiae extraction

0	1	0
0	1	0
0	0	0

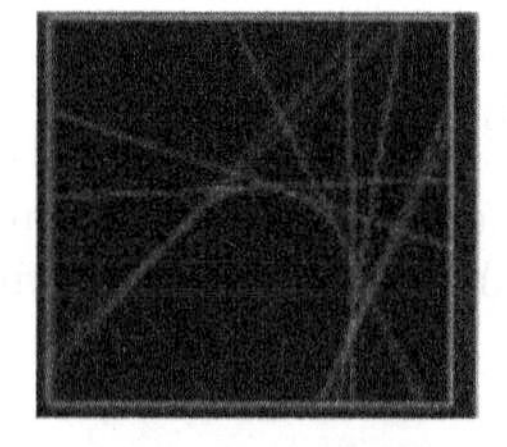

Fig. 3. a) CN(P)=1 Ridge Ending b) Ridge extraction using Hough Transformation

Step: 5 The extracted minutiae and ridge pixels are combined at feature level by taking the union of extracted feature sets of minutiae and ridges.

- The extracted minutiae feature set $M_i = \{ M_{i,1}, M_{i,2}, \ldots M_{i,m} \}$ where *m* is the number of minutiae in the fingerprint .
- The extracted ridge feature set $R_i = \{ R_{i,1}, R_{i,2}, \ldots R_{i,n} \}$ where *n* is the number of ridges in the fingerprint.

- As the ranges of extracted minutiae and ridge feature sets are incomparable, the range normalization is necessary before fusion to remap them to a common range. For example, the minutiae feature set may range from 1 to 1000 and the ridge feature set may range from 1 to 10. The Z-score normalization is used to convert the ranges to a distribution with the mean of 0 and standard deviation of 1.

$$X = \frac{n - \mu}{\sigma}$$

  Where, n is any raw score, and μ and σ are the mean and standard deviation of the stream specific ranges. The resulting normalized minutiae set $M'_i = \{ M'_{i,1}, M'_{i,2}, \ldots M'_{i,k} \}$ and normalized ridge set $R'_i = \{ R'_{i,1}, R'_{i,2}, \ldots R'_{i,k} \}$.

- The Fused feature set $F(M'_i, R'_i) = \bigcup_{j=1}^{k} (M'_{i,j}, R'_{i,j})$. Here, the union implies concatenation in which each element of the first vector $M'_{i,j}$ is concatenated with the feature of the second vector $R'_{i,j}$. The fused feature set $F_i = \{ F_{i,1}, F_{i,2}, \ldots F_{i,k} \}$

Step: 6 In this work, it is assumed that the minutiae feature extraction algorithm often misses genuine minutiae and does not find any false minutiae. So, taking the union of minutiae and ridge points not only recovers missing minutiae and ridges but also increases the finger area represented by the consolidated template.

Step: 7 Generating Match Score

Let the minutiae and ridge feature vectors of the database (input) and the query fingerprint images be $\{M_i, R_i\}$ and $\{M_j, R_j\}$ which are obtained at two different time instances i and j. The corresponding fused feature vectors are denoted as F_i and F_j, respectively. Let S_M be the match score generated by comparing M_i with M_j using Euclidean distance.

$$S_M = \sum_{r=1\ldots k} (M_{i,r} - M_{j,r})^2$$

Let S_R be the match score generated by comparing R_i with R_j.

$$S_R = \frac{No.of\ lines\ matched}{MAX(R_i, R_j)}$$

Where R_i and R_j represents the total number of straight lines detected in the input image and total number of lines detected in the query template image respectively. The fused match score S_{fused} is calculated using the sum rule.

$$S_{fused} = \frac{(S_M + S_R)}{2}$$

3 Experimental Results

The proposed multi representation system uses fingerprint database FVC2004 for performance analysis in which the fingerprint images are acquired using an optical sensor and contains 100 subjects with 8 images for each subject. The experiments were conducted on the sample sizes of 10, 20, 30, 40 and 50 fingerprint images. The present study analyses the performance of Closed-Set fingerprint identification system in which a person is identified from a group of known (registered) people. The performance is usually measured in terms of False Acceptance Rate (FAR), False Rejection Rate (FRR) and Equal Error Rate (EER). EER is obtained by averaging the closet False Acceptance Rate (FAR) and False Rejection Rate (FRR). Figure 1 shows the performance gain attained using feature level fusion in comparison with minutiae-based and ridge-based fingerprint identification systems.

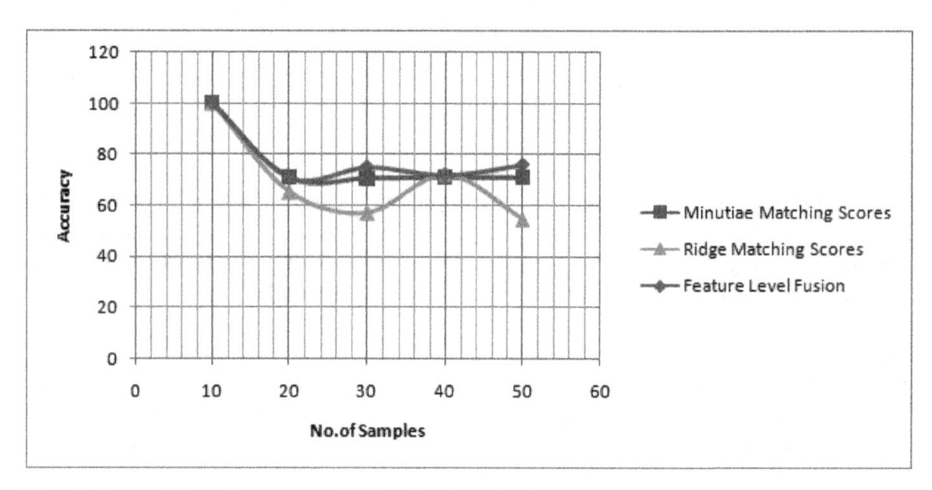

Fig. 4. Recognition Accuracy of Minutiae-based, ridge-based and feature level fusion of Fingerprint Identification Systems

Table 1. FAR, FRR and Accuracy of Fingerprint matching methods and Feature Fusion

Database	Matching Method	FAR (%)	FRR (%)	Accuracy
FVC 2004	Minutiae-based	5.35	4.87	93.5
	Ridge-based	6.71	5.52	94.6
	Feature Fusion	1.2	1.6	96.66

The Feature level fusion is adopted to enhance the accuracy of fingerprint identification. Finally the experimental results shows that the accuracy rate of identification of the proposed system using FVC2004 is 96.66% which is much better than the minutiae-based (93.5% for FVC2004) or the ridge-based fingerprint identification system (94.6% for FVC2004) shown in *Table 1*. Also, there has been a substantial improvement in Genuine Accept Rate (GAR) at very low FAR values thus underscoring the importance of feature level fusion.

4 Conclusions

A multi-representation system involves different types of processing and matching on the same image. The feature level fusion employed in the multibiometric systems can be used either for verification or for indexing in an identification system. The proposed multi-representation fingerprint identification system has shown the substantial improvement in the matching accuracy superior than either of minutiae-based or ridge-based recognition systems. This system is cost effective as it does not require the use of multiple scanners and the user need not interact with different scanners without affecting throughput and user convenience. The results indicates that feature level fusion boost up the classification effectively, and is suitable for integration with other biometric features of palm print, iris, hand and face. In the future work, more experimental test will be performed on the FVC2000, FVC2002 databases and wavelet based feature fusion techniques are used for feature reduction as concatenating uncorrelated feature vectors results in the large dimensionality.

References

[1] Ross, A.A., Nandakumar, K., Jain, A.K.: Handbook of Multibiometrics. Springer, New York (2006)
[2] Lee, H.C., Gaesslen, R.E.: Advances in Fingerprint Technology. Elsevier, New York (2001)
[3] Fierrez-Aguilar, J., Ortega-Garcia, J., Gonzalez-Rodriguez, J.: Fusion strategies in Biometric Multimodal Verification. In: Proceedings of International Conference on Multimedia and Expo, ICME 2003 (2003)
[4] Hong, L., Jain, A.: Integrating Faces and Fingerprints for Personal Identification. IEEE Transactions on Pattern Analysis and Machine Intelligence 20(12), 1295–1307 (1998)
[5] Ross, A., Govindarajan, R.: Feature Level Fusion Using Hand and Face Biometrics. In: Proceedings of SPIE Conference on Biometric Technology for Human Identification II, Orlando, USA, pp. 196–204 (March 2005)
[6] Singh, S., Gyaourova, G., Pavlidis, I.: Infrared and visible image fusion for face recognition. In: SPIE Defense and Security Symposium, pp.585–596 (2004)
[7] Zhou, X., Bhanu, B.: Feature fusion of face and Gait for Human Recognition at a distance in video. In: International Conference on Pattern Recognition, Hongkong (2006)
[8] Ross, A., Jain, A.K.: Information fusion in biometrics. Pattern Recognition Letters 24, 2115–2125 (2003)
[9] Gudavalli, M., Viswanadha Raju, S., Vinaya Babu, A., Srinivasa Kumar, D.: MultiModal Biometrics- Sources, Architecture & Fusion Techniques: An Overview. In: IEEE-International Symposium on Biometrics and Security Technologies (ISBAST 2012), Taipei, Taiwan, March 26-29 (2012)

Segment Based Image Retrieval Using HSV Color Space and Moment

R. Tamilkodi[1], R.A. Karthika[2], G. RoslineNesaKumari[2], and S.Maruthuperumal[3]

[1] MCA, GIET, Rajahmundry, A.P, India
[2] CSE, Saveetha School of Engg., Chennai, T.N, India
[3] CSE&IT, GIET, Rajahmundry, A.P, India
tamil_kodiin@yahoo.co.in,
{karthika78,rosemaruthu,maruthumail}@gmail.com

Abstract. This paper proposed a color image retrieval method based on the primitives of color space and moments (HSVCSM).The proposed HSVCSM analyses the visual properties of the HSV (Hue, Saturation, Value) color space and human visual system. This method is effectively generating the color histogram for CBIR applications. HSVCSM introduced color features in HSV space to quantize the color space into 15 non uniform bins to calculate color spatial feature. Proposed HSVCSM divides the query image into three segments, in which color moments for all segments are extracted like Mean, Variance and Skewness and clustered into three classes for HSV channels. Performance of HSVCSM is evaluated on the basis of precision and recall. Experimental results show that the proposed method HSVCSM for image retrieval is more precise, well-organized and quite comprehensible in spite of the accessibility of the existing retrieving algorithms.

Keywords: Content based image retrieval, feature extraction, Hue Saturation Value, color spatial feature, Histogram, Color Moments.

1 Introduction

CBIR is a system which uses pictorial contents, generally known as features, to search images from huge scale image databases according to users' requests in the form of a query image [1]. Traditional CBIR systems use low level features like color, texture, shape and spatial location of objects to index and retrieve images from databases [2]. Even though research on image retrieval has grown exponentially, particularly in the last few years, it appears that less than 20% were concerned with applications or real world systems. Though various combinations of contents and their possible descriptions have been tried, it is increasingly evident that a system cannot cater to the needs of a general database. Hence it is more relevant to build image retrieval systems that are specialized to domains. The selection of appropriate features for CBIR and annotation systems remain is largely ad-hoc [3, 4, 5]. In CBIR systems, image processing techniques are used to extract visual features such as color, texture and shape from images. Therefore, images are represented as a vector of extracted visual features instead of just pure textual annotations. Color, which represents physical

S.C. Satapathy et al. (eds.), *Emerging ICT for Bridging the Future – Volume 1*,
Advances in Intelligent Systems and Computing 337, DOI: 10.1007/978-3-319-13728-5_27

quantities of objects, is an important attribute for image matching and retrieval. Reasons for its development are that in many large image databases, traditional methods of image indexing have proven to be insufficient, laborious and extremely time consuming [6].These old methods of image indexing, ranging from storing an image in the database and associating it with a keyword or number, to associating it with a categorized description, have become obsolete. Early systems existed already in the beginning of the 1980s [7], the majority would recall systems such as IBM's QBIC (Query by Image Content) as the start of content based image retrieval [8, 9]. QBIC supports users to retrieve image by color, shape and texture. QBIC provides several query methods Simple Query Multi-Feature Query Multi-Pass Query. Few of the techniques have used global color and texture features [8, 10, 11] whereas few others have used local color and texture features [12, 13, 14]. The latter approach segments the image into regions based on color features. For color feature extraction the HSV color space is used that is quite similar to the way in which the colors are defined as human perception, which is not always possible in the case of RGB color space. This method shows that the use of the HSV color space instead of the RGB space can further improve the performance of auto correlograms in CBIR systems, making it more robust to varying illumination conditions. To use the HSV space in CBIR application, it must be quantized into several bins. J.RSmith [15] designed a quantization scheme to produce166 colors. Li Guohui [16] designed a non-uniform quantization method to produce 72 colors. Zhang Lei [17] designed a quantization method to produce 36 bins. Along with the analysis of color features in HSV space, the quantization of the color space into 15 non uniform colors is presented. The focus of HSVCSM method is to make efficient retrieval of a set of images that are similar to a given query image.

The rest of the paper is structured as follows. First, a brief review of HSV Color Space is provided. Then the concept of a proposed method is explained and the principles of the algorithm are obtained. Next, the Experiment result is presented. Later, the system is evaluated against other CBIR approaches. Finally, some conclusions are drawn.

2 HSV Color Space

An image is a collection of pixels each having a particular combination of Hue, Saturation and Intensity Value in the HSV color space. A three dimensional representation of the space is a hexacone [18], where the central vertical axis represents intensity, I (often denoted by V for Intensity Value).Hue, H is defined as an angle in the range $[0,2\pi]$ relative to the red axis with red at angle 0,green at $2\pi/3$,blue at $4\pi/3$ and red again at 2π. Saturation, S is measured as a radial distance from the central axis with value between 0 at the centre to 1 at the outer surface. For S-0, as one move higher along the intensity axis, one goes from black to white through various shades of gray. on the other hand, for a given intensity and hue, if the saturation is changed from 0 to 1, the perceived color changes from a shade of gray to the most pure from the color represented by its hue. When saturation is near 0, all pixels of an image, even with different hues, look alike and as we increase the saturation towards 1, they tend to get separated out and are visually perceived as the true colors represented by their hues .

3 Proposed HSVCSM System

The proposed HSVCSM system is based on the primitive of color space and moments. The proposed CBIR system is shown in Figure.1 and it's having various steps to retrieve the images from the data base. Initially the query image is converted from RGB to HSV.

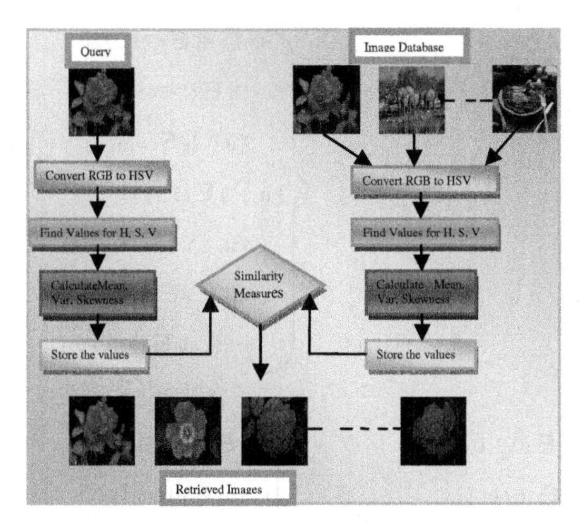

Fig. 1. Block Diagram of Proposed HSVCSM Method

The HVS stands for the Hue, Saturation and Value provides the perception representation according with human visual feature. The formula that transfers from RGB to HSV is defined as below:

$$H = \begin{cases} 60\left(0 + \frac{G - B}{C_{max} - C_{min}}\right), if\ C_{max} = R \\ 60\left(2 + \frac{B - R}{C_{max} - C_{min}}\right), if\ C_{max} = G \\ 60\left(4 + \frac{R - G}{C_{max} - C_{min}}\right), if\ C_{max} = B \end{cases} \tag{1}$$

$$S = \frac{C_{max} - C_{min}}{C_{min}} \tag{2}$$

$$V = C_{max} \tag{3}$$

Where C_{max}= Max(R, G, B) and
C_{min} = Min(R, G, B)

The R, G, B represent red, green and blue components respectively with value between 0-255. In order to obtain the value of H from 0o to 360 o, the value of S and V from 0 to 1, the present HSVCSM do execute the following formula:

H= ((H/255*360) mod 360
V= V/255
S= S/255

The quantization of the number of colors into several bins is done in order to decrease the number of colors; the proposed HSVCSM introduced the non-uniform scheme to quantize into 15 bins. These are H: 8bins; S: 3 bins and V: 3 bins.

$$H = \begin{cases} 0 \text{ if } & h \in (330, 22) \\ 1 \text{ if } & h \in (23, 45) \\ 2 \text{ if } & h \in (46, 70) \\ 3 \text{ if } & h \in (71, 155) \\ 4 \text{ if } & h \in (156, 186) \\ 5 \text{ if } & h \in (187, 278) \\ 6 \text{ if } & h \in (279, 329) \end{cases}$$

$$S = \begin{cases} 0 & \text{if } s \in (0.2, 0.65) \\ 1 & \text{if } s \in (0.66, 1) \end{cases} \qquad V = \begin{cases} 0 & \text{if } s \in (0.2, 0.65) \\ 1 & \text{if } s \in (0.66, 1) \end{cases}$$

In color quantization for each image in the database, colors in HSV model are quantized, to make later design easier. The proposed HSVCSM method used HSV color space, because it is natural and is approximately perceptually uniform. The proposed method is calculated the moments from these HSV values. Color moments have been positively used in content based image retrieval systems, retrieval of images only containing the objects of user's interest. Because most information can be captured by lower order moments, i.e. the first order moment, the second order and the third order moments and Skewness as shown in Equation (4), (5) & (6) . Color moments can be effectively used as color features. If the value of the i-th color channel at the j-th image pixel is pij, then the color moments are defined as:

$$\mu_i = \frac{1}{n}\sum_{j=1}^{n} p_{ij} \tag{4}$$

$$\sigma_i = \sqrt{\frac{1}{n}\sum_{j=1}^{n}(P_{ij} - \mu_i)2} \tag{5}$$

$$S_i = \sqrt[s]{\frac{1}{N}\sum_{j=1}^{N}\left(p_{ij} - \mu_i\right)^3} \tag{6}$$

When a user query is submitted for similarity matching, the step of analysis and feature selection is repeated as performed with image database. The value of query image is compared with the values of different images stored in database. The images having closest values compared to query image color values are extracted from database by this result.

4 Experimental Results

The proposed system experienced with Wang Database contains 1000 images from the Corel image database. The database consists of different categories such as Buses, Dinosaurs, Flowers, Building, Elephants, Mountains, Food, African people, Beaches and Horses. All the categories are used for retrieval. These images are stored with size 256×256 and each image is represented with RGB color space. Some of the sample Wang Data base images are shown in Figure 2. Figure 3 is the query image of the proposed system HSVCSM. Retrieved images based on query image Bus are shown in Figure 4. In order to measure retrieval effectiveness for an image retrieval system, precision and recall values are used. Table1 is summarizing the Precision and Recall results of the proposed HSVCSM. Recall measures the ability of the system to retrieve all the models that are relevant, while precision measures the ability of the system to retrieve only models that are relevant. The HSV values has a high recall and precision of retrieval, and is effectively used in content based image retrieval systems. They are defined as

Precision = (The Number of retrieved images that are relevant) / (The number of retrieved images)

Recall = (The Number of retrieved images that are relevant) / (The Total number of relevant images)

Fig. 2. Sample of WANG Image database

Fig. 3. Query Image: Rose

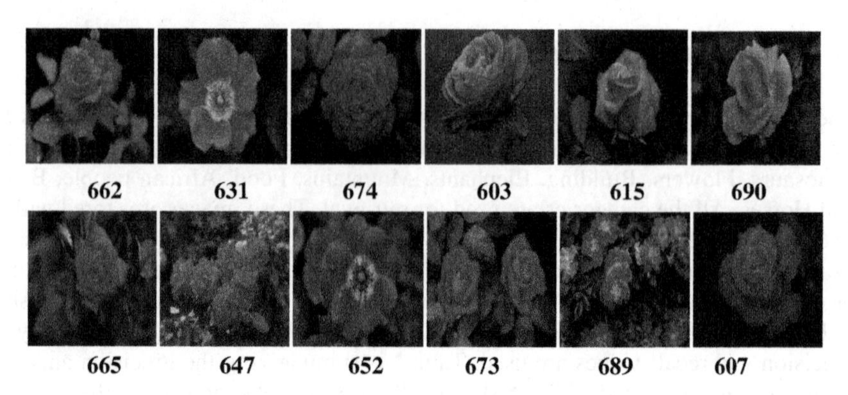

Fig. 4. Retrieved images based on query image Bus

Table 1. Summarize the Precision and Recall results of the proposed HSVCSM

Category of Images	Average Precision	Average Recall
Buses	0.846	0.733
Building	0.828	0.585
Flowers	0.917	0.647
Elephants	0.923	0.533
Mountains	0.811	0.732
Dinosaurs	0.923	0.706
Food	0.931	0.600
Beaches	0.892	0.805
African people	0.828	0.706
Horses	0.951	0.848
Average	0.885	0.690

Table 2. Comparison table of proposed method with other method

Category of Images	RGB		HSV		Proposed HSVCSM	
	Precision	Recall	Precision	Recall	Precision	Recall
Bus	0.757	0.622	0.771	0.600	0.846	0.733
Building	0.913	0.512	0.857	0.439	0.828	0.585
Flower	0.900	0.529	0.611	0.324	0.917	0.647
Elephant	0.774	0.533	0.778	0.467	0.923	0.533
Mountain	0.744	0.707	0.692	0.659	0.811	0.732
Dinosaur	0.857	0.706	0.793	0.676	0.923	0.706
Food	0.806	0.644	0.882	0.667	0.931	0.600
Beaches	0.706	0.585	0.781	0.610	0.892	0.805
African people	0.793	0.676	0.875	0.618	0.828	0.706
Horses	0.771	0.587	0.676	0.500	0.951	0.848
Average (%)	0.802	0.610	0.772	0.556	0.885	0.690

The comparison of the proposed system with RGB and HSV color space are shown in Table 2. The Figure 5 and Figure 6are showing the graph diagram of comparison of other methods with Precision and Recall. The proposed HSVCSM method proves clearly that image retrieval is more precise, well-organized and quite comprehensible in spite of the accessibility of the existing retrieving algorithms. In the present method color index from different regions of the image significantly increases the discriminating power compared to the color moment with HSV color values are extracted from the entire image.

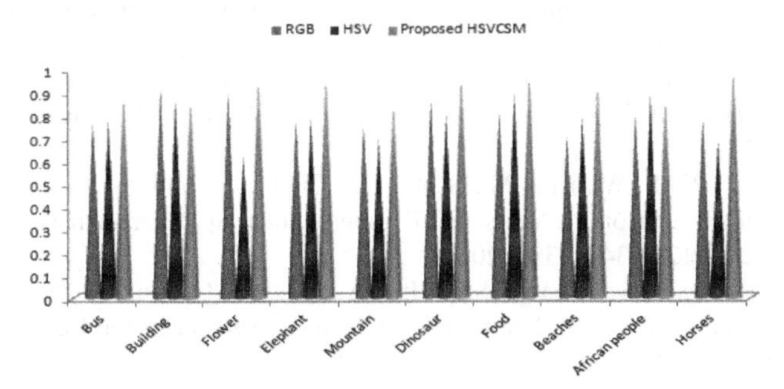

Fig. 5. Comparison of Precision

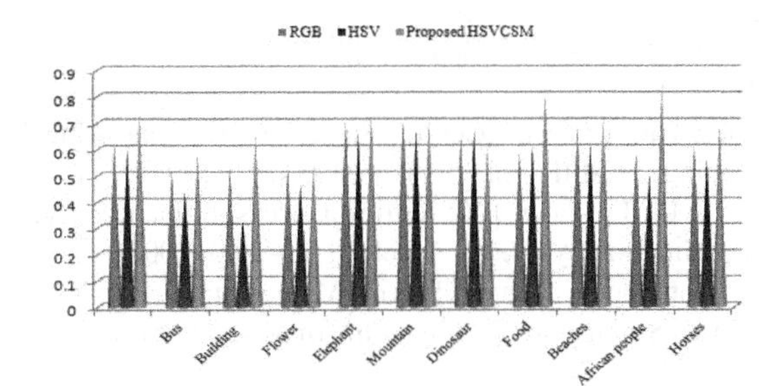

Fig. 6. Comparison of Recall

5 Conclusion

The proposed HSVCSM introduced color features in HSV space to quantize the color space into 15 non uniform bins to calculate color spatial feature. Proposed HSVCSM divides the query image into three segments, in which color moments for all segments are extracted like Mean, Variance and Skewness and clustered into three classes for HSV channels. The retrieval efficiency of the color descriptors is experimented by means of recall and precision. In most of the images categories color moment's shows better performance than the RGB and HVS technique. When robustness to radiance condition changes is an issue, HSV color space has been proven to be a good choice to work with. In this paper, a CBIR method has been proposed which uses the color moments and the proposed method HSVCSM gives higher precision and recall.

An experimental comparison of a number of descriptors for content based image retrieval is performed. The CBIR using HSV color space system handovers each pixel of image to a quantized color and using the quantized color code to compare the images of database.

References

1. James, I.S.P.: Face Image retrieval with HSV color space using clustering techniques. The SIJ Transactions on Computer Science Engineering & its Applications (CSEA) 1(1) (March-April 2013)
2. Shrivastava, N., Tyagi, V.: Content based image retrieval based on relative locations of multiple regions of interest using selective regions matching. Information Sciences 259, 212–224 (2014)
3. Smeulders, A.W.M., Worring, M., Santini, S., Gupta, A., Jain, R.: Content-Based Image Retrieval at the end of the years. IEEE Transactions on Pattern Analysis and Machine Intelligence 22(12), 1349–1380 (2000)
4. Rui, Y., Huang, T.S., Chang, S.F.: Image retrieval: Current Techniques, promising Directions and open issues. Journal of Communications and Image Representation 10(1), 39–62 (1999)
5. Das, M., Manmatha, R., Riseman, E.M.: Indexing flower patent images using domain knowledge. IEEE Intelligent Systems, 24–33 (September/October 1999)
6. Jain, M., Dong, J., Tang, R.: Combining color, texture and region with objects of user's interest for CBIR. IEEE (2007), doi:10.1109/SNPD.2007.104
7. Chang, N.-S., Fu, K.-S.: Query by Pictorial. IEEE Transactions on Software Engineering, 519–524 (1980)
8. Niblack, W.: The QBIC Project: Querying Images by Content using Color, Texture and Shape. In: Proceedings of SPIE, vol. 1908, pp. 173–187 (1993)
9. Datta, R., Joshi, D., Li, J., Wang, J.Z.: Retrieval: Ideas, Influences and Trends of the New Age. ACM Computing Surveys 40(2), Article 5, 5.1–5.60 (2008)
10. Pentland, A.: Photobook: Content-based Manipulation of Image Databases. In: Proceedings of SPIE Storage and Retrieval for Image and Video Databases II, pp. 34–47 (1994)
11. Stricker, M., Orengo, M.: Similarity of Color Images. In: Proceedings / Conference Na Proceedings of SPIE Storage and Retrieval for Image and Video Databases, pp. 381–392 (1995)
12. Natsev, A., Rastogi, R., Shim, K.: WALRUS: A Similarity Retrieval Algorithm for Image Databases. ACM SIGMOD Record 28(2), 395–406 (1999)
13. Li, J., Wang, J.Z., Wiederhold, G.: IRM: Integrated Region Matching for Image Retrieval. In: Proceedings of the Eighth ACM International Conference on Multimedia, pp. 147–156 (2000)
14. Chen, Y., Wang, J.Z.: A Region-based Fuzzy Feature Matching Approach to Content-based Image Retrieval. IEEE Transactions Pattern Analysis and Machine Intelligence 24(9), 1252–1267 (2002)
15. Smith, J.R.: Integrated spatial and feature image system: Retrieval, analysis and compression (Ph D Dissertation). Columbia University, New York (1997)
16. Guohui, L., Wei, L., Lihua, C.: An image Retrieval method based on color perceived feature. Journal of Image and Graphics 3 (1999)

17. Lei, Z., Fuzong, L., Bo, Z.: A CBIR based on color spatial feature, CBIR using color histogram processing an intelligent image database system. Journal of Theoretical and Applied Information Technology
18. Carson, C., Belongie, S., Greenspan, H., Malik, J.: Blobworld: Image Segmentation using Expectation-Maximization and its Application to Image Querying. IEEE Transactions Pattern Analysis and Machine Intelligence 24(8), 1026–1038 (2002)

Pedestrian with Direction Detection Using the Combination of Decision Tree Learning and SVM

G. Santoshi and S.R. Mishra

ANITS Engineering College, Visakhapatnam
{gsanthoshi,srmishra}.cse@anits.edu.in

Abstract. In the real world scenario of automatic navigation of a pedestrian on busy roads is still a challenging job to identify the pose and the direction of the pedestrian. To detect pedestrian with direction will involve large number of categories or class types. This paper proposes a combination of two techniques. The goal is to train the system on the bases of gradients, use the decision tree from that we can generate the candidate list (confusing pairs) with similar features. Taking the confusion matrix into consideration SVM is trained; this method will reduce the computational cost and generate appropriate results. The proposed work can be to classification and track the pedestrian direction on the still images.

Keywords: Decision tree, SVM, Confusing pairs.

1 Introduction

Pedestrian detection on the busy urban scenarios using vision based navigation form a dynamic environment should be detected classified and tracked the direction in order to reduce the accidents. High progress is made in the highway traffic situations and other pedestrian free environment. An environment with a large number of pedestrian moving- particularly in the interior of the city , still it is challenging to identify the direction to move that is to estimate the future states required for the decision-making and path planning.

Several surveys are made for the detection of human being .[2]The techniques are divided into top-down (e.g. comparing a projected 3D model of the human body with the current image) and bottom-up (e.g. body part detection and assembling). Due to different backgrounds, various style of clothing in appearance,different possible articulations, the presence of occluding accessories, frequent occlusion between pedestrians, different sizes, etc. still the pedestrian detection is an challenging job . In order to train the automatic automobile tracking system ,a standing person or slow moving person may be considered as an obstacle as most of the tracking system use the previous image to track the direction and cannot detect the direction for a standing person .

Recent works has shown that with the modern computer vision tools, visual environment modelling for direction by using multi-class SVM. [6] SVM(Support

S.C. Satapathy et al. (eds.), *Emerging ICT for Bridging the Future – Volume 1*,
Advances in Intelligent Systems and Computing 337, DOI: 10.1007/978-3-319-13728-5_28

Vector Machine) is a method used for binary classification(in which each object is classified into two classes).When dealing with multi-class classification ,in which each object is classified as one of *m*-classes where *m*>2, the problem is that the m-class must be decomposed into sub-classes to train. There are two methods to obtain the classification using SVM.1)We train m SVM classifiers each of which classifies a sample as pedestrian and non-pedestrian. Then train different SVM for pedestrian facing (N, E, S,W, NE,NW,SE,SW..etc).the number of SVM trained according to the permutation of the direction. The demerit is in terms of computational cost. The SVM has to train a large set of combinations. We train $m(m-1)/2$ class types, each of which classifies a sample example as N and S . The cost is terms of training to construct $m(m-1)/2$ classifiers.

We propose a novel method to decompose the number of classifiers. By using Decision tree for pre-processing of SVM. At first use a decision tree in which the pedestrian is represented as d-dimensional feature vectors. We go for multiple tree and get a set of nearest prototype **p** from each set of tree. All the nearest prototype in a candidate list of samples called confusing pairs. We then construct SVM classifiers for these pairs . In the testing phase when Y is give, we collect the top K-candidate list of Y and these confusing pairs are given as input to the SVM to get the appropriate results.

Taking the works into consideration, Our focus is purely on the static images, and integrating with dynamic navigation will certainly produce more accurate location and improve the performance of the described system. The proposed work is trained and tested on the INRIA People Dataset. The paper is organised as follows 2) Background work which explain about the generation of Feature vectors. Section 3) Explain the algorithm for decision tree learning process. 4) Explain about the SVM classifier.5) We demonstrated the experimental result and trace the pedestrian along with the direction in the form of bounding boxes and the direction labels. Final the paper contains conclusions with future works.

2 Proposed Work

The object detection method for classifying the pedestrian and direction is divided into two phases: learning and detection phase. This paper proposes a method in which the detection and training both follow the same processes. The learning phase a decision tree is applied which will help in training a multi class objects that is for a pedestrian with four direction and non-pedestrian for fixed size image regions; later on for the confusing matrix SVM is applied in order to reduce the cost of computation. Whereas detection phase classifies on a multi-scale scan and identifies the pedestrian along with direction label for the test images .

For achieving this HOG detection is considered as the background of the learning process except the use of SVM. As SVM is a binary classifier. The work extend with the decision tree learning algorithm followed by SVM classifiers.

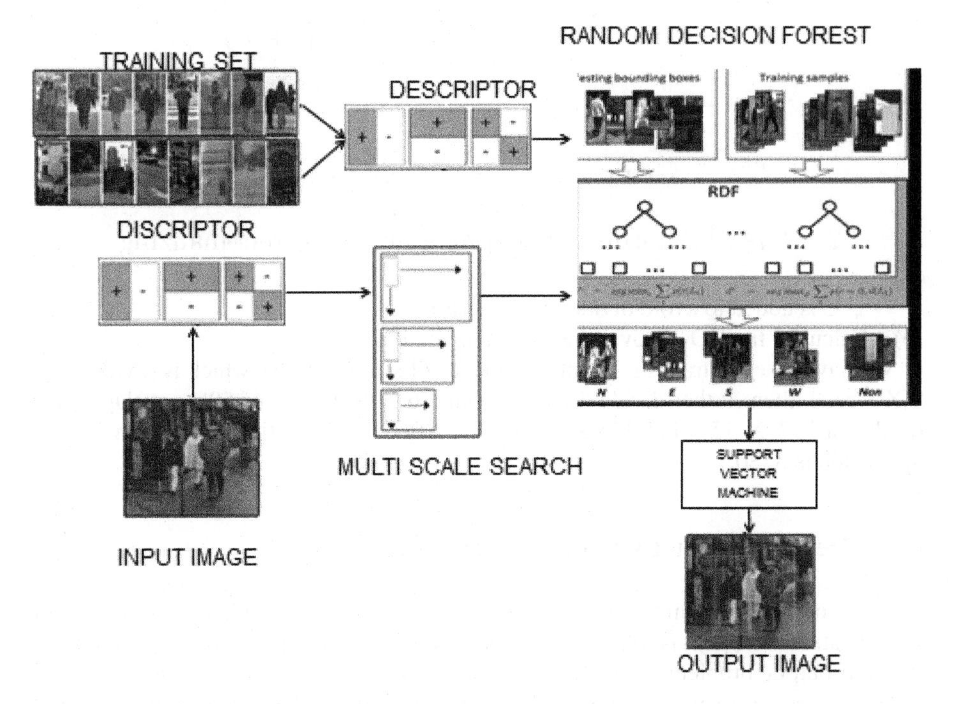

Fig. 1. Detection using the combination of decision tree and SVM

3 Background

The best way to predict pedestrian is by using Histogram of Oriented Gradient proposed by Dalal and Trigg's [1]. To achieve this , the Histogram of Oriented Gradient follow the following steps.

1) The input image is processed to ensure normalized color and gamma values.
2) Gradient computation is used to filter the color and intensity data of the image which uses both horizontal and vertical direction and the smooth the derived marks a Gaussian smoothing is applied.

$$[-1,0,1] \quad \text{and } [-1,0,1]T \tag{1}$$

3) Calculate the magnitude and orientation using formula

$$s = \sqrt{(s_x^2 + s_y^2)} \tag{2}$$

$$\theta = arctan(S^y/S^x) \tag{3}$$

4) Calculate cell histograms in a 9 bin orientation (0-180 degrees). The interpolated trilinearly method is applied to vote the spatial and orientation cell s which finds the difference of the nearest neighbouring bin and generate the ratio according to the nearest bin, so that the 9 bins are used properly.
5) Add a normalization over overlapping spatial blocks using four formula

$$L1 - norm : v \longrightarrow v/(||v||_1 + \epsilon) \tag{4}$$

$$L1 - sqrt : v \longrightarrow \sqrt{v/(||v||_1 + \epsilon)} \tag{5}$$

$$L2 - norm : v \longrightarrow v/\sqrt{||v||_2^2 + \epsilon^2} \tag{6}$$

$$L2 - hys : \text{L2-norm, plus clipping at .2 and renomalizing} \tag{7}$$

Є is added to avoid divide by zero error

6) Calculate the HOG's over the detection windows.

We have take's a training sample of images of size 64 X 128 which is divided into 8 X 8 cells, then further divide the image into 16 X 16 blocks of 50% overlap which results in 7 X 15 =105 blocks in total . After concatenation the feature vector dimension is 3780.

4 Decision Tree Learning Process

A decision tree learning process is a supervised learning process used for classification. The goal is to predict the target variables by learning the decision rules. A multi-output problem is a supervised learning problem with several output to predict. There are several outputs CART is the optimized version as it constructs binary tree using feature and threshold, which generates the largest information gain at each node.

1. Assume that training sample are expressed as d-dimensional feature vector, and we decide to grow t number of trees.
2. We divide each d-dimensional vector into t times e-dimensional sub-vectors so $t \times e \geq d$. If the equality holds, the t sub-vectors will not share any features; otherwise, they will overlap partially.
3. The training samples then split into t sets of sub-vectors and each set is input to a tree. When the tree-growing process has been completed, we store at each leaf the class types of the samples that have reached that leaf.
4. The smaller the leaves, and the greater the risk of losing critical candidates. One way to limit the growth of the tree is to stop splitting all nodes at level l, where $1 \leq l \leq e$, and count the root level as 1. However, it is unreasonable to require that all paths stop at the same level l, thereby generating leaves of various sizes. We limit the size of leaves to u=n/2^(l-1) as, the tree terminates at level l, then average leaf size of the tree will be u. A node contains less than u samples are cannot split further. Since the value of u depends on that of length, we write it as u(l). The optimal value of l is determined in the procedure for solving the candidate retrieval.
5. To retrieve candidates from multiple trees, we first grow t trees, in which the leaf sizes are bounded from above by u(l).

$$Candidate\_List(l,v)=\{C \in U^{t}_{i=1} Li : vote\_count(c) \geq v\} \tag{8}$$

Tao and Klitte[3] proposed a method in which the confusion matrixes for 4Dpedestrain random forest in which the author has given some experimental results of prediction for 4 directions(N,E,W,S) and the results shows that North and South, East and West are having the most confusing pairs. Therefore a SVM is trained only for the confusing pairs instead of complete set of input.

5 Support Vector Machine (SVM)

SVM is used to classify only the binary classifiers. For a large Classes of output data, we have to train the SVM using one-against-one or one against- many. The proposed method may have a m(m-1)/2 training set. For suppose we are having 5 classes, then the trained set is 10 SVM must be trained. If we use the decision tree we get a set of confusion pairs, which indicates how many conflicts may occur. As Tao and Klitte [3] has used a confusion matrix(Table-1) which results in a confusion between North and South, and the other set East and West have more confusion in generation the result.

So ,The proposed system used to train only 2 SVM instead of 10 SVM's. this way of computation reduces to time of training, and can obtain an accurate output as only decision tree will generate a confusion matrix and get an random outputs. These are the following steps

1) Given a set of class types, we define the train top-k candidates of a sample x as the top candidates which are sorted according to their distances from x.
2) We then collect the pairs (Ci, Cj), where Ci and Cj are, respectively, the ith and j^{th} candidates of x for $1 \leq i, j \leq k$.
3) Note different values of k can be chosen in the training and the testing phases from decision tree. We assume that k1 and k2 candidates are selected in the training and testing testing phases.
4) For each confusing pair (North,South), the samples of North and South are labeled, as 1 and –1. The task of the SVM method is to derive from the North- and South- training samples a decision function f(x) in the following form.

$$f(x)=\Sigma^{i}_{i=l}\ y_i\ \alpha_i K\,(s_i,\ x)\ +\beta \quad (9)$$

where s_i are support vectors, i α is the weight of s_i, β is the biased term, y_i is the label of s_i, i = 1, …., I, and K(·, ·) is a kernel function. The character x is classified as a North or South object, depending n whether f(x) is positive or negative; details are given in [2].
5) For our application, we adopt the kernel function.

$$K(s,x)=(<s,x>+1)^{\delta} \quad (10)$$

where <s , x> is the inner product of s and x, and δ is the degree of the polynomial kernel function. In the testing phase, for a character x, we first use the decision tree process to find k2 candidates for x.
6) We then compute the decision functions associated with all confusing pairs found within the decision tree by top-k2 candidates of x. If a confusing pair

(North, South) is found among the candidates, and x is classified as a north-type, then north scores one unit.

7) The candidate are generated with the highest score ranks , then with the second highest score ranks second, and so on. If two candidates have the same score, then their positions remain the same. We rearrange the involved candidates according to their assigned ranks.

To determine their values of the parameters δ, k1, k2, we used a set of samples (confusing matrix) in addition to the set used for training. This new set is called validation data. We use the training data of decision tree to construct SVM classifiers for various values of δ and k1 and the validation data to compute the accuracy rates of the classifiers and different set of parameters for each test sample. In so doing, we can find the best combination of values for the parameters .

6 Experimental Results

In this section we aim to prove that the proposed method, of combining decision tree with SVM is more efficient and reduce the computability time. We evaluated the INRIA dataset which consist of 2000 images. From which we have taken 250 images, 5 trees are generated for each categories. (N, S, E, W, Non-pedestrian). The experimental results in table 2 are used to show the confusion matrix for the pedestrian by using the Decision tree. The table with lighter (grey colour) boxes shows that how the pedestrian is predicted. For example, a pedestrian facing towards north is predicted 95% times as North and 5% as South, similarly if facing towards west is predicted as East(33%) and 60% as west. South is 61% predicted as North, which indicates that up to 93% the decision tree can predict the pedestrian correctly.

Processing time. The processing time for one bounding box when passed through a tree (*depth* < 15) is less than 5 ms on a PC defined by Core i7 1.9 Hz, and 2GB. As the processing of trees are independent calculations, *i.e.* the bounding box goes through each tree without interfering with the other trees and a parallel computing is possible.

Table 1. confusion matrix 4pedRD

	N	*E*	*S*	W
N	*0.95*	*0.00*	*0.05*	*0.00*
E	*0.06*	*0.78*	*0.11*	*0.06*
S	0.61	0.05	0.32	0.02
W	0.00	0.33	0.07	0.60

Processing time to train SVM for the confusion matrix take 15ms on the PC defined by Core i7 1.9Hz, and 2 GB from the confusion matrix we can see that south is predicted most of the times as North(61%) and West is predicted as East(33 %). So SVM is trained only two SVM (North and South, East and West).

We can see in the next case by using SVM to train the confusion matrix though it takes time to train but the accuracy for the prediction is more .We can see the Table 3 which shows that the accuracy increases to 5 %. The next table shows the comparative study of details of predicting the accurate values, the time to trace and train. The comparative study tells that by using the combination of the two classifier increases the accuracy and time of compilation.

Table 2. INRIA dataset (Number of class types=5, Number of prototypes=2000)

	Testing Accuracy	Computing Time(milliseconds)	Computing Speed(pedestrian /sec)
Decision tree	93.66%	5	5
SVM	98.9%	15	4
Decision tree+ SVM	98.00%	7	6

7 Conclusion

This paper proposes an innovative way to detect the pedestrian with direction for the still images .We have used two classifiers to train and test the images. The decision tree is to reduce the time of compilation and SVM to detect the accurate results. So by combining both the classifiers, there can be a good classifier in terms of reducing time, space, complexities and produce more accurate outputs. The work came be more apt is more number of prototype (N,S,E,W,NE,SE,NW,SW etc.) are taken into consideration and the expanding the algorithm for detecting the dynamic /moving objects will result in an efficient testing algorithm.

References

[1] Dalal, N., Triggs, B.: Histograms of Oriented Gradients for Human Detection. In: International Conference on Computer Vision & Pattern Recognition (IEEE), pp. 886–893 (June 2005)

[2] Gandhi, T., Trivedi, M.M.: Image Based Estimation of Pedestrian Orientation for Improving Path Prediction. In: 2008 IEEE Intelligent Vehicles Symposium, pp. 506–511 (2008)

[3] Tao, J., Klette, R.: Integrated Pedestrian and Direction Classification using a Random Decision Forest. In: The IEEE International Conference on Computer Vision (ICCV) Workshops, pp. 230–237 (2013)

[4] Sotelo, M.A., Parra, I., Fernandez, D., Naranjo, E.: Pedestrian Detection using SVM and Multi-feature Combination. In: Intelligent Transportation Systems Conference, ITSC 2006, pp. 103–108. IEEE (2006)

[5] Xu, F., Liu, X., Fujimura, K.: Pedestrian Detection and Tracking With Night Vision. IEEE Transactions on Intelligent Transportation Systems 6(1) (March 2005)

[6] Chang, F.: Techniques for Solving the Large-Scale Classification Problem in Chinese Handwriting Recognition. In: Doermann, D., Jaeger, S. (eds.) SACH 2006. LNCS, vol. 4768, pp. 161–169. Springer, Heidelberg (2008)

[7] Wilking, D., Röfer, T.: Realtime Object Recognition Using Decision Tree Learning. In: Nardi, D., Riedmiller, M., Sammut, C., Santos-Victor, J. (eds.) RoboCup 2004. LNCS (LNAI), vol. 3276, pp. 556–563. Springer, Heidelberg (2005)

Application of Radon Transform for Image Segmentation on Level Set Method Using HKFCM Algorithm

R. Nirmala Devi and T. Saikumar

Dept of EIE, Kakatiya Institue of Technology & Science(KITSW), Warangal, India
Dept of ECE, CMR Technical Campus, Hyderabad, India
tara.sai437@gmail.com, nimala123@yahoo.com

Abstract. In this Paper, HKFCM Clustering algorithm was used to generate an initial contour curve which overcomes leaking at the boundary during the curve propagation. Firstly, HKFCM algorithm computes the fuzzy membership values for each pixel. On the basis of HKFCM the edge indicator function was redefined. Using the edge indicator function the biomedical image was performed to extract the regions of interest for advance processing. Applying the radon transform on the output image will be shown with experiment demonstration. The above process of segmentation showed a considerable improvement in the evolution of the level set function.

Keywords: Image segmentation, HKFCM, level set method, Radon Transform.

1 Introduction

Image segmentation is plays an important role in the field of image understanding, image analysis, pattern identification. The foremost essential goal of the segmentation process is to partition an image into regions that are homogeneous (uniform) with respect to one or more self characteristics and features. Clustering has long been a popular approach to un tested pattern recognition. The fuzzy c-means (FCM)[1] algorithm, as a typical clustering algorithm, has been utilized in a wide range of engineering and scientific disciplines such as medicine imaging, bioinformatics, pattern recognition, and data mining. *Given* a data $X = \{x_i \ldots\ldots\ldots x_n\} \subset R^p$, the original FCM algorithm partitions X into c fuzzy subsets by minimizing the following objective function

$$J_m(U,V) \equiv \sum_{i-1}^{c} \cdot \sum_{k-1}^{n} u_{ik}^{m} \|x_i - v_i\|^2 \cdots \qquad \ldots\ldots\ldots (1)$$

Where c is the number of cluster and selected as a specified Value in the paper, n the number of data points, u_k , the member of x_k in class i , satisfying $\sum_{i-1}^{c} u_{ik}$, m the quantity controlling clustering fuzziness and v is set of control cluster centers or a prototypes $(v_i \in R^p)$. The function J_m is minimized by the famous alternate iterative algorithm. Since the original FCM uses the squared-norm to measure inner

S.C. Satapathy et al. (eds.), *Emerging ICT for Bridging the Future – Volume 1*,
Advances in Intelligent Systems and Computing 337, DOI: 10.1007/978-3-319-13728-5_29

product with an appropriate 'kernel' function, one similarity between prototypes and data points, it can only be effective in clustering 'spherical' clusters. And many algorithms are resulting from the FCM in order to cluster more general dataset. Most of those algorithms are realized by replacing the squared-norm in Eq (1) the object function of FCM with other similarity trial (metric) [1-2].

The level set method is [4-7] based on geometric deformable model, which translate the problem of evolution 2-D (3-D) close curve(surface) into the evolution of level set function in the space with higher dimension to obtain the advantage in managing the topology changing of the shape. The level set method has had great success in computer graphics and vision. Also, it has been widely used in medical imaging for segmentation and shape recovery [8-9]. However, there are some insufficiencies in traditional level set method.

2 Hyper Kernel Fuzzy c-Means Clustering (HKFCM):

Define a nonlinear map as $\phi : x \rightarrow \phi(x) \in F$,where $x \in X . X$ denotes the data space and F is the transformed feature space with higher even infinite dimensions. HKFCM minimized the following objective function:

$$J_m(U,V) \equiv \sum_{i-1}^{c} \cdot \sum_{k-1}^{n} u_{ik}^{m} \|\phi(x_i) - \phi(v_i)\|^2 \quad \ldots\ldots\ldots \quad (2.1)$$

Where

$$\|\phi(x_i) - \phi(v_i)\|^2 = K(x_k, x_k) + K(v_i, v_i) - 2K(x_k, v_i) \ldots\ldots \quad (2.2)$$

Where $K(x, y) = \phi(x)^T \phi(y)$ is an inner product of the kernel function. If we adopt the Gaussian function as a kernel function, $K(x, y) = \exp\left(-\|x - y\|^2 / 2\sigma^2\right)$, then $K(x, x) = 1$. according to Eq. (2.2), Eq. (2.1) can be rewritten as

$$J_m(U,V) \equiv 2 \sum_{i-1}^{c} \cdot \sum_{k-1}^{n} u_{ik}^{m} (1 - k(x_k, v_i)). \ldots\ldots\ldots \quad (2.3)$$

Minimizing Eq. (2.3) under the constraint of, $u_{ik}, m > 1$.

Here we now utilize the Gaussian kernel function for

Straightforwardness. If we use additional kernel functions, there will be corresponding modifications in Eq. (2.4) and (2.5).

We have

$$u_{ik} = \left[\frac{\left(1/\left(1 - K(x_k, v_i)\right)\right)^{1/(m-1)}}{\sum_{j=1}^{c} \left(1/\left(1 - K(x_k, v_i)\right)\right)^{1/(m-1)}} \right]^{\frac{1}{3}} \ldots\ldots\ldots \quad \ldots\ldots\ldots (2.4)$$

$$v_i = \frac{\sum_{k=1}^{n} u_{ik} K(x_k, v_i) x_k}{\sum_{k=1}^{n} u_{ik}^m K(x_k, v_i)} \quad \ldots\ldots\ldots\ldots \quad \ldots\ldots\ldots\ldots (2.5)$$

In fact, Eq.(2.2) can be analyzed as kernel-induced new metric in the data space, which is defined as the following

$$d(x, y) \underline{\underline{\Delta}} \|\phi(x) - \phi(y)\| = \sqrt{2(1 - K(x, y))} \ldots\ldots \quad \ldots (2.6)$$

And it can be proven that $d(x, y)$ is defined in Eq. (2.6) is a metric in the original space in case that $K(x, y)$ takes as the Gaussian kernel function. According to Eq. (2.5), the data point x_k is capable with an additional weight $K(x_k, v_i)$, which measures the similarity between x_k and v_i and when x_k is an outlier i.e., x_k is far from the other data points, then $K(x_k, v_i)$ will be very small, so the weighted sum of data points shall be more strong.

The full explanation of HKFCM algorithm is as follows:

HKFCM Algorithm:

Step 1: Select initial class prototype $\{v_i\}_{i=1}^{c}$.

Step 2: Update all memberships u_{ik} with Eq. (2.4).

Step 3: Obtain the prototype of clusters in the forms of weighted average with Eq. (2.5).

Step 4: Repeat step 2-3 till termination. The termination criterion is $\|V_{new} - V_{old}\| \leq \varepsilon$.

Where $\|.\|$ is the Euclidean norm. V is the vector of cluster centers ε is a small number that can be set by user (here ε=0.01).

3 The Level Set Method

The level set method was invented by Osher and Sethian [3] to hold the topology changes of curves. A simple representation is that when a surface intersects with the zero plane to give the curve when this surface changes, and the curve changes according with the surface changes. The heart of the level set method is the implicit representation of the interface. To get an equation describing varying of the curve or the front with time, we started with the zero level set function at the front as follows:

$$\phi(x, y, t) = 0, \text{ if } (x, y) \in 1 \ldots\ldots \quad .(3..1)$$

Then computed its derivative which is also equal to zero

$$\frac{\partial\phi}{\partial t}+\frac{\partial\phi}{\partial x}.\frac{\partial x}{\partial t}+\frac{\partial\phi}{\partial y}.\frac{\partial y}{\partial t}=0 \quad \ldots\ldots\ldots \quad .(3.2)$$

Converting the terms to the dot product form of the gradient vector and the x and y derivatives vector, we go

$$\frac{\partial\phi}{\partial t}+\left(\frac{\partial\phi}{\partial x}.\frac{\partial x}{\partial t}\right)\bullet\left(\frac{\partial\phi}{\partial y}.\frac{\partial y}{\partial t}\right)=0\ldots\ldots\ldots \quad .(3.3)$$

Multiplying and dividing by $\nabla\phi$ and taking the other part to be F the equation was gotten as follows:

$$\frac{\partial\phi}{\partial t}+F\left|\nabla\phi\right|=0\ldots\ldots\ldots\ldots \quad .(3.4)$$

According to literature [9]11], an energy function was defined:

$$E(\phi)=\mu E_{\text{int}}(\phi)+E_{ext}(\phi) \quad \ldots\ldots \quad .(3.5)$$

Where $E_{ext}(\phi)$ was called the external energy, and $E_{\text{int}}(\phi)$ was called the internal energy. These energy functions were represented as:

$$E_{\text{int}}(\phi)=\int_{\Omega}\frac{1}{2}(\nabla\phi-1)^{2}\,dxdy \quad \ldots\ldots\ldots\ldots \quad (3.6)$$

$$E_{ext}(\phi)=\lambda L_{g}(\phi)+vA_{g}(\phi)\ldots\ldots\ldots\ldots \quad (3.7)$$

$$L_{g}=\int_{\Omega}g\delta(\phi)\left|\nabla\phi\right|dxdy\ldots\ldots\ldots\ldots\ldots \quad (3.8)$$

$$A_{g}=\int_{\Omega}gH(-\phi)dxdy\ldots\ldots\ldots\ldots\ldots \quad (3.9)$$

$$g=\frac{1}{1+\left|\nabla G_{\sigma}*I\right|}\ldots\ldots\ldots\ldots\ldots\ldots\ldots\ldots \quad (3.10)$$

Where $L_{g}(\phi)$ was the length of zero level curve of ϕ; and A_{g} could be viewed as the weighted area; I was the image and g was the edge indicator function. In conventional(traditional) level set methods, it is numerically necessary to keep the evolving level set function close to a signed distance function[14][15]. Re-initialization, a technique for occasionally re-initializing the level set function to a signed distance function during the evolution, has been extensively used as a numerical remedy for maintaining stable curve evolution and ensuring desirable results.

From the practical viewpoints, the re-initialization process can be quite convoluted, expensive, and has subtle side effects [16]. In order to overcome the problem, Li et al [8] proposed a new variational level set formulation, which could be easily implemented by simple finite difference scheme, without the need of re-initialization. The details of the algorithm are in the literature [8]. However, because only the gradient information was imposed in the edge indicator function, Li's method has a little effect on the presence of fuzzy boundaries.

In the paper, a innovative method was proposed to modify the algorithm. The original image was partitioned into some sub images by HKFCM. The fuzzy boundary of each sub image was weighted by α, the edge indicator function was redefined:

$$g^{'} = g + \alpha \cdot g_2 \qquad (3.11)$$

Where $g_2 = \dfrac{1}{1+|\nabla G_\sigma * I_1|}$

I_1 Was the image after clustering. The iterative equation of level set functional was:

$$\frac{(\phi^{n+1} - \phi^n)}{\tau} = \mu\left[\Delta\phi - div\left(\frac{\nabla\phi}{|\nabla\phi|}\right)\right] + \lambda\delta(\phi)div\left(g^{'}\frac{\nabla\phi}{|\nabla\phi|}\right) + vg^{'}\delta(\phi) \ldots \quad (3.12)$$

Taking $g^{'} = g + \alpha \cdot g_2$ into 4.12

$$\phi^{n+1} = \phi^n + \tau\left\{\begin{matrix}\mu\left[\nabla\phi - div\left(\frac{\nabla\phi}{|\nabla\phi|}\right)\right]\lambda\delta(\phi)div\left(g\frac{\nabla\phi}{|\nabla\phi|}\right) \\ +vg\delta(\phi) + \alpha\begin{bmatrix}\lambda\delta(\phi)div\left(g_2\frac{\nabla\phi}{|\nabla\phi|}\right) + \\ vg_2(\phi)\end{bmatrix}\end{matrix}\right\} \ldots \quad \ldots (3.13)$$

Where $\alpha \in [0,1]$. When processing images of weak boundary or low contrasts, a bigger α was taken; otherwise, a smaller α was taken.

4 The Generation of Initial Contour Curve

On the basis of HKFCM clustering in image segmentation, the over segmentation usually exists. In this paper, the result of HKFCM was used as initial contour curve, and the automated initialization of deformation model was finished. For all the pixels in each cluster i.e. white matter, if 4 neighborhoods included the heterogeneous pixel, the pixel was regarded as candidate boundary point. So the algorithm of curve tracing [17] was proposed. The exterior boundary of the cluster was tracked in the candidate boundary points. Finally, the closed curve was obtained. The candidate boundary points, whose Euclidean distances to the origin coordinates were shortest, were chosen as initiation points of curve tracing. The steps of image segmentation with adapted level set method were as follows:

Step1. Set the number of clusters, then the original image was processed with HKFCM, and calculate the g_2.

Step2. Choose one cluster, evaluate the inside area with $-\rho$ and the outside area with $+\rho$, ρ is a plus constant. The boundary of the area is set to 0. The region of interest is defined initial contour.
Step3. Minimize the overall energy functional with 4.13 formula.

5 Radon Transform

The Radon transform of an image is the sum of the Radon transforms of each individual pixel.

The algorithm first divides pixels in the image into four subpixels and projects each subpixel separately, as shown in the following figure.

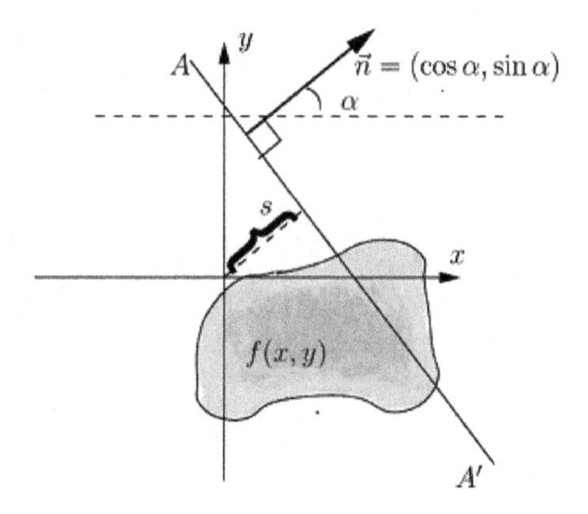

Fig. 1. Radon Transformation

Each subpixel's contribution is proportionally split into the two nearest bins, according to the distance between the projected location and the bin centers. If the subpixel projection hits the center point of a bin, the bin on the axes gets the full value of the subpixel, or one-fourth the value of the pixel. If the subpixel projection hits the border between two bins, the subpixel value is split evenly between the bins.

6 Experimental Results

The segmentation of image takes an important branch in the surgery navigation and tumor radiotherapy. However, due to medical imaging characteristics, the low contrast and fuzzy boundary is usually occurred in the images. In the experiment, the samples of images are taken from internet as shown in Figure 2a.

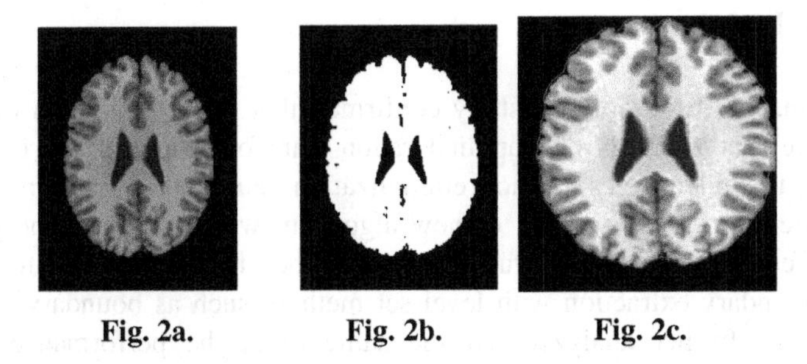

Fig. 2a. **Fig. 2b.** **Fig. 2c.**

Fig. 2a. are the original test images,
Fig. 2b. are the results of HKFCM clustering , to extracting the white matter.
Fig. 2c. are the results of final contour with proposed method.

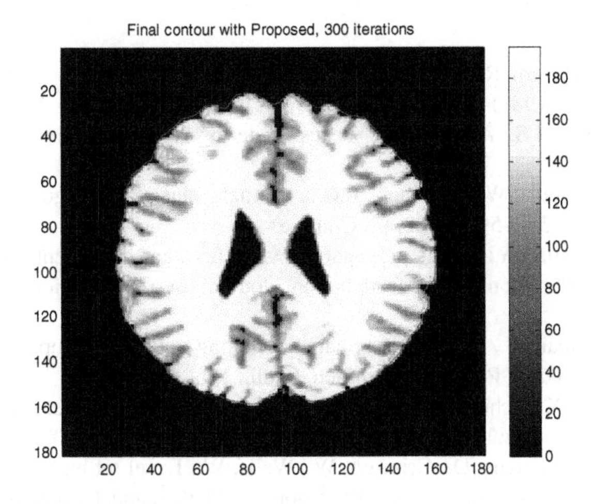

Fig. 3. Radon Transformed contour image

With the enhanced the approximate contour of white matter was got by HKFCM algorithm shown in Figure 2b. The snooping of regions else appear as a result of the in excess of segmentation.

7 Discussions

In this paper, we projected a new method to transform the algorithm. The original image was partitioned with HKFCM, and the controlled action of the edge indicator function was increased. The result of HKFCM segmentation was used to obtain the initial contour of level set method. With the new edge indicator function, results of image segmentation showed that the improved algorithm can exactly extract the corresponding region of interest. Under the same computing proposal, the average time cost was lower.

8 Conclusions

In conclusion, the results of this study confirmed that the combination of HKFCM with the level set methods and applied radon transformation to the images. The method has the advantages of no reinitialization, automation, and reducing the number of iterations. The validity of new algorithm was verified in the process of exacting different images. In the future research, the effect of priori information on the object boundary extraction with level set method, such as boundary, shape, and size, would be further analyzed. At the same time, the performance of image segmentation algorithms would be improved by reconstruction of classic velocity of level set method.

References

1. Bezdek, J.C.: Pattern Recognition with Fuzzy Objective Function Algorthims. Plenum Press, New York (1981)
2. Wu, K.L., Yang, M.S.: Alternative c-means clustering algorthims. Pattern Recognition 35, 2267–2278 (2002)
3. Zhang, L., Zhou, W.D., Jiao, L.C.: Kernel clustering algorthim. Chinese J. Computers 25(6), 587–590 (2002) (in Chinese)
4. Osher, S., Sethian, J.A.: Fronts propagating with curvature dependent speed: algorthim's based on the Hamilton-Jacobi formulation. Journal of Computational Physics, 12–49 (1988)
5. Malladi, R., Sethain, J., Vemuri, B.: Shape modelling with front propagation: A level set approach. IEEE Trans. Pattern Anal. Mach. Intell., 158–175 (1995)
6. Staib, L., Zeng, X., Schultz, R., Duncan, J.: Shape constraints in deformable models. In: Bankman, L. (ed.) Handbook of Medical Imaging, pp. 147–157 (2000)
7. Leventon, M., Faugeraus, O., Grimson, W., Wells, W.: Level set based segmentation with intensity and curvature priors. In: Workshop on Mathematical Methods in Biomedical Image Analysis Proceedings, pp. 4–11 (2000)
8. Paragios, D.R.: Geodesic active contours and level sets for the detection and tracking of moving objects. IEEE Transaction on pattern Analsis and Machin. In: telligence, 266–280 (2000)
9. Vese, L.A., Chan, T.F.: A multiphase level set frame wor for image segmentation using the mumford and shah model. International Journal of Computer Vision, 271–293 (2002)
10. Shi, Y., Karl, W.C.: Real-time tracking usin g level set. In: IEEE Computer Society Conference on Computer Vision and Pattern Recognition, pp. 34–42 (2005)
11. Li, C., Xu, C., Gui, C., et al.: Level set evolution without re-initialization:a new varitional formulation. In: IEEE Computer Society Conference on Computer Vision and pattern Recognition, pp.430–436 (2005)
12. Sethain, L.: Level set Methods and Fast Marching Methods. Cambridge University Press (1999)
13. Dunn, L.: A fuzzy relative of the ISODATA process and its use in detecting compact well-separated Clusters. J. Cybern., 32–57 (1973)
14. Bezedek, J.: A convergence thheorem for the fuzzy ISODATA clustering algorthims. IEEE Trans. Pattern Anal. Mach. Intell, 78–82 (1980)

15. Osher, S., Fedkiw, R.: Level set methods and Dynamic implicit surfaces, pp. 112–113. Springer (2002)
16. Peng, D., Merrimam, B., Osher, S., Zhao, H., Kang, M.: A PDE- based fast local level set method, J. Comp. Phys., 410–438 (1996)
17. Gomes, J., Faugeras, O.: Reconciling distance functions and Level Sets. J. Visiual Communic. And Imag. Representation, 209–223 (2000)
18. Mcinerney, T., Terzopouls, D.: Deformable models in medical image analysis: a survey. Medical Analysis, 91–108 (1996)
19. Zhang, D.-Q., Chen, S.-C.: Clustering in completed data usig Kernel-based fuzzy c-means algorthim. Neural Processing Letters 18(3), 155–162 (2003)

Performance Analysis of Filters to Wavelet for Noisy Remote Sensing Images

Narayan P. Bhosale, Ramesh R. Manza,
K.V. Kale, and S.C. Mehrotra

Geospatial Technology Laboratory, Dept. of Computer Science and IT,
Dr. Babasasaheb Marathwada University Aurangabad, MH, India
{narayanbhosale,manzaramesh,mehrotra.suresh15j}@gmail.com,
kvlale91@rediffmail.com

Abstract. In this paper, we have used Linear Imaging Self Scanning Sensor (LISS- III) remote sensing image data sets which are having four bands of Aurangabad region. For an empirical preprocessing work at lab an image is loaded and taken band image of spectral reflectance values and applied median 3x3, median 5x5, sharp 14, sharp 18, smooth 3x3, smooth 5x5 filters and the quality has been successfully measured. It gives better results than original noisy remote sensing image; therefore, the quality has been improved in all filters. Moreover to achieving high quality we have used multilevel 2D wavelet decomposion based on haar wavelet filter while applying various above filters on noisy remote sensing images, we can remove noise from remote sensing images at large level through multilevel 2D Wavelet decomposing based on haar wavelets over above filters has been proved successfully. Thus, this work plays significance important role in the domain of satellite image processing or remote sensing image analysis and its applications as a preprocessing work.

Keywords: RS image, Noise, Filter, Wavelet.

1 Introduction

The remote sensing Images are very useful for providing information about earth's surface by using the relative motion between antenna and its target. It has various applications such as remote sensing, resource monitoring, navigation, positioning and military commanding, high-resolution remote sensing for mapping, search and rescue, mine detection, surface surveillance and automatic target recognition [1], [2], [4], [6]. Noise will be unavoidable during image acquisition process and denoising is an essential step to improve the image quality. Image denoising involves the manipulation of the image data to produce a visually high quality image. Finding efficient image denoising methods is still valid challenge in image processing. Wavelet denoising attempts to remove the noise present in the imagery while preserving the image characteristics, regardless of its frequency content [7].

S.C. Satapathy et al. (eds.), *Emerging ICT for Bridging the Future – Volume 1*,
Advances in Intelligent Systems and Computing 337, DOI: 10.1007/978-3-319-13728-5_30

2 Denoising Remote Sensing Image

The main task of remote sensing image denoising is to remove noise when the edges are preserving with other relevant information. Noise is a common problem in the image processing task. If we consider the very high and accurate resolution image then also there is the chance of noise. The main purpose or the aim of image denoising is to recover the main image from the noisy image. Moreover, noise will exist in remote sensing then it may affect in an accuracy of classification results. Due to this reason one has to do preprocessing of their selected image and get confirmed denoised remote sensing data or images.

Noise is present in an image either additive or multiplicative form. Gaussian noise is most commonly known as additive white Gaussian noise which is evenly distributed over the signal. Each pixel in the noisy image is the sum of the true pixel value and random Gaussian distributed noise vale. Salt and pepper noise is represented as black and white dots in the images. This is caused due to errors in data transmission. Speckle noise is a multiplicative noise which occurs all coherent imaging systems like laser and Synthetic Aperture Radar imagery [7]. As we choose to use LISS-III remote sensing image to for our experimental work which as shown in Fig-1.

Fig. 1. LISS-III Remote Sensing Image of Aurangabad region [2], [4], [5]

The remote sensing analyst views an image only as "data". Noise is introduced into the data by the sensor. It is a variation in the sensor put that interfaces with our ability to extract scene information from an image. Image noise occurs in wide variety of form and often difficult to model; for these reasons, many noise reduction techniques are adhoc. It is beneficial to categorized noise type and denoising them at max level [11].

2.1 Data Used

Data type: Multi Spectral Radiometry High Resolution, Instruments: LISS-III, Resolution: 23.5 m, Satellites: IRS-P6,Dims: 1153 x 1153 x 4 [BSQ],Size: [Byte] ,354,913 bytes, File Type : TIFF, Sensor Type: LISS, Byte Order : Host (Intel), Projection : Geographic Lat/Lon, Pixel : 0.000225 Degrees Datum : WGS-84, Upper Left Corner: 1,1;Description: GEO-TIFF File Imported into ENVI [Sat Sep 06 19:01:32 2014].

2.2 Filters of Wavelet

Filter plays important role to denoising remote sensing Images. Wavelets are the latest research area in the field of image processing and enhancement. Wavelet analysis allows the use of long time intervals where we want more precise low frequency information, and shorter regions where we want high frequency information [3].

The various authors are addressed this issues with filters and shown their performance in their experimental works. And some of them used wavelet in their experimental work. Thus, we took new filters median 3x3, median 5x5, sharp 14, sharp 18, smooth 3x3, smooth 5x5 filters and apply the wavelet and interpreted the results.

2.3 Performance Measure

Image Quality is a characteristic of an image that measures the perceived image degradation Peak signal to noise ratio (PSNR) has used to measure the efficiency of the filters and wavelet.
This parameter carries the most significance as far as noise suppression is concerned.

PSNR is the peak signal to noise ratio in decibles (DB). The PSNR is measured in terms of bits per sample or bits per pixel. The image with 8 bits per pixel contains from 0 to 255. The greater PSNR value is the better the image quality and noise suppression [7], [8]. PSNR is the peak signal-to-noise ratio in decibels (dB). The PSNR is only meaningful for data encoded in terms of bits per sample, or bits per pixel. For example, an image with 8 bits per pixel contains integers from 0 to 255.

The mean square error (MSE) is the squared norm of the difference between the data and the approximation divided by the number of elements. MAXERR is the maximum absolute squared deviation of the data, X, from the approximation, XAPP.

L2RAT is the ratio of the squared norm of the signal or image approximation, XAPP, to the input signal or image, X.

Where-
X is a real-valued signal or image.
XAPP is a real-valued signal or image approximation with a size equal to that of the input data, X. BPS is the number of bits per sample in the data.
The following equation defines the PSNR in below equation (1):

$$20\ \log_{10}\left(\frac{2^B - 1}{\sqrt{MSE}}\right) \dots\dots\dots\dots \qquad \dots\dots(1)$$

Where MSE represents the mean square error and B represents the bits per sample.

The mean square error between a signal or image, X, and an approximation, Y, is the squared norm of the difference divided by the number of elements in the image as of equation (2).

$$\frac{||X - Y||^2}{N} \text{ ……… } \qquad \text{….(2)}$$

PSNR and MSE are used to the performance measure [2], [4], [6], [8].

3 Results Analysis

The experiments were conducted on Multi Spectral Radiometry High Resolution LISS-III, remote sensing image of Aurangabad region in at MNF noise levels combined with various Filters such as median 3x3, median 5x5, sharp 14, sharp 18, smooth 3x3, smooth 5x5. The proposed method is multilevel 2D Wavelet decomposing based on Haar wavelet compared with Filters [9], [10] based image denoising. Filter noising shown all statistical values & Fig.2 as shown in below Table 1. Filter performance denoising RS Images. The haar wavelet performance as shown in below Table 2 & Fig.3. And last given in Fig. 2. Remote Sensing Images with filters and Wavelet analysis

Table 1. Filter performance denoising RS Images

Image Names	**PSNR**	**MSE**	**Maxerr**	**L2rat**
Original Band	18.24840	9.732861	84	0.9540
MNF noisy	17.3431	1.1989	96	0.9462
median 3x3	22.4106	3.7327	98	0.9869
median 5x5	22.4106	3.7327	98	0.9869
sharap14	22.4106	3.7327	98	0.9869
sharap10	22.4106	3.7327	98	0.9869
sharap18	22.4106	3.7327	98	0.9869
smooth33	22.4106	3.7327	98	0.9869
smooth55	22.4106	3.7327	98	0.9869

Table 2. The haar wavelet perforamce

Image name	**PSNR**	**MSE**	**Maxerr**	**L2rat**
median 3x3	23.2411	3.082972	9.9	0.9892
median 3x3	23.2411	3.082972	9.9	0.9892
median 5x5	23.2411	3.082972	9.9	0.9892
sharap14	23.2411	3.082972	9.9	0.9892
sharap10	23.2411	3.082972	9.9	0.9892
sharap18	23.2411	3.082972	9.9	0.9892
smooth33	23.2411	3.082972	9.9	0.9892
smooth33	23.2411	3.082972	9.9	0.9892

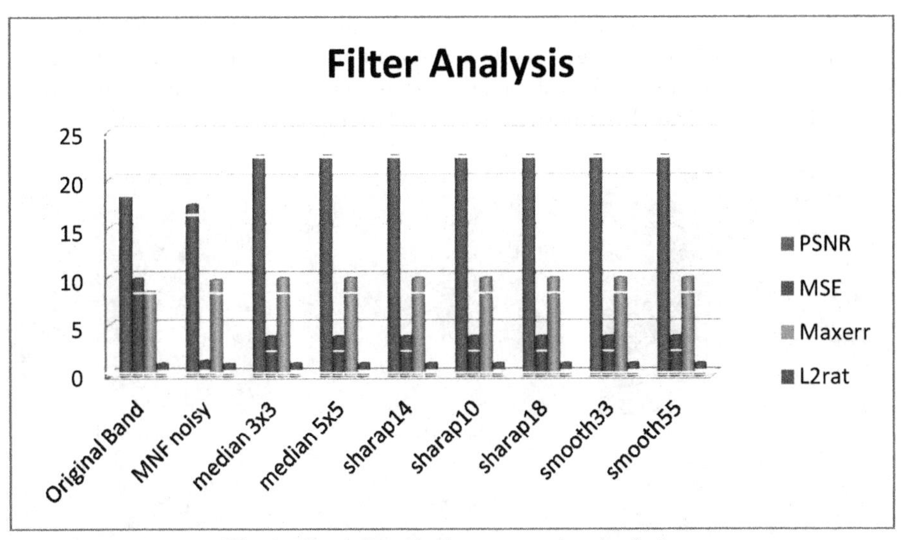

Fig. 2. Filters Graphical representation Analysis

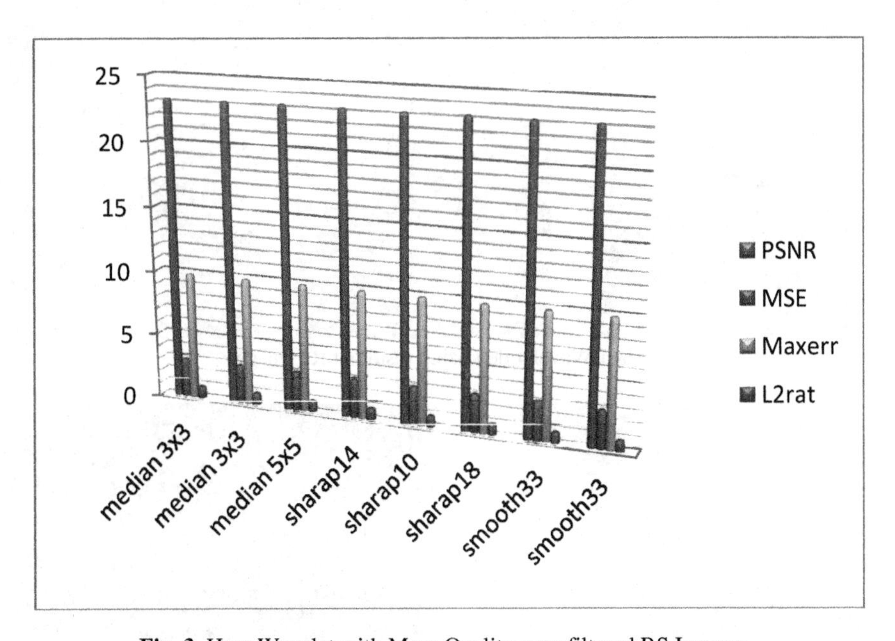

Fig. 3. Haar Wavelet with More Quality over filtered RS Images.

4 Remote Sensing Images with Performance of Filters and Wavelet

Image denoising has a significant role in the image processing. As the applications area of image denoising is more and big demand for effect denoising methods [11].

MNF is minimum noise function use MNF Rotation transforms to determine the inherent dimensionality of image data, to segregate noise in the data, and to reduce the computational requirements for subsequent processing, All experiments were carried using ENVI 4.4 and MATLAB R2012a were used for digital image processing [12].

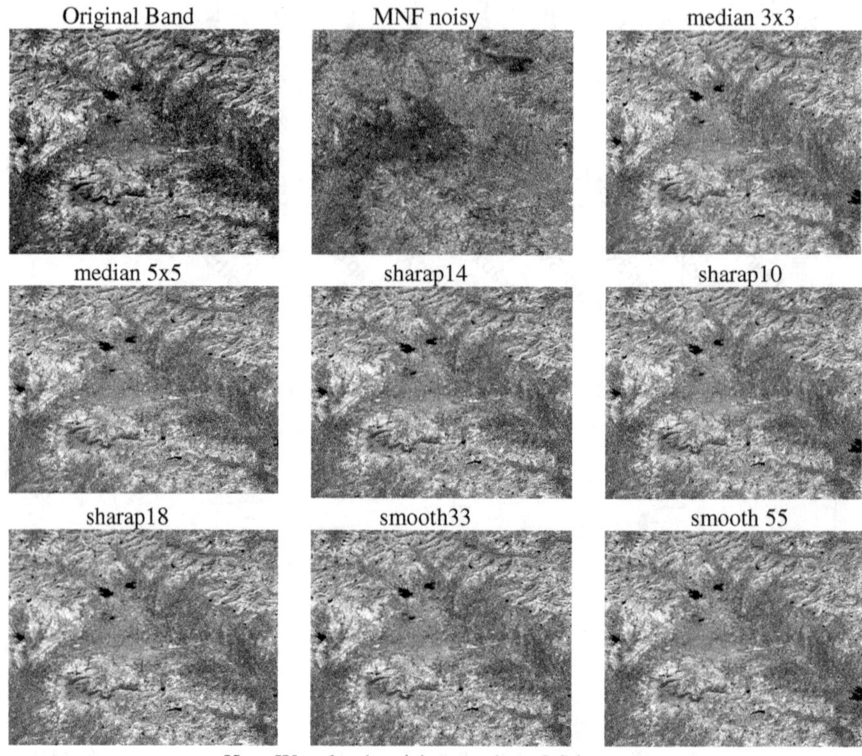

Haar Wavelet denoising resultant RS images

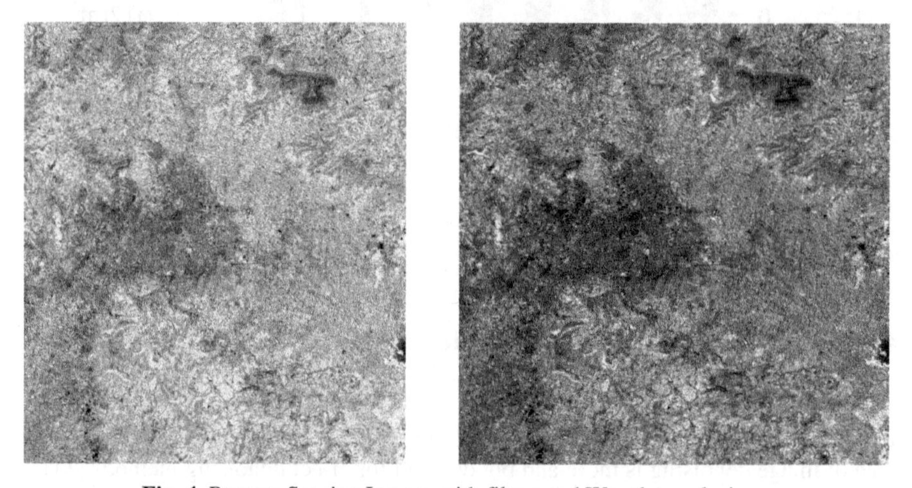

Fig. 4. Remote Sensing Images with filters and Wavelet analysis

5 Conclusion and Future Scope

In this paper, we have studied filters to wavelet performance for denoising the remote sensing images. Firstly, we have apply the MNF components have greatly increased noise fraction in our remote sensing image and more noise occur in band 4th band so that we applied median 3x3, median 5x5, sharp 14, sharp 18, smooth 3x3, smooth 5x5 filters on them. And quality has improved by haar wavelet over filters.

The results shows that the proposed multilevel 2D Wavelet decomposing based on haar wavelet transform is best than the traditional filter based denoising i.e. median 3x3, median 5x5, sharp 14, sharp 18, smooth 3x3, smooth 5x5 filters thus wavelet performance is a powerful noise tool for noise reduction. The future scope of this work will be to design robust wavelet filter to remove noise from remote sensing images.

Acknowledgements. The authors would like to thank for provide temporary access in Satellite Image Analysis Lab at CSRE, IIT Bombay & I am Specially thanks to UGC SAP (II) DRS Phase-II F.No-3-42/2009 (FIST) of Dept of Computer Science & IT, Dr. Babasaheb Ambedkar Marathwada University, Aurangabad, Maharashtra, India for providing Geospatial technology Lab to carry experimental work with data set.

References

1. Rajamani, A., Krishnaven, V.: Performance Analysis Survey of Various SAR Image Despeckling Techniques. International Journal of Computer Applications (0975 – 8887) 90(7) (March 2014)
2. Bhosale, N.P., Manza, R.R.: Analysis of Effect of Gaussian, Salt and Pepper Noise Removal from Noisy Remote Sensing Images. In: Second International Conference on Emerging Research in Computing, Information, Communication and Applications (ERCICA 2014). Elsevier (August 2014) ISBN: 9789351072607
3. Kaur, J.: Image Denoising For Speckle Noise Reduction In Ultrasound Images Using Dwt Technique. International Journal of Application or Innovation in Engineering & Management (IJAIEM) 2(6) (June 2013) ISSN 2319 – 4847
4. Bhosale, N.P., Manza, R.R.: Analysis of effect of noise removal filters on noisy remote sensing images. International Journal of Scientific & Engineering Research (IJSER) (10), 1511–1514 (2013)
5. Bhosale, N.P., Manza, R.R.: Effect of Poisson Noise on Remote Sensing Images and Noise Removal using Filters. IBMRD's Journal of Management & Research 3(2), 77–83 (2014)
6. Bhosale, N.P., Manza, R.R.: Image Denoising Based On Wavelet for Satellite Imagery: A Review. International Journal Of Modern Engineering Research (IJMER) (4), 63–68 (2014) ISSN: 2249–6645
7. Sulochana, S., Vidhya, R.: Image Denoising using Adaptive Thresholding in Framelet Transform Domain. International Journal of Advanced Computer Science and Applications(IJACSA) 3(9) (2012)
8. Bhosale, N.P., Manza, R.R.: A review on noise removal techniques from remote sensing images. In: National Conference, CMS 2012 (April 2013)

9. Kaur, G.: Image denosingusing wavelet transform and Various filters. Interantional Journal of Research in Computer Science 2(2), 15–21 (2012)
10. Subashini, P.: Image denoising based on Wavelet Analysis for satellite imagery. In: Wavelet Transform Book, Advance in Wavelet Theory and their Application (2012) ISBN:978-953-51-0494-0
11. Robert, A.: Schowengerdt: Remote Sensing models and methods for Image processing, 3rd edn. Acasemic Press, Elsevier (2007)
12. Kale, K., Manza, R., et al.: Understanding MATLAB, 1st edn. (March 2013)

Singer Identification Using MFCC and LPC Coefficients from Indian Video Songs

Tushar Ratanpara and Narendra Patel

C. U. Shah University, Wadhwan City, Gujrat, India
tushar.ratanpara@gmail.com

Abstract. Singer identification is one of the challenging tasks in Music information retrieval (MIR) category. Music of India generates 4-5% of net revenue for a movie. Indian video songs include variety of singers. The research presented in this paper is to identify singer using MFCC and LPC coefficients from Indian video songs. Initially Audio portion is extracted from Indian video songs. Audio portion is divided into segments. For each segment, 13 Mel-frequency cepstral coefficients (MFCC) and 13 linear predictive coding (LPC) coefficients are computed. Principal component analysis method is used to reduce the dimensionality of segments. Singer models are trained using Naive bayes classifier and back propagation algorithm using neural network. The proposed approach is tested using different combinations of coefficients with male and female Indian singers.

Keywords: Mel-frequency cepstral coefficients (MFCC), linear predictive coding (LPC), back propagation, music information retrieval (MIR).

1 Introduction

Indian video songs are sung by distinct singers. The listeners are attracted using the voice of singer. It is required to organize, extract and classify [1] singer's voice. Sometimes viewer is interested to listen or watch song of favorite singer only. So there is great need to develop a system for singer identification from Indian video songs.

Our proposed system can identify singer by extracting coefficients of audio part of Indian video songs. One of the usefulness of this system is famous Indian singer's video songs can be identified from large database. It can be useful to learn singer voice characteristics by listening songs of different genre. It can be useful in categorize unlabeled video songs and copy right protection. A significant amount of research has been performed on speaker identification from digitized speech for applications such as verification of identity. These systems for the most part use features similar to those used in speech recognition and speaker recognition. Many of these systems are trained on pristine data (without background noise) and performance [2] tends to degrade in noisy environments. Since they are trained on spoken data, they perform poorly to singing voice input. According to [3] Mel

S.C. Satapathy et al. (eds.), *Emerging ICT for Bridging the Future – Volume 1*,
Advances in Intelligent Systems and Computing 337, DOI: 10.1007/978-3-319-13728-5_31

Frequency Cepstral Coefficients (MFCCs) originally developed for automatic speech recognition applications can also be used for music modeling. Pitch and rhythm audio features are computed. MFCC feature vectors and neural networks are used to identify [4] singer identity. Classification of singers according to [5] based on voice type, and their voice quality. Classification is done using vector quantization and Gaussian mixture model. Singer's vibrato based octave frequency cepstral coefficients (OFCC) [6] were proposed which is used in singer identification. In [7] Hybrid selection method of audio descriptors is used in singer identification from north Indian classical music. Top few selected audio descriptors impact is more than other audio descriptors. Here Gaussian mixture model and hidden markov model are used for identification of a singer.

Rest of the paper is organized as follows: Section 2 describes proposed approach, Experimental results are explained in section 3 followed by conclusion in section 4.

2 Proposed Approach

The abstract model of our proposed system is given Figure 1. It contains six blocks. 1) Collect Indian video songs 2) Segmentation 3) Coefficients extraction 4) Singer model generation 5) Singer identification

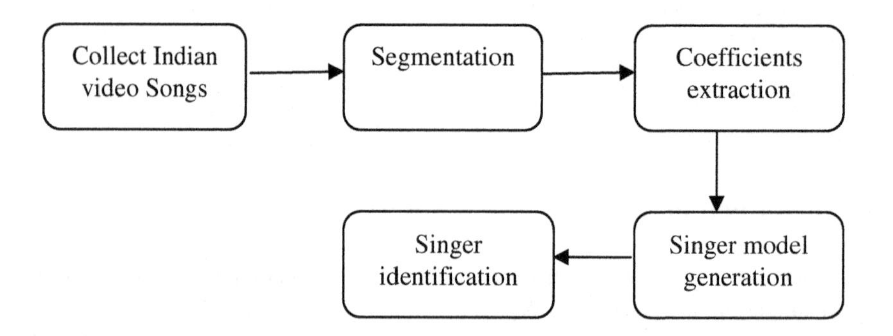

Fig. 1. Abstract model of our proposed system

Figure 2 represents algorithm steps of singer identification using MFCC and LPC coefficients and naïve bayes classifier and back propagation algorithm using neural net-work.

2.1 Collect Indian Video Songs

Four famous Indian singers are selected to create new dataset. Indian video songs are downloaded from different websites. Table 1 shows list of Indian singers used in our dataset. Audio portion is extracted from each Indian video song which is used in segmentation.

Algorithm Steps:

1. Collect N number of Indian video songs of M singers.
2. Separate audio portion from Indian video song.
3. Audio portion is divided into segments.
4. MFCC and LPC coefficients are calculated for each segment.
5. Dimension reduction using principal component analysis method
6. Training model is build using naïve bayes classifier and back propagation algorithm using neural network.
7. Model is tested using male and female singers of Indian video songs.
8. Singer is identified for a given Indian video song.

Fig. 2. Algorithm of singer identification

Table 1. List of Indian singers used in our dataset

ID	Singer Name	Gender
S1	Sonu nigam	MALE
S2	Sunidhi Chauhan	FEMALE
S3	Mohit Chauhan	MALE
S4	Atif Aslam	MALE

2.2 Segmentation

Audio portion of each video song is used for segmentation. Total number of segments is found using following equation

$$TNF = TS / 10 \tag{1}$$

Where TNF is total number of segments and TS total number of seconds of audio portion. First three minutes of audio portion is used to identify singer. So the value of TS is 360 and TNS is 36 used in our approach. Now coefficients are extracted for each segment which is explained in section 2.3.

2.3 Coefficients Extraction

Following two types of coefficients are considered for singer identification model.
1) MFCC 2) LPC.

2.3.1 MFCC

MFCC's are most useful coefficients used for speech recognition for some time because of their ability to represent speech amplitude spectrum in a compact form. Figure 3 shows the process of creating MFCC features [3]. Speech signal is divided into frames by applying a hamming windowing function at fixed intervals (typically 20 ms). Cepstral feature vector is generated using each frame.

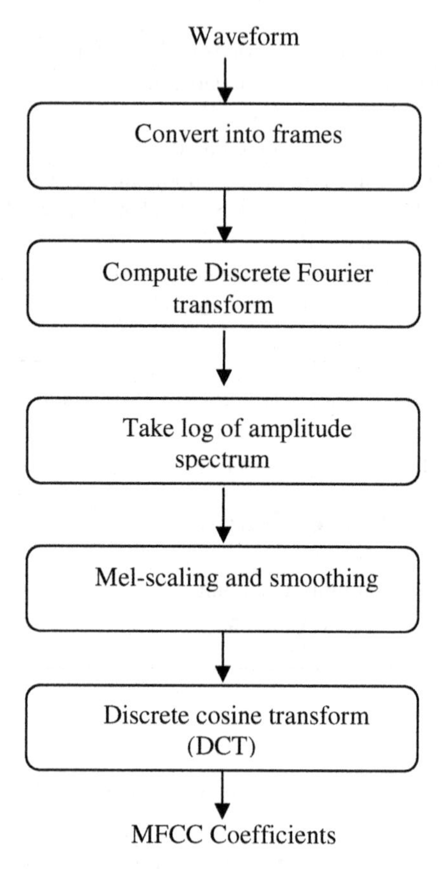

Fig. 3. The process of creating MFCC [3] features

In our proposed approach 13 MFCC coefficients are extracted for each segment and store it in array.

2.3.2 LPC

LPC [8] can provide a very accurate spectral representation for speech sound. LPC coefficients are computed by using following equation.

$$X(n) = -a(2)X(n-1) - a(3)X(n-2) \ldots\ldots -a(P+1)X(n-P) \tag{2}$$

Where p is the order of the prediction filter polynomial, a = [1 a (2) … A (P + 1)]. In our proposed approach 13 LPC coefficients are calculated. The value of P is 12 (12th order polynomial). Total 26 coefficients are extracted from each segment in our proposed approach. So now each video song contains matrix of 36 * 26 coefficients. It contains large number of coefficients so principal component method is used to reduce dimensionality of coefficients.

2.4 Singer Model Generation

Singer models are generated using naïve bayes classifier and back propagation algorithm using neural network. Naïve bayes classifier [9] is highly scalable which requires number of parameters. Maximum –likelihood training can be done by evaluating equation which takes linear time rather than expensive iterative approximation like other classifiers. Another singer model is based on back propagation algorithm feed forward neural network. Back propagation neural network [10] model is supervised learning model used in many applications.

2.5 Singer Identification

Singer identification is carried out using training models. Testing dataset is applied to input to training modes which leads to identification of a singer. Here k-fold cross-validation method [11] [12] is used. It is used for assessing how the result of a statistical analysis will generalize to an independent data set. To reduce variability, 10 rounds of cross-validation are performed using different partitions. After that validation results are averaged over the rounds.

3 Experimental Results

A novel music database is prepared for Indian Hindi video songs of a four famous Indian singers from Bollywood movies and albums which are publically available in CDs/DVDs. To maintain uniformity of our database, each Indian video song is converted to 256kbps bit rate. All the experiments have been performed in MATLAB, on a standard PC (Intel Core i3, 2.30 GHz, 4GB RAM). Dataset is divided into two parts 1) training dataset 2) testing dataset. 80% dataset is used for training purpose and remaining 20% dataset is used for testing purpose. Accuracy is found using equation 3.

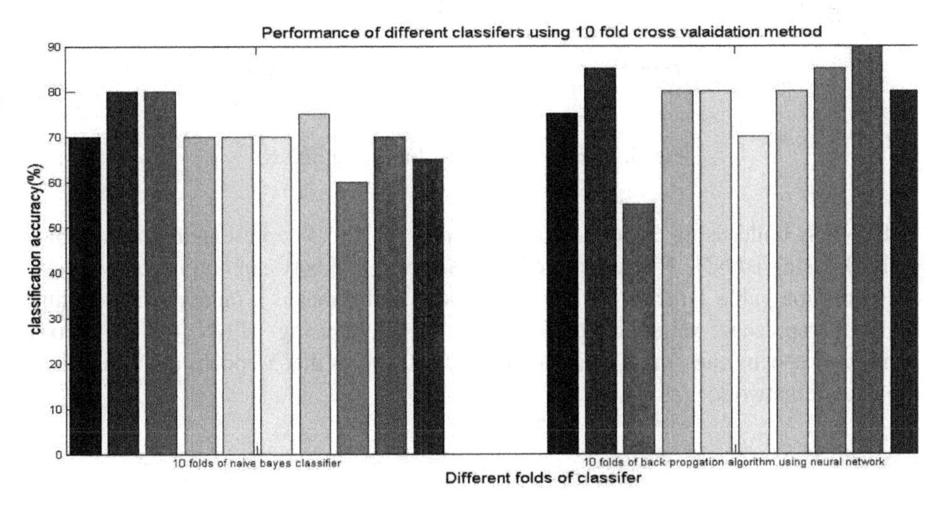

Fig. 4. Performance of different classifiers using 10 folds cross validation method

Accuracy (%) = total number of songs which is correctly identified / total number of songs used in testing dataset. (3)

To check accuracy of our proposed approach Performance of different classifier using 10 fold cross validation method is computed as shown in figure 4. It shows that back propagation neural network (BPNN) gives higher classification accuracy than the naive bayes classifier. 71% and 78% accuracy is achieved by Naïve bayes classifier and BPNN respectively.

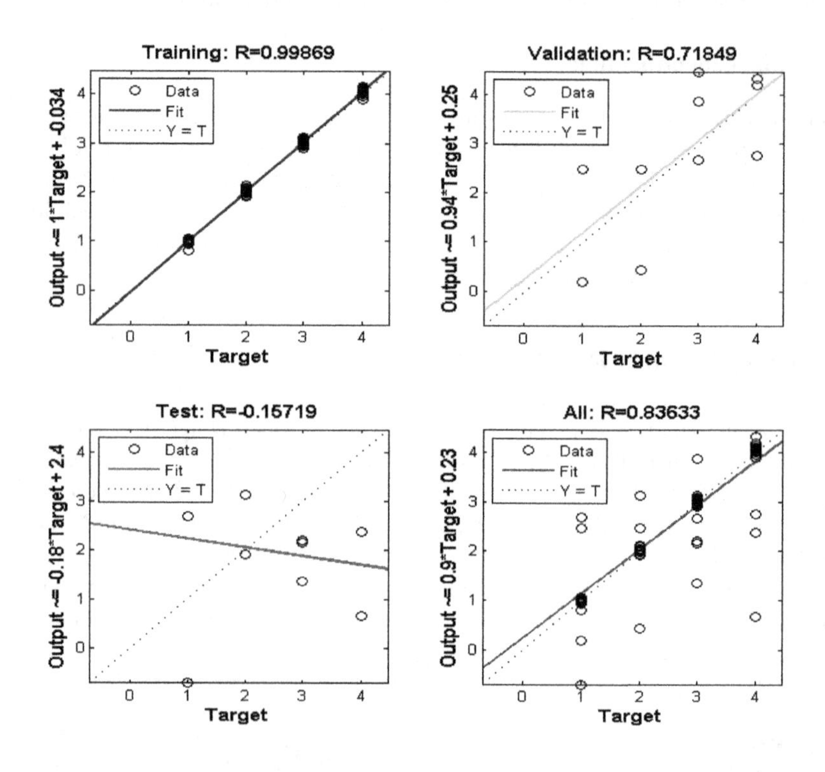

Fig. 5. Regression plot of 1 fold using BPNN

BPNN is build using 50 hidden layer neurons. Figure 5 represents regression plot of 1 fold using BPNN. Regression represents correlation between outputs and targets. A Regression value 1 means a close relationship and 0 means a random relationship. Figure 6 represents validation performance of 1 fold using BPNN. On the x-axis number of epochs and y-axis mean square error. It shows that 5 epochs are required to train neural network.

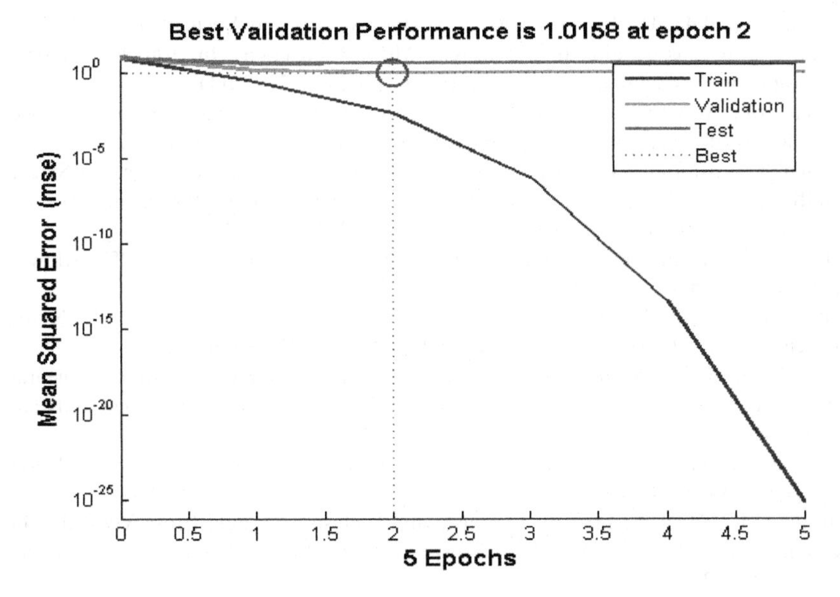

Fig. 6. Validation performance of 1 fold using BPNN

4 Conclusion

In this paper singer identification technique is proposed using MFCC and LPC coefficients from Indian video songs. The proposed scheme extracts 13 MFCC and 13 LPC features from collected Indian video songs. PCA is used to reduce dimensionality of coefficients. Then Naïve bayes classifier and BPNN are used to train the network using coefficients. Experimental result shows that BPNN gives higher classification accuracy than Naïve bayes classifier. Regression plot and validation performance curve is plotted to check performance of our proposed approach. BPNN is trained using 50 hidden layer neurons and gives best validation performance at epoch 2.

References

1. Regnier, L., Peeters, G.: Singer verification: singer model vs. song model. In: IEEE International Conference on Acoustics, Speech and Signal Processing, pp. 437–440 (2012)
2. Mammone, R., Zhang, X., Ramachandran, R.P.: Robust speaker recognition: A feature-based approach. IEEE Signal Processing Magazine 13, 58–71 (1996)
3. Logan, B.: Mel frequency cepstral coefficients for music modeling. In: International Symposium Music Information Retrieval (2000)
4. Doungpaisan, P.: Singer identification using time-frequency audio feature. In: Liu, D., Zhang, H., Polycarpou, M., Alippi, C., He, H. (eds.) ISNN 2011, Part II. LNCS, vol. 6676, pp. 486–495. Springer, Heidelberg (2011)

5. Maazouzi, F., Bahi, H.: Use of Gaussian Mixture Models and Vector Quantization for Singing Voice Classification in Commercial Music Productions. In: 10th International Symposium on Programming and Systems (ISPS), pp. 116–121 (2011)
6. Bartsch, M., Wakefield, G.: Singing voice identification using spectral envelope estimation. IEEE Transactions on Speech and Audio Processing 12, 100–109 (2004)
7. Deshmukh, S., Bhirud, S.G.: A Hybrid Selection Method of Audio Descriptors for Singer Identification in North Indian Classical Music. In: Fifth International Conference on Emerging Trends in Engineering and Technology, pp. 224–227 (2012)
8. Jackson, L.B.: Digital Filters and Signal Processing, 2nd edn., pp. 255–257. Kluwer Academic Publishers, Boston (1989)
9. Maniya, H., Hasan, M.: Comparative study of naïve bayes classifier and KNN for Tuberculosis. In: International Conference on Web Services Computing (2011)
10. Saduf, Wani, M.: Comparative study of back propagation learning algorithms for neural networks. International Journal of Advanced Research in Computer Science and Software Engineering 3 (2013)
11. Meijer, R., Goeman, J.: Efficient approximate k-fold and leave-one-out cross-validation for ridge regression. Biometrical Journal 55(2) (2013)
12. Rodríguez, J., Pérez, A., Lozano, J.: Sensitivity Analysis of k-Fold Cross Validation in Prediction Error Estimation. IEEE Transactions on Pattern Analysis and Machine Intelligence 32(2) (2010)

Role of Clinical Attributes in Automatic Classification of Mammograms

Aparna Bhale[1], Manish Joshi[2], and Yogita Patil[3]

[1] School of Computer Sciences, North Maharashtra University, Jalgaon, India
[2] Department of Computer Science and IT, Deogiri College, Aurangabad, India
{aparnakulkarnibhale,joshmanish,patilmyogita}@gmail.com

Abstract. It has been established that mammogram plays vital role in early detection of breast diseases. We also know that computational empowerment of mammogram facilitates relevant and significant information. Several research groups are exploring various aspects of mammograms in terms of feature selection to develop an effective automatic classification system.

Mammographic attributes including textural features, statistical features as well as structural features are used effectively to develop automatic classification systems. Several clinical trials explained that attributes of patient's clinical profile also plays an important role in determination of class of a breast tumor. However, usage of patients clinical attributes for automatic classification and results thereof are not reported in literature. None of the existing standard mammogram datasets provide such additional information about patients history attributes.

Our focus is to validate observations revealed by clinical trials using automatic classification techniques. We have developed a dataset of mammogram images along with significant attributes of patients clinical profile. In this paper, we discuss our experiments with standard mammogram datasets as well as with our extended, informative live data set. Appropriate features are extracted from mammograms to develop Support Vector Machine (SVM) classifier. The results obtained using mere mammographic features are compared with the results obtained using extended feature set which includes clinical attributes.

Keywords: Mammogram, SVM, Texture features, statistical features, Clinical attributes.

1 Introduction

In this paper we demonstrate how different features of mammograms and patients clinical features can be combined to obtain better classification of a breast tumor. Two parallel approaches of research associated with mammographical studies and its corresponding clinical observations are being explored in recent years. The first approach consists of clinical surveys (also called as clinical trials) where a sample of

S.C. Satapathy et al. (eds.), *Emerging ICT for Bridging the Future – Volume 1*,
Advances in Intelligent Systems and Computing 337, DOI: 10.1007/978-3-319-13728-5_32

adequate numbers of patients are surveyed to test certain hypothesis. Patients clinical profile attributes (clinical features) like 'age' is considered as the base for hypothesis and handful numbers of patients are surveyed to either accept or reject the hypothesis as a result of appropriate statistical analysis. Likewise age feature, several other features including gender, marital status, and age of menarche are interrogated and accordingly its significance is determined. Section 2 lists out various such recent papers.

The second approach uses principles of data mining and soft computing to determine class of a breast tumor. Data driven automatic classification based on the textural, structural, statistical features extracted from mammograms is the most common technique. Variations in feature selection as well as in algorithms of classifications can be easily noticed. A list of some of the recent research work is listed in section 2.

Most of the work reported for the second approach uses standard mammogram datasets. Clinical features of patients, although they are very significant, are not recorded in such datasets. Hence, these features are not at all considered during automatic breast tumor classification. In order to overcome this lacuna and revalidate the importance of clinical features using data mining approach we carried out certain experiments systematically. The details of our experiments are presented in section 3.

We have worked on mammogram images obtained from a standard mammogram dataset and live mammogram images obtained from a hospital. We have extracted texture features using Gabor filter and statistical features by gray level co-occurrence matrix. Support Vector Machine (SVM) is used to classify the mammograms as benign or malignant.

The paper is organized as follows. Related work is discussed in Section 2. Experimental details are elaborated in section 3 whereas results and observations are put forward in section 4. Conclusions are in section 5.

2 Related Work

Several evidences proved that certain clinical features can be decisive while determining the class of a breast tumor. For example use of contraceptive pills lead to Fibrocystic Changes (FCC), this is one of the subtypes of benign breast tumor. Despite of this fact, use of clinical features in automatic classification of breast tumor is not reported in literature. The main reason is the unavailability of a dataset having mammograms as well as corresponding patient's clinical features. In this section, we therefore decided to discuss recent research findings in automated breast tumor classification and observations of various clinical surveys (clinical trials).

Geethapriya Raghavan et al. [1] have developed and implemented pattern recognition techniques like support vector machine (SVM), kernel Fisher discriminant (KFD), relevance vector machine (RVM) and a multi resolution pattern recognition method using wavelet transforms to classify tumors as benign or malignant in statistical learning. S Mohan Kumar et al. [2] presented the classification of microcalcification in digital mammogram based on statistical features and SNE. They obtained accuracy for abnormal cases as 84% and for benign cases 94.67%. Ruey-Feng Chang et al. [3] used SVM to diagnose breast tumors on ultrasound and observed accuracy of 85.6%. Prabhakar Telagarapu et al. [4] shared an algorithm to reduce the speckle noise, feature extraction and classification methods for benign and malignant tumors. Their classifier gives 77% correct results. Xiangjun Shi et al. [5] presented a novel CAD system of automatic mass detection and classification of breast ultrasound images using fuzzy support vector machines. In this study they obtain SVM classifier's accuracy of 87.3% and Fuzzy SVM's classification result as 94.25%. Karmilasari et al. [6] and Osmar R. Zaiane et al. [7] classified tumors using SVM and association based rules to obtained results as 85% and 80% respectively. Mahmoud R. Hejazi et al. [8] proposed two segment classifications using region growing method and obtained 82.9% accuracy. The work of Tzikopoulos et al. [9] focuses on mammographic segmentation and classification. Several researchers proposed novel approaches for segmenting the lesions from the mammograms. Effective lesion segmentation ultimately contributes to the overall performance of automatic detection of microcalcification or masses. Xu et al. [10] enhanced the conventional watershed technique. Ashwini Kumar Mohanty et al.[21] proposed an algorithm using statistical techniques like GLCM, Intensity histogram. This algorithm is useful for medical image feature extraction.

Name, Age, gender, Age of Menarche, menstrual history, number of kids, breast feeding, dietary habits, alcohol intake, smoke, consumption of pregnancy prevention pills, family history of breast cancer are some of the clinical features that are used by many researchers[11,12] for breast tumor classification.

Jaree Thongkam et al. [13] used age, marital status as an one of the input attributes and used SVM and Weka 3.5.6 for evaluation of experimental results. Robert D. Rosenberg et al. [14] used various patient history attributes and reported that there exist a correlation between clinical attributes and tumor class. Linda Tintus et al. [15] made an attempt to evaluate the influence of menarcheal age to breast density. Many researchers studied various aspects which show relation of breast cancer to age [19], age of menarche, breast cancer history [16], dietary habits [17], and recall of early menstrual history [18].

In this work, we joined the mammographic and the clinical features to determine the combined effect of these features.

3 Experimental Details

In this section we elaborate the stepwise experimental process followed to obtain and compare the classification results. We have compared the results of two classifiers. We have taken into account the mammographic features for the classification of the first classifier. The results of second classifier are generated by using mammogram features augmented with clinical features. In first subsection we describe the details of the mammogram data sets used for the classification. How mammogram images are preprocessed to extract significant mammographic features is also depicted in this subsection. The description of the development of a data set of live mammograms and its usage to obtain various clinical attributes from patients' profile is presented in second subsection. The last subsection throws light on the approach used to compare the classifiers.

3.1 Mammographic Feature Extraction

For experimental purpose we have extracted mammographic features from a standard dataset of mammograms using Digital Database for Screening Mammography (DDSM). The details of feature extraction are as follows.

Image pre-processing is a necessary phase in order to improve image quality. Pre-processing is completed using morphological reconstruction operation. Gradient magnitude of image is pre-processed to remove the noise. Watershed transformation applied to refine gradient magnitude image. Marker Control Watershed Segmentation MCWS is used for deduction of over segmentation occurred by watershed. Images shown in figure 1 are obtained from live dataset. Figure 1 shows stages of segmentation:

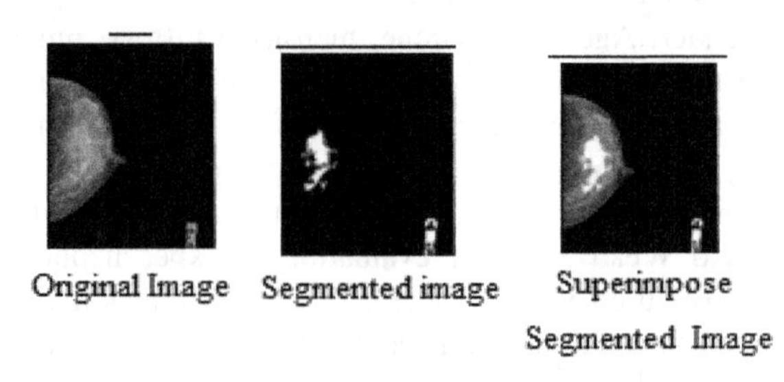

Fig. 1. Segmentation Stages

We have obtained Region of Interest (ROI) after segmentation by MCWS technique. Gabor filter is applied on ROI to extract features. ROI locates a lesion. Large number of statistical, textural features can be extracted. Appropriate features are selected that could significantly determine presence of a breast lesion and architectural distortions. By Gabor filter and GLCM (gray-level co-occurrence matrix) we extracted a few textural features from mammogram which are the quantitative measures of texture that are Contrast, energy, entropy, homogeneity. The significance of selected features used in our experiment is highlighted below.

1] Contrast: To distinguish an object from its surroundings, sufficient difference in signal contrast is required. Adequate signal-based object contrast is needed to detect a lesion for morphological analysis. Sensitivity of an imaging can be determined by reflection of image contrast [22].

2] Correlation: Correlation is used to measure noise from an image. The noise is usually described as the noise power spectrum (NPS). NPS is nothing but a frequency spectrum. The presence of noise at one small location in an image (like a pixel in a digital image) may influence the noise in adjacent locations and correlation signifies it. [22].

3] Energy: Spatial maps of signals represents assortment of interaction of energy with the body at a specific time to represent a biomedical image. A typical image is a graphical display of spatially and temporally dependent measurements of mass and energy. Energy is also known as uniformity [22].

4] Entropy: To characterize the texture of the input image, entropy is used. In case of tumor the edges of pathological lesions reveal much about their biological ferociousness and destructiveness. A benign lesion's margins are well defined but malignant lesion's margins are irregular. This is a biological reflection of entropy .Low entropy reflects normal, benign tissues but high entropy shows malignant tissues [22].

5] Kurtosis: Kurtosis depicts the shape of the PDF (probability density function). It is used to measure the peakness or flatness of an object/lesion [22].

6] Homogeneity: Homogeneity is used to measure the closeness of pixels/elements in the GLCM to the GLCM diagonal.

3.2 Clinical Feature Extraction

Research indicates that patients' history helps us to find out various aspects which are linked to breast cancer. We have developed a live dataset [20] which includes traditional mammogram also called as Screen Film mammograms (SFM) and digital mammograms (DM). While collecting images we gathered patients' information. We recorded patients' attributes and history information that is relevant to cancer. We have included following attributes in our data and augmented some of these clinical features during automatic classification. Several clinical surveys have underlined the significance of these features.

1] Age: Patient's age helps for prediction of breast tumor. Many researchers used age as one of the attributes for studies related to breast cancer [11, 13, 14, 19].

2] Menarcheal age: Late and early menarcheal age indicates hormonal imbalancing tendency in a patient. Many researchers have focused on this attribute and studied its correlation with breast cancer [11, 15, 16, 18].

3] Number of kids: Either having more numbers of kids or no kid at all, illustrates hormonal imbalance which may increase risk factor for breast cancer [11].

4] Number of abortions: It indicates disturbed/ irregular menstrual history of a patient. [13, 18].

5] Cancer History: This attribute helps to trace out genetic relation of a patient for chances of positive result for breast cancer detection. Researchers studied cancer history as an attribute for detection of not only breast cancer but for other types of cancers also [11, 12, 16].

6] Smoke: Study reveals that smoking habit may also lead to breast cancer [11].
7] Alcohol Consumption: Patient's habit of consuming alcohol is a significant attribute [11].

We have gathered information from patients as above. We observed that none of the patient from our dataset consumes alcohol and smoke. So we decided to exclude these attributes.

3.3 Comparison Strategy

The comparison of results of two classifiers is planned as follows. We developed two distinct classifiers namely "MammoFeature Classifier" and "MammoAndClinical feature Classifier". The first classifier works with only mammographic features whereas the second classifier operates upon clinical features augmented to mammographic features. Figure 2 and figure 3 correspond to these classifiers respectively.

MammoFeature Classifier is trained with standard data set (DDSM) mammogram images. This classifier is tested with two distinct datasets containing live mammogram images. The two data sets are 'Only_Mammo_Attrib_Dataset' and 'Mammo_Clinical_Attrib_Dataset'. The later data set contains those mammogram images for which clinical attributes are also available. The clinical attributes of patients are augmented with mammographic features. The Mammo_Clinical_Attrib_Dataset contains additional features and could not be tested on earlier classifier. Hence, "MammoAndClinical feature Classifier" is build and tested using 10 cross fold validation on Mammo_Clinical_Attrib_Dataset.

The results obtained and its comparison is tabled in the next section.

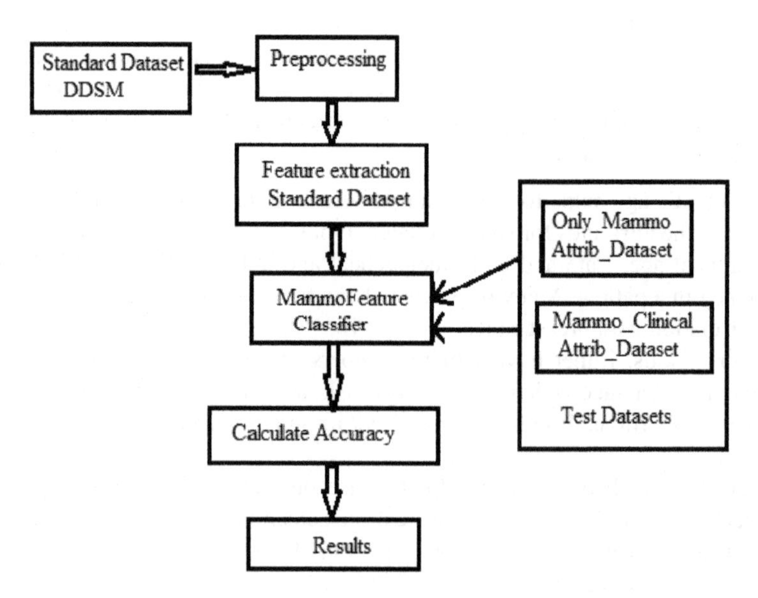

Fig. 2. Classifier that uses only mammographic features

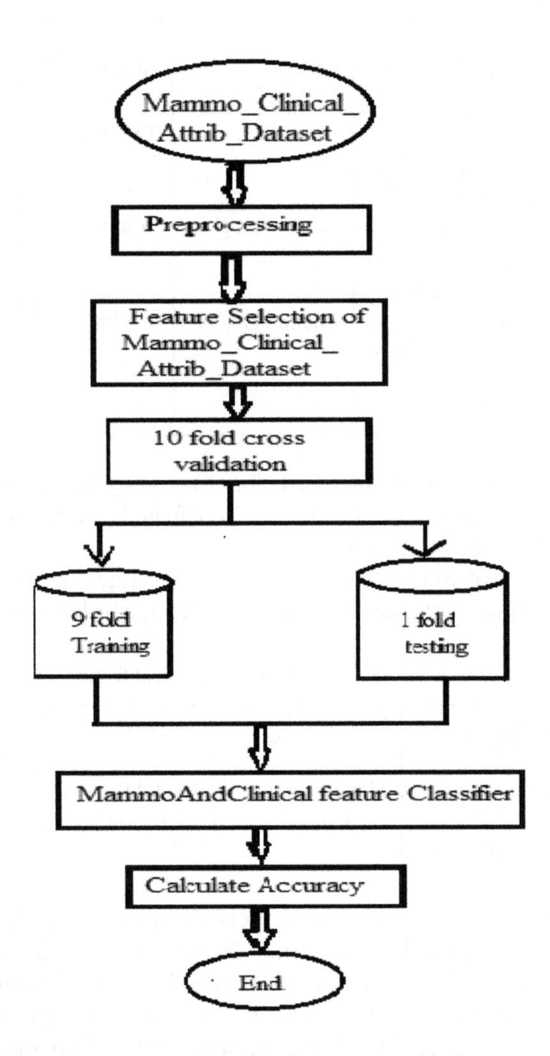

Fig. 3. Classifier that uses clinical and mammographic features

4 Results and Observations

We obtain classification accuracy of 'MammoFeature Classifier' while training and when it is tested with two datasets of live data. The results are shown in table 1.

Table 1. Results of MammoFeature Classifier:

Dataset	Features	Activity	Accuracy
Standard dataset DDSM	Mammographic	Training	79.72%
Live dataset-1 Only_Mammo_Attrib_Dataset	Mammographic	Testing	60.91%
Live dataset-2 Mammo_Clinical_Attrib_Dataset	Mammographic	Testing	52.17 %

The classification result obtained for Mammo_Clinical_Attrib_Dataset using MammoFeature Classifier is to be compared with result obtained by MammoAndClinical feature Classifier.

The MammoAndClinical feature Classifier generates result using 10 cross fold validation technique on the same 'Mammo_Clinical_Attrib_Dataset'.

Table 2. Results of MammoAndClinical feature Classifier

Dataset	Features	Activity	Accuracy
Live dataset Mammo_Clinical_Attrib_Dataset	Mammographic and Clinical	10 fold cross validation	69.56%

We can see that the classification result obtained for 'Mammo_Clinical_Attrib_Dataset' using MammoFeature Classifier is 52.17% and same dataset gives 69.56% accuracy for MammoAndClinical feature Classifier.

It shows that mammographic features if augmented with appropriate set of clinical features, can improve the overall classification result. Importance of such clinical features was studied and proved by various clinical surveys. This study authenticated the significance of clinical features with the help of automatic classification.

5 Conclusions

In this paper we discuss our experiments with standard mammogram datasets as well as with our extended, informative live data set. Using SVM we developed a model for classification. The results obtained using mere mammographic features is 52.17 % and whereas the classification accuracy of 69.56 % is obtained when clinical features are combined with mammographic features. It clearly shows that clinical attributes plays vital role in automatic classification of mammograms. We further propose to use these clinical features for sub-classification of benign breast tumors.

Acknowledgment. The authors wish to thank Dr. U. V. Takalkar [M.S. (Gen.Surg.) M.E.D.S F.U.I.C.C. (SWITZERLAND), MSSAT (USA),F.I.A.I.S.,MACG(USA), Chief surgeon and consultant : Kodlikeri Memorial Hospital, Director CIIGMA Group of Hospitals, Aurangabad] for their support and valuable guidance. Images were provided by Kodlikeri memorial Hospital, Aurangabad and Hedgewar Hospital, Aurangabad, Maharashtra, India.

References

[1] Raghavan, G.: Mammogram Analysis: Tumor Classification. EE 381K - Multidimensional Digital Signal Processing. Springer (2005)

[2] Mohan Kumar, S., et al.: Statistical Features Based Classification of Microcalcification in Digital Mammogram Using Stochastic Neighbors Embedding. International Journal of Advanced Information Science and Technology (IJAIST) 7(7) (November 2012) ISSN: 2319:2682

[3] Chang, R.F., et al.: Support Vector Machines for Diagnosis of Breast Tumors on US Images1. Academic Radiology 10(2) (February 2003)

[4] Telagarapu, P., et al.: Extraction to Classify the Benign and Malignant Tumors from Breast Ultrasound Images. International Journal of Engineering and Technology (IJET) 6(1) (February-March 2014) ISSN : 0975-4024

[5] Shi, X., Cheng, H.D., et al.: Mass Detection and Classificatio. In: Breast Ultrasound Images Using Fuzzy SVM, Utah State University 401B. Old Mail Hill, Logan UT 84322, USA

[6] Karmilasari, Widodo, S., Lussiana, E.T.P., Hermita, M., Mawadah,L.: Classification of Mammogram Images using Support Vector Machine, `http://saki.siit.tu.ac.th/acis2013/uploads_final/25.../ACIS%202013-0025.pdf`

[7] Zaiane, O.R., Antonie, M.L., Coma, A.: Mammography Classification By an Association Rule based Classifier. In: MDM/KDD 2002: International Workshop on multimedia Data Mining (with ACM SIGKDD) (2002)

[8] Hejazi, M.R., Ho, Y.-S.: Automated Detection of Tumors in Mammograms Using Two Segments for Classification. In: Ho, Y.-S., Kim, H.-J. (eds.) PCM 2005. LNCS, vol. 3767, pp. 910–921. Springer, Heidelberg (2005)

[9] Tzikopoulos, S.D., Mavroforakis, M.E., Georgiou, H.V., Dimitropoulos, N., Theodoridis, S.: A fully automated scheme for mammographic segmentation and classification based on breast density and asymmetry. Computer Methods and Programs in Biomedicine 102(1), 47–63 (2011)

[10] Xu, S., Liu, H., Song, E.: Marker-controlled watershed for lesion segmentation in mammograms. Journal of Digital Imaging 24, 754–763 (2011), `http://dx.doi.org/10.1007/s10278-011-9365-2`, doi:10.1007/s10278-011-9365-2

[11] Friedenreich, C.M., et al.: Risk factors for benign proliferative breast disease. International Journal of Epidemiology 29, 637–644 (2000)

[12] Yang, W.T., Tse, G.M.K.: Sonographic, Mammographic, and Histopathologic Correlation of Symptomatic Ductal Carcinoma In Situ. AJR 182 (January 2004)

[13] Thongkam, J., et al.: Toward breast cancer survivability prediction models through improving training space. Elsevier Ltd. (2009), doi:10.1016/j.eswa.2009.04.067

[14] Rosenberg, R.D.: Performance Benchmarks for Screening Mammography1. Radiology 241(1) (October 2006)

[15] Tintus-Ernstoff, L., et al.: Breast cancer risk factors in relation to breast density (United Sates). Springer Science + Business media B.V. (2006), doi:10.1007/s10552-006-0071-1

[16] Artmann, L.C.H., et al.: Efficacy of Bilateral Prophylactic Mastectomy in Women With A Family History of Breast Cancer. The New England Journal of Medicine 340 (January 14, 1999)

[17] Robsahm, T.E., et al.: Breast cancer incidence in food- vs. non-food-producing areas in Norway: possible beneficial effects of World War II. British Journal of Cancer 86, 362–366 (2002), doi:10.1038/sj/bjc/6600084

[18] Must, A., et al.: Recall of Early Menstrual History and Menarcheal Body Size: After 30 Years,How Well Do Women Remember? American Journal of Epidemiology 155(7)

[19] Roberts, M., et al.: Risk of breast cancer in women with history of benign disease of the breast. British Medical Journal 288 (January 28, 1984)

[20] Bhale, A., Joshi, M., et al.: Development of a Standard Multimodal Mammographic Dataset. In: Proceedings of National Conference on Advances in Computing (NCAC 2013), March 5-6, North Maharashtra University, Jalgaon (2013) ISBN: 978-81-910591-7-5

[21] Mohanty, A.K., et al.: An improved data mining technique for classification and detection of breast cancer from mammograms. Springer- Verlag London Limited (2012), doi:10.1007/s00521-012-834-4

[22] Nick Bryan, R.: Introduction to the Science of Medical Imaging. Cambridge University Press, doi:http://dx.doi.org/10.1017/CBO9780511994685 ISBN: 9780511994685

Quantifying Poka-Yoke in HQLS: A New Approach for High Quality in Large Scale Software Development

K.K. Baseer[1], A. Rama Mohan Reddy[2], and C. Shoba Bindu[3]

[1] JNIAS-JNTUA, Anantapuramu, India
[2] Dept. of CSE, SVU College of Engineering, SV University, Tirupati, India
[3] Dept. of CSE, JNTUACE, JNT University Anantapur, Anantapuramu, India
{kkbasheer.ap,ramamohansvu,shobabindhu}@gmail.com

Abstract. Improving performance of software, web sites and services is a holy grail of software industry. A new approach is the implementation of Poka-Yoke method in software performance engineering is proposed. Poka-Yoke is a mistake proofing technique used in product design. We are proposing HQLS: a new approach for high quality in the large scale software development in this paper. The effectiveness of Poka-Yoke in software development was evaluated using a case study: product redesign mini-project given to six groups of students with this both quantitative and qualitative evaluation was done. Our proposed mistake proofing technique for software development using Poka-Yoke evaluation demonstrated the usability goals. The results showed that implementing Poka-Yoke technique improves the software development process. Improved UGAM and IOI scores showed linearity and justified Poka-Yoke implementation. Our findings recommend usage of techniques for mistake proofing for overall software performance. The main aim is to reduce errors in software development process.

Keywords: Performance Issue, User Experience, UGAM Score, Mistake Proofing Solution, Poka Yoke Principles, Large Scale Software Development, Quality, Detection.

1 Introduction

Software architecture has emerged as an important discipline in the creation of complex software systems. A model of software architecture that consists of three components: {Elements, Form, and Rationale}, i.e., software architecture is a set of architectural elements that have a particular form [1], [6]. Consider some examples discussed by Neil [13] for better understand the objective of the paper. The paper mainly focuses on Quality attributes of run time-performance and user-usability categories.

Example-1: Consider any building; the usability of the building is affected by its layout. The architecture specifies structural elements that affect the reliability and strength of the building. This is also true of software. The architecture of a software

S.C. Satapathy et al. (eds.), *Emerging ICT for Bridging the Future – Volume 1*,
Advances in Intelligent Systems and Computing 337, DOI: 10.1007/978-3-319-13728-5_33

system affects the externally visible characteristics of software. **Impact:** The architecture of a system can have a significant impact on the characteristics of the system.

Example-2: An architecture designed was for: four simultaneous transactions were allowed. Further architectural decisions resulted in database queries and updates that were very expensive. **Impact:** Strong impact on performance, and through it, on usability.

Example-3: Fault-tolerant applications: an online banking system must correctly abort partially completed money the transfer if the communication link drops before the transfer is completed. **Impact:** These "user-visible ways" are often the quality attributes of the system. However, one cannot fully validate whether the characteristics of the system meet the system's quality attribute requirements until the entire system has been constructed – far too late to easily change the architecture if needed.

Quality measurements are performed primarily on the architectural description. Quality attributes can be addressed through architectural styles and patterns. Solutions for example 1, 2, and 3 discussed in [6], [3]. Different architectural styles address different sets of quality attributes and to varying degrees. The specification of quality attributes, therefore, affects the architectural style of the system. Not all quality attributes are addressed by architectural design. Different existing approaches were discussed in [3] to achieve quality attributes (nonfunctional requirements) knowledge and tools needed to analyze the behavior of a system with respect to some quality attributes. Usability and performance both have non architectural aspects. In terms of levels of systems knowledge, the identification of quality attributes is equivalent to the source level (level 0) of systems knowledge.

Mukesh Jain proposed a scheme and procedure to prevent and/or detect performance issues in time for the product engineering team to take action and fix them and prevent them from happening [2].

Software must possess the qualities like Safety, Reliability, Availability, Cost, Maintainability, Performance or Response, Time, Energy consumption [4]. Usability is important only to increase the speed and accuracy of the range of tasks carried out by a range of users of a system, but also to ensure the safety of the user. Productivity is also imperative where the software is used to control dangerous processes. Different authors defined different definitions for usability [4] and by combining all, Usability is composed of different attributes: Learnability, Efficiency, Understandability, Operability, Memorability, Effectiveness, Error avoidance, Error handling, Satisfaction, Complexity, Ease of use.

Improving performance of software, web sites and services is a holy grail of software industry. A new approach for implementation of Poka-Yoke method to achieve High Quality in Large scale (HQLS) Software Development.

1.1 Poka-Yoke

Poka-Yoke (ポカヨケ) [Poka yoke] is a Japanese term that means "mistake-proofing"[8]. A Poka-Yoke (PY) is any mechanism in a lean manufacturing process that helps an equipment operator avoid (*yokeru*) mistakes (*Poka*). Its purpose is to eliminate product defects by preventing, correcting, or drawing attention to human errors as they occur. For more information about Poka-Yoke see [5], [7], [8], [9], [10].

The field of Poka-Yoke has a large amount of literature in product design methods, software testing techniques, and management (M.Dudek Burlikowska et al, 2009; Lawrence P Chao et al, 2003; Harry Robinson, 1997, etc). These proposals are excellent demonstrations of how Poka-Yoke design methodology can result in improved user-experience design and service performance with fewer defects in their respective domains. Unfortunately, major gaps between Poka-Yoke and SE continue to exist in academics, literature, and industrial practice.

The next section gives an overview of related work / the state of the art in Poka-Yoke. Sections 3 describe the proposal for HQLS: a new approach for high quality in large scale software development. Sections 4 describe case study that evaluated the architecture for HQLS and their findings. Section 5 draws conclusions from the study.

2 State of the Art

In recent years, research on applying Poka Yoke in software has received much attention [11], [8], [9], [12] and different approaches talk in [5], [27].

3 Proposal for HQLS

We propose a new model for large scale [14], [15], [16], [17], [20] software development for Products and Services with high quality expectations. It would be based on investing up front in the Software Architecture of the system, designing with the software product monitoring and alerting logic in place, end-to-end user experience, experimentation and quality of service based on Poka-Yoke principles [5, 7]. The basic idea behind developing this new model is to have high quality software products and services that can be developed faster, cheaper and in a better way, it can scale with demand in various scenarios, can deliver an outstanding user experience and be failing safe for SDLC bottlenecks which arise in both conventional and the agile software development shown in figure 1. The proposed model has the following areas:

- Get the right Software Architecture in place
- Ensure high quality software is developed
- It is based on POKA-YOKE principles

We are going to examine each of the phases in the software development life cycle and find out opportunities for improvements. We will explore how things can be planned and designed up front to avoid discovering issues late in the cycle. One of the major challenges in the software is quality, to understand this better we will find opportunities to inject product monitoring at the right place to capture the user experience, product quality and help us in alerting at the right time. We are going to leverage the principles of Poka Yoke (primarily used in production to make the process mistake-proof, this prevents people from making mistakes and if mistakes are made, they are caught early in the cycle) [18], [19]. Provide an emphasis on investing in the Software Architecture before the design is started. While developing Software Architecture, adequate attention is to be provided with the Software Reliability, Scale, User experience and making it fail safe based on Poka Yoke Principles.

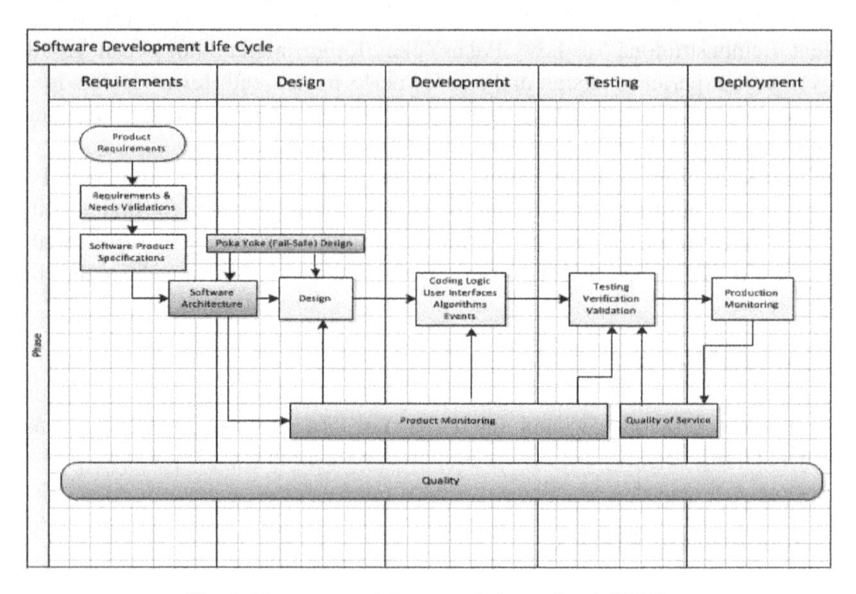

Fig. 1. The proposed framework for updated SDLC

4 Case Studies

Six groups of students were assigned the task of redesigning of PS (Personal Secretary/Persona) product using Poka-Yoke recommendations. PS is a proprietary product of Rajesh Kulkarni (RK) et. al [23], [24], [25], [26].

The reason for selecting PS was the objectives and tools used for better software performance were same. We derived our inspiration for better software from [20], [21], [22] and so we selected their works and tools and are furthering with our idea.

The following table 1 is a Personas (a fictitious representation of target user of the large scale product) P1:

Table 1. Persona of KKB

Name	Kamalapuram Khaja Baseer (KKB)
Background	He has insurance policies of LIC (5 NO), TATA AIG (2NO) and RELIANCE (1 NO). He also does online trading on a small scale, but not so frequently.
Sites-used	Google, gmail, blogger, youtube, rediffmail, rediff money, yahoomail, yahoogroups,wordpress,linkedin,facebook,irctc,makemytrip,cleartrip,yatra,slide share,way2sms,flipkart,sbicard.com,onlinesbi.com,pnb internet banking, naukri, member of many online groups.

Table 1. (*Continued*)

Email ids being used by the persona	Gmail, rediffmail, yahoomail.
Blog ids being used by the persona	blogger, wordpress.
Other site ids being used by the persona	irctc, makemytrip, facebook, linkedin, slideshare, flipcart, sciencedirect, Elsevier id for online research paper submission, sbicard.com, onlinesbicard.com, pnb, hdfc for online banking.
ATM/DEBIT/CREDIT cards being used	SBI (2 no), SBH, PNB, VISA, HDFC.
Awareness	He is aware of SDLC, agile, user centered design life cycle but not practically implemented. Lacks industry exposure.
Goals: End Goal:	i. Hassle free storage of all ids and password. He need not remember any id or password once created. He need not store passwords and IDs in mobile, mail, and in diaries or write on a paper and keep that paper in the wallet. He should be able to extract ids and passwords at will at any time and with just a click. The stored location/application again should not have the feature which makes him remember another password and id. ii. With just a click all his mark memos, certificates, experience certificates, appointment orders should be printed and updated resume and that too in order. iii. With just a click all the documents related to applying for a loan should be printed and that too in order: form 16s, IT returns, ID proof etc. iv. Alerts of various payments on mobile: loan EMIs, credit cards and insurance premiums etc.
Experience Goal	Secure and hassle free online transactions, removing the overhead of carrying all the ids, passwords, deadlines all the time. Just in click retrieval of documents, deadlines and alerts.
Life Goal	He wants to patent the above app as he has a profound belief in the utility of the above product and as well this will accelerate his Ph.D thesis acceptance.
Technology Expertise	Competent Performer.
Ideation of the Product	i. The software can be a web app or a mobile app. ii. The point of entry or access to the software should not be through user id and password. iii. The point of entry or access to the software can be through iris or thumb or voice. iv. The main objective of the software should be category wise storage of all ids and passwords of the persona created above. v. The secondary objective is category wise storage of all documents. vi. Retrieval of ids and passwords with print and mail option. vii. Retrieval of documents with prints and mail option. viii. Feature of adding and updating of IDs and passwords. ix. Feature of adding and updating documents.
Expected deliverables	i. Wireframes and mock-ups (optional, not mandatory). ii. Screens unit tested with sample data. iii. Not required: database, server etc. iv. Coding and documentation. v. Satisfactory, Learnability, usability factors and user experience.

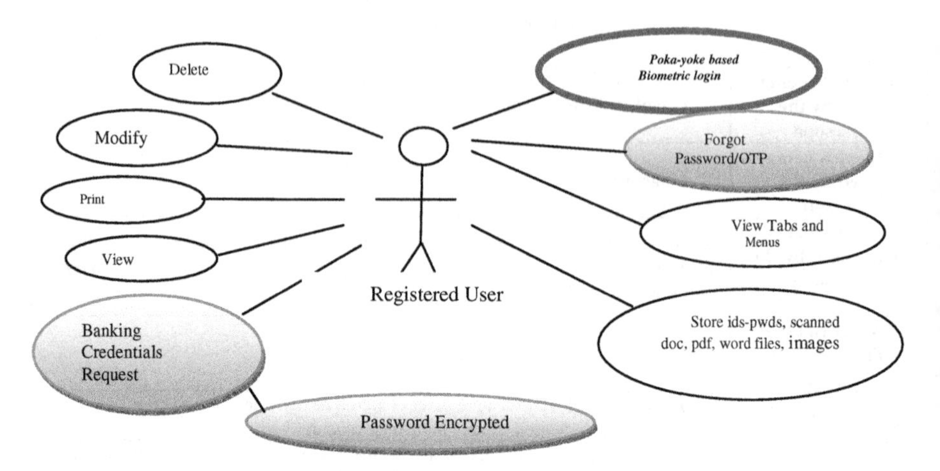

Fig. 2. Use case for Redesign of PS with poka-yoke

4.1 PS Implementation with PY

Figure 2 shows the use case for redesign of PS with PY. System Architecture of redesign PS with PY shown in figure 3, PS is aimed towards the online user who wants: Secure and hassle free online transactions, removing the overhead of carrying all the ids, passwords, deadlines all the time, just in click retrieval of documents, and alerts. PS should be user-friendly, 'quick to learn' and reliable software for the above purpose. PS is intended to be a stand-alone product and should not depend on the availability of other software. It should run on both UNIX and Windows based platform.

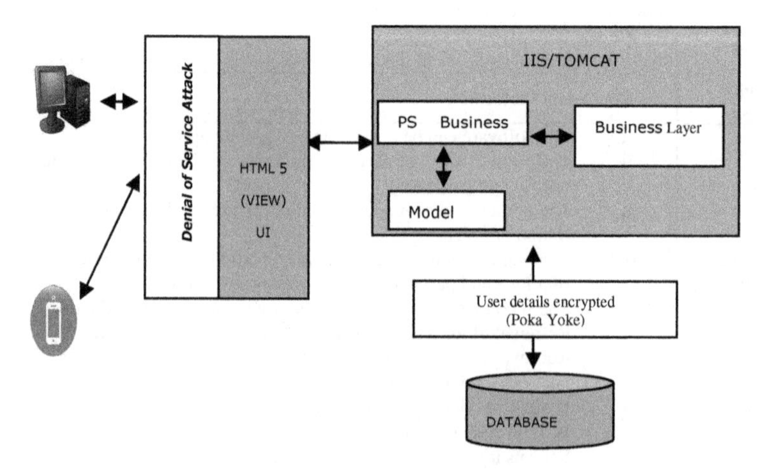

Fig. 3. System Architecture of Redesign PS with PY

4.2 Quantifying Poka-Yoke Approach in PS

Usability Goals Achievement Metric (UGAM) proposed by [22] is a product metric that measures the quality of user experience. UGAM scores of the initial PS product were taken as the base. UGAM calculations were recalculated for redesigning PS product. The improved results validated our Poka-Yoke hypothesis. UGAM scores of Rajesh Kulkarni et al derived from [24] and our recalculated UGAM scores are depicted in figure 4. It shows the detailed UGAM calculation. Now the average weight assigned is 2.8 which are in the range between 2.4 to 3.4. As per UGT rule, the goal setting has been balanced. The respective graph is shown in figure 4. IOI score indicates the Index of Integration [22] as shown in Table 2, which derives IOI calculations. Figure 5 indicates the correlation between UGAM and IOI scores.

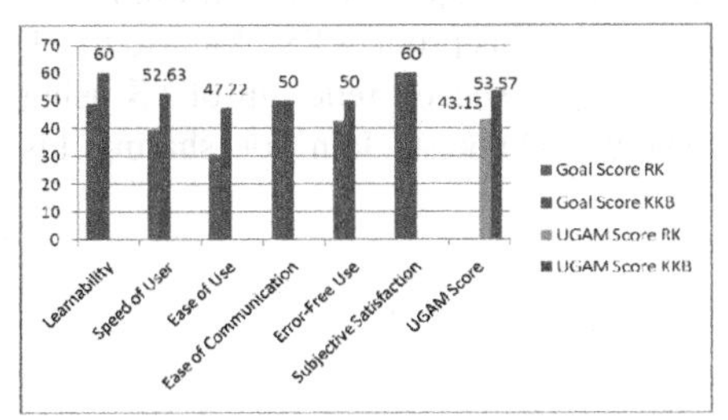

Fig. 4. Comparison of KKB UGAM score with RK UGAM score

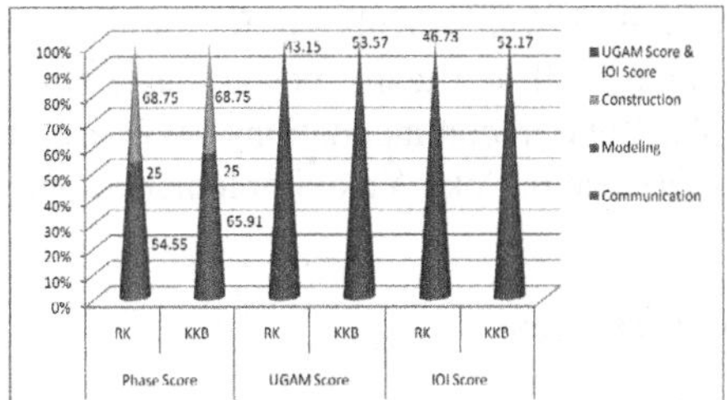

Fig. 5. Comparisons of RK and KKB UGAM & IOI Score

Table 2. KKB IOI Calculation

Phases	HCI activities	Recommended weights	weights	Activity Score	Phase Score	IOI Score
Communication	Contextual User Studies/modeling	3 to 4	3	75	65.91	52.17391
	Ideation with the multidisciplinary team	2	4	75		
	Product defn/IA/Wireframes	1 to 3	3	50		
	Usability evaluation, refinement	1 to 3	1	50		
Modeling	Detailed UI prototyping	4 to 5	4	50	25	
	Usability Evaluation, refinement	4 to 5	4	25		
Construction	Dev Support reviews by Usabilty team	3	3	75	68.75	
	Usability Evaluation(Summative)	1 to 3	1	50		

5 Conclusion

In this paper we proposed a new model for high quality in the large scale software development and focused on qualitative evaluation to achieve usability goals. UGAM and IOI scores showed improvement validating our hypothesis of Poka-Yoke implementations. Our findings recommend usage of techniques for mistake proofing for overall software performance. Quantitative validation of Poka-Yoke approach using Regression Coefficients and Analysis of Variance was discussed in previous work. We are planning come out with a poka-yoke index tool on the lines of UGAM/IOI tools which are indicators and verify how far a software development process has adopted our proposed model for mistake proofing.

Acknowledgement. We thank Dr. Anirudha Joshi who designed the UGAM tool. We thank Prof. Rajesh Kulkarni for allowing us to evaluate his product PS. We also thank Asmita Patil and team; postgraduate students who worked on redesign of PS using Poka-Yoke recommendations. We are indebted to Mukesh Jain for sharing his experience in Poka-Yoke implementation.

References

1. Perry, D.E., Wolf, A.L.: Foundations for the Study of Software Architecture. ACM SIGSOFT, Software Engineering Notes 17(4), 40–52 (1992)
2. Jain, M.: Performance Testing and Improvements using Six Sigma– Five steps for Faster Pages on Web and Mobile. In: Paper Excerpt from the Conference Proceedings of 28th Annual Software Quality, October 18-19, pp. 1–13 (2010)
3. Baseer, K.K., Reddy, A.R.M., Shoba Bindu, C.: A Survey of Synergistic Relationship: Scenarios, Quality Attributes, Patterns, Decisions, Reasoning Framework. International Journal of Web Technology 01(02), 74–88 (2012) ISSN: 2278-2389
4. Sai Aparna, S., Baseer, K.K.: A Systematic Review on Measuring and Evaluating Web Usability in Model Driven Web Development. International Journal of Engineering Development and Research (IJEDR) Conference Proceeding (NCETSE-2014), 171–180 (2014) ISSN: 2321-9939
5. Baseer, K.K., Reddy, A.R.M., Shoba Bindu, C.: HQLS-PY: A new Framework to achieve High Quality in Large Scale Software Product Development using Poka-Yoke Principles. International Journal of Engineering Development and Research (IJEDR) Conference Proceeding (NCETSE 2014), 164–170 (2014) ISSN: 2321-9939
6. Reddy, A.R.M.: Programming Methodologies and Software Architectures: Programming Methods, Process Model for Software Architecture, Evaluation of Software Architectures using integrated AHP-GP Model. Thesis, Sri Venkateswara University, Tirupati, A.P. (July 2007)
7. Poka-Yoke Guidelines, `http://pokayokeguide.com/`
8. Robinson, H.: Using Poka-Yoke Techniques for Early Defect Detection. In: Sixth International Conference on Software Testing Analysis and Review (STAR 1997), pp. 1–12 (1997)

9. Gordon Schulmeyer, G.: The Development of Zero Defect Computer Software. Submitted for the Shingo Prize for Excellence in Manufacturing Research Prepared for PYXIS Systems International, Incorporated, pp. 1–8 (December 1, 1991)
10. Chao, L.P., Ishii, K.: Design Process Error Proofing. Failure Modes and Effects Analysis of the Design Process: Draft of a Technical Paper Prepared for Stanford ME317 dfM Course, pp. 1–10 (March 2003)
11. Pan, W.: Applying Complex Network Theory to Software Structure Analysis. World Academy of Science, Engineering and Technology 60, 1636–1642 (2011)
12. Beizer, B.: Software Testing Techniques, 2nd edn., p. 3. Van Nostrand Reinhold (1990)
13. Harrison, N.: Improving Quality Attributes of Software Systems Through Software Architecture Patterns. Thesis: University of Groningen, the Netherlands (2011)
14. Gumuskaya, H.: Core Issues Affecting Software Architecture in Enterprise Projects. Proceedings of World Academy of Science, Engineering and Technology 9, 32–37 (2005)
15. Abdelmoez, W.M., Jalali, A.H., Shaik, K., Menzies, T., Ammar, H.H.: Using Software Architecture Risk Assessment for Product Line Architectures. In: Proceedings of International Conference on Communication, Computer and Power (ICCCP 2009), Muscat, February 15-18, pp. 1–8 (2009)
16. Dey, P.P.: Strongly Adequate Software Architecture. World Academy of Science, Engineering and Technology 60, 366–369 (2011)
17. Meedeniya, I.: Robust Optimization of Automotive Software Architecture. In: Proceedings of Auto CRC Technical Conference (2011)
18. Dudek-Burlikowska, M., Szewieczek, D.: The Poka-Yoke Method as an Improving Quality Tool of Operations in the Process. Journal of Achievements in Material and Manufacturing 36(1), 95–102 (2009)
19. Shingo, S.: Zero Quality Control: Source Inspection and the Poka-yoke System, p. 45. Productivity Press
20. Jain, M.: Delivering Successful Projects with TSP(SM) and Six Sigma: A Practical Guide to Implementing Team Software Process (SM), Kindle Edition. CRC Press (2006) ISBN 1420061437
21. Seffah, A., Donyaee, M., Kline, R.B., Padda, H.K.: Usability measurement and metrics: A consolidated model. Software Quality Journal 14, 159–178 (2006)
22. Joshi, A., et al.: Measuring Effectiveness of HCI Integration in Software development processes. The Journal of Systems and Software JSS-8496, 1–14 (2010)
23. Kulkarni, R., Padmanabham, P.: TEIM-The Evolved Integrated Model of SE and HCI. UNIASCIT 2(3), 301–304 (2012) ISSN 2250-0987, Impact Factor: 1
24. Kulkarni, R., Padmanabham, P.: Validating Utility Of TEIM: A Comparative Analysis. IJACSA 4(1) (January 2013) U.S, ISSN : 2158-107X(Print), Impact Factor: 1.187
25. Kulkarni, R., Padmanabham, P.: Varsha Sagare: Usability Evaluation of PS Using SUMI (Software Usability Measurement Inventory). In: IEEE International Symposium on Knowledge-Intensive Software Engineering (KISE 2013), Co-located with ICACCI 2013, pp. 22–25 (August 2013)
26. Kulkarni, R., Padmanabham, P., Namose, M.S.: Improving Software Quality Attributes of PS Using Stylecop. Global Journal of Computer Science and Technology (GJCST) 13(8), 20–26 (2013)
27. Baseer, K.K., Reddy, A.R.M., Shoba Bindu, C.: Quantitative Validation of Poka-Yoke approach using Regression Coefficients and Analysis of Variance. In: 3rd International Conference on Eco-friendly Computing and Communication Systems (ICECCS 2014), December 18-21, NITK Surathkal, Proceedings in IEEE Xplore Digital Library (accepted paper, 2014)

SIRIUS-WUEP: A Heuristic-Based Framework for Measuring and Evaluating Web Usability in Model-Driven Web Development

S. Sai Aparna and K.K. Baseer

SreeVidyanikethan Engineering College, Department of IT, Tirupati, India
{aparnasamudrala15,kkbasheer.ap}@gmail.com

Abstract. Now-a-days websites providing all kinds of services to the users, this role of importance of the web in our society has led to a tremendous growth of websites. Websites are now generally considered the most effective and efficient marketing channel. Usability plays an important role in the development of successful websites, in order to make usable designs expert recommendations are considered as guidelines. There is a lack of empirically validated usability evaluation methods that be applied to models in model-driven web development. To evaluate these models WUEP (Web Usability Evaluation Process) method is proposed. And also presents operationalization and empirical validation of WUEP into another method: WebML. The evaluation methods WUEP and HE (Heuristic Evaluation) were evaluated from the viewpoint of novice inspectors to know the effectiveness, efficiency, perceived ease of use and satisfaction of evaluation methods. Inspectors were satisfied when applying WUEP; it is easier to use than HE.Usability measurement is considered as part of the development process stands out amongthe expert's recommendations. Here we are defining Sirius, to perform expert evaluations an evaluation framework is proposed.

Keywords: Model driven web development, usability inspection, WebML, Heuristic Evaluation, usability metric.

1 Introduction

One of the most important quality factors for web applications is Usability. Usability is defined by ISO as "the capability of software product to be understood, learned, applied and attractive to the end user, when used under particular conditions". Characteristics of usability are Learnability, Understandability and Operability[2] [16]. To integrate usability into conventional software engineering practice and to operationalize usability is difficult. Usability can be distinguished into 2 approaches like bottom up defines ease of use and top down defines ability to use a product for its intended purpose.

Web applications success or failure determined with the ease or difficulty experienced by users [5] [14]. Development of more usable web applications needs to develop more number of usability evaluation methods. To detect and reduce the

S.C. Satapathy et al. (eds.), *Emerging ICT for Bridging the Future – Volume 1*,
Advances in Intelligent Systems and Computing 337, DOI: 10.1007/978-3-319-13728-5_34

usability problems, theevaluation methods should be incorporated in the early stages of the software development life cycle, by this we can improve the user experience and decrease maintenance costs. Here we are presenting the validation of evaluation methods, to evaluate analysis and design models of a web application under development.

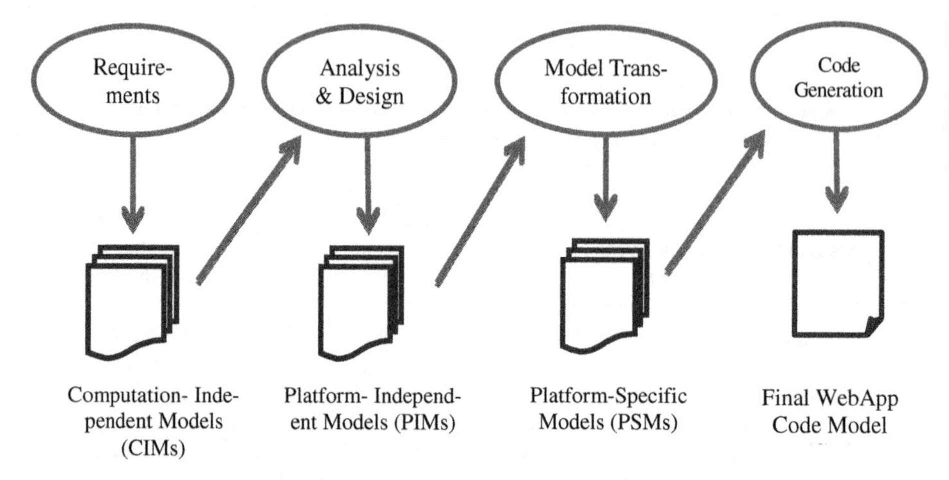

Fig. 1. Overview of a generic Model-Driven Web Development process (courtesy: [1])

To solve usability issues, a usability inspection method (WUEP) is proposed, which can be integrated into different (MDWD) Model-Driven Web Development processes. The peculiarity of these development processes is that Platform-Independent Models (PIMs) and Platform-Specific Models (PSMs) are built to show the different views of a web application (navigation, presentation and content) finally, the source code (Code Model - CM) is obtained from these models using model-to-text automatic transformations.

2 Literature Survey

The literature survey was done to know the importance of usability evaluation in web application development.To reduce the quality problems evaluation of usability attributes proposed before going to the deployment stage. After identifying the requirements, it focuses on the inspection of quality measurements during the system design. In a MDWD process, where abstract models are essential things for a system development, in this usability can be evaluated during the design stage of implementation .To evaluate the proposal this work introduces an empirical study. The main goal is to evaluate whether the evaluated value is concerned with the understandability value.

Much of the web development process has not originated the advantage of evaluation methods at design stage. Usability evaluation, execution of these design models is very difficult, but we can perform these evaluation methods in the "MDWD". It provides guidelines to solve usability problems prior to the source code generation.

3 WUEP(Web Usability Evaluation Process) Method

WUEP, a usability evaluation method that offers a generic process for evaluating the web application usability, which is developed in a model-driven web development environment. WUEP employs a web usability model as its principal input artifact which breaks down usability into sub-characteristics, attributes and measures.

3.1 Instantiation of WUEP in the WebML method

WUEP can be instantiated in the Web Modeling Language method (WebML) to evaluate the web applications. This method is supported by the Web Ratio tool, which offers the edition and compilation of the models proposed by the method [7] [11]. WebML is a domain-specific language; it consists of Data Model and Hypertext Models for specifying the structure of Web applications and the overall presentation of their contents in one or more hypertexts [9]. Hypertext Model consists of containers, session units, data units, service units and links, and it shows how the data is assembled, interconnected and presented web application. Synchronization should be done between database and models to export and import the data between them.

1. **Establishment of the evaluation requirements stage**: The evaluation designer defines the scope of the evaluation by (a) The purpose of the evaluation (b) Specifying the evaluation profiles. (c) Selecting the web artifacts (models) to be evaluated (d) selecting usability attributes. Domain model and Hypertext models are designed in order to develop Book Store web application usingaWUEP method in Model Driven Web Development Environment.
2. **Specification of the evaluation stage**: The evaluation designer operationalizes the measures associated with the selected attributes of Book Store web application in order to be evaluated. In addition, for each measure thresholds are establishedbased on their scale type and guidelines related to each measure.Based on these thresholds severity of the usability problem is classified into 3 types are low, medium, and critical.
3. **Design of the evaluation stage**: Evaluation plan is elaborated and template for the usability report is defined.
4. **Execution of the evaluation stage**: Measures are applied to the models or web artifacts and detect usability problems prior to the generation of source code based on the thresholds established for each measure.
5. **Analysis of changes stage:** To correct the affected artifacts from a specific stage of the Web development process, the Web developer analyzes all the usability problems in order to propose changes.

4 Empirical Validation

A Web Store for book e-commerce, it was developed through the use of the WebML (Ceri et al. 2000) by the Web Ratio Company located in Milano (Italy). This method is full supported by the Web Ratio Tool Personal Edition. Two different functional Web Store application (Book search and Book shopping) as object O1, two artifacts: a Hypertext model (HM) and a Final User Interface (FUI) [13]. Perceived ease of use and perceived satisfaction of use were calculated from a five-point Likert-scale questionnaire [8].

Table 1. Final results of Evaluation methods validation

Characteristics	(WUEP)	(HE)
Perceived Ease of Use (PEU)	4.2	3.38
Perceived Satisfaction of Use(PSU)	4.3	3.63
Effectiveness (%)	50	33.04
Efficiency (probs/min)	0.65	0.05

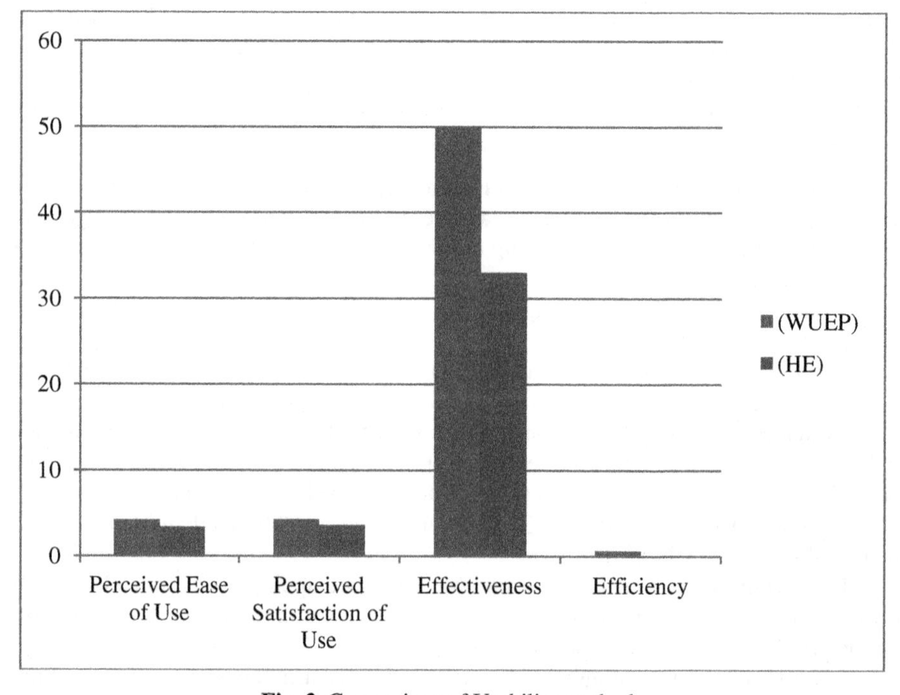

Fig. 2. Comparison of Usability methods

Table 1 summarizes the values obtained by conducting the experiment. The results shown by graph, WUEP is better in all the characteristics when compared to HE. The WUEP evaluation method is validated and satisfied the conditions effectiveness, efficiency, perceived ease of use and perceived satisfaction of use.The results are graphically represented in the figure 2.

5 Sirius Evaluation

Sirius framework an expert's evaluation phase, by considering only the aspects and criteria. Usability level of website is influenced by many factors such as the context of use, equipment of target groups of users, different services and tasks etc. Figure 3 shows the pillars of evaluation [12].

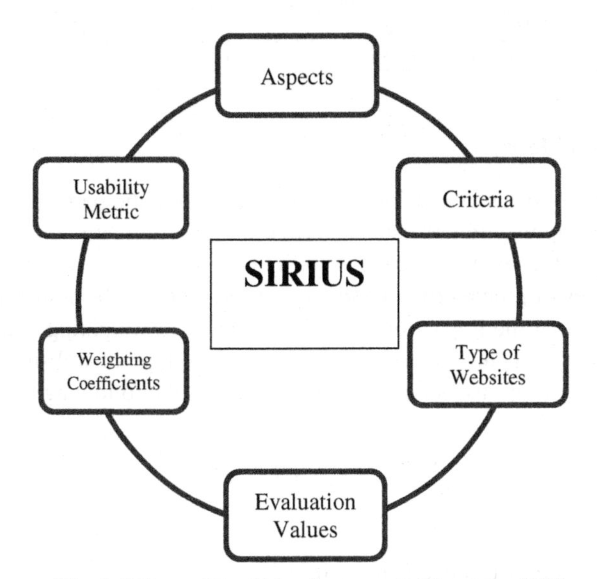

Fig. 3. Pillars of the Sirius framework (Courtesy: [12])

Sirius Aspects: The aspects which are defined in a Sirius evaluation framework are [12]: General Aspects (GA), Identity and Information (II), Structure and Navigation (SN), Labeling (LB), Layout of the page (LY), Comprehensibility and ease of Interaction (CI), Control and Feedback (CF), Multimedia Elements (ME), Search (SE), Help (HE).

5.1 Evaluation Values of the Criteria

Each aspect is divided into set of criteria, the scope of criteria compliance can be one of the following: Table 2 shows the values used for compliance/non-compliance of a page criterion.

1. Global: Must be globally compliant through the whole site.
2. Page: Must be compliant in each page of the site.

Table 2. Sirius evaluation values for page criteria

Evaluation Value	Definition	Numerical Value
0…10	0:Not compliant at all, 10: Fully compliant	0,1,2,…10
NWS	Not compliant in the whole website.	0
NML	Not compliant in the main links.	2.5
NHP	Not upheld in the home page.	5
NSP	Not compliant in one or more subpages.	7.5
YES	Fully compliant	10
NA	Criterion not applicable in the site.	-

5.2 Evaluation Metric

The final step in the Sirius evaluation framework is to find the percentage of usability for particular web application or website. After all the criteriaare assessed by experts defined to evaluate the site the level of usability is computed. In order to measure the usability the values of global criteria and page criteria have to be assigned by experts. In order to reflect that the impact of criteria on usability a weighting is applied to the value of each criterion is not the same regardless of the type of website. The formula used to compute the usability percentage of a website is the following:

$$PU=\frac{\sum_{1}^{nec}(\text{wci}*svi)}{\sum_{1}^{nec}(wci*10)}*100 \quad (1)$$

Where:

- nec: The number of evaluated criteria.
- sv: Sirius value.
- wc: Weighting coefficient.

This is computed as follows:

$$\text{wc}_{\text{i}}=\frac{rvi}{\sum_{j=1}^{nec}rvj} \quad (2)$$

Where rv: Relevance value for a given criterion

Sirius evaluation framework is applied to the Book Store web application to measure the percentage of usability. Evaluator examines and assigns values to each of the Sirius criteria. Based on the metric the usability level of the site is finally computed by aggregating aspect values.The final usability percentage of Book Store web application is obtained 90.35% as shown in Table 3.

Table 3. Final results of the Sirius Evaluation for Book Store Web Application

Aspects & Criteria	Sirius Value (SV)	Relevance Value (RV)	Weighing Coefficient(WC)	WC*10	SV*WC	Usability (%)
General Aspects (Criteria:10)	80	40	0.13	1.3	1.2	
Identity & Information (Criteria:7)	55	19	0.084	0.84	0.64	
Structure & Navigation(Criteria:14)	117	41	0.182	1.82	1.60	
Labeling (Criteria:6)	60	18	0.08	0.8	0.8	
Layout of the page (Criteria:10)	87	28	0.124	1.24	1.09	**90.356%**
Comprehensibility & ease of Interaction (Criteria:7)	55	20	0.08	0.8	0.82	
Control & Feedback (Criteria:10)	80	25	0.1	1.1	1.1	
Multimedia Elements (Criteria:6)	40	11	0.04	0.48	0.48	
Search(Criteria:8)	70	20	0.08	0.8	0.8	
Help(Criteria:5)	25	13	0.057	0.57	0.37	
Sum		225	1	10	9.035	

6 Conclusion

Usability evaluation methods are empirically validated for better results, thus validation is done by comparing the two evaluation methods used to develop web applications in model driven web development. The results of empirical validation shows that the WUEP evaluation method is better than HE with respect to effectiveness, efficiency, perceived ease of use and perceived satisfaction of use of evaluation methods. Finally, to show the Percentage of usability, Sirius framework is used. Based on the aspects and criteria of the web application percentage of usability is calculated.

References

1. Fernandez, A., Abrahão, S., Insfran, E., Matera, M.: Usability Inspection in Model-driven Web development: Empirical Validation in WebML. In: Moreira, A., Schätz, B., Gray, J., Vallecillo, A., Clarke, P. (eds.) MODELS 2013. LNCS, vol. 8107, pp. 740–756. Springer, Heidelberg (2013)
2. Carvajal, L., et al.: Usability through Software Design. IEEE Transactions on Software Engineering 39(11), 1582–1596 (2013)

3. Bevan, N.: Measuring usability as quality of use. Software Quality Journal 4(2), 115–150 (1995)
4. Bertoa, M.F., Vallecillo, A.: Quality Attributes for COTS Components. Journal of System and Software 1(2), 55–65 (2002)
5. Offutt, J.: Quality Attributes of Web Software Applications. IEEE Software Special Issue on Software Engineering for Internet Software 19(2), 25–32 (2002)
6. Matera, M.: Web Usability: Principles and Evaluation Methods. In: Web Engineering, pp. 143–180. Springer (2006)
7. Comai, S., Matera, M., Maurino, A.: A Model and an XSL Framework for Analyzing the Quality of WebML Conceptual Schemas. In: Olivé, À., Yoshikawa, M., Yu, E.S.K. (eds.) ER 2003. LNCS, vol. 2784, pp. 339–350. Springer, Heidelberg (2003)
8. Fernandez, A., Abrahãoa, S., Insfrana, E.: Empirical Validation of a Usability Inspection Method for Model-Driven Web Development. Journal of Systems and Software 86(1), 161–186 (2013)
9. Fraternali, P., Matera, M., Maurino, A.: WQA: an XSL Framework for Analyzing the Quality of Web Applications. In: IWWOST 2002, Malaga, Spain (June 2002)
10. Molina, F., Toval, A.: Integrating usability requirements that can be evaluated in design time into Model Driven Engineering of Web Information Systems. Advances in Engineering Software 40, 1306–1317 (2009)
11. Ceri, S., Fraternali, P., Bongio, A.: Web Modeling Language (WebML): a modeling language for designing Web sites. Computer Networks 33(1-6), 137–157 (2000)
12. Torrente, M.C.S., Prieto, A.B.M., Gutiérrez, D.A., de Sagastegui, M.E.A.: Sirius: A heuristic-based framework for measuring web usability adapted to the type of website. The Journal of Systems and Software 86(3), 649–663 (2013)
13. Hartson, H.R., Andre, T.S., Williges, R.C.: Criteria for Evaluating Usability Evaluation Methods. International Journal of Human–Computer Interaction 15(1), 145–181 (2000)
14. Tian, J.: Quality-Evaluation Models and Measurements. IEEE Computer Society 21(3), 740–7459 (2004)
15. SaiAparna, S., Baser, K.K.: A Systematic Review on Measuring and Evaluating Web Usability in Model Driven Web Development. International Journal of Engineering Development and Research-Conference Proceeding of NCETSE 2014 2(4), 164–170 (2014)
16. Baseer, K.K., Reddy, A.R.M., Shoba Bindu, C.: HQLS:PY: A New Framework to Achieve High Quality in Large Scale Software Product Development using POKA-YOKE Principles. International Journal of Engineering Development and Research-Conference Proceeding of NCETSE 2014 2(4), 164–170 (2014)

Implementation of Secure Biometric Fuzzy Vault Using Personal Image Identification

Sarika Khandelwal[1] and P.C. Gupta[2]

[1] Mewar university, Gangrar, Chittorgarh,Rajasthan, India
sarikakhandelwal@gmail.com
[2] Kota University, Kota, Rajasthan, India
pc.gupta26@gmail.com

Abstract. Biometric is proved to be an exceptional tool for identifying an individual. Security of biometric template is the most challenging aspect of biometric identification system. Storing the biometric template in the database increases the chance of compromising it which may lead to misuse of the individual identity. This paper proposes a novel and computationally simpler approach to store a biometric sample in the form of template by using cryptographic salts. Use of Personal Image Identification (PII) makes the proposed algorithm more robust and adds another level of security. The saltcrypted templates are created and stored instead of storing the actual sample behaving as a fuzzy vault. The algorithm has been analytically proved computationally simple compared to the existing template security mechanisms. The fuzzy structure of saltcrypted template is entirely dependent on user interaction through PII. Actual template is not stored at any point of time which adds new dimension to the security and hence to individual identity.

1 Introduction

The major problem with password based verification system is that, passwords or PINS can be stolen or lost[1].The suggested way for individual identity verification which seems to be robust and always available is use of biometric traits such as fingerprint, iris etc. Since the biometric templates are stored in the database, security of biometric template is major area of concern. Biometric template stolen once simply means that an individual's identity is stolen, as you cannot change this identity like passwords. This paper is an attempt to store a biometric in the form of template which is more secure and cannot be broken easily even though one gets an access to the template. A novel and relatively less complex approach to secure a biometric template without storing them in a database are proposed.

Fuzzy fingerprint vault is a secured construct used to store a critical data like secure key with fingerprint data. The secure template that is generated from a biometric sample is dependent on the attributes selected by the user. The two user selected attribute which are used in this paper are secret key and personal image. Along with that unique user id is also provided to the user which may be public.

S.C. Satapathy et al. (eds.), *Emerging ICT for Bridging the Future – Volume 1*,
Advances in Intelligent Systems and Computing 337, DOI: 10.1007/978-3-319-13728-5_35

Storing a biometric sample in the secured template form will avoid loss of privacy which could be there if the samples are stored in the database. Use of personal image identification along with biometric sample and secret key has made a system more robust in terms of security of a template.

Fuzzy vault binds biometric features and a secret key together without storing any of them. Thus it adds extra noise in the key as well as biometric sample and creates a fuzzy template for storage. At the time of verification, if both the saltcrypted stored template and query template are matched, then only the key can be released for further authentication. This work presents a novel approach to secure a biometric template using cryptographically salted fuzzy vault. The secured template that is generated using this cryptographically salted fuzzy system is dependent on the user defined personal images as well the secret key provided by the user.

2 Approaches to Secure Biometric Templates

Different strategies that are available to secure biometric template are generally based on cryptographic key binding/key generation mode. It includes transformations like salting or bio-hashing, cryptographic framework like Fuzzy vault, fuzzy commitment, secure sketches, fuzzy extractor etc.

A. Transformation: To secure a template it can be transformed into another form using either invertible or non-invertible transformations. Some of such transformations are salting or bio-hashing.

Salting: It is a template protection scheme in which template is converted or transformed into a different form using user specific key[2]. The random multi-space quantization technique proposed by Teoh et al. [3] is good example of salting. Salting can be done by extracting most distinguishing features of a biometric template and then obtained vectors can be projected in randomly selected orthogonal direction. This random projection vectors serves the basis of salting [4]. Another approach is noninvertible transform in which, the template is transformed into some other form using a key. Ratha et al [5] have proposed a method for noninvertible transformation of fingerprints.

B. Fuzzy vault: Fuzzy vault is biometric construct used to bind key as well as template together in a single framework. In order to secure a template using fuzzy vault, a polynomial is evaluated using secret key and some identifying points say minutia points in fingerprint templates are added to it to form a fuzzy vault. Some chaff points are also added to enhance the security. The security of fuzzy vault is based on infeasibility of polynomial reconstruction problem [6]. V.Evelyn Brindha[7] has proposed a robust fuzzy vault scheme in which fingerprints and palm prints are combined together to enhance the security of the template.Some results using fuzzy fingerprint vault have been reported [8-13].However, the major problems with all these approaches are that these do not consider all possible issues of fingerprint alignment, verification accuracy etc. Some of the difficulty and importance of alignment problem related to rotation in fuzzy fingerprint vault is explained by P. Zhang[14]. Chung and Moon [10-12] proposed the approach to solve the auto-alignment problem in the fuzzy fingerprint vault using the idea of the geometric hashing [15].Yang and Verbauwhede [16] has used the concept of automatic alignment of two fingerprints of fuzzy vault using the idea of reference minutia. Jin Zhe[17] has proposed protected template scheme which is alignment free. In his work, each minutia is decomposed into four minutiae triplets. From these triplets a geometric feature is extracted to construct a fingerprint template.The experimental result shows that it is computationally hard to retrieve minutia information even when both protected template and random matrix are known. Besides that, the scheme is free from alignment and light in complexity.

Another problem that is reported in literature with fuzzy vault is that, Fuzzy vault is susceptible to correlation attack. That is two fuzzy vault created using same fingerprints can be correlated to reveal fingerprint minutiae hidden in the vault.

3 Salt Cryptography

The purpose of salt is to produce a large set of keys corresponding to a given password among which one is selected as a random. Salt need not to keep secret, it should only be random. Its only purpose is to inflate the potential number of combinations for each individual password in order to exponentially increase the effort required to crack it. However, a salt has only little impact when an individual password is attacked with brute force.

Salt can also be added to make it more difficult for an attacker to break into a system if an attacker does not know the password and trying to guess it with a brute force attack. Then every password he tries has to be tried with each salt value. If the salt has one bit this makes the encryption twice as hard to break in this way, and if the salt has two bit this makes it four times harder to break. If the salt is 32 bits long for instance there will be as many as 2 ^ 32 keys for each password from which we can imagine how difficult to crack passwords with encryption that uses a 32 bit salt.[18]

4 Personal Image Identification (PII)

PII is commonly used identification mechanism based on the image selected by the user at the time of enrolment. If the same image is selected by the user at the time of verification then it proves that the user is genuine [19].This PII can be used to create transaction specific password or it can be combined with other biometric verification system to make it more secure. The basic application of Personal image identification (PII) is to provide enriched security to the ATM system. To use PII in authentication, user has to select the personal image out of the given N number of images. At the time of verification the same has to be selected by the user.PII adds second level of security to the existing identification system.

In this paper PII is used to generate random salt that has to be added to the biometric sample. Here user is asked to register three personal images out of 16 available images at the time of enrolment. Based on these images and the password provided by the user, salt is generated which can be used for further processing to create the template.

Figure 1 gives the details of the process used for registration of PII and its subsequent processing.

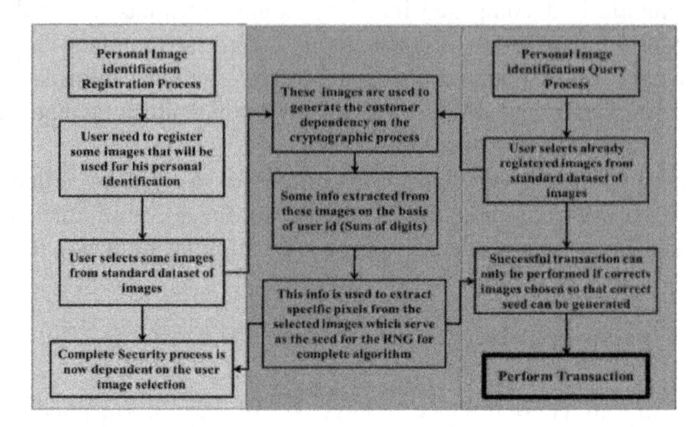

Fig. 1. Block schema of the PII registration and query process

5 Proposed Algorithm

This paper proposes a method to create a vault using salted cryptography where biometric traits i.e. fingerprint is stored in the form of transformed template. At the time of verification if the stored template and input template are matched then only key is released. The main idea behind the proposed method is not to store the actual template in the system. The algorithm for proposed verification system has two phases. Registration phase and matching phase. Figure 2 shows the phases of proposed algorithm.

5.1 Steps for Phase I (Registration/Enrollment Phase)

1. Input: Two fingerprint samples are merged together say f1 and f2 ,User ID and corresponding secret key for one time registration.
2. Personal image registration: User is asked to select the personal image from available images. Here three images are selected by the user from available set.
3. User ID of the customer is used to locate and extract the PII seed of the PN sequence generators from the pixels of PII images. Here we have used sum of digits of user ID to extract PII seed from the registered personal images.
4. Salt is generated using this seed for PN sequence generation after resetting the generator state.
5. Individual templates are combined and mixed with salt. The resultant saltcrypted template is stored. So no original templates are being stored.
6. Key taken while registration is to be embedded to the salted templates such that this embedding is guided by a pixel moving salt generated taking PII seed.

7. Now f3=f1 XOR f2 and then f4=f3 XOR salt, this f4 will be stored.
8. Embed the secret key inside the salted template.
9. Store the embedded salted fingerprints.

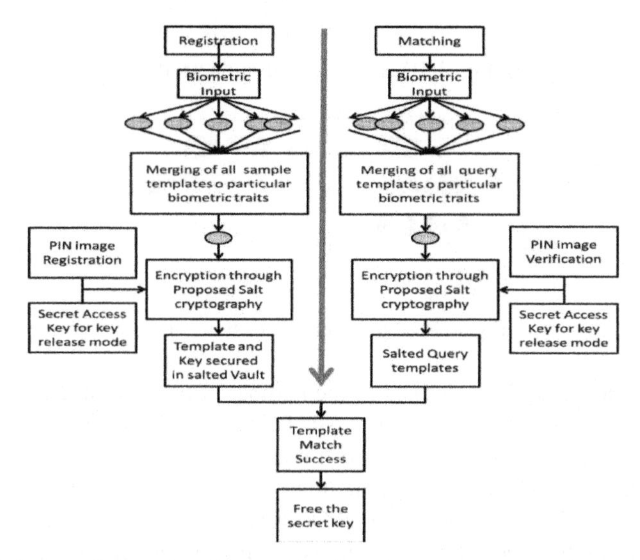

Fig. 2. Proposed algorithm (Phase –I Registration, Phase-II Matching phase)

5.2 Steps for Phase II(Matching phase)

1. During Verification phase biometric samples are taken. This is similar to enrolment phase.
2. User is asked to enter the user ID and to select the personal images.
3. Applying the same procedure as that of enrolment, the template is created.
4. This created saltcrypted template T' is matched with stored template T.

6 Simulation Results

The dataset used for fingerprint biometric is downloaded from NIST [9]. Any real time images can also be used. There is no restriction of the image size. But currently both the sample and query images must have same sizes. The sample fingerprint images are shown in figure. User need not to give the samples and queries in the same order. Any finger sample can be taken in any order. The algorithm has been simulated in MATLAB v12. Figure 3 shows the user ID and key registration window. The Personal images are selected through PII selection window. The user has to select three images out of the available 16 images. Personal image selection window is shown in figure 4.

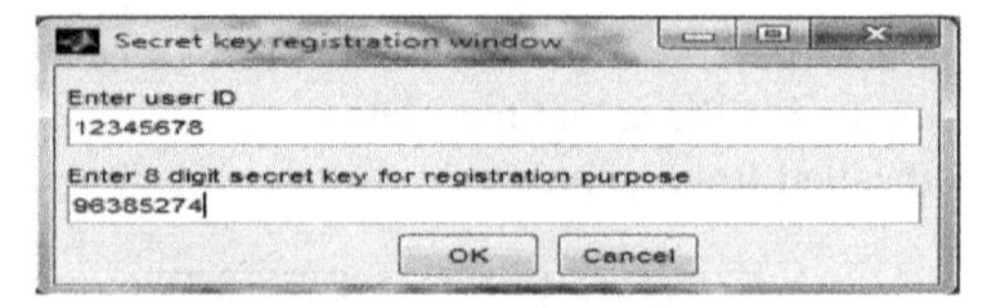

Fig. 3. ID and key registration window

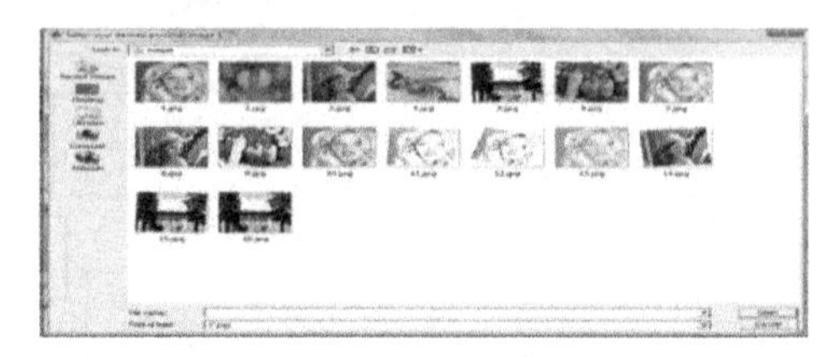

Fig. 4. Personal Image Selection Window

False acceptance rate : It is defined as the number of times a fingerprint is taken as a match even it is of some other person. FAR=False accepted queries/Total Queries
False rejection rate : It is the number of times a fingerprint of the genuine person is declared unmatched. FRR is acceptable up to some extent but FAR is not acceptable at all. FRR=False rejection queries/Total Queries.

The algorithm has been tested on variety of fingerprints images (more than 1000). FAR is 0.003% which may be due to bad quality of fingerprint image. FRR is a bit high 0.02% which is acceptable.

Table 1 shows the comparison of FAR and FRR with the existing approaches and proposed approach.Fig. 5(a) and 5(b) shows salted merged template and embedded saltcrypted template respectively.

Table 1. Comparison of the FAR and FRR with recent approaches

Approach	FAR	FRR
[20]	0.0016	0.05
[21]	0	0.21
[22,23]	7.1	7.15
[24]	4	2
Proposed	0.003	0.2

Fig.5.(a). **Fig. 5**(b).

Fig. 5. (a) Salted and merged template
Fig. 5.(b) Embedded saltcrypted template.

7 Analysis

First of all the major advantage of the algorithm needs discussion which is the complexity to get the actual template from the salted template.

1. Fingerprint image size = m*n,Salt1 size = m*n,Salt2 Size = bin_key_length= L
2. Complexity to know exact salts= 2m*n+L
3. Complexity of whole approach = O(m*n); If m==n then $O(n^2)$
4. Complexity of Embedding and extraction O(L)
5. Complexity to find exact embedding locations

a. No of images in PII database = P,Concurrent selection = k = 3,No of digits in user ID = N
b. Complexity of extracting user dependent info from the PII images = $k*10^N$= seed generation complexity.

The complexities of RSA, AES and other cryptography methods used in literature fuzzy vault are much larger than the proposed algorithm. The mathematical process is also easier than the existing process to be implemented in hardware and low cost devices. The combination of PII image and salted templates make the exact PN sequences determination impossible.

8 Conclusion and Future Work

The proposed algorithm has shown advantage in terms of complexity. The approach has very less complexity. On the other hand the seed generation and localization of the secret key in the salted templates is very hard to do if exact secret keys and the generation algorithms are unknown, so intrusion is almost impossible. Other advantage is that,nowhere actual fingerprint are stored. Only the salted templates are kept in storage with embedded key. Even the matching process is independent of the actual templates. User need to give the fingerprint template at run time, it is stored nowhere. The comparison of the stored saltcrypted template and query saltcrypted template is done and the embedded key is released only if they match.

Future work lies in eliminating need of the user to remember the three personal images which seems to be difficult. Another can be in improvement of the algorithm to identify the images taken in different seasons for enrolment and verification. This approach is simplest and can be applied to any other biometric modality. We are currently evaluating this algorithm on other modalities. Results are in pipeline for publishing.

References

1. Jain, A., Ross, A., Uludag, U.: Biometric template security: Challenges and solutions. In: Proceedings of European Signal Processing Conference (EUSIPCO), pp. 469–472 (2005)
2. Jain, A.K., Nandakumar, K., Nagar, A.: Review Article Template Security. EURASIP Journal on Advances in Signal Processing 2008, Article ID 579416, 17 (2008), doi:10.1155/2008/579416

3. Teoh, A.B.J., Goh, A., Ngo, D.C.L.: Random multispace quantization as an analytic mechanism for BioHashing of biometric and random identity inputs. IEEE Transactions on Pattern Analysis and Machine Intelligence 28(12), 1892–1901 (2006)
4. Belhumeur, P.N., Hespanha, J.P., Kriegman, D.J.: Eigenfacesversusfisherfaces: recognition using class specific linear projection. IEEE Transactions on Pattern Analysis and MachineIntelligence 9(7), 711–720 (1997)
5. Ratha, N.K., Chikkerur, S., Connell, J.H., Bolle, R.M.: Generating Cancelable Fingerprint Templates. IEEE Transactionson Pattern Analysis and Machine Intelligence 29(4), 561–572 (2007)
6. Juels, A., Sudan, M.: A Fuzzy Vault Scheme. In: Proceedings of IEEE International Symposium on Information Theory, vol. 6(3), p. 408 (2002)
7. Brindha, E.: Biometric Template Security using Fuzzy Vault. In: IEEE 15th International Symposium on Consumer Electronics (2011)
8. Clancy, T., et al.: Secure Smartcard-based Fingerprint Authentication. In: Proc. of ACM SIGMM Multim., Biom. Met. & App., pp. 45–52 (2003)
9. Uludag, U., Pankanti, S., Jain, A.K.: Fuzzy vault for fingerprints. In: Kanade, T., Jain, A., Ratha, N.K. (eds.) AVBPA 2005. LNCS, vol. 3546, pp. 310–319. Springer, Heidelberg (2005)
10. Nandakumar, et al.: Fingerprint-based Fuzzy Vault: Implementation and Performance. EEE Transactions on Information Forensics and Security 2(4), 744–757 (2007)
11. Yang, S., Verbauwhede, I.: Automatic Secure Fingerprint Verification System Based on Fuzzy Vault Scheme. In: Proc. of IEEE International Conference on Acoustics, Speech, and Signal Processing, vol. 5, pp. 609–612 (2005)
12. Chung, Y., Moon, D.-s., Lee, S.-J., Jung, S.-H., Kim, T., Ahn, D.: Automatic Alignment of Fingerprint Features for Fuzzy Fingerprint Vault. In: Feng, D., Lin, D., Yung, M. (eds.) CISC 2005. LNCS, vol. 3822, pp. 358–369. Springer, Heidelberg (2005)
13. Moon, D., et al.: Configurable Fuzzy Fingerprint Vault for Match-on-Card System. IEICE Electron Express 6(14), 993–999 (2009)
14. Zhang, P., Hu, J., Li, C., Bennamound, M., Bhagavatulae, V.: A Pitfall in Fingerprint Bio-Cryptographic Key Generation. In: Computers and Security. Elsevier (2011)
15. Wolfson, H., Rigoutsos, I.: Geometric Hashing: an Overview. IEEE Computational Science and Engineering 4, 10–21 (1997)
16. Yang, S., Verbauwhede, I.: Automatic Secure Fingerprint Verification System Based on Fuzzy Vault Scheme. In: Proc. of IEEE International Conference on Acoustics, Speech, and Signal Processing, vol. 5, pp. 609–612 (2005)
17. Zhe, J.: Fingerprint Template Protection with Minutia Vicinity Decomposition. ©2011 IEEE (2011) 9781- 4577-1359-0/11/$26.00
18. Sharma, N., Rathi, R., Jain, V., Saifi, M.W.: A novel technique for secure information transmission in videos using salt cryptography. In: 2012 Nirma University International Conference on Engineering (NUiCONE), December 6-8, pp. 1–6 (2012)
19. Santhi, B., Ramkumar, K.: Novel hybrid Technology in ATM security using Biometrics. Journal of Theoretical and Applied Information Technology 37(2) ISSN: 1992- 8645
20. Moon, D.-s., Lee, S.-J., Jung, S.-H., Chung, Y., Park, M., Yi, O.: Fingerprint Template Protection Using Fuzzy Vault. In: Gervasi, O., Gavrilova, M.L. (eds.) ICCSA 2007, Part III. LNCS, vol. 4707, pp. 1141–1151. Springer, Heidelberg (2007)
21. Uludag, U., Pankanti, S., Jain, A.: Fuzzy Vault for Fingerprints. In: Kanade, T., Jain, A., Ratha, N.K. (eds.) AVBPA 2005. LNCS, vol. 3546, pp. 310–319. Springer, Heidelberg (2005)

22. Long, T.B., Thai, L.H., Hanh, T.: Multimodal Biometric Person Authentication Using Fingerprint, Face Features. In: Anthony, P., Ishizuka, M., Lukose, D. (eds.) PRICAI 2012. LNCS, vol. 7458, pp. 613–624. Springer, Heidelberg (2012)
23. Qader, H.A., Ramli, A.R., Al-Haddad, S.: Fingerprint Recognition Using Zernike Moments. The International Arab Journal of Information Technology 4(4) (October 2007)
24. Shrivastava, R., Thakur, S.: Performance Analysis of Fingerprint Based Biometric Authentication System using RSA. Engineering Universe for Scientific Research and Management 6(2) (February 2014)

Robust Pattern Recognition Algorithm for Artifacts Elimination in Digital Radiography Images

Igor Belykh

St. Petersburg State Polytechnical University, ul. Polytechnicheskaya 29,
St. Petersburg 195251, Russia
igor.belyh@avalon.ru

Abstract. In projection radiography stations image quality is enhanced by using anti-scatter grids that improve image contrast but form specific patterns that may be visible or may cause Moiré effect when digital image is resized on a diagnostic monitor. In this paper a robust, efficient and fully automated grid pattern recognition and elimination algorithm is proposed which is still an actual problem especially in computer aided diagnosis. The pattern recognition is based on statistical approach in both spatial and frequency domains and provides features extracted for the elimination stage. The pattern suppression is based on a 2-D filter approach preserving diagnostic quality of the image. Experimental results and advantages over existing approaches are discussed.

Keywords: pattern recognition, feature extraction, image processing, digital radiography.

1 Introduction

In modern radiography the reliable diagnosis depends on quality of digital X-ray images. One of the factors that affect image quality is the absence or minimal amount of different type of noise and artifacts. Anti-scatter grids are used in radiology stations [5] for image contrast enhancement at acquisition stage but leave visible artifacts that may cause problems while digital image representation at diagnostic stage.

The use of anti-scatter grids is based on assumptive decomposition of X-ray radiation propagated through the object of interest into a primary and a secondary component. The primary component is formed by the electron beams with small deviations from initial straight line trajectories while secondary component is formed by scattered electrons which are deflected (or even reflected) by object interior. Scattered component degrades image contrast. Anti-scatter grids are designed to enhance image contrast by means of scattered radiation partial absorption and are located between the object and the receiving device, i.e. computed radiography (CR) plate [4] or direct radiography (DR) digital panel [3], and can be oriented vertically or horizontally.

Grids can be stationary or moving during the exposure time, and can be linear or cross and also parallel or focused. Currently stationary linear grids [5] are still the most usable. Primary X-ray beams are mostly transmitted through the grid interspace

S.C. Satapathy et al. (eds.), *Emerging ICT for Bridging the Future – Volume 1*,
Advances in Intelligent Systems and Computing 337, DOI: 10.1007/978-3-319-13728-5_36

material while scattered beams along with some primary ones are absorbed by grid lead stripes. Different grid structure and design were described in details in [1] including the ranges for main characteristics such as grid ratio and grid frequency measured in lines per inch (LPI) or per mm. The advantage of the grid use is the higher image contrast but the disadvantages are: the need of higher X-ray dose (up to 3 times) and the visible thin line artifacts that may cause a known Moiré pattern [4] when digital image is resized for display on a diagnostic monitor (Fig. 1).

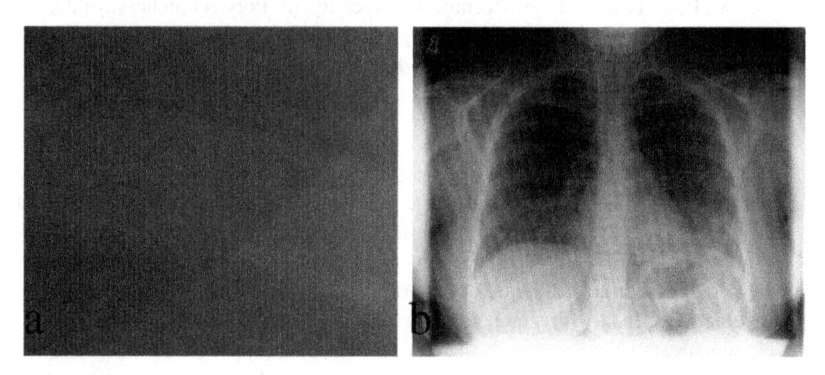

Fig. 1. Line artifacts and Moiré pattern on magnified (a) and minified (b) CR image fragments

During the last 10-15 years there were many efforts in studying the grid artifacts and developing different methods for their detection and suppression. Most of grid detection methods are based on a simple threshold analysis of the maximum in 1-D Fourier spectrum of image rows/columns providing a grid frequency [5, 6]. Those algorithms may fail on detecting the grids with high frequencies. Also detection algorithm is supposed to provide accurate and comprehensive information for the suppression algorithm, i.e. not only a grid frequency but some other features that are needed for automated filter design and tuning for the fine elimination of the artifacts preserving the diagnostic quality of the image. The suppression algorithms based on different type of filters either in spatial [2, 4, 5, 6] or in frequency domains [3] in 1-D or in 2-D [6] are reviewed in [1]. That review can be summarized that there is still a need for robust and reliable grid pattern recognition method followed by its elimination that will improve but not degrade the diagnostic quality of digital X-ray image.

2 Grid Pattern Recognition

2.1 Grid Pattern Formation

Grid pattern is formed in radiographic image in two steps described in [1]. First, a periodical structure of radiation intensity is formed behind the grid device and then it is digitized by CR scanners or registered by DR digital detectors. Digital image profile orthogonal to grid lines contains a periodic grid function which is close but not equal to a harmonic function and that is why grid spectral maxima are usually a bit dispersed (Fig. 2).

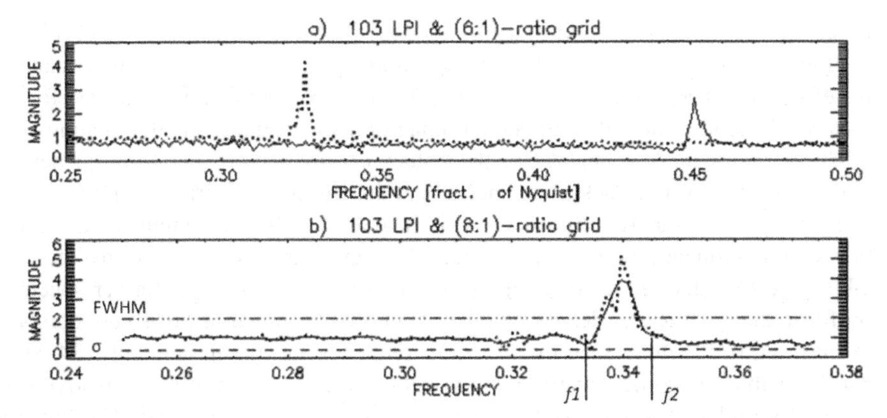

Fig. 2. Fragment of spectra for different grid LPI and ratio (a, b) scanned with different sampling frequency: (a) f_s = 5.85 – dotted, and f_s = 8.7 – solid; (b) f_s = 5.85 – dotted, smoothed spectrum - solid, σ-level – dashed, half-maximum-level – dotted-dashed, *f1* and *f2* – extracted frequency values of grid maximum width.

The periodic grid pattern formation details related to: (a) grid frequencies f_g possible aliasing for some sampling frequencies f_s that do not satisfy the Nyquist theorem in CR/DR; (b) grid spectral maxima dispersion due to modulation transfer functions of CR scanners or DR digital detectors; (c) grid spectral maxima migration due to possible inclination of grids from ideal position while acquisition; and (d) grid pattern intensity variation in different image parts due to different absorption of radiation by different body parts/tissues are analyzed in [1] and are shown in Fig. 2. Moreover, grid designs by different manufacturers give another degree of freedom for grid pattern features.

All those reasons mentioned above require a reliable recognition algorithm invariant to grid inclinations and able to track grid pattern artifacts in images with different body parts and tissues acquired with grids of different design.

2.2 Pattern Recognition and Feature Extraction

Since grids require the increase of X-ray dose in some studies (e.g. in ICU – intensive care units) they are not used, so there is also a need to detect whether grid is present or not in acquired image. In this paper the proposed model for grid pattern recognition in digital X-ray image emulates the work of "eye and brain processor" to reveal a regular structure in 2-D random intensity field. This problem can be divided into two tasks: a) reveal the periodicity in 1-D; and b) track it in orthogonal dimension. The first task can be solved by calculating an autocorrelation function (ACF) for each image row x_i and detecting and comparing the dominant periods. The second task can be can be done by calculating the cross-correlation function (CCF) for the pair of neighbor image rows x_i and x_{i+1} followed by analysis of dominating periods. If the majority of periods are the same then a decision that grid pattern is detected is made. This two-step process can be repeated for image columns to check what type of grid design or orientation was used. Thus the pattern recognition algorithm is proposed based on non-transformed and transformed structural features extraction and selection [8] followed by their classification with maximal descriptor dimensionality reduction for efficiency. The descriptors are formed by the following features required for grid pattern detection and its further elimination: the values of dominating periods/frequencies in ACF and CCF, the values of grid pattern period/frequency and magnitude M_g, level of background noise, and the width of spectral maximum.

Feature extraction algorithm starts with the intensity shift (DC) removal and reduction of the influence of intensity trends of image ultra large objects by means of taking the first derivatives of image profiles (row or column). A Fourier spectra of profile x_i and profile x_{i+1} derivative taken from image shown in Fig.1b are compared in Figs. 3a, b. It can be seen that the grid frequency maximum at about 0.34 can be hardly visible in profile spectrum and is better detectable in profile derivative spectrum. More reliable results can be achieved if the spectra of ACF and CCF functions of profile derivatives are compared (Fig. 3c, d), where the peak corresponding to grid frequency is even dominating. The frequencies lower than 0.01 are not shown in all charts due to the big magnitudes related to image large objects in chart on Fig. 3a.

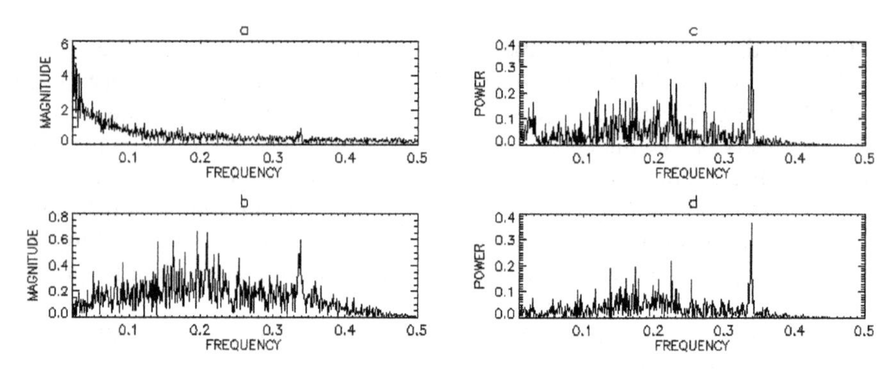

Fig. 3. Fragments of 1-D Fourier spectra of: (a) - profile x_i, (b) - profile x_{i+1} derivative, (c) - ACF of profile x_i derivative, (d) - CCF of profiles x_i and x_{i+1} derivatives

This statistical approach in both spatial and frequency domains can be further optimized using a known [7] Khinchin-Kolmogorov theorem and substituting a profile derivative ACF spectrum with its derivative power spectrum and two profiles derivatives CCF spectrum with their derivatives cross-spectrum.

Grid pattern frequency f_g (or its aliased harmonic f_g^*) [1] and its magnitude are taken from each profile power spectrum. The level of background noise around each maximum can be estimated as a standard deviation σ of the right half (high frequency part) of each spectrum. The width of detected maximum can be obtained in two substeps (Fig. 2b): (1) smoothing of each profile spectrum by a gentle lowpass or morphological filter (solid line); (2) taking two end values $f1$ and $f2$ of the line segment equal to full width at half maximum (FWHM) projected down to the level of σ within the angle with the top at the maximum.

Thus an iterative transformational feature extraction algorithm can be built in several steps in spatial and frequency domains:

(1) take first two image neighbor profiles x_1 and x_2;
(2) get their first derivatives, i.e. their differentials dx_1 and dx_2;
(3) calculate power spectrum of dx_1 and cross-spectrum for dx_1 and dx_2;
(4) pick up frequencies of dominating peaks in power spectrum of dx_1 and cross-spectrum for dx_1 and dx_2;
(5) pick up magnitudes from profile spectrum at frequencies picked up at step (4);
(6) pick up the grid peak frequency bandwidth;
(7) repeat steps 1-6 for the next pair of profiles and so on for the whole image.

Finally a 6-element grid pattern descriptor is obtained for each image profile.

The classification process of pattern recognition is optimized by means of reduction the descriptor dimensionality down to one, i.e. the frequencies extracted at step 4 need to coincide within the predefined frequency interval df. The coincidence of the majority of those frequencies corresponds to the solution of initially formulated the two-task problem of a 2-D pattern recognition making the algorithm statistically robust and efficient. A final decision is based on preset *threshold* for the amount of counted frequency coincidences.

In order to increase the classification algorithm efficiency a neighbor profiles for analysis can be taken with a small distance *delta* between them, e.g. take every 2nd, 5th or 10th profile. For regular 5 Mpix CR image the predefined parameters for the feature extraction and classification algorithm with their experimentally obtained values are: *delta=5, df=0.005, threshold=50%*. Those presets were tested on a representative number of images and can be used as default values to make the algorithm fully automated.

The features required for grid pattern elimination are averaged overall number of classified descriptors of image profiles except bandwidth limits that are taken as minimal value of all *f1* values and maximal value of all *f2* values. Finally, the extracted features form a 5-element descriptor which can be stored in image file DICOM auxiliary tags [8].

3 Grid Pattern Elimination

As mentioned earlier, lowpass filters blur an image and remove high frequencies that provide the presence of image normal background. The removal of that background makes an image look artificial and not acceptable by radiologists for clinical diagnosis. Another point is that linear grid patterns are narrow banded and so the band stop filters are the most appropriate for their suppression by definition. The problem here is a ringing effect especially for narrow banded filters applied to images containing sharp edges and/or spikes. The applied research of the influence of bandstop filter Gibbs oscillations δ on image quality is one the subjects of this article. A variety of approaches in digital filter transfer function (TF) design and implementation is known, including bandstop filters. One of the best methods for medical image processing is a known Potter310 smoothing window [4]. This filter has very smooth TF combined with powerful attenuation and computational efficiency due to relatively short operator length N. The only its disadvantage is a gentle slope of TF that might not be acceptable if very narrow bandstop filtering is needed (Fig. 4, dashed line). This disadvantage can be overcome with a filter based on Kaiser transfer function [7]. Theoretical difference of those filters bandwith is revealed by the fact that Potter filter is based on smoothing of a trapezoidal window while Kaiser function approximates a rectangular window. The remarkable feature of Kaiser filter is that beside filter length parameter there is another parameter $A = -20\lg(\delta)$ that controls the shape of smoothing window and what is very important the level of Gibbs events that cause periodic oscillations of intensity around narrow bandwidth limited by filter stop and pass frequencies (Fig. 4, dotted line). The design of Kaiser TF is described in [1] and its advantage over Potter approach is shown in Fig. 4 by solid line. The level of oscillations can be controlled by means of A and N value variations. If δ is set to a constant value then Kaiser window coefficients will be a function of one parameter N, which will affect filter attenuation power and will depend on "grid-to-noise ratio" defined by detector as $20\lg(\frac{M_g}{\sigma})$. The optimal value for Gibbs events level was estimated experimentally and is equal to $\delta = 0.001$, i.e. $A = 60$ dB.

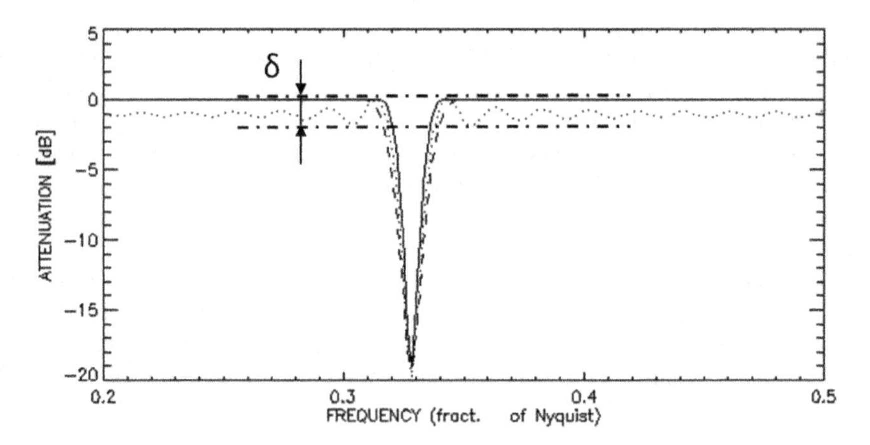

Fig. 4. Fragments of Potter (dashed) with N=64 and Kaiser (solid) transfer function with different A and N values: A= 20 dB, N=50 – dotted line; A= 60 dB, N=116 – solid line; δ - denotes Gibbs events full amplitude

The designed Kaiser bandstop filter was tested on a set of CR images (Fig. 5a). The analysis of image fragment with label "R" and numbers (Fig. 5b) similar to represented in [2] showed that due to the physics of radiographic acquisition all objects are projected onto X-ray receiving device and then digitized, so there are no ideally sharp edges in any registered image object. Even artificial objects located on receiver, such as, have steep slopes in intensity as shown in Fig. 5c, and so the chances of appearance of strong filter ringing are very small. After recognition step followed by filtering there are neither visible grid pattern artifacts nor filter ringing effects noticeable neither in filtered magnified image fragment nor in an image profile fragment or its spectrum and neither at the edges nor at few visible spikes (Fig. 5 b, e, f).

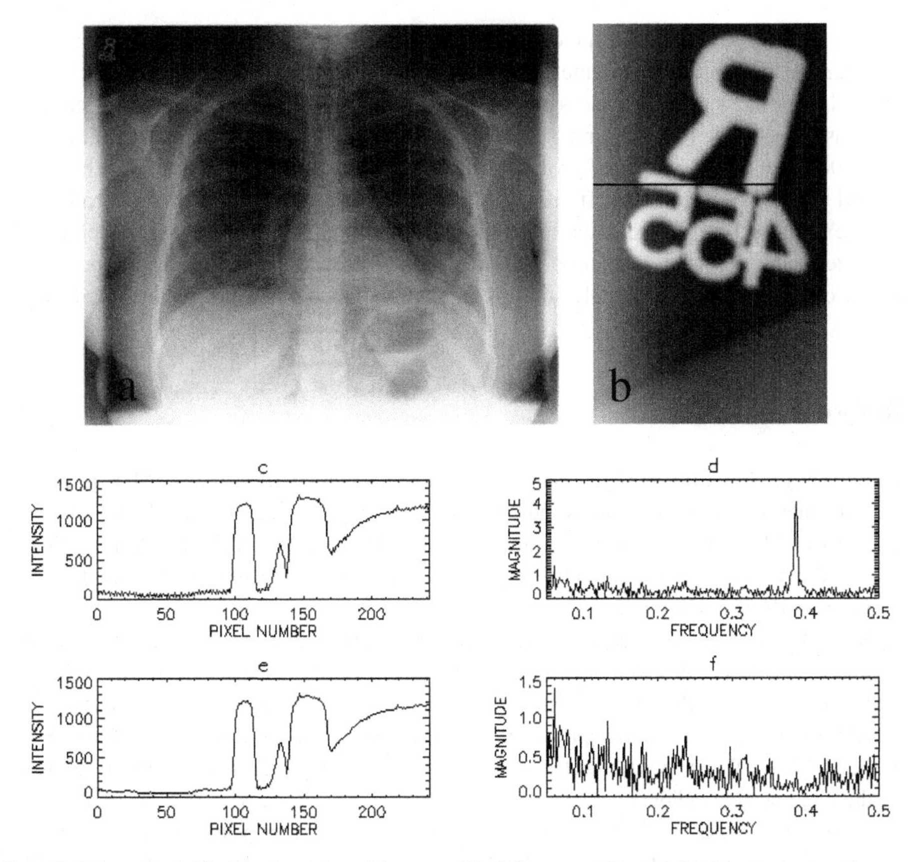

Fig. 5. Filtered minified image (a) and its magnified fragment (b), original and processed profiles (c, e) and their spectra (d, f). The profile position is shown by solid black line across the image fragment in (b).

A 2-D filter is recommended to be applied for pattern elimination and for the cross grid it is mandatory. For linear parallel grid pattern the filter can be applied in 1-D for the efficiency. The diagnostic quality of 1-D and 2-D filtered images was not degraded as confirmed by radiological analysis based on clinically established metrics.

4 Results and Conclusions

The implemented grid pattern recognition and elimination algorithm was tested on a representative set of 72 CR images of two sizes (1Mpix and 5Mpix) acquired using grids with three different frequencies (85, 103, 152 LPI) and four different ratios (6, 8, 10 and 12) and scanned with three different pixel sizes (0.171, 0.115 and 0.097 mm).

There was no failure of grid recognition algorithm and processed images revealed an effective grid pattern elimination preserving acceptable image diagnostic quality as evaluated by radiologists based on a standard clinical metrics.

The advantages of the proposed algorithm over existing approaches are that it is invariant to possible grid inclinations at acquisition stage and it is also stable in different image parts (body parts/tissues). The proposed grid pattern elimination method demonstrated very reliable and accurate grid pattern filtering preserving clinically valuable high frequency background structure without ringing and image blur revealed in [2]. The algorithm efficiency is less than 2 sec per image on a modern PC.

The proposed fully automated and efficient algorithm is recommended for practical use due to the described theoretical and applied advantages over existing approaches and it can be built into CR/DR stations or PACS systems and computer aided diagnostic (CAD) systems.

References

1. Belykh, I.: Grid artifacts suppression in computed radiographic images. WASET. Intl. J. of Computer, Information, Systems and Control Engineering 8(8), 1280–1283 (2014)
2. Lin, C.Y., Lee, W.J., Chen, S.J., Tsai, C.H., Lee, J.H., Chang, C.H., Ching, Y.T.: A Study of Grid Artifacts Formation and Elimination in Computed Radiographic Images. J. Digit. Imaging 19, 351–361 (2006)
3. Kim, D.S., Lee, S.: Grid artifact reduction for direct digital radiography detectors based on rotated stationary grids with homomorphic filtering. Med. Phys. 40(6), 061905-1–14 (2013)
4. Belykh, I.N., Cornelius, C.W.: Antiscatter stationary grid artifacts automated detection and removal in projection radiography images. In: Proc. SPIE, vol. 4322, pp. 1162–1166 (2001)
5. Barski, L.L., Wang, X.: Characterization, detection and suppression of stationary grids in digital projection radiography imagery. In: Proc. SPIE, vol. 3658, pp. 502–519 (1999)
6. Sasada, R., Yamada, M., Hara, S., Takeo, H.: Stationary grid pattern removal using 2-dimensional technique for Moiré-free radiographic image display. In: Proc. SPIE, vol. 5029, pp. 688–698 (2003)
7. Hamming, R.W.: Digital filters. Prentice-Hall, Englewood Cliffs (1989)
8. Meyer-Baese, A., Schmid, V.J.: Pattern Recognition and Signal Analysis in Medical Imaging. Elsevier Academic Press (2014)

Hull Detection from Handwritten Digit Image

Sriraman Kothuri and Mattupalli Komal Teja

Department of CSE, Vignan's Lara Institute of Technology & Science,
Vadlamudi, Guntur District, Andhra Pradesh, India
ksriraman@hotmail.com, komalteja@gmail.com

Abstract. In this paper we proposed a novel algorithm for recognizing hulls in a hand written digits. Those hull regions are detected in order to find out in a digit of user drawn. To achieve that it is necessary to follow the three steps. Those are Pre-processing, Boundary Extraction and at last apply the Hull Detection system in a way to attain the most relevant results. The detection of Hull Regions is mainly aim to intend the increase of machine learning capability in detection of characters or digits. This provides an artificial intelligence to the system in away in can identify the digits or characters easily. This can also be extended in order to get to detect the hull regions and their intensities in Black Holes in Space Exploration.

Keywords: Chain code, Machine Learning, Hull Regions, Hull Recognition System, SASK algorithm.

1 Introduction

When we saw a digit it is very easy to recognize it. Using the same ideology in the Machine Learning Process i.e., to impart the intelligence to the system by maintaining appropriate training data set. So, that when a user interacts with the system using GUI to draw an input digit which will identify the digit just like Humans. While doing the research on Digit Recognition Using Freeman Chain Code [1] we came across to know about the hull regions in some digits like 0, 2, 4, 6, 8 etc. As these Hull Regions are very useful in many Image Processing applications that's the reason why to extend our project for the Hull Detection also. For this we have proposed a new algorithm i.e., SASK to extract the hull region for the given input image.

The process of detecting is not only mainly focused for the digits can also be applied same for the alphabets also. It is easy to identify the hull which is resides in a digit by humans and for the computer it must have capability to identify what actually the input is. To improve the computer visionary in a way that it can have identification regarding the input given by the user.

S.C. Satapathy et al. (eds.), *Emerging ICT for Bridging the Future – Volume 1*,
Advances in Intelligent Systems and Computing 337, DOI: 10.1007/978-3-319-13728-5_37

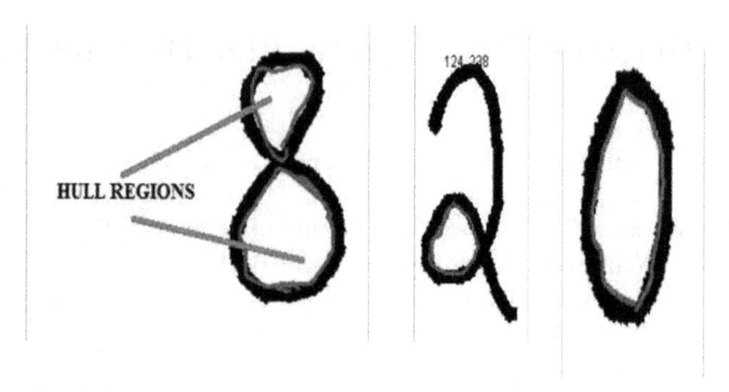

Fig. 1. Highlighted portions shows the hull regions for 8, 2, 0 digits

Hull is a region which is surrounded by data but does not contain any data inside it. Hull regions for the digits 8, 2, 0 is shown in the below figure and we can observe two hull regions for digit 8 and a single hull region for both digits 2 and 0. By using the SASK algorithm these hulls are identified easily and from there we can easily locate where the region exactly having the hull.

2 Overall Design

2.1 Input Data

User gives the input digit by using the GUI which is designed by using the concept of swings in JAVA that will read by computer. The user draws the digit on canvas with the help of a mouse. In order to accept the digit it meant to read the digit in order to recognize it first.

2.2 Preprocessing

The input data will be saved in the image format with .(PNG) extension. This input image will be same as whatever the user draws using the canvas. It will become as the input.

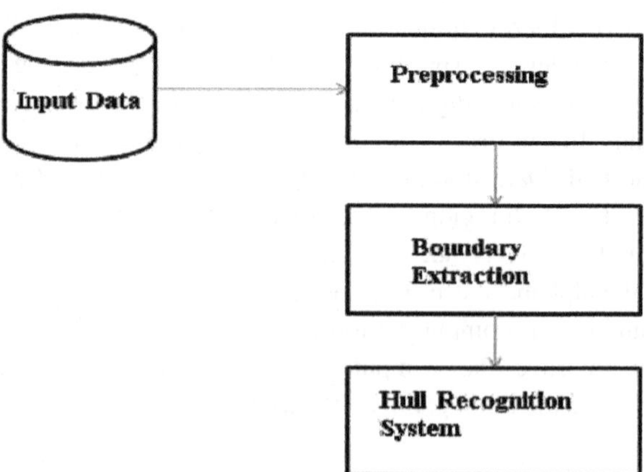

Fig. 2. Overall Design of Hull Recognition System

2.3 Boundary Extraction

After the preprocessing step, our aim is to detect the boundary of the acquired image. In order to get the boundary, we need to check each and every pixel properties till now most of the papers [1][2] that we have seen they checked only the 8-connectivity of the particular pixel. 8-connectivity means if a pixel contains 8 surrounded pixels then it does not belong to boundary pixel but by checking only 8-connectivity we have a disadvantage I.e., we can't get effective boundary of an image so in order to get exact boundary we have also checked 4-connectivity of every pixel.

(X-1,Y-1)	(X-1,Y)	(X-1,Y+1)
(X,Y-1)	Pixel at (X,Y)	(X,Y+1)
(X+1,Y-1)	(X+1,Y)	(X+1,Y+1)

Fig. 3. Connectivity of a pixel

	(X-1,Y)	
(X,Y-1)	Pixel at (X,Y)	(X,Y+1)
	(X+1,Y)	

Fig. 4. Connectivity of a pixel

a) Algorithm for Boundary Extraction:

Input: Source Image

Output: Image with only boundary

Step 0: Take the image as input file

Step 1: Scan the image all the pixels from left to right and top to bottom as follows. Assume Image left, top corner starts from 0,0. Then

```
for ( i=0 to image width )
 {
for (j=0 to image height )
 {
Pixel = getImagepixel (i,j)
if ( pixel is presented at location (i,j) ) then
 {
if(check_8-connectivity_ pixel ( i,j) or check_4-connecitivity_pixel(i,j))
then
 {
Just remove that pixel at location(i,j)
 }
 }
 }
 }
```

The time complexity of the algorithm is O(n) . The output for the above algorithm is shown in the below Figure 5.

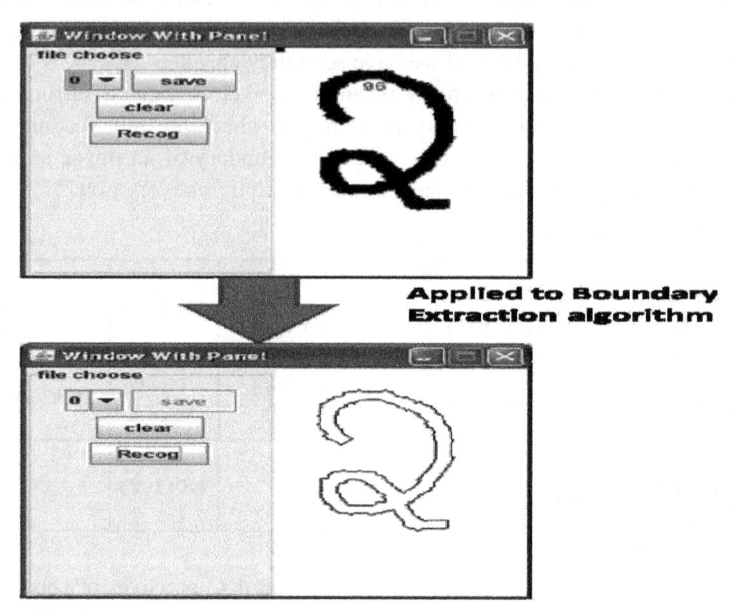

Fig. 5. Shows the extraction of boundary for the drawn input digit 2

2.4 Hull Recognition System

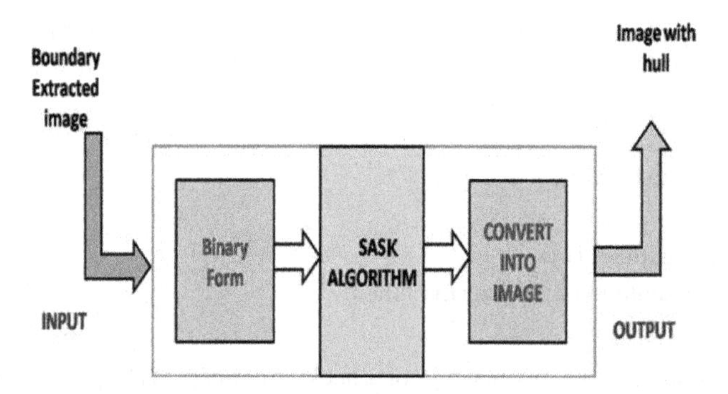

Fig. 6. Hull Recognition System Design

In this step we need to convert the boundary extracted image in to the binary format, to convert the boundary image into the binary form we need to follow the below steps. Below figure 7 shows the Binary output.

- Just check if the pixel is white then mark it as 0.
- If it is not a white pixel mark it as 1.

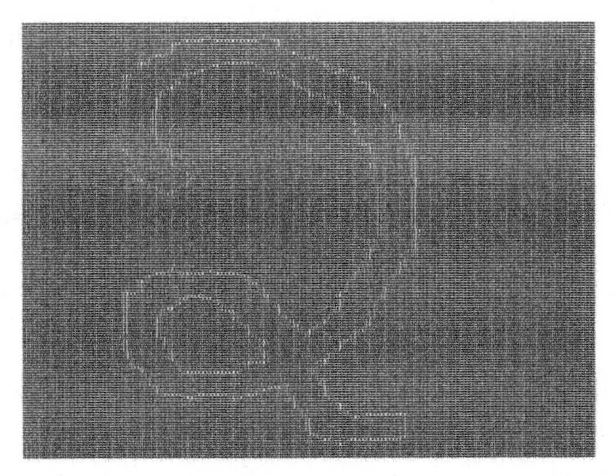

Fig. 7. Binary form of the boundary extracted image

After getting the binary form of the image we have to apply the binary output to hull recognition algorithm i.e. SASK-Algorithm.

A. SASK-ALGORITHM

For the binary output it first checks for the 1 marked pixel from left to right whenever it appears mark it as any number here we consider it as 2(not exact 2 any number other than 0 and 1) it checks for the upward pixel if it is present it moves the position to that pixel if it is not present it rotates 45^0 angle clockwise direction and for every 1 pixel we have to check it from upwards only(no need of exact upward any direction but the same direction follows for every pixel) and checks again and this process repeats until we reach to the starting pixel and finally all these 2's becomes the outer boundary of the Hull and the 1's present inside this outer boundary forms the Hull Region.

2.5 SASK-Algorithm (Hull Recognition)

Input: Binary form of Boundary Image data

Output: Hull region

Step 0: Take input as binary file

Step 1: Store binary file data as 2-d array

a[ ][ ]=binary file data.

Step 2: Scan array from left to right and from top to bottom manner. Check if 1 is present or not and if it is the first 1 that is appeared while scanning from Left to Right and Top to Bottom then Make x=row position of fist pixel y= column position of first pixel.

Step 3: check a[x][y]=1 && x,y not equal the positions got at step 2 (first time ignore) then

```
{

if check a[x-1][y]=1 then set x=x-1 , a[x][y]=2;

else if check a[x-1][y+1]=1 then set x=x-1, y=y+1 , a[x][y]=2;

else if check a[x][y+1]=1 then set y=y+1 , a[x][y]=2;
else if check a[x+1][y+1]=1 then set x=x+1, y=y+1, a[x][y]=2;

else if check a[x+1][y]=1 then set x=x+1 , a[x][y]=2;
else if check a[x+1][y-1]=1 then set x=x+1,y=y-1, a[x][y]=2;

else if check a[x][y-1]=1 then set y=y-1 , a[x][y]=2;

else if check a[x-1][y-1]=1 then set x=x-1,y=y-1, a[x][y]=2;

}
```

Step 4: Repeat step 3 until we reach the same 1 position that we get in step-2 in the beginning.

Step 5: Now we get all the outer region of the border pixel gets 2 (any value and here we consider it as 2) value and remaining inner regions not changed they have same 1 value

Step 6 : Restore the same above array a[ ] [ ] data to image file as by giving 2- color codes

```
if a [x][y]!=0 && a[x][y]!=1 set pixel color = red;

else if a[x][y]=1

set pixel color = black ;
```

Step 7 : End

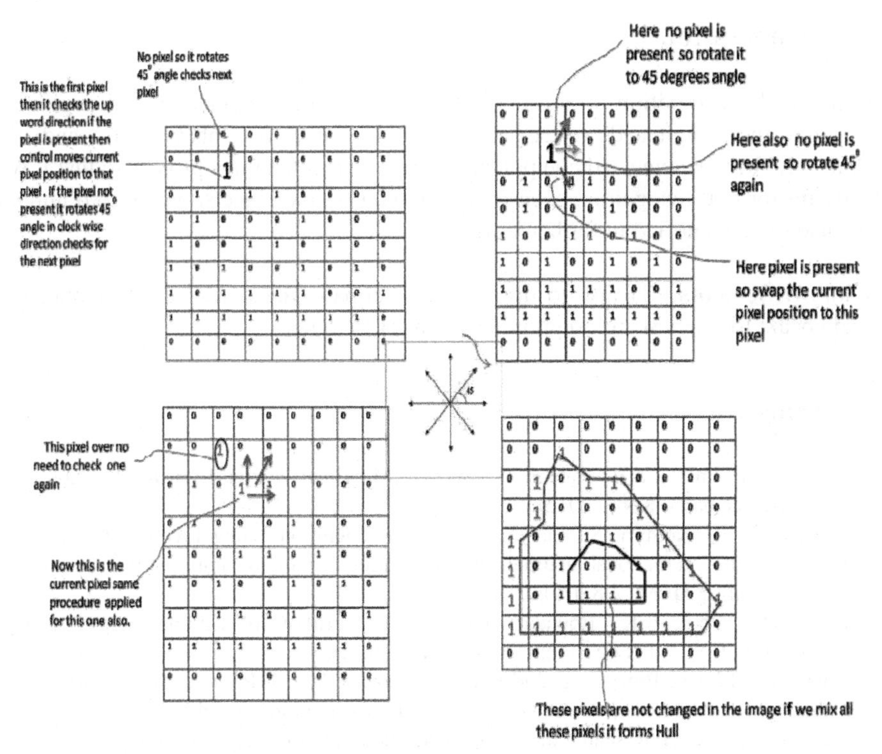

Fig. 8. SASK Algorithm Overall Procedure

In this entire algorithm time complexity for finding the outer region it needs < n times and to restore to image it needs same complexity of n^2 so the time complexity is O(n). For the SASK algorithm it takes time complexity of O(n) So, After applying SASK algorithm here, in the Figure 8 the outer boundary of the digit 2 is represented in red color and the hull portion is represented in black color.

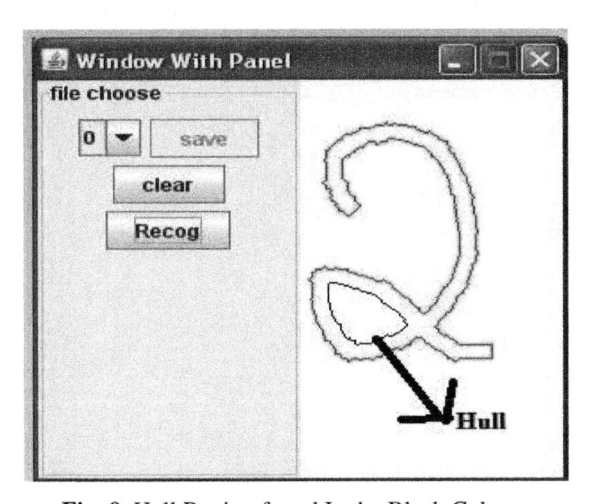

Fig. 9. Hull Region found In the Black Color

3 Conclusion

By this Hull Recognition SASK algorithm, System will recognize the hull region for the given input digit and we can also apply the same for alphabets also and it is applicable for the connected components and we can extend this project so that it will be useful for hull recognition even in the Irregular shape objects, the layered hulls, if any also to be identified in further work.

Acknowledgments. Thanks to my students Pulipati Annapurna & Karampudi siva naadh baazi for the research support and data set training.

References

1. Bernard, M., Fromont, E., Habard, A., Sebban, M.: Hand Written Digit Recognition using Edit-Distance based KNN (2012)
2. Annapurna, P., Kothuri, S., Lukka, S.: Digit Recognition Using Freeman Chain Code. Published in International Journal of Application or Innovation in Engineering and Management (IJAIEM) 2(8) (August 2013)
3. Park, S.C., Choi, B.K.: Boundary Extraction Algorithm for Cutting Area Detection. Published in ElSEVIER on Computer Aided Design (2000)
4. Mich Digital Image Processing, 2nd edn. Gonzalez and Woods Pearson Publications
5. Chaudhari, P.P., Sardhe, K.: Handwritten Digits Recognition Special point. International Journal of Advanced Research in Computer Science and Software Engineering
6. Choudhary, S., Patnaik, T., Kumar, B.: Curved Text Detection Techniques - A Survey. International Journal of Engineering and Innovative Technology (IJEIT) 2(7) (January 2013)
7. Labusch, K., Barth, E.: Simple Method for High-Performance Digit Recognition Based on Sparse Coding
8. Zekovich, S., Tuba, M.: Hu Moments Based Handwritten Digits Recognition Algorithm, Recent Advances in Knowledge Engineering and Systems Science

Learning Approach for Offline Signature Verification Using Vector Quantization Technique

Aarti Chugh[1], Charu Jain[1], Priti Singh[2], and Preeti Rana[3]

[1] Department of Computer Science, Amity University Gurgaon
{achugh,cjain}@ggn.amity.edu
[2] Department of Electronics and Communication, Amity University Gurgaon
psingh@ggn.amity.edu
[3] Department of Computer Science, JKP Polytechnic, Sonepat, India
rana.preeti21@gmail.com

Abstract. Signature is a behavioral trait of an individual and forms a special class of handwriting in which legible letters or words may not be exhibited. Signature Verification System (SVS) can be classified as either offline or online. [1] In this paper, we used vector quantization technique for signature verification. The data is captured at a later time by using an optical scanner to convert the image into a bit pattern. The features thus extracted are said to be static. Our system is designed using cluster based features which are modeled using vector quantization as its density matching property provides improved results compared to statistical techniques. The classification ratio achieved using Vector Quantization is 67%.

Keywords: Off-line Signature Verification, Feature Extraction, Vector Quantization.

1 Introduction

Signature has been a distinguishing feature for person identification through ages. Approaches to signature verification fall into two categories according to the acquisition of the data: On-line and Off-line. On-line data records the motion of the stylus while the signature is produced, and includes location, and possibly velocity, acceleration and pen pressure, as functions of time [1, 2]. Online systems use this information captured during acquisition. Off-line data is a 2-D image of the signature. Processing Off-line is complex due to the absence of stable dynamic characteristics. Difficulty also lies in the fact that it is hard to segment signature strokes due to highly stylish and unconventional writing styles. The non-repetitive nature of variation of the signatures, because of age, illness, geographic location and perhaps to some extent the emotional state of the person, accentuates the problem. All these coupled together cause large intra-personal variation. A robust system has to be designed which should not only be able to consider these factors but also detect various types of forgeries [3]. It should have an acceptable trade-off between a low False Acceptance Rate (FAR) and a low False Rejection Rate (FRR). Numerous approaches have been proposed for

S.C. Satapathy et al. (eds.), *Emerging ICT for Bridging the Future – Volume 1*,
Advances in Intelligent Systems and Computing 337, DOI: 10.1007/978-3-319-13728-5_38

Handwritten Signature Identification, Recognition and Authentication systems. Many research works on signature verification have been reported. In March, 2007, Debnath Bhattacharyya, Samir Kumar Bandyopadhyay and Poulami Das [4] have proposed a new recognition technique; an Artificial Neural Network is trained to identify patterns among different supplied handwriting samples [4, 5]. Another important method for offline signature verification is the application of Hidden Markov's rule. Justino, Bortolozzi and Sabourin proposed an off-line signature verification system using Hidden Markov Model [5]. Hidden Markov Model (HMM) is one of the most widely used models for sequence analysis in signature verification. A well chosen set of feature vectors for HMM could lead to the design of an efficient signature verification system. Application of Support Vector machine (SVM) at the time of signature verification is also a new dimension in this field. Emre Ozgunduz, Tulin _enturk and M. Elif Karslıgil has proposed an algorithmic approach according to which off-line signature verification and recognition can be done by Support Vector Machine [7, 9]. In this paper we deal with Offline signature Verification System.

Vector Quantization is a clustering technique mainly used for speech recognition and lossy image compression [23]. Here we use vector quantization in training our system. The main advantage of VQ in pattern recognition is its low computational burden when compared with other techniques such as dynamic time warping (DTW) and hidden markov model (HMM). It works by dividing a large set of points (vectors) into groups having approximately the same number of points closest to them. This feature set is based on Vector Quantization.

2 Methodology

This section describes the methodology behind signature verification system development. It starts with data acquisition through scanner followed by pre-processing, feature extraction, training and at last classification (Fig. 1).

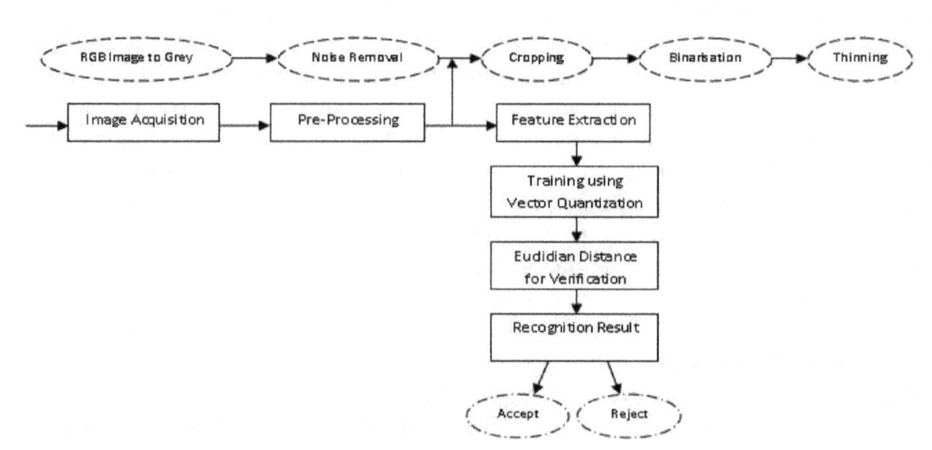

Fig. 1. Block Diagram of Proposed Signature Verification System

2.1 Image Acquisition

Handwritten signature images are collected from different individuals. The system has been tested for its accuracy and effectiveness on data from 25 users with 10 specimens of each making up a total of 250 signatures. The proposed verification algorithm is tested on both genuine and forged signature sample counterparts. So we developed a signature database which consists of signatures from all the age groups. Our database is also language independent and also it consists of signatures done with different pens with different colors. 10 users were asked to provide genuine signatures, 5 were asked to do skilled forgeries, 5 provide casual forgeries and 5 did random forgeries. A scanner is set to 300-dpi resolution in 256 grey levels and then signatures are digitized. For further working we cut and pasted scanned images to rectangular area of 3 x 4 cm or 200 x 200 pixels and were each saved separately in files.

2.2 Pre-processing

The pre-processing step is normally applied to make signatures ready for further working. It is also applied to improve the efficiency and performance of the SVS. Signatures are scanned in color. We have assumed that background is white and signature is done with any colored pen. Following are the preprocessing steps that are implemented:-

(a) RGB to grayscale image conversion: - For signature verification, color of signature has no importance. Only the form of the two signatures must be compared. Hence, all the scanned images are converted to grayscale images. This conversion makes future coding easier.
(b) Noise Removal: - This step involves removing noise from images. Noises are the disturbances (like spurious pixels) that can be attached to the image during scanning time. These noises are not a part of signature and should be removed. In this paper, we have assumed that background is white and signature is made from black points. The points which are far from black points are considered noises. The noise removal module returns a noise free image of the same size as the input image. For noise removal, a median filter is used followed by the mean filter.
(c) Image Cropping: - In this step, the Region of Interest (ROI) is determined using auto cropping approach. ROI is the signature object itself from which unwanted region apart from signature is removed. Using auto cropping, we segment the signature smoothly.
(d) Grayscale image to binary image: - Binarizing is a process of converting the image into black and white image, so that further processing could be easy.
(e) Thinning: - Thinning is a process that deletes the dark points and transforms the pattern into a "thin" line drawing called skeleton. Image thinning reduces a connected region in the image to a smaller size and minimum cross-sectional width character [24]. The minimum cross-sectional width could be one character in which case it would be a stick figure representation. The thinned pattern must preserve the basic structure of the original pattern and

the connectedness. Thinning [10] plays an important role in digital image processing and pattern recognition, since the outcome of thinning can largely determine the effectiveness and efficiency of extracting the distinctive features from the patterns.

2.3 Feature Extraction

Decision of authenticated signatures is usually based on local or global features extracted from signature under processing. A feature is called global if it is extracted from the whole signature and is called local if it is extracted from each sample point [12]. The global features can be extracted easily and are robust to noise. But they only deliver limited information for signature verification. On the other hand, local features provide rich descriptions of writing shapes and are powerful for discriminating writers, but the extraction of reliable local features is still a hard problem.

This is the most vital and difficult stage of any off-line signature verification system. The accuracy of the system depends mainly on the effectiveness of the signature features use in the system. Here we extracted Skewness, Kurtosis, Signature area (Signature Occupancy Ratio, height, width, horizontal projection, vertical projection, and width to height Ratio (Aspect Ratio), centre of gravity, density of thinned image, density of smoothed image and normalized area of black pixels. These features are robust and work well with vector quantization technique used in training phase.

2.4 Training Phase and Verification

In order to analyze the discriminatory power of each individual feature, we have used vector quantization technique.

Vector quantization is a statistical clustering technique which allows the modeling of probability density functions by the distribution of prototype vectors. Vector quantization is presented as a process of redundancy removal that makes effective use of four interrelated properties of vector parameters: linear dependency (correlation), nonlinear dependency, shape of the probability density function (pdf), and vector dimensionality itself. It is based on competitive learning technique and hence is related to self-organizing feature map. It follows the principle of dividing a large set of input points or vectors into groups having approximately the same number of points closest to them and those groups are called as "code book" vectors [15]. In Vector quantization, input is given to input layer and this is used for training our network. In training phase, the output units are used for decision and assigning class to which the input vector belongs. After our network is trained, Vector quantization can classify the input vector by assigning it to appropriate class. Hence, it is a form of supervised learning because each output unit has a known class. If the given input matches our target output, it is considered as accepted otherwise rejected.

In vector quantization, the vector X is mapped onto another real-valued, discrete-amplitude, N dimensional vector Y. We say that X is quantized as Y, and Y is the quantized value of X. We write $Y = Q(X)$ where Q is the quantization operator [28].

First, the signature image to be verified is partitioned into a set of blocks represented by feature vectors X. Y={y_i;i=1,2,…,N} is the set of reproduction vectors which is called a codebook. Each K-dimensional input feature vector X=(x_1,x_2,…,x_k) is then compared with all the codewords in the codebook, and quantized to the closest codeword or best match. A reference vector Y is considered to best match a feature vector X=(x_1,x_2,…,x_k) in the sense that appropriately defined distortion measure such as the squared error $||X - Y||^2$ is minimal. This distortion measure is actually Euclidean distance. Each time an input X is presented, we first make an ordering of the elements of a set of distortions E_x = {||X - Y]],i =l;.. ;N} and then determine the adjustment of reference vector Y. For the given training set and initial codebook, the block performs an iterative process.

Verification is a process that decides whether a signature is genuine or forged [21]. The reference feature vector containing the values of 13 global features are given as input to the system and the comparison is based on the assumption that the values of the feature sets extracted from genuine signatures are more stable than the signatures that are forged. That is, the intra-personal variations are smaller than inter-personal variations. The neural network techniques that are used for verification is VQ.

The verification process consists of two parts: - comparison and decision making. The varying values of vigilance parameters are given to the system to gain maximum efficiency so that more and more forgeries can be detected.

3 Results

The efficiency and execution time of VQ is computed on above defined database. The main advantage of using Vector Quantization is that the efficiency does not vary with the varying value of learning rate ρ. For our system ρ=0.9. Following setup is done in Matlab to implement vector quantization.

Table 1. Results of Simulation using Matlab

Simulation Parameters	Results
No. of layers	2
No. of input units	13
No. of output units	2
Learning rate	(0.1-0.9)
Training algorithm	VQ
Initial weights	Randomized
Initial biases	Randomized
No. of signatures used for training	50
No. of tested signatures	200
No. of recognized signatures	154
No. of iterations performed	34

Having worked on Vector quantization, the efficiency achieved is 67% and execution time required is 11 seconds. The results of our simulation for forged and genuine signatures are as shown in the Table 1. The system is robust; it rejected all the random forgeries but 5 false signatures are accepted in case of casual forgeries leading to a FAR of 1%. Out of the 100 genuine signatures that were fed in, 5 were rejected as forgeries. This yielded a FRR of 5%. Also out of 50 skilled forgeries fed into the system, 5 signatures were accepted. This gave us a FAR of 2.5%

Table 2. The value of FAR & FRR at Different Samples

Nature of Signature	Samples	FAR	FRR
Genuine	100	....	5%
Random	50	0%	....
Casual	50	1%	....
Skilled	50	2.5%	....

4 Conclusion

Vector Quantization has been used for signature verification. They have adaptive nature of supervised learning by example in solving problems. This feature makes them computational models very appealing for a wide variety of application domains including pattern recognition. The distortion measure between trained signature set and test signature is done using Euclidean distance. The system has been tested for its accuracy and effectiveness on data from 25 users with 10 specimens of each making up a total of 250 signatures. The classification ratio achieved by the system using Vector quantization is 67%. The FAR and FRR computed are 2.5% and 5% respectively. The system can be further designed using local features classification.

References

1. Jain, C., Singh, P., Chugh, A.: An Offline Signature Verification using Adaptive Resonance Theory 1 (ART1). International Journal of Computer Applications (0975 – 8887) 94(2), 8–11 (2014)
2. Jain, C., Singh, P., Chugh, A.: An optimal approach to offline signature verification using GMM. In: Proc. of the International Conference on Science and Engineering (ICSE 2011), pp. 102–106 (June 2011)
3. Jain, C., Singh, P., Chugh, A.: Performance Considerations in Implementing Offline Signature Verification System. International Journal of Computer Applications (0975 – 8887) 46(11) (May 2012)
4. Ferrer, M.A., Alonso, J.B., Travieso, C.M.: Offline Geometric Parameters for Automatic Signature Verification Using Fixed- Point Arithmetic. IEEE Transactions on Pattern Analysis and Machine Intelligence 27(6), 993–997 (2005)

5. Bowyer, K., Govindaraju, V., Ratha, N.: Introduction to the special issue on recent advances in biometric systems. IEEE Transactions on Systems, Man and Cybernetics 37(5), 1091–1095 (2007)
6. Zhang, D., Campbell, J., Maltoni, D., Bolle, R.: Special issue on biometric systems. IEEE Transactions on Systems, Manand Cybernetics 35(3), 273–275 (2005)
7. Prabhakar, S., Kittler, J., Maltoni, D., O'Gorman, L., Tan, T.: Introduction to the special issue on biometrics: progress and directions. Pattern Analysis and Machine Intelligence 29(4), 513–516 (2007)
8. Vargas, J.F., Ferrer, M.A., Travieso, C.M., Alonso, J.B.: Off-line signature verification based on grey level information using texture features. Pattern Recognition(Elsevier) 44, 375–385 (2011)
9. Hanmandlu, M., Yusof, M.H.M., Madasu, V.K.: Off-line signature verification and forgery detection using fuzzy modeling. Pattern Recognition (Elesevier) 38, 341–356 (2005)
10. Uppalapati, D.: Thesis on Integration of Offline and Online Signature Verification systems. Department of Computer Science and Engineering, I.I.T., Kanpur (July 2007)
11. Aykanat, C.: Offline Signature Recognition and Verification using Neural Network. In: Proceedings of the 19th International Symposium on Computer and Information Sciences, pp. 373–380. Springer, Heidelberg (2004)
12. Kisku, D.R., Gupta, P., Sing, J.K.: Offline Signature Identification by Fusion of Multiple Classifiers using Statistical Learning Theory. Proceeding of International Journal of Security and It's Applications 4(3), 35–44 (2010)
13. Baltzakisa, H., Papamarkos, N.: A new signature verification technique based on a two-stage neural network classifier. Engineering Applications of Artificial Intelligence 14, 95–103 (2001)
14. Özgündüz, E., Şentürkand, T., Elif Karslıgil, M.: Offline Signature Verification And Recognition By Support Vector Machine, Antalya, Turkey, pp. 113–116 (September 2005)
15. Kalera, M.K.: Offline Signature Verification And Identification Using Distance Statistics. International Journal of Pattern Recognition and Artificial Intelligence 18(7), 1339–1360 (2004)
16. Karouni, A., Daya, B., Bahlak, S.: Offline signature recognition using neural networks approach. Procedia Computer Science 3, 155–161 (2011)
17. Ahmad, S.M.S., Shakil, A., Faudzi, M.A., Anwar, R.M., Balbed, M.A.M.: A Hybrid Statistical Modeling, Normalization and Inferencing Techniques of an Off-line Signature Verification System. In: World Congress on Computer Science and Information Engineering (2009)
18. Nguyen, V., Blumenstein, M., Leedham, G.: Global Features for the Off-Line Signature Verification Problem. In: IEEE International Conference on Document Analysis and Recognition (2009)
19. Bansal, A., Garg, D., Gupta, A.: A Pattern Matching Classifier for Offline Signature Verification. In: IEEE Computer Society First International Conference on Emerging Trends in Engineering and Technology (2008)
20. Kiani, V., Pourreza, R., Pourreza, H.R.: Offline Signature Verification Using Local Radon Transform and SVM. International Journal of Image Processing 3(5), 184–194 (2009)
21. Das, M.T., Dulger, L.C.: Signature verification (SV) toolbox: - Application of PSO-NN. Engineering Applications of Artificial Intelligence 22(4), 688–694 (2009)
22. Maya, V., Karki, K., Indira, S., Selvi, S.: Off-Line Signature Recognition and Verification using Neural Network. In: International Conference on Computational Intelligence and Multimedia Applications, pp. 307–312 (December 2007)

23. Shukla, N., Shandilya, M.: Invariant Features Comparison in Hidden Markov Model and SIFT for Offline Handwritten Signature Database. International Journal of Computer Applications 2(7), 975–995 (2010)
24. Wen, J., Fang, B., Tang, Y.Y., Zhang, T.P.: Model-based signature verification with rotation invariant features. Pattern Recognition 42(7), 1458–1466 (2009)
25. Fahmy, M.M.M.: Online handwritten signature verification system based on DWT features extraction and neural network classification. Ain Shams Engineering Journal 1(1), 59–70 (2010)
26. Batista, L., Granger, E., Sabourin, R.: Dynamic selection of generative–discriminative ensembles for off-line signature verification. Pattern Recognition 45(4), 1326–1340 (2012)
27. Radhika, K.R., Venkatesha, M.K., Sekhar, G.N.: Signature authentication based on subpattern analysis. Applied Soft Computing 11(3), 3218–3228 (2011)
28. Samuel, D., Samuel, I.: Novel feature extraction technique for off-line signature verification system. International Journal of Engineering Science and Technology 2(7), 3137–3143 (2010)
29. Deepa, S.N.: Introduction to Neural Network. Tata McGraw Hill (2006)

Fuzzified Inverse S-Transform for Identification of Industrial Nonlinear Loads

Srikanth Pullabhatla[1] and Chiranjib Koley[2]

[1] Dept. of Electrical & Inst.,
EDRC-Kolkata, L&T ECC, W.B.
[2] Dept. of Electrical Engg.,
NIT Durgapur, W.B.
srikanth.srikki@gmail.com, chiranjib_k@yahoo.com

Abstract. In this paper a modified inverse Stockwell transform has been proposed for the identification of industrial nonlinear loads. The proposed method is based on the maximum values of unfiltered inverse Stockwell Transform termed as MUNIST. It is a well known fact that Stockwell transform technique produces time-frequency representation. As the proposed MUNIST technique is obtained from inverse operation of time-frequency data it gives only time resolution. MUNIST technique found to provide unique signatures for accurate identification of the industrial loads. Later the results obtained using the proposed technique has been used as input to the fuzzy decision box. Using the fuzzy logic design automatic identification of different nonlinear loads has been carried out efficiently and accurately. Current measurements of different industrial loads have been used as the input for the proposed MUNIST algorithm. The results obtained using the proposed technique, have been compared with the existing techniques to show its efficacy.

Keywords: Fuzzy, harmonics, nonlinear load, power quality, signal processing.

1 Introduction

Identifying the type of industrial load has become a challenge to the present day engineers as these inject various harmonics into the power system [1-2]. Further, there is a need for various techniques for identification and analysis of harmonic loads problems. A large number of methods for identifying the power quality problems based on signal processing techniques like wavelet transform (WT), S-transform (ST), Gabor-Wigner transform (GWT) etc. have been proposed [3-7]. Using these techniques researchers are successful in identifying the power quality problems viz. sag, swell, transient, harmonics. However, identification of the source of such power quality disturbances has not been dealt extensively. Load identification and its analysis using Short time Fourier transforms (STFT) and wavelet have been reported in the literature [8-11]. However, automatic identification of industrial nonlinear loads using recent advances in signal processing has not been explored and the present work is a step ahead in this direction.

S.C. Satapathy et al. (eds.), *Emerging ICT for Bridging the Future – Volume 1*,
Advances in Intelligent Systems and Computing 337, DOI: 10.1007/978-3-319-13728-5_39

In this paper ST has been used as the base tool. ST is an advanced time-frequency technique which is much more superior to STFT and WT. Its unique and dynamic Gaussian window dilation with change in signal frequency provides a strong platform for time-frequency resolution (TFR). Unlike the previous TFR techniques it overcomes the disadvantages of poor resolution of both time and frequency. It has been widely used by the researchers in various fields. ST technique can be used in both continuous and discrete transformation. Continuous ST (CST) is used for time-frequency visualization whereas discrete ST (DST) is used for automatic identification. DST is only a time representation replica of the input signal. It is generally used as an input to the decision box to automatically identify the type of signal. However, few disadvantages of DST are lack of uniqueness and time complexity. Such pitfalls in DST have lead the authors to propose the modified inverse ST method.

The proposed methodology is not the exact inverse as defined by Stockwell. However, the proposed technique is based on maximum values of unfiltered inverse ST termed as MUNIST and here it is used for identification of non-linear loads [12-14]. MUNIST gives information about magnitude variation with time like DST but maintains uniqueness. Furthermore, the time taken by the algorithm is much less compared to DST technique. The MUNIST amplitude versus time plot is also useful for signatory identification. Whereas, the features extracted from such plot is used for automatic identification and forms the main base of this paper. The proposed method has not yet been utilized as a tool for identification of unknown nonlinear loads. Furthermore, to automatically identify the non-linear loads fuzzy logic decisions have been used [15-17]. The fuzzy logic decision box is fed with the MUNIST magnitude values. It is well known that study on such loads requires highly accurate and practical data which is bit difficult. Hence, a simulation of the equivalent circuits of the industrial loads has been adopted in the present work to get data of the current waveforms. All the test signals are obtained using MATLAB®/SIMULINK 7.6 version [18]. Further sections discuss about the methodology and process of identification of nonlinear loads.

2 Proposed Methodology

The methodology proposed in this paper is based on the inverse properties of Stockwell transform (ST). The theory of ST is well known and hence its background information has not been provided. However, it can be observed that the local time-frequency spectra $S(\tau, f)$ obtained with the generalized S-Transform can easily be back-transformed since the S-Transform windows satisfy the condition shown in eqn. (1) [12, 13]

$$\int_{-\infty}^{+\infty} w(\tau - t, f).dt = 1 \tag{1}$$

This ensures that the time averaging of the S-spectrum $S(\tau, f)$ yields the spectrum $U(f)$ as shown in eqn. (2).

$$\int_{-\infty}^{+\infty} S(\tau,f)d\tau = \int_{-\infty}^{+\infty} u(t)e^{-i2\pi ft}\int_{-\infty}^{+\infty} w(\tau-t,f)d\tau dt = U(f) \quad (2)$$

It means that the S-Transform is exactly invertible with one inverse Fourier Transform and is a simple unfiltered inverse S-Transform (IST) [12, 13]. Such an operation can be handled in MATLAB using the IFFT function [18]. Hence, IST can be represented as shown in eqn. (3).

$$IST = IFT\left[S(\tau,f)\right] \quad (3)$$

where, IFT is the inverse Fourier transform.

The result in eqn. (3) gives a matrix of the same order as that of $S(\tau, f)$. Now, if a plot between the absolute value of such a matrix and number of time samples is taken it will be a cluster of dilated and translated Gaussian windows. Subsequently, a plot with maximum value of each Gaussian window versus time samples can be obtained. This methodology has been used for identification of nonlinear loads in the present paper. As the maximum values of the Gaussian windows obtained from the unfiltered inverse ST are used to determine the load signatures the method is henceforth termed as MUNIST (Maximum values of unfiltered inverse ST) [13-14]. Finally, the proposed MUNIST technique can be represented as eqn. (4).

$$M(t) = \max\left(IFT\left[S(\tau,f)\right]\right) \quad (4)$$

In eqn. (10) the term 'max' refers to maximum value of the inner matrix and $M(t)$ is the MUNIST matrix obtained in time domain. To demonstrate MUNIST a test signal as shown in Fig.1 is considered. The signal in Fig. 1 is a swell event i.e. an increase in the amplitude of a particular signal for a short time interval. The parameters of the test signal i.e. voltage and frequency are 1p.u. and 50Hz respectively. Time interval of the signal is 0.5s with sampling frequency 1 kHz. The duration 't' of swell is considered for t = 0.2s to 0.28s. The signal is now processed using the proposed MUNIST technique and its output is shown in Fig. 1(b). It is seen from Fig. 1(b) that the Gaussian window behaves according to the signal amplitude. The number of Gaussian windows depends on the sampling frequency and number of samples. In the present case this number is 250, which is half the number of samples.

From Fig. 1(b) it is visualized that Gaussian window changes its amplitude depending on the swell signal from interval 0.2 to 0.28s as shown between dotted lines. In Fig. 1(b) crystal clear information is not available as the Gaussian windows present over the complete duration of the plot are more due to more number of samples. Hence, the plot between maximum values of the windows and the time related to MUNIST plot provides a distinguishable signature and is as shown in Fig. 1(c). Though similar signatures are also be obtained using DST. However, DST has some disadvantages which is can be clarified from the results obtained in the subsequent section.

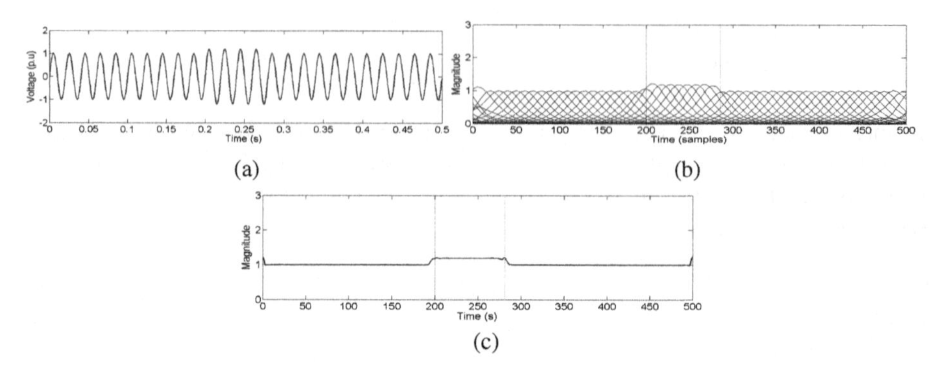

Fig. 1. (a) Test signal considered (b) Magnitude vs. time plot obtained from the inverse of ST (c) Plot of maximum value vs. time of the Gaussian windows shown in (b)

3 Case Studies and Discussion

Different types of industrial loads that pollute the ideal power signal with harmonic content have been simulated. However, only one case study has been considered in this paper. The current waveforms obtained from the simulation have been considered for identification of loads. The CST, DST and MUNIST techniques are used to for identification of nonlinear loads. All the signals are of 1p.u magnitude and have a total time period t = 1s. However, the analysis has been done on the waveform within 0.9 to 1s. The data is sampled at 1kHz to obtain 1000 samples for input.

In the results obtained Fig 2(a) shows the current waveform obtained. Fig. 2(b) is the contour plot of the test signal plotted in Fig. 2(a) using CST and they are named as CONPLOT. By searching through the rows and columns of the matrix the maximum value of the matrix is obtained and that maximum value will be present in r^{th} row and c^{th} column. The values of the rth row and c^{th} column thus obtained are used for analysis using Discrete ST technique. Fig. 2(c) is the plot of DST obtained by plotting all the values of cth column versus frequency and is named as FPLOT and 2(d) represents the plot of ST amplitude variation of the rth row versus time obtained from the ST technique and is labeled as TPLOT [5-6]. Fig. 2(e) represents the plots obtained from the proposed methodology MUNIST discussed in section 2 and is termed as MUPLOT. Next subsection discuss about the case study considered for identification.

3.1 AC Voltage Regulator

The main application of AC regulators is to control polyphase induction motor, tap changing, variable heat equipments etc. Using a regulator the supply voltage is served as a controlled input to the load and thus the voltage is non-sinusoidal. The nature of the current is also non-sinusoidal and its phase depends on the type of load used. Consequently, it affects and pollutes the supply current. The investigation and identification of such loads is of vast importance. Based on the theory of operation of a bidirectional ac regulator a simulation model has been developed. Fig. 2 shows the waveforms and results obtained from the analysis of the current of a full wave ac voltage regulator of 50Hz frequency. Fig. 2(a) is the current waveform obtained with time interval t=0.9 to 1s. It is seen that the signal is not sinusoidal and resembles a cut version of a sinusoidal waveform.

The CONPLOT obtained from the CST is shown in Fig. 2(b). It is seen that there are contours in top and bottom of the plot. The bottom most strip at a value of 0.05 corresponds to the fundamental frequency of the signal in the ST domain. Again the contours are seen in the frequency range of 0.2 to 0.4 in the plot. The plot is not exactly clear about the frequencies present. However, the plot is used to show the range of frequencies at every time instant and it serves a signature for each type of waveform.

FPLOT which is the plot between the column with maximum value of ST magnitude and frequency is shown in Fig. 2(c). This plot is used to identify the dominating frequencies present in a particular signal. The frequency corresponding to the peak values i.e. peak values sorted in decreasing order are of importance. It is seen here that there are nearly three peaks in Fig. 2(c). The fundamental frequency component of 50 Hz has a value of 0.05 in the ST domain frequency axis. The consecutive peaks have a value of 0.025, and 0.035 which indicates the presence of 5th and 7th harmonics respectively [1]. Compared to the fundamental these are of lower magnitude which shows the amount of harmonic present in the signal.

DST plot between the row with maximum value of ST magnitude and time samples (named as TPLOT) is shown in Fig. 2(d). The plot is just to show how the maximum value of DST changes for a signal with typical variations [4]. Here the DST plot has gradually reached a value of 0.45 from an initial value of 0.4. A dip is seen at the first and last 150 samples of the plot. This plot is useful to know the rms behaviour of the signal. However, visually it is not of major importance. It would be useful if the magnitude variation is dynamic without sacrificing the uniqueness. Such problems are the main roots for developing the MUNIST technique.

The proposed MUNIST technique is used to obtain a signature in time domain and is as shown in Fig. 2(e). The signature may not be clear due to more number of cycles considered for analysis and high compression ratio of the image. However, from a mere observation it can be recognized that the MUNIST plot is giving a signature with an oscillatory magnitude at regular intervals. As the input waveform is of 50Hz for 0.1s there will be 5 cycles with a time period of 0.02s per cycle. In Fig. 2(e) there are two oscillations per cycle and in total 100 oscillations for the input waveform with t=1s. The obtained MUPLOT shown in Fig. 2(e) is having real time behaviour and also serves as a signatory identification. Besides these features the main advantage lies in computing and plotting time of the proposed MUNIST technique.

Similar analysis has been done on various loads and the results obtained using MUNIST technique are highly discriminative for each type of load. Due to space limitation all the results have not been displayed. However, the comparison for generation of MUNIST plot for various loads has been provided in Table 1. Though the signatures obtained are unique and the highest amplitude variation is dynamic, visual identification is less accurate. This is due to repetitive checking of the

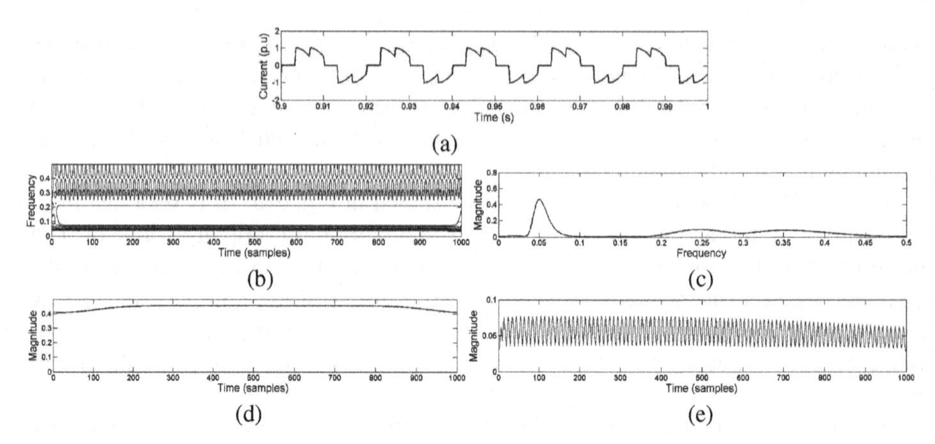

Fig. 2. Analysis of current from an ac regulator. (a) & (b) current waveforms obtained from the AC regulator, (c) CONPLOT, (d) FPLOT, (e) TPLOT, (f) MUPLOT of the current waveform shown in (a).

Table 1. Comparison of Computing Times of MUNIST and DST Algorithm

Load	Average time taken for 25 runs of MUNIST algorithm (sec)	Average time taken for 25 runs of DST algorithm (sec)
PWM Inverter fed load (PWM)	0.45	8.3
Cycloconverter fed load with an output frequency of f/5 (CYCLOCON)	0.43	8.9
ASD load (ASD)	0.42	8.4
Traction load with two stage converter control (DC TRAC)	0.41	7.86
AC Regulator fed load (AC REG)	0.4	8

signatures which may lead to confusion. The proposed technique is more effective with an automatic machine which can identify the non-linear load by its own. Hence, a fuzzy logic based expert system is designed which gives the fuzzy decisions about the type of load by taking few parameters extracted from the proposed MUNIST technique. The detail about parameter extraction and fuzzy logic design is provided in further sections.

4 Parameter Extraction

In order to identify various nonlinear loads parameters are to be extracted. These parameters may or may not be related with the current waveforms recorded at the busbar. Here, in this case the extracted parameters are derived from the MUNIST values obtained from various case studies. It has been observed from all case studies that there is variation in the behavior, maximum value and the mean values in MUPLOT. It is useful if the designed parameters are based on these variations. Hence, the parameters designed based on the MUNIST magnitude values and are presented in Table 2.

Table 2. Parameters extracted from the MUNIST algorithm results

Parameter	Description of the parameter
$MUNIST_{\text{MAG-SINE}}$	Array of values obtained from the algorithm by taking the ideal sine wave of 1p.u, 50Hz as input. Here its value is 0.00868.
$MUNIST_{\text{MAG-LOAD}}$	Array of values obtained from the MUNIST algorithm by taking the non-linear load waveform of 1p.u amplitude as input.
MM	600*(Avg. of [$MUNIST_{MAG-SINE}$ - $MUNIST_{MAG-LOAD}$]

In calculating MM the value 600 is considered as the voltage rating of all the nonlinear loads is 600V. Based on the above defined factors the identification of non-linear loads is carried out and the fuzzy decision box is designed. The typical values of the above mentioned factors for each type of non-linear loads are tabulated and presented in Table 3. It is clearly visible that the values obtained from the defined or the extracted parameters are not overlapped for any of the considered case studies. The case studies other than discussed in section 3 are also included to check the robustness of the proposed technique. Now based on these values the rule sets and fuzzy variables have been designed. The design and results obtained from the fuzzy decisions is discussed in the next section.

Table 3. Parameters extracted from the MUNIST algorithm results

Load type	MM
PWM	5 to 0
CYCLOCON	0 and -3
ASD	-3 to -10
DC TRAC	-10 to -20
AC REG	< -20

5 Fuzzy Logic Design and Results

The broad structure of the FIS or the fuzzy decision box design is shown in Fig. 3 and the complete rule base generated based on the requirement of identification of type of load is shown in Table 4 [17]. Now, based on the FIS structure shown in Fig. 3 and the rules in Table 4 the results have been obtained and tabulated in Table 5. The values of MM are obtained by taking the average of 25 simulations of each case

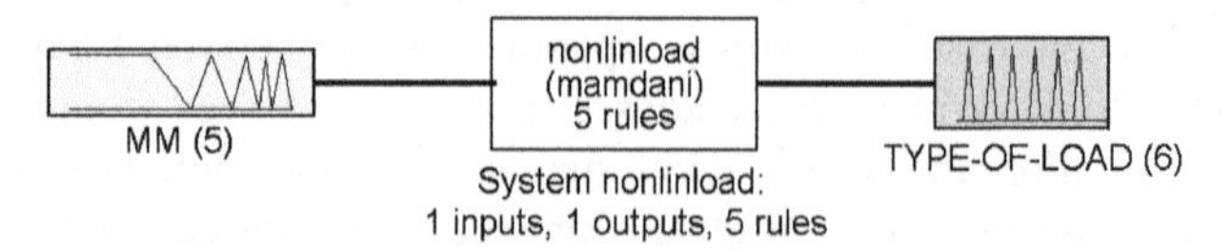

Fig. 3. A high level diagram of the FIS named "nonlinload". Inputs and their membership functions shown in the left side of the FIS structural characteristics, while outputs and their membership functions are shown on the right side.

Table 4. Set of rules defined in the FIS rule base for identifying the type of load

S. No.	Rules
1	If (MM is mf5) then (TYPE-OF-LOAD is PWM) (1)
2	If (MM is mf4) then (TYPE-OF-LOAD is CYCLOCON) (1)
3	If (MM is mf3) then (TYPE-OF-LOAD is ASD) (1)
4	If (MM is mf2) then (TYPE-OF-LOAD is DCTRAC) (1)
5	If (MM is mf1) then (TYPE-OF-LOAD is ACREG) (1)

Table 5. Extracted parameters values obtained from simulations of various case studies

Load type	MM (Average of 25 simulations)	Count of correct Fuzzy Decisions for 25 simulations	Accuracy (%)
PWM	1.8	23	92
CYCLOCON	-2.7	25	100
ASD	-6.84	24	96
DC TRAC	-10.38	24	96
AC REG	-27.54	25	100

study. Various samples of 1sec interval (i.e. for t=0 to t=1s OR t=0.2s to t=1.2s etc.) are fed to the get the values of MM. As the signal remains same for every 1s interval all the MM values obtained are nearly same. Such simulation is carried out to test the robustness of the technique for repeated applications. All the cases mentioned in Table 3 have been simulated and are identified. It is seen that the proposed MUNIST methodology with the addition of the fuzzy decision box is 96.8% accurate in identifying the type of nonlinear load. Thus the blend of MUNIST and FL serves as a robust platform for identifying various types of nonlinear loads.

6 Conclusions

A complete analysis of different types of harmonic loads using the ST has been carried out. The ST technique gives a complete idea about the time and frequency properties of the loads considered. It is seen that the behaviour of the typical loads considered is distinctive. The nature of harmonics present in current is completely different. The technique is effectively used to provide unique signatures for different harmonic loads. However, the DST is lacking uniqueness and the magnitude variations cannot be used for load identification. In order to provide better uniqueness a new method has also been proposed. The method is based on inverse of ST. The proposed method is termed as MUNIST and is only in time domain. The amplitude variations are very clear compared to the DST. In order to reduce the misinterpretation and increase the accuracy of the proposed technique a fuzzy decision box is incorporated. The implementation provided a huge for automatic identification of the non-linear loads. It is found that the fuzzy decisions are 100% accurate in identifying the unknown nonlinear loads. It is envisaged that the exact identification of nonlinear loads shall facilitate in taking up proper remedial action for the harmonic mitigation and will prove to be an effective tool for non-linear loads identification.

References

[1] Arrilaga, J., Watson, N.R.: Power System Harmonics, 2nd edn. John Wiley and Sons, England (2003)

[2] Singh, M.D., Khanchandani, K.B.: Power Electronics, 2nd edn. Tata McGraw-Hill Publishing Company Limited, INDIA (2009)

[3] Santoso, S., Powers, E.J., Grady, W.M., Hofmann, P.: Power quality assessment via wavelet transform analysis. IEEE Trans. on Power Delivery 18(2), 924–930 (1996)

[4] Stockwell, R.G., Mansinha, L., Lowe, R.P.: Localization of the complex spectrum: The S transform. IEEE Trans. Signal Processing 44(4), 998–1001 (1996)

[5] Dash, P.K., Panigrahi, B.K., Panda, G.: Power quality analysis using S–Transform. IEEE Trans. on Power Delivery 18(2), 406–411 (2003)

[6] Lee, I.W.C., Dash, P.K.: S-Transform-based intelligent system for classification of power quality disturbance signals. IEEE Trans. on Ind. Elect. 50(4), 800–805 (2003)

[7] Cho, S.H., Jang, G., Kwon, S.H.: Time-Frequency analysis of power-quality disturbances via the gabor–wigner transform. IEEE Trans. on Power Delivery 25(1), 494–499 (2010)

[8] Sharma, V.K., Doja, M.N., Ibraheem, I., Khan, M.A.: Power quality assessment and harmonic, comparison of typical non-linear electronic loads. Int. Conf. on Industrial Technology, 729–734 (2000)

[9] Aintablian, H.O., Hill, H.W.: Harmonic currents generated by personal computers and their effects on the distribution system neutral current, pp. 1483–1489. IEEE (1993)

[10] Larsson, E.O.A., Lundmark, C.M., Bollen, M.H.J.: Measurement of current taken by fluorescent lights in the frequency range 2 - 150 kHz, pp. 1–6. IEEE (2006)

[11] Rönnberg, S.K., Wahlberg, M., Bollen, M.H.J., Lundmark, M.: Equipment currents in the frequency range 9-95 kHz, measured in a realistic environment, pp. 1–8. IEEE (2008)

[12] Simon, C., Ventosa, S., Schimmel, M., Heldring, A., Dañobeitia, J.J., Gallart, J., Mànuel, A.: The S-Transform and its inverses: side effects of discretizing and filtering. IEEE Trans. Signal Processing 55(10), 4928–4937 (2007)

[13] Srikanth, P., Chandel, A.K., Naik, K.A.: Identification of Power Quality Events Using Inverse Properties of S Transform. IEEE PES Power System Conference and Exposition, 1–7 (2011)

[14] Srikanth, P., Chandel, A.K., Naik, K.: Inverse S-Transform based decision tree for power system faults identification. Telkomnika Journal of Electrical Engineering 9(1), 99–106 (2011)
[15] Banshwar, A., Chandel, A.K.: Identification of Harmonic Sources Using Fuzzy Logic. In: IEEE PEDES, pp. 1–6 (2010)
[16] Ibrahim, W.R.A., Morcos, M.M.: Preliminary application of an adaptive fuzzy system for power quality diagnostics. Proc. of IEEE Power Eng. Rev. 20, 55–58 (2000)
[17] Fuzzy logic toolbox user"s guide by The Mathworks, Inc.
[18] MATLAB/SIMULINK 7.6.

Bandwidth Allocation Scheme in Wimax Using Fuzzy Logic

Akashdeep

UIET, Panjab University Chandigarh
akashdeep@pu.ac.in

Abstract. WiMAX referes to IEEE 802.16 standard for metropolitan area networks that has inherent support for variety of real and non real applications. The quality of service mechanism in WiMAX has been left as open issues for vendors. This paper proposes a bandwidth allocation scheme for WIMAX networks using fuzzy logic concepts. The system works as adaptive technology in granting bandwidth to all traffic classes and helps to satisfy quality of service requirements for all service classes. The results demonstrate that proposed system was able to fulfill requirements of all classes and avoid starvation of low priority classes.

Keywords: Fuzzy Logic, WiMAX, Quality of service, bandwidth allocation.

1 Introduction

WiMAX stands for World Wide Interoperability for Metropolitan Area Networks which is popularized and licensed by WiMAX Forum in compliance with IEEE 802.16 standard[1]. Traffic in WiMAX network is categorized into five different service classes namely UGS(unsolicited grant service), ertPS(extended real time polling service), rtPS(real time polling service), nrtPS(non real time polling service) and BE(best effoert). IEEE 802.16 standard specifies only priority to these classes and does not specify any fixed mechanism for allocation of resources to these. Equipment manufacturers are free to design and implement their own algorithms.[2] Increasing number of multimedia applications in today's environment makes process of resource allocation a very complex and tedious process as real time applications are always hungry for more and more resources. This puts lots of pressure on scheduler serving these service classes and maintaining relatively good quality of service levels gets more difficult with rise in number of packets in network. This paper proposes a bandwidth allocation scheme using fuzzy logic concepts. Uncertainty principles of fuzzy logic have been utilized to work according to the changes in traffic patterns of incoming traffic. The scheme adapts itself to these changes so that appreciable performance level can be maintained for all service classes.

Use of fuzzy logic in bandwidth allocation process has got popularized very shortly as number of papers in this direction is still limited. Few of these studies are available at [4]-[10]. Fuzzy logic has been employed by Tarek Bchini et. al. [3] and Jaraiz Simon et. al[4] in handover algorithms. Use of fuzzy logic for implementing inter-class scheduler for 802.16 networks had been done by Yaseer Sadri et al.[6]. Authors

S.C. Satapathy et al. (eds.), *Emerging ICT for Bridging the Future – Volume 1*,
Advances in Intelligent Systems and Computing 337, DOI: 10.1007/978-3-319-13728-5_40

had defined fuzzy term sets according to two variables dq_{rt} which means latency for real time applications and tq_{nrt} meaning throughput for non real time applications. Shuaibu et al.[5] has developed intelligent call admission control (CAC) in admitting traffics into WiMAX. Mohammed Alsahag et al.[7] had utilized uncertainty principles of fuzzy logic to modify deficit round robin algorithm to work dynamically on the basis of approaching deadlines. The Fuzzy based scheduler dynamically updates bandwidth requirement by different service classes according to their priorities, latency and throughput by adjusting the weights of respective flows. Similar studies were also given by Hedayati[8], Seo[9] and Akashdeep[10] et al.

2 Proposed Model

The bandwidth allocation process in WIMAX is shown in Figure 1. Different SS connected to BS request resources from BS and these requests are classified into different queues by classifier of IEEE 8021.6. The scheduler at BS listens to these request and allocates resources to these request made by SS on per connection ID basis by taking into consideration available resources and request made by that particular connection. Presently IEEE 802.16 does not specify any algorithms for resource allocation and therefore a number of schedulers are available in literature. Real time classes have high priority as a result of which low priority non real time classes tend to suffer increased delays in WFQ algorithm. This can be improved by devising a suitable policy that could adapt itself to changing requirements of incoming traffic. The proposed system is motivated by theories of fuzzy logic where fuzzy logic can work to serves queues belonging to different scheduling services. The designed system works as component of base station and works on two input and one output variables. The input variables are taken as follows:- Latency of real time

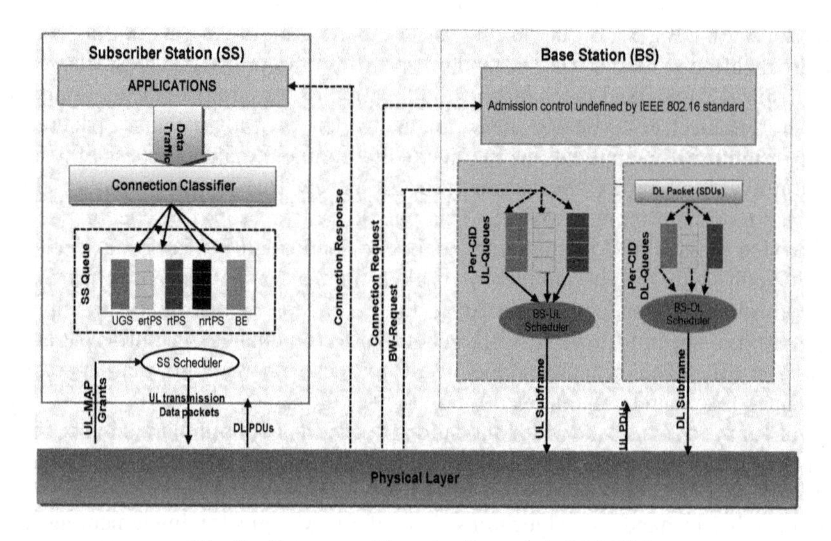

Fig. 1. - Resource Allocation Process in WIMAX

applications and throughput for non real traffic are taken as input variable. The output of fuzzy system is taken as weight of queues serving real time traffic. Membership functions for these variables have been defined utilizing knowledge of domain expert as shown in figures 2-4.

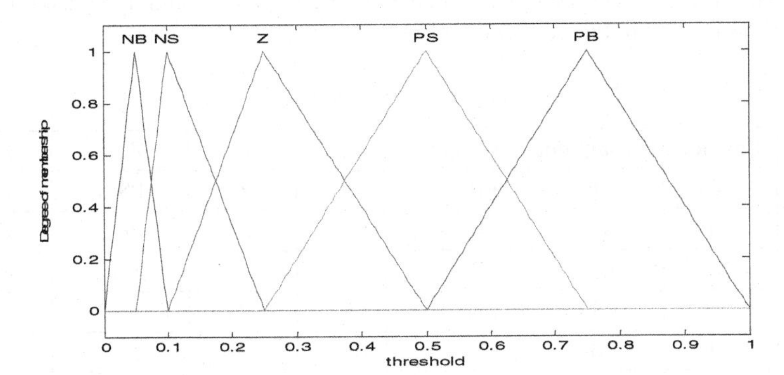

Fig. 2. Fuzzy membership diagram for input variable throughput

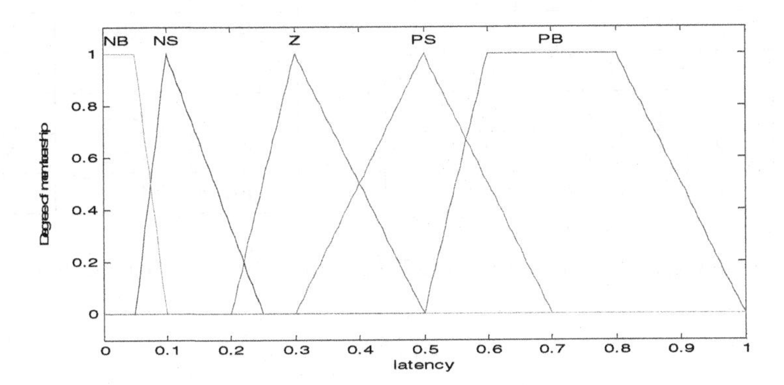

Fig. 3. Fuzzy membership diagram for input variable latency

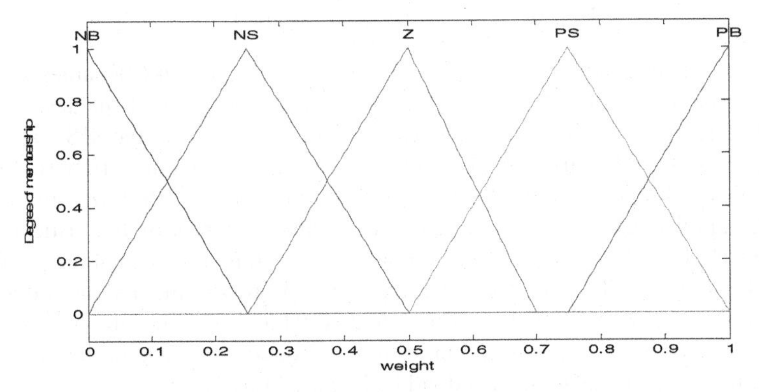

Fig. 4. Fuzzy membership diagram for output variable weight

Five different linguistic levels are defined for input and output variables .The membership function are defined as NB(Negative Big), NS(negative small), Z(Zero), PS (Positive small) and PB(positive big). The nature of variables forces their dynamics to range between 0 and 1. The rule base consists of 25 rules as depicted in table I. The rule base has been defined considering the nature and dynamism of input traffic and is considered to be sufficiently large.

Table 1. Fuzzy Rule-base

S. No	Latency	Throughput	Weight
1.	NB	NB	PB
2.	NB	NS	PB
3.	NB	Z	PB
4.	NB	PS	PB
5.	NB	PB	PB
6.	NS	NB	PS
7.	NS	NS	PS
8.	NS	Z	PS
9.	NS	PS	PS
10.	NS	PB	PS
11.	Z	NB	Z
12.	Z	NS	Z
13.	Z	Z	Z
14.	Z	PS	Z
15.	Z	PB	Z
16.	PS	NB	NS
17.	PS	NS	NS
18.	PS	Z	NS
19.	PS	PS	NS
20.	PS	PB	NS
21.	PB	NB	NB
22.	PB	NS	NB
23.	PB	Z	NB
24.	PB	PS	NB
25.	PB	PB	NB

The initial weight for any flow (i) is calculated from the following equation

$$w_i = \frac{R_{min(i)}}{\sum_{i=0}^{n} R_{\min(i)}} \tag{1}$$

where $R_{min(i)}$ is the minimum reserved rate for flow(i).

$$\sum_{i=0}^{n} w_i = 1 \qquad 0.001 \le w_i \le 1 \tag{2}$$

All flows shall satisfy the constraint of equation 2. Equation (2)enables system to allocate minimum value of bandwidth to all flows as weights of queues cannot be zero. Whenever new bandwidth request is received by BS, the BS calls fuzzy inference system. The fuzzy system reads values of two input variables, fuzzifies these values and inputs it to the fuzzy scheduler component at BS. Fuzzy reasoning is thereafter applied using fuzzy rulebase and a value in terms of linguistic levels is outputted. At last, de-fuzzification of output value is done to get final crisp value for weight. De-fuzzification is performed using centre of gravity method and inference is applied using Mamdami's method. The outputted value is taken as the weight for real time traffic. The bandwidth allocation to different queues is made on basis of weight assigned to that queue using the bandwidth allocation formula

$$\text{Allocated Bandwidth} = R_{max} \times \frac{w_i}{\sum_{i=0}^{n} w_i} \tag{3}$$

3 Results and Discussion

The proposed solution has been tested on Qualnet Simulator. Experiments are conducted to check whether proposed system was able to provide desired quality of service levels to traffic classes. Performance is studied on the basis of parameters like delay, throughput and jitter. Simulation is aimed at making sure that proposed scheme is able to provide a relative good QoS levels to all traffic classes. Two different experiments are conducted to validate scheduler performance.

First experiment was conducted by increasing the number of UGS connections in scenario while keeping number of rtPS, ertPS and nrtPS connections fixed. Scenario consists of one BS and 110 SSs. For experimentation 10 ertPS, 10 rtPS connections, 25 nrtPS connections and 25 BE connections were established, and number of active UGS connections were varied from varies from 10 to 40. Figure 5 plots average delay incurred by real time classes consisting of nrtPS, BE with increase in number of UGS connections. The variation in delay of UGS connections is minimal as scheduler makes fixed allocations to it. Delay for ertPS and rtPS classes shows an increase when number of UGS connections approaches above 30 which is understandable as more resources are consumed to satisfy high priority traffic class UGS.

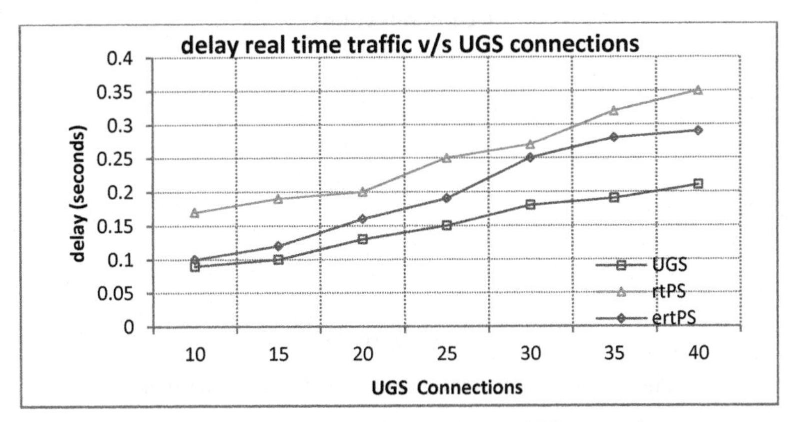

Fig. 5. Delay of UGS, ertPS and rtPS wrt UGS connections

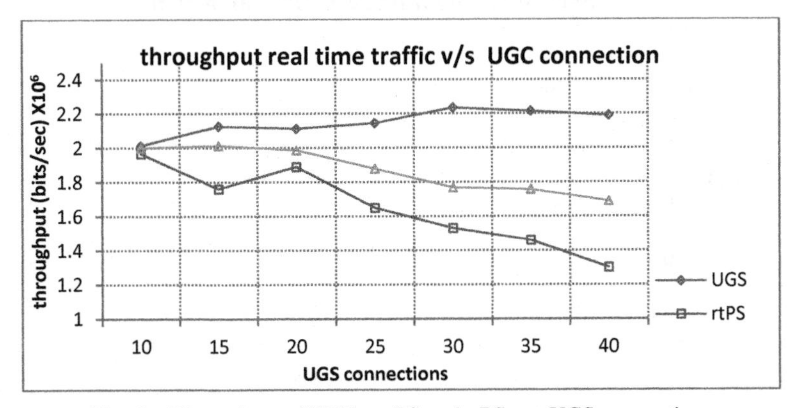

Fig. 6. - Throughput of UGS, ertPS and rtPS wrt UGS connections

Figure 6 and 7 show plot of throughput for real and non real time classes with variations in UGS connections. The graph shows that throughput obtained for UGS is maximum. This is expected because of periodic grants made to this class by scheduler. Throughput for ertPS and rtPS was quite good until number of UGS connection increased beyond a limit after which throughput for nrtPS started to decrease but throughput for ertPS still remained competitive enough. Fig. 7 shows throughput of nrtPS and of BE connections. nrtPS throughput shows small oscillations and scheduler provides minimum bandwidth to achieve minimum threshold levels for nrtPS class. The BE service enjoys same throughput as enjoyed by nrtPS service for small number of active connections, since scheduler could have allocated the residual bandwidth to this service. When number of connections is high, the throughput for BE decreases as this is least prioritized class out of all classes. Throughput of both non real time classes was good in the start as there were enough resources available but increased amount of UGS led to degradation of performance levels of these classes.

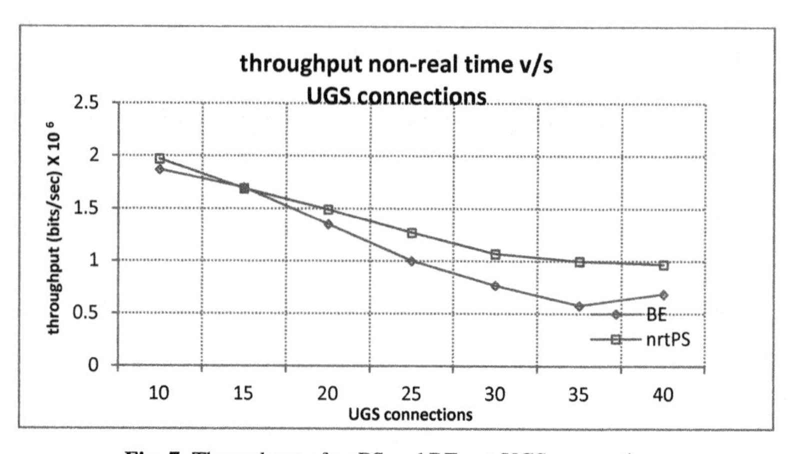

Fig. 7. Throughput of nrtPS and BE wrt UGS connections

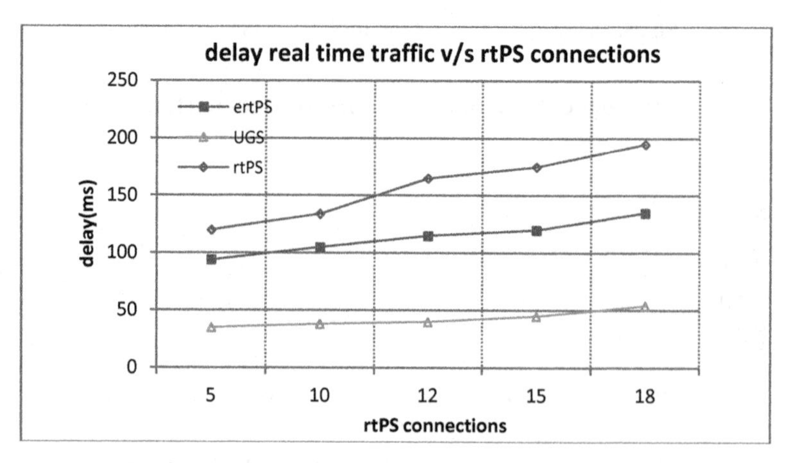

Fig. 8. Delay of UGS, ertPS and rtPS wrt rtPS connections

Another experiment was designed to study effect of varying traffic of another real time class namely rtPS on performance of other traffic classes. Effect of rtPS on network performance is studied separately from UGS as rtPS service normally consists of video traffic which has very bursty nature and can put scheduler under tremendous load. The designed scenario consists of one BS and 103 SS having 10 UGS and ertPS connections. Number of nrtPS and BE connections were 25 and 40 respectively while rtPS connections were varied from 5 to 18. Performance is measured by measuring delay across real time classes and throughput of non real time classes as given in figure 8 and 9 respectively. The figures show that performance of both real and non real time classes was not much affected by increase in rtPS traffic.

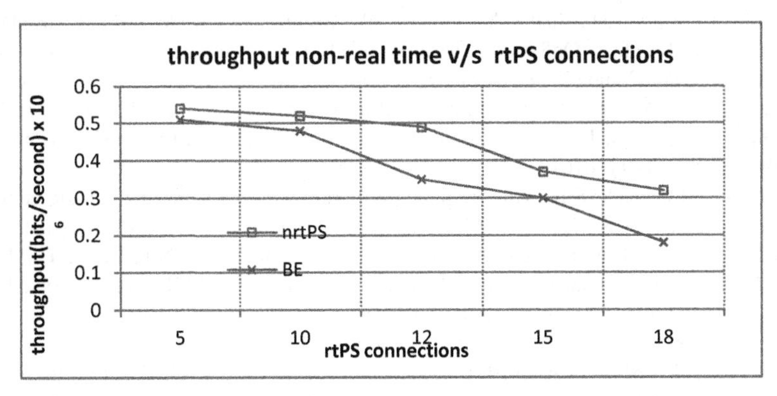

Fig. 9. Throughput of non real time wrt rtPS connections

This is because fuzzy logic adapts itself to traffic variations. Delays of UGS and ertPS are almost bounded. Throughput of both non real time classes was good in the start as there were enough resources available but increased amount of rtPS packet led to degradation of performance levels of these classes. BE class performance degraded quickly as it has no qos requirements.

4 Conclusion and Future Scope

The above study proposed a bandwidth allocation scheme utilizing fuzzy logic for allocation of resources in WiMAX networks. The proposed scheme soaks the transitions in incoming traffic and mould the allocation policy accordingly. The scheme is implemented as component of base station and is an adaptive approach for scheduling resources. Results indicate that system was able to provide desired quality of service levels to all traffic classes. Increase in relative increase of traffic of UGS was also not able to deteriorate system performance. The system can further be improved by incorporating more input parameters to fuzzy system and widening prospect of decision variables. One such parameter can be the length of queues of various traffic classes where by the scheduler can base its decision on the instantaneous amount of traffic in queues of SS.

References

1. IEEE, Draft. IEEE standard for local and metropolitan area networks. 727 Corrigendum to IEEE standard for local and metropolitan area networks—Part 16: 728 Air interface for fixed broadband wireless access systems (Corigendum to IEEE Std 729 802.16- 2004). IEEE Std P80216/Cor1/D2. 730 (2005)
2. IEEE, Draft. IEEE standard for local and metropolitan area networks. 731 Corrigendum to IEEE standard for local and metropolitan area networks – 732 Advanced air interface. IEEE P80216m/D10, 1–1132 (2010)
3. Bchini, T., Tabbane, N., Tabbane, S., Chaput, E., Beylot, A.: Fuzzy logic based layers 2 and 3 handovers in IEEE 802.16e network. J. Com. Comm. 33, 2224–2245 (2010)
4. Simon, J., Maria, D., Juan, A., Gomez, P., Miguel, A., Rodriguez, A.: Embedded intelligence for fast QoS-based vertical handoff in heterogeneous wireless access networks. J. Per Comp (2014), http://dx.doi.org/10.1016/j.pmcj
5. Shuaibu, D.S., Yusof, S.K., Fiscal, N., Ariffin, S.H.S., Rashid, R.A., Latiff, N.M.: Fuzzy Logic Partition-Based Call Admission Control for Mobile WiMAX. ISRN Comm and Netw. 171760, 1–9 (2010)
6. Sadri, Y., Mohamadi, S.K.: An intelligent scheduling system using fuzzy logic controller for management of services in WiMAX networks. J. Sup Com. 64, 849–861 (2013)
7. Mohammed, A., Borhanuddin, A., Noordin, A.M., Mohamad, N.K., Fair, H.: Uplink bandwidth allocation and latency guarantee for mobile WiMAX using fuzzy adaptive deficit round robin. J. Net. Com. Appl (2013), http://dx.doi.org/10.1016/j.jnca.2013.04.004i
8. Hedayati, F.K., Masoumzadeh, S.S., Khorsandi, S.: SAFS: A self adaptive fuzzy based scheduler for real time services in WiMAX system. In: 2012 9th International Conference on Communications (COMM), June 21-23, pp. 247–250 (2012)
9. Seo, S.S., Kang, J.M., Agoulmine, N., Strassner, J., Hong, J.W.-K.: FAST: A fuzzy-based adaptive scheduling technique for IEEE 802.16 networks. In: 2011 IFIP/IEEE International Symposium on Integrated Network anagement (IM), May 23-27, pp. 201–208 (2011)
10. Deep, A., Kahlon, K.S.: An Adaptive Weight Calculation based Bandwidth Allocation Scheme for IEEE 802.16 Networks. J. Emer. Tech. Web Inte. 6(1), 142–147 (2014)

GMM Based Indexing and Retrieval of Music Using MFCC and MPEG-7 Features

R. Thiruvengatanadhan, P. Dhanalakshmi, and S. Palanivel

Dept. of Computer Science and Engineering, Annamalai University, Chidambaram, Tamil Nadu, India
{thiruvengatanadhan01,abidhana01}@gmail.com, spal_yughu@yahoo.com

Abstract. Audio which includes voice, music, and various kinds of environmental sounds, is an important type of media, and also a significant part of video. The digital music databases in place these days, people begin to realize the importance of effectively managing music databases relying on music content analysis. The goal of music indexing and retrieval system is to provide the user with capabilities to index and retrieve the music data in an efficient manner. For efficient music retrieval, some sort of music similarity measure is desirable. In this paper, we propose a method for indexing and retrieval of the classified music using Mel-Frequency Cepstral Coefficients (MFCC) and MPEG-7 features. Music clip extraction, feature extraction, creation of an index and retrieval of the query clip are the major issues in automatic audio indexing and retrieval. Indexing is done for all the music audio clips using Gaussian mixture model (GMM) models, based on the features extracted. For retrieval, the probability that the query feature vector belongs to each of the Gaussian is computed. The average Probability density function is computed for each of the Gaussians and the retrieval is based on the highest probability.

Keywords: Acoustic Feature Extraction, Indexing and Retrieval, Mel-Frequency Cepstral Coefficients (MFCCs), MPEG-7 and Gaussian mixture model (GMM).

1 Introduction

Audio content analysis and description have been a very active research and development topic. Retrieval of multimedia data is different from retrieval of structured data. A very popular means of music retrieval is to annotate the media with text, and use text-based database management systems to perform the retrieval [18]. However, text-based annotation has significant drawbacks when confronted with large volumes of music data. An annotation can then become significantly labor intensive. Content-based retrieval systems describe audio data by their content rather than text [8].

Music information retrieval (MIR) has a rapid growth in the field of audio processing. Search engines like Google facilitate in audio retrieval for music fans and the reputation of music has also grown exponentially. The World Wide Web massively contributes to

S.C. Satapathy et al. (eds.), *Emerging ICT for Bridging the Future – Volume 1*,
Advances in Intelligent Systems and Computing 337, DOI: 10.1007/978-3-319-13728-5_41

this [14]. A retrieval strategy based on audio content that follow the query-by-example paradigm, gives an audio recording as input, and automatically retrieves documents from a given music collection containing the music as a part or similar to it. [9]. Figure 1 shows the proposed method for Music indexing and retrieval system.

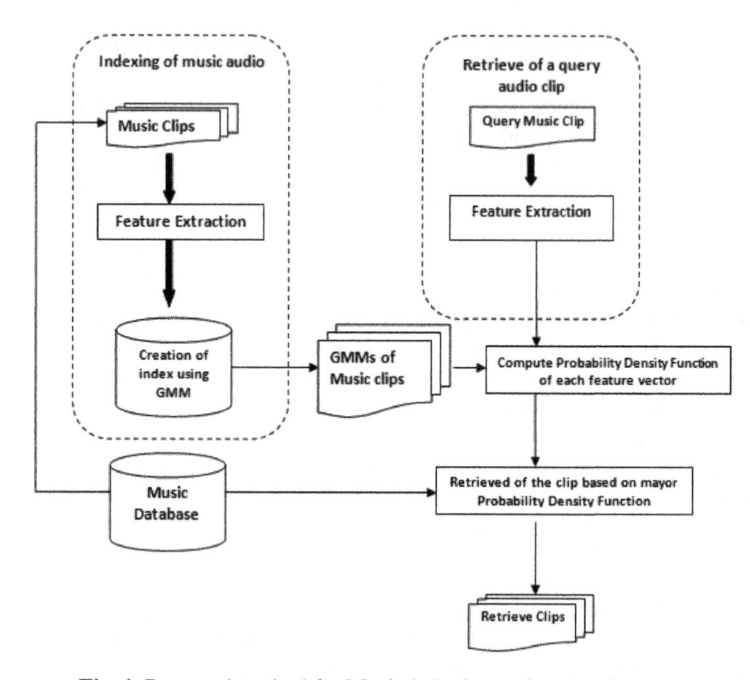

Fig. 1. Proposed method for Music indexing and retrieval system

2 Acoustic Feature Extraction

Music follows certain rules which are highly structured and it provides strong regularities, whereas music signals are random and chaotic [1]. In this work, acoustic features, namely MFCC and MPEG-7 features that robustly capture and reveal musically meaningful information concealed in the audio waveform are extracted.

2.1 Timbre

The timbre differentiates the sound of the same note played by two different instruments. The timbre is defined as everything about a sound and it is neither loud nor pitch [2]. MFCC has been extracted as timbre feature.

2.1.1 Mel-Frequency Cepstral Coefficients

Mel-frequency cepstral coefficients (MFCCs) are extensively used in music information retrieval algorithms. The Mel-frequency cepstrum has proven to be highly effective in recognizing structure of music signals and in modeling the subjective pitch and frequency content of audio signals [3]. MFCCs are based on the known variation of the human ears critical bandwidths with frequency, filters spaced linearly at low frequencies and logarithmically at high frequencies to capture the phonetically important characteristics of speech and audio [4]. The audio signals were segmented and windowed into short frames. Magnitude spectrum was computed for each of these frames using fast Fourier transform (FFT) and converted into a set of Mel scale filter bank outputs. Logarithm was applied to the filter bank outputs followed by discrete cosine trans-formation to obtain the MFCCs [5].

2.2 MPEG-7 Audio Descriptors

The Moving Picture Experts Group (MPEG) is a working group of ISO/IEC in charge of the development of standards for digitally coded representation of audio and video [9]. The MPEG-7 low-level audio descriptors are of general importance in describing audio. The MPEG-7 basic spectral audio descriptors are:

2.2.1.1 Audio Spectrum Centroid

The audio spectrum centroid (ASC) gives the centre of gravity of a log-frequency power spectrum. It describes the center-of-gravity of a log-frequency power spectrum [10]. The spectral centroid is calculated as the weighted mean of the frequencies present in the signal, which is determined using a Fourier transform, with their magnitudes as the weights

$$centroid = \frac{\sum_{n=0}^{N-1} f(n)x(n)}{\sum_{n=0}^{N-1} x(n)} \tag{1}$$

Where *x(n)* represents the weighted frequency value, or magnitude, of bin number n, and f(n) represents the center frequency of that bin.

2.2.1.2 Audio Spectrum Spread

The audio spectrum spread (ASS) is another simple measure of the spectral shape. The spectral spread, also called instantaneous bandwidth. In MPEG-7, it is defined as the second central moment of the log-frequency spectrum. For a given signal frame, the ASS feature is extracted by taking the root-mean-square (RMS) deviation of the spectrum from its centroid ASC:

$$ASS = \sqrt{\frac{\sum_{k'=0}^{(N_{FT}/2)-K_{low}} [log_2\left(\frac{f'(k')}{1000}\right) - ASC]^2 P'(k')}{\sum_{k'=0}^{(N_{FT}/2)-K_{low}} P'(k')}} \tag{2}$$

Where the modified power spectrum coefficients P'(k') and the corresponding frequencies f'(k') are calculated in the same way as for the ASC descriptor [8].

2.2.1.3 Audio Spectrum Flatness

The audio spectrum flatness (ASF) reflects the flatness properties of the power spectrum [8]. More precisely, for a given signal frame, it consists of a series of values, each one expressing the deviation of the signal's power spectrum from a flat shape inside a predefined frequency band. The first step of the ASF extraction is the calculation of the power spectrum of each signal frame as specified in Equation (3). In this case, the power coefficients *P(k)* are obtained from non-overlapping frames. For each band b, a spectral flatness coefficient is then estimated as the ratio between the geometric mean and the arithmetic mean of the spectral power coefficients within this band:

$$ASF(b) = \frac{\sqrt[hiK'_b - loK'_b + 1]{\prod_{k'=loK'_b}^{hiK'_b} P_g(k')}}{\frac{1}{hiK'_b - loK'_b + 1}\sum_{k'=loK'_b}^{hiK'_b} P_g(k')} \quad (1 \le b \le B) \tag{3}$$

For all bands under the edge of 1 kHz, the power coefficients are averaged in the normal way. In that case, for each band b, we have $P_g(k') = P(k')$ between $k' = loK'_b = loK_b$ *and* $k' = hiK'_b = hiK_b$.

2.2.1.4 Zero Crossing Rate

The zero crossing rate (ZCR) is commonly used in characterizing audio signals. The ZCR is computed by counting the number of times that the audio waveform crosses the zero axis [8]. In this expression, each pair of samples is checked to determine where zero crossings occur and then the average is computed over N consecutive samples.

$$Zm = \sum_n |sgn\,[x(n)] - sgn\,[x(n-1)]|w(m-n) \tag{4}$$

Where the *sgn* function is

$$sgn[x(m)] = \begin{cases} 1, x(m) \ge 0 \\ -1, x(m) < 0 \end{cases} \tag{5}$$

And *x(n)* is the time domain signal for frame *m*.

3 Techniques for Music Indexing

In this section, Gaussian mixture models (GMM) are used to index music clips acoustic features namely MFCC and MPEG-7 are extracted from music audio clips. Index is created using GMMs. Retrieved is made depending on the maximum probability density function.

3.1 Gaussian Mixture Models (GMM)

The probability distribution of feature vectors is modeled by parametric or non parametric methods. Models which assume the shape of probability density function are termed parametric [6]. In non parametric modeling, minimal or no assumptions are made regarding the probability distribution of feature vectors. In this section, we briefly review Gaussian mixture model (GMM), for audio classification. The basis for using GMM is that the distribution of feature vectors extracted from a class can be modeled by a mixture of Gaussian densities [7]. For a D dimensional feature vector x, the mixture density function for category s is defined as

$$p\left(\frac{x}{\lambda^s}\right) = \sum_{i=1}^{M} \alpha_i^s f_i^s(x) \tag{6}$$

The mixture density function is a weighted linear combination of M component uni-modal Gaussian densities $f_i^s(.)$.Each Gaussian density function $f_i^s(.)$ is parameterized by the mean vector μ_i^s and the covariance matrix Σ_i^s using

$$f_i^s(x) = \frac{1}{\sqrt{(2\pi)^d |\Sigma_i^s|}} exp\left(-\frac{1}{2}(x-\mu_i^s)^T (\Sigma_i^s)^{-1}(x-\mu_i^s)\right) \quad (7)$$

Where $(\Sigma_i^s)^{-1}$ and $|\Sigma_i^s|$ denote the inverse and determinant of the covariance matrix Σ_i^s, respectively. The mixture weights $(\alpha_1^s, \alpha_2^s, \ldots, \alpha_M^s)$ satisfy the constraint $\sum_{i=1}^M \alpha_s^i = 1$. Collectively, the parameters of the model λs are denoted as $\lambda^s = \{\alpha_s^i, \mu_i^s, \Sigma_i^s\}$, i=1, 2,…M. The number of mixture components is chosen empirically for a given data set. The parameters of GMM are estimated using the iterative expectation-maximization algorithm [17].

4 Proposed Method for Indexing and Retrieval of Music Clip

4.1 Creation of Index Using GMM

1. Collect 100 music m_1, m_2,…m_{100} from Television broadcast music channels (Tamil).
2. Extract music audio clips of 50 seconds duration each from each music audio.
3. Extract 13 dimensional MFCC feature from 100 music audio clips.
4. Fit Gaussian mixture models for each music audio clip using EM algorithm.
5. Repeat steps 3 and 4 for extracting features using MPEG-7.

4.2 Retrieval of Music for a Given Query Clip

1. For a given query of 10 seconds duration extract MFCCs.
2. Compute the probability density function of the feature vectors with query music clip to all hundred GMM in the index database.
3. The maximum probability density function corresponds to the query audio.
4. Retrieve ranked list of music audio clips in descending order.
5. Repeat steps 1 to 4 for extracting features using MPEG-7.

5 Experimental Results

For music audio, experiments are conducted to study the performance of the retrieval algorithms in terms of the performance measures. Experiments are conducted for indexing music using the television broadcast music audio data collected from Tamil channels. Music audio of 2 hours duration is recorded from broadcast audio using a TV tuner card. A total data set of 100 different music clips, each of 1minute duration is extracted from the 2 hour music, which is sampled at 22 kHz and encoded by 16-bit. In our study, we have used fixed duration music clips of 50 second duration for indexing a music database and 10 second duration for query to retrieve music.

5.1 Acoustic Feature Extraction during Indexing

For each of the music audio clips of 50 seconds duration, MFCC and MPEG-7 features are extracted as described in Section 2. A frame size of 20 ms and a frame shift of 10 ms are used. Thereby 13 MFCC features are extracted for each music audio clip of 50 seconds. Hence, 5000 × 13 feature vectors are arrived at for each of the 50 second clip and this procedure is repeated for all 100 clips. Similarly, Experiments are conducted to extract feature for each clip, we are extracted MPEG-7 of 5000 × 4 feature vectors. Same procedure is repeated for all 100 clips.

5.2 Creation of Index

GMM is used to create the index for the music clips as described in Section 3.1. In our experiment, the index is created for first 50 seconds music clips using a Gaussian with 2, 3 and 5 components respectively.

5.3 Acoustic Feature Extraction during Retrieval

For a given query music audio clips of 10 seconds duration, a frame size of 20 ms and a frame shift of 10 ms are used. Thereby 13 MFCC features are extracted for each music audio clip of 10 seconds. Hence, 1000 × 13 feature vectors is arrived at for each of the 10 seconds clip Similarly, Experiments are conducted to extract feature for each clip, we are extracted MPEG-7 of 1000 × 4 feature vectors is arrived at for each of the 10 seconds clip.

5.4 Retrieval of a Clip Using Index

For retrieval, the remaining 10 second in a music clip is used as query. For every frame in the query the probability density function that the query feature vector belongs to the first Gaussian is computed. The same process is repeated for all the feature vectors. Retrieval is based on the maximum probability density function. Fig. 2 shows Percentage of retrieval in top ranked list.

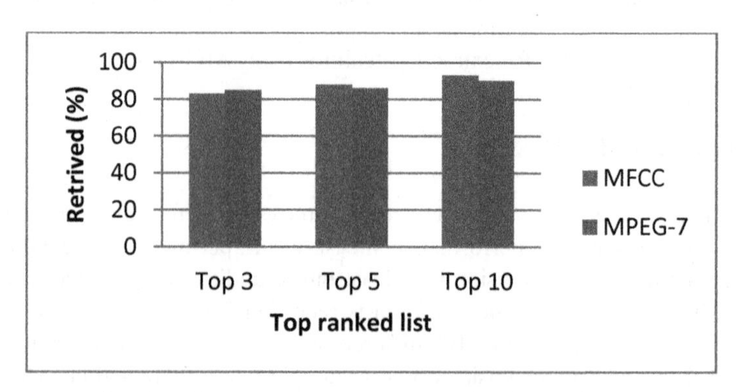

Fig. 2. Accuracy of retrieval in percentage of correct audio clips retrieved in the top n ranked list of music

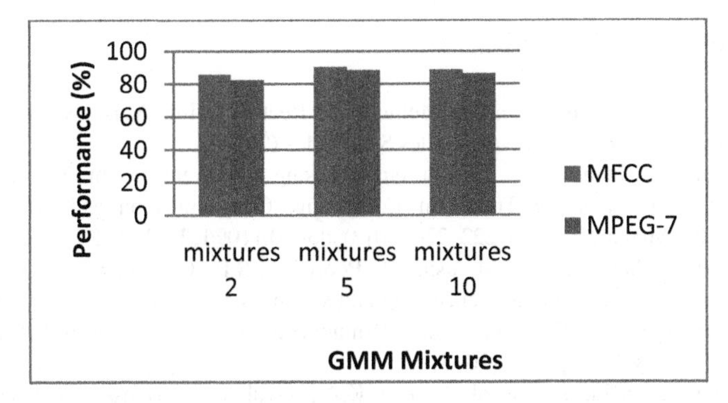

Fig. 3. Shows the performance of GMM for different mixtures

The distributions of the acoustic features are captured using GMM. We have chosen a mixture of 2, 5 and 10 mixture models. The index to which the music sample belongs is decided based on the highest output. The performance of GMM for different mixtures is shown in Fig. 3. Experiments show that when the mixtures were increased from 5 to 10 there was no considerable increase in the performance. With GMM, the best performance was achieved with 5 Gaussian mixtures. Table 1 shows average No. of clips retrieved for a given query.

Table 1. Average No. of clips retrieved

No. of Clips used for Index Creation	Average No. of clips retrieved for a given query	
	MFCC	MPEG-7
50	5.3	4.9
100	6.7	5.8
200	7.9	8.0

6 Summary

In this work, methods are proposed for Music indexing and retrieval. In the Music indexing for music audio clips using MFCC and MPEG-7 features extracted. For all the music audio clips fit Gaussian mixture model based on the features extracted. The average probability density function is computed for each of the Gaussians and the retrieval is based on the highest probability density function. The query feature vectors were tested and the retrieval performance was studied. Performance of music audio indexing system was evaluated for 100 clips, and the method achieves about 89.0% accuracy rate and a rate of average number of clips retrieved for each query. These clips are the highly similar clips retrieved. The method is sensitive to the duration of the query clip and shows the probability that the query is in the top 5 ranked list of retrieval music clip is high.

References

[1] Grosche, P.M.: Signal Processing Methods for Beat Tracking, Music Segmentation and Audio Retrieval. Thesis Universität des Saarlandes (2012)

[2] Kumar, R.C.P., Chandy, D.A.: Audio retrieval using timbral feature. In: 2013 International Conference on Emerging Trends in Computing, Communication and Nanotechnology (ICE-CCN), March 25-26, pp. 222–226 (2013), doi:10.1109/ICE-CCN.2013.6528497

[3] Nagavi, T.C., Anusha, S.B., Monisha, P., Poornima, S.P.: Content based audio retrieval with MFCC feature extraction, clustering and sort-merge techniques Computing. In: 2013 Fourth International Conference on Communications and Networking Technologies (ICCCNT), July 4-6, pp. 1–6 (2013)

[4] Kinnunen, T., Saeidi, R., Sedlak, F., Lee, K.A., Sandberg, J., Hansson-Sandsten, M., Li, H.: Low-Variance Multitaper MFCC Features: A Case Study in Robust Speaker Verification. IEEE Transactions on Audio, Speech, and Language Processing 20(7), 1990–2001 (2012)

[5] Krishna Kishore, K.V., Krishna Satish, P.: Emotion recognition in speech using MFCC and wavelet features. In: 2013 IEEE 3rd International Advance Computing Conference (IACC), February 22-23, pp. 842–847 (2013)

[6] Zanoni, M., Ciminieri, D., Sarti, A., Tubaro, S.: Searching for dominant high-level features for Music Information Retrieval. In: 2012 Proceedings of the 20th European Signal Processing Conference (EUSIPCO), August 27-31, pp. 2025–2029 (2012)

[7] Tang, H., Chu, S.M., Hasegawa-Johnson, M., Huang, T.S.: Partially Supervised Speaker Clustering. IEEE Transactions on Pattern Analysis and Machine Intelligence 34(5), 959–971 (2012)

[8] Kim, H.-G., Moreau, N., Sikora, T.: MPEG-7 Audio and Beyond Audio Content Indexing andRetrieval. John wiley & sons Ltd (2004)

[9] Lidy, T.: Evaluation of New Audio Features and Their Utilization in Novel Music Retrieval Applications, Thesis (December 2006)

[10] Raś, Z.W., Zhang, X.: Multi-hierarchical Music Automatic Indexing and Retrieval System (2008)

[11] Goto, M.: A real-time music-scene description system: Predominant-f0 estimation for detecting melody and bass lines in real-world audio signals. Speech Communication 43, 311–329 (2004)

[12] Redner, R.A., Walker, H.F.: Mixture densities, maximum likelihood and the EM algorithm. SIAM Review 26, 195–239 (1984)

[13] Chechik, G., Ie, E., Rehn, M.: Large-Scale Content-Based Audio Retrieval from Text Queries. In: MIR 2008, October 30–31, pp. 105–112 (2008)

Heart Disease Prediction System Using Data Mining Technique by Fuzzy K-NN Approach

V. Krishnaiah[1], G. Narsimha[2], and N. Subhash Chandra[3]

[1] CVR College of Engineering, Mangalpally (V), Ibrahimpatnam (M), R.R. Dist, TS, India
varkala.krish@gmail.com
[2] JNTUH College of Engineering, Kondagattu, Karimnagar (D), TS, India
narsimha06@gmail.com
[3] Vignana Bharathi Institute of Technology, Aushapur (V), Ghatkesar (M), R.R. Dist., TS, India
subhashchandra_n@yahoo.co.in

Abstract. Data mining technique in the history of medical data found with enormous investigations resulted that the prediction of heart disease is very important in medical science. The data from medical history has been found as heterogeneous data and it seems that the various forms of data should be interpreted to predict the heart disease of a patient. Various techniques in Data Mining have been applied to predict the patients of heart disease. But, the uncertainty in data was not removed with the techniques available in data mining. To remove uncertainty, it has been made an attempt by introducing fuzziness in the measured data. A membership function was designed and incorporated with the measured value to remove uncertainty. Further, an attempt was made to classify the patients based on the attributes collected from medical field. Minimum distance K-NN classifier was incorporated to classify the data among various groups. It was found that Fuzzy K-NN classifier suits well as compared with other classifiers of parametric techniques.

Keywords: Fuzzy K-NN Classifier, Membership function, Cleveland Heart Disease Data Base, Statlog Heart Disease Database,Data Mining,Heart disease.

1 Introduction

The diagnosis of diseases is a vital and intricate job in medicine. The recognition of heart disease from diverse features or signs is a multi-layered problem that is not free from false assumptions and is frequently accompanied by impulsive effects. The health care industry collects huge amount of health care data which unfortunately are "not mined" to discover hidden info for effective decision making.

Number test should be performed to diagnosis the heart disease and it forms various investigations with the data available and it is varied from patient to patient for health care organisations, data mining techniques are applied to predict the Heart Disease and useful in preparing knowledge base for further treatment.

It was found that more than one in three adults are found diseased because of heart problems as per the reports of the World Health Organisation (W.H.O.). It was found

S.C. Satapathy et al. (eds.), *Emerging ICT for Bridging the Future – Volume 1*,
Advances in Intelligent Systems and Computing 337, DOI: 10.1007/978-3-319-13728-5_42

that Heart diseases are compared in most of the developed countries are found more than that of the third world countries (under developing countries). These disorders of the system are termed cardiovascular diseases. Cardiovascular diseases cause an estimated 17 million people worldwide each year, majority of these are due to heart attacks and strokes. The term Cardio Vascular Diseases (CVD) refer to diseases of the heart and blood vessels and includes conditions such as coronary heart disease which is also known as ischaemic (heart disease), Cerebrovascular disease (stroke), heart failure, Rheumatic heart disease and hypertension (high blood pressure). In India, about 25% of deaths at the age group of 25 - 69 years occur because of heart disease. In urban areas, 32.8% deaths occur because of heart ailments, while this percentage in rural areas is about 22.9%.

Most of the authors in data mining classifications techniques proposed for the prediction of heart disease, but the prediction system did not considered the uncertainty in the data measure. So, to remove the ambiguity and uncertainty, we made an experiment with fuzzy approach by introducing a membership function to the classifier. The fuzzy *K-NN classifier* results show promising in nature for removing the redundancy of data and to improve the accuracy of classifier as compared with other classifiers of supervised and un-supervised learning methods in data mining.

The term, Heart disease encompasses the diverse diseases that affect the heart. Heart disease is the major cause of casualties in the United States, England, Canada and Wales as in 2007. Heart disease kills one person every 34 seconds in the United States [13].Coronary heart disease; Cardiomyopathy and Cardiovascular diseases are some categories of heart diseases. The term, "cardiovascular disease" includes a wide range of conditions that affect the heart and the blood vessels and the manner in which blood is pumped and circulated through the body. Cardio Vascular Disease (CVD) results in severe illness, disability and death [14]. Narrowing of the coronary arteries results in the reduction of blood and oxygen supply to the heart and leads to the Coronary Heart Disease (CHD). Myocardial infarctions, generally known as a heart attack and angina pectoris or chest pain are encompassed in the CHD. A sudden blockage of a coronary artery, generally, due to a blood clot that results a heart attack. The chest pains arise when the receiving of the blood by the heart muscles is inadequate [15].

High blood pressure, coronary artery disease, valvular heart disease, stroke or rheumatic fever/rheumatic, heart diseases are the various forms of cardiovascular disease. The World Health Organization (WHO) has estimated that 12 million of deaths occurred worldwide every year due to the cardiovascular diseases. Half of the deaths in the United States and other developed countries occur due to cardio vascular diseases. It is also the chief reason of the deaths in numerous developing countries. On the whole, it is regarded the primary reason behind the deaths in adults [16].

A number of factors have been shown that increases the risk of Heart disease [17]:

i. Family history
ii. Smoking
iii. Poor diet
iv. High blood pressure

v. High blood cholesterol
vi. Obesity
vii. Physical inactivity

2 Review of the Related Literature

Various works in literature related with heart disease diagnosis using data mining techniques have provoked the present work. Many authors applied different data mining techniques for diagnosis and achieved different probabilities for various methods. A few of the works are discussed beneath:

A new technique to extend the multi-parametric feature with linear and nonlinear characteristics of Heart Rate Variability (HRV) was proposed by Heon Gyu Lee et. al. [1]. The techniques of Statistical and classification were utilized to develop the multi-parametric feature of the HRV. They have also been assessed the linear and the non-linear properties of the HRV for three recumbent positions, to be accurate the supine, left side and right side location. Many experiments have been conducted on linear and nonlinear characteristics of the HRV indices to assess a number of classifiers, e.g., Bayesian classifiers [4], Classification based on Multiple Association Rules (CMAR) [3], Decision Tree (C4.5) [5] and Support Vector Machine (SVM) [2]. The SVM surmounted the other classifiers.

A representation of Intelligent Heart Disease Prediction System (IHDPS) is developed by using data mining techniques i.e., Decision Trees, Naïve Bayes and Neural Network was projected by Sellappan Palaniappan et. al. [6]. Each method has its own power to get suitable results. The hidden patterns and relationships among them have been used to construct this system. The IHDPS is web-based, user-friendly, scalable, reliable and expandable.

The prediction of Heart disease, Blood Pressure and Sugar with the aid of neural networks was proposed by Niti Guru et. al. [7]. Experiments were passed out on a sample database of patients' records. The Neural Network is tested and trained with 13 input variables such as Age, Blood Pressure, the reports of Angiography etc., The supervised network has been suggested for diagnosis of heart disease. Training was passed out with the aid of back-propagation algorithm. The mysterious data was fed at any time by the doctor, the method identified that the unknown data from the comparisons with the trained data and generated a list of probable diseases that the patient is prone to heart disease.

The inhibited problem to identify and predict the association rules for the heart disease according to Carlos Ordonez presented in "Improving Heart Disease Prediction Using Constrained Association Rules," [8]. The assessed set of data encompassed medical records of people having heart disease with attributes for risk factors, heart perfusion measurements and artery narrowing. Three constraints were introduced to reduce the number of patterns, they are as follows:

1. The attributes have to appear on one side of the rule only.
2. It separates the attributes into uninteresting groups.
3. The ultimate constraint restricts the number of attributes in a rule.

Besides decreasing the running time as per the experiments illustrated the constraints of exposed rules have been extremely decreased the number. The presence or absence of heart disease in four specific heart arteries have been anticipated two groups of rules.

Data mining methods may help the cardiologists in the predication of the survival of patients and the practices adapted consequently. The work of Franck Le Duff et. al. [9] might be executed for each medical procedure or medical problem and it would be feasible to make a wise decision tree fast with the data of a service or a physician. Comparison of traditional analysis and data mining analysis have been illustrated the contribution of the data mining method in the sorting of variables and concluded the significance or the effect of the data and variables on the condition of the present study. A chief drawback of the process was knowledge acquisition and the need to collect adequate data to create an appropriate model.

A novel heuristic for able computation of sparse kernel in SUPANOVA was projected by Boleslaw Szymanski et. al. [10]. It was applied to a standard Boston housing market dataset and to the discovery of the heart diseases in the population generally major topic of striking with the aid of a novel, non-invasive measurement of the heart activities on the basis of attractive field generated by the human heart. 83.7% predictions on the results were correct, in this manner outperforming the results obtained through Support Vector Machine and equivalent kernels. The spline kernel yielded good results equally on the standard Boston housing market dataset.

Latha Parthiban et. al.[11] projected a move towards on the basis of coactive neuro-fuzzy inference system (CANFIS) for the prediction of heart disease. The CANFIS model diagnosed the occurrence of disease by integrating the neural network adaptive capabilities and the fuzzy logic qualitative approach and further integrating with genetic algorithm. On the basis of the training performances and classification accuracies, the performances of the CANFIS model were evaluated. The CANFIS model is shown the potential in the prediction of the heart disease as illustrated by the results.

Kiyong Noh et. al.[12] has been placed forth a classification method for the extraction of multi parametric features by assessing HRV from ECG, the data pre-processing and the heart disease pattern. The proficient FP-growth method was the foundation of this method which is an associative. They accessible a rule cohesion measure that allows a strong press on pruning patterns in the pattern of generating method as the volume of patterns created could probably be huge. The several rules and pruning, biased confidence (or cohesion measure) and dataset consisting of 670 participants, circulated into two groups, namely normal people and patients with coronary artery disease, were employed to take out the experiment for the associative classifier.

3 Proposed Prediction System

Many hospitals manage the clinical data built with health care information systems, as system updates a data on regular basis extracting the information based on decision levels is a little difficulty for proper prediction. The main objective of this research work is to build "Predicting the heart disease patients with fuzzy approach" the diagnosis of the heart disease based on historical data. Most of the authors made an attempt to predict the heart disease with the build in classifiers available in data mining, but the uncertainty in the collection of the data is not removed and found which is to be ignored in most of the existing systems. To remove the uncertainty in the data, we have tried to propose a Fuzzy approach by introducing an exponential membership function with Standard Deviation and Mean calculated for the attributes measured (i.e. table.1) of each person and the resultant membership value is multiplied with the attributes to remove uncertainty. The data source is collected from Cleveland Heart Disease data Base [18] and Statlog Heart disease database [19]. Cleveland Heart Disease database consists of 303 records and Statlog Heart disease database consists of 270 records. Combined data bases of 550 records are chosen for our prediction system.

The data set consists of 3 types of attributes, i.e. Input, Key and Predictable attributes which are listed below:

3.1 Input Attributes

Table 1. Description of 13 input attributes

S. No.	Attribute	Description	Values
1	Age	Age in years	Continuous
2	Sex	Male or Female	1=male,0=female
3	Cp	Chest pain type	1 = typical type 1 2 = typical type angina 3 = non-angina pain 4 = asymptomatic
4	thestbps	Resting blood pressure	Continuous value in mm hg
5	chol	Serum cholesterol	Continuous value in mm/dl
6	Restecg	Resting electrographic results	0 = normal 1 = having_ST_T wave abnormal 2 = left ventricular hypertrophy
7	fbs	Fasting blood sugar	1 ≥ 120 mg/dl 0 ≤ 120 mg/dl
8	thalach	Maximum heart rate achieved	Continuous value
9	exang	Exercise induced angina	0= no, 1 = yes
10	oldpeak	ST depression induced by exercise relative to rest	Continuous value
11	solpe	Slope of the peak exercise ST segment	1 = un sloping 2 = flat 3 = down sloping
12	Ca	Number of major vessels colored by floursopy	0-3 value
13	thal	Defect type	3 = normal 6 = fixed 7 = reversible defect

3.2 Key Attribute

Patient Id: Patient's Identification Number

3.3 Predictable Attribute

Diagnosis: Value 1= <50% (no heart disease)

Value 0= >50 % (has heart disease)

It has been found that there is an uncertainty, in the measured attributes, i.e. cp, thestbps, chol, restecg, fbs, thelach. To remove the uncertainty, it has been multiplied these attributes with a membership value given by the membership function.

Fuzzy sets were introduced by Zadesh in 1965 [21]. Since that time researchers have found numerous ways to utilize this theory to generalize the existing techniques and to develop new algorithms in pattern recognition and decision analysis [22]-[23]. In [22] Bezdek suggests an interesting and useful algorithms that could result from the allocation of fuzzy class membership to the input vectors, thus affording fuzzy decisions based fuzzy labels. This work is concerned with incorporating fuzzy set methods into the classical K-NN decision rule. In particular, a *fuzzy K-NN* algorithm is developed utilizing fuzzy class memberships of the sample sets and thus producing a fuzzy classification rule. An assigning method of fuzzy membership for the training sets is proposed and their advantages and disadvantages are discussed. Results of both the 'crisp' (that based on traditional set theory) and *fuzzy K-NN* rule are compared on two data sets and the fuzzy algorithm is shown to dominate its crisp counterpart by having lower error rates and by producing membership values that serve as a confidence measure in the classification.

The *fuzzy K-nearest neighbor* algorithm assigns class membership to a sample vector rather than assigning the vector to particular class. The advantage is that the algorithm makes no arbitrary assignments. In addition, the vector's membership values should provide a level of assurance to accompany the resultant classification.

The basis of the algorithm is to assign membership as a function of the vector's distance from its *K-nearest neighbors* and those memberships of their neighbors in the possible classes. Beyond obtaining these K samples, the procedures differ considerably. James M. Keller [24] has proposed Fuzzy *K-Nearest Neighbour Algorithm* and a brief description of algorithm is mentioned below:

Let W=(x_1, x_2... x_n) be the set of n labelled samples. Also let u_i(x) be the assigned membership of vector x (to be computed), and u_{ij} be the membership in the ith class of the jth vector of the labelled sample set. The algorithm is as the following;

Table 2. A Fuzzy K-NN Classifier Algorithm

```
BEGIN
  Input 'x', of an unknown classification.
  Set K, 1 ≤ K ≤ n.
  Initialize i=1.
  DO UNTILE ('x' is found by K-nearest neighbours)
    Calculate distance between 'x' to'x_i'
     IF (I ≤ K) THEN
        Include 'x_i' in the set of K-nearest neighbours
     ELSE IF ('x_i' closer to x than any previous nearest neighbour) THEN
       Delete the farthest of the K-nearest neighbours
       Include 'x_i' in the set of K-nearest neighbours.
     END IF
  END DO UNTIL
  Initialize i=1.
  DO UNTIL ('x' assigned membership in all classes)
    Calculate u_i(x) from (1).
     Increment i.
  END DO UNTIL
 END
```

$$u_i(x) = \frac{\sum_{i=1}^{k} u_{ij}\left(\frac{1}{\|x-x_j\|^{2/(m-1)}}\right)}{\sum_{j=1}^{k}\left(\frac{1}{\|x-x_j\|^{2/(m-1)}}\right)} \tag{1}$$

We made an attempt in choosing the correct members hip function by measuring error rate and it was found successful with exponential membership function.

$$u_i = e^{-\left(\frac{X-\mu}{2\sigma^2}\right)}$$

$$\text{Where } \mu = \sum_{j=1}^{13} \mu_j \text{ and } \sigma = \sqrt{\frac{\sum_{i,j}^{13}(x_i-\mu_j)^2}{13}} \tag{2}$$

3.4 Fuzzy Based Approach

Nowadays, let us deal by "fuzzy logic" in medicine in wide intellect. In the medicine, particularly, in oriental medicine, nearly all medical concepts are fuzzy. The unfocused nature of medical concepts and their associations requires the use of "fuzzy logic". It defines incorrect medical entities as fuzzy sets and provides a linguistic approach with an brilliant estimate to texts. "Fuzzy logic" presents analysis methods able of sketch estimated inferences.

4 Results

Our dataset consists of total 550 records in Heart disease database. Initially dataset contained some fields, in which some value in the records was missing. These were identified and replaced with most appropriate values using Replace Missing Values filter from Weka 3.6.6. The Replace Missing Values filter scans all records & replaces missing values with mean mode method. This process is known as Data Pre-Processing.

A confusion matrix is obtained to calculate the accuracy of classification. A confusion matrix shows how many instances have been assigned to each class. In our experiment we have two classes and therefore we have a 2x2 confusion matrix.

Table 3. Class a = YES (has heart disease) Class b= NO (No heart disease)

	a(Has heart disease)	**b(No heart disease)**
a(Has heart disease)	TP	FN
b(No heart disease)	FP	TN

The data set consisting of 550 records have been divided into 25 classes where each class consists of 22 records. To predict the correctness of the system, we have divided the dataset in to equal amount of training and testing sets. To classify the system, a symbolic learning approach by interval method, i.e.[μ-σ , μ-σ] has been chosen with an importance of the measured certainty of thirteen attributes mentioned above in Table 1. The classification is based on the values falling in that interval. We have constructed a confusion matrix with the predictable attribute mentioned for diagnosis resulted in Table 4.

Table 4. Confusion matrix obtained for *Fuzzy K-NN* method with 13 attributes

	A	**B**
A	123	5
B	8	146

A *Fuzzy K-NN* symbolic classifier for different values of k is shown in Table 6. To show the difference between the performances of two classifiers, it is visualized in Figure 1.

Table 5. Overall Accuracy obtained for dataset with varying *k* (*K-NN method*)

K	Accuracy
1	0.55
3	0.48
5	0.45
7	0.44
9	0.42

Table 6. Overall Accuracy obtained for dataset with varying *k (Fuzzy K-NN method)*

K	Accuracy
1	0.91
3	0.87
5	0.84
7	0.81
9	0.80

To evaluate the performance of our approach, we have calculated the class-wise accuracy, precision and recall to illustrate the importance of 13 attributes obtained without redundancy. It was shown in Figure 2, Figure 3 and Figure 4.

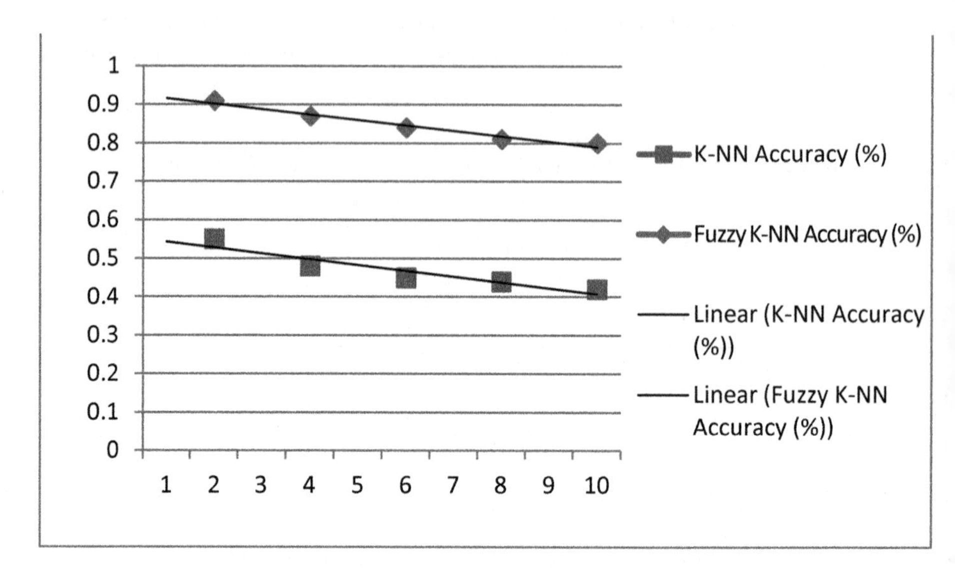

Fig. 1. Comparision of accuracy between *K-NN* and *Fuzzy K-NN* classifier

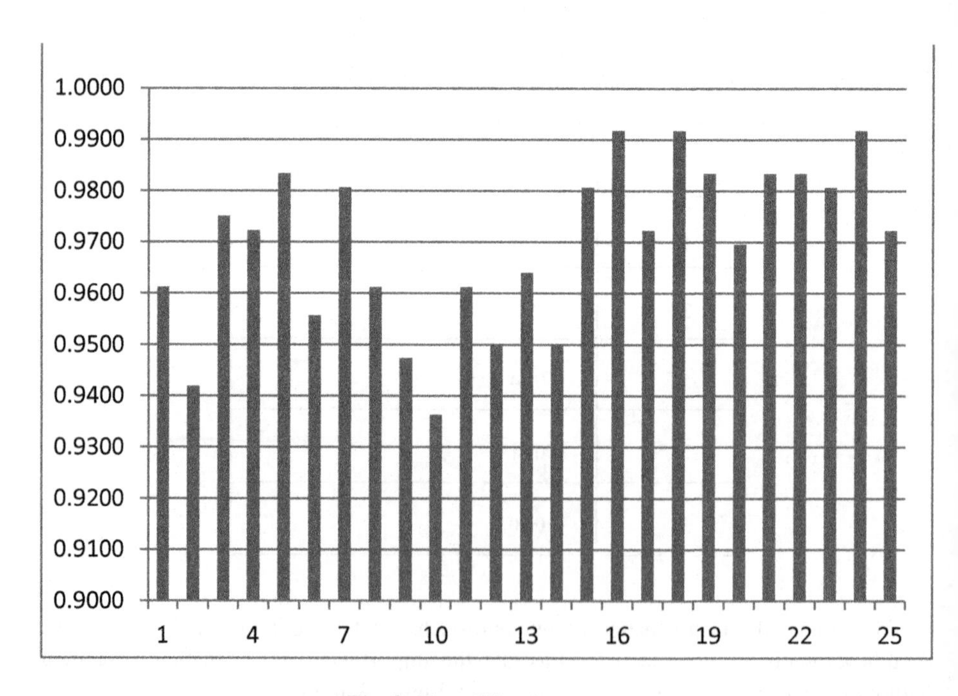

Fig. 2. Class-Wise Accuracy

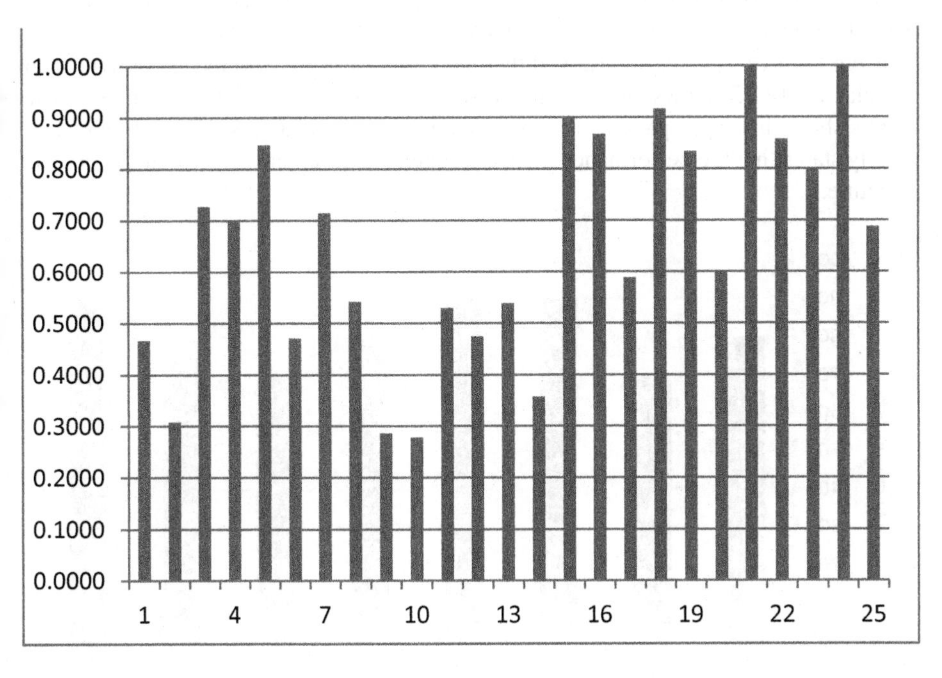

Fig. 3. Class-Wise Precision

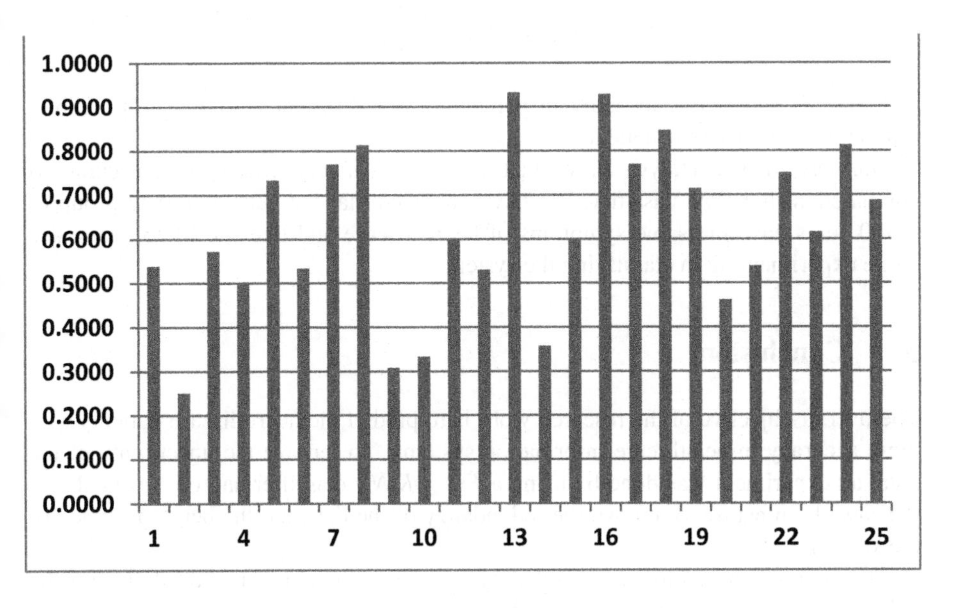

Fig. 4. Class-Wise Recall

It was observed that from Figure.1 the accuracy of predicting a heart disease patient with the 13 attributes mentioned above is reached up to 98%.

From Figures 3, 4 it was observed that more precision and less recall configuration for classes 17-25 shows that healthy cells are removed from those patients. The results show that those patients are more suitable for heart disease. From Figure.1, it clearly states that the performance measure evaluated based on 13 attributes is precise in nature.

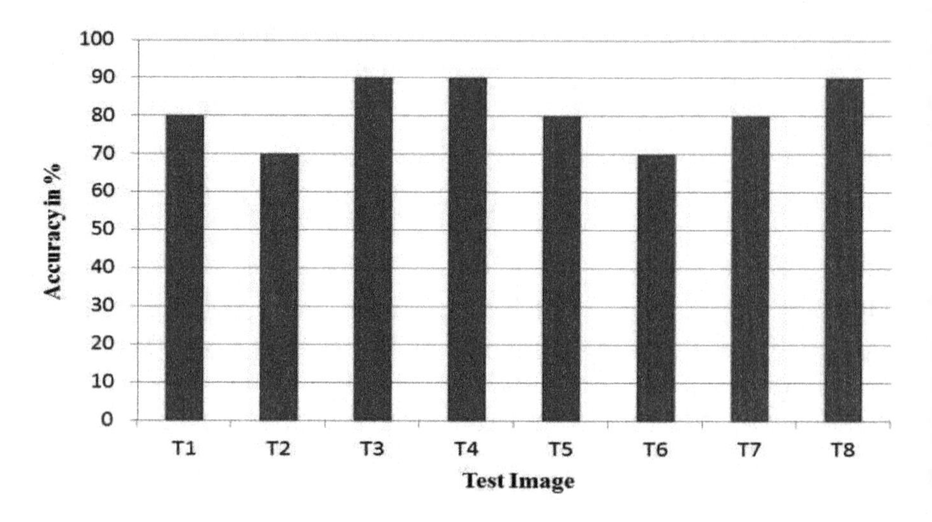

Fig. 5. Accuracy chart

Figure 5. shows the accuracy in percentage for different 8 sample test samples randomly taken from the database.

Our performance analysis shows that the *fuzzy K-NN* classifier is more accurate as compared with *K-NN* classifier. We have observed that the persons in age group of 50-60 years have the same symptoms of heart disease and like for 40-45 years also while experimenting in classifying the system.

5 Conclusion

The overall objective of the research work is to predict the heart disease patients with more accuracy in health care information systems. To remove the uncertainty of the data, an experiment was done by using a *Fuzzy K-NN* classifier and our result shows the capable in nature to remove the redundancy of the data and the better accuracy of the system.

The system can be further expanded by increasing the number of attributes for the existing system. The classifier can be tested further with the unstructured data available in health care industry data base and it can be structured with the *fuzzy K-NN* classifier.

References

[1] Lee, H.G., Noh, K.Y., Ryu, K.H.: Mining biosignal data: Coronary artery disease diagnosis using linear and nonlinear features of HRV. In: Washio, T., Zhou, Z.-H., Huang, J.Z., Hu, X., Li, J., Xie, C., He, J., Zou, D., Li, K.-C., Freire, M.M. (eds.) PAKDD 2007. LNCS (LNAI), vol. 4819, pp. 218–228. Springer, Heidelberg (2007)
[2] Cristianini, N., Shawe-Taylor, J.: An introduction to Support Vector Machines. Cambridge University Press, Cambridge (2000)
[3] Li, W., Han, J., Pei, J.: CMAR: Accurate and Efficient Classification Based on Multiple Association Rules. In: Proc. of 2001 Interna'l Conference on Data Mining (2001)
[4] Chen, J., Greiner, R.: Comparing Bayesian Network Classifiers. In: Proc. of UAI-1999, pp. 101–108 (1999)
[5] Quinlan, J.: C4.5: Programs for Machine Learning. Morgan Kaufmann, San Mateo (1993)
[6] Palaniappan, S., Awang, R.: Intelligent Heart Disease Prediction System Using Data Mining Techniques. IJCSNS International Journal of Computer Science and Network Security 8(8) (August 2008)
[7] Guru, N., Dahiya, A., Rajpal, N.: Decision Support System for Heart Disease Diagnosis Using Neural Network. Delhi Business Review 8(1) (January - June 2007)
[8] [8] Ordonez, C.: Improving Heart Disease Prediction Using Constrained Association Rules. Seminar Presentation at University of Tokyo (2004)
[9] Duff, F.L., Munteanb, C., Cuggiaa, M., Mabob, P.: Predicting Survival Causes After Out of Hospital Cardiac Arrest using Data Mining Method. Studies in Health Technology and Informatics 107(Pt. 2), 1256–1259 (2004)
[10] Szymanski, B., Han, L., Embrechts, M., Ross, A., Sternickel, K., Zhu, L.: Using Efficient Supanova Kernel For Heart Disease Diagnosis. In: Proc. ANNIE 2006, Intelligent Engineering Systems Through Artificial Neural Networks, vol. 16, pp. 305–310 (2006)
[11] Parthibanand, L., Subramanian, R.: Intelligent Heart Disease Prediction System using CANFIS and Genetic Algorithm. International Journal of Biological, Biomedical and Medical Sciences 3(3) (2008)
[12] Noh, K., Lee, H.G., Shon, H.-S., Lee, B.J., Ryu, K.H.: Associative Classification Approach for Diagnosing Cardiovascular Disease, vol. 345, pp. 721–727. Springer (2006)
[13] Hospitalization for Heart Attack, Stroke, or Congestive Hear Failure among Persons with Diabetes, Special report, New Mexico (2001 – 2003)
[14] Heart disease, from `http://en.wikipedia.org/wiki/Heart_disease`
[15] Chen, J., Greiner, R.: Comparing Bayesian Network Classifiers. Proc. of UAI-99, 101–108 (1999)
[16] Heart Disease, from `http://chinese-school.netfirms.com/heart-disease-causes.html`
[17] Lemke, F., Mueller, J.-A.: Medical data analysis using self-organizing data mining technologies. Systems Analysis Modelling Simulation 43(10), 1399–1408 (2003)
[18] Statlog database, `http://archive.ics.uci.edu/ml/machine-learningdatabases/statlog/heart/`
[19] Statlog database, `http://archive.ics.uci.edu/ml/machine-learning-atabases/statlog/heart/`

[20] Singh, Y., Chauhan, A.S.: Neural Networks in data mining. Journal of Theoretical and Applied Information Technology, JATIT (2005 - 2009)
[21] Zadeh, L.A., Bezdck, J.C.: Fuzzy Sets, Inf. Control Pattern Recognition with Fuzzy Objective Function Algorithm, vol. 8, pp. 338–353. Plenum Press, New York (1965)
[22] Kandel, A.: Fuzzy Techniques in Pattern recognition. John Wiley, New York (1982)
[23] Keller, J.M., et al.: A Fuzzy K-Nearest Neighbor algorithm. IEE Transaction on Systems,Man,Cybernetics, SMC 15(4), 580–585 (1985)

Applying NLP Techniques to Semantic Document Retrieval Application for Personal Desktop

D.S.R. Naveenkumar[1], M. Kranthi Kiran[1], K. Thammi Reddy[2], and V. Sreenivas Raju[1]

[1] Department of Computer Science and Engineering,
Anil Neerukonda Institute of Technology and Sciences,
Sangivalasa, Bheemunipatnam [M], Visakhapatnam
naveen.dannina@gmail.com,
{mkranthikiran.cse,vsraju.cse}@anits.edu.in
[2] Department of Computer Science and Engineering,
GITAM University, Visakhapatnam
thammireddy@gitam.edu

Abstract. Information retrieval in a semantic desktop environment is an important aspect that is to be stressed on. There are many works in the past which proposed many improved information retrieval techniques in this environment. But in most of them there is one thing that is lacking, is the involvement of user in the retrieval process. In other words it should be an interactive information retrieval system. This paper proposes an interaction based information retrieval system which interacts with the user to find out the hints and the suggestions from his/her side to get the best of the results which might satisfy the user.

Keywords: Semantic Desktop, Natural Language Processing techniques, Ontology.

1 Introduction

The Semantic Desktop [1][2] paradigm stems from the Semantic Web movement [3] and aims at applying the developed technologies of Semantic Web to the desktop computing.

Semantic Desktop is a device where a user can store and manage all the digital information such as documents, multimedia, and messages. Ontology [4] is a data model that represents knowledge as a set of concepts within a domain and the relationships between these concepts. Ontologies were developed to make sharing of knowledge easier and to reuse them with the aid of semantic languages such as RDFS [5] and OWL [6]. RDF is used for representing entities, which are referred to, by their unique identifiers or URIs, and a binary relationship among those entities. RDF Schema is a set of classes with certain properties using the RDF extensible knowledge representation language, which provides the basic elements for the description of ontologies. Ontologies will allow the user to express personal mental models and

S.C. Satapathy et al. (eds.), *Emerging ICT for Bridging the Future – Volume 1*,
Advances in Intelligent Systems and Computing 337, DOI: 10.1007/978-3-319-13728-5_43

form the semantic glue, interconnecting the information and systems. Semantic Desktop mainly aims at managing personal information on desktop computers.

In order to implement semantic desktop retrieval, a certain information retrieval model is required. Earlier, the associative retrieval techniques [7] and vector space semantic models [8] are applied for information retrieval on the Semantic Desktop to improve higher retrieval performance results based on the user query. But the problem with the associative retrieval technique is that the information retrieved was not relevant every time. Both techniques use some algorithms and retrieves documents by the systems own assumption. This paper proposes the incorporation of Natural Language Processing Techniques (NLP) [9] to improve the accuracy of information retrieval and retrieval performance, when compared to existing approaches by involving the user to interact with the system for producing the exact document which the user needs.

2 Related Work

There are several techniques like associative retrieval and vector space semantic model techniques, but the two main problems associated with them are the low retrieval performance and irrelevant file extraction. Since ontology is a new technology, the research work has been still going on. Considering the drawbacks of existing research work, we have developed a system to overcome those limitations. The following are the background works related to associative retrieval techniques and semantic vector retrieval model.

2.1 Background Works Related to Associative Retrieval Techniques

Crestani [10] has understood that the **associative retrieval** is a form of information retrieval which tries to find out the relevant information by retrieving information that is by some means **associated** with information which has already been known to be relevant. The information items which are associated might be documents or part of the documents or concepts, etc. 1960s, has seen a boost in the idea of associative retrieval [11] when the researches in the field of information retrieval have tried to increase the performance of retrieval by using the concept of associations between the documents and index terms. Semantic Similarity of Concepts and Text-based Similarity of Documents are calculated.

Semantic enrichment of resources has seen some better search results and the resources are covered with less semantic information which presents a major issue in realizing the vision of search on the Semantic Desktop. In order to address this problem Peter Scheir, Chiara Ghidini [12] have investigated on how to improve retrieval performance where the annotation of resources are very thin with semantic information. They suggested that employing techniques of associative information retrieval [7] will find relevant material, which was not annotated with the concepts which are used in a query. The drawback of the associative retrieval system for the Semantic Desktop is providing irrelevant documents because of unnecessary associations.

2.2 Background Works Related to Semantic Vector Retrieval Model

Semantic vector retrieval (SVR) model [8] for the desktop documents is fully based on the ontology. In the Semantic Vector model, words and concepts are represented by high-dimensional vectors in a mathematical space. To measure the high semantic similarity or relevance, any of the two vectors with a close distance in that particular space is taken into consideration. SVR basically draws on some thinking of traditional vector space model, in which some useful improvements are made based on the specific features of semantic information expression. In semantic vector space model, the document characteristic item sequence is not represented by the keywords as usual but the concepts that are extracted from documents. These concepts contain rich meaning in the ontology [4]. At the same time, for each concept in the concept space, there is a corresponding list to describe. The list represents a vector in the property space. In the semantic model, the weight of an item is not only related to the frequency of a keyword, but also the description of corresponding concept involved in the document. In this, the comparability and relativity between two concepts are fully taken into account.

3 Proposed Work

In this paper we proposed Natural language processing techniques mainly for two purposes which are,

1. Effective information retrieval for the user query.
2. To make the Desktop system intelligent. So that the system may interact with the user for exact search whenever ambiguities occur.

3.1 Ontology Creation

In our system the ontology acts as a information carrier. First we are creating a documents ontology model by using apache Jena API [13], where jena is a open source framework. The metadata includes the path of the file, author name, title, subject and context. The metadata of every document (we mainly considered PDF documents) is included into the ontology. The sample structure of the ontology is as follows:

```
  <xmlns:rdf="http://www.w3.org/1999/02/22-rdf-syntax-
ns#"
  xmlns:owl="http://www.w3.org/2002/07/owl#"
  xmlns:j.0="http://semanticdesktop/ontology#leo"
  xmlns:xsd="http://www.w3.org/2001/XMLSchema#"
  xmlns:rdfs="http://www.w3.org/2000/01/rdf-schema#"
  xmlns:j.1="http://semanticdesktop/ontology#" >
  <rdf:Description
rdf:about="http://semanticdesktop/ontology#leo saurman">
```

```
  <j.1:hasAuthor
rdf:resource="http://semanticdesktop/ontology#semantic
desktop Using  natural language processing techniques"/>
  </rdf:Description>
  <rdf:Description
rdf:about="http://semanticdesktop/ontologyAuthor">
  <rdf:type
rdf:resource="http://semanticdesktop/ontology#leo
saurman"/>
  <j.1:hasAuthor
rdf:resource="http://semanticdesktop/ontologyTitle"/>
  <rdfs:subClassOf
rdf:resource="http://semanticdesktop/ontologyTitle"/>
  <rdf:type
rdf:resource="http://www.w3.org/2002/07/owl#Class"/>
  </rdf:Description>
  <rdf:Description
rdf:about="http://semanticdesktop/ontologyTitle">
  <rdf:type
rdf:resource="http://www.w3.org/2002/07/owl#Class"/>
  </rdf:Description>
  <rdf:Description
rdf:about="http://semanticdesktop/ontology#hasAuthor">
  <rdf:type
rdf:resource="http://www.w3.org/2002/07/owl#ObjectPropert
y"/>
  </rdf:Description>
```


3.2 Applying NLP Techniques

The semantic document retrieval application acts as an interface between the system and the user. If the user enters queries, then the query undergoes the pre-processing with NLP techniques.

3.2.1 Tokenization

The task of chopping it up into pieces is tokenization, called tokens, at the same time eliminating certain characters, such as punctuations. These tokens are often loosely referred to as terms or words. A *token* is an instance of characters sequence in some particular documents that are grouped together as a useful semantic unit for processing. In this the query is taken as a sentence and it will chop up into pieces i.e. tokens and it will also remove unnecessary punctuations. We are using document preprocessor to tokenize the sentence. The sentence S is divided into tokens with words as Sentence=Word1 | Word2 | Word3 | Word4 | Word5.

3.2.2 Stop Words Removal

In our query there are some words which may not effect the search i.e., having little value in searching a document. So, to remove them we are giving a pre-defined less value words called stop words they are compared with the query and excluded from the vocabulary and then the remaining keywords will be processed. After removing those stop words, the sentence will be SW(Sentence)=Word1 | Word3 | Word5 ,where Word2 and Word4 are stop words and hence they are excluded from the sentence. Here the SW (sentence) represents Stop Words of a sentence.

3.2.3 Lower Case

In this technique all the keywords which are retrieved from the query, and the keywords which are retrieved from the ontologies are converted into lower case to generalize matching. The keywords are converted in to lower case to generalize the matching between the keywords from the user query and the keywords from the ontology. LC(Sentence)= LC(word1) | LC(word3) | LC(word5). Here LC represents Lower Case.

3.2.4 Stemming

Stemming terms is a linguistic process that attempts to determine the base (lemma) of each word in a text. Its aim is to diminish a word to its root, so that the key words in a query or text document are represented by their roots instead of the original words. Below given is an example which represents that, for all keywords car, cars….etc, the root word is car.

Ex – car, cars, car's, cars' → car

For every keyword which is converted in to lower case will undergo stemming process. Where the original word is changed to the root of that word and then sent to the dictionary. The words are stemmed to eliminate morphological variation. After that the meaning is added and stores the keywords with new meanings.

M(Sentence) = M(word1) | M(Word3) |M(word5),Where M represents Meaning.

3.2.5 Word Sense Disambiguation

Word Sense Disambiguation [12] is the process of identifying which sense of a word (i.e. meaning) is used in a sentence, when the word has multiple meanings. For example let us take two sentences.

Sentence1: This story book is really very interesting.
Sentence2: Please book my ticket as early as possible.

Here in the above two sentences both are having Book as a keyword, but the sense is different. The sense of a word is identified by using the meanings of the corresponding keywords of the particular sentence.

In our proposal, we apply this word sense disambiguation on the metadata of a document which contains some attributes of that document. We actually apply this concept to understand the relationship between these attributes of a particular document. This understanding can be helpful in removing the ambiguity as explained in the further sections.

Algorithm:

1. Let S be the user entered query string to be searched for retrieving a document based on its name.
2. Apply string tokenizing to the given sentence S, then it will be ST(S)
3. Remove Stop words from ST(S), then it will be SW(S).
4. Convert the SW(S) into small letters, then it will be LC(S)
5. Append meaning to LC(S), then it will be M(S)
6. If string is equal to subject, then it retrieves the path of the file. i.e., it checks the metadata whether it is author, context or subject .If it matches exactly with any of them it retrieves the document.
7. If there is an ambiguity, that is if the string matches with more than one context or subject or author, then our system involves the user to enter another search string which may be the another part of metadata of the document to be retrieved.
8. It searches the entire ontology with the combination of these two strings as meta-data of the same document and retrieves the relevant results.

4 Results

For each and every document, its meta data is included into the documents ontology which has been created as explained in section 3.1. An ontology file stores the metadata of the document and the ontology file is saved with an extension of .owl. Fig 1 depicts how the metadata of every document was stored. Here we have stored two files which is having the same context but with different subjects. Fig 2 depicts the searching of a document by the user. Fig 3 shows how a user searches the document with the same context but with different subjects which leads to ambiguity in document retrieval, in such cases our system involves the user by posing questions for getting additional information to retrieve the exact result the user needs. Basing on the information given by user the system will retrieve a particular document based on user query.

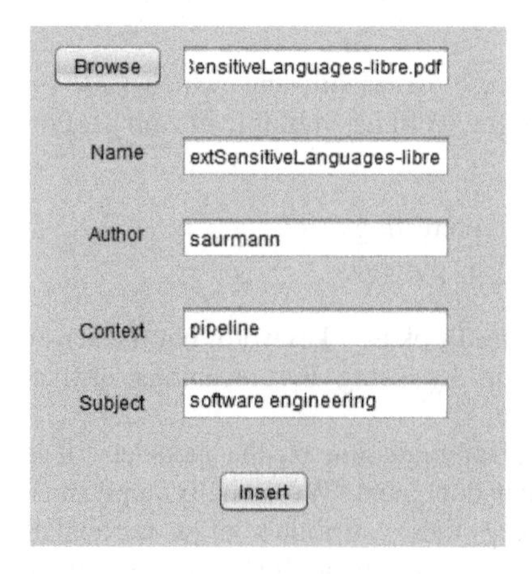

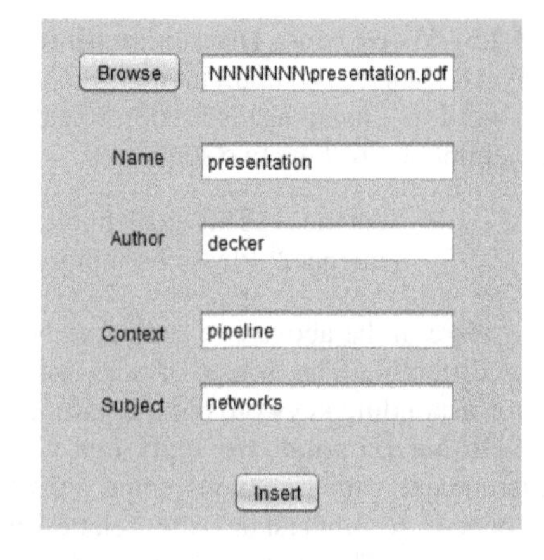

Fig. 1. Ontology Creation

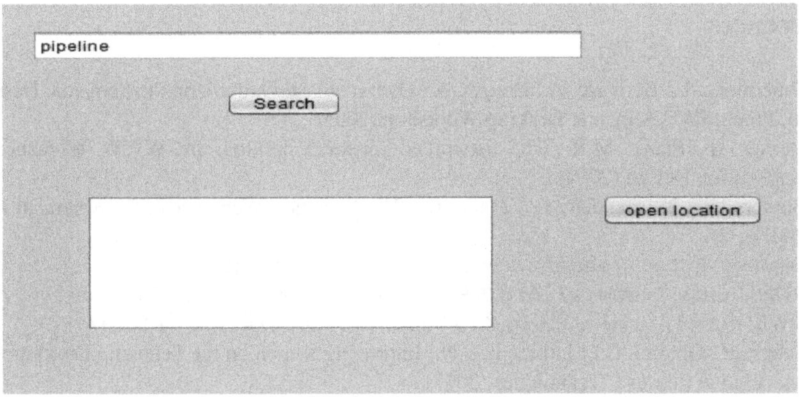

pipeline

Search

open location

Fig. 2. Showing the entered search string

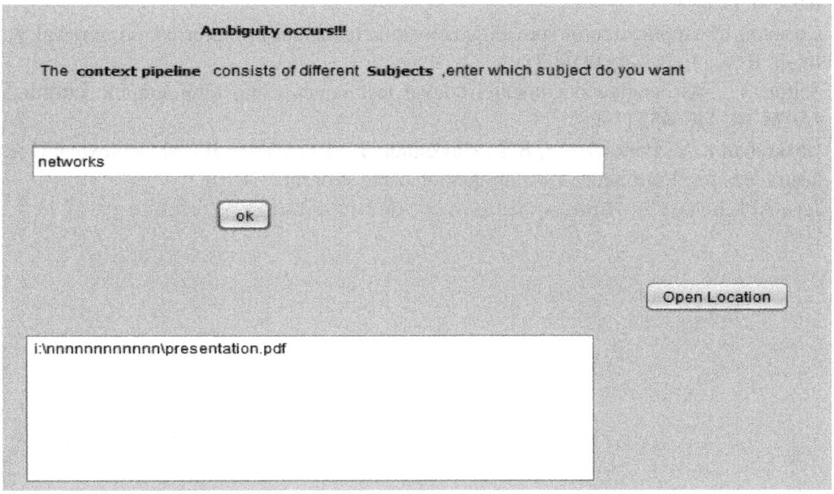

Ambiguity occurs!!!

The context pipeline consists of different Subjects ,enter which subject do you want

networks

ok

Open Location

i:\nnnnnnnnnnnn\presentation.pdf

Fig. 3. Shows the systems reaction which poses a question to the user when ambiguity occurs and retrieves exact document upon further clarification from the user

5 Conclusion

To overcome the problem of not involving the user in searching the desired data in the desktop environment this paper proposed an approach that the system was made interactive with the user in searching process. Overall, the application of Natural language processing techniques to information retrieval on the Semantic document retrieval application is an adequate strategy. As our system constantly takes the help of user for better results with a higher increase of retrieval performance and exact document retrieval, which will ultimately satisfy the user.

References

1. Sauermann, L., Bernardi, A., Dengel, A.: Overview and Outlook on the Semantic Desktop. In: Proc. ISWC Semantic Desktop Workshop (2005)
2. Decker, S., Frank, M.R.: The networked semantic desktop. In: WWW Workshop on Application Design (2004)
3. Berners-Lee, T., Hendler, J., Lassila, O.: The Semantic Web. Scientific American (May 2001)
4. Ontology, http://semanticweb.org/wiki/Ontology
5. RDFS, http://www.w3.org/TR/rdf-schema/#ch_introduction
6. OWL, http://www.w3.org/TR/2004/REC-owl-ref-20040210/
7. Scheir, P., Ghidini, C., Lindstaedt, S.N.: Improving Search on the Semantic Desktop using Associative Retrieval Techniques (2007)
8. Li, S.: A Semantic Vector Retrieval Model for Desktop Documents (2009)
9. Vallez, M., Pedraza-Jimenez, R.: Natural language processing in textual information retrieval (2007)
10. Crestani, F.: Application of spreading activation techniques in information retrieval. Artif. Intell. Rev. 11, 453–482 (1997)
11. Salton, G.: Associative document retrieval techniques using bibliographic information. JACM 10, 440–457 (1963)
12. Chakrabarty, A., Purkayastha, B.S., Gavshinde, L.: Knowledge-Based Contextual Overlap keen Ideas for Word Sense Disambiguation using Wordnet (2010)
13. Jena API, http://jena.apache.org/documentation/ontology/

An Adaptive Edge-Preserving Image Denoising Using Epsilon-Median Filtering in Tetrolet Domain

Paras Jain and Vipin Tyagi

Jaypee University of Engineering and Technology, Raghogarh – Guna (MP) – 473226, India
{parasbe05,dr.vipin.tyagi}@gmail.com

Abstract. Image denoising is a well-studied problem in the field of image processing and computer vision. It is a challenge to important image features, such as edges, corners, etc., during the denoising process. Wavelet transform provides a suitable basis for suppressing noisy signals from the image. This paper presents a novel edge-preserving image denoising technique based on tetrolet transform to preserve edges. Experimental results, compared to other approaches, demonstrate that the proposed method is suitable especially for the natural images corrupted by Gaussian noise.

Keywords: Wavelet, Tetrolet transform, Adaptive epsilon-median filtering.

1 Introduction

An image is often corrupted by noise during the process of acquisition and transmission. The goal of a denoising method is to reduce the noise while retaining as much as possible the important signal features. Several approaches [1-3] for noise reduction have been proposed in last few decades, many of them are based on linear spatial domain filters. Linear spatial filters (for example, Gaussian filters [2]) are simple and easily implementable and usually smooth the data to reduce noise effects; however, they can also result in blurring of important image structures such as edges [1].

To improve the linear filters, filtering based on non-linear edge-preserving methods is suggested [4]. These non-linear edge-preserving filtering methods can preserve the important image features while suppressing the undesirable noise. Denoising by wavelet methods has received much interest in edge-preserving image denoising problems. With wavelet transform, one can decompose the image signal into multiple subbands (a smooth subband at coarsest scale and three detail subbands at all resolution scales) of wavelet coefficients. For most signals, energy mainly distributes in the smooth subband and energy in the detail subbands is clustered on a few large wavelet coefficients, corresponding to the edge structure of original signal. In contrast, noise energy spreads over both the smooth subband and the detail subbands. Thus, noise can be suppressed through the shrinkaging of the small coefficients. By estimating an appropriate threshold, large coefficients attributed to signals details are preserved and small coefficients mostly contain noise are thresholded. Lastly, the thresholded coefficients are transformed back to the original domain to reconstruct the image.

S.C. Satapathy et al. (eds.), *Emerging ICT for Bridging the Future – Volume 1*,
Advances in Intelligent Systems and Computing 337, DOI: 10.1007/978-3-319-13728-5_44

Numerous approaches use the concept of wavelet thresholding for image denoising [5-16]. This paper describes a new image denoising method using tetrolet transforms (Haar-type wavelet transforms).

2 Proposed Technique

The multiscale decomposition of image into the wavelet coefficients is not adaptive i.e. local structures of image are not taken into account during decomposition. Although this issue is resolved by an adaptive Haar wavelet transform (also called tetrolet transform) proposed by Krommweh [17] but this tetrolet system is suitable only for sparse image representation due to its non-redundant nature, while for image denoising redundant information is helpful. Thus, it is very much needed to exploit redundancy by a denoising method based on tetrolet transform.

Inspired by an adaptive tetrolet decomposition algorithm proposed by Krommweh [17] and the $\in$-median filtering method [19], we propose a novel adaptive edge-preserving image denoising method. [Fig. 1].

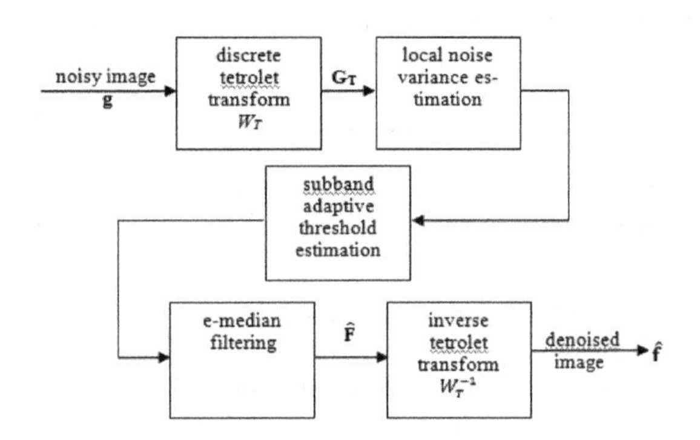

Fig. 1. Proposed denoising method

Let $\mathbf{f} = (f[x,y])_{x,y=1}^{N}$, $N = 2^K$, $K \in$ N denote $N{\times}N$ original image to be recovered. During transmission, the image **f** is corrupted by independent and identically distributed (i.i.d) zero mean white Gaussian noise according to :

$$\mathbf{g} = \mathbf{f} + \mathbf{n} \tag{1}$$

where **n** represents the noise and **g** the observed image.

Initially, the input image **g** corrupted by additive Gaussian noise, is decomposed into tetrolet coefficients through a tetrolet transform [17] W_T, expressed as:

$$\mathbf{G}_T = W_T(\mathbf{g}) \tag{2}$$

When applying the tetrolet transform to an image $\mathbf{g} = (g[x,y])_{x,y=1}^{N}$ with $N = 2^K$, $K \in$ N, we divide an image into 4×4 blocks. Then each block is covered with any four free tetrominoes. These four tetrominoes, which form the adaptive basis, are denoted by $\{I_0, I_1, I_2, I_3\}$ and the four indices in each tetromino subset I_v are mapped to a unique order {0, 1, 2, 3} by applying a bijective mapping L. On the basis of these definitions, for each tetromino subset I_v, Krommweh [17] defined following discrete basis functions:

$$\phi_{I_v}[x',y'] := \begin{cases} 1/2, & (x',y') \in I_v \\ 0, & \text{otherwise} \end{cases} \tag{3}$$

$$\psi_{I_v}^{l}[x',y'] := \begin{cases} \epsilon[l, L(x',y')], & (x',y') \in I_v \\ 0, & \text{otherwise} \end{cases} \tag{4}$$

for $l = 1, 2, 3$. Due to the underlying tetromino support, ψ_v^l are called tetrolets, and ϕ_v the corresponding scaling function. The function values $\epsilon[l, L(x', y')]$ in the tetrolet definition come from the Haar wavelet transform matrix:

$$\mathrm{W} = (\epsilon[m,n])_{m,n=0}^{3} = \frac{1}{2}\begin{pmatrix} 1 & 1 & 1 & 1 \\ 1 & 1 & -1 & -1 \\ 1 & -1 & 1 & -1 \\ 1 & -1 & -1 & 1 \end{pmatrix} \tag{5}$$

The above defined tetrolet is used as basis for the discrete tetrolet transform W_T, which will produce the multiscale tetrolet decomposition $\mathbf{G}_\mathrm{T}$ of the noisy image $\mathbf{g}$ as $\mathbf{G}_\mathrm{T} = \{\mathbf{g}^J; \mathbf{w}_1^j; \mathbf{w}_2^j; \mathbf{w}_3^j, j = J, J\text{-}1,\ldots..,1\}$, where j indicates the decomposition level (or resolution scale) of tetrolet transform and J is the largest scale in the decomposition.

Let the detail subband S at scale j, i.e., $S = \mathbf{w}_l^j$, $l = 1, 2, 3, j = 1,\ldots..,J$.

Note that the additive noise model for image $\mathbf{g}$ in spatial domain in Eq. (1) is also applicable for the subband $\mathbf{w}_l^j$ in wavelet domain. Thus, we have

$$\mathbf{w}_l^j = w_l^j + n_l^j, \tag{6}$$

where the noisy subband $\mathbf{w}_l^j$ is obtained after introducing the noise n_l^j in its noiseless counterpart w_l^j. Let $y_k \in \mathbf{w}_l^j$, $w_k \in w_l^j$ and $n_k \in n_l^j$. Then,

$$y_k = w_k + n_k, \qquad k = 1\ldots\text{no. of tetrolet coeffs.} \tag{7}$$

where y_k is the noisy observation of w_k and n_k is the noise sample. The objective of the proposed adaptive e-median (epsilon-median) filtering method (3×3 window size) is to obtain $\widehat{w}_k$, the estimate of w_k, from its noisy observation y_k. The e-median filter can be defined as:

$$\widehat{w}_k = y_k^m + X(y_k - y_k^m)$$
$$X(x) = \begin{cases} x, & |x| > \lambda_l^j \\ 0, & \text{otherwise} \end{cases} \tag{8}$$

where y_k represents the degraded data, y_k^m represents the median filtered data, and λ_l^j represents the threshold value [18]. The e-median filter preserves edges while removing noise [18, 19]. The threshold λ_l^j on a given subband $\mathbf{w}_l^j$ is computed by using the BayesShrink [7, 8] as threshold estimation criterion, expressed as

$$\lambda_l^j = \frac{\hat{\sigma}_{noise,j}^2}{\hat{\sigma}_{signal,l}^j} \tag{9}$$

where $\hat{\sigma}_{noise,j}^2$ is the local noise variance which can be estimated from the diagonal detail coefficients at the same scale j as the subband under consideration (that is, the coefficients in $\mathbf{w}_3^j$ subband).

$$\hat{\sigma}_{noise,j} = \frac{median(|y_i|)}{0.6745}, y_i \in \mathbf{w}_3^j \tag{10}$$

The expression on RHS of Eq. (10) is a robust median estimator used to estimate noise variance from diagonal detail coefficients at the finest scale (i.e. coefficients in subband $\mathbf{w}_3^1$) [5]. The term $\hat{\sigma}_{signal,l}^j$ in Eq. (9) is the local estimated signal deviation on the subband under consideration, estimated as:

$$\hat{\sigma}_{signal,l}^j = \sqrt{max\left(\hat{\sigma}_y^2 - \hat{\sigma}_{noise}^2, 0\right)} \tag{11}$$

where $\hat{\sigma}_y^2 = \frac{1}{N_s}\sum_{k=1}^{N_s} y_k^2$ and N_s is the number of tetrolet coefficients y_k on the subband under consideration.

The threshold computed in (9) will be used to obtain the elements $\hat{w}_k$ of $\hat{w}_l^j$ (the estimate of noiseless counterpart w_l^j of the given noisy subband $\mathbf{w}_l^j$) according to (8). The above procedure will be used to obtain the estimates of noiseless counterparts for all the detail subbands at each level and the final thresholded result $\hat{F}$ of complete image in wavelet domain is:

$$\hat{F} = \left\{\mathbf{g}^J;\ \hat{w}_1^j;\ \hat{w}_2^j; \hat{w}_3^j, j = J, J-1, \dots, 1\right\} \tag{12}$$

After completion of above stage, we get the thresholded subbands of tetrolet coefficients. Now an inverse discrete tetrolet transform is applied:

$$\hat{\mathbf{f}}_{rec} = W_T^{-1}(\hat{\mathbf{F}}) \tag{13}$$

where W_T^{-1} denotes the inverse discrete tetrolet transform, and $\hat{\mathbf{f}}_{rec}$ is the reconstructed image. For the reconstruction of the image, we use the low-pass coefficients from the coarsest level and the thresholded tetrolet coefficients from all levels obtained in above stage as usual. Additionally, the information about the respective covering in each level and block is used. We now apply the mechanism of tetrolet decomposition process elaborated earlier in reverse order. The whole procedure is iteratively executed for each admissible tetromino covering. As there are 117 possible tetromino coverings that can be used for partitioning image blocks, so many samples (instead of one) are obtained for a pixel value. This situation leads to redundancy in samples for the image pixels. The average of all the collected samples is taken to obtain the desired denoised version of a noisy pixel.

3 Experimental Results

The proposed image denoising method is applied to several test images corrupted by simulated additive Gaussian white noise at six different power level $\sigma \in [10, 15, 20, 25, 30]$ and iterated over ten different noise realizations for each standard deviation and the results are averaged over these ten runs. The test set comprises images from the standard gray-scale image dataset [20], as well as images such as *Lena*, *cameraman* and *peppers*. We estimate a set of parameters used by the proposed technique elaborated in Section 2: number of decomposition levels J, number of admissible tetromino coverings *num_cov*. In proposed method, wavelet coefficients are obtained using Haar-type wavelets (tetrolets) and the results are obtained using four decomposition levels (i.e. $J = 4$). Therefore, for all other wavelet-based methods (SURELET, Bayes, Bivariate), the Haar (Daubechies-1) wavelet with four decomposition levels are used for fair comparison.

The denoising results are obtained by averaging the collected samples over number of tetromino coverings used. Different values of *num_cov* are considered in the experiments. It can be observed that redundancy improves the denoising performance by a huge factor. The denoising performance improves as more and more tetromino coverings are averaged. Though, the performance improves rapidly at the beginning but gets saturated after a certain point. The primary reason for this is the duplication in the generated tetrolet coefficients. To assess the denoising effectiveness, the proposed method is compared to state-of-the-art methods. Namely, SURELET [12], Bivariate [11] and Bayes [8]. PSNR (in dB) and SSIM [21] values of the denoised images relative to their original images are shown in Table 1. The results obtained by the proposed method reveal significant gain in comparison to other wavelet-based methods, specially considering Bivariate and Bayes methods.

Fig. 2. Denoising results for test images using proposed approach

Table 1. PSNR and SSIM results for a set of test images and noise levels

	PSNR results				SSIM results			
	SURELET method	Bivariate method	Bayes method	Proposed method	SURELET method	Bivariate method	Bayes method	Proposed method
Lena								
$\sigma = 10$	**30.94**	30.02	29.97	30.45	0.90	0.85	0.88	**0.91**
$\sigma = 15$	28.50	27.12	27.30	**28.78**	0.85	0.76	0.82	**0.87**
$\sigma = 20$	26.99	25.14	25.96	**27.25**	0.80	0.67	0.77	**0.83**
$\sigma = 25$	26.07	23.84	23.75	**26.31**	0.76	0.60	0.65	**0.78**
$\sigma = 30$	25.04	23.10	23.09	**25.54**	0.72	0.54	0.62	**0.74**
$\sigma = 35$	24.42	22.20	22.51	**24.73**	0.68	0.48	0.56	**0.70**
Cameraman								
$\sigma = 10$	**32.35**	30.96	30.91	32.01	**0.90**	0.80	0.82	**0.90**
$\sigma = 15$	29.79	27.93	28.12	**30.57**	**0.85**	0.67	0.75	**0.85**
$\sigma = 20$	28.27	25.88	26.56	**28.62**	0.80	0.57	0.69	**0.81**
$\sigma = 25$	26.85	24.35	25.25	**27.24**	**0.76**	0.50	0.65	**0.76**
$\sigma = 30$	25.66	23.29	24.16	**26.18**	**0.72**	0.44	0.62	**0.72**
$\sigma = 35$	24.63	21.75	23.38	**25.03**	0.68	0.40	0.59	**0.70**
Barbara								
$\sigma = 10$	**30.30**	29.24	29.56	29.42	**0.91**	0.87	0.89	**0.91**
$\sigma = 15$	26.84	26.43	27.12	**28.15**	0.85	0.77	0.83	**0.87**
$\sigma = 20$	26.19	24.40	24.93	**26.61**	0.80	0.69	0.75	**0.83**
$\sigma = 25$	25.25	23.26	23.96	**25.78**	0.76	0.62	0.71	**0.78**
$\sigma = 30$	24.24	22.02	22.52	**24.45**	0.73	0.55	0.62	**0.74**
$\sigma = 35$	23.58	21.42	22.14	**23.72**	0.69	0.49	0.60	**0.70**
Man								
$\sigma = 10$	29.74	29.17	28.90	**30.81**	**0.89**	0.87	0.88	**0.89**
$\sigma = 15$	27.29	26.37	26.63	**28.15**	0.83	0.78	0.81	**0.84**
$\sigma = 20$	25.83	24.57	24.98	**26.70**	0.78	0.73	0.75	**0.80**
$\sigma = 25$	24.98	23.19	23.72	**25.24**	0.74	0.62	0.68	**0.75**
$\sigma = 30$	23.97	22.50	22.55	**24.35**	0.69	0.55	0.61	**0.71**
$\sigma = 35$	**23.80**	22.68	22.44	23.26	0.65	0.50	0.56	**0.67**
Boat								
$\sigma = 10$	30.68	29.96	29.76	**30.87**	0.90	0.85	0.87	**0.91**
$\sigma = 15$	28.07	27.12	27.16	**28.48**	0.85	0.75	0.80	**0.86**
$\sigma = 20$	26.51	25.00	24.99	**27.18**	0.79	0.66	0.68	**0.81**
$\sigma = 25$	25.34	23.90	24.07	**25.52**	0.74	0.58	0.66	**0.76**
$\sigma = 30$	24.61	22.53	23.07	**25.03**	0.70	0.52	0.62	**0.72**
$\sigma = 35$	23.95	21.85	21.95	**24.42**	0.66	0.47	0.52	**0.68**
Peppers								
$\sigma = 10$	**30.53**	29.62	29.58	30.06	**0.91**	0.86	0.88	**0.91**
$\sigma = 15$	28.01	26.63	27.04	**28.29**	0.86	0.77	0.82	**0.88**
$\sigma = 20$	26.58	24.70	24.96	**26.83**	0.82	0.68	0.75	**0.83**
$\sigma = 25$	25.24	23.27	23.99	**25.58**	**0.78**	0.62	0.72	**0.78**
$\sigma = 30$	24.50	22.41	22.57	**24.89**	**0.74**	0.56	0.63	**0.74**
$\sigma = 35$	23.86	21.13	21.59	**24.28**	0.71	0.51	0.58	**0.72**

4 Conclusions

In this paper, an adaptive edge-preserving image denoising method in tetrolet domain using an epsilon-median filtering is presented. The proposed approach has several desirable features. First, redundancy is exploited in the method to achieve significant gain in denoising performance. Second, estimating the term noise variance, used in the computation of thresholds, locally at each resolution scale makes it more beneficial as it takes the noise strength at that scale into consideration. Third, thresholding is done in subband-dependent manner which can suppress noise while preserving finer details and geometrical structures in the original image. The method denoises square natural gray-scale images with dimensions in the exponential order of two. If image is not a square then it has to be extended to make it a suitable input image. Both the quantitative and qualitative analysis of the results indicates that the proposed method effectively suppresses Gaussian noise without eliminating the important image details. Experiments demonstrate that the proposed method produces better results when compared with results obtained from other wavelet-based denoising methods.

References

1. Gonzalez, R.C., Woods, R.E.: Digital image processing. Prentice-Hall, Upper Saddle River (2008)
2. Shapiro, L., Stockman, G.: Computer Vision. Prentice Hall (2001)
3. Jain, P., Tyagi, V.: Spatial and frequency domain filters for restoration of noisy images. IETE Journal of Education 54(2), 108–116 (2013)
4. Jain, P., Tyagi, V.: A survey of edge-preserving image denoising methods. Information Systems Frontier, 1–12 (2014), doi:10.1007/s10796-014-9527-0
5. Donoho, D.L., Johnstone, I.M.: Ideal spatial adaptation via wavelet shrinkage. Biometrika 81, 425–455 (1994)
6. Donoho, D.L., Johnstone, I.M.: Adapting to unknown smoothness via wavelet shrinkage. Journal of the American Statistical Association 90(432), 1200–1224 (1995)
7. Chipman, H., Kolaczyk, E., McCulloc, R.: Adaptive Bayesian wavelet shrinkage. Journal of the American Statistical Association 440(92), 1413–1421 (1997)
8. Chang, S., Yu, B., Vetterli, M.: Adaptive wavelet thresholding for image denoising and compression. IEEE Transactions on Image Processing 9(9), 1532–1546 (2000)
9. Donoho, D.L.: De-noising by soft-thresholding. IEEE Transactions on Information Theory 41(3), 613–627 (1995)
10. Antoniadis, A., Fan, J.: Regularization of wavelet approximations. Journal of the American Statistical Association 96(455), 939–967 (2001)
11. Sendur, L., Selesnick, I.W.: Bivariate shrinkage with local variance estimation. IEEE Signal Processing Letters 9(12), 438–441 (2002)
12. Blu, T., Luisier, F.: The SURE-LET approach to image denoising. IEEE Transactions on Image Processing 16(11), 2778–2786 (2007)
13. Luo, G.: Fast wavelet image denoising based on local variance and edge analysis. International Journal of Intelligent Technology 1(2), 165–175 (2006)
14. Chang, S., Yu, B., Vetterli, M.: Spatially adaptive wavelet thresholding based on context modeling for image denoising. IEEE Transactions on Image Processing 9(9), 1522–1531 (2000)
15. Silva, R.D., Minetto, R., Schwartz, W.R., Pedrini, H.: Adaptive edge-preserving image denoising using wavelet transforms. Pattern Analysis and Applications. Springer (2012), doi:10.1007/s10044-012-0266-x
16. Jain, P., Tyagi, V.: An adaptive edge-preserving image denoising technique using tetrolet transforms. The Visual Computer (2014), doi:10.1007/s00371-014-0993-7

17. Krommweh, J.: Tetrolet transform: A new adaptive Haar wavelet algorithm for sparse image representation. Journal of Visual Communication Image Representation 21(4), 364–374 (2010)
18. Haseyama, M., Takezawa, M., Kondo, K., Kitajima, H.: An image restoration method using IFS. In: Proceedings IEEE International Conference on Image Processing, vol. 3, pp. 774–777 (2000)
19. Koç, S., Ergelebi, E.: Image restoration by lifting-based wavelet domain E-median filter. ETRI Journal 28(1), 51–58 (2006)
20. `http://decsai.ugr.es/cvg/CG/base.htm`
21. Wang, Z., Bovik, A., Sheikh, H., Simoncelli, E.: Image quality assessment: from error visibility to structural similarity. IEEE Transactions on Image Processing 13(4), 600–612 (2004)

A Comparison of Buyer-Seller Watermarking Protocol (BSWP) Based on Discrete Cosine Transform (DCT) and Discrete Wavelet Transform (DWT)

Ashwani Kumar[1], S.P. Ghrera[2], and Vipin Tyagi[3]

[1] Jaipur-National-University, Computer Science & Engineering, Jagatpura, Jaipur, R.J
[2] Jaypee University of Information Technology, Computer Science & Engineering, Waknaghat, Solan, H.P
[3] Jaypee University of Engineering & Technology, Computer Science & Engineering, Raghogarh, Guna, M.P
{Ashwani.Kumarcse,Dr.Vipin.Tyagi}@gmail.com,
sp.ghrera@juit.ac.in

Abstract. Buyer-Seller watermarking protocol (BSWT) is used to preserve the rights for the buyer and the seller. Frequency domain watermarking embedding that is DCT and DWT can affect the robustness and imperceptibility of watermarking algorithm. This paper studies the comparison of both domain which is DCT and DWT and concludes which one is better on the bases of some parameters. Digital watermarking is a key technology to embed information as unperceivable signals in digital contents. Buyer-seller watermarking protocols based on Discrete Cosine Transform (DCT) and Discrete Wavelet Transform (DWT) integrate digital watermarking algorithm and cryptography techniques for copyright protection. In this paper we have shown the comparison of these two, buyer-seller watermarking protocol based on Discrete Cosine Transform (DCT) and Discrete Wavelet Transform (DWT). These two protocols use Public Key Infrastructure (PKI), arbitrator and watermarking certificate authority (WCA) for better security. This paper shows results of watermark image quality based on peak signal-to-noise ratio (PSNR) mean square error (MSE) and similarity factor (SF).

1 Introduction

Digital multimedia has become innovative in the field of internet application and data securities because copyright protection and data integrity detection has become a vast concern. This technology works well and uses suitable tools to identify the source of the content, creator of the content, owner of the content, distributor of the content, or authorized consumer of a digital content such as document, image, video and audios. Digital watermarking techniques can be used to detect a document or image is illegally distributed or modified [1]. These arises an important problem concerning copyright protection of digital production [2]. The progressively growth of computer networks increased the use of multimedia data via the Internet have resulted in fast

S.C. Satapathy et al. (eds.), *Emerging ICT for Bridging the Future – Volume 1*,
Advances in Intelligent Systems and Computing 337, DOI: 10.1007/978-3-319-13728-5_45

and convenient exchange of digital information. In digital watermarking an imperceptible signal, i.e. watermark is embedded into the cover object which identifies the ownership of the digital content.

Copyright marking [3] is a relatively new technique for hiding information in multimedia content with the aim of tracing any traitor who redistributes the content illegally. Digital watermarking [4] has been proposed, complementing encryption techniques, to establish and prove ownership rights by embedding the seller's information in the redistributed content. There are number of watermarking protocols have been proposed in [5] to identify the illegal distribution of digital content. The buyer-seller watermarking protocol is one that combines encryption, digital watermarking, and other techniques to ensure rights protection for both the buyer and the seller in e-commerce. The classical methods are to modify the least significant bits (LSB) of specific pixels of the host image based on the watermark bits [6].For frequency domain, the main concept to insert a watermark into frequency coefficients of the transformed image using the discrete cosine transform (DCT), the discrete wavelet transform (DWT) [7], or other kind of transforms techniques [6,8]. In general, a buyer-seller watermarking scheme for traitor tracing includes three steps. First, a seller embeds a watermark that identifies the buyer into a digital product, second, when a pirated copy is found the seller will detect the watermark from the pirated copy. At last, once the watermark of a specific buyer is identified, the seller will take the case to a court [9]. The buyer-seller watermarking protocol is one that combines encryption, digital watermarking in order to ensure rights protection and security for both the buyer and the seller simultaneously in to the network.

In section 2 we have discussed about previous work i.e. related work. In section, 3 & 4 we have shown an overview about the watermarking protocol using DCT and DWT. Section 5 & 6 contain result analysis and conclusion.

2 Related Work

There are many watermarking protocols that have been proposed using cryptography and digital watermarking techniques. The intuitive idea of watermark-based fingerprinting has been implemented by a number of schemes using cryptographic techniques before the customer's right problem was first identified in [10]. Recent researches show that a secure watermarking protocol is protecting the participants' digital content during transaction using digital watermarking technique and a public key cryptosystem. The first known buyer-seller watermark protocol was introduced by Memon et al. [11], and it was improved by Ju et al. [12]. Since the first introduction of the concept several alternative design solutions have been proposed in [10, 13, 14]. Memon and Wong proposed a buyer-seller watermarking protocol in [12] to deal with the customer's right problem, but also introduced a new issue, the unbinding problem, in their solution. Our propose work compare buyer-seller watermarking protocol based on discrete cosine transform (DCT) and discrete wavelet transform (DWT) to fulfill the design requirements, different from the predecessors, our approach makes improvements on the many aspects such as anonymous communication between buyer and seller [9] it support multi-transaction and dispute resolution and avoid double watermark insertion. In the buyer seller watermarking protocol based on DCT and DWT we use robust watermark technique proposed by L Qiao [15] with the RSA cryptosystem [16, 17].

3 Buyer Seller Watermarking Protocol (BSWP) Using DCT

In this section we have used DCT to provide more security for the buyer and the seller during the transmission of digital content same as we do with DWT. In this section we have defined define the role and notations which are given in table-2.

In this, there are four roles i.e. one is buyer, second is a seller, third is WCA device and fourth is DCT [18]. The seller provides the watermark embedding operation and sells the watermarked product to the buyer. The WCA device is integrated into the seller's computer system and it will generate the watermark with the help of DCT for the buyer. We have assume that every seller in transaction has unique water marking embedded function algorithm in their software. In this protocol we use watermarking embedding with discrete cosine transform (DCT), and arbiter (ARB) and watermarking certificate authority (WCA). The discrete cosine transform (DCT) is a technique for converting a signal into elementary frequency components. It uses a transformation function which transforms the representation of data from space domain to frequency domain.

Here we have shown the details of possible transactions in the buyer seller watermarking protocol with DCT [18]. Step-by-step procedures of the Transactions are given below:

1. When B wants to purchase a digital content from S, after negotiation B will place the purchase

   order (PO) by encrypting SB, ESB (PO), along with digitally signing the PB using SB, DSSB (PB).
2. S checks the authenticity by analyzing the digital signature using B's public key and decrypts the PO using PB. S places request for valid digital watermark (WR), buyer contact information (BI) such as IP address and other details to WCA, along with encrypted PB using SS key, ESS(WR+BI+PB), and with digitally signing the PB using SS, DSSS(PB).
3. Forward the request for watermark to the DCT.
4. Receiving the request for generating the watermark W from WCA, to DCT will do so after checking the digital signature. It generates a watermark with the help of DCT which is DCTWID and sends it to a WCA device.
5. Upon receiving the request for generating the watermark W from S, WCA will do so after checking the digital signature. Watermark W directly depends on B's and S's public keys and Product-Copy ID number (PB, PS and PCID). Following conditions are satisfied when the watermark W is generated:

$$\Gamma\,(PB, PS\ PCID) \rightarrow W \quad (1)$$

$$f\,(W,SB) \rightarrow PS \quad (2)$$

$$\mu(X'\Phi\,(W, PB)) \rightarrow X' \quad (3)$$

$$\mu'\,(X'\copyright\,(PB)) \rightarrow W \quad (4)$$

$$\mu''(X'\varnothing\,(SB)) \rightarrow W' \quad (5)$$

Where Γ, f, μ, μ', and μ" are homomorphism functions to generate or check the digital content or watermark. A privacy homomorphism with respect to the binary operator Θ is applied to extract the watermark W from any original copy of the watermark embedded digital content X'. This homomorphism binary operator Θ and function μ' is only known to the WCA.

$$\delta\,(W, S_S)\_WID\text{-}S \quad (6)$$

$$\delta'(WIS\text{-}S, S_S)\_W' \quad (7)$$

$$\sigma\,(W', S_B)\_WID\text{-}B \quad (8)$$

$$\sigma\,(WID\text{-}B, S_B)\_W \quad (9)$$

6. Generated watermark, W, is encrypted, EPS(W), by WCA is send to S along with digitally signing the WID-S and PS, DSSC(WID S+PB).
7. The encrypted Watermark Identification number for B, EPS(WID-B), and DSSC(PB) is send to B. This will help B to check the originality of the purchased digital content in a later stage of the transaction.
8. Upon receiving (EPS (W), DSSC (WID-S+PB)) from WCA, S checks the authenticity by decrypting the digital signature using PC to get WID-S and PB. The equation (6) is applied to generate WID-S to crosscheck, with the one that is received from WCA, originality of the watermark. S decrypts, DSB (EPS (W)), to get the watermark W, which will be embedded using the secrets homomorphism function μ.
9. Digitally signed PB, DSSS (PB), and encrypted watermarked digital content, EPB (μ (XΦ (W, PB)) or X', is now forwarded to B.
10. Upon receiving (EPB (μ (XΦ (W, PB)), DSSS (PB)) from S, B checks the authenticity by analyzing the digital signature. W' is generated from X¢ and SB from the equation (5) and WID-B is from the equation (8) to crosscheck, with the one that is received from WCA, the originality of the digital content purchased.

4 Buyer Seller Watermarking Protocol (BSWP) Using DWT

Discrete wavelet transform (DWT) based watermarking techniques are gaining more popularity because DWT has a number of advantages over other transform. It contains progressive low bit-rate transmission and quality scalability. The buyer seller watermarking protocol based on DWT uses robust watermark technique proposed by L Qiao [15] with the RSA cryptosystem [16]. For that there are four roles i.e. one is buyer, another is a seller, and other is a WCA device, the fourth is DWT. Wavelets are obtained from a signal prototype wavelet y (t) called mother wavelet by dilations and shifting. Equation (c) shows the general form of a discrete wavelet transform.

Here we have shown the details of possible transactions in the buyer seller watermarking protocol with DWT [9]. Step-by-step procedures of the Transactions are given below:

1. When B wants to purchase a digital content from S, after negotiation B will place the purchase order (PO) by encrypting SB, ESB(PO), along with digitally signing the PB using SB, DSSB(PB).
2. S checks the authenticity by analyzing the digital signature using B's public key and decrypts the PO using PB. S places request for valid digital watermark (WR), buyer contact information (BI) such as IP address and other details to WCA, along with encrypted PB using SS key, ESS(WR+BI+PB), and with digitally signing the PB using SS, DSSS(PB).
3. Forward the request for watermark to the DWT.
4. Receiving the request for generating the watermark W from WCA, to DWT will do so after checking the digital signature. It generates a watermark with the help of DWT which is DWTWID and sends it to a WCA device.
5. Upon receiving the request for generating the watermark W from S, WCA will do so after checking the digital signature. Watermark W directly depends on B's and S's public keys and Product-Copy ID number (PB, PS and PCID). Following conditions are satisfied when the watermark W is generated:

$$\Gamma\,(PB, PS\ PCID) \rightarrow W \quad (1)$$

$$f\,(W,SB) \rightarrow PS \quad (2)$$

$$\mu(X'\Phi\,(W, PB)) \rightarrow X' \quad (3)$$

$$\mu'\,(X'©\,(PB)) \rightarrow W \quad (4)$$

$$\mu''(X'\emptyset\,(SB)) \rightarrow W' \quad (5)$$

Where Γ, f, μ, μ', and μ" are homomorphism functions to generate or check the digital content or watermark. A privacy homomorphism with respect to the binary operator Θ is applied to extract the watermark W from any original copy of the watermark embedded digital content X'. This homomorphism binary operator Θ and function μ' is only known to the WCA.

$$\delta\,(W, S_S)\_WID\text{-}S \quad (6)$$

$$\delta'(WIS\text{-}S, S_S)\_W' \quad (7)$$

$$\sigma\,(W', S_B)\_WID\text{-}B \quad (8)$$

$$\sigma\,(WID\text{-}B, S_B)\_W \quad (9)$$

6. Generated watermark, W, is encrypted, EPS(W), by WCA is send to S along with digitally signing the WID-S and PS, DSSC(WID S+PB).

7. The encrypted Watermark Identification number for B, EPS (WID-B), and DSSC (PB) are sending to B. This will help B to check the originality of the purchased digital content in a later stage of the transaction.
8. Upon receiving (EPS (W), DSSC (WID-S+PB)) from WCA, S checks the authenticity by decrypting the digital signature using PC to get WID-S and PB. The equation (6) is applied to generate WID-S to crosscheck, with the one that is received from CA, originality of the watermark. S decrypts, DSB (EPS (W)), to get the watermark W, which will be embedded using the secrets homomorphism function μ.
9. Digitally signed PB, DSSS (PB), and encrypted watermarked digital content, EPB (μ (XΦ (W, PB)) or X', is now forwarded to B.
10. Upon receiving (EPB (μ (XΦ (W, PB)), DSSS (PB)) from S, B checks the authenticity by analyzing the digital signature. W' is generated from X¢ and SB from the equation (5) and WID-B is from the equation (8) to crosscheck-with the one that is received from CA, the originality of the digital content purchased.

5 Result Analysis of These Two Protocols

In this section, we have study the effect of DWT and DCT upon the buyer seller watermarking protocol. Buyer seller watermarking protocol based on DWT is better and gives more secure watermark insertion and extraction algorithm as we describe above it uses wavelet decomposition and sub-band distribution. Many researchers have used Lena image as the original image. We have also use a Lena image of size 256×256 as a test image. The researcher has applied some types of attacks on the Lena image after watermark embedding to prove the quality of their proposed work. We have choose some previous work [19],[20],[21],[22] to obtain robustness results. For that we first defined some parameter to measure the quality of image.

5.1 Parameter Used

For analysis the comparison of DWT & DCT domain we have used PSNR, MSE and Similarity factor (SF) measurements.

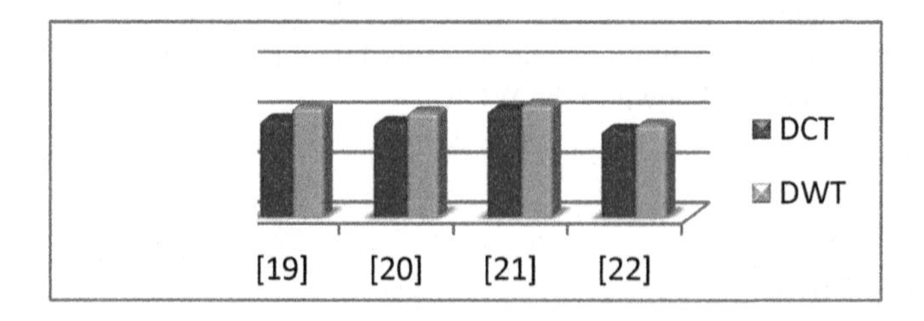

Fig. 1. Comparison of PSNR value in DCT & DWT domain

Figure 1 shows the comparison of PSNR value in DCT & DWT domain. The DWT embedding domain is more robust than DCT embedding domain and it also contain high capacity. In the brief DWT based watermarking is better than the DCT domain watermarking for embedding the watermark into the buyer seller watermarking protocol.

The value of PSNR, MSE & SF is given in Table no. 3.

Previous Work	Discrete Cosine Transform(DCT)			Discrete Wavelet Transform(DWT)		
	PSNR	MSE	SF	PSNR	MSE	SF
[19]	39.393	227.748	0.423	44.876	221.999	0.323
[20]	38.765	228.180	0.499	42.675	225.093	0.423
[21]	43.897	223.139	0.325	45.987	219.368	0.291
[22]	37.675	229.113	0.538	37.654	229.748	0.536

6 Conclusion

This paper compares Buyer Seller Watermarking Protocol based on Discrete Cosine Transform (DCT) and Discrete Wavelet Transform (DWT). These protocols use arbitrator, public key infrastructure and watermark certificate authority (WCA) for providing the efficient security for the buyer and seller. It is clarify that the DWT is better than DCT in terms of imperceptibility, robustness and capacity. Discrete wavelet transform (DWT) based watermarking techniques are gaining more popularity because DWT has a number of advantages over other transform. It contains progressive low bit-rate transmission and quality scalability. DCT and DWT based buyer seller watermarking protocol is use to fulfill the design requirements, different from the previous, these approach makes huge change on the many aspects such as anonymous communication between buyer and seller it support multi-transaction and dispute resolution and avoid double watermark insertion. This paper has shown the various result which obtained by calculating PSNR, MSE and SF measurement. In future work will study the affect of others optimization techniques and watermarking requirements with some testing experiments.

References

1. Shih, F., Wu, Y.-T.: Information Hiding by Digital Watermarking. Information Hiding and Applications (2009)
2. I Hartung, F., Kuter, M.: Multimedia Watermarking Techniques. Proceeding of the IEEE (1999)
3. Mintzer, F., Braudaway, G.W.: If one watermark is good, are more better. In: Proceedings of the IEEE International Conference on Acoustics, Speech, and Signal Processing (ICASSP 1999), Phoenix, Ariz, USA (1999)

4. Cox, I., Kilian, J., Leighton, T., Shamoon, T.: Secure spread spectrum watermarking for multimedia. IEEE Transactions on Image Processing (1997)
5. Wong, P.W., Memon, N.: Secret and public key image watermarking schemes for image authentication and ownership verification. IEEE Transactions on Image Processing (2001)
6. Wang, F.-H., Pan, J.-S., Jain, L.: Intelligent Techniques, Innovations in Digital Watermarking Techniques. Studies in Computational Intelligence. Springer, Heidelberg (2009)
7. Vetterli, M.A.K.: Wavelets and Subband Coding. Prentice Hall (1995)
8. S. J. a. N. B. Hingoliwala H.A.: An image comperession by using haar wavelet transform. Advances in Computer Vision and Information Technology (2008)
9. Kumar, A., Tyagi, V., Ansari, M.D., Kumar, K.: A Practical Buyer-Seller Watermarking Protocol based on Discrete Wavelet Transform. International Journal of Computer Applications (2011)
10. Choi, J.-G., Sakurai, K., Park, J.-H.: Does it need trusted third party? Design of buyer-seller watermarking protocol without trusted third party. In: Zhou, J., Yung, M., Han, Y. (eds.) ACNS 2003. LNCS, vol. 2846, pp. 265–279. Springer, Heidelberg (2003)
11. Memon, N.D., Wong, P.W.: A buyer-seller watermarking protocol. IEEE Transactions on Image Processing (2001)
12. Ju, H.S., Kim, H.-J., Lee, D.-H., Lim, J.-I.: An anonymous buyer-seller watermarking protocol with anonymity control. In: Lee, P.J., Lim, C.H. (eds.) ICISC 2002. LNCS, vol. 2587, pp. 421–432. Springer, Heidelberg (2003)
13. Lei, C.-L., Yu, P.-L., Tsai, P.-L., Chan, M.-H.: An efficient and anonymous buyer-seller watermarking protocol. Transactions on Image Processing (2004)
14. Zhang, J., Kou, W., Fan, K.: Secure buyer-seller watermarking protocol. IEE Proceedings Information Security (2006)
15. Qiao, L., Nahrstedlt, K.: Watermarking Schemes and Protocols for Protecting Rightful Ownership and Customer's Right. Journal of Visual Communication and Image Representation (1998)
16. Rivest, R.L., Shamir, A., Adleman, L.: A Method for Obtaining Digital Signatures and Public Key Cryptosystems. Communications of ACM (1978)
17. Kutter, M., Bhattacharjee, S.K., Ebrahimi, T.: Towards Second Generation Watermarking Scheme Image Processing (1999)
18. Kumar, A., Ansari, M.D., Ali, J., Kumar, K.: A New Buyer-Seller Watermarking Protocol with Discrete Cosine Transform. In: Das, V.V., Stephen, J., Chaba, Y. (eds.) CNC 2011. CCIS, vol. 142, pp. 468–471. Springer, Heidelberg (2011)
19. Shieh, C.-S., Huang, H.-C., Wang, F.-H., et al.: Genetic watermarking based on transform-domain techniques. Pattern Recognition (2004)
20. Promcharoen, S., Rangsanseri, Y.: Genetic Watermarking with Block-Based DCT Clustering
21. Deng, M., Weng, L., Preneel, B.: Anonymous buyer-seller watermarking protocol with additive homomorphism. In: Proc. of International Conference on Signal Processing and Multimedia Applications (2008)
22. Hameed, K., Mumtaz, A., et al.: Digital Image Watermarking in the Wavelet Transform Domain. World Academy of Science, Engineering and Technology 13 (2006)

Log Analysis Based Intrusion Prediction System

Rakesh P. Menon

TIFAC Core in Cyber Security
Amrita Viswa Vidyapeetham, Coimbatore, India
rakeshmenon090689@gmail.com

Abstract. This paper proposes an intrusion prediction system in which network log file entries are used for the prediction of attacks. Good filtering and classification techniques helps to process huge amount of data and find patterns of anomalies pertaining to network attacks. The techniques used in this paper are Naive Bayes and Adaboost Cost Sensitive Learning algorithms. The network log files obtained from network devices like IDS, Firewalls etc. are collected, normalized and correlated with the help of Alienvault SIEM and the fields which are important for classification are extracted. Next, the training data is classified with the help of Naive Bayes and misclassified entries are passed on to Cost Sensitive variant of Adaboost by which the classification rate is improved. Now with the help of this train data the system creates an attack model with the help of which it predicts whether an attack is about to happen or not.

Keywords: Intrusion Prediction, Alienvault, SIEM, Naive Bayes, IDS, Adaboost.

1 Introduction

In the recent years, the widespread use of internet and the introduction of network enabled applications have increased the likelihood of cyber-attacks and malicious activities on network devices. Thus, it has become very much necessary for computer and network attacks to be detected timely. The voluminous amount of data and the complexity of network attacks make it harder for manual analysis and timely detection of attacks. Machine Learning helps to process huge amount of data and find patterns of anomalies pertaining to network attacks.

Intrusion detection involves identifying the malicious behavior of a packet that targets a network or a device and its resources. With existing techniques, log collection and analysis using attack signatures; attacks are being identified only after they have occurred. By the time the attacks are identified it might have caused enough damage before any remediation could be done. So it would be beneficial if the attack can be predicted probabilistically, in order to thwart the incoming attack. Intrusion prediction involves a sequence of actions performed in order to recognize an attack strategy. In this paper, Naive Bayes classifier is used for the classification of log data to detect and predict attacks and Adaboost Cost Sensitive technique is used to improve the accuracy. A computer or a network of computers connected to Internet is monitored and the log files are collected. The log files are preprocessed to extract features and identify if each data packet is anomalous or normal. This forms the

S.C. Satapathy et al. (eds.), *Emerging ICT for Bridging the Future – Volume 1,*
Advances in Intelligent Systems and Computing 337, DOI: 10.1007/978-3-319-13728-5_46

training data and it contains all data which includes the normal, and anomalous packets involved in simple DDOS attack, SYN flood attack etc. First, Naive Bayes is used for classifying the given data set. Then the misclassified instances are passed through Adaboost cost sensitive AdaC2 for improving the classification accuracy. This serves as reference and is known as the train data. The test data is then passed through the system for prediction based on the patterns in the training data and then the performance of the proposed system is evaluated.

The remainder of this paper is organized as follows. In Section 2 details regarding the background and related works are given. In Section 3 the framework for the proposed system is given. Section 4 covers the implementation of the proposed system. Chapter 5 presents the conclusion and future work.

2 Background and Related Works

The work by Dewan.Md.Farid, Mohammad.Z.Rahman and Chowdhury.M.Rahman [1] proposed a new learning algorithm for Adaptive Intrusion Detection system based on Boosting and Naïve Bayesian classifier. During each round, the proposed framework generates probabilities for the input data using the naïve bayesian classifier and updates the weights of the same taking into account the rate of the misclassified instances. Also this system tackles the problem of classifying large datasets, so that detection rate (DR) is improved and false positives (FP) are reduced. The work by Mrutyunjaya Panda and Manas Ranjan Patra [2] proposed a framework of Network Intrusion Detection System (NIDS) based on Naive Bayes (NB) algorithm. Patterns, for the data sets given as input to the system, are generated by the proposed framework. With the built patterns, the framework detects attacks in the datasets using the Naive Bayes classifier algorithm. Compared to the other approaches, the proposed approach achieves higher detection rate, consumes less time and costs less. However, it generates somewhat more false positives. Then the work was continued in this direction in order to build an efficient intrusion detection model. The work by Andrew K.C.Wong, Mohamed S. Kamel, Yanmin Sun and Yang Wang [3] proposed Adaptive Boosting (Adaboost) algorithm as a successful meta-technique for improving classification accuracy. In this three cost sensitive boosting strategies are being analyzed after a thorough analysis of the advantages and shortcomings of Adaboost algorithm as a method to tackle the class imbalance problem. The three cost sensitive variants of adaboost algorithm are developed by introducing cost items into Adaboost framework. The boosting algorithms are also being studied by analyzing their weighting strategies for different samples where the class imbalance problem prevails. The various Cost Sensitive (CS) variants include First Cost Sensitive variant of Adaboost (AdaC1), Second Cost Sensitive variant of Adaboost (AdaC2) and Third Cost Sensitive variant of Adaboost (AdaC3) out of which AdaC2 is the most effective. The work by Cristina Abad *et al* [4] argues that by correlating data among different logs, the accuracy of the intrusion detection system can be improved. It shows how different attacks are being logged in different network devices and argues that when, only a single log is analyzed, some attacks cannot be detected correctly. Experimental results were presented using anomaly detection for the virus yaha and through the use of data mining tools and correlation between logs from syslog, firewall, DNS, Mail etc., effectiveness of the intrusion detection system have been improved and false positives reduced. The work by Natesan P, P. Balasubramanie and G. Gowrison [5] proposed a system in which an Adaboost algorithm with single weak classifier for network intrusion detection system is used. The classifiers Naive Bayes, Decision Tree and Naive Bayes are being used as a weak classifier. A benchmark data set is being used in the experiments so as to demonstrate that the classification accuracy of weak classification algorithms can be greatly improved by boosting algorithm. The result of this paper concludes with a survey which predicts that by combining these algorithms there is an increase in the attack detection rate, as opposed to the normal case. The paper by Chotirat Ann Ratanamahatana Dimitrios Gunopulos describes regarding the best features that can be used by the Naive Bayesian classifier and also have explained more on Selected Naive Bayes and Augmented Naive Bayes. The Feature Selection uses various methods like the Information Gain, Gain ratio. Gain ratio provides a better classification for attributes even with less value.

3 Intrusion Prediction System

With the use of Intrusion Detection Systems [8], attacks are being detected only after they have occurred or currently the attack is taking place. By the time the attacks are identified it might have caused enough damage before any remediation could be done. Keeping in mind these limitations, as well as others like huge amount of data to be processed and time taken, the Intrusion Prediction System (IPS) [7] using Naive Bayesian classifier and Cost Sensitive Adaboost has been proposed. The anomaly based IPS can probabilistically predict whether an attack is going to happen or not, that is whether an incoming log belongs to an attack packet or a normal packet. Fig. 1 shows the architecture of the proposed Intrusion Prediction System. The intrusion prediction system collects logs from several devices such as IDS, Firewalls, Syslog, and DNS etc. and then applies several methods to develop a model based on which the prediction can be done.

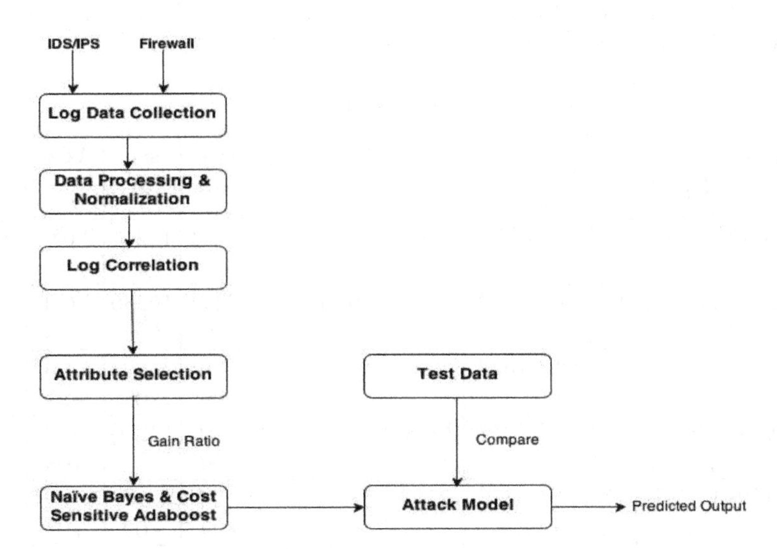

Fig. 1. Architecture of Intrusion Prediction System

3.1 Data Processing and Normalization

It involves removing the noisy and inconsistent data from dataset and to make the dataset suitable for mining. Intrusion detection dataset may contain large number of attributes, many of which are irrelevant for the classification of the input data or they may be redundant. Also, with the use of all the attributes for classification may increase the overall complexity of model, increase the time required for processing and decrease the detection accuracy of the intrusion detection algorithm. So only those attributes which are relevant for the proper classification of the input data are chosen. Normalization is done in order to convert logs from different devices which are in different into a single format. Consider an example, in which we have to correlate events recorded by a website and a network sniffer. The website will record the events in W3C format and the network sniffer records the events in LibPcap format. So we need a common format for all the logs so that further processing can be done easily. This is done with the help of normalization [9], which involves finding fields which are common among logs of different formats.

3.2 Log Correlation

Log correlation [6] is an important step which helps in more effective detection and also helps in making prediction of future attacks. The logs from different devices like IDS (Snort), Firewall (Nessus), Syslog are taken and after passing them through the first phase are converted into a single format. Then correlation is done to make the classification more effective, thereby reducing false negatives and false positives. Correlating multiple logs [10] enables an IPS to make a better decision for predicting an oncoming attack. Alienvault SIEM takes logs, which are in different format, normalizes and then correlates them and gives the output in which the logs are in a common format. The log collection and log normalization are done by the alienvault sensor and log correlation by alienvault server. The normalized and correlated output is shown in Table 1.

Table 1. Data Set

pkt time	ip from	ip to	port from	port to	len	flag	seq from	seq to	ACK	WIN
0.26745	172.17.128.10	172.17.128.17	3006	1214	0	F	13211141	13211141	2.65E+08	8575
0.267545	172.17.128.87	172.17.128.17	1214	3005	0	A	13211142	16286	2.65E+08	8745
0.275855	172.17.128.15	172.17.128.88	1022	22	20	P	3.43E+09	3.43E+09	3.14E+09	24820
0.280674	172.17.128.14	172.17.128.16	2157	33388	0	A	2.25E+09	17520	2.65E+08	2345
0.281179	172.17.128.12	172.17.128.16	3338	2157	1176	P	2.25E+09	2.25E+09	8.41E+08	8760

3.3 Attribute Selection

The correlated output contains attributes of which only some are required for classification. That is, only the attributes which gives maximum information needs to be selected. For this attribute selection is done. Gain ratio is used for attribute selection which gives more accurate attributes for classification and prediction. Gain ratio takes the number and size of branches into account when choosing an attribute. Using the Information Gain, attributes with many values are chosen over those attributes with few values. Gain Ratio of each of the attributes is given in Table 2.

Table 2. Gain Ratio of Attributes

Packet Attributes	Gain Ratio
pkt_time	0
length	0.127
flag	0.160
ip_from	0.205
port_from	0.283
port_to	0.283
seq_from	0.283
seq_to	0.283
ACK	0.339
ip_to	0.736
WIN	1

3.4 Classification and Boosting

In this system, classification of the input data set is done using Naive Bayes algorithm, followed by Cost sensitive adaboost for the purpose improving the accuracy of the train data. The train data is given as input to the naïve bayes algorithm and it classifies the given data with good computational efficiency and less error rate compared to other classification algorithms. If we have sample set $X = x_1, x_2...x_n$ with measured values $A = a_1, a_2...a_n$ to be classified in m classes. Then according to naive bayesian algorithm,

$$P(c_i/X) = \frac{P(X/c_i) \times P(c_i)}{P(X)}$$

Where $P(c_i/X)$ is the Posterior Probability which is the probability that an attack occurs on some data; $P(X/c_i)$ is the Likelihood which is the probability that given data exhibits the characteristics of attack; $P(c_i)$ is the class Prior Probability for the attack class which is probability for that type of attack to occur; P(x) is the Predictor Prior Probability also known as the evidence. In this paper, Adaboost algorithm is used as a strong classifier after using naive bayes as weak classifier. This is not much effective in case of imbalanced data sets, so we use cost sensitive adaboost in such cases.

In case of imbalanced datasets there will be an imbalance in the collected data set that is, higher number of normal data and lower number of anomaly. The classification can be increased by including a Cost Matrix and thereby making it cost sensitive adaboost. To differentiate samples (log fields), a cost item is being associated with each sample. Finally, the one having higher cost gives the correctly classified values.

3.5 Attack Model and Prediction

Next, an attack model is created based on the train data. For example, in case of DDOS attack, the attacker is targeting a system through a specific port, say 80, its window size is unique that is the packets that it sends all have same window size, say 8745, length of the attack packet may be more, say 1460. So this forms the model for predicting the attack. The test data is then compared with the model in order to classify the packet as anomalous or normal.

4 Implementation

Various attack scenarios like simple DDOS, SYN flood etc. are simulated. Dataset consists of logs which are generated by multiple devices like IDS, IPS, and Firewalls etc. The dataset are to be classified as either Anomalous or Normal. Our motive is to extract features from the log data which best represents a particular class. Logs from multiple devices are collected by the sensors of SIEM which are installed on multiple networks. These sensors help in collecting and normalizing the huge amount of log data. This processed log data is then sent to the SIEM server for correlation. After correlation, attribute selection is done. This set of features forms train data set. The SIEM component helps in correlating the log data which are collected from multiple devices. This is done by calculating the reliability associated with each log data and increasing the reliability value if the same log data is detected as anomalous by multiple devices. A low reliability parameter value is given to each of the log data at first and its value is incremented each time the same log data is detected as anomalous by a device. Higher value of reliability parameter implies the log data is anomalous. OSSIMs directives or rules are defined using xml 1.0. The correlation engine reads all the directives on startup in order to match individual rules or events. In order to differentiate between attack and normal log data, the risk value associated with each event is calculated. High risk value means the event is anomalous and vice versa. But this may be a false alarm, so correlation rule is applied continuously in order to improve the detection rate. Thus the number of false positives and false negatives are reduced with the help of SIEM correlation rules and risk value.

$$Riskvalue = \frac{\text{Assetvalue} \times \text{Priority} \times \text{Reliability}}{10}$$

Now the correlated log data forms the train data set which consists of all the features. From this only those features which best represents the log data is selected with the help of gain ratio. For selecting the best attribute with the help of gain ratio we used rapid miner tool and this forms the test dataset. This is given as input for naïve bayes classification algorithm which classifies the given log data as anomalous or normal based on the previous instances. In case class imbalance problem occurs, cost sensitive adaboost (AdaC2) is used in order to solve the problem and also to improve the classification accuracy. This is done with either rapid miner tool or Weka. Thus a new log data can be classified as either anomalous or normal and thus an attack model was created with the help of which we will be able to predict future attacks. The Prediction result obtained by giving the train and test data set to Naïve Bayes is given in Table 3.

Table 3. Prediction Results

Confid-ence (Yes)	Confid-ence (No)	Predicti-on (Attack)	ip from	ip to	port from	port to	seq from	seq to	ACK	WIN
0	1	No	172.17.128.15	172.17.128.92	8076	3336	27337407	27337502	32452511	10136
1	0	Yes	172.17.128.13	172.17.128.17	3336	8076	27337407	1488	32452511	1189
1	0	Yes	172.17.128.14	172.17.128.17	1214	4425	16610071	16610108	40797710	8298
1	0	Yes	172.17.128.14	172.17.128.17	3006	1214	13211141	13211141	26470488	8745
0	1	No	172.17.128.15	172.17.128.88	1022	22	34250753	34250773	31395515	24820

5 Conclusion and Future Works

Network log files are essential for analysis in finding as any issues in complex computer systems. The packets of a network may be normal, anomalous, appear to be normal but anomalous, or appear to be anomalous but normal. The training data which are collected from multiple devices contains all data which includes the normal and anomalous packets involved DDOS, SYN flood etc. The logs are then normalized and correlated with the help of Alienvault SIEM in order to gain more information for a certain type of attack, which might be impossible to detect or predict with the use of log from a single device only. The correlated log files are then preprocessed to extract features and from those, some features that can give maximum idea about the data are selected. This is done by selecting those features whose gain ratio is more compared to others. First, Naive Bayes is used for classifying the given data set. Then the misclassified instances are passed through Cost Sensitive Adaboost for improving the classification accuracy. This forms the train data. The test data is then passed through the system for prediction.

The future work can be extended to include more number of network devices such as DNS, Syslog, IPS, mail log etc. This would be helpful in providing better correlation results leading to reduction of false positives and false negatives. Next, more number of attacks such as CSS, botnet infections, other variants of DOS can be included.

References

1. Farid, D.M., Rahman, M.Z., Rahman, C.M.: Adaptive Intrusion Detection based on Boosting and Naive Bayesian Classifier. International Journal of Computer Applications 24(3) (2011)
2. Panda, M., Patra, M.R.: Network Intrusion Detection using Naive Bayes. International Journal of Computer Science and Network Security 7(12), 258–263 (2007)
3. Sun, Y., et al.: Cost-sensitive boosting for classification of imbalanced data. Pattern Recognition 40(12), 3358–3378 (2007)
4. Abad, C., et al.: Log correlation for intrusion detection: A proof of concept. In: 19th Annual IEEE Conference on Computer Security Applications (2003)
5. Natesan, P., Balasubramanie, P., Gowrison, G.: Improving the Attack Detection Rate in Network Intrusion Detection using Adaboost Algorithm. Journal of Computer Science 8(7), 1041 (2012)
6. Forte, D.V.: The Art of log correlation-Tools and Techniques for Correlating Events and Log Files. Computer Fraud & Security 2004(8), 15–17 (2004)
7. Kannadiga, P., Zulkernine, M., Haque, A.: E-NIPS: An event-based network intrusion prediction system. In: Garay, J.A., Lenstra, A.K., Mambo, M., Peralta, R. (eds.) ISC 2007. LNCS, vol. 4779, pp. 37–52. Springer, Heidelberg (2007)
8. Sendi, A.S., Dagenais, M., Jabbarifar, M., Couture, M.: Real Time Intrusion Prediction based on Optimized Alerts with Hidden Markov Model. Journal of Networks 7(2), 311–321 (2012)
9. Kruegel, C., Tóth, T., Kerer, C.: Decentralized event correlation for intrusion detection. In: Kim, K.-c. (ed.) ICISC 2001. LNCS, vol. 2288, p. 114. Springer, Heidelberg (2002)
10. Wang, L., Liu, A., Jajodia, S.: Using Attack Graphs for Correlating, Hypothesizing and Predicting Intrusion Alerts. Computer Communications 29(15), 2917–2933 (2006)

Multi-objective k-Center Sum Clustering Problem

Soumen Atta* and Priya Ranjan Sinha Mahapatra

Department of Computer Science and Engineering
University of Kalyani, Nadia, W.B.
{soumen.atta,priya}@klyuniv.ac.in

Abstract. Given a set P of n objects in two dimensional plane and a positive integer k ($\leq n$), we have considered the problem of partitioning P into k clusters of circular shape so as to minimize the following two objectives: (i) the sum of radii of these k circular clusters and (ii) the number of points of P covered by more than one circular cluster. The *NSGA-II* based *multi-objective genetic algorithm* (*MOGA*) has been proposed to solve this problem.

Keywords: k-center sum problem, Clustering problem, Multi-objective optimization, NSGA-II, Facility location problem.

1 Introduction

Clustering is the process of partitioning a set of objects into some small finite groups based on some similarity measurements among the objects. Details on different clustering algorithms can be found in [1,2]. The similarity among these objects is often measured as the distance among them. Different types of distance metrics such as Euclidean, squared Euclidean, Manhattan distance can be used. The objective function which are based on any distance metric is optimized to make clusters of the given objects. One of the well-known clustering problem where the maximum radius of the clusters is minimized is known as *k-center problem* [3,4]. The *k-center problem* is known to be NP-hard [3]. Sometimes *k-center problem* suffers from *dissection effect* for which objects which should lie in the same cluster are placed in the different cluster [5,6,7]. This *dissection effect* can be avoided if we minimize the sum of radii or the sum of diameters of these clusters [6,7]. When the objective is to minimize the sum of cluster radii the problem is known as *Minimum Sum Radii* (*MSR*) problem [6] whereas when the objective is to minimize the sum of cluster diameter the problem is known as *Minimum Sum Diameters* (*MSD*) problem [6]. *MSR* problem is also known as *k-center sum problem* (or *k-median problem*) in facility location [8]. We can relate clustering problem to the facility location problem as both the problems are concerned with minimizing the sum of distances of demand points

* Corresponding author.

S.C. Satapathy et al. (eds.), *Emerging ICT for Bridging the Future – Volume 1*,
Advances in Intelligent Systems and Computing 337, DOI: 10.1007/978-3-319-13728-5_47

from their cluster centers. In facility location problem it is desirable to find the best positions to install facilities such that customers can get facilities efficiently. These positions where facilities are installed are analogous to cluster centers in clustering problem. *Range* of a facility refers to the distance to which the facility can deliver its service. A customer can get the service of a facility if and only if the customer is within the *range* (or *coverage area*) of the facility. There are some facilities (e.g hospital, ambulance, fire station etc.) for which it is mandatory to facilitate each customer. Customers are considered as points in facility location model and the *range* of facilities can also be thought of as geometric objects such as a circle, a rectangle, a square etc. such that the facilities are installed at the center of the respective geometric objects [8]. Observe that the *cost* for installing facility may be taken as proportional to the range of that facility. In this model, the objective is to identify the center of geometric objects of given type such that together they cover all input points, representing customers, and the sum of areas (or perimeter) is minimized. This problem is referred as *full enclosing problem* in computational geometry [8,9] as well as *covering location problem* in *locational science* [10] and found to be NP-hard [11]. This problem is also NP-hard even if the geometric object used for covering is a circle [11], a square [12], a rectangle [13], a rectangular annulus [14] etc. Genetic algorithm [15] based approach to solve the different variations of this problem can be found in [16]. In another case, it is sometime not possible to provide services to all customers due to financial constraints. In this case, the objective is to facilitate the maximum number of customers using limited facilities. These problems are well studied in facility location as well as in computational geometry [17,18]. Interested reader may look into the papers [19,20] and the references therein for further studies although our work does not consider this model.

The k*-median problem* is known to be NP-hard relative to both the Euclidean and the rectilinear metrics [21]. This problem is known as *Euclidean* k*-median problem* when the distance metric is Euclidean. In some cases facilities may be negatively interfere with each other. Therefore, the area (or perimeter) of overlapped region between each pair of facilities must be minimized. This optimization criterion ensures that the number of customers lying in the overlapped region is minimum. In this paper we have considered two objectives for the *Euclidean* k*-center sum problem* (also known as *Euclidean* k*-median problem*). The objectives are to minimize the sum of the distances from the customers to their nearest facilities and to minimize the sum of the number of customers served by more than one facility. This problem is a multi-objective optimization problem.

Multi-objective optimization problem is an optimization problem with several objective functions without or with some constraint functions. *Multi-objective optimization (MOO)* [22] can deal with several objective functions. These functions may be conflicting in nature. In case of *MOO* the final solution set contains a number of Pareto-optimal solutions. The *Non-dominated Sorting Algorithm-II (NSGA-II)* [23] is a popular elitist multi-objective GA. In this work we have used *NSGA-II*. The framework of *NSGA-II* is shown in the Fig. 1.

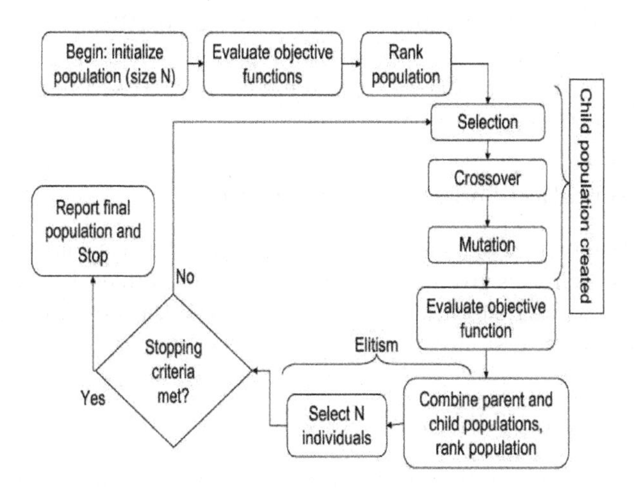

Fig. 1. Framework for NSGA-II

The organization of the paper is as follows: in *Section 2* we define the problem. In *Section 3* we propose multi-objective GA technique for the *multi-objective k-center sum clustering problem.* Results are discussed in *Section 4*. *Section 5* concludes the paper.

2 Problem Definitions

In this work we have considered the following problem: Given a set P of n points in a plane and a positive integer number k ($\leq n$), find a partition of P into k clusters viz. $C_1, C_2, \ldots, C_k$ of circular shape with their respective cluster radii as $r_1, r_2, \ldots, r_k$ such that $C_1 \cup C_2 \cdots \cup C_k = P$ with the following two objectives: (i) $\min \sum_{i=1}^{k} r_i^2$ and (ii) $\min |(C_i \cap C_j)|$, $\forall i, j = 1, 2, 3, \ldots, k$ and $i \neq j$, where $|X|$ represents the cardinality of the set X.

We call this problem as *multi-objective k-center sum clustering problem.* This problem is related with the facility location model in the following way: if we install facility with circular *range* at each of the cluster centers then any point (which is analogous to a customer here) inside cluster C_i can get the service from the facility installed at the center of C_i. In some cases facilities may negatively interfere with each other. Then the overlapping regions for more than one facility are devoid of any service from any such interfering facilities. So it would be very much desire to minimize the number of points (customers) within those regions. Different variations of the second objective function may also be considered. The area of overlapping region among all C_i's can be considered as the second objective function.

3 Proposed Multi-objective GA Technique

In this section we propose *NSGA-II* based multi-objective GA technique to solve *multi-objective k-center sum clustering problem.*

3.1 Representation and Encoding Scheme

To find the placement of k circles we need to identify the position of these circles. The position of a circle is given by its center coordinate values. Each chromosome is of length $2k$. So we take the first two consecutive values from chromosome to identify the center of first circles and so on.

3.2 Population Initialization

At first we create the initial population. If k denotes the number of circles then we create real-coded chromosome of length $2k$. To restrict our solution space, we find the minimum and the maximum coordinate values of the given point set and generate the real number to form chromosomes in this range. An user given parameter determines the *population size.*

3.3 Evaluation of Objective Functions

According to the problem definition the number of objective functions is two here. First objective function determines the sum of the radius of the k-circles and the second objective function determines the number of overlapping points. Here we want to minimize both these objective functions. Basically input points are clustered into k clusters. We take one input point and then find its nearest neighbor among the k centers represented by chromosome under consideration. In this way all the input points are clustered into k clusters. Here the shape of the cluster is circle. So we need to find out the radius of each circle now. Before that we update the chromosome as follows: we take the mean of the coordinate values of all the points in a particular cluster and then update the corresponding cluster center. We take the distance of the farthest point among the points in associated cluster to the center of a circle as the radius. The sum of such radii for k circles is taken as the value of the first objective function for the chromosome under consideration. Then we find the distance of each input point from all the k centers. If these distances are less than its corresponding radius then we consider the point is enclosed by that circle. If a point is enclosed by more than one circle then we say that the point is overlapped. We keep a count on the number of circles enclosing a particular point. The sum of such counts for all the input points is taken as the value of the second objective function for the chromosome under consideration.

3.4 Selection

The individuals are selected by using a *binary tournament selection* with *crowed-comparison-operator* to form the matting pool as done in *NSGA-II*. The details of the selection process can be found in [23].

In order to explore the solution space, we use two genetic operators, crossover and mutation. The operators are applied iteratively in every generation with their corresponding probabilities.

3.5 Crossover

The original NSGA-II algorithm uses *Simulated Binary Crossover (SBX)*. Real-coded GAs use *SBX* [24], [25] operator for crossover. Here we have set the *Crossover probability* (μ_c) to 0.9.

3.6 Mutation

Real-coded GAs use *polynomial mutation* [24], [26]. Here we have set the *Mutation probability* (μ_p) to $\frac{1}{n}$, where n is the number of decision variables. Here the number of decision variables is $2k$, where k is the number of circles.

3.7 Elitism

After the crossover and mutation intermediate population is generated by combining the population of parents and offspring of the current generation. So, the population size is two times the initial population. Top p solutions from the combined population are propagated to the next generation based on non-dominated sorting and crowding distance operator.

4 Results and Discussion

In this section we have discussed some experimental results. We have used some randomly generated input data sets of points and also some standard input data sets of points. Here we have always selected the first solution in rank 1 from the non-dominated front. The values of different parameters of genetic algorithm such as *population size*, *number of generations*, *crossover probability* and *mutation probability* are taken as 50, 50, 0.9 and $\frac{1}{2k}$ (k is the number of circle) respectively.

The result obtained using the proposed technique for the standard data set of 250 points [27,28,29] is shown in Fig. 2, where $k = 5$. Here the five circles enclose all the input points and the number of overlapping points is very less.

We have got the result for the standard data set of 300 points [28] as shown in Fig. 3, where $k = 6$. In this case, we have got five disjoint circles enclosing all the input points. Therefore, the number of overlapping points is zero here. If we want to install negatively interfering facilities such as mobile tower, then

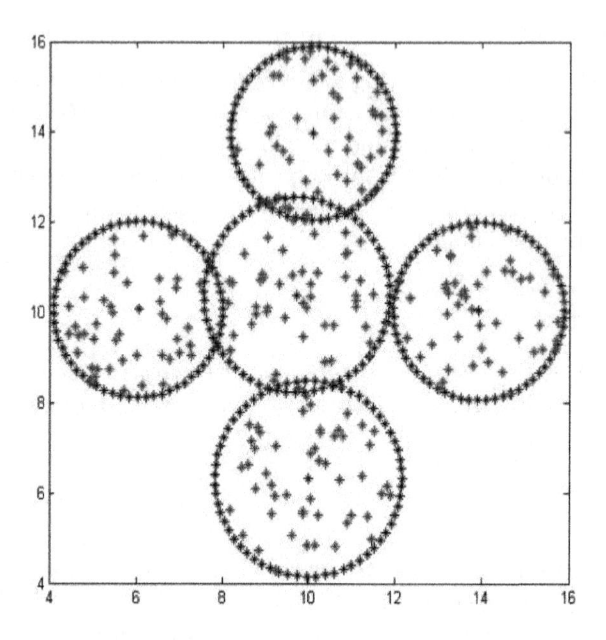

Fig. 2. Result for 250 points

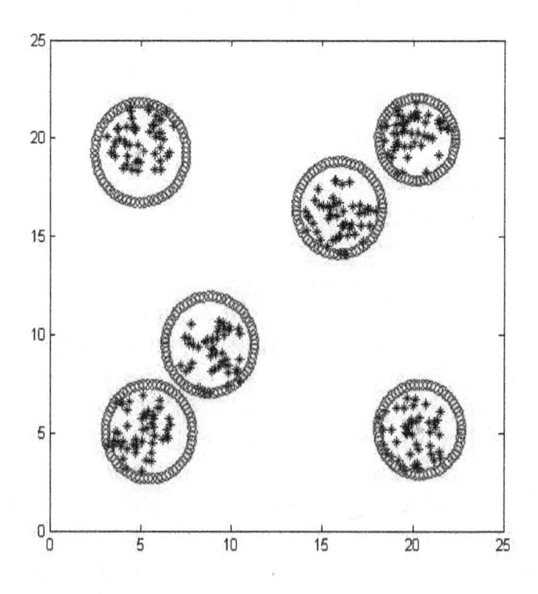

Fig. 3. Result for 300 points

this type of result is best suited. Here we can install each mobile tower at the center of each such circle.

For randomly generated input set of 50 points, the result is shown in Fig. 4, where $k = 5$. In this case also the number overlapping points is zero.

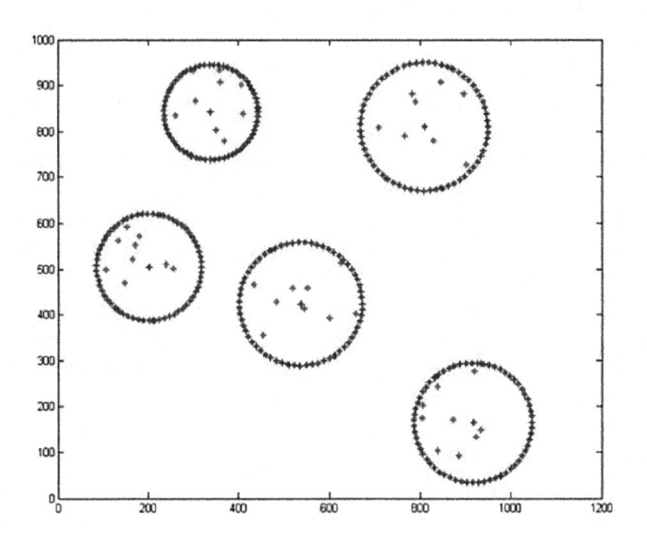

Fig. 4. Result for 50 points

5 Conclusion

In this work we have considered the *multi-objective k-center sum clustering problem*. If we consider the overlapping area as our second objective then the resulting clusters would be more disjoint. If we consider the number of points in the overlapping region as our second objective then there may be certain scenarios when overlapped regions contain no points but the area of the overlapping region will be large. In this paper we have considered that the center of a cluster can be any point in the plane. Another variation of the problem are there where the cluster center must be a point of the given input point set.

References

1. Xu, R., II Wunsch, D.: Survey of clustering algorithms. IEEE Transactions on Neural Networks 16(3), 645–678 (2005)
2. Xu, R., Wunsch, D.: Clustering. Wiley-IEEE Press (2009)
3. Hochbaum, D.S., Shmoys, D.B.: A best possible approximation algorithm for the k-center problem. Math. Oper. Res. 10, 180–184 (1985)
4. Dyer, M.E., Frieze, A.M.: A Simple Heuristic for the P-centre Problem. Oper. Res. Lett. 3(6), 285–288 (1985)

5. Hansen, P., Jaumard, B.: Minimum sum of diameters clustering. Journal of Classification 4(2), 215–226
6. Behsaz, B., Salavatipour, M.R.: On Minimum Sum of Radii and Diameters Clustering. In: Fomin, F.V., Kaski, P. (eds.) SWAT 2012. LNCS, vol. 7357, pp. 71–82. Springer, Heidelberg (2012)
7. Charikar, M., Panigrahy, R.: Clustering to Minimize the Sum of Cluster Diameters. In: Proceedings of the Thirty-Third Annual ACM Symposium on Theory of Computing, STOC 2001, pp. 1–10. ACM, New York (2001)
8. Hamacher, H.W., Drezner, Z.: Facility location: applications and theory. Springer (2002)
9. Ranjan, P., Mahapatra, S.: Studies on Variations of Enclosing Problem using Rectangular Objects. PhD thesis, University of Kalyani (2012)
10. Churchand, R.L., ReVelle, C.S.: Theoretical and Computational Links between the p-Median, Location Set-covering, and the Maximal Covering Location Problem. Geographical Analysis 8(4), 406–415 (1976)
11. Alt, H., Arkin, E.M., Brönnimann, H., Erickson, J., Fekete, S.P., Knauer, C., Lenchner, J., Mitchell, J.S.B., Whittlesey, K.: Minimum-cost Coverage of Point Sets by Disks. In: Proceedings of the Twenty-Second Annual Symposium on Computational Geometry, SCG 2006, pp. 449–458. ACM, New York (2006)
12. Sharir, M., Welzl, E.: Rectilinear and Polygonal p-Piercing and p-Center Problems.. In: Symposium on Computational Geometry, pp. 122–132 (1996)
13. Mukherjee, M., Chakraborty, K.: A polynomial-time optimization algorithm for a rectilinear partitioning problem with applications in VLSI design automation. Inf. Process. Lett. 83(1), 41–48 (2002)
14. Mukherjee, J., Ranjan, P., Mahapatra, S., Karmakar, A., Das, S.: Minimum-width rectangular annulus. Theor. Comput. Sci. 508, 74–80 (2013)
15. Goldberg, D.E.: Genetic Algorithms in Search, Optimization and Machine Learning. Addison-Wesley, Reading (1989)
16. Atta, S., Mahapatra, P.R.S.: Genetic Algorithm Based Approaches to Install Different Types of Facilities. In: Satapathy, S.C., Avadahani, P.S., Udgata, S.K., Lakshminarayana, S. (eds.) ICT and Critical Infrastructure: Proceedings of the 48th Annual Convention of CSI - Volume I. AISC, vol. 248, pp. 195–203. Springer, Heidelberg (2014), http://dx.doi.org/10.1007/978-3-319-03107-1_23
17. Ranjan, P., Mahapatra, S., Goswami, P.P., Das, S.: Covering Points by Isothetic Unit Squares. In: Proceeding of the 19th Canadian Conference on Computational Geometry (CCCG 2007), pp. 169–172 (2007)
18. Ranjan, P., Mahapatra, S., Goswami, P.P., Das, S.: Maximal Covering by Two Isothetic Unit Squares. In: Proceeding of the 20th Canadian Conference on Computational Geometry (CCCG 2008), pp. 103–106 (2008)
19. Mahapatra, P.R.S., Karmakar, A., Das, S., Goswami, P.P.: k-enclosing axis-parallel square. In: Murgante, B., Gervasi, O., Iglesias, A., Taniar, D., Apduhan, B.O. (eds.) ICCSA 2011, Part III. LNCS, vol. 6784, pp. 84–93. Springer, Heidelberg (2011)
20. Atta, S., Ranjan, P., Mahapatra, S.: Genetic Algorithm based Approach for Serving Maximum Number of Customers Using Limited Resources, 10th edn., pp. 492–497. Elsevier (2013)
21. Megiddo, N., Supowit, K.J.: On the Complexity of Some Common Geometric Location Problems.. SIAM J. Comput. 13(1), 182–196 (1984)
22. Deb, K.: Multi-objective optimization using evolutionary algorithms. John Wiley & Sons, Chichester (2001)

23. Deb, K., Pratap, A., Agarwal, S., Meyarivan, T.: A fast and elitist multiobjective genetic algorithm: NSGA-II. IEEE Transactions on Evolutionary Computation 6(2), 182–197 (2002)
24. Agrawal, R.B., Deb, K., Agrawal, R.B.: Simulated binary crossover for continuous search space (1994)
25. Beyer, H.-G., Deb, K.: On Self-adaptive Features in Real-parameter Evolutionary Algorithms. Trans. Evol. Comp 5(3), 250–270 (2001)
26. Raghuwanshi, M.M., Kakde, O.G.: Survey on multiobjective evolutionary and real coded genetic algorithms. In: Proceedings of the 8th Asia Pacific Symposium on Intelligent and Evolutionary Systems, pp. 150–161 (2004)
27. Bandyopadhyay, S., Maulik, U.: Nonparametric genetic clustering: comparison of validity indices. IEEE Transactions on Systems, Man, and Cybernetics, Part C: Applications and Reviews 31(1), 120–125 (2001)
28. Bandyopadhyay, S., Maulik, U.: Genetic clustering for automatic evolution of clusters and application to image classification.. Pattern Recognition 35(6), 1197–1208 (2002)
29. Bandyopadhyay, S., Pal, S.K.: Classification and learning using genetic algorithms: applications in bioinformatics and web intelligence. Springer (2007)

Reliability Aware Load Balancing Algorithm for Content Delivery Network

Punit Gupta, Mayank Kumar Goyal, and Nikhil Gupta

Department of Computer Science Engineering,
Jaypee University of Information Technology,
Himachal Pradesh, India
{punitg07,mayankrkgit,candid.nikhil}@gmail.com

Abstract. With increasing use of internet and data sharing over internet network traffic over internet has increased beyond a limit, which has also increased the number of request made for a resource over a server. So to maintain the Quality of service even if the request made become large CDN (Content Delivery Network) is used. Main goal of CDN is to balance the load over the servers. But with the increase of load over a server even after balancing decreases the reliability of server with increase in fault rate and processing time. Proposals made for CDN do not take into consideration real time faults accuring in a server with time. To overcome this fault and reliability aware load balancing algorithm for CDN is proposed in this paper to increase the scalability and reliability of CDN.

Keywords: CDN, QoS, Reliability, Load balancing, Fault rate, Network load, System load.

1 Introduction

With the rapid development of Internet, network traffic is exploding. There is now an upsurge in demand for content services. Therefore to provide uninterrupted services to users, Internet Service Providers should scale their infrastructure. A network of few servers is not sufficient to handle the large number of requests. As the request over servers is rapidly increasing its management is also getting tougher. This fast growing traffic over servers is triggering many problems. Content Delivery Networks is among one of the best methods to cope up with the increasing demands. A Content Delivery Network (CDN) is emerging as an effective solution to support the load of fast growing web applications by adopting a distributed network of servers [8]. CDN has been widely accepted as a method of circulating large amount of content to the user's .By making several redundant copies of content CDN can solve even high congestion issues occurred due to unexpectedly high request rate from clients to an extent. CDN consists of a backend server which consists of new data to be spread, together with more than one stand-in servers. Only static data is stored upon the stand-in servers, while dynamic information is stored on few backend servers. The backend server also updates the stand-in servers regularly. Therefore CDN maintains accessibility, while

S.C. Satapathy et al. (eds.), *Emerging ICT for Bridging the Future – Volume 1*,
Advances in Intelligent Systems and Computing 337, DOI: 10.1007/978-3-319-13728-5_48

preserving correctness by keeping the redundant copies of data over multiple servers. It involves use of the techniques like load balancing, request routing and maintaining multiple copies of resources. There are many issues and parameters which restricts the performance of CDN such as issue of load balancing, cost, request traffic, response time. Many proposals based on Cost [1-3] have been proposed to balance load based on Cost, and few others based on Response time and load on server [4-6]. CDN has even been discussed on the basis of Energy consumption and data transfer rate [3,7, 9]. These proposals take into consideration energy consumed by the server and data transfer rate in the server. So the primary issue that persists in CDN is load balancing which can be defined as (number of requests fulfilled by a data center and queue length of a data center).

In this paper we have tried to resolve the problem of load balancing in CDN and to overcome the drawbacks of proposed studies and to design a generalized load balancing algorithm for CDN networks. This paper is divided into 5 sections: Introduction, Related Work, Proposed Model, Experimental result and Conclusion. Section 2 related work discusses the survey of proposed model for load balancing in CDN and their drawbacks. In section 3 we present the proposed models for request scheduling. This section discusses the proposal to overcome the drawbacks.

2 Related Work

Taking into consideration load balancing algorithm proposed for CDN and other distributed system like grid computing and cloud computing. Many load balancing algorithms have been proposed in content delivery networks based on load, cost, response time, QoS, response time and energy.

Some of the similar load balancing algorithm based on cost was proposed by Chrysa Papagianni [1], in which a hierarchical framework is proposed which is further evaluated towards an efficient and scalable of content distribution over a multi provider networked cloud environment, where inter and intra cloud communication resources are simultaneously considered along with traditional cloud computing resources. The performance of this proposed framework is accessed via simulation and modeling, while appropriate metrics are defined associated with and reflecting the interests of different key players.

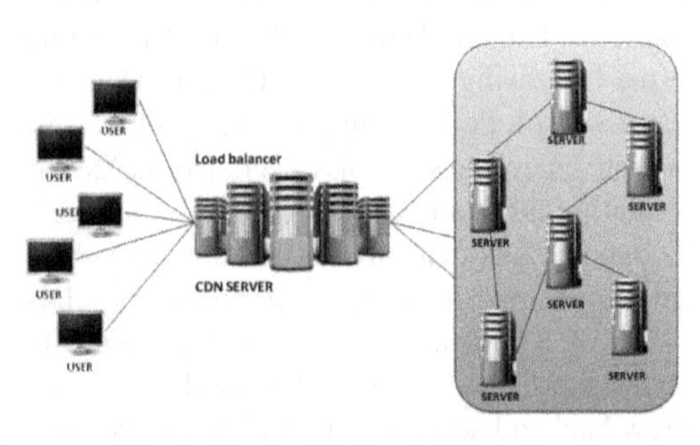

Fig. 1. Content Delivery Networks

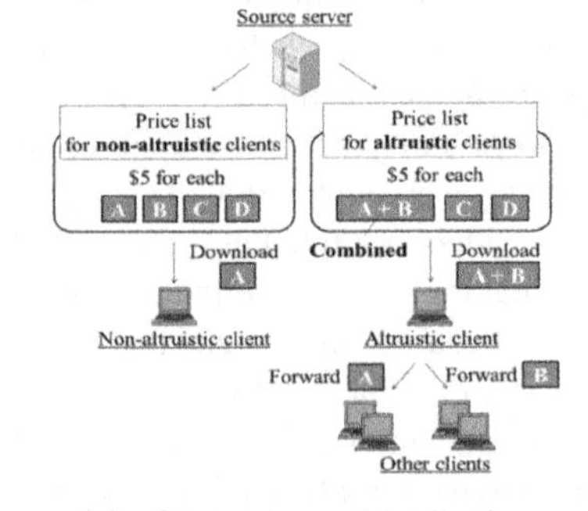

Fig. 2. Cost based Distribution

Similar cost based algorithm was proposed by Naoya Maki [3] which propose a periodic combined-content distribution mechanism to increase the gain in traffic localization. As shown in figure 2, this proposed mechanism automatically optimizes the distribution period by using how long we can expect the previous downloaded combined-content to localize traffic. Another algorithm based on linear program formulation was proposed by Derek Leong [2] takes into account various costs and constraints associated with content dissemination from the origin that is server to nodes that is storage, and the eventual fetching of content from storage nodes by end users.

Vimal Mathew [7] giving new dimension to CDN proposed an energy aware load balancing algorithm which is an optimal offline and online algorithm and can be used to extract energy savings both at the level of local load balancing at the data center and global load balancing across data centers. S.Manfredi [10] also proposed an algorithm to improve overall throughput and response time and hence the performance. It proposes a highly dynamic distributed strategy based on the periodical exchange of information about the status of the nodes, in terms of load. Also they study the scalability of this algorithm and performed a comparative evaluation of its performance with respect to its best solutions.

3 Proposed Work

In all the papers which we discussed previously, Fault aware load balancing in Content delivery network (FLB) has been simulated and modeled taking in consideration the parameters like cost, load, request time, QoS, throughput. But none of them took fault as a parameter while considering load balancing for CDN. So our algorithm along with the factors discussed previously will also take fault as a parameter for load balancing in CDN. Therefore the factors on which our algorithm is based on are Network Load, Fault Rate, Queue length, and Response Time. These parameters can be defined as:

a) **System Load:** Total number of MIPS (Million of Instruction per Second) under use.
b) **Fault rate:** Number of faults over a period of time.
c) **Queue Length**: Maximum size of request queue length a server can maintain and fulfill.
d) **Response time:** Time taken to start fulfilling a request.
e) **Network load**: Total bandwidth of server out of total under utilization.

Proposed algorithm is divided into three phases:

a) Initialization
b) Load balancing
c) Updating

A. Initialization

In this phase fitness value for a datacenter is initialized with default values of all the parameters explained above. All the parameters are checked and updated periodically. Initially fault rate and network load is zero where as Queue length and response time are based on the datacenter properties. For example if a datacenter can maintain a queue length of 100 request and response time on an average for any request be 10 milliseconds. Based on these values fitness values is calculated.

When a new datacenter is introduced in CDN it is initialized with default values and fitness value is calculated and updated after equal interval of time. Initial parameters are defined as:

Fault_Ini: Initial fault rate.
Quel_Ini: Initial Queue length based on datacenter.
Resp_Ini: Initial response time based on datacenter.
Netload_Ini: Initial network load.

B. Load Balancing

In this phase when the server queue is full and no more request can be queued other request are dropped or rejected because they cannot be fulfilled. So to overcome this replica of the data been requested is made on other server i.e. we require to find a datacenter which can fulfill the request. Here we can classify the servers into two categories as hot spot and cold spot. Hot spots are those servers which are over loaded with requests and have most of the MIPS and network bandwidth under utilization. Cold spots are those datacenters which have low request rate and can accommodate more requests .In other words datacenters with low MIPS and network bandwidth under utilization i.e. load network and processing load.

Load balancing is required to stop server to become hot spot. Whenever load balancing is called we need to find best fit server based on following parameters.

1) **Fault rate :** It is directly propositional to the load on the server that can be network load which leads to network failure and system load which leads to system failure on the other had degradation in QoS (quality of service) provided by the server.

$$\Lambda(t) = F(N\_Load\_new, S\_load\_new) \quad (1)$$

$\Lambda(t)$: fault over a time t.
N_load : network load.
S_load : system load.

Equation 1 defines that fault rate at particular instance of time is functionally and directly proportional to system load and network load.

Λ: fault rate

$$\Lambda = \sum \text{total number of fault / per hour;} \quad (2)$$

2) **Response time**: This can be defined as the time taken to start processing a request i.e. difference between the time request was submitted and the time server started processing the request. It is directly propositional to system load. As the CPU utilization of server increases response time increases. So the server which needs to be selected should have least average response time.

Resp = Response time

3) **Queue length:** Every server has a fixed request queue length which can be fulfilled without request failure. So we need to select a server for load balancing which have a sufficient largest free queue size to accommodate new requests without failure.

To balance the load we need to take all the above parameters into consideration and calculate fitness value for each server over which load can be balanced to provide better QoS and increase the reliability of the overall system by balancing the load and reducing failures.

Fitness value for a server can be determined as:

fval (s): Fitness value of datacenter s

$$\text{Fval}(s) = \left(\alpha 1 * \frac{1}{\lambda}\right) + \left(\alpha 2 * \frac{1}{\text{Resp}}\right) \quad (3)$$

server_id= min (fval1, fval2...., fvaln); (4)

Server for load balancing will be selected based on the one with highest fitness value i.e. least fault rate, lease network load, least system load, and largest free queue length. This approach helps in maintaining skewness and increase reliability and decrease fault rate.

C. Fitness Updating

This phase includes updating the value of current network load, system load, fault rate, queue length of datacenter. This phase is repeated after an equal interval of time to get the updated current status of the servers. Initially all the parameters are initialized with default values in which fault rate λ (t) is initially zero, network load in also zero and system load is also taken zero. Queue length of a server in always initially zero because there are no request made for that server.

```
ʎ (t) _Initial = 0      \\ Initial fault rate
N_load_initial = 0   \\Initial network load.
S_load_initia = 0     \\Initial system load.
Q_len_initial = 0     \\Initial queue length
Res_Initial = 0;        \\Initial server response time
```

For calculating new fitness value we need to find changes in the parameters. Let Si be the server, ʎ (t)_new, N_load_new, S_load_new, Q_len_new , Res_new are new fault rate over a time 't' ,new network load, system load, queue length and response time correspondingly. Let new fitness value be fvalt_new (Si) of server i.

$$ʎ(t) = F(N\_Load\_new, S\_load\_new) \tag{5}$$

$$Fval\_new(s) = \left(\alpha 1 * \frac{1}{ʎ\_new}\right) + \left(\alpha 2 * \frac{1}{Resp\_new}\right) \tag{6}$$

$$server\_id = \min(fval1\_new, fval2\_new \ldots, fvaln\_new) \tag{7}$$

Whenever a fitness value is upgraded next request is always diverted to server with highest fitness value based on updated fitness values giver equation 4.

4 Experimental Results

In this for simulation GridSim API [10] is used. GridSim API basically support scheduling and load balancing in parallel and distributed environment. Load balancing, fault in server and server request queue feature of GridSim are used to simulate CDN.

Initially GridSim do not support failure in servers. In this implementation we have introduces fault aware scheduling in GridSim to study the performance of CDN in fault aware environment. To study the performance and compare the improvement we have referred queue length based load balancing proposal [11] by S manfredi. In this paper author has taken queue length as a basis for load balancing.

Based on this attribute we have computed the result to study the problem. For this we have considered 3 servers S1, S2, S3 with queue length, fault rate and processing rate i.e. is the number of requests that can be parallel processed shown in table 1.

With 60 user requests which are bifurcated as 50 request for server1, 10 request for server2 and server3 is free shown in table 2. So the total number of failure that occurred is shown in table 3 using queue length based load balancing (QLBLB).

On the other hand using proposed algorithm selects the server with sufficient free queue length and least fault rate which decreases overall failure in the system and output is shown in table 4.

Table 1. Servers Parameters

Server Name	Queue length	Fault rate	Processing rate
Server1	10	0.143	10
Server2	50	0.125	50
Server3	50	0.5	50

Table 2. Request rate

Server Name	Request
Server1	50
Server2	10
Server3	0

Table 3. Failure Count with QLBLB

Server Name	Failure count
Server1	1
Server2	1
Server3	11

Table 4. Failure Count with FLB

Server Name	Failure count
Server1	1
Server2	5
Server3	0

Table 5. Configuration Details

Server Name	Queue length	Fault rate	Processing rate
Server1	100	0.143	30
Server2	250	0.125	90
Server3	250	0.5	90

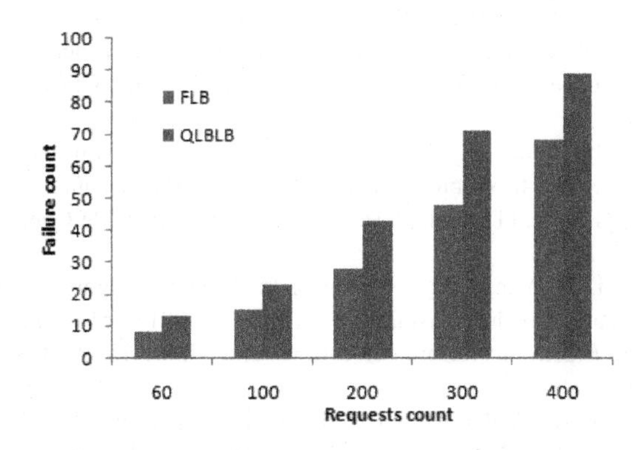

Fig. 3. Comparison of FLB and QLBLB with various requests counts

Table 6. Failure rates

	Request count			
	60	100	200	300
Failure count of Proposed (FLB)	8	15	28	48
Failure count of QLBLB	13	23	43	71

Figure 3 shows improvement in fault rate with increase in requests. This shows the performance of QLBLB and proposed fault aware load balancing algorithm. Proposed algorithm is also tested for 200, 300, 400, 500 user requests and 3 datacenters with following configuration shown in table 5. Table6 shown the over reduce in failure rate and increase in reliability as the request count increases. So as the result shown increase in overall system reliability with the use of fault based load balancing algorithm.

5 Conclusion

In this paper different type of Load balancing algorithm have been discussed with their drawbacks in CDN. To overcome the drawbacks an efficient fault aware load balancing algorithm is proposed which perform better than other load balancing algorithm proposed for CDN in fault aware environment. For future work this algorithm may be compared with other proposals and study the improvement in the QoS.

References

[1] Chrysa, P., Leivadeas, A., Papavassiliou, S.: A cloud-oriented content delivery network paradigm: modeling and assessment. IEEE Transactions on Dependable and Secure Computing 10(5), 287–300 (2013)

[2] Derek, L., Ho, T., Cathey, R.: Optimal content delivery with network coding. In: 43rd Annual Conference on Information Sciences and Systems, CISS 2009, pp. 414–419. IEEE (2009)

[3] Naoya, M., Shinkuma, R., Mori, T., Kamiyama, N., Kawahara, R.: A periodic combined-content distribution mechanism in peer-assisted content delivery networks. In: Proceedings of the ITU Kaleidoscope: Building Sustainable Communities (K-2013), pp. 1–8. IEEE (2013)

[4] Xueying, J., Li, S., Yang, Y.: Research of load balance algorithm based on resource status for streaming media transmission network. In: 2013 3rd International Conference on Consumer Electronics, Communications and Networks (CECNet), pp. 503–507. IEEE (2013)

[5] Li, L., Xiaozhen, M., Yulan, H.: CDN cloud: A novel scheme for combining CDN and cloud computing. In: 2013 International Conference on Measurement, Information and Control (ICMIC), vol. 1, pp. 687–690. IEEE (2013)

[6] TaeYeon, K., Song, H.: Hierarchical Load Balancing for Distributed Content Delivery Network. In: 2012 14th International Conference on Advanced Communication Technology (ICACT), pp. 810–813. IEEE (2012)

[7] Vimal, M., Sitaraman, R.K., Shenoy, P.: Energy-aware load balancing in content delivery networks. In: Proceedings IEEE INFOCOM, pp. 954–962. IEEE (2012)

[8] Naoya, M., Nishio, T., Shinkuma, R., Takahashi, T., Mori, T., Kamiyama, N., Kawahara, R.: Expected traffic reduction by content-oriented incentive in peer-assisted content delivery networks. In: 2013 International Conference on Information Networking (ICOIN), pp. 450–455. IEEE (2013)

[9] Sabato, M., Oliviero, F., Romano, S.P.: Optimised balancing algorithm for content delivery networks. IET Communications 6(7), 733–739 (2012)

[10] Rajkumar, B., Murshed, M.: Gridsim: A toolkit for the modeling and simulation of distributed resource management and scheduling for grid computing. Concurrency and computation: Practice and Experience 14(13-15), 1175–1220 (2002)

[11] Sabato, M., Oliviero, F., Romano, S.P.: A distributed control law for load balancing in content delivery networks. IEEE/ACM Transactions on Networking (TON) 21(1), 55–68 (2013)

SQL Injection Detection and Correction Using Machine Learning Techniques

Garima Singh[1], Dev Kant[2], Unique Gangwar[3], and Akhilesh Pratap Singh[4]

[1] Department of CSE, Jaypee University of Information Technology,
Waknaghat, Solan, H.P, India
garimasingh2841@gmail.com
[2] Department of CSE, Indian Institute of Information Technology Allahabad
Deoghat, Jhalwa, Allahabad, U.P, India
devkantg@gmail.com
[3] Department of CSE, Jaypee Institute of Information Technology,
Noida, U.P, India
unique.gangwar@gmail.com
[4] Department of CSE, Kamla Nehru Institute of Technology,
Sultanpur, U.P, India
akhilesh.chauhan88@gmail.com

Abstract. SQL is a database language which is used to interact with the database. SQL is a language with the help of which database could be created, modified and deleted. Nowadays every organization used to have their own databases which may keep important information which should not be shared publicly. The SQL injection technique is now one of the most common attacks on the Internet. This paper is all about SQL injection, SQL injection attacks, and more important, how to detect and correct SQL injection. This paper proposes an algorithm to detect not only the SQL injection attack but also detects unauthorized user by maintaining an audit record using machine learning technique (clustering).

Keywords: SQL Injection Technique, SQL Injection Attack, Detect & Correct SQL Injection, Clustering, Audit Record.

1 Introduction

SQL Injection is one of the most effective methods for stealing the data from the backend, with the help of these attacks hackers can get access to the database and steal sensitive information. According to the "Open Web Application Security Project", injection attack is a technique used in hacking or cracking to access information or unauthorized activity [1]. Now a day's most web applications are being hacked using SQL Injection method [2] Till now many approaches have been proposed for detecting the injecting query but practical implementation is not possible. As we have three main technique i.e. Prevention, detection and correction. So prevention is a little bit complication because none of the proposed method provides the exact solution for preventing injected query [2, 3, 4], there are many additional methods through which injection is possible. This paper presents a

S.C. Satapathy et al. (eds.), *Emerging ICT for Bridging the Future – Volume 1*,
Advances in Intelligent Systems and Computing 337, DOI: 10.1007/978-3-319-13728-5_49

technique for the detection and correction of different type of SQL attacks. In the proposed method an audit record should be maintained for each transaction and with the help of that record, it could be checked that which kind of data is being accessed by the transaction, which type of resources are being used and whether they have privilege to query such data and with the help of clustering technique clusters are made for attacks and if attributes of record falls under attack clusters then that transaction must be blocked and verified and if negative result comes then that query must not be executed. In order to solve these problems, researchers have developed various detection [8-10] and Prevention techniques [11-14]. This paper has been organized in the following way; section 2 describes the basic idea of SQL Injection attack (SQLIAs). The proposed algorithm is defined in Section 3. In section 4, the experimental outcomes are introduced and discussed. Finally section 5 concludes the complete paper.

2 Types of Attacks

There are various methods of SQL Injection attacks which may be performed either together or sequentially [5, 6].

Tautologies: Tautology attack injects SQL tokens to the conditional query statement which is always evaluated as true [18][22][23]. Example;
"SELECT * FROM employee_record WHERE emp_id = '102' and Password ='aa' OR '1'='1'"

Invalid/Logically Incorrect Queries: when a query is rejected, an error message is returned from the database [10]. In the given example an attacker makes a type mismatch error by injecting the following text into the pin input field:
1) Original URL:
http://www .archeive.polimLitieventil?id _ nav=8864
2) SQL Injection:
http://www.archeive.polimi.itleventil?id nav=8864'
3) Error message showed:
SELECT emp_name FROM Employee WHERE id =8864\'

Union Query: With the help of union query attacker can append injected query with the original query by the word UNION and then can extract information about other tables from the database [10][18][23].

Piggy-Backed Queries: In this type of attack, unauthorized user exploit database with the help of query delimiter, such as ";", by appending an extra query with the original SQL query [10][19][20][22].

Stored Procedure: This is also a part of the database that the programmer can set an extra abstraction layer on the database [7].

Inference: With the help of inference attack, intruders can change the behavior of a database or web application. [12].

3 Proposed Work

The proposed approach provides a model for the purpose of understanding the behavior of the object. Thus by observing the behavior of the object it could be verified that whether the user is authorized or not. An audit record is a fundamental tool for detecting the behavior of an object. Record of ongoing activity of users must be maintained and can be used for determined that whether the transaction is valid or not.

3.1 Audit Record Must Contain the Following Fields

Subject: A subject is an end user but might also be process acting on behalf of users or group of users. All activity arises through SQL commands issued by subjects.

Action: Action involves operations performed by the subject (SQL Command) on or with an object (Database); for example, login, read, write, perform I/O, execute.

Object: Receptors of actions. The example involves a database.

Exception Condition: Defines which, if any, exception condition is raised on return.

Resource Usage: A list of quantitative elements in which each element gives the amount used of some resource.

Time-Stamp: It's a unique time and date stamp used for identifying when the action took place.

3.2 Measures Must Be Used for Unauthorized Detection

Login and Session Activity: Login frequency by day and time, Time since last login, Elapsed time per session, session resource utilization, password failures at login.

Command or Program Execution Activity: Execution frequency, Program resource utilization, Execution denials.

File Access Activity: It involves Read, write, create, delete frequency, records read and written, failure count for read, write, create and delete.

3.3 These Are the Various Measures Which Can Be Verified in the Audit Record with the Help of Following Clustering Technique

The following steps of the algorithm are:

Step1-To check the behavior of an object or transaction, an audit record must be maintained by host agent to record the ongoing activity of an object or transaction.

Step2- The attributes of audit record are clustered with the help of k-mean clustering technique for detecting attacks by central manager.

Step3- Now, with the help of one-rule classification attacks are classified into classes to reduce computational time.

Step4- Repeat step2 and step3 until a sufficient number of samples are available.

Step5- Tabulate the various values of the decision criterion and choose the best policy.

The proposed model is based on Intrusion Detection System (IDS) to detect SQL Injection attacks (SQLTAs), which involves a machine learning technique. As the query parser generates the original query and clustering must be done of the injected query and there are various attributes have been defined on the basis of which behavior of the query must be judged. Figure 1 shows the proposed architecture.

3.4 The Algorithm Works on the Basis of the Following Three Main Components Are as Follows

Host Agent Module: Its purpose is to collect data on the above defined measures of host and transmit these to the central manager.

Monitor Agent Module: Its work is to check Execution frequency, program resource utilization and reports the results to the central manager.

Central Manager Module: It receives reports from the above two modules and processes and correlates these reports to detect unauthorized access.

The agent collects each audit record produced by the native audit collection module. A filter is implemented that retains only those commands that are of security interest. Records are then formatted in a standardized format as the host audit record (HAR). After this, a template driven logic module analysis the SQL commands or unauthorized commands for suspicious activity. At the lowest stage, the agent scans for interest. Records are then formatted in a standardized format as the host audit record (HAR).

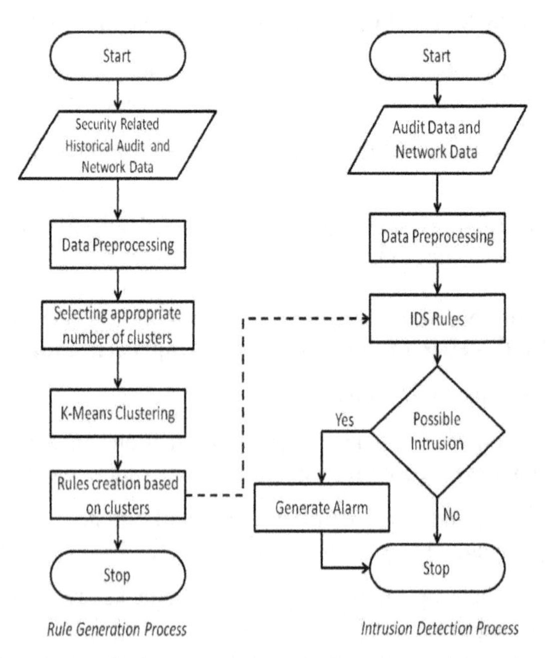

Fig. 1. Flow chart of Rule creation through clustering and Intrusion detection

After this, a template driven logic module analysis the SQL commands or unauthorized commands for suspicious activity. At the lowest stage, the agent scans for suspected events that are of interest independent of any past events. Examples include failed record accesses, accessing system files, and changing a file's access control.

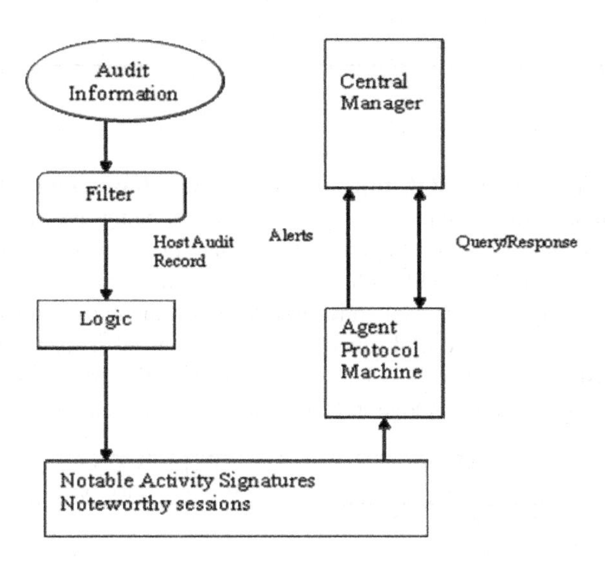

Fig. 2. Architecture

At the next higher level, the agent looks for anomalous behavior of an individual user based on a historical profile of that user, such as the number of programs accessed or executed, number of records accessed, authorized for accessing the records etc. When suspicious activity is detected, an alert is sent to the central manager. The central manager module includes an expert system that can draw inferences from received data. The manager can also query individual systems for copies of HARs to correlate with those of other agents.

4 Result and Analysis

4.1 Result

The experiment has done to evaluate the detection rate for SQL Injection; we have used 3 different types of web applications which is the same method as SQLCheck [15] and AMNESIA [16].

Table 1. Experiment Results

Audit Record	Proposed Algorithm		SQL Check [11]		AMNESIA[12]	
	Attack\ Detection	Detection Rate(%)	Attack\ Detection	Detection Rate(%)	Attack\ Detection	Detection Rate(%)
Employee	257/257	100	3937/3937	100	280/280	100
Record	298/298	100	3473/3473	100	182/182	100
Library Portal	374/374	100	3685/3685	100	140/140	100

4.2 Analysis

This paper proposes a new SQL injection attack detection method that utilizes both Static and Dynamic Analysis. An audit record is being used to compare the attack frequency and detection frequency to calculate the detection rate. As shown below in table 1 there is no difference in the detection rate compared to other researchers. This is because the SQL Injection detection methods compare the static SQL queries with the dynamic SQL queries for detection rather than using machine learning or statistical methods. It results in a high detection rate. As a result, the detection rate cannot be used to judge the efficiency of the detection.

Detection/Prevention Method	Tautologies	Illegal/Incorrect Queries	Union Queries	Piggy-Backed Queries	Stored Procedures	Inference	Alternate Encodings
SQLCheck[11]	✓	✓	✓	✓	×		
IDS	o	o	o	o	o	o	o
SQLGuard	✓	✓	✓	✓	×		
AMNESIA[12]	✓	✓	✓	✓	×		
Security Gateway	■	■	■	■	■	■	■
Proposed Method	✓	✓	✓	✓	✓	✓	✓

Symbols:

✓ This symbol defines that detection and prevention is possible.

× Defines that detection and prevention is impossible.

o Defines that detection and prevention are partially possible.

▪ Defines that there is no relation.

5 Conclusions

In this paper, the various types of SQL injection attacks have been organized. Then we investigated the SQL injection detection and prevention techniques. After that we have proposed an algorithm to detect the SQL Injection Attack. As Prevention is little bit complicated, so focus is on detection and correction of unauthorized access or transaction. In the future work we will separate techniques which have been implemented as tools then compare effectiveness, efficiency, stability, flexibility and performance of tools to show the strength and weakness of the tool.

References

[1] Open Web Application Security Project, Top Web application vulnerabilities for (2010), http://www.owasp.org/index.php/
[2] William, G.J., Fond, H., Orso, A.: Projecting applications using positive tainting and syntax Member. IEEE Comptter Society 34(I) (January - February 2008)
[3] Halfond, W.G., Viegas, T., Orso, A.: A Classification of SQL injection Attacks and Counter measures. In: Proc. of the Intl. Symposium on Secure Software Engineering (March 2006)
[4] Tajpour, A., Masrom, S.M.: SQL Injection Detection and Prevention Techniques. Proc. International Journal of Advancements in Computing Technology 3(7) (August 2011)
[5] Kindy, D.A., Pathan, A.-S.K.: A Survey On SQL Injection: Vulnerabilities, Attacks And Prevention Techniques. In: IEEE 15th International Symposium on Consumer Electronics (2011)
[6] Kavi, S.B., Bisht, P., Madhusudan, P.: CANDID: Preventing SQL injection attacks using Dynamic candidate Evaluations. ACM, Alexandria (2007)
[7] Cova, M., Balzarotti, D.: Swaddler: An approach for the Anomaly-based Detection of state violations in web applications. In: Proceedings of the 10th International Symposium on Recent Advances in Intrusion Detection, Queensland, Australia, pp. 63–86 (September 7, 2007)
[8] Wei, K., Muthuprasanna, M., Kothari, S.: Preventing SQL Injection Attacks in Stored Procedures. In: Proceedings of the 2006 Australian Software Engineering Conference (ASWEC). IEEE (2006)
[9] Kruegel, C., Vigna, G.: Anomaly Detection of Web based Attacks, CCS (2003)
[10] McDonald, S.: SQL Injection: Modes of attack, defense, and why it matters. White paper, Government Security.org (April 2002)
[11] Bertino, E., Kamara, A., Early, J.P.: Profiling Database Application to Detect SQL Injection Attacks (2007)
[12] Spett, K.: Blind sql injection. White paper, SPI Dynamics, Inc. (2003), http://www.spidynamics.com/whitepapers/BlindSQLInjection.pdf
[13] Ezumalai, R., Aghila, G.: Combinatorial Approach for Preventing SQL Injection Attacks. In: International Advance Computing Conference (IACC 2009). IEEE (2009)
[14] Muthuprasanna, M., Wei, K., Kothari, S.: Eliminating SQL Injection Attacks- A Transparent Defense Mechanism. In: Eight IEEE International Symposium on Web Site Evolution(WSE 2006) (2006)
[15] Halfond, W.G., Orso, A.: AMNESIA: Analysis and Monitoring for Neutralizing SQL-Injection Attacks. In: Proceedings of the 20th IEEE/ACM International Conference on Automated Software Engineering, pp. 174–183 (2005)
[16] Buehrer, G., Weide, B.W., Sivilotti, P.A.: Using Parse Tree Validation to Prevent SQL Injection Attacks. In: Proceedings of the 5th International Workshop on Software Engineering and Middleware, pp. 105–113 (2005)
[17] Kim, J.-G.: Injection Attack Detection using the Removal of SQL Query Attribute Values, pp. 26–29. IEEE (2011)
[18] Anley, C.: Advanced SQL Injection In SQL Server Applications. White paper. Next Generation Security Software Ltd (2002)
[19] Bouma, F.: Stored Procedures are Bad, O'kay?Technical report, Asp.Net Weblogs (November 2003), http://weblogs.asp.net/fbouma/archive/2003/11/18/38178.aspx

[20] Fayo, E.M.: Advanced SQL Injection in Oracle Databases. Technical report, Argeniss Information Security, Black Hat Briefings, Black Hat USA (2005)

[21] Finnigan, P.: SQL Injection and Oracle - Parts 1 & 2. Technical Report, Security Focus (November 2002), http://securityfocus.com/infocus/1644, http://securityfocus.com/infocus/1646

[22] Howard, M., LeBlanc, D.: Writing Secure Code, 2nd edn. Microsoft Press, Redmond (2003)

[23] Labs, S.: SQL Injection. White paper. SPI Dynamics, Inc. (2002), http://www.spidynamics.com/assets/documents/WhitepaperSQLInjection.pdf

An Enhanced Ontology Based Measure of Similarity between Words and Semantic Similarity Search

M. Uma Devi and G. Meera Gandhi

Faculty of Computer Science,
Sathyabama University, Chennai, Tamil Nadu, India
{umadevi.as2006,drmeeragandhii}@gmail.com

Abstract. Measures of Semantic Similarity of two sets of words that describe two entities is an important problem in Web Mining. Semantic Similarity measures are used in various applications in Information Retrieval (IR) , Natural Language Processing (NLP) such as Word Sense Disambiguation (WSD), synonym extraction, query expansion and automatic thesauri extraction. The Computer being a syntactic machine, it cannot understand the semantics. Ontology is the explicit specialization of concepts, attributes and the relationships between them. It is for providing relevant and accurate information to the users for a particular domain. A new Semantic Similarity measure based on the domain Ontology is proposed here. It brings out a more accurate relationship between the two words The main purpose of finding Semantic Similarity is to enhance the integration and retrieval of resources in a more meaningful and accurate way. The performance analysis in terms of Precision and Recall for Traditional Search and Semantic Similarity Search is done. The Precision value of Semantic Similarity Search is high compared with the Traditional Search. This paper focuses on the approaches that differentiates the Semantic Similarity Research from other related areas.

Keywords : Semantic Similarity, Information Retrieval, Semantics, Semantic Web, Ontology, Text Snippets.

1 Introduction

Semantic Similarity is the Semantic relatedness of two words where a set of documents or terms within term lists are assigned a metric based on the likeness of their meaning / Semantic content. The study of Semantic Similarity between words is an important problem in the domain of Web Mining and various applications in Information Retrieval (IR) , Natural Language Processing (NLP). A Web Search Engine is a software code that is designed to search for information on the World Wide Web. Search results are generally presented in a line of results often referred to as Search Engine Results Pages (SERP's) [6]. This is convenient for human users to read and view but difficult for computers to understand. The Semantic Web offers an approach in which computers can use symbols with well-defined, machine-interpretable Semantics to share knowledge. Semantic Search uses Semantics or the

S.C. Satapathy et al. (eds.), *Emerging ICT for Bridging the Future – Volume 1*,
Advances in Intelligent Systems and Computing 337, DOI: 10.1007/978-3-319-13728-5_50

Science of meaning in language, to produce highly relevant search results [5], [6]. Semantic Similarity Search uses the Semantic Similarity measure which gives the accurate relationship between the two words.

The Role of Semantic Web

The Semantic Web is a framework that allows computers to publish, share, and reuse data and knowledge on the Web and across application, enterprise, and community boundaries [2]. The Semantic Web is an extension of the current Web which offers Web page documents as well as the relationships among resources denoting real-world objects.

When Berner-Lee described his vision in 2001, the development of the Semantic Web made use of two existing technologies: Extensible Markup Language (XML) and the Resource Description Framework (RDF) [2]. The Semantic Web is the second-generation WWW, enriched by machine - processable information which supports the user in his tasks. Berners-Lee suggested a layer structure for the Semantic Web [1] , [2].

RDF- (Resource Description Framework)

RDF is a data-model where each resource can be described by means of a triple <Subject, Predicate, Object> [2] , [3] . The Subject must always be a resource, which can be uniquely identified through an URI (Universal Resource Identifier). The object can be a string or another resource. The Predicate describes the relation between them. Fig. 1. depicts the author of a certain Book that is modeled in RDF

```
AuthorOf ('http://www.w3.org/employee/id1321','http: //www.books.org/ISBN0621')
```

Fig. 1. RDF of a Book

There are different formal languages with which the RDF-model can be serialized and the most appropriate for the Semantic Web is XML. **Fig. 2** depicts XML expression of the book RDF.

```
<rdf : Description
rdf : about: "http://www.w3.org/employee
e/id1321">
<authorOf rdf :
resource="http://www.books.org/ISBN0
621"/>
</rdf: Description>
```

Fig. 2. XML expression of book RDF

Semantic Similarity Search

Semantic Web Search involves the same set of high-level tasks: discovering and revisiting online documents, processing users' queries, and ordering search results. An advantage of machine-readable metadata such as Semantic mark-up is that the Search Engines can use it to infer additional Semantic relations; then, they can apply the so-called Semantic Search [6] , [14]. A new generation of intelligent Search Engines incorporates Web Semantics and uses more advanced search techniques based on concepts of Semantic Similarity.

The remaining part of the paper is laid out as follows. The next section , Section 2 provides an overview of the different methods used in Semantic Similarity measures. The various ways to find Semantic Similarity of words has been described in Section 3. Section 4 follows up this discussion by describing the investigated Similarity measures. Section 5 provides the details of our experimental evaluation on the query-query Similarity task. Finally, in Section 6 Conclusions and directions of Future Work has been wrapped up.

This paper is concerned with the discussion of new infrastructures, architectures, algorithms, etc., whose goal is to Search the Semantic Web on Similarity based approach.

2 Materials and Methods

2.1 Similarity Using Page Counts and Text Snippets

Page count of a query gives the number of pages that contain the query words. For example, the page count of the query "apple" AND "computer" in Google is 288,000,000, whereas the same for "Mango" AND "computer" is only 3,590,000. The more than 80 times more numerous page counts for "apple" AND "computer" indicate that "apple" is more Semantically Similar to "computer" than is " Mango" [9], [10], [12].

A brief window of text extracted by a Search Engine around the query term in a document is a Snippet. It provides useful information regarding the local context of the query term. Semantic Similarity measures defined over Snippets have been used in query expansion. The main aim of query expansion (also known as query augmentation) is to add new meaningful terms to the initial query. It proposes a method that considers both page counts and lexical syntactic patterns extracted from snippets.

2.1.1 Similarity Measures Using Page Count

The Semantic Similarity between the two words X and Y is modeled as a function Sim (X, Y) that returns a value in range [0: 1]. The popular Co-Occurrence measures; Jaccard, Overlap (Simpson), Dice, and Point wise Mutual Information (PMI), have been dealt in the equations (1),(2),(3) & (4) that are used to compute Semantic Similarity using page counts.

WebJaccard (X,Y) is defined as,

$$\text{WebJaccard (X,Y)} = \begin{cases} 0, & \text{if } H(X \cap Y) \leq c, \\ \frac{H(X \cap Y)}{H(X) + H(Y) - H(X \cap Y)}, & \text{otherwise.} \end{cases} \quad (1)$$

WebOverlap (X,Y) is defined as,

$$\text{WebOverlap (X,Y)} = \begin{cases} 0, & \text{if } H(X \cap Y) \leq c, \\ \frac{H(X \cap Y)}{\min(H(X), H(Y))}, & \text{otherwise} \end{cases} \quad (2)$$

WebDice (X,Q) is defined as

$$\text{WebDice (X,Q)} = \begin{cases} 0, & \text{if } H(X \cap Y) \leq c, \\ \frac{2H(X \cap Y)}{H(X) + H(Y)}, & \text{otherwise} \end{cases} \quad (3)$$

WebPMI(X,Q) is defined as,

$$\text{WebPMI(X,Y)} = \begin{cases} 0, & \text{if } H(X \cap Y) \leq c, \\ \log\left(\frac{\frac{H(X \cap Y)}{N}}{\frac{H(X)}{N}\frac{H(Y)}{N}}\right) & \text{otherwise.} \end{cases} \quad (4)$$

Example

Google returns 12,300,000 as the page count for "car" AND "automobile," whereas the same is 59,000,000 for "car" AND "apple." Automobile is more Semantically Similar to "car" than "apple". Page counts for the query "car" AND "apple" are more than four times greater than those for the query "car" AND "automobile."

To get an accurate Semantic Similarity we must consider the page counts not just for the query X AND Y, but also for the individual words X and Y to assess Semantic Similarity between X and Y.

2.1.2 Similarity Measures Using Text Snippets

The lexical syntactic patterns extracted from the text Snippets are used to compute the Semantic Similarity between words.

Example

The **Jaguar is the largest cat** in Western Hemisphere and can subdue larger prey than can the Puma

The pattern for the above Snippet is "**X is the largest Y**", where the two words "Jaguar" and "cat" are replaced by two variables X and Y. The extracted set of patterns are clustered to identify the different patterns that describe the same semantic relation [11] , [13] , [14]. A sequential pattern clustering algorithm was proposed to identify different lexical patterns that describe the same Semantic relation.

2.2 Similarity Measures for Short Segments of Text

The tasks like query reformulation (query-query Similarity), sponsored search (query/and keyword Similarity), and image retrieval (query-image caption Similarity) require the Similarity computing between two very short segments of text. If the query and document do not have any terms in common, then they receive a very low Similarity score, regardless of how topically related they actually are [8]. This is known as the vocabulary mismatch problem.

Example
"UAE" and "United Arab Emirates" are semantically equivalent, yet they share no terms in common.

2.2.1 Representation

The short segments of Text is represented in the following 3 ways

1. **Surface Representation**
2. **Stemmed Representation**
3. **Expanded Representation**

Surface Representation
Surface Representation is a Sparse representation. However, it is very high quality because no automatic or manual transformations (such as stemming) have been done to alter it. This transformations enhance the representation, but it is possible that they introduce noise.

Stemmed Representation
Stemming is one of the most obvious ways to generalize (normalize) text. For this reason, Stemming is commonly used in Information Retrieval systems as a rudimentary device to overcome the vocabulary mismatch problem. Various Stemmers exist, including Rule-based Stemmers [7] , [8] and Statistical Stemmers [4].

Although Stemming can significantly improve matching coverage, it also introduces noise, which can lead to poor matches. Using the Porter stemmer, both "marine vegetation" and "marinated vegetables" stem to "marin veget", which is undesirable.

Expanded Representation
Although stemming helps to overcome the vocabulary mismatch problem to a certain extent, it does not handle the contextual problem. It fails to discern the difference between the meaning of "bank" in "Bank of America" and "river bank". Therefore, it is desirable to build representations for the short text segments that include contextually relevant information.

One approach is to enrich the representation using an external source of information related to the query terms. Possible sources of such information include Web (or other) search results returned by issuing the short text segment as a query, relevant Wikipedia articles, and, if the short text segment is a query, query reformulation logs.

2.2.2 Similarity Measures

Three methods for measuring the similarity between short segments of text are as follows.

1. **Lexical**
2. **Probabilistic**
3. **Hybrid**

Lexical

The most basic similarity measures are purely lexical. That is, they rely solely on matching the terms present in the surface representations [16] , [17] . Given two short segments of text, *Q* and *C*, treating *Q* as the query and *C* as the candidate, to measure the similarity the following lexical matching criteria is defined

Exact – *Q* and *C* are lexically equivalent. (*Q*: "seattle mariners tickets", *C*: "seattle mariners tickets")

- Phrase – *C* is a substring of *Q*. (*Q*: "seattle mariners tickets", *C*: "seattle mariners")
- Subset – The terms in *C* are a subset of the terms in *Q*. (*Q*: "seattle mariners tickets", *C*: "tickets seattle")

Lexical measures are binary. That is, two segments of text either match are deemed "similar" or they do not. Exact matches are very high Precision (excellent matches), yet very low Recall since they miss a lot of relevant material. Therefore, the matches generated using these Lexical rules will be having high Precision but poor Recall.

Probabilistic

Lexical matching alone is not enough to produce a large number of relevant matches. In order to improve Recall, expanded text representations can be used. To do so, use the language modeling framework to model query and candidate texts.

To estimate unigram language models for the query (θQ) and each candidate (θC),use the negative KL-divergence between the query and candidate model. The ranking function has been derived in the Equation (5).

$$-\mathrm{KL}(\theta Q,\theta c) = \mathrm{H}(\theta Q)-\mathrm{CE}(\theta Q,\theta c) \equiv \sum_{w \in v} p(w/\theta Q)\ \log p(w/\theta c) \quad (5)$$

Where V is the Vocabulary, H is Entropy, CE is Cross Entropy, and ≡ denotes Rank Equivalence.

To estimate a Query model use the Surface Representation. It has been estimated using the Equation (6).

$$\mathrm{p}\,(w/\theta Q) = \frac{tf\ w,QS}{|QS|} \quad (6)$$

Where QS denotes the Query Surface Representation, tfw , QS is the number of times *w* occurs in the representation, and |QS| is the total number of terms in QS.

To estimate a query model for an expanded query use the following Equation (7).

$$p(w/\theta Q) = \frac{tf\ w,QE + \mu Q\ p(w/c)}{|QE| + \mu Q} \quad (7)$$

Where QE is the Query Expanded Representation, and µQ is a Smoothing parameter.

The Candidate Mode l(Qc) is estimated as follows using the Equation (8).

$$p(w/\theta C) = \frac{tf\ w,CE + \mu Q\ p(w/c)}{|CE| + \mu C} \quad (8)$$

Where CE is the Candidate Expanded representation, and µC is a Smoothing parameter.

Hybrid

It is often interested in taking the matches generated by several different Similarity measures and combining them. It is called as hybrid technique. Given two or more lists of matches, according to some pre-defined ordering (denoted ">") of the lists are stacked , to form a combined list. For example, when the given match lists A and B, and ordering A > B, form the hybrid list AB, which represent that the list B is appended to the end of list A. Since the same match may occur in more than one set of results, duplicates from the combined list must be removed.

An accuracy of Similarity measures can be improved by using a Machine Learning approach [15] . It has been done through Web Relevance Similarity Measure. It is an extension of the Web-based similarity kernel function by using a better term weighting scheme. The Web-based Similarity kernel function uses Term and Document Frequencies (TDF) to measure the importance of the terms in the expanded representation of the input text segment [15]. A word that appears in the beginning of a document is typically more important and relevant to the topic of the document.

3 An Approach to Measure Similarity Using Ontology

The following subsection describe the complete details of our approach to compute Semantic Similarity using Ontology.

Algorithm for finding Semantic Similarity in Entity Description in Set Spreading

Step 1 : Bag of Word (BOW) is created for the terms.
Step 2 : Include the terms that are related to the original terms in BOW by referring to Ontology.
Step 3 : Compute Similarity between original two Entity Description (EDs).
Step 4 : Spread the two EDs if none of the termination conditions are not met else go to Step7.
Step 5 : Compute Cosine Similarity between the two extended EDs.
Step 6 : Go to Step4.
Step 7 : Calculate the mean of the similarity values computed in all the iterations
Step 8 : Return the mean Similarity value.

The steps in the above algorithm is depicted in Fig 3.

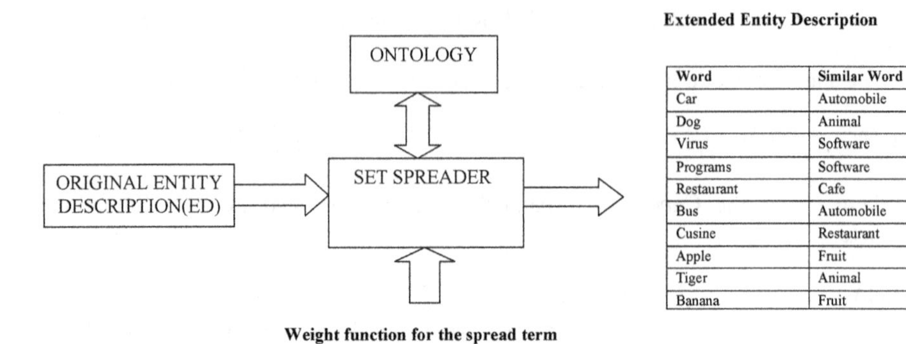

Word	Similar Word
Car	Automobile
Dog	Animal
Virus	Software
Programs	Software
Restaurant	Cafe
Bus	Automobile
Cusine	Restaurant
Apple	Fruit
Tiger	Animal
Banana	Fruit

Fig. 3. Set Spreading Process

4 Result and Discussion

To evaluate the performance of the system, a sample of 10 words have been taken and Semantic Similarity Searching has been done using Ontology. The performance measures of Precision and Recall have been depicted in the Equations (9) & (10).

Recall:
Measure of how much relevant information the system has extracted (coverage of the system).

$$\text{Recall} = \frac{\{Relevant\ Documents\} \cup \{\ Retrieved\ Documents\}}{\{Retrieved\ Documents\}} \tag{9}$$

Precision:
Measure of how much of the information the system returns is correct (accuracy).

$$\text{Precision} = \frac{\{Relevant\ Documents\} \cap \{Retrieved\ Documents\}}{\{Relevant\ Documents\}} \tag{10}$$

Although a small fraction of the result had negative and similar Performance with a Traditional Search Engine for the result of Semantic Similarity analysis, a marked improvement in performance has been observed for most of the words. Table - 1 depicts the Comparison of Traditional Search and Semantic Similarity Search and its performance is shown in Fig. 4, Fig. 5 and Fig. 6.

Table 1. Comparison of Traditional and Semantic Similarity Search

Input Word	Traditional Search		Semantic Similarity Search	
	Recall	Precision	Recall	Precision
Car	41	68%	50	87%
Dog	43	62%	55	86%
Virus	45	68%	56	78%
Program	48	75%	58	87%
Restaurant	49	64%	60	75%
Bus	51	58%	61	89%
Cusine	48	63%	57	88%
Apple	45	69%	59	83%
Tiger	49	77%	63	91%
Banana	48	68%	63	85%

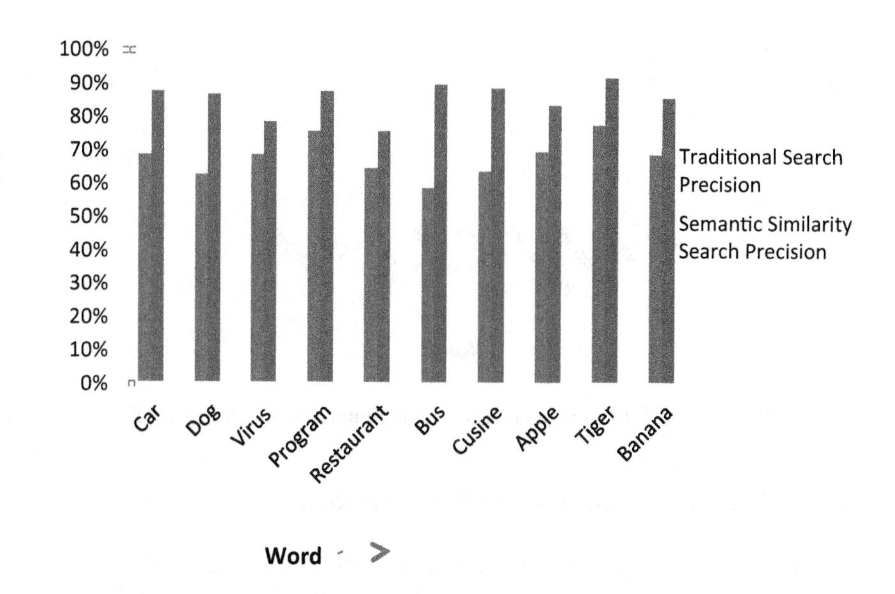

Fig. 4. Precision value of Semantic Similarity Search and Traditional Search

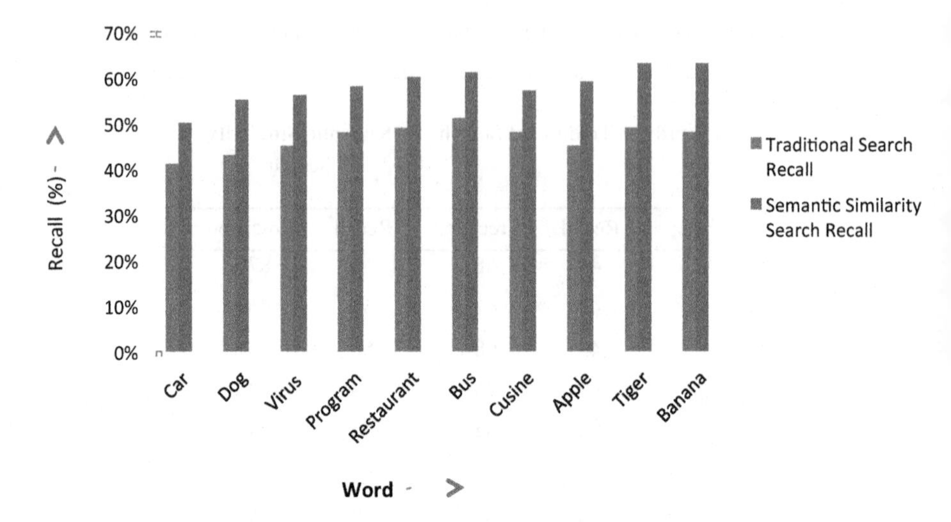

Fig. 5. Recall value of Semantic Similarity Search and Traditional Search

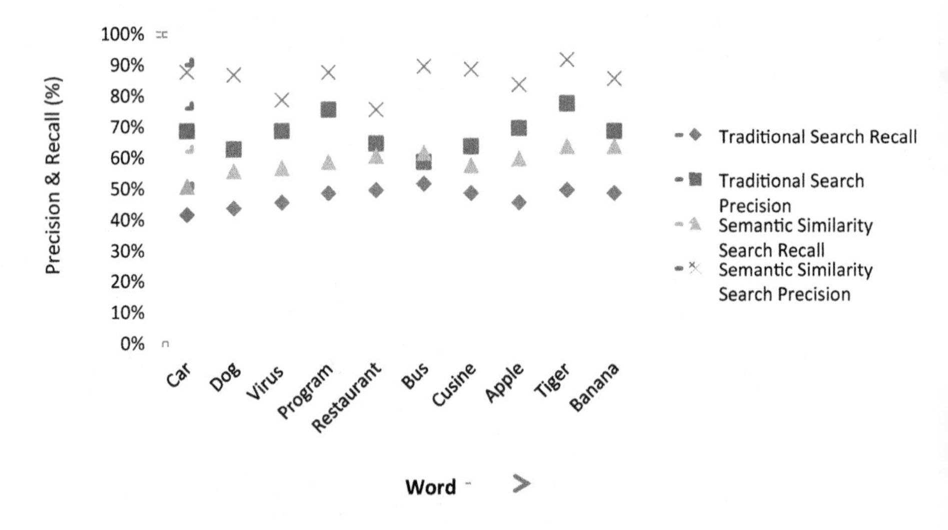

Fig. 6. Precision & Recall value of Semantic Similarity Search and Traditional Search

5 Conclusion and Future Enhancement

Semantic Similarity assessment is a crucial component embedded in many applications framed in the artificial intelligence research area. After the detailed analysis of the Fig. 4,5,6 we can infer that there is higher value of average Precision and Recall for the Semantic Search system when compared to Traditional Search. It depicts that our system has a better performance and accuracy in retrieving the results than the generic Search Engines. This paper provides a detailed and different method of Semantic

Similarity measures that can be used to estimate the resemblance between terms and Ontology based Semantic Similarity Search. They use synonyms in most cases, but they sometimes use Semantically Similar concepts with different rule structures. For example, Amazon uses the concept region for shipping destinations, but Powells.com uses country in every shipping rate rules. Country is not the synonym of region, but is Semantically Similar to region. Therefore, use a Semantic Similarity measure, in addition to synonyms in order to increase the recall rate.

Our ongoing research work is to use Semantic Similarity measures in query expansion, where a user query is modified using synonymous words to improve the relevancy of the search. In order to improve the performance of the search result (i.e.) to increase the Recall and Precision rate Ontology based Semantic Similarity can be used.

References

[1] Tjoa, A.M., Andjomshoaa, A., Shayeganfar, F., Wagner, R.: Semantic Web Challenges and New requirements. In: Proceedings of the 16th International workshop on Database and Expert Systems Applications (DEXA 2005). IEEE (2005)

[2] Berners-Lee, T., Hendler, J., Lassila, O.: The Semantic Web. In: Proceedings of Scientific American (2001)

[3] Bollegala, D., Matsuo, Y., Ishizuka, M.: A Web Search Engine-Based Approach to Measure Semantic Similarity between Words. IEEE Transactions on Knowledge and Data Engineering 23(7) (July 2011)

[4] McCrae, J., Campaña, J.R., Cimiano, P.: An Architecture for Cross Language Semantic Data Querying. In: WWW 2010, Raleigh, North Carolina (April 2010)

[5] Thirunarayan, K.: On Embedding Machine Processable Semantics into Documents. IEEE Transactions on knowledge and data engineering17(7) (July 2005)

[6] Ding, L., Finin, T., Joshi, A., Peng, Y., Pan, R., Reddivari, P.: Search on the Semantic Web. Published by the IEEE Computer Society 0018-9162/05/ © 2005 IEEE (2005)

[7] Metzler, D., Bernstein, Y., Croft, W.B., Moffat, A., Zobel, J.: Similarity measures for tracking information flow. In: Proceedings of CIKM 2005, pp. 517–524 (2005)

[8] Metzler, D., Dumais, S.T., Meek, C.: Similarity measures for short segments of text. In: Amati, G., Carpineto, C., Romano, G. (eds.) ECiR 2007. LNCS, vol. 4425, pp. 16–27. Springer, Heidelberg (2007)

[9] Ng, H., Lim, C., Foo, S.: A case study on interannotator agreement for word sense disambiguation. In: SIGLEX 1999, pp. 9–13 (1999)

[10] Resnik, P.: Using Information Content to Evaluate Semantic Similarity in a Taxonomy. In: Proc. 14th Int'l Joint Conf. Artificial Intelligence, pp. 448–453 (1995)

[11] Richard Benjamins, V., Contreras, J., Corcho, O., Gómez-Pérez, A.: Six challenges for the semantic web. In: KR 2002 Semantic Web Workshop (2002)

[12] Sahami, M., Heilman, T.: A web-based kernel function for measuring the similarity of short text snippets. In: Proceedings of WWW 2006, pp. 377–386 (2006)

[13] Park, S., Kang, J.: Using Rule Ontology in Repeated Rule Acquisition from Similar Web Sites. IEEE Transactions on Knowledge and Data Engineering 24(6) (2012)

[14] Stumme, G., Hotho, A., Berendt, B.: Semantic Web Mining — A Survey. In: ECML/PKDD Conference (2004)
[15] W.-T. Yih, C.: Meek.: Improving Similarity Measures for Short Segments of Text. Association for the Advancement of Artificial Intelligence (2007)
[16] Lau, R.Y.K., Song, D., Li, Y.: Towards A Fuzzy Domain Ontology Extraction Method For Adaptive E-Learning. IEEE Transactions on Knowledge and Data Engineering 21(6), 800–813 (2009)
[17] Corley, C., Mihalcea, R.: Measuring the Semantic Similarity of Texts. In: Proceedings of the ACL Workshop on Empirical Modeling of Semantic Equivalence and Entailment, Ann Arbor, pp. 13–18. Association for Computational Linguistics (June 2005)

Intelligent Traffic Monitoring Using Internet of Things (IoT) with Semantic Web

Manuj Darbari, Diwakar Yagyasen, and Anurag Tiwari

Department of Computer Science, Babu Banarasi Das University,
Lucknow, Uttar Pradesh, India
manuj_darbari@acm.org, manujdarbari@ieee.org

Abstract. The sudden rise in population has brought a very heavy demand of vehicles, hence the need to control them. The paper highlights the issue of Intelligent Traffic monitoring system using the technologies like IoT, MultiAgent system and Semantic Web. The paper links IoT sensors using Zigbee protocol and traffic movements are continously monitored by control centre using Granular classification in Ontology.

Keywords: IoT, OWL Urban Traffic System (UTS), Granular Computing, Ubiquotous Computing.

1 Introduction

With the rise in population there is sudden super in travel needs of people, the impact of spur leads to traffic congestion which is probably the most perennial problems faces by various countries. As per the latest survey on Urban Traffic Mobility (2009) traffic congestion results in loss of 7.3 million hours of productivity valued at 62 billion(In India).

Since mobility is the basic need of survival congestion cannot be eliminated completely but necessary steps can be taken to ease out the traffic conditions. There is significant research going in this direction but unfortunately none of the models developed so far eliminate the problem of congestion

The previous model like PLOTS, ATLAS and TRANST developed for vehicular studies only considers limited micro mobility, involving restricted vehicle movements but no attention was paid on micro mobility and its interaction. Thus there exists a basic need of development of framework which can cater the micro-mobility aspect focusing on the behavioural aspect of commuting. This framework will be able to support the realistic behavioural aspect of urban traffic congestion.

To develop close to real time simulation of urban Mobility we have developed the concept of Granular computing , Internet of Things (IoT) and Multi Agent Group Behaviour based on People -Machine Interface.[16,17]

1.1 Granular Computing

"Granulation of an Object 'A' leads to the collection of granules 'A' with the granule being a clump of point (objects) drawn together by indistinguishability, similarity,

S.C. Satapathy et al. (eds.), *Emerging ICT for Bridging the Future – Volume 1*,
Advances in Intelligent Systems and Computing 337, DOI: 10.1007/978-3-319-13728-5_51

proximity or functionality" was stated by Zadeh[1] . The main purpose of granularization is to describe important or interesting patterns in data[2]. In general it aims to discover meaningful structure in a particular data set. The concept of granularization started for Fuzzy classification of data. The use of granularization has promoted the necessary flexibility to represent vague and imprecise linguistic term supporting meaningful refinement of the data set. Granularization provides generalization as well as specialization of the data set using the concept of linguistic modifiers. It works on two basic pattern analysis and dissimilarity pattern. There are various algorithms defined for cluster analysis which are categorised as:

1. Hierarchical cluster analysis
2. Objective function based cluster analysis

The extension of granular computing into cluster formation relates to rough set theory proposed by Pawtak in 1982. The main concept of rough set is based on an assumption that every object in the entire universe is associated with some information, which helps us in forming clusters. These clusters have the property of indiscernible (similar). A single granule is a set of similar objects that have the very similar characteristic. For example group of cars with similar Horse power. But while forming a granule there can some confusion in classifying an object one into particular granule hence there exists vagueness (or imprecision). It is here the importance of mathematical foundation named rough set is applied which can be defined by means of topological operations, interior and closure called approximations.

Consider a Universe U and X be set of approximations in terms of granule of knowledge then according to rough set theory, the set approximations by lower approximation and upper approximation can be shown by the figure 1.

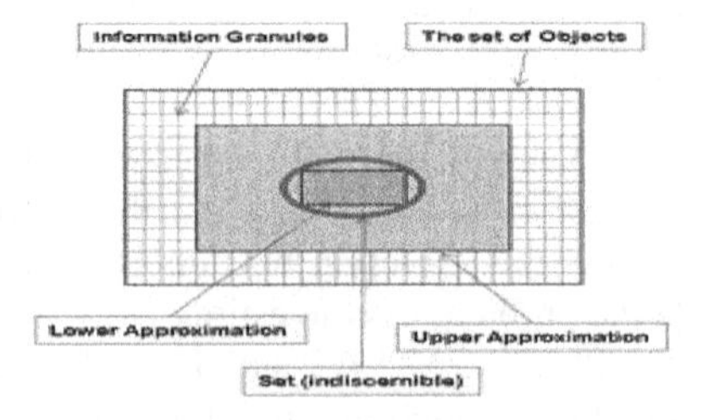

Fig. 1. Approximations in Rough Set (Adapted from Andrzej Skowron and James F. Peters, Handbook of Granular Computing, pp 29) Let X be the set of attributes B by constructing $\underline{B}$ and $\overline{B}$, approximation of X denoted by $\underline{B}X$ and $\overline{B}X$

1.2 Internet of Things

The concept of IoT is surrounded by SMART OBJECTS which uses Web service as a medium to communicate between heterogeneous environments. XML is customised for this purpose accommodating special tags for sensor network and spatial locations, EEML (Extended Environments Markup Language).

The basic idea of developing IOT framework is to make the system self reliant and at the same time linked with web service for receiving the signals. IOT is derived from the concept of Autonomic Computing developed by IBM research. The word autonomic forms the Central Nervous System acting as Central Unit of self control in human being. SMART objects can automatically detect the fault and recover from the fault. And if necessary these objects can synchronize themselves in order to improve the overall performance of the system.

System reliability and robustness also plays a significant role as IOT support self detection and there are various devices Raspberry pie, Yellow Pic and XBee following ZigBee protocol standard developed by Zigbee Alliance. It is based on IEEE 802.15.4 and has frequency range from 915 MHz & 2.4 GHz spectrum. Network topologies used in ZigBee is same as wireless topology like Star, Tree or Mesh.

The concept of Internet of Things(IoT) started in 2003 for tracing the flow of goods in supply Chain. IoT acts as interfere between the real world object and electronic devices often called as Context of Things. Ubiquitous framework helps in achieving many tasks relating to physical world. Significant effort in the area of combining sensor information into community information is taking place by maintaining high value of semantic integration and information entropy. Internet of Thing is an evolving internet which ranges for object collaboration which could be as small as pen to as big as a car.

The mix of cloud services, Web services and Sensor Network from Internet of Things (IoT); any real world change could be linked to global network infrastructure and the objects sets themselves by the help of interpretable communication protocol, thus creating a virtual intelligent background for any industry or human services.

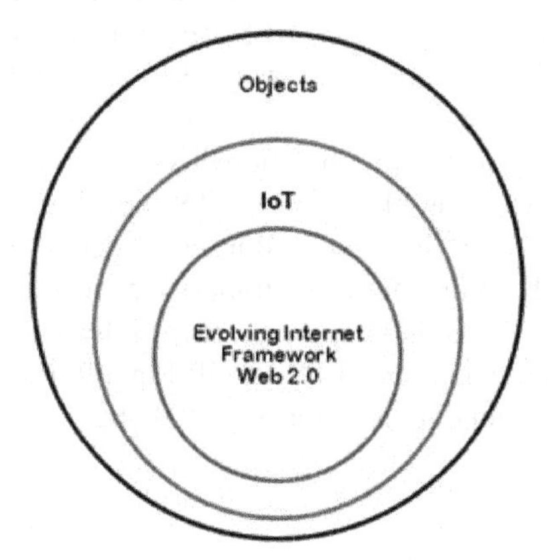

Fig. 2. IoT Framework

1.3 Collaborative Agents

Collaboration amongst agent started from the concept of Multi Agent system. Multi Agent system provides a structured method of solving complex problem. If we have to collaborate multiple heterogeneous activities we use co-operative multi Agent system. It involves learning by many Agents simultaneously. There are various learning schemes being used in Multi Agent Learning: Supervised, unsupervised and evolutionary learning. These learning are applied on teams for collaboration performance. It could be a homogeneous team learning and heterogeneous team learning. Homogeneous Team belongs to group of Agents of same cluster. We Authors will be using the combination of Heterogeneous and Homogeneous Agents combination of Heterogeneous and Homogeneous Agents in modeling Urban Traffic system for collaborative work support.

In order to coordinate this pervasive computing environment we use the concept of amorphous agents which is a nature inspired computing. In order to control the complexity of the system these amorphous agents will be linked with Amorphous clusters. Amorphous clusters provide coherent behavior from the cooperation perspective. Formally it represents a system of irregularly placed interacting elements. In our case each XBee sensor network are programmed identically with separate storage facility of local storage. The addition of amorphous cluster leads to creation of imperfect knowledge which is being achieved by Rough Set theory.

2 Literature Survey

The paper by Li. M. and Chong[3] on "Agent Oriented Urban Traffic Simulation" describes the use of interaction agent in controlling and management of Urban Traffic system. The shortcomings with paper was improper connectivity between multiagent and object modeling. Zhao & Xin Chen[4] proposed "Intelligent Cooperation Algorithm" highlighting the use of generic reinforcement learning which will be used by authors in maintaining the heterogeneity. DESIRE[5] developed by Frances, Nick and Jain provides high-level modeling framework enabling both the specification and implementation of systems engineer view and systems designer view to explicitly and precisely specify both the ingredients functionality. The paper by R.J.F. Rossetti and S. Bampi[6] shows a synthesis of complex UTS into three types of flows: Vehicle flow, Information flow and Decision flow. Bargiela A. et al.[9] highlights the issue of Data granulation through optimization of similarity measure [2,9,15], also highlights the "Integration of heterogeneous traffic and travel information through a combined Internet and Mobile communication". EPSRC GR Final Report[18]. We are motivated by the papers from Bargiela in transforming IoT based Traffic Information into granules.

3 Proposed Framework

The Combination Internet of Things (IoT)[10,11,12] and Granular computing provides a holistic view on Heterogeneous Control System. For IoT technology we will be using Zigbee Protocol in determining the concentration of Traffic Linking of Traffic Light Cluster and Road Information is done by Web 2.0 services. The two clusters are controlled by the help of control centre (Fig. 2).

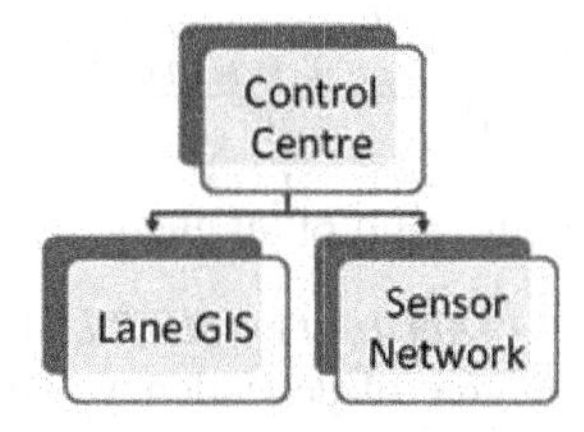

Fig. 3. Basic Building Block of Control Centre

For Modeling the Urban Traffic system we will be using Granular Ontology language to provide efficient flow of information. Each vehicle is assumed to be smart having Zigbee standard of transmitting the signal.

Figure 3 shows the schematic arrangement of sensor Network. Each crossing is supported by transreceiver attached to it which send/receives signal from the nearest ZigBee Transreceiver Tower. All these towers are connected to the Control Centre. The Control centre has all the GIS related information covering the main road and lane by-passes. Consider a situation where there is heavy traffic between Crossing I and II. There too possible solutions to the problem either timing of the traffic light should be altered or the traffic needs to be diverted to the lane bypass. The signal is given by the help of VMS to the drivers the control centre, which find out the best suitable. The entire process of information works in real time and in collaborative mode. The signals generated are analyzed by the control centre machines; secondly it provides an online access to the commuters. In Task-Driven granularization we are able to derive application-specific information.

Based on the concept of Rough Set Theory application specific categorization can be dynamically created.

A rough set according to definition is characterized as:

$$_{B}X \frac{|\underline{BX}|}{|\overline{BX}|} \tag{1}$$

which is called as Accuracy of Approximation where |X| denotes cardinality of X. Let us consider a table of some set of readings relating ZigBee signals:

Table 1. Various Attributes of Traffic Management System

LaneGIS	Sensor Tx	Jam detected	TLC	VMS
e_1-e_4	TX_4-TX_3	Yes	Change Timing	Activate VMS
e_1-e_2	TX_1-TX_2	No	No Change	No Signal
e_3-e_4	TX_3-TX_4	No	No Change	No Signal
e_3-e_4	TX_3-TX_4	Yes	Change Timing	Activate VMS
e_4-e_5	TX_4-TX_5	No	No Change	No Signal

From the table we find that Activate VMS and change in Timing is totally dependent on Jam Detected with a degree K=1. It is represented as $K_{\text{Jam Detected}}$.

$$\left\{\begin{matrix} Activate\ VMS, \\ Change\ in\ Light\ Timing\ (TLC) \end{matrix}\right\} \frac{\left| POS_{JamDetected} \{ActivateVMS\ TLC\} \right|}{|U|} \quad (2)$$

The coefficient P_k represents the ratio of all elements of the universe, which can be classified to block of partition $U/_{\{Actuvate\ VMS, TLC\}}$ with all the attributes of Jam Detected and this is called as degree of dependency. There exists discernibility relation amongst all the attributes of Traffic Management. A pair of (Universal set and Traffic Management Attribute) is defined as approximation space. Equivalence classes of Traffic management Attribute are called Granules.

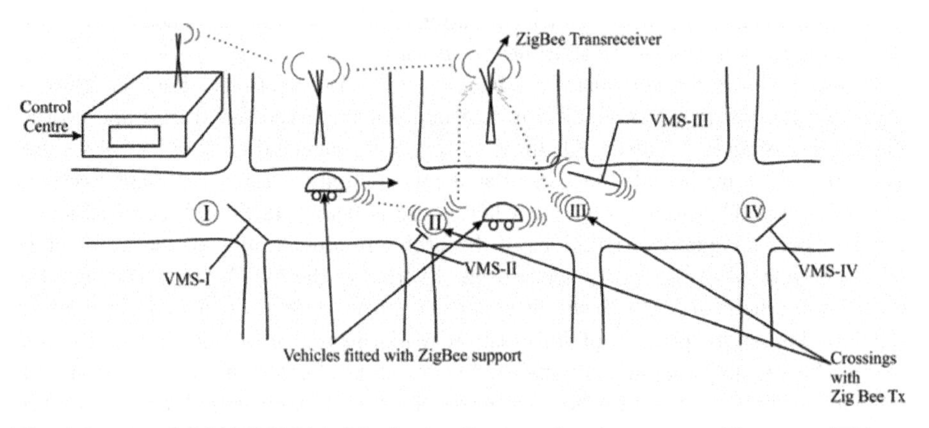

Fig. 4. Layout of SMART Vehicle Monitoring System using the concept of Internet of Things (IoT)

Figure 4 shows the formation of granules based on the Task-Driven information in real time scenario. The Separation of Task Driven distance functions between patterns of control information of variety of signals like (Car, Zigbee Tx) as shown in equation 3.

$$d(CarCar_{46}, ZigbeeTx) = \sqrt{\sum_{i=1}^{n}(Car_{46}, ZigbeeTx_{II})^2} \quad (3)$$

The dynamic information generated by the Clusters are made machine understandable by the help of OWL. To make Ubiquitous devices collaborate intelligently, we represent the framework using knowledge representation and semantics supporting a more closely knitted and dynamic environment. The basic RDF structure is shown in figure 5.

The Granular Classification in Ontology can be realized by OWL. The category of two granules which are possible are: (1) Lane GIS Granule (2) Sensor Granule (3) Control Granule.

In order to transmit the information in Ubiquitous mode and make it Machine understandable we write a OWL which could possibly link the granules for faster and

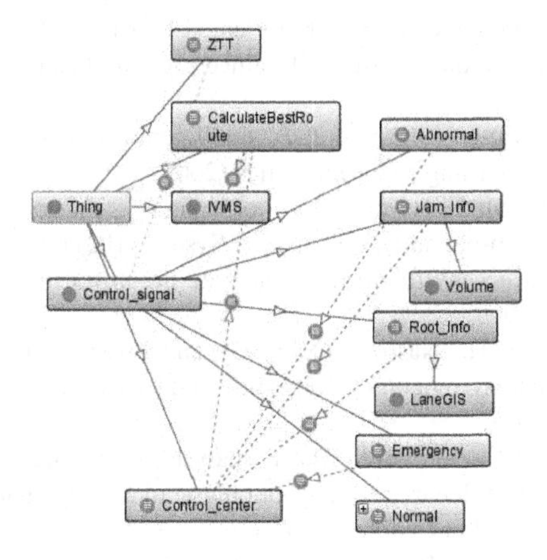

Fig. 5. RDF of Traffic Monitoring Framework

accurate processing along with latest information about traffic conditions. A sample OWL which links ZigBee sensors and Control Centre is written as:

```
< owl: Class rdf: ID = 'ZigBee Sensor'>
<rdfs: subclassof>
<owl: Class rdf: ID = 'Control Centre'/>
</rdfs: subclassof>
</owl: Class>
```

The above sample code provides information that set of all individuals of Class ZigBee Sensor which is a subset of the set of all individuals in the Class Control Centre.

4 Conclusion

The paper focuses on a framework consisting of Ubiquitous environment supported by IoT classified into Granules for clustering dynamic information about urban traffic. For making the entire process more machine understandable we used the concept of OWL as it helps the commuters in providing intelligent search[13,14] and correct information.

References

1. Zadeh, L.A.: Toward a theory of fuzzy information granulation and its centrality in human reasoning and fuzzy logic. Fuzzy Sets and Systems (1997)
2. Bargida, A., Pedrycz, W.: A model of granular data: a design problem with Tchebysher FCM. Soft Computing (2005)

3. Li, M., Chong, K.W., Hallan, C.S.: Agent –Oriented Urban Traffic Simulation. In: The1st International Conferences on Industrial Engineering Application and Practice, China (1996)
4. Yang, Z.-S., Chen, X., Tang, Y.-S., Sun, J.-P.: Intelligent cooperation control of urban traffic networks. Machine Learning and Cybernetics (2005)
5. Prithivi, J., Nick, F.: DESIRE: A Framework for designing and specification of interacting compounds. Journal of Co-operative Information Systems (III) (1997)
6. Rossetti, B.S.: A software environment to integrate urban traffic simulation tasks. Communications of ACM (1999)
7. Krishapuram, R., Frigui, H., Nasrovi, O.: Fuzzy and Possibilistic shell clustering algorithms and their application to boundary detection and surface approximation. IEEE Trans. Fuzzy System 3(1) (1995)
8. Hoppner, F., Klawonn, F., et al.: Fuzzy Clster Analysis, Viky, England (1999)
9. Bargiela, A., Pedrycz, W.: Recursive information granules: Aggregation and interpretation issues. IEEE Trans. Syst. Man Cyborn. (2003)
10. Uckelmann, D., Harison, M., Michahelles, F.: An Architectural Approach Towards the Future Internet of Things. In: Architecting the Internet of Things. Springer, Heidelberg (2011)
11. Vemesan, O., Friess, P.: Pan European Cross Fertilisation and Integration of IoT Programmes and Initiative through National Value Creation networks. In: Business-Information Systems, Inernet of Things European Research Cluster (IERC), 3rd edition of Cluster Book (2012)
12. Felix, A.A., Taofiki, A.A., Adetkunbo, S.: A conceptual framework for an Ontology-Based Examination system. International Journal of Advanced Computer Science and Applications 2(5) (2011)
13. Yagyasen, D., Darbari, M., Shukla, P.K., Singh, V.K.: Diversity and Convergence Issues in Evolutionary Multiobjective: Optimizationtion to Agriculture Science. In: International Conference on Agricultural and Natural Resources Engineering (ICANRE 2013). Elsevier (2013)
14. Yagyasen, D., Darbari, M., Ahmed, H.: Transforming Non-Living to Living: A Case on Changing Business Environment. In: International Conference on Agricultural and Natural Resources Engineering (ICANRE 2013). Elsevier (2013)
15. Siddiqui, I.A., Darbari, M., Bansal, S.: Application of Activity Theory and Particle Swarm Optimization Technique in Cooperative Software Development. International Review on Computers & Software 7(5), 2126–2130 (2012)
16. Dhanda, N., Darbari, M., Ahuja, N.J.: Development of Multi Agent Activity Theory e-Learning (MATeL) Framework Focusing on Indian Scenario. International Review on Computers & Software 7(4), 1624–1628 (2013)
17. Darbari, M., Dhanda, N.: Applying Constraints in Model Driven Knowledge Representation Framework. International Journal of Hybrid Information Technology 3(3), 15–22 (2010)
18. EPSRC GR Final Report (2011)

Development of a Real-Time Lane Departure Warning System for Driver Assistance

N. Balaji[1], K. Babulu[2], and M. Hema[1]

[1] Department of ECE, JNTUK UCEV, Vizianagaram, A.P., India
narayanamb@rediffmail.com,
hema_asrith@yahoo.co.in
[2] Department of ECE, JNTUK UCEK, Kakinada, A.P., India
kapbbl@gmail.com

Abstract. According to World Health Organization's (WHO) global status report on road safety, the number of road accidents per day is increasing drastically. Majority of these accidents occur due to violation of the safety norms by the drivers. The main objective of this paper is to develop a real-time lane departure warning system based on Beagle board. The proposed system avoids road accidents by giving warning to the driver when the vehicle begins move out of its lane. The developed system is also useful to minimize accidents by addressing the main causes due to drivers error and drowsiness.

The proposed system runs on an embedded operating system called angstrom. The system architecture deals with the development of device driver for the USB web camera on OMAP3530 processor in standard Linux kernel and installation of Open CV package to merge USB web cameras with the navigation system.

Keywords: Beagle board, Open CV, Canny edge detection, Hough transform, RANSAC algorithm, Vanishing point.

1 Introduction

Driver Assistance System (DAS) is a kind of active system that is used to help the driver in driving process. Lane departure warning (LDW) is a driver warning system designed to help the driver in reducing the number of unintended lane departures [1], [2]. Many single vehicle roadway departure crashes take place in light traffic situations and good weather conditions. Such crashes are often due to drivers' inattention or drowsiness. Drowsiness is frequently reported during night-time driving and in monotonous driving conditions.

National highway traffic safety administration (NHTSA) data indicates that in recent years there have been about 100,000 crashes annually in which police cited driver drowsiness, resulting in about 1,500 fatalities. The goal of a lane departure warning (LDW) is to warn drivers if they are unintentionally drifting out of their lanes. LDW relies on the detection of the vehicle position with respect to the road lane markings in order to detect a lane departure. The vehicle position is evaluated by a camera system

S.C. Satapathy et al. (eds.), *Emerging ICT for Bridging the Future – Volume 1*,
Advances in Intelligent Systems and Computing 337, DOI: 10.1007/978-3-319-13728-5_52

mounted behind the windshield. The proposed driver assistance system based on an embedded processor called beagle board shown in Fig.1. Also the system is useful in the case of overspeeding of vehicle, as it identifies accident prone areas like bends, corners and school zone and it alters the driver to maintain safe speeds at those areas. The alerts will be given in the form of both image display and audio prompts.

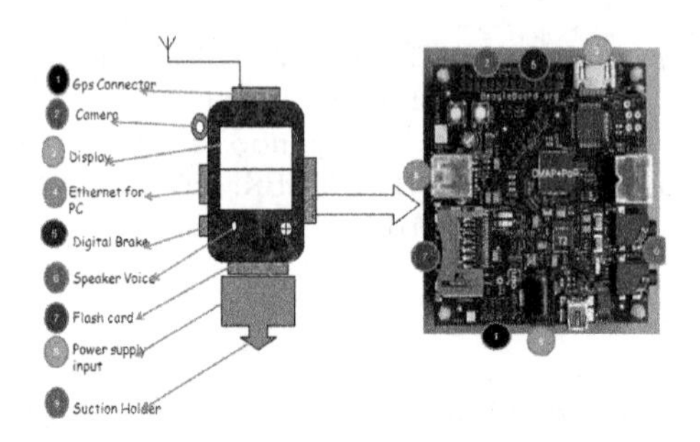

Fig. 1. Driver Assistance System with interface of components

This system is specially designed for safety and fleet management. The system will be mounted in front of the driver, which will capture the road images and display them on the screen.

The most lane detection methods are based on edge detection techniques used in image processing [2], [3]. After the edge detection step, the edge-based methods organize the detected edges into meaningful structure like lane markings or fit a lane model to the detected edges. Most of the edge-based methods use straight lines to model the lane boundaries. Hough Transform is one of the most common techniques used for lane detection [4], [5], [6]. Hough Transform is a method for detecting lines, curves and ellipses, but in the lane detection it is preferred for the lane detection capability. It is mostly employed after an edge detection step on grayscale images. The images are first converted into grayscale using only the R and G channels of the color image. The grayscale image is passed through very low threshold canny edge detection filters [7].

For line sequence estimation RANSAC algorithm [8] and error functions can be used. The RANSAC algorithm partitions the data set into inliers (the largest consensus set) and outliers (the rest of the data set), and also delivers an estimate of the model. RANSAC uses as small an initial data set as feasible and enlarges this set with consistent data when possible. The vanishing points are elements of great interest in the computer vision field, since they are the main source of information for structured environments. Vanishing point is a point in the image plane on which images of parallel lines of the scenario seem to converge [9].

2 Block Diagram of the System and Hardware Description

The Block diagram of the proposed DAS is shown in Fig. 2. In the block diagram input is a video signal that comes through the camera, which is interfaced to the beagle board. A Secured Digital (SD) card is used for the mounting operating system kernel image called uImage file and root file system of Angstrom operating system.

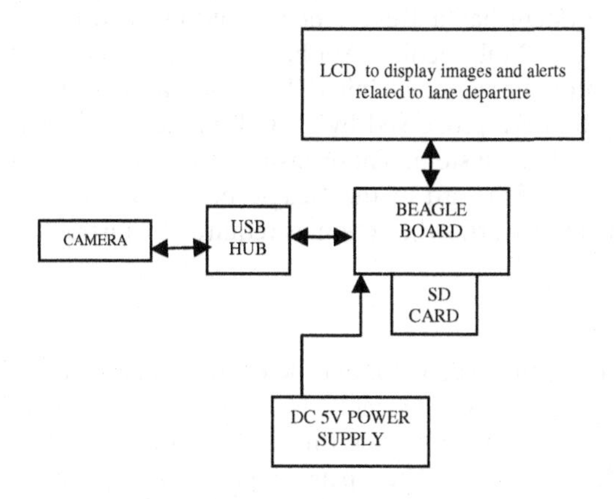

Fig. 2. Block diagram of the Driver Assistance System

2.1 Beagle Board

The Beagle Board is a low-cost, fan-less single-board computer based on TI's OMAP3 device family. It uses a TI OMAP3530 processor as shown in Fig. 3.

Fig. 3. Photograph of the Beagle board

2.2 Porting Angstrom Operating System

Angstrom is complete Linux distribution includes the kernel, a base file system, basic tools and even a package manager to install software from a repository. It is optimized for low-power microprocessors like the one in Beagle Board and intends to be small and basic system to modify on needs. It uses the Open Embedded (OE) platform, a tool chain that makes cross-compiling and deploying packages easy for embedded platforms. Linux installed on the personal computer was used to build the kernel and the same with only required features was built on the Beagle Board.

3 Development of Lane Departure Detection Application

The Driver Assistance System has a Beagle board interfaced with Camera. Beagle board has Open source Multimedia Application Processor (OMAP 3530). OMAP3530 acts as Central Processing Unit (CPU) for tracking system. All operations of the tracking system are processed by OMAP processor. OMAP requires instructions to operate the whole system. These instructions are provided to processor by writing the software into SD card of the Beagle board. It reads the software instruction by instruction and performs the action as required by instruction.

3.1 Open CV Package

OpenCV as a platform to develop a code for lane detection in real time [10]. The code is then implement on the beagle board. OpenCV is an open source computer vision library. The library is written in C and C++ and runs under various operating systems like Linux, Windows and Mac OS X [11]. The OpenCV has features of Image data manipulation, basic image processing, structural analysis and various dynamic data structure.

4 Main Stages of the System

The main stages of the lane departure warning system are represented in flow chart Fig. 4. The image sequences are captured from a camera which is mounted on the front of the vehicle. First it checks the camera availability, if camera is available then it will read frame by frame from the video. If camera is not available, it cannot open the video file. In the case of a moving vehicle, the lane recognition process must be repeated continuously on a sequence of frames. The scene image can be divided into two regions in the vertical orientation i.e. the sky region and the road region. The road region needs to be processed. Then the lane marking pixels can be detected only on the region below the dividing line using both the edge and color features.

4.1 Color Conversion

Color image are converted into grayscale. However, the processing of grayscale images becomes minimal as compared to a color image. This function transforms a 24-bit, three-channel, color image to an 8- bit, single-channel grayscale image. First, the color of the pavement is black or gray, so the intensity of the pavement in road image is weak.

For each color pixel (R, G, B), the gray pixel P can be computed by

$$P = 0.3 * R + 0.59 * G + 0.11 * B. \quad (1)$$

4.2 Canny Edge Detection

Lane boundaries are defined by sharp contrast between the road surface and painted lines or some type of non pavement surface. These sharp contrasts are edges in the image. Therefore edge detectors are very important in determining the location of lane boundaries. The Canny edge detector is an edge detector operator that uses multistage algorithm to detect wide range of edges in image.

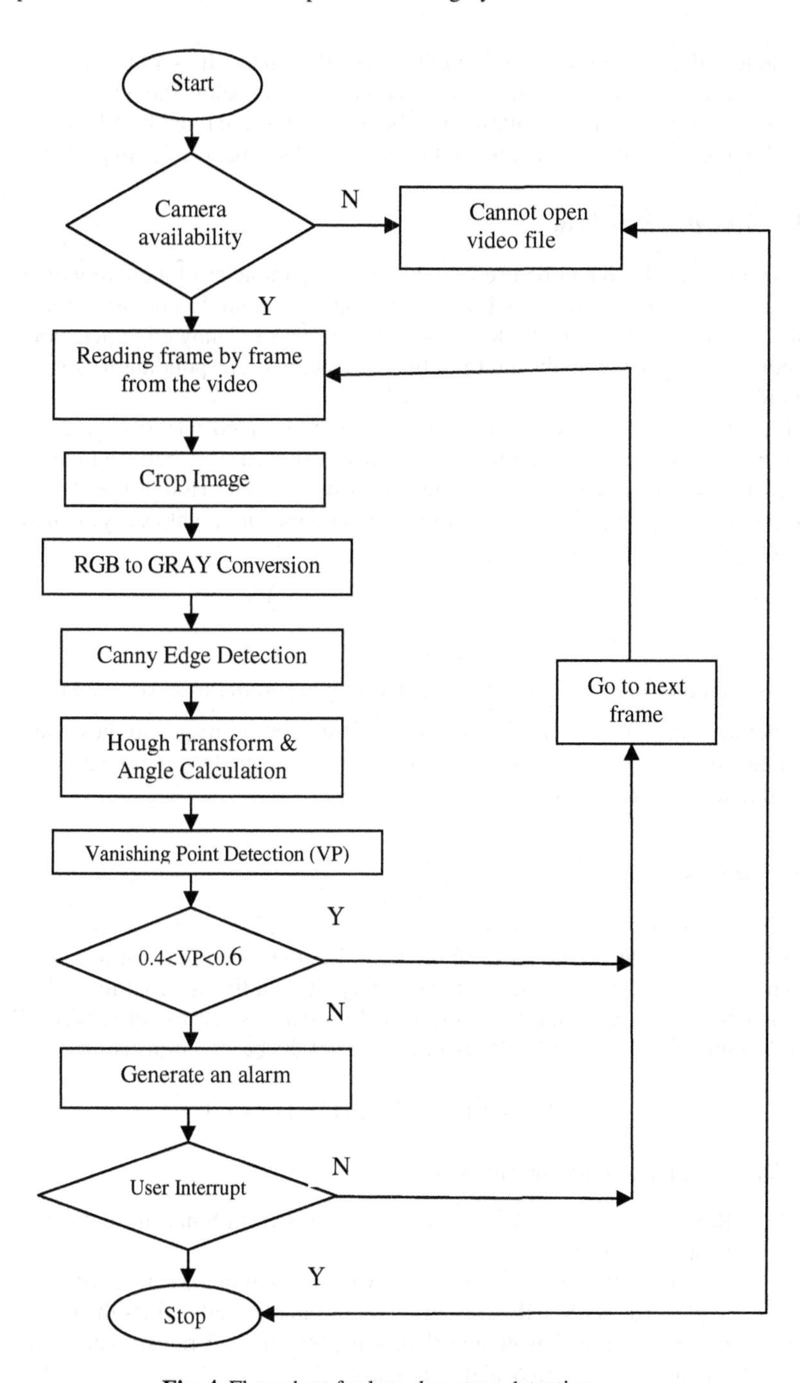

Fig. 4. Flow chart for lane departure detection

The Canny edge detection function finds edges in the input image and marks them in the output map edges using the canny algorithm. The Canny edge detection uses two thresholds. If a pixel gradients is higher than the upper threshold (threshold 1), the pixel is accepted as an edge. If a pixel gradient value is below the threshold (threshold 2), then it is rejected. The smallest value between threshold1 and threshold2 is used for edge linking. The largest value is used to find initial segments of strong edges.

4.3 Hough Transform

The Hough transform can be used to identify the parameter of a curve which best fits a set of given line points. This line description is commonly obtained from a feature detecting operator such as the Roberts Cross, Sobel or Canny edge detector and may be noisy, i.e. it may contain multiple line fragments corresponding to a single whole feature.

The lines that got by the Hough transform are divided into two groups by if the slope is greater than 0. As the linear road model is used, it is believed that the slopes of left boundaries and right boundaries are different. The Hough transform give out two points of the line: the start point (x1, y1) and the end point (x2, y2) and the slope *tan p* is given by equation (2).

$$\tan p = \frac{(y_2 - y_1)}{(x_2 - x_1)} \tag{2}$$

It is worth noting that if $|\tan P| < 0.1$, the line is parallel to *x-axes*. When the line is parallel to x-axes it is mostly because the obstruction on image, so these lines will be ignored. At last if $\tan P > 0$, it will be stored in the queue left line otherwise it will be stored in the queue right line.

4.4 RANSAC Algorithm

RANSAC (random sample consensus) algorithm which randomly chooses minimal subset of data points necessary to fit model. It has the inliers and outliers.

The inliers i.e., points which approximately fitted a line and outliers which cannot be the line. It can put some threshold based on that consensus set is there. We have take N-samples for the number of iterations to get the best consensus set.

$$N = \log(1-p)/\log\left(1-(1-e)^s\right) \tag{3}$$

The steps involved in the algorithm are

(i) Randomly select a sample of *s* data points from S and instantiate the model from this subset.

(ii) Determine the set of data points S_i which is within a distance threshold *t* of the model. The set S_i is the consensus set of samples and defines the inliers of S.

(iii) If the subset of S_i is greater than some threshold *T*, re-estimate the model using all the points in S_i and terminate

(iv) If the size of S_i is less than *T*, select a new subset and repeat the above.

(v) After *N* trials the largest consensus set S_i is selected, and the model is re-estimated using all the points in the subset S_i.

4.5 Vanishing Point

Vanishing point is point at which two parallel lines are converge. Lane markings are parallel to each other. From the perspective of the driver the lane markings converge at a point, which is referred as vanishing point. Line segments extracted from whose extensions intersect the vanishing point are likely to define a lane division. For the calculation of vanishing point it is required to approximate line segments. For each line segment l_i, we recomputed the normalized homogeneous representations of the coincident infinite line, l_i, the endpoints, e_i^1 and e_i^2, and the midpoint, m_i.

$$e_i^1 = \left(x_i^1, y_i^1, 1\right)^T \quad (4)$$

$$e_i^2 = \left(x_i^2, y_i^2, 1\right)^T \quad (5)$$

$$m_i = \left(\frac{\left(x_i^1 + x_i^2\right)}{2}, \frac{\left(y_i^1 + y_i^2\right)}{2}, 1\right)^T \quad (6)$$

$$l_i^1 = e_i^1 \times e_i^2 \quad (7)$$

By using all the above equations the vanishing points are calculated efficiently. The RANSAC algorithm is applied several times on the data to get the best consensus set of lane estimation to locate the single vanishing point.

If the detected vanishing point lies in the range of 0.4 to 0.6, alarm will not be generated then it will go to the next frame. In case if the vanishing point lies in that range alarm will be generated so that driver will be alerted by an alarm.

5 Results and Discussions

A mobile operating system "Angstrom" is ported on to the Beagle board [12]. A Linux kernel image is created using Linux kernel 2.6.34 which is compatible with OMAP [13]. Fig. 5 shows the hardware for the lane departure detection system.

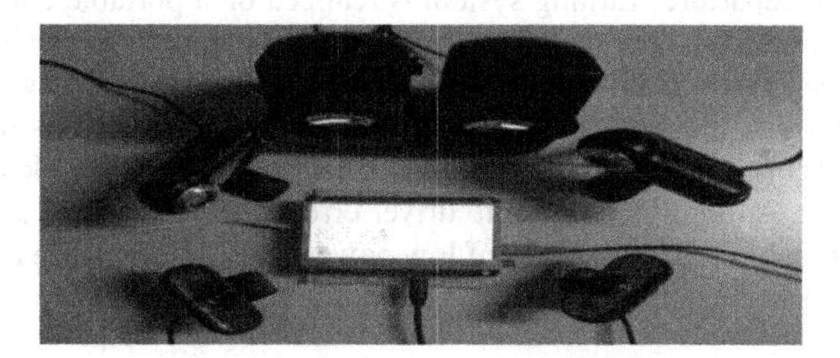

Fig. 5. Hardware for th-e lane departure detection system

The video or scene is captured through a USB web camera interfaced to the beagle board. When the driver is driving the vehicle on road, the lanes will be detected as shown in Fig. 6 and Fig. 7.The Fig. 6 shows the lane detection. The lanes are detected in red color. The orientation of lane marks changes is small and smoother along the lane. Lane marks are parallel to left and right from the center of the lane.

The g++ compiler is. The application code used to compile which creates the object files linking the OpenCV libraries and header files. Fig.7 shows lane departure warning system. Whenever the vehicle crosses the lane it warns the driver about the departure of lanes by an buzzing an alarm.

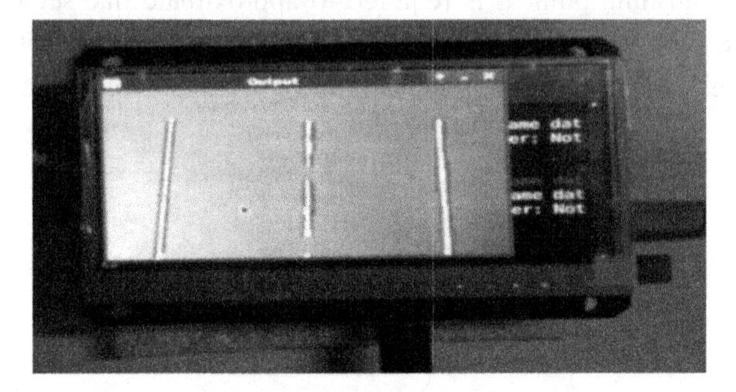

Fig. 6. Lane Detection

Fig. 7. Lane departure warning system

6 Conclusion

A real time lane departure warning system is realized on a portable embedded system with open source Linux operating system for improving vehicle driving safety. The lane departure system contains latest embedded processor interfaced with camera and a complex algorithm embedded in the system continuously monitors the lane departure. Lane departure warning systems are designed to minimize accidents by addressing the main causes of collisions due to driver error, distractions and drowsiness. The developed system has the advantages of low power, high performance and easy installation. Also the system achieved high detection rate for lane departure.

References

1. Soetedjo, A., Yamada, K.: Traffic sign classification using ring partitioned method. IEICE Transactions on Fundamentals of Electronics, Communications and Computer Sciences E88-A(9), 2419–2426 (2005)

2. Piccioli, G., De Micheli, E., Campani, M.: A robust method for road sign detection and recognition. In: Eklundh, J.-O. (ed.) ECCV 1994. LNCS, vol. 800, pp. 493–500. Springer, Heidelberg (1994)
3. Acharya, T., Ray, A.: Image processing: Principles and Applications. Wiley, New York (2005)
4. Hough, P.V.: Method and means for recognizing complex patterns (1962)
5. Li, Q., Zheng, N., Cheng, H.: Springrobot:a prototype autonomous vehicle and its algorithms for lane detection. IEEE Transactions on Intelligent Transportation Systems 5(4), 300–308 (2004)
6. Yu, B., Jain, A.K.: Lane boundary detection using a multiresolution hough transform. In: International Conference on Image Processing, vol. 2, pp. 748–751 (October 1997)
7. Wang, Y., Teoh, E.K., Shen, D.: Lane detection using b-snake. In: International Conference on Information, Intelligence, and Systems, pp.438–443 (1999)
8. Hartely, R., Zisserman, A.: Multiple view geometry in computer vision, Cambridge (2008)
9. Nieto, M., Salgado, L.: Plane rectification through robust vanishing point tracking using the Expectation-Maximization algorithm. In: IEEE International Conference on Image Processing (ICIP 2010), pp. 1901–1904 (September 2010)
10. Bradski, G.R., Kaehler, A.: Learning OpenCV-computer vision with the OpenCV library. O'Reilly (October 2008)
11. Driks, B., Schimek, M.H., Verkuil, H.: video for Linux two API Specification (1999-2006)
12. http://www.angstromdistribution.org
13. USB implementation forum, universal serial bus device class definition for video devices, http://www.usb.org/developers/devclass_docs

Visual Simulation Application for Hardware In-Loop Simulation (HILS) of Aerospace Vehicles

Pramod Kumar Jha[2], Chander Shekhar[1], and L. Sobhan Kumar[2]

[1] Terminal Ballistics Research Laboratory (TBRL), DRDO, Chandigarh, India-160030
chander_shekhar@tbrl.drdo.in
[2] Research Centre Imarat (RCI), DRDO, Hyderabad, India-500069
jha@rcilab.in

Abstract. Visual simulation which is a basic form of virtual reality (VR), uses 3-D images and models of real world which can be explored interactively by the user in a 3-dimensional realistic immersive environment. Some of the well known applications of VR and visual simulation are in the areas of Flight simulator and interactive games. VR is now extensively being used for many practical and scientific applications including mining, oil & gas exploration, automobile design, business training, medical science, robotics to name a few. Advanced Simulation centre (ASC) at RCI is primarily involved in carrying out Hardware In-loop Simulation (HILS) of various aerospace vehicles (ASVs). HILS is the only tool for carrying out complete validation and performance analysis of embedded systems, hardware, control & guidance software of missile systems in an integrated manner subjecting it to trajectory dynamics. HILS generates lot of data which needs to be analyzed thoroughly before the test launch. It's a quite tedious task to go through and analyze such a huge data manually. Modern developments in the field of VR based man machine interactions has made it easy to facilitate such tasks enormously to analyze and understand in a more convenient and collaborative manner. In our work, we have exploited the strengths of this technology for analysis purpose with fault injection & detection during the HILS of ASV. A 3-channel distributed rendering based immersive projection system has been established where visual application runs on a high end graphics cluster.The application is capable to visualize complete trajectory dynamics of an ASV from lift-off to impact point as per mission sequence and makes simulation engineer able to walkthrough inside the vechicle, visualize effects and troubleshoot the defects caused due to failures in the hardware and software. It also enhances post flight analysis capability. This paper focuses on the visual simulation system and its application in HILS of Aerospace vehicle.

Keywords: 3-D Graphical Modeling, Hardware In-Loop Simulation (HILS), Virtual Reality, Visual Simulation.

1 Introduction

In the last decade, lot of advancement has happened in the fields of Computer Graphics, Virtual Reality (VR), Computer based Modelling & Simulation. This ad-

S.C. Satapathy et al. (eds.), *Emerging ICT for Bridging the Future – Volume 1*,
Advances in Intelligent Systems and Computing 337, DOI: 10.1007/978-3-319-13728-5_53

vancement has led to the growth of Visualization Simulation technology and VR technology to be used as high-tech means for development of various applications of interest to the society such as visual simulation of automobile engine [1], process manufacturing [3], game instruction and Training [4], Shipborne Crane Control [5], Aircraft & space flight training [6], investigation on control methods and path planning in lunar exploration [8], flight simulations [11, 12]. In this technology users can participate in the process of simulation actively through various human computer interfaces.

This technology has also spread its footprints even in the military & defense sector. In [7] author discusses visual simulation of anti-torpedo system while in [9] this technology has been used to simulate Unmanned Aerial Vehicle (UAV) digital tactical battlefield. This paper emphasis on use of virtual reality technology in HILS of aerospace vehicle (ASV). The paper first discusses the visual system setup and later its application in visual simulation of ASV.

The paper has been organized in six sections. Section 2 is the simulation system setup. In Section 3, Principle of HILS for ASVs and mathematical modeling has been described. Section 4 describes 3-D graphical modeling process & visual simulation based on Vega Prime®. Section 5 of the paper discusses visual simulation results. Section 6 summarizes conclusion and future scope.

2 Visual System Setup

Taking inspiration from the pioneering work done in wide applications of the Visual Simulation &VR technology [1, 2, 3-9, 11,12] in different fields including defense sector, the idea of setting up a Visual simulation centre for HILS was conceived at Research Centre Imarat (RCI). It was proposed that system will be capable to render the optimized 3-D ASV models. The graphical distributed rendering and generation of proper blended image will be done at high end graphical workstations cluster unit. The basic system setup of the VR facility conceived at RCI has been shown in Fig. 1.

This facility consists of a cluster of 3 ultra-high end workstations with dual GPU cards (Image generators). After rendering and electronic blending at these workstations, operation output is fed to the stereoscopic 3-D projection system through Barco's Magik box which is an easy-to-use very high RGB bandwidth interface to connect with image generators. The projection system consists of three 3-chip DLP™ (Digital Light Processing) projectors and each projector is having SXGA+ resolution (1400*1050) to increase total field of view (FOV). In addition, the projectors chosen are capable to up-convert the input low refresh rate to good stereo rates higher than 110 Hz for better stereoscopic visualization thereby giving enough bandwidth for image generators to generate more detailed information instead of wasting this bandwidth for very high refresh rates. A high end seamless semi rigid large screen with FOV of 80 degree has been deployed in the above setup to give an immersive feeling. After optical blending at projectors, the whole scene (combined scene from 3 projectors) is displayed on seamless 3-D stereoscopic display system. The projectors has been placed on rear side of screen to avoid any kind of distracting shadows on the screen due to presence of users or any other object in the path between projectors and screen. Rear projection also helps in more collaborative development approach. For display alignment on screen, the system is equipped with laser array. The observer will wear light weight infitec glasses and will get immersive feel/involvement while sitting in front of the screen. The user controls right from touch based control panel.

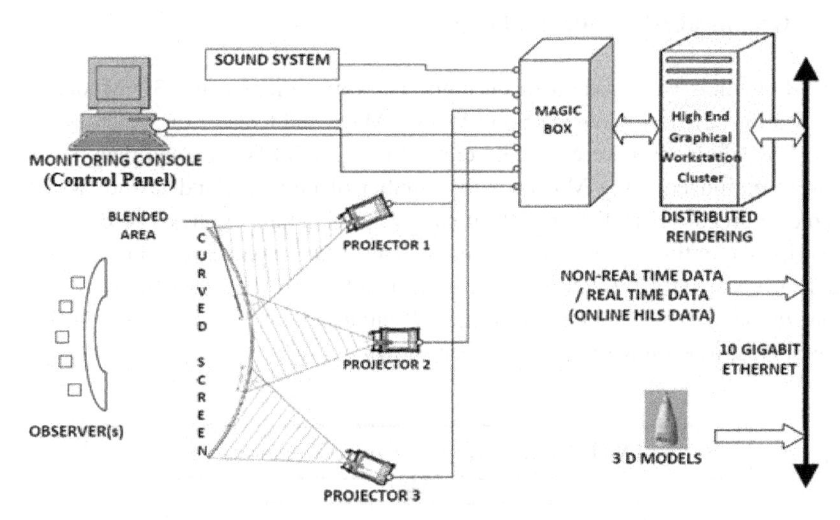

Fig. 1. Visual system design & setup at RCI, Hyderabad

3 Principle of HILS of ASV and Mathematical Modelling

HILS is the only validation tool available to the designer to validate the embedded software and hardware under the trajectory dynamics before the actual test launch of the ASV. Six-Degrees of Freedom (6-DOF) model is the plant of the HILS test bed and generates three rotational and three translational acceleration. The output of (6-DOF) model is fed to simulated navigation system which generates the current attitude of the ASV. The control guidance algorithm which is resident in onboard computer generates the error based on desired trajectory. The control system generates the steering command based on this error and steers the ASV to its desired path. In our present scheme, the trajectory data of the ASV and deflection commands of the actuator and fins is fed to the Image generator using a high speed Gigabit Ethernet. HILS uses rapid prototyping model and to start with simulation is carried out with simulated systems and it is replaced with their respective hardware one at a time. The performances of all subsystems are studied under various nominal as well as off-nominal conditions and based on their performance they are cleared for the flight trial.

4 3-D Modelling and Visual Simulation Based on Vega Prime

The emphasis of kinematics simulation is generation of simulation results. It serves design optimization and scene driving. However, visual simulation places more weight on more rendering and driving of 3D graphics [8]. This section will introduce technique of visual simulation of ASV, that is, visual model building, LOD, DOF and its implementation.

4.1 Graphical 3-D Model Building

The 3-D models were generated on high end workstations using 3D Modelling and visualization software tools viz Solid Works, Maya and Creator on windows XP platform. The 3-D models were initially created using Solid Works, and then these models were optimized using Maya for the number of polygons and texture was added. Degree of freedom (DOF) and different levels of details (LODs) were added after importing the optimized models to Creator. This 3-D Graphical data-model with DOF and LODs not only provides precise object definition, but also supports real-time visualization [10] and interaction in the VR environment. The process followed during graphical modelling process has been shown in Fig 2.

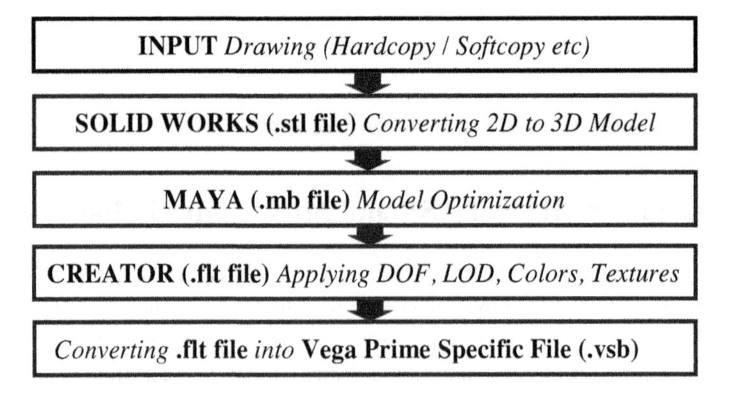

Fig. 2. 3-D Graphical modelling process followed

Multigen® Creator [13] uses hierarchical data for different level of details for same model making the model most suitable for real time visual simulation. When doing walkthrough, low level details model is used while when we want to simulate flight scenario of ASV same model with optimum high levels details is useful as it reduces the undesired rendering time. DOF are added to object nodes in the Creator in local coordinate system and all its decedent nodes inherit transformations from it. Vega prime does the necessary traversal walks through the scene graph to get to the specified DOF node. Following above 3-D Graphical modeling process, profile models of two different sections of an ASV are shown in Fig 3. Comparison of polygon counts and size variation during the 3-D Graphical Modelling process for the two sections of the ASV shown in Fig. 3 has been given in Table 1.

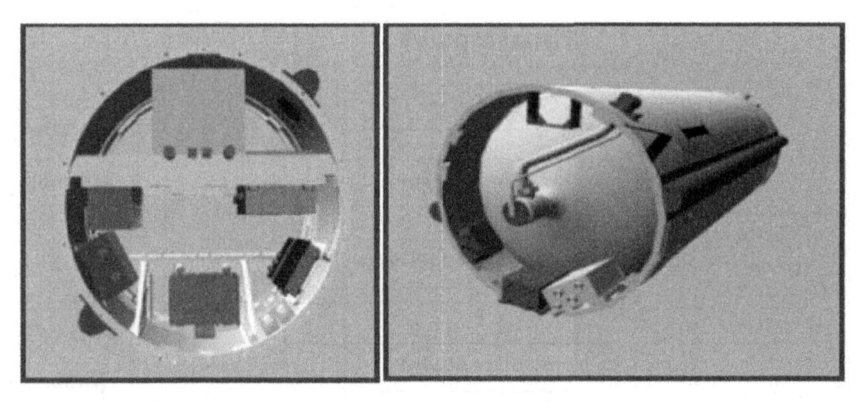

Fig. 3. Profile 3-D Graphical models of Section-2 and Secton-5 of an aerospace vehicle (ASV)

Table 1. Polygon Count and Size Comparison during 3-D Model Development Process

Parameter	Section No.	Solid Works (.stl)	Maya (.mb)	Creator (.flt)
Polygon Count	Section 2	1,05,134	50, 922	37, 217
	Section 5	1,12,729	38, 180	26, 017
Size	Section 2	28.7 MB	9.61 MB	2.30 MB
	Section 5	30.9 MB	9.36 MB	1.39 MB

4.2 Vega Prime Based Visual Simulation for HILS of Aerospace Vehicle

Vega Prime is cross-platform application software for visual simulation with application programming interface (API) that is an extension of Vega Scene Graph with multi-threaded mode. Vega prime provides good user interface. This interface and additional modules with powerful functions such as special effects (e.g. Explosion, flames, debris etc) and Distributed rendering [14]. Being products from same company i.e. MultiGen-Paradigm (Now Presagis) Vega Prime and Creator provide adequate support to each other. Because of flexible expansibility of development for real-time rendering it is used for visual simulation. The generated 3D model runs on Vega Prime on high end workstations with Windows XP environment. VS.NET 2003 was used for programming purpose with Vega Prime. The basic visualization block diagram has been given in Fig. 4.

The instructor module is the main controlling unit of the application from where various modes of operation will be initiated. From graphical user interface (GUI) of this module user can select the different modes such as walkthrough, Prelaunch, autolaunch and sequence of operations. Various effects like smoke, fire, explosion etc. that occur during aerospace vehicle launch will also be shown by this module. The HILS setup provides mission data for the flight vehicle during the online mode of operation in the launch scenario or may be provided through recorded file for further visual simulation. In the application, provision has been provided to record certain portion of aerospace vehicle's flight and may be played later to analyze specific part.

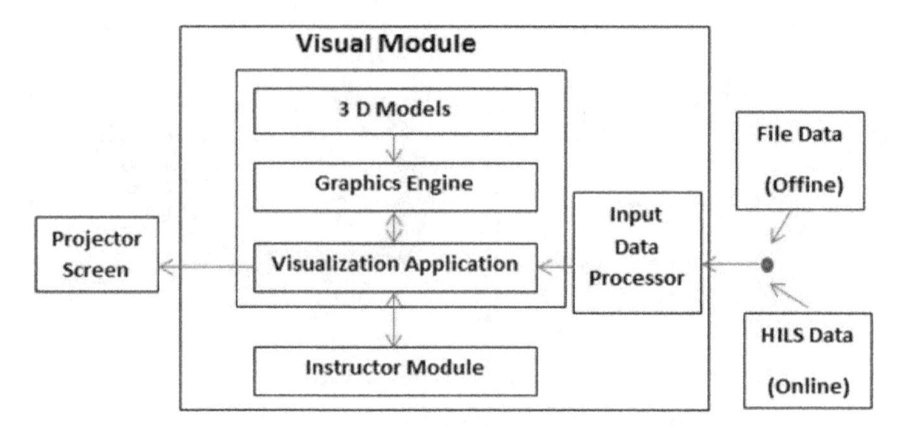

Fig. 4. Basic visualization block diagram

5 Visual Simulation Results and Discussion

During the 3-D graphical modelling process, it was found by observing Table 1 that solid works generates models with very high polygon counts and obviously with more file size while models optimization can be easily done with Maya and Creator software which as a result reduces number of polygons and file size without much affecting the shape of the object. In Creator as it uses hierarchical data structure we can easily define LOD and DOF.

Fig. 5. Visual simulation seen captured during aerospace vehicle (ASV) launch

Visual simulation was implemented by using VS.NET 2003 and Vega Prime. The visual system setup shown in Section 1, proved good enough to provide good immersive feel of the simulation. The user is able to have walkthrough of these models with appropriate zooming and see inside of them as well as can have nice interaction with these models. Observer is able to see complete trajectory dynamics of ASV from lift-off to impact point along with actuator, engine, fin and ASV body movements on the HILS data. Fig 5 shows snapshot of visual simulation of ASV during launch stage.

Various test runs show that the system performs distinctively in interaction and real-time. The added facility of record and replay makes it easy for the simulation engineer to analyze the specific recorded part of interest. This application of visual simulation in HILS make simulation engineer able to visualize effects and trouble-shoot the defects caused due to failures in the hardware and software.

6 Conclusions and Future Scope

In this paper, process and flow of visual simulation for HILS of an ASV was demonstrated. Interactions between the users and the virtual environment are based on the Vega Prime and VS.NET 2003. It proved good real-time immersive visual simulation. The said application is capable to enhance post flight analysis capability and flight failures can be visualized. Further enhancements of this visual simulation will make simulation engineer more capable to visualize effects and sorting out the defects caused due to failures in the hardware and software.

Visual simulation and Virtual Reality has the ability to provide heightened level of understanding to quickly focus on issues, reach strategic goals and consensus. Its use in complex technological projects could be very helpful in reducing the work cycle and hence reducing the overall cost & efforts substantially. This technology has ability to provide heightened level of understanding to quickly focus on issues, reach strategic goals and consensus. Besides being truly effective method for training & understanding, by its use various risk analysis and linked technological challenges can be accessed in advance and efficient work direction can be set.

Acknowledgements. Authors would like to thank Director, RCI for his kind support. Authors would also like to express their sincere thanks to Dr SK Chaudhuri, former Director, RCI for his efforts in establishing this facility. Our Sincere thanks to Shri KR Tolia, Scientist 'E' for his sincere contributions in making this system operational. Authors also express their gratitude to all the officers and Staffs of our group who contributed directly or indirectly for this paper.

References

1. Liang, S., Xianguang, L., Yebin, B.: Research on key Techniques of Visual Simulation of Automobile Engine based on Virtual Reality Technology. In: Third International Conference on Measuring Technology and Mechatronics Automation, pp. 14–17 (2011)

2. Huang, S., Xu, Y., Pang, Y., Wang, Z.: Visual Simulation of Unmanned Surface Vehicle in MFC framework based on Vega, pp. 166–170. IEEE (2011)
3. Lu, S., Yue, H.: Real-time Data Driven Visual Simulation of Process Manufacturing: A Case Study. In: Chinese Control and Decision Conference (CCDC), pp. 1806–1809. IEEE (2011)
4. Zyda, M.: From visual Simulation to Virtual Reality to Games, pp. 25–32. IEEE Computer Society (2005)
5. Li, P., Wang, C.: Design and implementation of Visual Simulation System for Shipborne Crane Control. In: Chinese Control and Decision Conference (CCDC), pp. 5482–5486 (2009)
6. Carey, P.M., Eng, C., Eng, P.: Visual Simulation for Aircraft and Space Flight Trainers. The Institution of Electronic and Radio Engineers, pp. 61–69 (1966)
7. K. Hong, H. Liang, C. Yan: Visual simulation of Anti-Torpedo System, pp. 243–246. IEEE (2009)
8. Meng, Y., Wang, Y., Xie, Y., Zhou, J.: Visual Simulation of Wheel-terrain Interaction of Lunar Rover based on Creator/Vega. In: The 1st International Conference on Information Science and Engineering (ICISE), pp. 5291–5294. IEEE (2009)
9. Shi, G., Gao, X.: Unmanned Aerial Vehicle Digital Visual Simulation System Load and Missile Modules Simulation Software Design. In: International Conference on Measuring Technology and Mechatronics Automation, pp. 1063–1066 (2010)
10. Li, Z.-L., Wang, W.-F., Zhao, Z.: A Modified Visual Simulation Method Based on LOD. In: 2nd International Conference on Industrial Mechatronics and Automation (ICIMA), pp. 58–61 (2010)
11. Li, Z., Zhang, S., Cui, J., Gong, J., Yu, F.: Visual simulation of flight attitude based on MFC and Vega. In: International Forum on Information Technology and Applications, pp. 132–133 (2010)
12. Hu, X., Bo, S., Huiqin, Z., Bing, X., Hao, W., Ge, H.: Visual Simulation System for Flight Simulation Based on OSG. In: ICALIP, pp. 562–566. IEEE (2010)
13. Creator Modelling Guide. Technical Manual, MultiGen-Paradigm Inc., Version 3.7 (2007)
14. Vega Prime Programmer's Guide. Technical Manual, MultiGen-Paradigm Inc., Version 2.2 (2007)

Multiprotocol Label Switching Feedback Protocol for Per Hop Based Feedback Mechanism in MPLS Network

Ankur Dumka and Hardwari Lal Mandoria

[1] University of Petroleum and Energy Studies, Dehradun, India
adumka@ddn.upes.ac.in
[2] College of Technology, G.B. Pant University, Pantnagar, India
drmandoria@gmail.com

Abstract. Multiprotocol feedback (MFB) protocol is suggested in this paper which suggest per hop base feedback mechanism in MPLS network including piggy backing in the algorithm, thus reduces the time needed for error detection and correction and also efficient for bandwidth utilization in the MPLS network. This is achieved by including a packet of 7 bit which carry three different types for correction and detection. Thus increases efficiency of MPLS network and also provide efficient method to find the error detection and correction and efficient for bandwidth utilization.

Keywords: MPLS, LDP, RSVP, LER, LSR.

1 Introduction

Multiprotocol label switching (MPLS) is a label switching technology that combines the traffic engineering capability of ATM with the flexibility and scalability of IP. MPLS provides the ability to establish connection-oriented paths over a connectionless IP network, and facilitates a mechanism to engineer network traffic pattern independently of routing tables. MPLS technology offers many services, including layer 2 and layer 3 VPN services, traffic engineering, and resiliency.

RFC 3031 describes the multiprotocol label switching architecture. The term Multiprotocol indicate that an MPLS architecture can transport payloads from many different protocols (IPv4, IPv6, Ethernet, ATM, Frame Relay, etc.), whereas label switching describes that an MPLS domain switches, rather than routes, packets in the service provider core.

MPLS allows routers to forward traffic based in a simple label embedded in the packet header. An MPLS router examines the label to determine the next-hop for the packet. This simplifies the forwarding process and separates it from the routing protocol, which determines the route that traffic will take across the network. Thus, MPLS sets up a specific path for a sequence of packets. Each packet is identified by a label inserted in the packet and forwarding occurs based on this label.

S.C. Satapathy et al. (eds.), *Emerging ICT for Bridging the Future – Volume 1*,
Advances in Intelligent Systems and Computing 337, DOI: 10.1007/978-3-319-13728-5_54

Label switching was initially considered an improvement over IP packet switching as it involves a simple lookup. However with the advances in hardware technology, MPLS for Layer 3 forwarding alone has become obsolete in recent years.

The MPLS label binding table lookup process is simpler. The table only contains the forwarding information associated with an exact match rather than a longest match so the forwarding table can be smaller than routing table. The modes forward traffic using a predetermined label sent down a preselected path and replaced at each hop. So they can decide much more quickly where to send the packets next.

In a MPLS network there are customer edge routers resides on the customer premises. The customer edge routers access to the service provider network over a link to one or more provider edge routers. The end user typically operates these devices. The customer edge routers are unaware of tunneling protocols or VPN services that are provided by the service provider. The provider edge routers has at least one interface that is connected to customer edge router and at least one edge connect to providers core routers. Provider routers are located in the provider core network. The provider router support the service providers bandwidth and switching requirements over a geographically dispersed area. Label edge router resides in the boundary between the MPLS domain and customer domain. It is similar to provider edge router. Non-MPLS traffic enters the MPLS domain through ingress label edge router (iLER) and egress label edge router (eLER) removes the label from MPLS packet and forward the unlabeled packet to the customer edge router. Label switch router resides within the MPLS domain. They are same as provider routers and they replace the incoming label with an outgoing label and forward the packet to the next hop router.

Label switch path (LSP) is defined as a sequence of labels and label actions performed on MPLS routers to forward the packets from point A to point B, using label switching. A label switch path always starts from an ingress label edge router and ends at an egress label edge router. An LSP is thus an end-to-end, unidirectional path that can carry traffic from one end router to other end router of MPLS network. The encapsulation and forwarding of packets using label is also referred to as tunneling; as such, LSPs are often called as tunnels. Tunnels must be established prior to the arrival of data packets. Label negotiation and distribution protocols are used to build the tunnels with negotiated label values.

Forwarding equivalence class (FEC) is a group of IP packets forwarded in the same manner, over the same path, and with the same forwarding treatment. In IP-based routing, forwarding equivalence class lookup is done at each hop.[5] Forwarding equivalence class allows for the classification of packets into groups based on common criteria. In MPLS FEC can be defined based on destination IP prefixes, and other administrative criteria. In MPLS based forwarding, FEC lookup is done only at the ingress LER on incoming data packets. The FEC lookup determines the next hop label switch router and the label the source router pushes onto the packet. The label switch router then simply perform swap operations based on the previously determined label values.

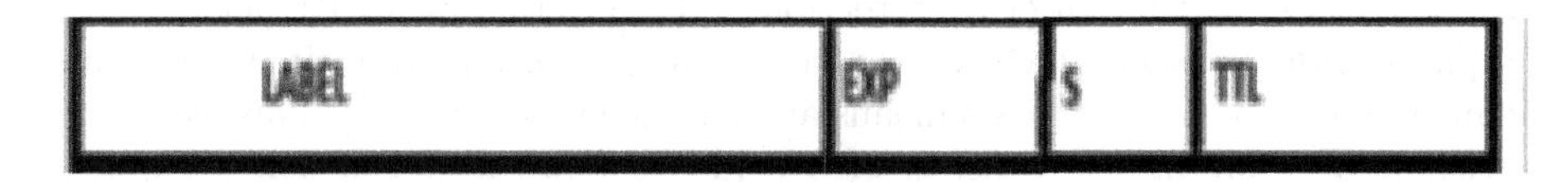

Fig. 1. MPLS HEADER

The structure of MPLS header, which exist specifically for MPLS is shown below. Each MPLS header is a fixed length of 4 byte (32 bits) in size and contain following fields: Label : MPLS label is of size 20 bits, which contain MPLS label value EXP : Experimental bits S : Bottom of stack of size 1 bit flag, used to indicate bottom of MPLS stack. TTL : Time to Live of size 8 bits, used for loop prevention by limiting packet lifetime in MPLS header. The label field contains the value information of the label. Label can be given values ranging from 0 to 1,048,575. Of which 0-15 labels are reserved for special use labels. 16-31 reserved for future use , 32-1023 reserved for static label switch paths for forwarding of packets, 1024-2047 reserved for future use, 2,048- 18,431 statically assigned for services, 18,432- 32,767 reserved for future use, 32,768 131,071 dynamically assigned by MPLS protocols and 131,072- 1,048,575 are reserved for future use. Out of these 20 labels we can use one bit label as flag for hop by hop error detection to reduce the time required for detecting any error. [6]

The MPLS experimental bits are used to carry the class of services information of the transported data traffic over the path of an LSP. This information, inside the label header, allow label switch routers to apply different Quality of Services treatments to various types of MPLS encapsulation traffic.

The bottom of stack is only one bit in length and set to 1 at the bottom header to indicate the end of the stack.[6] This bottom header resides closest to the customer payload; the remaining header, with the stack bit set to 0, serve to transport the payload to the tail end provider routers.

The 8 bit MPLS header TTL field is decremented by 1 at each IP hop. If the TTL value drops to 0, the packet is discarded and is not transmitted to the next-hop.[7]

2 Structure of Feedback Packet

In this algorithm we had used 7 bit label for hop by hop feedback mechanism and piggy backing for performing per hop based error detection for error detection and correction with efficient bandwidth utilization. In this approach, since label switch path is unidirectional, so there should exist two label switch path from source to destination and for label switch path we have to use an interior gateway protocol and label distribution protocol or resource reservation protocol[2]. Once the label switch path is being set from source to destination router the packets send from source to destination and if there is any error due to checksum error or timeout then that error being detected at the destination label edge router as there is only switching take place at label switch router or intermediary routers.

So, to reduces the time in error detection and correction in MPLS network we proposes a for a packet which sends to the sender after a certain time period as defined by user at per hop basis means at each receiving router[1]. This feedback packet is of 7 bit in size consists of 3 fields that are feedback field of 1 bit, send sequence number of 3 bits and receive sequence number of 3 bits and this feedback packet is attached with the packet destination router to the source router. If there is no packet transferring from destination router to source router then FB feedback packet is send from destination to source router independently checking for error.

Table 1. FB Feedback packet

1 bit	3 bit	3 bit
Feedback bit	Send sequence number	Receive sequence number

7 bits feedback packets information is shown in Table 1 In this 7 bit packet feedback bit of 1 bit can contain 0 if a packet is not received and 1 if the packets received. If the packet is received at the destination then next 3 bit will give the send packet sequence number and receive bit sequence number to test whether the packet received is in order or not and the source router can understand that packet received at destination next hop is of correct order or not and if packet is not received at destination next hop then after a particular time the destination next hop send FB feedback packet with feedback bit as 0 and in the send sequence bit the last sequence number what it received from the source and in receive sequence number the last sequence number that it send. This type of packet can be used as feedback when there is no packet need to be transmitted from destination next hop router to source router as the label switch path between source routers to destination next hop router is unidirectional. The next packet FP feedback packet can be used in feedback mechanism when there is some packet need to be send from destination next hop router to source router. Thus, this frame attach with the packet coming from destination next hop router to source router as piggy back with same 7 bits being used as FB frame with packet attached at the end of this feedback frame.

Table 2. FB Feedback packet

1 bit	3 bit	3 bit	
Feedback Bit	Send sequence number	Receive sequence number	Message

Thus by using this technique, we can implement per hop based error detection rather than end to end signaling for error detection and correction. Thus save the time needed for error detection and correction.

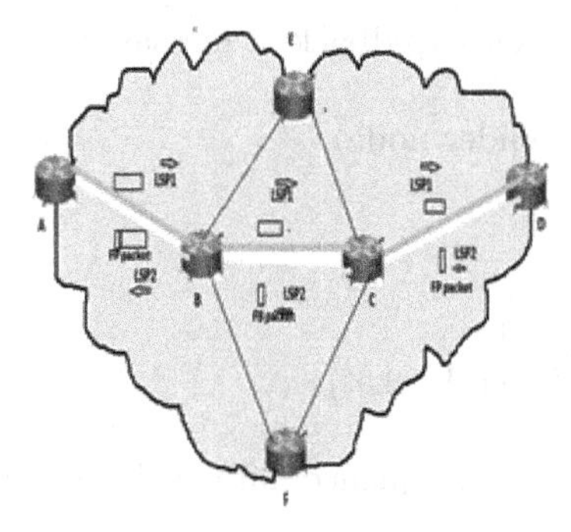

Fig. 2. MPLS network with FB and FP packets

3 Demonstration Model of MPLS Feedback Algorithm

Here in the above figure, label switch path (LSP1) is created from router A to router D via router B and router C. In the same way label switch path (LSP2) is created from router D to router A via router C and router B. Now router A wants to send packet 1 to router D. In this case as packet reaches to router B, then router B redistribute to synchronize label switch path 1 and label switch path 2 and router B then send acknowledged packet back to router A sending acknowledge of the packet 1. Since , there is a packet meant to send from router B to router A, so this feedback packet is attached with this packet thus save bandwidth utilization and resource utilization and FB packet is send from router B to router A. In the same way packet 1 reaches from router B to router C. Feedback is send from router C to router B through label switch path 2 and since there is no packet to be send from router C to router B so FB feedback packet is send from router C to router B giving feedback of packet 1 from router B to router C.

4 Algorithm for MPLS Feedback Protocol

The algorithm for the following feedback mechanism:
Variable Frame arrive successfully, variable
Enum checksum and timeout is event type
While (forever and repeat timer(time))

If (variable ==frame arrival successful send sequence number==receive sequence number)
Redistribute LSP1 feedback to LSP2 and viceversa
If(packet sending to sender node)

Then feedback bit=1
Send Sequence number++;
Receive sequence number++;

Packet= PG feedback Packet (feedback + message)

If (no packet sending to sender node)

Then feedback bit =1;
Send sequence number++;
Receive sequence number++;
Packet= FB Packet (feedback Packet);

If(variable==event type —— sequence number != receive sequence number)

Then feedback bit=0;
If (packet sending to sender node)
Packet = PG packet (feedback + message);
Else
Packet= FB packet (feedback frame);

This algorithm is applicable in MPLS network, protocol like LDP or RSVP is used to find out the path taken by the label within the MPLS network termed as label switch path.[4] Since LSP is unidirectional path from one provider edge router to another thus for two way communication 2 LSPs need to be established by LDP or RSVP protocol. But if there is event generated because of checksum error or timeout then that error can be detected at the edges routers as there is only switching occur at the label switch routers. Thus this will take time as error detection is done at edge routers and retransmitted to ingress edge routers.

Thus this protocol proposes an algorithm for per hop basis error detection in MPLS network thus help in reducing the time taken to find the error detection and correction. In this algorithm, since packets get numbered and send according to their numbering if variable arrive successful at the next hop router and the send packet sequence is same as required by the destination router keeping track of the sequence, then feedback bit will set to 1 indicating for successful receiving of packet from sender router and receiver sequence number gets incremented by 1 and the feedback packet sends to the sender through label switch path 2 indicating packet arrive successful. If there is some message that need to be send from label switch path 2 to label switch path1 then this packet is attached with the packet sending from label switch path2 as piggy backing, thus also reduces the bandwidth utilization this frame is termed as FP packet. If there is no packets need to be send from label switch path 2 to label switch path1 then frame will be send individual as feedback frame to sender router, this packet is termed as FB packet.

If there is some event generated because of checksum error or time out or the sender frame sequence number is not the same as that of receiver sequence number then feedback packet will contain 0 as feedback bit and sender sequence number and receiver sequence number will remain the same. As this feedback packet reached to the sender, the sender resends the same sequence number packet to the next hop router through label switch path 2. If there is any packet need to be send from label switch path 2 to sender router then this feedback packet will be attached with the same packet as piggy backing and send in form of FP packet thus reduces bandwidth otherwise this feedback packet will be send individually as FB feedback packet. This process keeps on repeating at repeated time period set for the following protocol and all these processes keep on repeating for the lifetime till there are packets to be send.

5 Conclusion

The algorithm discussed in this paper is used to find error detection and correction on per hop basis thus reduces the time taken to find an error in MPLS network by passing a new packet of 7 bit in the alternative label switch path. Also using piggy backing in the same algorithm is used for better bandwidth utilization thus prevail resource for other packets and improve the time taken by protocol to detect error and thus enhance its performance than other protocols which work on end to end signaling.

References

[1] Qui, Y., Zhu, H.B., Zhou, Y., Gu, J.: A Research of MPLS- based Network Fault Recovery. In: Third International Conference on Intelligent Networks & Intelligent Systems. IEEE (2010)

[2] Xiao, Y., Yang, L., Li, Y., Qin, Z.: A loopback Detection Mechanism for MPLS failure. In: Proceding of IC-NIDC 2010. IEEE (2010)

[3] Xie, W., Huang, S., Gu, W.: An improved Ring Protection Method in MPLS-TP Network. In: Proceeding of IC-NIDC 2010 (2010)

[4] Porwal, M.K., Yadav, A., Charhate, S.V.: Traffic Analysis of MPLS and Non MPLS Network including MPLS signaling Protocols and Traffic distribution in OSPF and MPLS. In: IEEE 2008 (2008)

[5] Oomen, B.J., Misra, S., Granmo, O.C.: Routing Bandwidth- Gauranteed Path in MPLS Traffic Engineering: A Multiple Race Track Learning Approach. IEEE Transaction on Computers 56 (July 2007)

[6] Gupta, A., Kumar, A., Rastogi, R.: Exploring the Trade-off Between Label Size and Stack Depth in MPLS routing. In: IEEE 2003 (2003)

[7] Zhang, Z., Shao, X., Ding, W.: MPLS ATCC: An Active Traffic and Congestion Control Mechanism in MPLS. In: IEEE 2001 (2001)

[illegible]

6 Conclusion

[illegible]

References

[illegible]

Towards Improving Automated Evaluation of Java Program

Aditya Patel, Dhaval Panchal, and Manan Shah

AES Institute of Computer Studies, Ahmedabad University, Ahmedabad, India
adityabpatel@yahoo.com, {pdp3323,mhshah92}@gmail.com

Abstract. E-learning is gaining widespread use and is being used as importance method of education, particularly in higher education. Manual program evaluation by the instructor or expert is a time consuming process and subject to manual variations and errors. The automatic assessment and grading of student's answers/assignments plays an important role in improving e-learning and providing relief to instructors from the time consuming and lengthy task of manually evaluating student's assignments/computer programs and grading them. To address this problem, we have proposed a model for automated evaluation and grading of Java programs submitted by students during term work submission or practical examination. This research work focuses on automated evaluation of Java program using various parameters like number of compilation errors, correct output, lines of code, use of coding and naming conventions, cyclomatic, time and space complexity. The experimental results obtained from the initial prototype implementation are encouraging and validates the effectiveness of the proposed model.

Keywords: E-learning, Automated Evaluation, Java program grading, cyclomatic complexity, time and space complexity.

1 Introduction

In today's educational environment, information and communication technologies for E-learning are being widely used. E-learning management systems like Moodle, online courses, discussion forums, E-classrooms and E-assessment tools are playing an important role in ICT enabled teaching learning environment. As part of the e-learning tools, online exams are widely used to test students' skills through using computers and network services. Online exams reduce the overall expenses of processing exams especially in saving papers, storage, and materials costs [1]. The assessment of student's program is a necessary part of most programming courses, as it provides a way to see if learning goals are being met. Unfortunately, assessing student code manually can be difficult and time-consuming for the instructors/faculty members. In addition, it is difficult to judge the correctness of student code without spending a large amount of time understanding it. Also, the evaluator may need to write test cases/code to evaluate or assign grade [2].

S.C. Satapathy et al. (eds.), *Emerging ICT for Bridging the Future – Volume 1*,
Advances in Intelligent Systems and Computing 337, DOI: 10.1007/978-3-319-13728-5_55

2 Problem Definition and Motivation

In current educational environment, large number of students enrolls for courses offered by all the Universities and Colleges offering graduates/post-graduate programs. In most of the courses related to programming and technology, there is a practical component involving programs to be written by students to evaluate their practical solution development skills. During the practical exams of such courses, students are required to submit their programming solution to the respective faculty member/evaluator [3]. Manual program evaluation by the faculty is a time consuming process and subject to manual variations and errors. Also, in case of practical programming assignment submissions, it creates a burden on the faculty/evaluator and they would be busy most of time in testing and grading work at the expense of time spent with the students [4]. There are various issues related to automated program evaluation like checking multiple evaluation parameters - number of errors, correct output, lines of code, use of coding and naming conventions, cyclomatic, time and space complexity. It may happen that the program is giving correct output, but the program design or efficiency is not good. In academic curriculum of most of the Indian and foreign universities, Java language is being increasing used for teaching object oriented programming paradigms and programming techniques in general in engineering, computer science and IT based courses. This problem faced in academic environment motivated us to research on automated evaluation model for Java programs submitted by the students.

3 Literature Survey and Related Work

We have surveyed the research work that has been done in broad area of automated evaluation of programs. Several research projects have been undertaken which focuses on C and C++ and their structured or semi-structured automated evaluation. Ikdam Alhami and Izzat Alsmadi [5] have developed a tool which automatically grades students C++ programs. Concepts or code from Students' answers are first parsed. Key abstractions and keywords are extracted from students' assignments and compared with typical or expected answers. Weights are given to code keywords by the instructor based on their value and importance in the overall answer. In [6], cyclomatic complexity based program evaluation is discussed. It is used to measure the amount of decision logic in a software module. Cyclomatic complexity is based entirely on the structure of software's control flow graph. Each flow graph consists of nodes and edges. The node represents computational statements or expressions, and the edges represent transfer of control between nodes. Automated evaluation in E-learning domain is also being applied to evaluate UML use case diagrams submitted by students [7].

3.1 Automated Program Code Testing

In the paper [8], a web based application is made in which proposed question structure allows entering full program code and hiding some parts that must be re-written by the students to avoid hard coding. The system generates results on the basis of functional outputs only. It checks whether the student's code contains allowable or disallowable keywords. The system doesn't check the correctness of the program when the program code doesn't generate exact output.

3.2 Automatic Code Homework Grading Based on Concept Extraction

In this work [9], Ikdam Alhami, and Izzat Alsmadi have developed a tool which automatically grades students C++ programs. It compares students program with the expected answers submitted by the evaluator and gives weights or grades depending on the number and the type of keywords found.

3.3 The BOSS Online Submission and Assessment System

Boss is an online submission system which provides feature for course management and automated testing for Java [10]. BOSS is a tool for the assessment of programming assignments, which supports a variety of assessment styles and strategies, and provides maximum support to both teachers and students. The process of marking a programming assignment includes three principle components: correctness - relates to the extent to which a program's functionality matches that of its specification, style - describes those attributes of a program that a student's submission is expected to display, authenticity – covers verification of the student's identity and checks for plagiarism. The web tool BOSS uses both black box testing which defines input and output as data files, and a test is constructed by specifying the content of the expected output file (or files) for given input data files [11][12].

3.4 Automatic Evaluation System with a Web Interface [13]

This work describes scheme and implementation for automatic evaluation of C programming assignments. The scheme was motivated by the need to handle programming assignments with large cohorts of students (approximately eight hundred students). The paper explores some of the limits of what can be automated and how this can impact on the design of assignments. White box approach is being chosen which is more general than the Black box approach to testing programs. It considers the correctness of the solution, use of algorithm and suitable data structures, and performance of the program in terms of efficiency and space utilization. It uses XML format for specification of the questions and all other grading details from the web interface [13].

4 Proposed Model for Automated Evaluation of Java Program

4.1 Proposed Program Testing Approach

The proposed approach uses both Black Box testing and White Box testing. In Black Box testing, a program is treated as a single entity and using particular set of inputs and outputs, the overall program is tested. In White box testing, the structure, programming logic as well as behavior of a particular program is tested. In an environment where the students are learning to program, testing the final output is not feasible because conformance to an overall input/output is particularly hard to achieve. As a part of White box testing, the time, space and cyclomatic complexity of the student's code and the evaluator's model solution will be calculated and compared. The system executes the model program and students program with a specific input and will calculate the execution time of program. Now, the execution time of both the programs is compared and the grades are then awarded accordingly.

It is very important to test Java program on the basis of their execution times and space complexity in order to determine efficiency of the program. As a part of White box testing, this testing will be made by comparing the student's code with the model solution submitted by the evaluator on the basis of space and time complexity. The system executes the model program with a specific input and will calculate the execution time. Now the system will then execute the Student's program using the same input value, and will determines the execution time and then compare it with the execution time of model program and the grades are then given accordingly. The free memory and total memory on Java Virtual Machine by using Runtime class in Java. If the student has used the memory efficiently, then the memory requirement for student's program will not much vary from that of the model program [14].

4.2 Proposed Model for Automated Program Evaluation

The proposed architecture for the automated evaluation system is shown in Figure 1. In this system the whole process of automatic evaluation begins with the evaluator's submission of the problem statement, model solution of the problem stated in the assignment question, and grading schemes. The student's program files will be collected according to the predefined format (e.g. User names and roll no.). The submissions of student's programs (.java files) are stored in a server side database. The submissions will be accepted only if proper filenames, class names and method signatures are used as specified by the evaluator. This operation's will be performed by the file Basic Validate module.

The next module is the server side compilation module. After the students submit their work, the program is compiled through JVM (Java Virtual Machine) on the successful compilation of the program the system proceeds to the next stage.

After the Compilation module the next module which will be executed is the Program Testing module. This module includes both Black box testing and White box Testing. The student's program will be tested on the basis of Black box testing initially. In this module, the JUnit Test framework is used in which student code will be tested with the set of input, output and test data parameters specified by the evaluator. It uses the testing procedures provided by the evaluator. If the results of testing the program on user defined inputs are positive, then the next phase of testing is white box testing which includes testing the correctness and efficiency of the program on the basis of parameters such as execution time and space complexity. During this testing, parameters of the user program are compared against parameters of the model

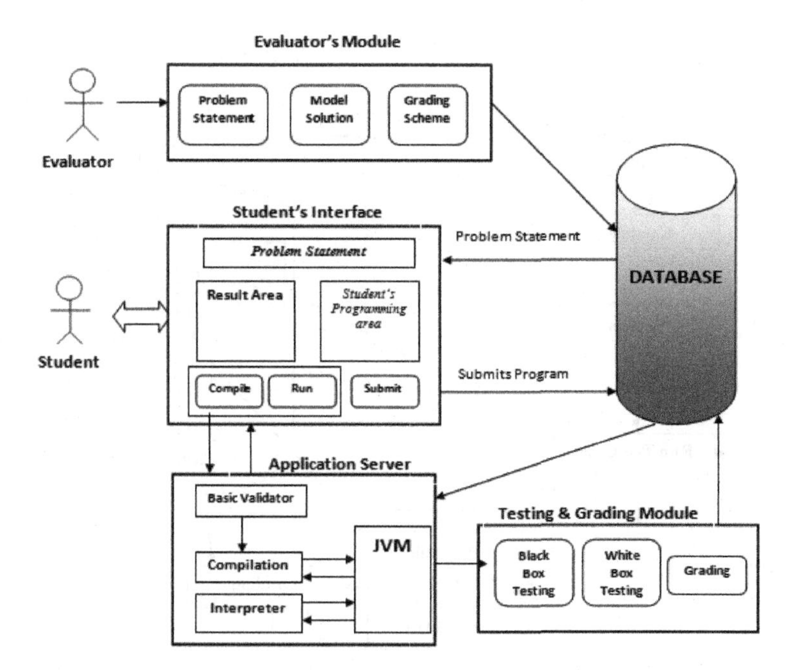

Fig. 1. Automated Program Evaluation System Architecture

solutions submitted by the evaluator. The execution time of the student program as well as the evaluator's program is determined for a particular number of inputs. The execution time of the student's program and the evaluator's program are compared and to analyze the time efficiency of program and grading is done accordingly. If the test for space and time complexity goes well then the next phase is measuring the cyclomatic complexity of code on the basis of some of the characteristics of the program such as, total program length, total number of branching or conditional statements, total number of go to, continue and break statements, total number of looping statements, etc [15]. After the system generates automated grading of the Program, it is shown to the evaluator. The evaluator can also manually check the code and override the grades generated by the system.

5 Proposed Algorithm for Automated Evaluation of Java Program

5.1 Explanation of the Algorithm

The students submitted Java programs will be fetched from the database. The fetched code is first compiled through the JVM (Java Virtual Machine) to check if there exists any compilation error or not. If the compilation error exits then system will not proceed further and the system will evaluate based on the comparison of keywords and similarity with the model solution. If the program compiles without any error, then the set of

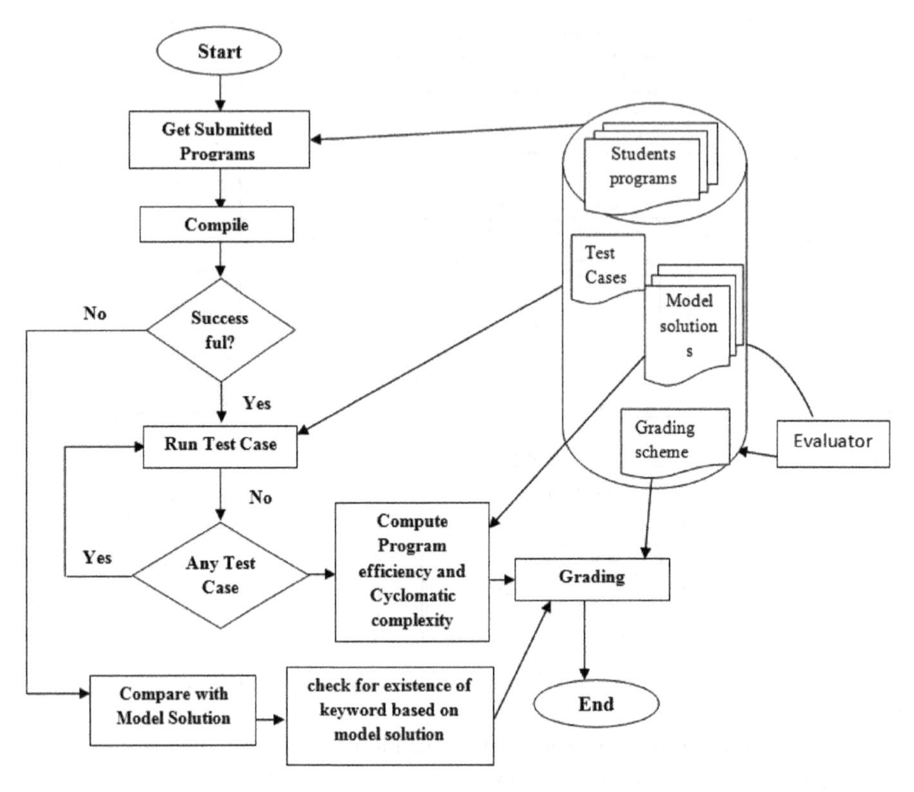

Fig. 2. Algorithm for Automated Evaluation of Java Program

JUnit test cases submitted by the evaluator will be conducted on the code to check whether the code generates the expected output when particular user defined inputs are given. If the results of testing the program on user defined inputs are positive, then the next phase of testing is the testing of programs on the basis of characteristics such as Space, Time and Cyclomatic complexity, and comparing its values with the values specified in evaluator's model solution and the grades are than given accordingly. Finally, the grades that are assigned to student during evaluation as per the grading scheme specified by the evaluator will be displayed to the student.

6 Prototype Implementation and Experiment Results

The proposed model is under implementation in our computer lab consisting of multiple windows machines running Windows 7/Ubuntu linux 9.10 and Windows server running the application server connected with LAN network. The proposed model and algorithm is implemented in Java server side programs deployed in Apache Tomcat Application server 6.

Technology Used: JDK 1.7, Java EE 1.7 (Java Servlet, JSP)
Student Operating System: Ubuntu linux 9.10, Windows XP
Java Application Server: Apache Tomcat 6
Database: MySQL Server 5

6.1 Experiment Result and Discussion

After the initial prototype implementation, we have performed basic experiments with the system on automated evaluation of Java programs by conducting practical exam of 50 random sample of students. The faculty expert submitted the question with model answer and 3 unit test cases to the students. The students submitted their answers through the system and their answers were evaluated by the System. The automated results of system evaluation were compared with the evaluation of faculty expert to validate the effectiveness of the system in automated grading of Java programs. The results obtained are presented in below table.

Table 1. Comparison of Student grades

Evaluation mode of Student Programs	Manual Grade assigned by Faculty Expert	Grade assigned by Automated Evaluation System
Grading of Student Program		
A Grade	12	14
B Grade	25	23
C Grade	8	8
Fail	5	5

The above experiment result indicates that the proposed model is effectively able to perform automated evaluation of Java program with considerable accuracy while compared it with manual evaluation by faculty expert.

7 Conclusions and Future Work

The proposed model we have illustrated our system with its usability and explained the details of the automatic evaluation process with an example in a systematic way. Our system includes each possible ways for testing the programs i.e. testing the programs on basis of user defined set of inputs and outputs, testing the programs on time complexity, space usage and cyclomatic complexity of the student's program. Evaluator's gets benefit because they do not have to spend hours for grading the programs. Currently we have proposed a system that supports a single programming language i.e. the 'Java' language, but the system can be extended to test programming assignments in other popular languages such as C, C++, etc. The Code Plagiarism tool can be used in this system which looks for possible plagiarism locally among students code. It can also compare each student code and search through Google search engine to find exact matches for lines or paragraphs in the code.

References

1. Alhami, I., Alsmadi, I.: Automatic Code Homework Grading Based on Concept Extraction (2011)
2. Ihantola, P., Ahoniemi, T., Karavirta, V., Seppälä, O.: Review of recent systems for automatic assessment of programming assignments. In: Proceedings of the 10th Koli Calling International Conference on Computing Education Research, Koli Calling 2010, pp. 86–93. ACM, New York (2010)
3. Edwards, S.H.: Improving student performance by evaluating how well students test their own programs. Journal on Educational Resources in Computing (JERIC) 3(3), 1 (2003)
4. Mandal, A.K., Mandal, C., Reade, C.: A System For Automatic Evaluation of Programs For Correctness and Performance (2010)
5. Eason, G., Noble, B., Sneddon, I.N.: On certain integrals of Lipschitz-Hankel type involving products of Bessel functions. Phil. Trans. Roy. Soc. London A247, 529–551 (1955)
6. Watson, A.H., McCape, T.J.: Stuructured Testing: A Testing Methodology Using the Cyclomatic Complexity Metric, pp. 500–535. National Institute of Standard and Technology Special Publication (September 1996)
7. Vachharajani, V., Pareek, J., Gulabani, S.: Effective Label Matching for Automatic Evaluation of Use–Case Diagrams. In: 2012 IEEE Fourth International Conference on Technology for Education (T4E), pp. 172–175. IEEE (July 2012)
8. Drasutis, S., Motekaitytė, V., Noreika, A.: A Method for Automated Program Code Testing. Informatics in Education 9(2), 199–208 (2010)
9. Alhami, I., Alsmadi, I.: Automatic Code Homework Grading Based on Concept Extraction. International Journal of Software Engineering and Its Applications 5(4) (2011)
10. Symeonidis, P.: Automated Assessment of Java Programming Coursework for Computer Science Education.The University of Nottingham School of Computer science and Information Technology (2006)
11. Higgins, C.A., Gray, G., Symeonidis, P., Tsintsifas, A.: Automated assessment and experiences of teaching programming. Journal on Educational Resources in Computing (JERIC) 5(3), 5 (2005)
12. Joy, M., Griffiths, N., Boyatt, R.: The BOSS Online Submission and Assessment System. Preprint of ACM JERIC 5(3) (September 2005)
13. Mandal, C., Reade, C.M.P., Sinha, V.L.: An Automatic Evaluation System with a Web Interface (2011)
14. Truong, N., Roe, P., Bancroft, P.: Static analysis of students' Java programs. In: Proceedings of the Sixth Australasian Conference on Computing Education, vol. 30, pp. 317–325. Australian Computer Society, Inc. (January 2004)
15. Watson, A.H., McCape, T.J.: Stuructured Testing: A Testing Methodology Using the Cyclomatic Complexity Metric, pp. 500–535. National Institute of Standard and Technology Special Publication (September 1996)

Fault Tolerant Scheduling - Dual Redundancy in an Automotive Cruise Control System

Manne Lakshmisowjanya, Annam Swetha, and V. Radhamani Pillay

Amrita Vishwa Vidyapeetham (University),
Coimbatore, India
{manne.lakshmi448,swethaannam4}@gmail.com,
vr_pillay@cb.amrita.edu

Abstract. Safety Critical real time systems are required to meet high reliability requirements, stringent deadlines and temporal demands. Such demands are met with fault tolerant mechanisms for applications like automotive, space and avionics systems. For such safety critical systems, to ensure the success of systems, various redundancy schemes are built into hard real-time systems. In this paper, a dual redundant scheme with active hot standby system has been employed in a Cruise Control System. A framework based on a paradigm for fault tolerance to provide adaptive fault tolerance scheduling of tasks in a DAG of the CCS is proposed. The scheme when implemented gives an efficient offline task scheduling, adaptive online dynamic reconfiguration of resources for single point of failure and guarantees functional and timing correctness of essential tasks. Efficient use of the redundant resources under fault free conditions and fail safe mechanism for fault ensures full functionality and enhanced performance. The comparative evaluation with a typical traditional dual system with performance metric highlights the enhanced performance and the importance of this work for the automotive industry.

Keywords: Cruise Control System; dual redundancy scheme; hard real time system; Fault tolerant system.

1 Introduction

The emerging current trends ensure sophisticated real-time applications meet high computational demands. Safety critical system requires high dependability to avert catastrophic consequences. One method usually employed uses either hardware or software redundancy or both to provide fault tolerance. Typically hardware redundancy is used to provide fault tolerance for permanent faults in the system. The requirements of higher computational capability and power and the complexities due to the synchronization and timing constraints on a multiprocessor platform makes the fault tolerant scheduling [1, 2, 3]. The real-time system is vulnerable to process failures and if a time-critical task is under execution, when a fault occurs this can lead to permanent failure. Hardware and software redundancy play a major role in fault tolerance in safety critical system. The model proposed is based on one such

S.C. Satapathy et al. (eds.), *Emerging ICT for Bridging the Future – Volume 1*,
Advances in Intelligent Systems and Computing 337, DOI: 10.1007/978-3-319-13728-5_56

traditional dual redundant system for adaptive fault tolerance to maintain full functionality of system even under fault scenario. This scheme is implemented using the TORSCHE scheduling tool box and the performance evaluated.

Rest of the paper is organised as follows, literature review backround study of cruise system, task scheduling, faults and fault tolerance aspects are discussed in Section 2. The description of system model, in Section 3. its implementation with simulation environment is explained in Section 4. Section 5 deals with conclusion and future scope of the work.

2 Literature Review and Background

Several approaches have been developed to schedule task systems on multiprocessor processor platforms. Liu and Layland [4] proved that Rate Monotonic (RM) scheduling is optimal among fixed priority algorithm and along with that utilization bound for RM for periodic tasks are derived, later this bound was improved by Bini et al[5]. Multiprocessor real-time scheduling theory has its origins in the late 1969s. Liu [6] inclined the course of research in this area for two decades. Graham[7] in 1976 adding extra processors or relaxing constraints implies reducing execution time requirements of task set that is optimal scheduled on a multiprocessor platform increases length of schedule. In depth survey of multiprocessor real-time scheduling is found in [8]. Partitioned approach has its origin, with a fixed allocation of tasks to processors were preferred compared to global approach during 1980s and 1990s.

A symbolic approach for reliability analysis focusing on permanent faults was presented by Glab et al. Spatial redundancy is considered and duplicating the same task to multiple Processing Elements (PE), with permanent faults are regarded in [9]. The design optimization of fault-tolerance systems using both spatial and temporal redundancy is presented by Izosimov et al[10]. A. Avizienis incorporated masking with practical techniques of error detection, fault diagnosis, and then into concept of fault-tolerant systems [11]. Later Ghosh[12] analyzed a fault tolerant scheduling algorithms for multiprocessor. The enhanced fault tolerance in static real-time scheduling was proposed by Oh and Son[13]. Further fault tolerance scheme dealt by Krishna and Shin[14] with the concept of fast recovery of tasks from failure. The fault tolerance enhances the performance and provides effective utilization of additional resources of the system. A prototype of a Cruise Control System has been developed in the year 1987, in European union EUREKA program.Main aim was to reduce the workload of driver and provides him more comfort by automatically adjusting the speed of the vehicle[15].

Safety Critical Systems

Typically real-time systems are represented by a real-time task model that gives information about workload of the real-time application, system timing constraints, resource model that describes availability of resources for real-time applications and scheduling algorithm that defines resource allocation to applications at all times. To ensure the functionality of safety critical system even during occurrence of fault by operating in different modes where certain non-critical tasks are dropped, a new way of fault tolerant scheduling in safety critical multiprocessors where three task level criticality has been explored to schedule tasks in multiple processors is explained in [16, 17, 18].

Fault Tolerance and Real-time Scheduling

In computer dependent world reliability and availability have become increasingly important. Failure in system occurs due to environment, system operations, and software or hardware faults. To increase reliability one of the method is fault tolerance which uses additional components (redundancy) hardware, software or both to bypass the effects of failure. The advantage of hot standby with dual redundancy scheme is, it has very short or no outage time and does not require recovery of application. To guarantee deadlines for hard real-time system hardware and software redundancy can be used by [19]. Predictability is one of the primary issues in real-time systems. Schedulability analysis or feasibility checking of the tasks of a real-time system has to be done to predict whether the tasks will meet their timing constraints. Static table-driven approaches performs static schedulability analysis and the resulting schedule (or table, as it is usually called) is used at run time to decide *when* a task must begin execution. The above mentioned scheduling approaches are selected for implementation for fault free and fault conditions respectively. Based on number of processors required by the parallel task model, each subtask in the graph can be executed sequentially or parallel. Existing parallel task model in practice is fork-join model used in OpenMP [20] Directed Acyclic Graph (DAG), is a combination of subtasks and directed edges, representing the precedence constraints and flow of execution of sub tasks In this paper, scheme for task allocation and real-time scheduling using table driven cyclic execution and at the same time dual redundancy scheme with fault tolerance with programming flexibility is proposed.

Case Study: Cruise Control System (CCS)

Cruise control system can be characterized as an embedded real time system having a number of sensors, actuators and processors. The functionality of such system is based on the tasks done by sensors, actuators and controllers that interact among each other in real-time.

Normal tasks:

- Sensors processes.
- Computing control values.
- Actuator process and Updating parameters.

Monitoring functions:

- User interface monitoring.
- Brake monitoring.
- Accelerator monitoring.
- Engine monitoring.
- Monitoring speed using wheel revolution sensor.

Control functions:

- To maintain a constant speed comparing current speed and desired speed.
- Updating parameters.

Actuating functions:

- Based on signal after controller action, throttle actuator is controlled to maintain a desired speed.

3 Approach

Assumptions

In this paper a real-time CCS with dual redundancy is considered.

1. Problem of concurrency control of tasks are not taken into account. System can tolerate one processor failure.
2. Maximum time required for re-allocating the tasks due to failure is greater than minimum time interval between the faults.
3. Only periodic tasks scheduling is attempted.
4. Time period (T_i) of all tasks are assumed to be same as deadlines (D_i) of corresponding tasks.
5. To enable the detection of processor failures with a bounded latency a suitable watchdog mechanism is assumed.

In CCS taskset independent tasks are scheduled in different threads (based on DAG) which reduce the execution time of the process.

Dual Redundant Scheme (DRS)

The fault tolerant scheme with dual redundancy scheme employs two independent, identical processors for performing the same computations as shown Fig.2. By specified switching mechanism, error detection is done by a synchronized internal hardware and hardware disables that system if there is failure. If the primary processor fails then the corresponding redundant processor takes over the operation into hot standby. Such an implementation ensures the system continues to function without any break in the computations in case of failure and thus meeting the real time constraints essential for such critical missions.

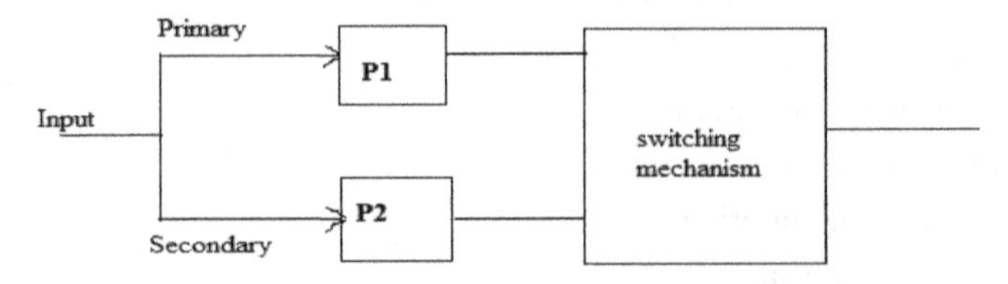

Fig. 1. Dual Redundancy Scheme

Model Schematic

Table 1. Task set – ACC

SI.No	Tasks	C_i (time units)	T_i (time units)
1.	Speed monitoring	3	15
2.	Acceleration monitoring	2	10
3.	CCS clutch monitoring	2	10
4.	Brake monitoring	3	15
5.	Proximity sensor monitoring	2	15
6.	Computing control values	10	55
7.	Actuating the throttle valves	5	30
8.	CCS measurement update	10	15
9.	Vehicle control Update	2	60
10.	Fwheel Speed update	2	80
11.	Fwheel brake update	2	80
12.	Rwheel Speed update	2	80
13.	Rwheel brake update	2	80
14.	Driver display update	2	80

Task attributes of Cruise Control System (CCS) task set is given in Table 1.The task set contains periodic task T1-T13. The periodic tasks T1 – T5 are sensing tasks and are classified as non-critical sensing tasks that are parallel real-time tasks. Tasks T6, T7 are tasks that compute control values and perform actuating functions are classified under critical tasks and sensing tasks are predecessor tasks to T6, and is the predecessor task to T7. Tasks from T8 to T14 are update tasks that are classified as non-critical tasks, which improve the performance of CCS. Fig 3 represent the fault tolerant model with Dual Redundancy Scheme and with a Global Clock which provides synchronization between the two processors. A signal is sent by both the processors to master node periodically which acts as a health check signal for master node

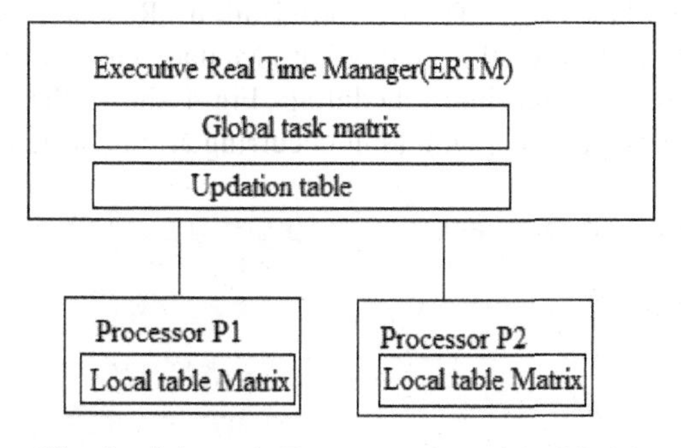

Fig. 2. Schematic Representation of the Model

Executive Real Time Manager (ERTM) – The ERTM maintains two global task matrix and updation table matrix and acts as master node along with that this node monitors the health status of the processors. If master node does not receive a signal from processor indicates the permanent failure of the processor. Global task matrix maintains the task attributes of CCS, nature and criticality of the tasks. And updation table indicates the level of criticality and remaining utilization of the processor. It is updated based on the information from each processor.

Each processor contains a local table matrix where the tasks to be scheduled in that processor are allocated.

Methodology for Task allocation using Dual Redundancy Scheme:

In DRS all tasks in the task set are allocated and executed in both processors and each processor will execute the tasks using DAG, parallel task model.contolling task which is dependent on all sensing tasks is through inter process communication. This system functions even during the failure of one processor and provides parallelism for sensing tasks in the task set.

Algorithm for task allocation in Dual Redundancy Scheme:

Input: is the periodic task set of CCS stored in ERTM.
Output: Dual Redundancy Scheme in normal mode.

1. *for* i= 1to n do
2. Schedule the tasks in both processors,
3. *end*
4. *for* each clock pulse do
5. Signal from both processors trigger the master node
6. Update ERTM
7. *End*

One evaluation metric for the performance is the total execution time [21] of algorithm.

4 Implementation and Simulation

The Dual Redundancy Scheme proposed in section IV is analyzed and simulation is carried out in Matlab with Time Optimization Resource and SCHEduling (TORSCHE) tool box.The model is entered into Matlab. The work load is assumed to be 70%. Fig 3 shows the real-time scheduling. Fig 4 shows the scheduling without fault. Fig 5 shows the scheduling with fault occurring at time instant 20 in the P2.

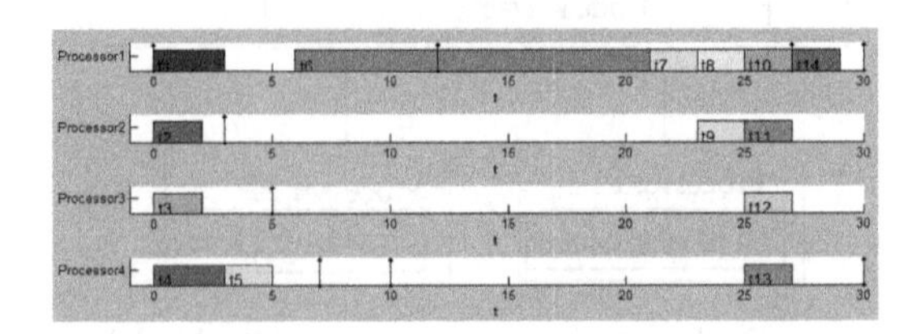

Fig. 3. Real time Scheduling

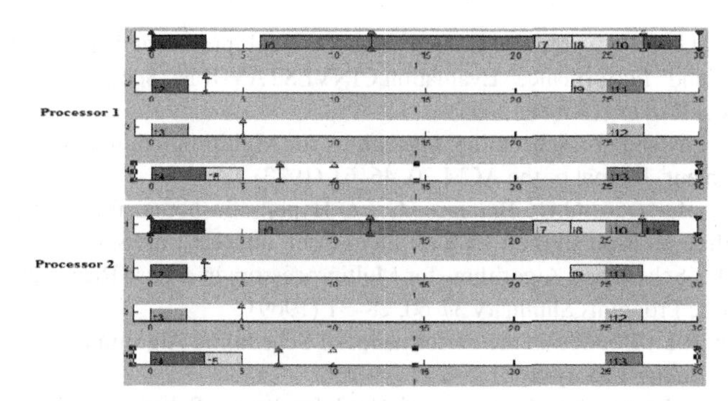

Fig. 4. Scheduling under normal mode

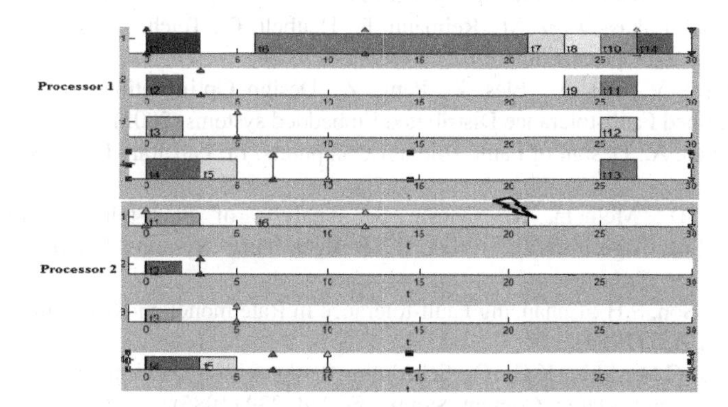

Fig. 5. Scheduling under fault mode

The execution time of the process has been found to be 29 time units under fault free mode.

5 Conclusion

In this paper a scheme has been implemented for fault tolerance scheduling for given CCS. Along with schedulability guarantees and flexibility in fault tolerance strategy provides safe implementation even under fault conditions. Under fault condition, effective fault tolerance mechanism upto one processor fault, and to ensure optimum system performance it can be extended to m processor which can withstand m-1 processor faults. The proposed scheme can be beneficial for future design that include new features such as possibility for execution of advanced algorithms with higher computational requirements.

References

[1] Isovic, D.: Design of Real-time Systems. Malardalen University, Swedan (1998)

[2] Luo, J., Jha, N.K.: Power-conscious Joint Scheduling of Periodic Task Graphs and Aperiodic Tasks in Distributed Real-time Embedded Systems. In: Proceedings of the 2000 IEEE/ACM International Conference on Computer-aided Design (2000)

[3] Zapata, O.U.P., Alvarez, P.M.: EDF and RM Multiprocessor Scheduling Algorithms: Survey and Performance Evaluation.CINVESTAV-IPN Seccion de Computacion. Technical Report (October 2005)
[4] Liu, C., Layland, J.: Scheduling Algorithms for Multiprogramming in a hard real-time environment. Journal of the ACM 30, 46–61 (1973)
[5] Bini, E., Buttazzo, G.C., Buttazzo, G.: A Hyperbolic bound for the Rate monotonic Algorithm. In: Proc. of the 13th Euromicro Conf. on Real-time Sys., p. 59 (2001)
[6] Liu, C.L.: Scheduling Algorithms for Multiprocessors in a Hard Real-time environment. JPL Space Programs Summary 37-60, 28–31 (1969)
[7] Graham, R.L.: Bounds on the Performance of Scheduling Algorithms. In: Computer and Job Scheduling Theory, pp. 165–227 (1976)
[8] Davis, R., Burns, A.: A survey of Hard Real-time Scheduling for Multiprocessor systems. ACM Computing Surveys (2011)
[9] Glab, M., Lukasiewycz, M., Reimann, F., Haubelt, C., Teich, J.: Symbolic Reliability Analysis and Optimization of ECU Networks (2008)
[10] Izosimov, V., Pop, P., Eles, P., Peng, Z.: Design Optimization of Time and Cost-constrained Fault-tolerance Distributed Embedded systems (2005)
[11] Avizienis, A.: Design of Fault-Tolerant Computers. In: Fall Joint Computer Conference (1967)
[12] Mosse, D., Melhem, R., Ghosh, S.: Analysis of Fault-tolerant Multiprocessor Scheduling Algorithm. In: Proc. IEEE Real Time Systems Symp., pp. 129–139 (December 1981)
[13] Oh, Y., Son, S.H.: Enhancing Fault-tolerance in Rate monotonic Scheduling. *Real-Time Systems 7(3) (1994)*
[14] Krishna, C.M., Shin, K.G.: On Scheduling Tasks with a Quick recovery from Failure. In: Proc. Fault-tolerant Comput. Symp., pp. 234–239 (1985)
[15] Björnander, S.: Technical Report, Adaptive Cruise Controllers – A Literature review. Mälardalen University, Sweden (2008)
[16] Pillay, R., Punnekkat, S., Chandran, S.K.: An improved redundancy scheme for optimal utilization of onboard Computers. In: IEEE INDICON 2009, India (2009)
[17] Pillay, R., Chandran, S.K., Punnekkat, S.: Optimizing Resources in Real-time Scheduling for Fault tolerant processors. In: IEEE International Conference on Parallel, Distributed and Grid Computing (PDGC 2010), Solan India (October 2010)
[18] Chandran, S.K., Pillay, R., Dobrin, R., Punnekkat, S.: Efficient scheduling with Adaptive Fault tolerance in heterogeneous multiprocessor systems. In: International Conference on Computer and Electrical Engineering (ICCEE) Chengdu, China (November 2010)
[19] Oh, Y., Son, S.H.: An algorithm for Real-time Fault-tolerant scheduling in Multiprocessor system. In: Proc. Euromicro Workshop on Real-Time Systems, vol. 20, pp. 190–195 (1992)
[20] Openmp, `http://www.openmp.org`
[21] Birolini, A.: Reliability Engineering –Theory and Practice. Springer, Heidelberg (2004)

A Simple Mathematical Model for Performance Evaluation of Finite Buffer Size Nodes in Non- Saturated IEEE 802.11 DCF in Ad Hoc Networks

Neeraj Gupta[1] and C.S. Rai[2]

[1] K.R. Mangalam University,
Gurgaon, India
[2] USICT, GSSIndraprastha University
Dwarka, Delhi, India
neerajgupta37@rediffmail.com

Abstract. Analytical models help in predicting the change in results once the input parameters of the networks are changed. A lot of modeling work has already been carried out to evaluate the performance of IEEE 802.11 DCF for both saturated networks and non-saturated networks. Most of work considering the arbitrary buffer size nodes in non-saturated conditions uses complex mathematically approach. In the current paper, we present a flexible and practical transform-free approach for evaluating the performance of IEEE 802.11 MAC protocol. The simplicity and fastness of our algorithm is due to the simple closed form expressions to determine idle probability of buffer and blocking probability of buffer with size K assuming each node as a discrete M/G/1/K queue. The proposed model gives accurate results which are validated through ns2 simulations.

Keywords: IEEE 802.11, Backoff, service time, variance, collision probability, buffer.

1 Introduction

The IEEE 802.11 [1], [2] WLAN, both infrastructure based and Infrastructure less, has been widely accepted and deployed for the indoor and short range communications. In order to evaluate key parameters of network various analytical models has already been proposed. Depending on the traffic arrival rate, the analytical models can be classified as saturated networks and non-saturated networks. The saturated networks are based on the assumption that there is always the availability of packet whenever a station wants to transmit [3], [4], [5], [6], [7]. However, in real condition the availability of packet cannot always be assured leading to bursty networks. This leads to development of non-saturated analytical models [8], [9], [10], [11], [12]. Most of the analytical models proposed in case of non-saturated networks involve the assumption of either small or infinite buffers [13], [14], [15], [16]. Such assumptions are neither realistic nor very useful. The extension of these models, capturing influence of arbitrary buffer size is a challenging task. The generalized cases for

S.C. Satapathy et al. (eds.), *Emerging ICT for Bridging the Future – Volume 1*,
Advances in Intelligent Systems and Computing 337, DOI: 10.1007/978-3-319-13728-5_57

non-saturated networks with nodes having arbitrary finite buffers size are investigated in [17], [18], [19], [20]. The key challenge while designing such models is determination of MAC layer packet service time. While most of the model uses the complex concept of probability generating function (PGF) and classical queuing model and inverse transform methods to solve for service time. The current paper presents a simplified and fast method for non-saturated model taking into consideration the arbitrary buffer size. The performance descriptors in this paper are evaluated using M/G/1/K queuing model and fixed point analysis of single cell IEEE802.11WLAN.

The paper is organized as follows: In section 2, discusses assumptions, terminologies and fixed point formulation model for non-saturated networks. Section 3 presents our simplified model for non-saturated model taking into consideration arbitrary buffer size. The comparison of models in terms of complexity and accuracy are presented in section 4. Section 5 provides validation of our model via ns2 simulations. Finally, section 6 concludes this paper.

2 Assumptions and Terminologies

It is assumed that 1) the channels is error free and the packets are lost only due to collision 2) All nodes reside in one-hop network and there are no hidden or exposed terminals 3) the key assumption is collision-error probability of a transmitted packet is constant and independent of the transmission error or number of collisions suffered in the past 4) all stations relay the packet of same size and at same rate. The packet inter-arrival rate is defined in terms of slots and is independent and identically distributed. The backoff period is divided into discrete slots. Whenever the medium is sensed to be idle the backoff is decremented slot by slot. When the backoff counter reaches zero the process for packet transmission begins at next time slot. The transmission process of packet is said to be *finished* when either packet is transferred successfully or it is dropped once the retry limit is reached. The total amount of time required when packet becomes head-of-line packet till the transmission period is finished is referred to as *service time* of MAC layer. The station enters the idle state when there are no more packets to be transmitted and backoff counter is initialized to its minimum value.

Our analytically model is based on the work proposed by [10], [20]. Let p and τ represents the collision probability and attempt rate per slot respectively. Whenever the buffer of node is empty we have

$$\tau = (1 - p_0)\, \tau_s \tag{1}$$

where, $p_0$and τ_s represents the probability of empty buffer and attempt rate per slot for saturated network. Assuming decoupling approximation the collision probability can be expressed as

$$p = \Gamma(\tau) \triangleq 1 - (1 - \tau)^{n-1} \tag{2}$$

In order to solve (1) and (2), we need to determine the values $p_0$and τ_s. The value of τ_s can be determined by equation (3) [5]. The problem is to determine the value of p_0.

$$\tau_s = \frac{1+p_s+p_s^2+\cdots+p_s^{M-1}}{b_0+p_s b_1+p_s^2 b_2+\cdots+p_s^{M-1} b_{M-1}} \quad (3)$$

3 Computations of Service Time and Probability of Empty Buffer

We are assuming M/G/1/K queuing model where K represents the maximum amount of buffers space to store the packets awaiting their transmission. The packet arrival process to queue is assumed to be Poisson distributed with parameter λ. In order to approximate service time we will use the approach developed in [20]. Let Y^c denote the MAC service time. Backoff decrement process is represented using random variable X. The random variable Ω is used to mathematical represent each decrement of slot. Since a station must complete backoff count of X before the transmission is finished, the service time is given by

$$Y^c = \sum_{i=1}^{X} \Omega \quad (4)$$

The mean value of X, Ω and Y^c are represented as $\bar{X}$, $\bar{\Omega}$ and $\overline{Y^c}$ respectively. From (4) and [21] we have

$$\overline{Y^c} = \bar{X} . \bar{\Omega} \quad (5)$$

In (4), $\bar{X}$ and $\bar{\Omega}$ is calculated as

$$\bar{X} = b_0 + pb_1 + p^2 b_2 + \cdots + p^{M-1} b_{M-1} \quad (6)$$

$$\bar{\Omega} = \sigma + P_s T_s + P_c T_c \quad (7)$$

In equation (6), b_k represents the mean backoff duration at the kth attempt of packet. The terms P_b, Ps and P_c in equation (7) denotes the probability of busy slots, the probability of successful transmission and probability of collision respectively by any of n contending stations. The symbol σ represents slot time and is equal 20 μs, T_s and T_c represents the time spent in slots for successful transmission and unsuccessful transmission of packet respectively.

Let M represent the maximum retry limit before the packet is dropped. The value of contention count is doubled till maximum value is reached, $W_k = 2^{m'} W$, where m' is contention window increasing factor. Once the maximum value is attained, the contention window remains at W_m until it is reset either due to successful transmission or short retry limit is reached. The probability that transmission process has finished at j^{th} stage is represented by $\delta(p, j)$.

$$\delta(p, j) = \begin{cases} (1-p)p^j, & j = 0, \ldots, M-2 \\ p^j, & j = M-1 \end{cases} \quad (8)$$

From [21], the probability generating function random variable X and Ω denoted as $\hat{X}(p, z)$ and $\hat{\Omega}(p, z)$ are expressed as

$$\hat{X}(p,z) = \sum_{i=0}^{M-1}\left[\delta(p,i)\prod_{k=0}^{i}\hat{\eta}_k(z)\right]$$

$$\hat{\Omega}(p,z) = (1-P_b)z^{\sigma} + P_s z^{T_s+\sigma} + P_c z^{T_c+\sigma}$$

$$\widehat{\eta_k}(z) = \begin{cases} \frac{1}{W_k}\left[\sum_{i=0}^{W_k-1} z^i\right], & k = 0,..,m-1 \\ \frac{1}{W_m}\left[\sum_{i=0}^{W_m-1} z^i\right], & k = m,..,M-1 \end{cases} \quad (9)$$

Let V(X) and V(Ω) denote the variance of X and Ω can be calculated as

$$V(X) = \left(\frac{d^2\hat{X}(p,z)}{dz^2}\bigg|z=1\right) + \bar{X} - \bar{X}^2$$

$$V(\Omega) = \left(\frac{d^2\hat{\Omega}(p,z)}{dz^2}\bigg|z=1\right) + \bar{\Omega} - \bar{\Omega}^2 \quad (10)$$

From (6), (7), (10) and [21] the variance of service time can be computed as

$$V(Y^c) = V(\textstyle\sum_{i=1}^{X}\Omega) = \bar{X}.V(\Omega) + V(X).\bar{\Omega}^2 \quad (11)$$

From (5) and (11) squared coefficient of variation of the service time (S^2) can be derived as

$$S^2 = \frac{V(Y^c)}{\overline{Y^c}^2} = \frac{\bar{X}.V(\Omega) + V(X).\bar{\Omega}^2}{(\bar{X}.\bar{\Omega})^2} \quad (12)$$

In queuing theory one of the important parameter is traffic intensity, ρ, and can be expressed in terms of p.

$$\rho = \lambda.\overline{Y^c} = \lambda\,\bar{X}\,\bar{\Omega} \quad (13)$$

where λ is arrival rate for Poisson arrival process and in our case it is represented as packet arrival per slot.

In order to directly calculate the value of p_0, unconditional probability that there is no packet in the queue (either being served or being held after service), we will use the equation (13) defined in [22].

$$p_0 = \frac{(\rho-1)}{\rho\left(2\frac{\sqrt{\rho e^{-S^2}}S^2 - \sqrt{\rho e^{-S^2}} + K + 1}{2 + \sqrt{\rho e^{-S^2}}S^2 - \sqrt{\rho e^{-S^2}}}\right) - 1} \quad (14)$$

The blocking probability, p_K, of finite queue of size K is give as

$$p_K = \frac{\rho\left(\frac{\sqrt{\rho e^{-S^2}}S^2 - \sqrt{\rho e^{-S^2}} + 2K}{2 + \sqrt{\rho e^{-S^2}}S^2 - \sqrt{\rho e^{-S^2}}}\right)(\rho - 1)}{\rho\left(2\frac{\sqrt{\rho e^{-S^2}}S^2 - \sqrt{\rho e^{-S^2}} + K + 1}{2 + \sqrt{\rho e^{-S^2}}S^2 - \sqrt{\rho e^{-S^2}}}\right) - 1} \quad (15)$$

Substituting (14) in (1) and we can easily calculate the value of p and τ.

4 Validations and Simulation

To validate the results obtained through our model we simulate single hop network for basic mechanism using ns-2 [23]. The parameters for model comparison are presented in Table I. The packet size is 1024 bytes at MAC layer which is fixed. The throughput of the system is calculated using the expression (16) [23]

$$S = \frac{P_S L}{P_S T_S + (1 - P_{idle} - P_S)T_C + P_{idle}\sigma} \quad (16)$$

The expression for calculating T_s is same as in [1]. Once the packet has suffered a collision, the station will defer for a period of EIFS (extended interframe space). Considering basic mechanism, the time for collision, T_c, can be expressed as

$$T_c = \text{DATA} + \text{EIFS} + \text{DIFS} \quad (17)$$

Table 1. IEEE 802.11b Parameters

Parameters	Value
SIFS	10us
DIFS	50us
Slot time Duration	20us
Propagation delay	2 us
Physical layer Header	192 bits / 1Mbps
Mac Header	224 bits / 11Mbps
RTS	160 bits / 1 Mbps
CTS	112 bits / 1Mbps
ACK	112 bits / 1Mbps
Minimum Contention Window Size	32
Maximum Contention Window Size	1024
Maximum retry Limit	7

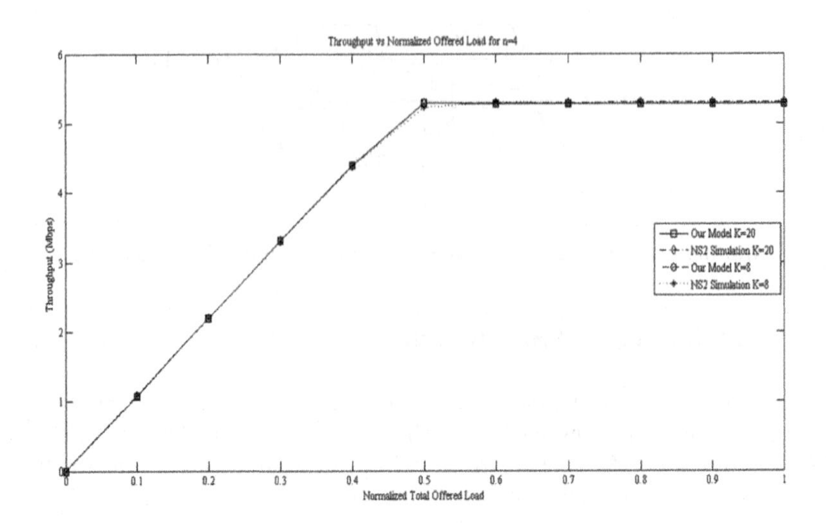

Fig. 1. System Throughput vs Normalized Offered Load for four number of stations. The graph is plotted when K = 8 and K=20 respectively.

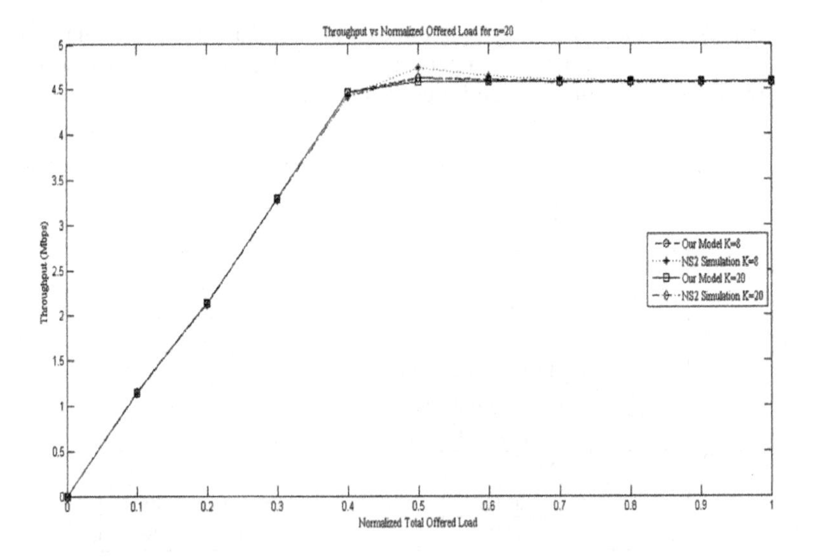

Fig. 2. System Throughput vs Normalized Offered Load for number of stations are twenty. The graph is plotted when K = 8 and K=20 respectively.

To validate our analytical model, we compare the analytical results with that of ns-2 simulation. During simulation, the offered traffic arrives at each station following a Poisson manner. Fig. 1 and Fig. 2 show the results for ad hoc network with sending stations number being four and twenty respectively. With buffer size of eight and twenty, we compare simulation results for the throughput with results obtained from our analysis. Initially, the throughput of the system increases with the increase in offered load. At higher load value network converges to saturated conditions leading the graph to go flat. It can be observed from both the graph that there is no substantially gain in throughput even if the buffer is increased from eight to twenty. It also shows that the results of our model are close to the simulation results.

5 Conclusion

The current paper presents an analytical model for performance evaluation of IEEE 802.11 DCF utilizing arbitrary buffer size. This provides a simple transform-free approach solution for comprehensive M/G/1/K queuing analysis of IEEE 802.11for non-saturated networks. The use of simple closed form expressions to determine idle probability of buffer is much simple in comparison to other work present in literature. It can be seen that increasing the buffer size of node does not necessarily increase the throughput of system. The simulation results proved that our mathematical approach provides correct results.

References

1. IEEE 802.11: IEEE standard for wireless LAN Wireless LAN Medium Access Control (MAC) and Physical Layer (PHY) Specifications.: Higher-Speed Physical Layer Extension in the 2.4 GHz Band. IEEE standard 802.11b (1999)
2. IEEE 802.11, Part II: Wireless LAN Medium Access Control (MAC) and Physical Layer (PHY) Specifications (August 1999)
3. Bianchi, G.: Performance Analysis of the IEEE 802.11 Distributed Coordination Function. IEEE Journals on Selected Areas in Communication 18(3), 535–547 (2000)
4. Chatzimisios, P., Vitsas, V., Boucouvalas, A.C.: Performance Analysis of IEEE 802.11 MAC Protocol for Wireless LANS. International Journal of Communication Systems 18, 545–569 (2005)
5. Kumar, A., Altman, E., Miorandi, D., Goyal, M.: New Insights from a Fixed Point Analysis of Single Cell IEEE 802.11 WLANs. IEEE/ACM Transaction on Networking 15(3), 588–601 (2007)
6. Cali, F., Conti, M., Gregori, E.: IEEE 802.11 protocol: design and perfromance evaluation of an adaptive backoff mechanism. IEEE Journal on Selected Area in Communications 18(9), 1774–1780 (2000)
7. Daneshgaran, F., Laddomada, M., Mesiti, F., Mondin, M.: Saturation Throughput Analysis of IEEE 802.11 in Presence of Non-ideal Transmission Channel and Caputre Effects. IEEE Transactions on Communications 56(7), 1178–1188 (2007)
8. Liaw, Y.S., Dadej, A., Jayasuriya, A.: Perfromance Analysis of IEEE 802.11 underLimited Load. In: IEEE Perth Asia-Pacific Conference on Communications, pp. 759–763 (2005)
9. Malone, D., Duffy, K., Leith, D.: Modelling the 802.11 Distributed Coordination Function in Nonsaturated Heterogenous Conditions. IEEE/ACM Transaction on Mobile Computing 15(1), 159–172 (2007)
10. Zhao, Q., Tsang, D.H.K., Sakurai, T.: A Simple and Approximate Model for Non- Saturated IEEE 802.11 DCF. IEEE Transaction on Mobile Computing 8(11), 1539–1553 (2009)

11. Senthilkumar, D., Krishan, A.: Nonsaturation Throughput Enchancement of IEEE 802.11b Distributed Coordination Function for Heterogenous Traffic under Noisy Environment. International Journal of Automation and Computing 7(1), 545–569 (2010)
12. Daneshgaran, F., Laddomada, M., Mesiti, F., Mondin, M.: Unsaturated Throughput Analysis of IEEE 802.11 in Presence of Non-ideal Transmission Channel and Caputre Effects. IEEE Transactions on Wireless Communications 7(4), 1276–1286 (2008)
13. Huang, K.D., Duffy, K.R.: On Buffering Hypothesis in 802.11 Analytic Models. IEEE Communications Letters 13(5), 312–314 (2009)
14. Duffy, K., Ganesh, A.J.: Modelling the Impact of Buffering on 802.11. IEEE Communication Letters 11(2), 219–221 (2007)
15. Dong, W., Zhang, W., Chen, X., Wei, G.: A New Load Equation for 802.11 MAC Performance Evaluation Under Non-Saturated Conditions. In: First IEEE International Conference on Communication in China: Wireless Communication Systems, pp. 482–486. IEEE (2012)
16. Gupta, N., Rai, C.S.: Performance Evaluation of IEEE802.11 DCF in Single Jop Ad Hoc Networks. Wireless Personal Communication (July 2014), doi:10.1007/s11277-014-1979-05
17. Garetto, M., Chiasserini, C.-F.: Performance Analysis of 802.11 WLANs under Sporadic Traffic. In: Boutaba, R., Almeroth, K.C., Puigjaner, R., Shen, S., Black, J.P. (eds.) NETWORKING 2005. LNCS, vol. 3462, pp. 1343–1347. Springer, Heidelberg (2005)
18. Ozdemir, M., McDonald, A.B.: On the Performance of Ad Hoc Wireless LANs: A Practical Queuing Theortical Model. Performace Evaluation 63(11), 1126–1157 (2006)
19. Zhai, H., Kwon, Y., Fang, Y.: Performance Analysis of IEEE 802.11 MAC Protocols in Wireless LANs. Wireless Communication and Mobile Computing 4, 917–931 (2004)
20. Zhao, Q., Tsang, D.H.K., Sakurai, T.: Modelling Nonsturated IEEE 802.11 DCF Networks Utilizing an Arbitrary Buffer Size. IEEE Transaction on Mobile Computing 10(9), 1248–1263 (2011)
21. Tijms, H.C.: A First Course in Stochastic Models, pp. 434–437. John Wiley & Sons Ltd. (2003)
22. Smith, J.M.: Propoties and Performance Modelling of Finite Buffer M/G/1/K Neworks. Journal of Computer & Operations Research 38(4), 740–754 (2011)
23. Gupta, N., Rai, C.S.: New Analytical Model for Non-Saturation Throughput Analysis of IEEE 802.11 DCF. In: International Conference on Advances in Communication, Network and Computing, pp. 66–76. Elsevier Science Ltd. (2014)
24. The VINT Project. The ns Mannual (May 2010)

An Efficient Algorithm and a Generic Approach to Reduce Page Fault Rate and Access Time Costs

Riaz Shaik and M. Momin Pasha

Department of Computer Science and Engineering, K L University, Vijayawada, India
shaikriaz@kluniversity.in,
mominmunna.cs@gmail.com

Abstract. Memory management systems are generally custom-made where the replacement algorithm deals only with the number of memory hits. So, to reduce the page fault rate, we are proposing a page replacement algorithm called Farthest Page Replacement (FPR) algorithm. More the page faults lesser the performance. However, there are many other parameters which are to be considered like access time costs in the aspects of time and energy. Access time is the time required by a processor to access data or to write data from and to memory. Higher performance can be achieved with a reduction of access time costs. So, to decrease the access time and energy consumption, we are proposing an approach for selecting the page to be replaced.

Keywords: Page replacement, Fault Rate, Access Time Cost, Memory Addresses.

1 Introduction

The time a program or device takes to locate a single piece of information and make it available to the computer for processing is called access time cost. The term is applied to both random access memory (RAM) access and to hard disk and CD-ROM access. The access time for disk drives includes the time it actually takes for the read/write head to locate a sector on the disk (called the seek time). It depends on how far away the head is from the desired data. Access time is also frequently used to describe the speed of disk drives. It is composed of a few independently measurable elements that are added together to get a single value when evaluating the performance of a storage device. Access time costs are considered while measuring overall performance of a system.

In the following fig.2, access time of A is more than access time of B. It is because B is located nearer than A.

Access Time Costs (A) > Access Time Costs (B)

S.C. Satapathy et al. (eds.), *Emerging ICT for Bridging the Future – Volume 1*,
Advances in Intelligent Systems and Computing 337, DOI: 10.1007/978-3-319-13728-5_58

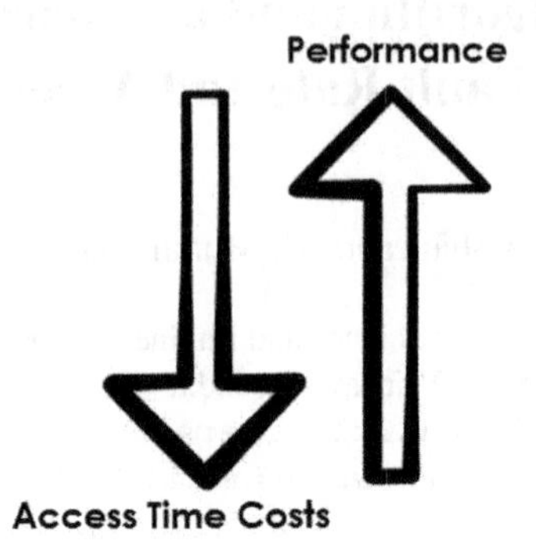

Fig. 1. Lesser the Access Time Costs higher the performance

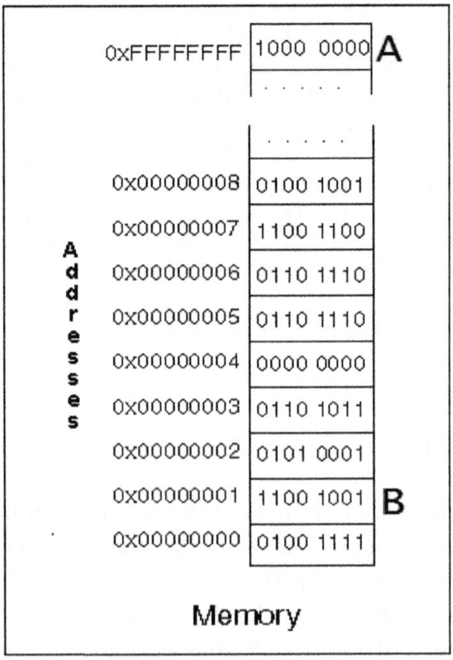

Fig. 2. Access Time Costs of different addresses

Main memory is the important and limited resources of a computer. As it is limited, it must be utilized efficiently to get better performance. Now one has to design capable approaches to handle main memory effectively. A page fault occurs when a program requests an address on a page that is not in the current set of memory resident pages. What happens when a page fault occurs is that the thread that experienced the page fault is put into a Wait state while the operating system finds the specific page on disk and restores it to physical memory. The OS then locates a copy of the desired page on the page file, and copies the page from disk into a free page in RAM. Once the copy has completed successfully, the OS allows the program thread to continue on. If the RAM does not have any empty frame then page replacement[1] is done. And the replacement is done based upon different approaches. With this, the overall objective is to decrease the total page fault rate[2] and thereby improve the performance.

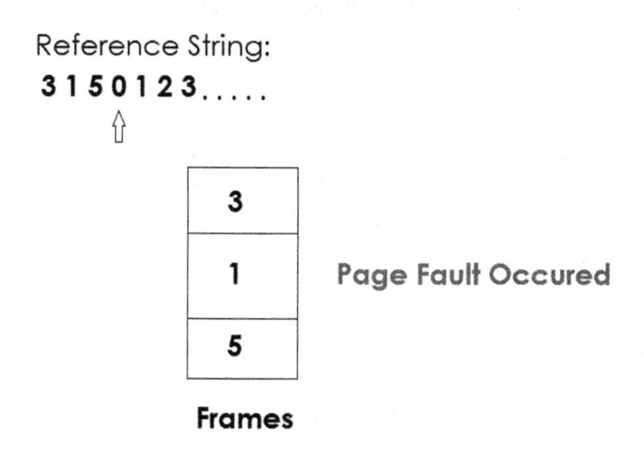

Fig. 3. Occurrence of Page Fault

2 Related Work

In a computer operating system that uses paging for virtual memory management, page replacement algorithms decide which memory pages to be replaced when a new page request causes page fault. There are a variety of page replacement algorithms. The simplest page-replacement algorithm is a FIFO algorithm. The first-in, first-out (FIFO) page replacement algorithm is a low-overhead algorithm which replaces the first page entered into the frame. The other one is least recently used page (LRU) replacement algorithm. LRU[3] works on the idea that pages that have been most heavily used in the past few instructions are most likely to be used heavily in the next few instructions too. In this manner, there are various algorithms for page replacement. But as of now the finest among them is optimal page replacement algorithm which has the least page fault rate.

In this algorithm the page that will not be used for the longest period of time will be replaced. For example, let us consider s reference string

1 2 5 8 3 2 5 1 5 3

1	1	1	8	3	3	3	3	3
	2	2	2	2	2	2	1	1
		5	5	5	5	5	5	5

Initially when the memory is empty the requests up to the frame size can be served. From then, the reference to the page 8 replaces page 1 because 1 will not be used until reference 8. Whereas page 2 will be used at 6 and page 5 at 7.The reference to page 3 replaces page 8 as page 8 is no longer used again in the string. This process continues until the last reference. But the disadvantage of the optimal page replacement policy is that it is very difficult to implement. In this paper, we are proposing a simple page replacement algorithm which can be implemented easily and also yields the same number of page faults like optimal page replacement. However, if we have a reference strings like this

9 1 7 3 8 2 4 6 0

in which no page is referred more than once, all the page replacement algorithms[6] would yield the same number of page faults[7] which results in performance degradation. Considering such cases, to improve the performance, we are proposing an approach.

3 Paper Contribution

3.1 Farthest Page Replacement Algorithm

Farthest Page replacement (FPR) algorithm is the simplest page replacement algorithm. A FPR replacement algorithm associates with address of the each page. When a page must be replaced, the farthest page is chosen. For our example reference string, our three frames are initially empty.

7 0 1 2 0 3 0 4 2 3 0 3 2 1 2 0 1 7 0 1

7	7	7	2	2	3	3	4	2	3	3	3	2	2	2	2	2	7	7	7
	0	0	0	0	0	0	0	0	0	0	0	0	0	0	0	0	0	0	0
		1	1	1	1	1	1	1	1	1	1	1	1	1	1	1	1	1	1

The first three references (7, 0, 1) cause page faults and are brought into these empty frames. The next reference (2) replaces page 7, because page 7 was farthest. Since 0 is the next reference and 0 is already in memory, we have no fault for this reference. The first reference to 3 results in replacement of page 2, since it is now farthest in line. Because of this replacement, the next reference, to 4, will fault. Page 3 is then replaced by page 4. This process continues and there are 9 faults altogether. For the same reference string FIFO[5] would result in 15 page faults, LRU[4] would yield 12 page faults and Optimal Page Replacement algorithm would generate 9 page faults. Thus the number of page faults generated for this string by our algorithm is the least page fault rate of all the existing algorithms.

3.1.1 Pseudo Code

```
/*This pseudo-code is to find the farthest
page among the pages present in the frame*/

for(j=1; j < Frame_Size; j++)
If (frames[j] > FARTHEST)
   FARTHEST=j;

/*replacing the farthest page with the new page
which caused page fault*/

frames[FARTHEST]=New_Page;
```

3.2 An Approach to Reduce the Access Time Costs Basing on Addresses of the Pages

Although there are many page replacement algorithms to decrease the page fault rate, it is well known fact that page fault rate is completely based on references [8]. There may be cases where the best algorithm also yields the same number of page faults as the worst algorithm. At that time, fault rate is unavoidable. So, in such cases, to improve the performance, we are proposing an approach which reduces the access time cost. The main principle of our approach is

Replace the page which will be nearest of the available pages.

i.e. whenever a page fault occur, compare the addresses of the pages which are already present in the frame and find out the nearest among them. Select that page as victim and replace it with the new page which caused page fault. For example, let us consider the same string we have taken earlier

9 1 7 3 8 2 4 6 0

9	9	9	9	9	9	9	9	9
	1	1	3	8	8	8	8	8
		7	7	7	2	4	6	0

For this kind of references any algorithm would yield same number of page faults. But by following our approach performance can be improved. The main reason for selecting nearest page as victim page[9] is, if the same page is referred again, access time cost will be less and thus performance will be improved.

3.2.1 Application

These cases will occur in all the algorithms. In LRU page replacement algorithm what if a page fault occurred and all the pages in the frame are referred same number of times. And in optimal page replacement algorithm consider a scenario where none of the pages in the frame are going to be referred again. In such cases FIFO will be implemented by default. Instead of that, if our approach is implemented we can minimize the access time and thus advance the performance.

3.2.2 Pseudo Code

```
for(i=0;i<No_of_Pages;i++)
{
/*Checking wether requesting page is in frame*/
If (Inframe(New_Page)=FALSE)
{
/*page fault occured*/
Optimal(Pages,Frame);
/*inplace of optimal any other algorithm can be
used*/
}
Else
{
/*when there is no redundancy*/
/*or for any default case*/
for(j=1;j<Frame_size;j++)
/*finding the nearest page*/
if(frames[j]<NEAREST)
NEAREST=j;
/*replacing the nearest page*/
frames[NEAREST]=pages[i];
}
```

4 Results

The results of various page replacement algorithms are noted and given below. Fig.4, Fig,5 and Fig.6 show that FPR works better than LRU and FIFO page replacement algorithms.

In Figure 7 it is proved that, FPR results in same number of page faults that optimal would yield. Thus FPR can be said as the finest page replacement algorithm.

```
Farthest Page Replacement
Reference string:
7 0 1 2 0 3 0 4 2 3 0 3 2 1 2 0 1 7 0 1
Addresses of pages :
 64914 64916 64918 64920 64922 64924 64926 64928 64930 64932
64934 64936 64938 64940 64942 64944 64946 64948 64950 64952

2          0          1
3          0          1
4          0          1
2          0          1
3          0          1
2          0          1
7          0          1

Number of page faults : 9_
```

Fig. 4. FPR Algorithm

```
LRU PAGE REPLACEMENT ALGORITHM
REFERENCE STRING :
 7 0 1 2 0 3 0 4 2 3 0 3 2 1 2 0 1 7 0 1
No. of frames : 3
        7
        7       1
        7       1       2
        0       1       2
        0       3       2
        0       3       4
        0       2       4
        3       2       4
        3       2       0
        3       2       1
        0       2       1
        0       7       1

The no of page faults is 12
```

Fig. 5. LRU algorithm

```
2       0       1
2       3       1
2       3       0
2       3       0
4       3       0
4       2       0
4       2       3
4       2       3
0       2       3
0       2       3
0       2       3
0       1       3
0       1       2
0       1       2
0       1       2
0       1       2
7       1       2
7       0       2
7       0       1
Page Faults=15
FIFO PAGE REPLACEMENT ALGORITHM
REFERENCE STRING:
7 0 1 2 0 3 0 4 2 3 0 3 2 1 2 0 1 7 0 1
No. of frames : 3
```

Fig. 6. FIFO algorithm

```
OPTIMAL PAGE REPLACEMENT
Reference String:
7 0 1 2 0 3 0 4 2 3 0 3 2 1 2 0 1 7 0 1
Number of frames : 3

7
7       0
7       0       1
2       0       1
2       0       3
2       4       3
2       0       3
2       0       1
7       0       1
page faults:9
```

Fig. 7. Optimal algorithm

4.1 Comparitive Analysis of FPR

For the sample string

Table 1. Comparison of Page Replacement Algorithms

7 0 1 2 0 3 0 4 2 3 0 3 2 1 2 0 1 7 0 1

PAGE REPLACEMENT ALGORITHM	NO. OF PAGE FAULTS
FIFO	15
LRU	12
OPTIMAL	9
FPR	9

In order to evaluate the performance of FPR algorithm, we have experimented many reference strings on different existing algorithms and plotted the average of page fault rates. We observed that FPR has got less number of page faults than many other existing page replacement algorithms and same page fault rate with optimal page replacement algorithm which is considered as the best page replacement algorithm. The results are plotted in Figure 8

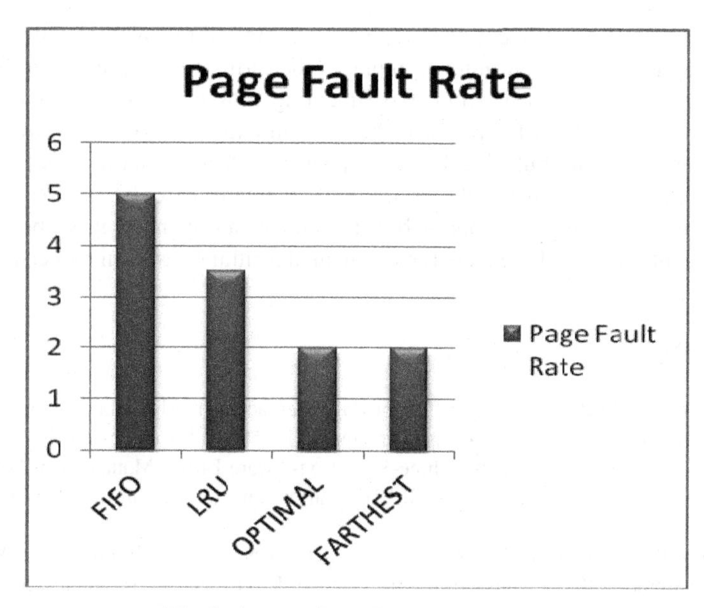

Fig. 8. Average Page Faults Comparison

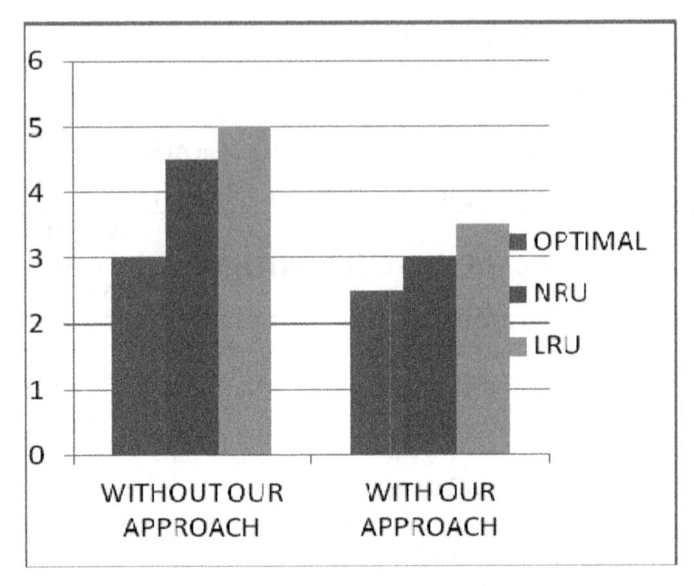

Fig. 9. Access Time Costs Comparison

In Figure 9, a graph is plotted for access time costs of existing algorithms with and without our approach applied on them. It is observed that, access time cost has been reduced when our approach is also applied to the page replacement algorithms.

5 Conclusion

This paper presents an efficient page replacement algorithm which is called Farthest Page replacement (FPR) algorithm. It replaces the farthest page present in the frame with the new page reference. After experimenting FPR on many strings of various types, it is observed that FPR performs better than existing algorithms [by reducing not only page fault rate but also the time spent on selecting victim page]. We have also proposed an approach which can be used in default cases of the existing algorithms. By implementing our approach in the above said manner, it is observed that the total execution time of the page replacement algorithms has been reduced.

References

1. Lin, M., Chen, S., Zhou, Z.: An Efficient Page Replacement Algorithm for NAND Flash Memory. IEEE Transactions on Consumer Electronics 59(4) (November 2013)
2. Jin, R., Cho, H.-J., Chung, T.-S.: Three-State Log-Aware Buffer Management Scheme for Flash-Based Consumer Electronics. IEEE Transactions on Consumer Electronics 59(4) (November 2013)
3. Li, Z., Jin, P., Su, X., Cui, K., Yue, L.: CCF-LRU: A New Buffer Replacement Algorithm for Flash Memory. IEEE Transactions on Consumer Electronics 55(3) (August 2009)

4. Jung, H., Shim, H., Park, S., Kang, S., Cha, J.: LRU-WSR: Integration of LRU and Writes Sequence Reordering for Flash Memory. IEEE Transactions on Consumer Electronics 54(3) (August 2008)
5. Seo, D., Shin, D.: Recently-Evicted-First Buffer Replacement Policy for Flash Storage Devices. IEEE Transactions on Consumer Electronics 54(3) (August 2008)
6. Silberschatz, A., Galvin, P.B., Gagne, G.: Operating System Concepts. Wse Wiley, ISBN 9788126520510
7. Stallings, W.: Operating Systems Internals And Design Principles. Pearson Prentice HallTM, ISBN 1098765432
8. Tanenbaum, A.S.: Operating Systems: Design And Implementation. Prentice-Hall (1987) ISBN 0136373313
9. Tanenbaum, A.S.: Modern Operating Systems. Prentice-Hall (1992) ISBN 0135957524

A Cryptography Scheme through Cascaded Session Based Symmetric Keys for Ubiquitous Computing

Manas Paul[1] and J.K. Mandal[2]

[1] Dept. of Computer Application, JIS College of Engg.,
Kalyani, West Bengal, India
manaspaul@rediffmail.com

[2] Dept. of Computer Science & Engg., Kalyani University,
Kalyani, West Bengal, India
jkmandal@rediffmail.com

Abstract. A cryptography scheme through cascaded implementation of six sessions based symmetric key techniques, which are available in literature, has been proposed. The scheme may introduce new dimension to ensure data security at maximum possible level based on available infrastructure and is suitable for the security of the system in the paradigm of unify computing. The scheme is idle to trade-off between security and performance of light weight devices having very low processing capabilities or limited computing power.

1 Introduction

With the fast evolution of digital data exchange of any file format, it is essential to protect the confidential data from unauthorized access. Cryptography is one of the important tools that provide data security. As a result researchers are working in the field of cryptography to enhance the security further[1, 2, 3, 4].

Six sessions based independent symmetric key cryptographic techniques, termed as MBBOT[5], MBBST[6], MSBBOT[7], SMBBOT[8], PCT[9] and SBSKCT[10] (say T_1, T_2, T_3, T_4, T_5 and T_6 respectively) have been integrated to generate the proposed cascaded scheme where the number of cascading stages n is selected randomly. The plain text is considered as a binary bit stream and the output stream of the nth stage is the cipher text. At each stage, a technique is chosen randomly from among six techniques with or without repetition of the same. Repetition of same technique in consecutive cascading stages is not allowed. No technique be implemented more than m number of times where $m < n$. It may so happen that one or more out of the six available techniques for cascading not implement at all.

Section 2 explains the proposed scheme. Section 3 deals with the generated results and analysis of the scheme. Conclusions are drawn in section 4.

2 The Scheme

The plaintext is the input stream of the first encryption technique of the sequence and the output stream generated from the nth stage of cascading is the cipher text.

S.C. Satapathy et al. (eds.), *Emerging ICT for Bridging the Future – Volume 1*,
Advances in Intelligent Systems and Computing 337, DOI: 10.1007/978-3-319-13728-5_59

Whenever a technique is selected for encryption, a sub-key is also generated associated with that technique. A session key K is generated for the proposed scheme which contains the value of n, the order of encryption techniques for n cascading stages and the information of n number of sub-keys. During decryption, the cipher text is considered as binary bit stream and passes through each of n number of decryption techniques in exactly the reverse order of the sequence followed during encryption. The final output stream generated from the nth stage of cascading reproduced the plain text. At any intermediate stages of this approach, the output stream of the technique of that stage is the input stream to the next cascading stage.

Section 2.1 describes the encryption process of the proposed cascaded implementation and that of decryption process of the same is described in section 2.2. Generation of session key is discussed in section 2.3.

2.1 Encryptor Module

The proposed scheme has n number of cascaded stages where n is a finite random integer. In maiden stage any one out of six techniques can be chosen in 6 ways and then for remaining cascading (n-1) stages any one out of five (since consecutive repetition of same technique is not allowed) techniques can be chosen in 5 ways. So there are $6*5^{n-1}$ ways to choose a cascading sequence. Now, $6*5^{n-1} = 1.2*5^n$ means that the formation of session key is order of 5^n ways. It indicates that the key space of the session key is very large. By notation, the sequence of encryption techniques for n cascading stages is represented as $E_iE_jE_k$...... $E_uE_vE_w$ where $i{\neq}j$, $j{\neq}k$, ... $u{\neq}v$, $v{\neq}w$. At each stage, the input binary stream is chopped into variable size of blocks and a corresponding sub-key is generated. S-box and P-Box operations are performed in the corresponding stages. The encryption algorithm is described as follows:

Algorithm:

Input: *Source stream i.e. plaintext.*

Output: *Encrypted stream i.e. ciphertext.*

Method: *The process takes binary stream and generates encrypted bit stream through a combination of S-Box and P-Box operations.*

STEP 1: *The input stream, say* C_0, *is taken as a stream with finite number of binary bits.*

STEP 2: *Obtain the number of cascaded stages, say n, randomly.*

STEP 3: *Set* $i=0$ *and initialize* $S_0 = E_0$ *(i.e. Null).*

STEP 4: *The sub-key* K_{i+1} *is generated using the binary bit stream* C_i.

STEP 5: *Select* $S_{i+1} \in \{E_1, E_2, E_3, E_4, E_5, E_6\}$ *randomly in such a way that* $S_{i+1} \neq S_i$.

STEP 6: *The bit stream* C_i *is encrypted into* C_{i+1} *using the encryption technique* S_{i+1} *and the key* K_{i+1}.

STEP 7: *Set* $i=i+1$. *If* $i<n$ *then go to step 8 else go to Step 9.*

STEP 8: C_i *is the input stream for next cascading stage and go to Step 4.*

STEP 9: $C_i \equiv C_n$ *is the final output of the encryptor module i.e.* C_n *is the ciphertext*

2.2 Decryptor Module

Processing the information of the session key K, the order of the decryption which is exactly the reverse of the sequence followed during encryption i.e. $D_wD_vD_u\ldots\ldots D_kD_jD_i$ and the sub-keys in the order of K_n, K_{n-1}, ……., K_2, K_1 are identified. The first decryption technique D_w considers the input cipher text C_n as a binary bit stream. The final output stream generated from the final stage of cascading using the decryption technique D_i reproduced the plain text. At the intermediate stages of this approach, the output stream of any decryption technique is the input stream to the next cascading stage. Processing the corresponding sub-key information at each stage, the input binary stream is chopped into variable size of blocks. S-box and P-Box operations are performed in the corresponding stages. The decryption algorithm is described as follows:

Algorithm:

Input: *Encrypted stream i.e. ciphertext and the key K.*

Output: *Source stream i.e. plaintext.*

Method: *The process takes encrypted binary stream and generates decrypted bit stream i.e. the plain text through a combination of S-Box and P-Box operations.*

STEP 1: *The stream containing the information of the key K obtained.*

STEP 2: *The value of n (number of cascading stages) and the sub-keys K_1, K_2, …., K_n are extracted from the key K and used for decryption.*

STEP 3: *The order of decryption is fetched from the key K which is exactly reverse of the sequence of encryption and initialized into the variables S_1, S_2, ….., S_n (i.e. S_1 is the first decryption technique, S_2 is the second decryption technique and so on) where $S_i \in \{D_1, D_2, D_3, D_4, D_5, D_6\}$ $\forall i \in N$, the set of first n natural numbers.*

STEP 4: *The input stream, say C_n, is taken as a stream with finite number of binary bits.*

STEP 5: *Set $i = n$.*

STEP 6: *Input bit stream C_i is decrypted into C_{i-1} using the decryption technique S_{n-i+1} and the sub-key K_i.*

STEP 7: *Set $i=i-1$. If $i>0$ then go to step 8 else go to step 9.*

STEP 8: *C_i is the input stream for next cascading stage and goes to step 6.*

STEP 9: *$C_i \cong C_0$ is the final output of the decryptor module i.e. C_0 is the plaintext.*

2.3 Session Key Generator

The scheme has n number of cascading stages. At each stage, the input binary bit stream C_i passes through the key generator to generate the corresponding sub-key K_i where $i \in N$, the set of first n natural numbers. A session key K is generated for the proposed scheme and this key K contains the followings

i. *The value of n (number of cascading stages).*

ii. *The order of encryption techniques for n cascading stages (say $E_iE_jE_k$..... where i≠j, j≠k, ... and every $E_i \in \{E_1, E_2, E_3, E_4, E_5, E_6\}$).*

iii. *The information of n number of sub-keys (say K_1, K_2, K_3, ..., K_n) which are generated at the corresponding cascading stage of encryption using the input binary bit stream for that stage.*

The key space of the session key K is very large. The value of n can be represented by a character having ASCII value from 1 to 255. For each technique only three bits are required to store the technique index (the range 000 to 111 is sufficient to store 1 to 6). So (3*n) number of bits i.e. 3*n/8 number of characters are required to store the sequence of cryptographic techniques for n cascading stages. Each sub-key has a length of 32 to 48 characters. The length of n number of sub-keys is (32*n) to (48*n) number of characters i.e. on an average (4*32*n+2*48*n)/6 [weighted avg.] ≈ 38*n number of characters. So the proposed scheme has a session key K with length of [1+(3*n/8)+38*n] ≈ 39*n number of characters which conform a huge variability of the key space in terms of randomness. As a result of variability of time complexity for exhaustive key search is thirty nine times the number of cascading stages.

3 Results and Analysis

Results are computed on twenty files each of four different file types (.dll, .doc, .exe and .txt) with different file sizes varying from 0 to 6.3 MB (approx.) and exhaustive analysis have been made with a huge variability of cascading stages.

The encryption and decryption times taken are the differences between processor clock ticks at the starting of execution and at the end of execution respectively. The encryption and decryption times of the proposed approach are near equal to the cumulative sums of n number of encryption and decryption times respectively of the individual cryptographic techniques.

Avalanche, Strict avalanche and Bit independence values are generated for comparison between the source and encrypted bytes. The calculated values of all three tests are very high which may indicate good security of the proposed approach.

Spectrums of the frequency distribution of the encrypted characters generated using the proposed approach are analyzed and it is observed that characters with ASCII values ranging from 0 to 255 appeared all with near equal frequencies which may indicate that it is very hard to regenerate the original file for a cryptanalyst.

Chi-square value has been calculated from the character frequencies using "Pearsonian Chi-square" formula. The higher the Chi-square values the more deviation from the original message. The calculated Chi-square values for all the sample files using the proposed scheme are very large compare to tabulated one which may indicates that the degree of security of the proposed scheme using cascaded implementation is good. All eighty input streams are grouped into eight intervals depends on its sizes such that on an average ten input streams are there in each interval. The average Chi-square value of all input streams for each interval is given in table 1.

Table 1. Average Chi-square value of all input streams for each group interval

Group	Intervals	Average Chi-square values
1	(0 - 20) KB	105392868
2	(20 - 80) KB	449505660
3	80 KB - 0.5 MB	4140411940
4	(0.5 - 1.4) MB	125816009058
5	(1.4 - 2.6)MB	347313150591
6	(2.6 - 3.7)MB	812550544958
7	(3.7 - 4.7)MB	1555607712005
8	(4.7 - 6.3)MB	6263636864529

Since for any file the maximum number of possible characters is 256, so 255 degrees of freedom be maximum. The tabulated value of Chi-square is 310.46 for 255 degrees of freedom at 1% level of significance. The calculated average Chi-square values are very large than 310.46. Figure 1 shows the average Chi-square values (in logarithmic scale with base 100) of eight groups and the corresponding deviation of that from tabulated value for 255 degrees of freedom at 1% level of significance.

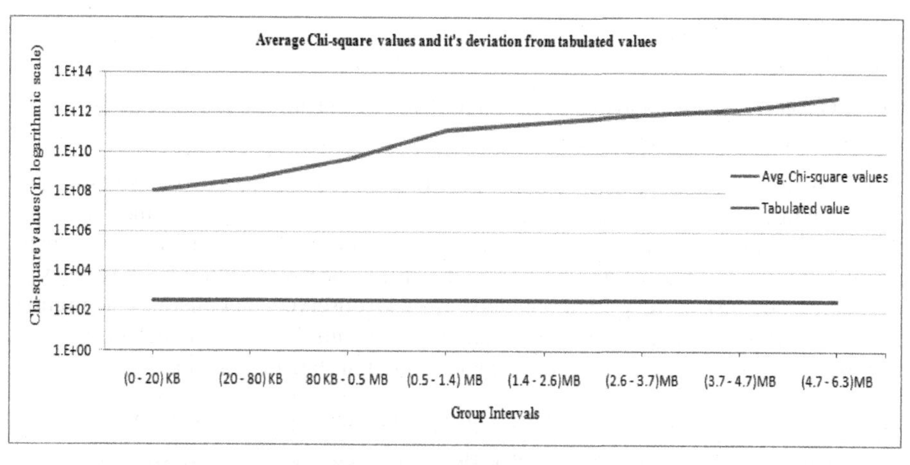

Fig. 1. Average Chi-square values (in logarithmic scale with base 100) of eight groups and the corresponding deviation of that from tabulated value for 255 degrees of freedom at 1% level of significance

The complexity of any symmetric key algorithm is generally compared with the Brute-force attack. Brute-force approach simply involves computing every possible key until an intelligible translation of ciphertext into plaintext is obtained. On average, half of all possible keys must be tried to achieve success. In section 2.3, the length of the session key of the proposed scheme has been discussed and it has approximately (39∗n) number of characters (not only alphabets, may be any character with ASCII value from 0 to 255). Therefore the number of alternate keys = 256^{39*n}. Since on an average half of all possible keys must be tried to achieve success, so total time required at 1 decryption / μs is $0.5*256^{39*n}$ μs = $2^{(312*n - 1)}$ μs. Analyzing the data, it may conclude that the proposed scheme is highly secured from Brute-force attack.

Let T be the average time in years required at 10^6 decryptions per μs for exhaustive search of the session key for the proposed scheme. If n, number of cascading stages, is plotted along X-axis and $\log_{10}T$ along Y-axis then the generated curve is a straight line which is shown in figure 2. Extending this straight line along positive X-axis, it may predict the required average time T in years for any large value of n. Since the slope value of that straight line is very high (approximately 94) and the value of T is plotted as $\log_{10}T$, so the value of T is increased sharply with the increase of n.

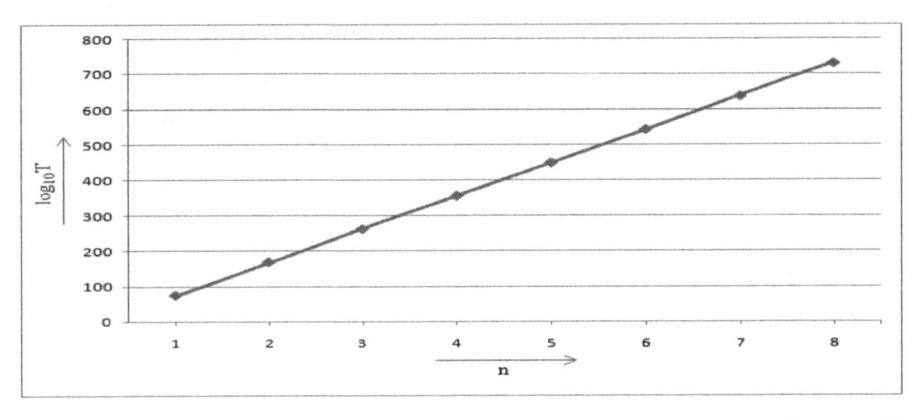

Fig. 2. Graphical representation of average time T in years (T in logarithmic scale as $\log_{10}T$) against n, number of cascading stages

Values of median, mode and standard deviation of four types of source streams and corresponding encrypted streams has been given in table 2. All three measures are different for source stream and encrypted stream. From this point of view it may conclude that the degree of non-homogeneity of the encrypted stream with respect to the source streams is very high. In table 2, it is observed that there is negligible correlation between the source stream and cipher stream. So it may conclude that the proposed scheme can effectively resist data correlation statistical attack.

Table 2. Some statistical measures of the source and the cipher stream using proposed scheme

Value of	Stream	IPDESIGN.dll	s20.doc	QlbOv.exe	s20.txt
Median (Character with ASCII value)	Source	101	121	134	76
	Encrypted	124	128	127	114
Mode (Character with ASCII value)	Source	0	0	255	32
	Encrypted	201	64	73	58
Standard Deviation	Source	52184.4	50230.4	101071.8	72812.7
	Encrypted	3797.2	3598.6	6596.2	4617.6
Correlation coefficient	Source & encrypted	0.005	0.033	–0.074	0.295

It is observed that the proposed scheme generates an entirely different cipher stream with the change of a single bit randomly in the key K. It is also noticed that the scheme totally fails to decrypt the cipher stream into plaintext with a slightly different secret key. From this point of view, it may be concluded that the proposed scheme is highly key sensitive.

Some of the salient features of the proposed scheme are summarized as follows:

- *Proposed scheme provides good degree of security. The number of alternate keys of the proposed scheme is approximately* 256^{39*n}. *It may be noted that the scheme is highly secured from Brute-force attack.*
- *The session key can be formed in order of* 5^n *ways which is a vast one and the length of session key is of approximately (39*n) number of characters where n is the number of cascading stages. It indicates that the key space of the session key is very large.*
- *The scheme has the flexibility to adopt the complexity based on infrastructure, resource and energy available for computing in any system. So the scheme is suitable for the security of the system in the paradigm of unify computing.*
- *The scheme may be ideal for trade-off between the security and performance of any light weight devices having very low processing capabilities or limited computing power.*
- *The proposed scheme generates an entirely different cipher stream with a small change in the key K and the scheme totally fails to decrypt the cipher stream with a slightly different secret key.*
- *All the measures like Measures of central tendency, dispersion and Chi-square values indicate that the degree of non-homogeneity of the encrypted stream with respect to the source stream is good.*
- *The cipher stream generated through proposed scheme is negligibly correlated with the source stream. Therefore the scheme may effectively resist data correlation statistical attack.*
- *The scheme can work with any size of source file. It doesn't require any sort of padding, which results in the same encrypted file size as the original file size. The scheme is very much useful for file transmission in any form.*

4 Conclusion

The approach of cascaded implementation is pretty simple logically. The analysis of results also indicates enhanced security of this approach. The real strength of proposed approach lies in the possible formation of a large key space. The key space increases drastically with allowing the much more cascading stages. The proposed scheme is highly secured from Brute-force attack. Other strength of this proposed scheme is the adoption of complexity based on infrastructure, resource and energy available for computing in a mesh or node. For any network having low energy, less number of cascading stages be chosen where as number of cascading stages be more for a fiber optic based network. So the scheme is very much suitable for the security of the system where unify computing is an essential component. One of the most important features of the proposed scheme is that the scheme is idle to trade-off between security and performance of any light weight devices having very low processing capabilities or limited computing power.

References

1. Navin, A.H., Oskuei, A.R., Khashandarag, A.S., Mirnia, M.: A Novel Approach Cryptography by using Residue Number System. In: 6th International Conference on Computer Sciences and Convergence Information Technology (ICCIT 2011), Seogwipo, South Korea, November 29-December 01, pp. 636–639 (2011)
2. Niemiec, M., Machowski, L.: A New Symmetric Block Cipher Based on Key-Dependent S-Boxes. In: 4th International Congress on Ultra Modern Telecommunications and Control Systems and Workshops (ICUMT 2012), St. Petersburg, Russia, October 3-5, pp. 474–478 (2012)
3. Verma, H.K., Singh, R.K.: Enhancement of RC6 Block Cipher Algorithm and Comparison with RC5 & RC6. In: 3rd IEEE International Advance Computing Conference (IACC 2013), Ghaziabad, Uttar Pradesh, India, February 22-23, pp. 556–561 (2013)
4. Mandal, B.K., Bhattacharyya, D., Bandyopadhyay, S.K.: Designing and Performance Analysis of a Proposed Symmetric Cryptography Algorithm. In: International Conference on Communication Systems and Network Technologies (CSNT 2013), Gwalior, India, April 6-8, pp. 453–461 (2013)
5. Paul, M., Mandal, J.K.: A Novel Session Based Bit Level Symmetric Key Cryptographic Technique to Enhance the Security of Network Based Transmission. In: RITS-International Conference on Advancements in Engineering & Management (RITS ICAEM 2012), February 28-29. Royal Institute of Technology & Science, Hyderabad (2012)
6. Paul, M., Mandal, J.K.: A Novel Generic Session Based Bit Level Encryption Technique to Enhance the Security of Network Based Transmission. In: International Conference on Computational Intelligence and Information Technology (ICCIIT 2012), INFO Institute of Engineering, Coimbatore, Tamilnadu, India, March 2-3, pp. NS043–NS049 (2012)
7. Paul, M., Mandal, J.K.: A Novel Generic Session Based Bit Level Cryptographic Technique based on Magic Square Concepts. In: International Conference on Global Innovations in Technology and Sciences (ICGITS 2013), SAINTGITS College of Engineering, Kottayam, Kerala, India, April 4-6, pp. 156–163 (2013)
8. Paul, M., Mandal, J.K.: A Novel Symmetric Key Cryptographic Technique at Bit Level Based on Spiral Matrix Concept. In: International Conference on Information Technology, Electronics and Communications (ICITEC-2013), International Academic and Industrial Research Solutions, Bangalore, India, March 30-31, pp. 6–11 (2013)
9. Paul, M., Mandal, J.K.: A Permutative Cipher Technique (PCT) to Enhance the Security of Network Based Transmission. In: 2nd National Conference on Computing for Nation Development, Bharati Vidyapeeth's Institute of Computer Applications and Management, New Delhi, India, February 08-09, pp. 197–202 (2008)
10. Paul, M., Mandal, J.K.: A Universal Session Based Bit Level Symmetric Key Cryptographic Technique to Enhance the Information Security. International Journal of Network Security & Its Application (IJNSA) 4(4), 123–136 (2012)

A Comparative Study of Performance Evaluation of Services in Cloud Computing

L. Aruna[1] and M. Aramudhan[2]

[1] Comp. Science, Periyar University, Salem, Tamilnadu
[2] Department of IT, PKIET, Karaikal, Tamilnadu
arunaloga@gmail.com, aranagai@yahoo.co.in

Abstract. Cloud Computing is a new service area in Information Technology environment with huge requirements on the shared information, infrastructure, software, resources, devices and services. Performance Evaluation is an important aspect of Cloud Computing environment. Efficient Performance evaluation technique is used to evaluate the all Performance activities, because Cloud users on demand basis in pay-as-you-go model. In this paper, we analyze and evaluate the cloud performance in different environment based on quality attributes, features, services, support specifications and access.

Keywords: Cloud Computing, SLA, Performance Evaluation, Cloud Computing Services, icloud, Egnyte, Google Apps Google, OpenDrive, Dropbox, Amazon Cloud drive and Load balancing.

1 Introduction

Cloud Computing is the most recent emerging paradigm promising to turn the vision of "computing utilities" into reality, it provides[2] a flexible and easy way to store and retrieve huge data without worrying about the hardware needed. However, this business clouds are designed to support and little information workloads that are terribly completely different from typical scientific computing workloads. As the number of users on cloud increases, [3]the existing resources decreases automatically which leads to the problem of delay between the users and the cloud service providers. The confluence of technological advances and business development in web services. Internet with broadband, computer systems and applications has created complete storm for cloud computing [1] during the past decade. Nowadays, cloud is the best solution for people who are looking for rapid implementation methods.

Thus, the load balancing concepts are also important and come into the analysis picture. The traffic over the network must be dealt[4] smartly such that the situation in which some nodes are overloaded and some other are under loaded should never arise. To overcome this situation, many load balancing algorithms are proposed by researchers, with their own pros and cons.

The ideas of each fact [6] are discussed and finally summarized as an overview [18].

S.C. Satapathy et al. (eds.), *Emerging ICT for Bridging the Future – Volume 1*,
Advances in Intelligent Systems and Computing 337, DOI: 10.1007/978-3-319-13728-5_60

1.1 Why Use Cloud Computing Services?

Cloud computing refers to employing a third-party network of remote servers hosted on the net to store and manage all of your information, instead of regionally. Simply put, cloud services offer you along with your own [8]Winchester drive within the cloud – or on the net. Such services became standard as a result of they are reasonable, convenient and supply ample cupboard space. However maybe the largest charm of such services is their accessibility, you will be able to access your documents, photos, videos and the other saved files from any device with net access. With cloud services you'll be able to connect reception, work or on the go via a portable computer, desktop, Smartphone or different hand-held device.

Some of the world's largest technical school firms have launched [17]cloud services, together with icloud, Egnyte, Opendrive, Dropbox, Apple, Amazon, Cloud Drive and Google Apps Google. [18]These technical school giants, along side some notable up-and-comers, offer many storage tier plans tailored for each customers and businesses. It ought to be noted that Microsoft Windows additionally offers cloud solutions. However in contrast to its competitors, Windows provides such at[12] low, restricted quantity of free storage (with no choice to upgrade) that it's troublesome to vie with the remainder of the services in our lineup.

The services we have a tendency to reviewed offer sturdy options and ample cupboard space thus you will [14]be able to access any file you saved within the cloud notwithstanding wherever you are. make sure to see out our articles on cloud services, further as our side-by-side comparisons and reviews of the highest services, together with iCloud, Egnyte , Dropbox, Amazon Cloud Drive [18]Cloud Services. What to appear for some firms might tout themselves as cloud services however square measure additional admire on-line backup or file sharing services. Whereas there is definitely some overlap between every of those services, cloud services square measure distinctive as a result of they permit you to look at, edit and share files saved within the cloud. With some services, you'll be able to [8] even synchronize your content across all of your computers and devices. Where as no two cloud services square measure identical, every of the services we have a tendency to reviewed offer a similar elementary options and practicality.

1.2 Service Level Agreements

A service level agreement (SLA) is a method to measure a cloud provider's comfort level with its service delivery platform. If one cloud supplier offers 99% computing handiness and another offers 100%, [16] it's an honest bet the latter could be a higher suited mission-critical applications. However ensure your provider's SLA has some teeth there to, within the unlikely event of a service outage, the cloud supplier ought to provide you with generous service credits reciprocally.

1.3 Features

The most vital issue to contemplate in a very cloud service is options, together with the kind of content you'll be able to store. [3]The simplest cloud computing services square measure those who enable you to transfer and save any kind of file you'd save on your native Winchester drive, from word documents to music files and everything in between. Some services even enable you to stay email, [5]contacts and your calendar within the cloud. Any cloud service you think about have to to additionally enable you to look at, edit and share your content in spite of what pc or device you're victimization.

Different options to appear for embrace automatic syncing of your files across all of your devices, and password-protected sharing and file secret writing to safe guard your content.

1.4 Mobile Access

Possibly one in every of the largest commerce points of cloud computing services is their wide-ranging access. Whether or not you're on your work pc at the workplace or reception on your iPad, cloud services enable you to [20]access your content anyplace, anytime and on anyone of your devices. Search for services that provide the best vary of mobile access, together with apps for standard smart phones and therefore the ability to log into your account from any mobile browser.

1.5 Ease of Use

Considering however usually you'll probably be accessing [18] your content within the cloud, it's vital to pick a cloud service that's intuitive and simple. The service's interface and tools ought to be straightforward to navigate and convenient to use.

1.6 Help and Support

Cloud suppliers are not illustrious to supply nice support, however that's commencing to modification. The first cloud adopters have typically been code developers and alternative techniques that don't want abundant hand-holding. Today, enterprise officers ought to grasp they'll contact somebody at their [13]cloud supplier after they expertise issues. Some cloud suppliers bundle in support services whereas others supply varied support tiers. The current utility nature of cloud platforms means it's easier than ever to provision and check infrastructure within the cloud. You will simply spin up infrastructure at multiple cloud suppliers to check your applications [16]and see if your service needs are met. Chances are high that you will realize the correct cloud for your organization. Getting facilitate after you would like it's crucial once victimization any kind of technology, together with cloud services. Obtainable support choices ought to embrace technical help via phone, email and live chat. The service have to additionally offer a knowledge base and user forums as resources.

1.7 Why Cloud Storage?

If you sign on for a cloud supplier your information is saved on a range of computers of those exact firms that house in information centers. Ideally, your information isn't solely saved on one pc however instead displayed equally in multiple locations. Cloud storage and on-line storage are often accustomed store your [7]most vital files that can't be reinstalled like photos, videos, business documents and additional. Betting on the scale of the documents it will take a short while till everything is uploaded to the cloud storage supplier of your alternative. at that time you'll be able[5] to use that service as an internet Winchester drive to access your files remotely from the online or synchronize them to different devices as an example i.e. mobile phone.

1.8 Cloud Storage for Business

Businesses will enjoy cloud storage and create their enterprise future proof. Several cloud storage services supply many on-line collaboration tools that companies will use to manage documents and spreadsheets. [15]Also, with cloud on-line storage businesses will share and synchronize file files across multiple devices and computers. Secure secret writing technology permits for safe usage and compliance with company regulation. Certify to see our secure cloud storage services to grasp that supplier's square measure safe for your business.

2 Related Work

A Comparative study of Performance Evaluation of Services in Cloud Computing based on the Cloud environments may be done in different categories. [8]The existing evaluation method is considered for performance evaluation and [1] Simulation technique. The Cloud Computing Performance evaluation studies related with the some criteria such as number of requests, waiting time in Queue, [15] load balancing for service level agreement, throughput, cost, response time for transactions, service delay and productivity are studied.

Performance evaluation based on the some specific services comparisons and review. Lot of services for cloud suppliers [4]and providers are offered in Cloud environment such as Scalability, Price comparison, Content, Features, Mobile Access, Help and Support and Software configuration related areas are identified.

Performance evaluation based on the different Cloud environments now a days, well known and large organizations like [20] Icloud, Egnyte, Google Apps, Opendrive, Dropbox, and Amazon Cloud drive are suitable for the valuation or may be based on different provides too. Performance evaluation in this paper is based on[23] combination of the above categories and it is performed with the help of http://cloud-services-review.toptenreviews.com.

3 Performance Evaluation in Cloud Computing Services Comparisons and Review Criteria

Cloud Computing Services and review must be compatible, high powerful and performance. We take this survey mechanism for the following area of tasks are identified. It is a series of criteria for performance evaluation [17]in Cloud Computing services of all factors are affecting, some of which will be identified in this paper.

The performance evaluation in cloud computing environments is consists of the three major categories are discussed and it will be analyzed. These are End User, [22]Cloud Supplier and Cloud Developer.

The performance Evaluation in cloud computing environment aspects of end user is likely to Safety measures, Origin, Time alone, High Accessibility, Reduced value and Convenience.

The second aspects of Performance evaluation is the Cloud Supplier is likely to administration resources, Outsourcing, [13]Resource utilization, metering, Providing resource, Meet user needs and Helpful Computing.

The another important aspects of [19]evaluation is Cloud Developer is likely to Measurability, Virtualization, Quickness and flexibility, Availability, Data supervision, Dependability and Programmability these criteria are under development.

The following Table.1 are listed out the [17] recent best cloud computing services comparisons and review.

Table 1. are listed out the performance evaluation in recent best cloud computing services comparisons and review

S. No	Criteria Type	Performance Evaluation in Best Cloud Computing Services					
		Icloud	Egnyte	Google Apps Google	OpenDrive	Dropbox	Amazon Cloud drive
1	Pros	Automatic Synchronization of data among devices.	Unlimited file storage with authentication Facilities.	Easy accessibility using internet.	Synchronization of elementary documents.	Easy to use of file sharing.	Hierarchical unlimited storage options. .
2	Cons	Services limited only to apple products.	Media streaming not possible	Lack of centralized apps for divergent data.	Lack of support for messaging and communications.	Lack of support for messaging and communications	Limited access in mobile device.
3	Ranking	97%	96.5%	93.8%	92.3%	89%	79.%8
4	Cost(Per month)	$1.66	$24.99	$5	$5	$9.99	$1.66
5	Storage Space allotment	15GB	150GB	Yes	100GB	100GB	20GB
	CONTENT						
6	Email	Yes	Yes	Yes	No	No	No
7	Contacts	Yes	Yes	Yes	No	No	No
	Calendar	Yes	Yes	Yes	No	No	No
9	Documents	Yes	Yes	Yes	Yes	Yes	Yes

Table 1. (*continued*)

10	Spreadsheets	Yes	Yes	Yes	Yes	Yes	Yes
11	Music	Yes	Yes	Yes	Yes	Yes	Yes
12	Photos	Yes	Yes	Yes	Yes	Yes	Yes
13	Videos	Yes	Yes	Yes	Yes	Yes	Yes
14	Slideshows	Yes	Yes	Yes	No	Yes	Yes
	FEATURES						
15	View Files	Yes	Yes	Yes	Yes	Yes	Yes
16	Edit Files	Yes	Yes	Yes	Yes	Yes	No
17	Automatic Sync	Yes	Yes	No	Yes	Yes	No
18	Password Protection	Yes	Yes	Yes	Yes	Yes	Yes
19	File Encryption	Yes	Yes	Yes	Yes	Yes	Yes
20	Media Streaming	Yes	No	No	Yes	No	Yes
21	Share Files	No	Yes	Yes	Yes	Yes	No
22	Desktop Applications /Access	No	Yes	No	Yes	Yes	No
	MOBILE ACCESS						
23	iphone	Yes	Yes	Yes	Yes	Yes	No
24	ipad	Yes	Yes	Yes	Yes	Yes	No
25	Android	No	Yes	Yes	No	Yes	Yes
26	Windows Mobile	No	Yes	Yes	No	No	No
27	BlackBerry	No	No	Yes	No	Yes	No
28	Mobile Browser	No	Yes	Yes	Yes	Yes	Yes
	HELP & SUPPORT						
29	Phone Support	Yes	Yes	Yes	No	Yes	Yes
30	Email Support	Yes	Yes	Yes	Yes	Yes	Yes
31	User Forums	Yes	Yes	Yes	Yes	Yes	Yes
32	Knowledgebase	Yes	Yes	Yes	Yes	Yes	Yes
33	Online Ticket Form	Yes	Yes	No	Yes	Yes	No
34	Live Chart	No	No	Yes	No	No	No
	SUPPORTED CONFIGURATIONS						
35	Windows	Yes	Yes	Yes	Yes	Yes	Yes
36	Windows Vista	Yes	Yes	Yes	Yes	Yes	Yes
37	Windows XP	Yes	Yes	Yes	Yes	Yes	Yes
38	Mac OS X	Yes	Yes	Yes	Yes	Yes	Yes
39	Linux	No	Yes	No	No	Yes	No
40	Ranking Mark	0-10 : Bad, 20-30 : Poor, 40-50 : Average, 60-80 : Good, 90-100 : Excellent.					

4 Conclusion

A Comparative Study of Performance Evaluation of services in cloud computing is an essential task of to identify the best services provider and to achieve maximum utilization of services and resources. In this paper, we have analyzed various performance evaluations in best cloud computing services comparisons and review criteria like Cloud Computing Services, Service Level Agreement(SLA), pros, cons, features, mobile access, ease of use, help and support, content, ranking, price comparison and supported configurations etc.. We analyzed major issues which must be taken into consideration of while best cloud computing services comparisons and reviews in any recent cloud computing.

References

[1] Khiyaita, A., Zbakh, M., El Bakkali, H., El Kettani, D.: Load balancing cloud computing: state of art. In: 2012 National Days of Network Security and Systems (JNS2), pp. 106–109. IEEE (2012)

[2] Menascé, D.A., Ngo, P.: Understanding cloud computing: Experimentation and capacity planning. In: Computer Measurement Group Conference (2009)

[3] Alonso-Calvo, R., Crespo, J., Garcia-Remesal, M., Anguita, A., Maojo, V.: On distributing load in cloud computing: A real application for very-large image datasets. Procedia Computer Science 1(1), 2669–2677 (2010)

[4] Chaczko, Z., Mahadevan, V., Aslanzadeh, S., Mcdermid, C.: Availability and load balancing in cloud computing. In: International Conference on Computer and Software Modeling, Singapore, vol. 14 (2011)

[5] Iosup, A., Ostermann, S., Yigitbasi, M.N., Prodan, R., Fahringer, T., Epema, D.H.: Performance analysis of cloud computing services for many-tasks scientific computing. IEEE Transactions on Parallel and Distributed Systems 22(6), 931–945 (2011)

[6] Iosup, A., Ostermann, S., Yigitbasi, M.N., Prodan, R., Fahringer, T., Epema, D.H.: Performance analysis of cloud computing services for many-tasks scientific computing. IEEE Transactions on Parallel and Distributed Systems 22(6), 931–945 (2011)

[7] Nine, Z., Sq, M., Azad, M., Kalam, A., Abdullah, S., Rahman, R.M.: Fuzzy logic based dynamic load balancing in virtualized data centers. In: 2013 IEEE International Conference on Fuzzy Systems (FUZZ), pp. 1–7. IEEE (2013)

[8] Kaur, J.: Comparison of load balancing algorithms in a cloud. International Journal of Engineering Research and Applications 2(3), 1169–1173 (2012)

[9] Gulati, A., Chopra, R.K.: Dynamic round robin for load balancing in a cloud computing (2013)

[10] Lin, C.-C., Liu, P., Wu, J.-J.: Energy-aware virtual machine dynamic provision and scheduling for cloud computing. In: 2011 IEEE International Conference on Cloud Computing (CLOUD), pp. 736–737. IEEE (2011)

[11] Mao, Y., Chen, X., Li, X.: Max–min task scheduling algorithm for load balance in cloud computing. In: Patnaik, S., Li, X. (eds.) Proceedings of International Conference on Computer Science and Information Technology. AISC, vol. 255, pp. 457–465. Springer, Heidelberg (2014)

[12] Randles, M., Lamb, D., Taleb-Bendiab, A.: A comparative study into distributed load balancing algorithms for cloud computing. In: 2010 IEEE 24th International Conference on Advanced Information Networking and Applications Workshops (WAINA), pp. 551–556. IEEE (2010)
[13] Kokilavani, T., Amalarethinam, D.: Load balanced min-min algorithm for static meta-task scheduling in grid computing. International Journal of Computer Applications 20(2) (2011)
[14] Vinothina, V., Shridaran, R., Ganpathi, P.: A survey on resource allocation strategies in cloud computing. International Journal of Advanced Computer Science and Applications 3(6), 97–104 (2012)
[15] Aruna, L., Aramudhan, M.: Text book for Fundamentals of Cloud Computing, ISBN No: 978-93-5137-266-0
[16] Aruna, L., Aramudhan, M.: A Novel Survey on SLA based Load leveling in Cloud Computing. International Journal of Research in Computer and Communication Technology 3(6) (June 2014) ISSN (Online) 2278- 5841and ISSN (Print) 2320- 5156
[17] Best cloud computing services comparisons review in 2014, http://cloud-services-review.toptenreviews.com
[18] Icloud, http://cloud-services-review.toptenreviews.com/icloud-review.html
[19] Egnyte, http://cloud-services-review.toptenreviews.com/egnyte-review.html
[20] Google Apps google, http://cloud-services-review.toptenreviews.com/google-apps-review.html
[21] Opendrive, http://cloud-services-review.toptenreviews.com/opendrive-review.html
[22] Dropbox, http://cloud-services-review.toptenreviews.com/dropbox-review.html
[23] Amazon cloud drive, http://cloud-services-review.toptenreviews.com/amazon-cloud-drive-review.html

A Case Study: Embedding ICT for Effective Classroom Teaching & Learning

Sandeep Vasant and Bipin Mehta

School of Computer Studies, Ahmedabad University, Gujarat, India
{sandeep.vasant,bv.mehta}@ahduni.edu.in

Abstract. Information and Communication Technology (ICT) has presence in all sectors and so as in Education too. Instructors, Learners, Administrators and Researchers of education field are thriving for innovative teaching pedagogy to make the learning experience effectual. In this paper we have discussed the use of current trends in ICT such as BYOD, LMS and learning through Cloud Computing. We have also summarized the responses of teachers and students on the campus through online survey. Analysis shows the significant improvement in class participation and examination result of the students.

Keywords: BYOD, ICT, LMS, SaaS, PaaS, IaaS, SCS, Moodle, MOOC, TED, NPTEL.

1 Introduction

In today's technological arena traditional "Chalk and Talk" teaching pedagogy is either completely replaced with interactive e-learning tools/software's or it is used in integration with ICT enabled tools/resources. ICT has touched almost all people and in all sectors at large and education is no exception in that. Researchers and Teachers are trying to find the ways since decade to make the teaching-learning interactive, effective and interesting. Use of ICT now a day is common amongst students and teachers. Hence effective use of e-learning tools is also common while teaching and learning.

For effective communication with students, University/Institutes uses the various communication technologies such as SMS, e-mail, Portal and hence use of smart devices such as Smart phone, tablets and laptops quickly become ubiquitous amongst students. As per author, college students are part of big chunk of the heaviest user of technology. Findings of this research show that embedding ICT into classroom will encourage students to learn through smart devices [1] [17, 18]. Only challenge is to provide the sate-of-the-art Infrastructure which turns the traditional classroom into smart class room. Majority of the universities/institutes has the classrooms equipped with audio and visual resources along with internet connectivity.

When teachers integrate the multimedia usage (audio, video, animation software's), it enhances the capability of students to learn and conceptualization of the topic which is being taught. Furthermore, Recording of such lectures/sessions will be useful for students either in their absentia or during their revision [2] [19 20 21].

S.C. Satapathy et al. (eds.), *Emerging ICT for Bridging the Future – Volume 1*,
Advances in Intelligent Systems and Computing 337, DOI: 10.1007/978-3-319-13728-5_61

In developed countries majority of the people are exposed to the use of ICT at fullest while developing country like India is still in the implementation phase of integrating ICT in education especially for effective and interactive classroom teaching [7]. Researcher in [3] has discussed the integration of ICT plays a major role to revive the economy of the country. Educational institutes foster and train students to use the ICT as vehicle to get prosper in their knowledge. Hence, use of ICT in classroom cannot be viewed as a burden. As per [6], it is teacher's conviction and approach influence successful integration of ICT into teaching-learning process. It is only the positive stance of teachers' for the effective use of ICT in education can truly provide interesting insight about the subject knowledge with its applicability.

2 Current Trends in ICT for Smart Class Room

- **BYOD**

BYOD stands for "Bring Your Own Device". Concept is to give freedom to people to allow bringing their own devices at work place or class room but in a controlled manner. Concept emerged by the introduction of iPhones. Senior executives wanted to use their iPhones rather than a Blackberry. BYOD devices typically have an access to organization's email, file servers, web servers and databases. BYOD is happening trend in Business and Education sector both. BYOD has certain benefits for educational institutes which are as follows [4][15][16],

- Reduced Hardware Costs, costs shifts to the user of the device
- Availability of information anytime, any places any where
- Seamless Communication
- Makes the most of Cloud Technologies
- Reduced Software Licensing Cost

However, while allowing BYOD in Education or at work place there are few things to consider which are, continuous monitoring for intruder, best security policy, ethical and staff issue, heterogeneity conflicts such as Android/IOS/Windows devices and their communication with your server [4][16].

Author in [5], studied on Schools going Mobile in Western Australian independent schools and come up with interesting findings about m-learning. He revels in his research that use of BYOD will motivate and encourage students to conceptualize their subject knowledge and improves the student learning but at the same time there will be a challenge to manage the technology carefully to see its ethical usage and staff roles to deploy such technology in the campus.

- **Cloud Computing**

Learning through the medium of Internet is contemporary now days and in near future Cloud Computing will hold a considerable stack in education. Through which, learners or instructors will have an advantage to complete their task with less cost by exploiting the available cloud based application in their day to day routing for teaching or learning. Cloud Computing enables user to consume the services on lease provided by the service provider such as Google or Amazon. To access the Cloud Computing services user just need a simple computer regardless of the back end platforms. There are various services which are available on Cloud are SaaS (Software as a Service), PaaS (Platform as a Service) or IaaS (Infrastructure as a Service). In this paper author has presented a Framework for Cloud base e-Learning and shown some interesting result that suggest that soon e-learning will certainly steer in a new era of Cloud Computing [8].

- **Learning Management System (LMS)**

Learning Management System is a Software Application developed for effective management of a course to provide an effective training to students. It is a smart tool which allows teacher or student to interact with each other. Colleges or Universities are using LMS for recording Attendance, Grades, Assignments, Quiz and many more day to day activities of students. Typically LMS works on intranet or internet depends on the policy of the University/College. Objective of implementing LMS is to facilitate students and teacher to access their work and interact from anywhere and anytime [9]. There are few popular LMS adopted by many universities such as Moodle (Modular Object Oriented e-learning), Blackboard, Piazza, Sakai etc [11, 12, 13, 14].

3 Embedding ICT for Class Room Teaching-Learning @School of Computer Studies, Ahmedabad University.

School of Computer Studies (SCS) is running Master of Computer Application course and is affiliated to Ahmedabad University in Gujarat, India. SCS has adopted the state of the art infrastructure to meet the requirement for contemporary education and innovative teaching – learning pedagogy. It is equipped with latest IT infrastructure and use the concept of BYOD and Cloud Campus through server virtualization. Each classroom is having audio visual resource with plug and play mobile devices. Figure – 1 shows the class room enabled with BYOD concept. Teachers and Students can bring their mobile devices and plug and play through Wi-Fi or LAN connection and perform their day to day college activities such as reading, completing lab assignments, watching educational video by connecting to a college server easily.

Fig. 1. BYOD Enabled Class Room@SCS

Figure-2 depicts the implementation strategy of Server Virtualization in the campus. One Physical Server is partitioned to run number of secure virtual server which caters the different kinds of user for different functionalities such as Mail, File and Database server. It also benefits in terms of scalability of users either increasing or decreasing. So in a single classroom different students may access different virtual servers under one roof of server virtualization [10]. State of the art IT infrastructure gives the flexibility with ease to teach and learn in the campus. It also motivates students to stay longer period of time in the campus which will increase their duration of learning and sharing the educational tasks with their peers.

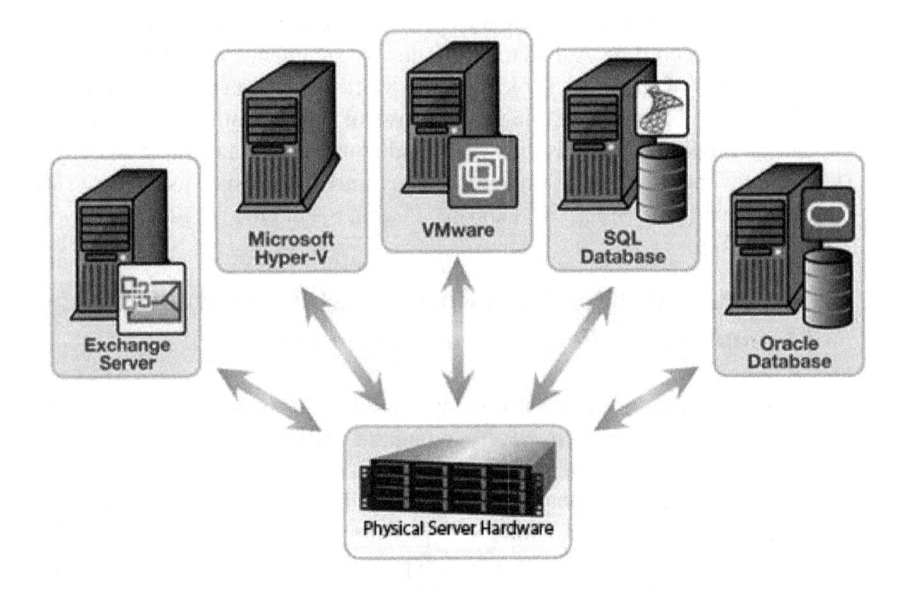

Fig. 2. Strategy for Server Virtualization (Source: x2technologysolutions)

Figure- 3 Shows the screenshots of the LMS - Moodle implemented on the college server along with (24*7) availability on intranet and internet both. Moodle (Modular Object Oriented Learning) is an open source project which facilitates teachers and students to interact for online course training and learning. Moodle is widely popular and in practice in many Indian and foreign universities/colleges. Moodle has salient features starting from sharing Class Notes, Power Point Presentations, Online Assignment Management, Forum Discussion, Multiple Choice Questions Quiz, Blogs, RSS feeds, Wikis, Embedding Audio & Video URLs, User Administration and Online Notice Board etc.

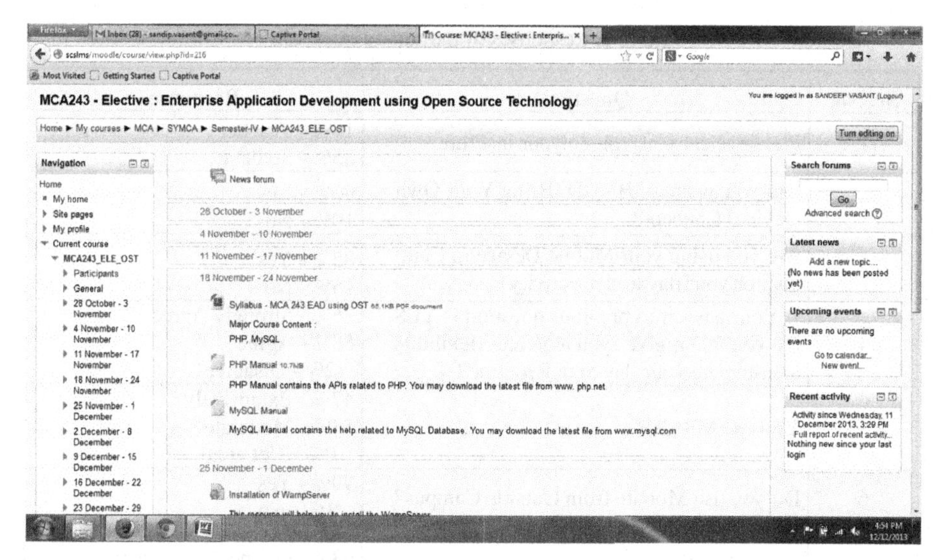

Fig. 3. Online Course with some activities created on Moodle for teacher student interaction

4 Results and Discussion of Online Survey of Teachers and Students on the Campus

Online survey is carried out to know the impact of embedding ICT in class room and campus for teaching and learning process. Responses of 18 teachers and nearly 150 post graduate students have been recorded. The questions and consolidated responses of students and teachers are shown in table-1.

Findings of this survey indicate that use of BYOD is very useful in classroom teaching and it gives the greater flexibility to students and teacher to complete their task with ease. It also reveals that e-learning tools like Moodle is being used extensively for Online Assignment, Quiz, forum discussion, blogs etc. which not only helps in increasing the participation in class discussions but also improves the final result of whole class. Out of 18 teachers 12 are using other e-learning tools such as MOOC, TED Talk and Videos from NPTEL to extend the knowledge beyond the curriculum. Whether Teacher or Student 99% of them are using college server from outside campus as well which helps them to study and interact with each other for doubt solution.

At last, overall result of this survey indicates that educational institute should leverage the ICT for effective teaching pedagogy. Embedding ICT in classroom learning has high potential and provides a greater clarity to students for understanding a particular topic.

Table 1. Responses of the survey questions

Sr. No.	Question	Response
1	Please select your role in your institute.	15% - Teacher 85% - Student
2	Are you aware of BYOD (Bring Your Own Device) Concept?	90% - Yes 10% - No
3	Are you using your Mobile Device in Campus for your day to day activity?	85% - Yes 15% - No
4	If Your answer to previous question is YES, Do you feel it gives you a greater flexibility to complete your day to day routine?	36% - Strongly Agree 55% - Agree 2% - Disagree
5	You use Moodle	47% - Extensively 53% - Moderate 0% - Not at all
6	Do you use Moodle from Outside Campus?	99% - Yes 1% - No
7	Which feature of Moodle from following, you are using for your course?	93% - Online Assignment 60% - Assignment Evaluation & Feedback 79% - Online Quiz 50% - Forum Discussion 20% - Blogs 4% - RSS Feeds 9% - Embedding Audio/Video 13% - Other
8	If You are a Teacher, do you see increase participation/interest of Students in Moodle based learning as compared to traditional teaching-learning	29% - Yes 6% - No 11% - Maybe
9	Whether you are a Teacher or Student, have you noticed significant improvement in result after using ICT based Teaching-Learning.	74% - Yes 3% - No 23% - May be
10	Do you use any other tools other than Moodle for Teaching - Learning	27% - MOOC 18% - TED Talk 18% - NPTEL 57% - Other

5 Conclusion

The objective of this cases study was to explore the perception of Teachers and Students about embeding ICT for effective teaching pedagogy in a classroom at School of Computer Studies, Ahmedabad University. Analysis of this sutdy finds the significant improvement in class participation and result both. Furthermore, findings of this study will be useful to other educational institutes or universities who are yet to embed ICT or are in the process of embeding ICT to their classrooms. Challenges could be to adopt the new technology, skilled staff and an extensive training to be given to the users. The success of this stretagy is highly depends on effective involvement of all the stack holders of the implementing insititute or university.

References

1. Junco, R., Merson, D., Salter, D.W.: The Effect of Gender, Ethnicity, and Income on College Students' Use of Communication Technologies. Cyberpsychology, Behavior, And Social Networking 13(6) (2010)
2. Zhang, D., Zhao, J.L., Zhou, L., Nunamaker, J.F.: Can e-learning replace classroom learning? Communications of the ACM 47(5) (2004)
3. Chandra, V.: An ICT Implementation Strategy for Primary Schools in Fiji. Changing education through ICT in developing countries. Aalborg University Press (2013)
4. Brooks, C.: BYOD (Bring your own device). Sovereign, Business Integration Group (2012)
5. Pegrum, M., Oakley, G., Faulkner, R.: Schools going mobile: A study of the adoption of mobile handheld technologies in Western Australian independent schools. Australasian Journal of Educational Technology 29(1) (2013)
6. Hashemi, B.: The investigation of factors affecting the adoption of ICTs among English language Teachers in ESL context. The International Journal of Language Learning and Applied Linguistics World 4(1) (2013)
7. Zander, P.-O.: Fundamentals of Education and ICT for Development, Changing education through ICT in developing countries. Aalborg University Press (2013)
8. Madan, D., Pant, A., Kumar, S., Arora, A.: E-learning based on Cloud Computing. International Journal of Advanced Research in Computer Science and Software Engineering 2(2) (2012)
9. Doctor, G., Bhavsar, S.: Exploring learning management systems (LMS): E-Learning Tools, CEPT University (2013)
10. Virtualization in Education, IBM Global Education, White Paper (2007)
11. Moodle, http://www.moodle.org
12. Blackboard, http://www.blackboard.com
13. Sakai, http://www.sakaiproject.org
14. Piazza, http://www.piazaa.com
15. Ifenthalera, D., Schweinbenzb, V.: The acceptance of Tablet-PCs in classroom instruction: The teachers' perspectives. Computers in Human Behavior 29(3), 525–534 (2013)
16. Sangani, K.: BYOD to the classroom. Engineering & Technology (2013)
17. Junco, R., Timm, D.: Using emerging technologies to enhance student engagement. New directions for student services issue #124. Jossey-Bass, San Francisco (2008)
18. Nelson Laird, T.F., Kuh, G.: Student experiences with information technology and their relationship to other aspects of student engagement. Research in Higher Education 46, 211–233 (2005)
19. Hiltz, S.R., Turoff, M.: What makes learning networks effective? Communications of ACM 45(4), 56–59 (2002)
20. Hiltz, S.R., Wellman, B.: Asynchronous learning networks as a virtual classroom. Communications of ACM 40(9), 44–49 (1997)
21. Latchman, H.A., Salzmann, C., Gillet, D., Bouzekri, H.: Information technology enhanced learning in distance and conventional education. IEEE Transactions on Education 42(4), 247–254 (1999)

Cloud Based Virtual Agriculture Marketing and Information System (C-VAMIS)

A. Satheesh[1], D. Christy Sujatha[1], T.K.S. Lakshmipriya[3], and D. Kumar[2]

[1] Department of Software Engineering,
[2] Department of Electronics and Communication Engineering
Periyar Maniammai University
Thanjavur, Tamilnadu, India
[3]Avinashilingam Institute for Home Science and Higher Education for Women University,
Coimbatore, Tamil Nadu, India
asatheesh@pmu.edu,
{christysujatha,tkslp.dr,kumar_durai}@gmail.com

Abstract. Today's agricultural marketing has to undergo a series of exchanges or transfers from one person to another before it reaches the consumer. The challenges of the traditional agriculture are addressed significantly by using information and communication technologies (ICT) that play an important role in uplifting the livelihoods of the rural poor. Our main objective of the proposed system is to provide an environment for the farmers that would facilitate their cultivation to deliver the Agricultural Products to the marketing place in time and enriches the farmers with up to date farming technology by coordinating all transactions in a *Cloud based E-commerce Environment.* The usage of Cloud platform will reduce the cost of maintenance in isolated environments and with the application of big data helps the activity of data analytics fast, which is utilized in cost effective manner and extract needed forecasting from very large volumes of data.

Keywords: Perishable Agricultural Products, E-Commerce Environment, Minimum Support Price Policy.

1 Introduction

Current literatures suggest that the agriculture is a source of livelihoods for 86% of rural people in India and it provides 1.3 billion jobs for small-scale farmers and landless workers [1]. Today's agricultural marketing has to undergo a series of exchanges or transfers from one person to another before it reaches the consumer. The productivity and income of the poor farmers have stagnated due to the middle men who have no hesitation in taking the advantage of the farmer's dependence upon them. The majority of the farming community are not getting upper bound yield, because of the appropriate and timely advice about the latest farming technology is not reaching the Farmers properly.

S.C. Satapathy et al. (eds.), *Emerging ICT for Bridging the Future – Volume 1*,
Advances in Intelligent Systems and Computing 337, DOI: 10.1007/978-3-319-13728-5_62

At present, Agricultural knowledge and information system has been addressed by the Extension Systems of State Departments of Agriculture, State Agricultural Universities (SAUs), KVKs, KISAN , NGOs, Private Extension Services through various extension approaches in transfer of technology. The purpose of these Centers is mainly to respond to issues raised by Farmers instantly in the local language, on continuous basis [4], [5].

Our proposed system eliminates the middlemen and arranges facilities for the farmers to sell their highly *Perishable Agricultural Products* (PAP) [2] (e.g., vegetables, fruits, plants, flowers) directly to the wholesalers at reasonable rates. Due to this scheme the Farmers, the Wholesalers and the consumers get the products in affordable prices. Using this system, the excess supply can be sent to other markets by correlating the available supply and demand and hence there is no wastage of the products. The *Centralized Cloud storage* maintains the Data base of the Farmers, Nodal Center, Transport, Wholesaler, Crops, Weed, Soil, Weather details and the Views of the Domain Experts. The *Cloud Service Provider* in our proposed system furnishes the farmers with the latest agricultural knowledge and information, product demand and weather forecasting from various sources for better farming and improved livelihoods. The usage Cloud platform [3][7] will reduce the cost of maintenance singly in isolated environments and with the application of big data will help in activity the data analytics fast. Big data technologies are recent technologies that help in developing generic architectures and it is utilized in cost effective manner and extract needed forecasting from very large volumes of data.

The novelty of the proposed system is that it introduces a new generation of Electronic Farmers and Wholesalers in Cloud based Virtual Agricultural Marketing. It provides virtually the traditional wholesale services, adding value to trading services due to Internet electronic commerce capabilities and enhanced market knowledge.

The objectives of our proposed system:

- To avoid the involvement of intermediate persons between farmers and Customers.
- To deliver the produce to the marketing place in time.
- To facilitate the consumers to get the product in affordable prices.
- To avoid the wastage of day to day production.
- To analyze the Product demand.
- To furnish the Farmers with up to date agricultural knowledge.
- To maintain all the required data in the Cloud storage.
- To improve the welfare of the Farmers by introducing various beneficial schemes

2 Proposed Methodology

The logical graph of the proposed Cloud based Virtual Agriculture Marketing and Information System is represented in Fig.1.

2.1 Architecture of the Proposed System

The proposed architecture consists of five main categories of actors: *Nodal Centre, Farmers, Customers, Transport and Cloud Agro Marketing and Information system.* These actors as related to PAP (Perishable agricultural Product) market participants are interpreted as follows:

The functions of the Nodal Centre is,

- It facilitates the *collection and selling* of Perishable Agricultural Products (PAP) to the Customers s with optimal market rate.
- It provides *hi-tech infrastructure* with cold storage facilities and services available to the Farmers with regard to store the products for direct marketing, and repacking.
- It provides adequate and cheap *transport facilities* which could enable the Farmers to take their surplus produce to the nodal center for marketing.
- It improves the efficiency in agricultural marketing through *regular training and other instructions* to reach the Farmers in their own language.
- It formulates the Farmers to overcome the new challenges in agricultural marketing by using *ICT as a vehicle* of extension.

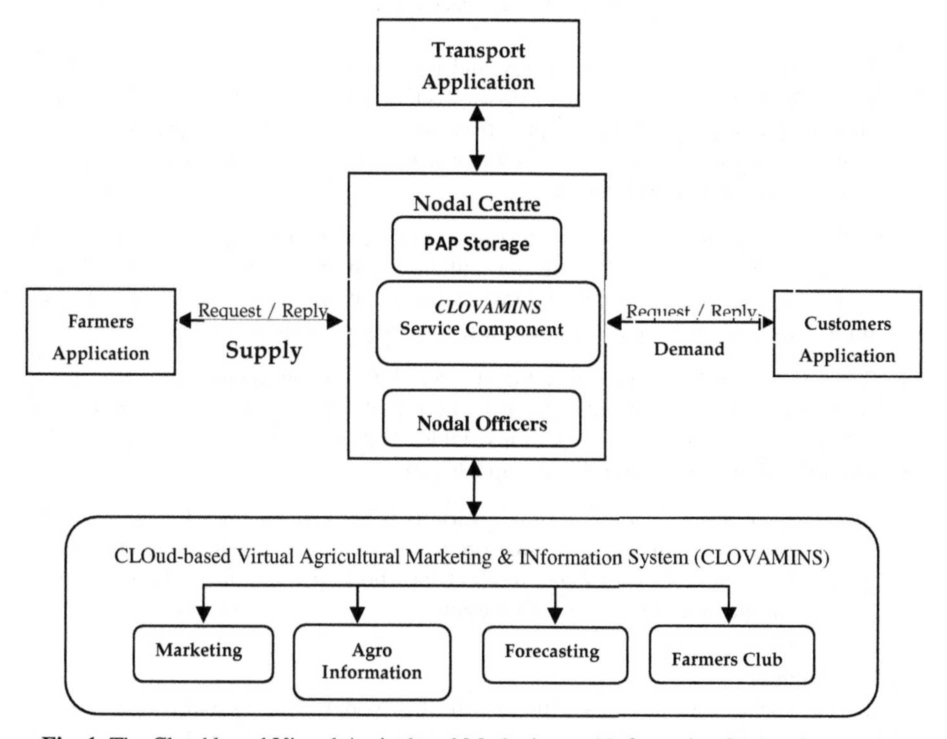

Fig. 1. The Cloud based Virtual Agricultural Marketing and Information System frame work

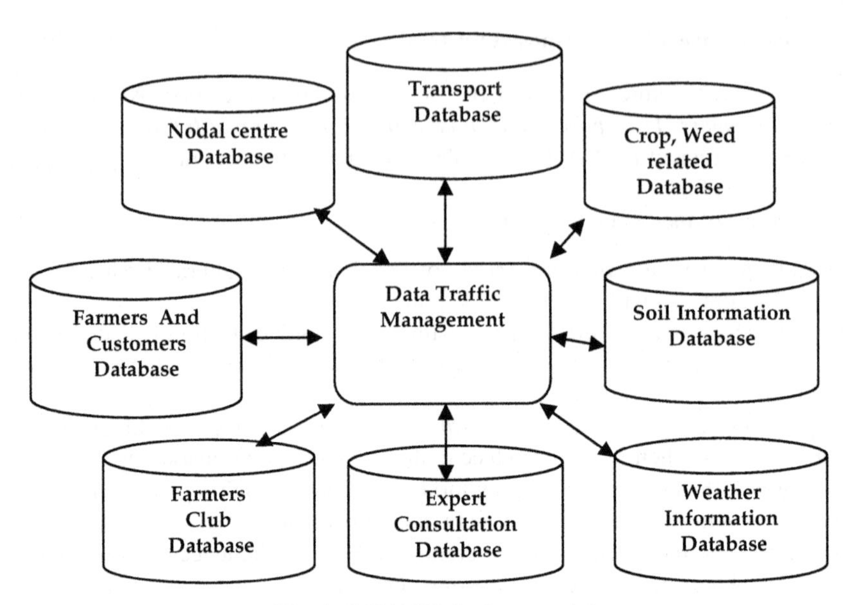

Fig. 2. C-VAMIS database modules

- It promotes ICT driven technology and *agricultural information dissemination* system for quick, effectual and cost-effective delivery of messages to all the stakeholders in agriculture. Keeping pace with the current knowledge distribution trends, it showcases current technologies, policies and other activities through print, electronic and web mode.
- It frequently delivers *voice messages* that cover different areas like soil management, crop management, horticultural crop management, plant protection practices, market rates, weather forecasts information, human and plants health information, employment opportunities, various government welfare schemes etc.
- The voice messages are customized according to the region and associated cooperative. The Farmers also can get solution of their *queries through a helpline* manned by the experts from various fields. Thus Farmers are updated with the latest developments, technologies, updates etc.
- It includes *Farmers' Club* that holds various beneficial schemes for the Farmers which includes Comprehensive *Insurance scheme* to ensure guaranteed damage cover, *Agriculture Loan scheme* to meet the short-term credit needs of Farmers for crop production and allied activities etc.

2.1.1 Farmers

The Farmers who want to connect them with this Nodal Center should register their details to get the benefits of this system. Farmers are provided with free AKASH tablets to communicate with the Nodal Centre. They should register by entering their personal details such as their Name, Address, Mobile Number, Bank account number, Aathar ID number as well as land details such as Land type, Land Size, Types of cultivation they used to do and product details like the production type and variety, the form of product packaging, the estimated production quantity, quality, period of availability and price, for each Farmers type. The Farmers can intimate their produce to the nodal center periodically as soon as their agricultural yields are cultivated. So that there is no need of going to the whole sale market to transform their product in to money. They also can pay attention to the voice messages with reference to the Agricultural information; can get solution of their queries.

2.1.2 Customers

The customers of the nodal centers are one who may be the Wholesaler *or Retailer.* They have to register their personal details such as their Name, Address, Phone number, Bank Account Number as well as requirement details such as types of marketing daily or weekly requirements. For safekeeping, at the outset they have to pay the caution deposit amount during the registration, which is a money back scheme while leaving the nodal centre. They should update their requirements report of agricultural produce to the nodal center in daily or weekly basis.

2.1.3 Transport System

The *transport* is considered as an important third party who can include any distinct market actor who performs specific value adding activities in the PAP supply chain. The transport vehicle is designed specifically to service the rural market and it provides adequate and cheap transport facilities which could enable the Farmers to take their surplus produce to the nodal center for marketing. It also focuses the provision of transport services to the Customers s to transport the yields to the market place. *Transport data* consists of structured information about the contact person, the number of trucks and containers, the characteristics of each vehicle (e.g., refrigerated or not), the ability to group loads, the estimated period of availability, and transportation expenses, for each transport firm.

2.1.4 Cloud Based Virtual Agricultural Marketing and Information System (CLOVAMINS)

CLOVAMINS system implements the cloud computing technology in the Agricultural sector. It monitors and fulfills end user requirements with a user-friendly and faster approach, and stores all relevant data in a centralized location. It connects all the Nodal Centers which are located at 50 Km from each other and can be shared among several Nodal Center users. Nodal Users access the server through a client application from their corresponding Nodal Centre. It has full protection with safeguards that prevent problems such as having multiple users trying to update the same piece of data at the same time. It also allocates the available resources effectively, such as memory, network bandwidth, and disk I/O, among the multiple users.

3 The Layers of the Proposed Architecture

3.1 End User Application

The ***end-user applications*** refer to the implementation of practices and functions for trading the agriculture products. These software applications are Cloud based Technology that are utilized by the Farmers, Wholesalers and Transport domain.

Customer Applications include their Demand, Advertisement and Forecaster modules. The Customer Demand application gives the opportunity for the wholesalers to express their long or short-term ordering preferences for PAP. The Customer Advertiser application is responsible for advertising their demand for perishables and posting their profile. The Customer Forecaster application offers a module for dynamic forecasting with the objective of estimating the demand of perishables that enables an efficient planning for ordering.

Farmer applications include Production Information, Farmers Advertiser and Production Forecaster modules. The Production Information application offers actual production information such as the type and variety of crop, cultivation methods, free text description or field images, product pictures. The Farmers Advertiser application is responsible for advertising of agricultural products through the provision of a rich set of templates, and posting their profiles. These templates include Farmers information such as the estimated production quantity, quality, period of availability and price. The Production Forecaster application is similar to the Customer Forecast application. It offers a module for dynamic forecasting for estimating the production of perishables of each particular provider.

Transport applications include Transport Advertiser and the Scheduling module. The Transport Advertiser application concerns the firms profile and capabilities. Scheduling application offers a module for analytical matching of different product loads and itinerary needs, for effective scheduling.

3.2 Supporting Services

Supporting services are based on Marketing and Information system components for access and service control. It provides Marketing components, Agricultural knowledge and information components, Forecasting component and Farmers association component.

Marketing Component

Using AKASH tablet any farmer can sell his produce through the web page present in the application. This webpage will let the Farmers to enter their personal and produce details, helping them to reach the Customers. The Customers also place their orders for their required items with their personal details. The details from the Farmers and Customers are collected and our proposed system computes the fulfillment of current requirements with the available products (Fig.3). If there is any requirement which has not been satisfied (or) any excess of product, can be connected with other nodal centers to check the availability (or) requirements of the product.

Agriculture Knowledge and Information

- Seasonal Cultivation
- Tillage
- Irrigation management
- Weed management
- Crop protection
- Soil management

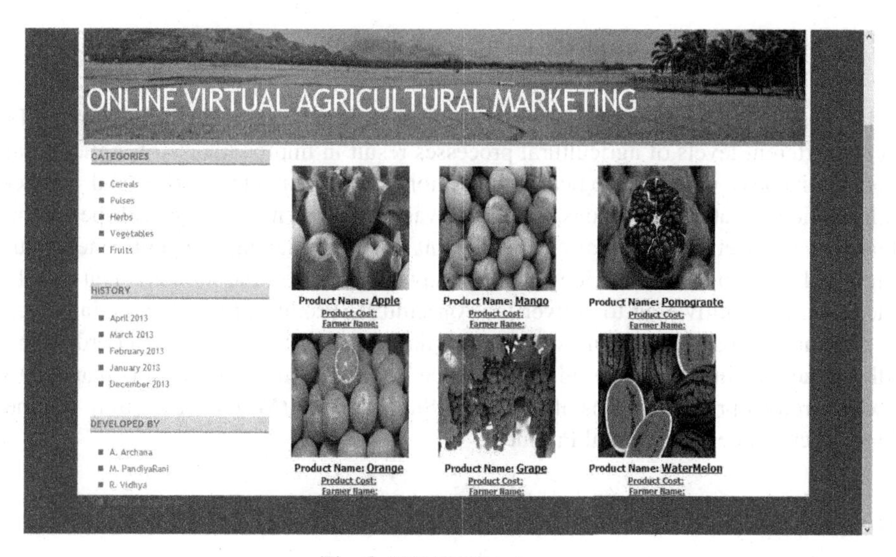

Fig. 3. C-VAMIS webpage

Forecasting

Weather forecasting is the application of science and technology to predict the state of the atmosphere for a given location. *Yield Demand Forecasting* involves techniques including both informal methods, such as educated guesses, and quantitative methods, such as the use of historical sales data or current data from test markets.

Farmers Association

Crop insurance is an insurance arrangement aiming at mitigating the financial losses suffered by the Farmers due to damage and destruction of their crops as a result of various production risks. It provides insurance coverage and financial support to the Farmers in the event of failure of any of the notified crop as a result of natural calamities, pests & diseases.

Crop Loan Farmers depend on loans for agricultural inputs, such as diesel for irrigation, seeds and fertilizers, whose expenses are rising. Crop Loans activities include, ploughing and preparing land for sowing, weeding, and transplantation where necessary, acquiring and applying inputs such as seeds, fertilizers, insecticides etc.

Crop Income Tax Farmers who operate their farming business through a corporation or other business entity must report income and expenses on the appropriate business tax return.

Cost of Cultivation and Cost of Production The cost of cultivation includes factor costs up to the stage of gathering the harvest and that cost of production includes factor costs up to the stage of marketing the produce. The Intelligent system calculates both the cost of cultivation and protection for the benefit of the farmers.

4 Conclusion

It has been acknowledged by various reports of government that application of ICTs at the different levels of agricultural processes result in improvements of agricultural competitiveness of technological information which includes price and market information, weather conditions, economic variables, communication with peers and business transactions etc., plays a signify cant role in achieving competitiveness. Our proposed system is to provide such an environment for the farmers that would facilitate their cultivation to deliver the Agricultural Products to the marketing place in time and enriches the farmers with up to date farming technology by coordinating all transactions in a Cloud based E-commerce Environment. In future, it is planned to fix the product price by means of Minimum Support Price (MSP) Policy [6] to sell the highly Perishable Agricultural Products.

References

1. Sinha, G.R.: ICT Enabled Agriculture Transforming India. CSI Communication, 27–28 (October 2013)
2. Costopoulou, C.I., Lambrou, M.A.: An architecture of Virtual Agricultural Market systems: The case of trading perishable agricultural products. Journal of Information Services and Use 20(1), 39–48 (2000)
3. Chaudhuri, S., Nath, B.: Application of Cloud Computing in Agricultural Sectors for Economic Development. JOMASS 1(2), 79–93 (2014)
4. http://www.cloudtweaks.com/2011/12/infographic-value-of-cloud-computing-servicesthroughthe-years/
5. http://www.portal.bsa.org/cloudscorecard (2012)
6. Alia, S.Z., Sidhub, R.S., Vatta, K.: Effectiveness of Minimum Support Price Policy for Paddy in India with a Case Study of Punjab. Agricultural Economics Research Review 25(2), 231–242 (2012)
7. Sujatha, D.C., Satheesh, A., Kumar, D., Manjula, S.: Smart Infrastructure at Home using Internet of Things. In: Satapathy, S.C., Avadahani, P.S., Udgata, S.K., Lakshminarayana, S. (eds.) ICT and Critical Infrastructure: Proceedings of the 48th Annual Convention of CSI - Volume II. AISC, vol. 249, pp. 627–634. Springer, Heidelberg (2014)

An Adaptive Approach of Tamil Character Recognition Using Deep Learning with Big Data-A Survey

R. Jagadeesh Kannan[1] and S. Subramanian[2]

[1] VIT University, Chennai
[2] R.M.D Engineering College, Chennai

Abstract. Deep learning is currently an extremely active research area in machine learning and pattern recognition society. It has gained huge successes in a broad area of applications such as speech recognition, computer vision, and natural language processing. With the sheer size of data available today, big data brings big opportunities and transformative potential for various sectors; on the other hand, it also presents unprecedented challenges to harnessing data and information. As the data keeps getting bigger, deep learning is coming to play a key role in providing big data predictive analytics solutions. This paper presents a brief overview of deep learning and highlight how it can be effectively applied for optical character recognition in Tamil language.

Keywords: Classifier Design and Evaluation, Feature Representation, Machine Learning, Neural Nets Models, Parallel Processing, Deep Learning, Big Data, GPGPU and Optical Character Recognition.

1 Introduction

Deep Learning is a new area of Machine Learning research, which has been introduced with the objective of moving Machine Learning closer to one of its original goals: Artificial Intelligence. In recent years, there's been resurgence in the field of Artificial Intelligence. It's spread beyond the academic world with major players like Google, Microsoft, and Face book creating their own research teams and making some impressive acquisitions.

Some this can be attributed to the abundance of raw data generated by social network users, much of which needs to be analyzed, as well as to the cheap computational power available via GPGPUs.

But beyond these phenomena, this resurgence has been powered in no small part by a new trend in AI, specifically in machine learning, known as "Deep Learning". In this paper, the key concepts and algorithms behind Deep Learning, beginning with the simplest unit of composition and how it is going to be applied for optical character recognition was introduced.

Big data the large volumes of data that are now produced in many fields can present problems in storage, transmission, and processing, but their analysis may yield useful information and useful insights.

In broad terms, the potential benefits of this Deep Learning System, as applied to big data, are in these areas:

S.C. Satapathy et al. (eds.), *Emerging ICT for Bridging the Future – Volume 1*,
Advances in Intelligent Systems and Computing 337, DOI: 10.1007/978-3-319-13728-5_63

Overcoming the problem of variety in big data. Harmonizing diverse kinds of knowledge, diverse formats for knowledge, and their diverse modes of processing, via a universal framework for the representation and processing of knowledge.

Interpretation of data: The Deep Learning system has strengths in areas such as pattern recognition, information retrieval, parsing and production of natural language, translation from one representation to another, several kinds of reasoning, planning and problem solving.

Velocity: Analysis of streaming data. The Deep Learning system lends itself to an incremental style, assimilating information as it is received, much as people do.

Volume: Making big data smaller. Reducing the size of big data via lossless compression can yield direct benefits in the storage, management, and transmission of data, and indirect benefits in several of the other areas discussed in this article.

Additional economies in the transmission of data: There is potential for additional economies in the transmission of data, potentially very substantial, by judicious separation of `encoding' and `grammar.

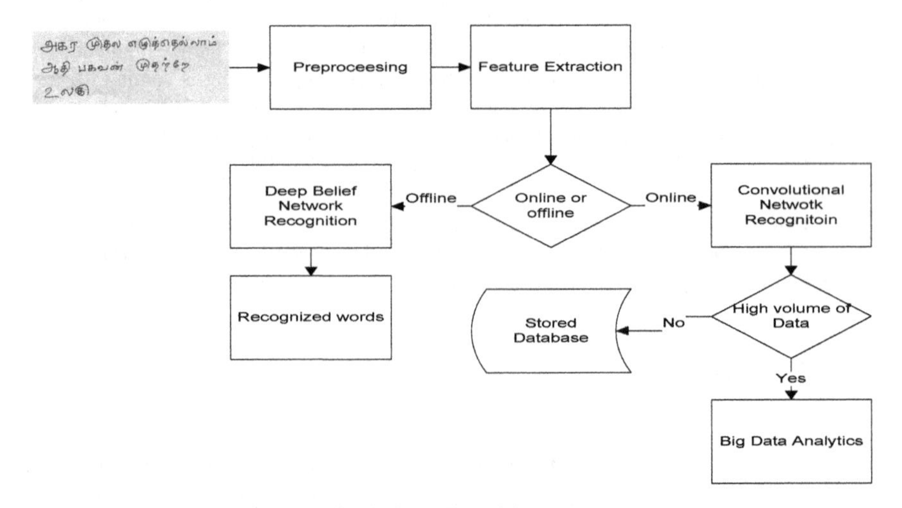

Fig. 1. Overall Architecture

1.1 Problem with Large Networks

A neural network can have more than one hidden layer: in that case, the higher layers are "building" new abstractions on top of previous layers. And as we mentioned before, you can often learn better in-practice with larger networks.

However, increasing the number of hidden layers leads to two known issues:

Vanishing gradients: As we add more and more hidden layers, back propagation becomes less and less useful in passing information to the lower layers. In effect, as information is passed back, the gradients begin to vanish and become small relative to the weights of the networks.

Over fitting: Perhaps the central problem in Machine Learning. Briefly, over fitting describes the phenomenon of fitting the training data too closely, maybe with hypotheses that are too complex. In such a case, your learner ends up fitting the training data really well, but will perform much, much more poorly on real examples.

2 Studies on Tamil Character Recognition

Siromoney et al. [1] described a method for recognition of machine printed Tamil characters using an encoded character string dictionary. The scheme employs string features extracted by the row- and column-wise scanning of character matrix. The features in each row (column) are encoded suitably depending upon the complexity of the script to be recognized.

Chinnuswamy and Krishnamoorthy [2] have proposed an approach for hand-printed Tamil character recognition. Here, the characters are assumed to be composed of line like elements, called primitives, satisfying certain relational constraints.

Suresh and Ganesan [3] have proposed an approach to use the fuzzy concept on handwritten Tamil characters to classify them as one among the prototype characters using a feature called distance from the frame and a suitable membership function. The unknown and prototype characters are preprocessed and considered for recognition.

Hewavitharana S and Fernando [6] designed a system to recognize the handwritten Tamil characters using a two stage classification approach, for a subset of the Tamil alphabet. In the first stage, an unknown character is pre-classified into one of the three groups: core, ascending and descending characters.

Bhattacharya et al. [5] proposed a two stage approach. In the first stage, an unsupervised clustering method was applied to create a smaller number of groups of handwritten Tamil character classes. In the second stage, a supervised classification technique was considered in each of these smaller groups for final recognition. The number of transitions and the chain code histogram are the features used in the first and second stage respectively.

An approach was proposed to recognize handwritten Tamil characters using neural network [6]. Fourier descriptor was used to recognize the characters. The system was trained using several different forms of handwriting provided by both male and female participants of different age groups.

The above literature survey indicated that the existing research works on handwritten Tamil character recognition were using only fuzzy approach, neural network and statistical approach. Recently the Deep Learning Technique has received attention for character recognition.

In this paper, the next section describes an Overview of Deep Learning and in Section 3, the preprocessing methods are explained. Sections 5 and 6 deals with Massive amount and Deep velocity of Data.

3 Character Recognition Using Deep Learning

Deep learning refers to a set of machine learning techniques that learn multiple levels of representations in deep architectures. In this section, a brief overview of two well-established deep architectures: deep belief networks (DBNs) and convolutional neural networks (CNNs) was presented.

3.1 Preprocessing

Data samples were collected from different writers on A4 sized documents. They were scanned using a flatbed scanner at a resolution of 300 dpi and stored as grayscale images. The raw input of the digitizer typically contains noise due to erratic hand movements and inaccuracies in digitization of the actual input. Original documents are often dirty due to smearing and smudging of text and aging [15]. In some cases, the documents are of very poor quality due to seeping of ink from the other side of the page and general degradation of the paper and ink. Preprocessing is concerned mainly with the reduction of these kinds of noise and variability in the input. The number and type of preprocessing algorithms employed on the scanned image depend on many factors such as paper quality, resolution of the scanned image, the amount of skew in the image and the layout of the text [16]. Some of the common operations performed prior to recognition are: Thresholding, the task of converting a gray-scale image into a binary black-white image, skeletonization, reducing the patterns to a thin line representation; line segmentation, the separation of individual lines of text; character segmentation, the isolation of individual characters and Normalization, converting the random sized image into a standard sized image [9].

It makes the subsequent phases of image processing like recognition of characters easier. Thinning is one of the preprocessing methods discussed in this paper. In thinning, the image regions are reduced to one-pixel width characters.

Thinning algorithms should perform thinning effectively by successive deletion of dark points (i.e. changing them to white points) along the edges of the pattern until it is thinned to a line.

An effective thinning algorithm is one that can ideally compress data, eliminate local noise without introducing distortions of its own. But the key goal is to retain significant features of the pattern. There are two types of thinning algorithms

1. Sequential thinning algorithms
2. Parallel thinning algorithms

In Sequential thinning algorithms, result of nth iteration depends on result of $(n-1)^{th}$ iteration as well as pixels already processed in the nth iteration.

In Parallel thinning algorithms, deletion of pixels in n^{th} iteration depends on the result of $(n-1)^{th}$ iteration.

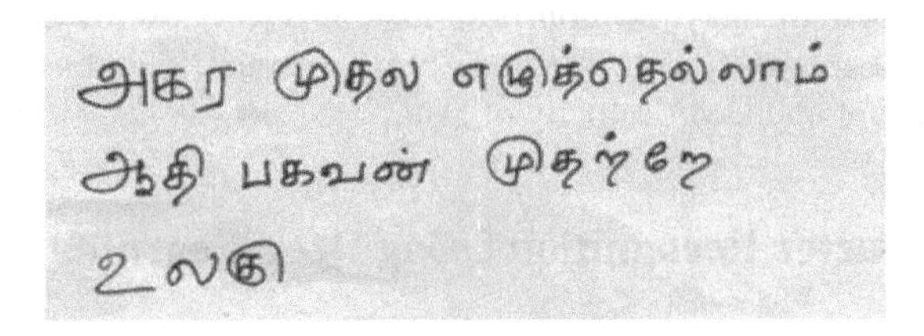

Fig. 2. Preprocessed Image of Tamil Characters

3.2 Feature Extraction

Feature extraction is the problem of extracting the information from the preprocessed data which is most relevant for classification purposes, in the sense of minimizing the within-class pattern variability while enhancing the between-class pattern variability. The Characters from the preprocessing stage are given as inputs to the feature extraction stage. The frame containing the normalized character is divided into several non-overlapping zones. The pixel density is calculated for each zone and used as a feature. Different sizes of zones were used in the work ranging from 2 9 2 to 8 9 8 pixels. When the zone size is 2 9 2 then there will be 256 different zones and 256 features. When the zone size is 4 9 4 then there will be 64 different zones and 64 features. When the zone size is 8 9 8 then there will be 16 different zones and 16 features. The features are extracted for the training set and test set.

It is well known that the performance of a handwritten character recognition system depends largely on the feature extraction phase. In this work, the standard 8-directions chain code is used to represent the closed outer contours of the characters as proposed by Freeman (1961).

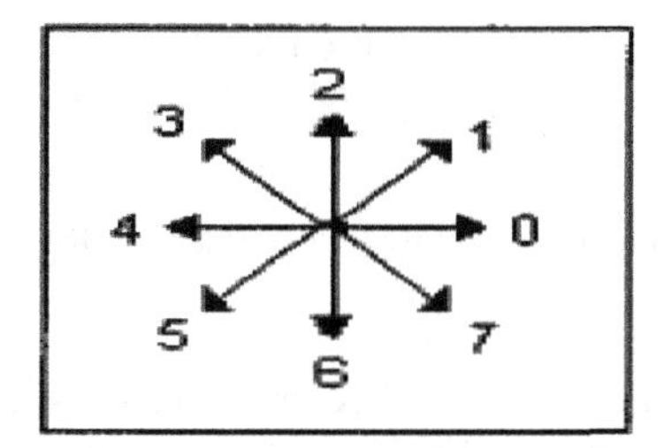

Fig. 3. Chain Code sequence

The extracted features are sequences of integers from 0 to 7. However, the chain code method has its own weaknesses. It is not capable of: (i) performing good corner detection; (ii) detecting sharp boundary changes, and thus not capable of capturing structural information. To solve this problem, a sequence of structures was assigned to represent the characters. The first step of the structure extraction is to determine all strokes. The strokes are separated when a significant change of a contour direction of a chain code occurs. In other words, if the difference (which is the minimum value of clockwise and counterclockwise change) between two successive chain code directions is not less than a preset threshold, then a new stroke is created. An example showing how to separate two successive strokes is illustrated in Figure 3. The chain code of the contour segment in the round corner rectangle box is "....556660007...." The biggest chain code difference is between the "6" and "0" in the middle. This difference is which is not less than our preset threshold value "2". Thus, we consider "....55666" as one stroke, and "0007...." as another stroke. After extracting all the strokes from chain codes, it is grouped. Each group is considered as a structure.

3.3 Deep Belief Networks

Conventional neural networks are prone to get trapped in local optima of a non-convex objective function, which often leads to poor performance [14]. Furthermore, they cannot take advantage of unlabeled data, which are often abundant and cheap to collect in Big Data. To alleviate these problems, a deep belief network (DBN) uses a deep architecture that is capable of learning feature representations from both the labeled and unlabeled data presented to it [21]. It incorporates both unsupervised pre-training and supervised fine-tuning strategies to construct the models: unsupervised stages intend to learn data distributions without using label information and supervised stages perform local search for fine tuning.

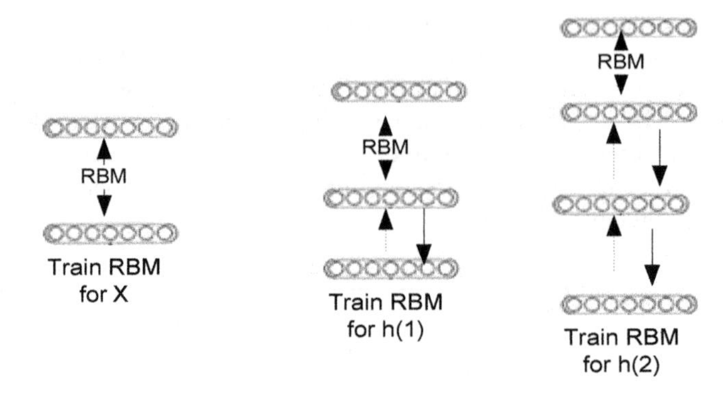

Fig. 4. Typical DBN architecture composed of a stack of Restricted Boltzmann Machines (RBMs)

Fig. 4 shows a typical DBN architecture, which is composed of a stack of Restricted Boltzmann Machines (RBMs) and/or one or more additional layers for discrimination tasks. RBMs are probabilistic generative models that learn a joint probability distribution of observed (training) data without using data labels. They can effectively utilize large amounts of unlabeled data for exploiting complex data structures. Once the structure of a DBN is determined, the goal for training is to learn the weights (and biases) between layers. This is conducted firstly by an unsupervised learning of RBMs. A typical RBM consists of two layers: nodes in one layer are fully connected to nodes in the other layer and there is no connection for nodes in the same layer. Consequently, each node is independent of other nodes in the same layer given all nodes in the other layer. This characteristic allows us to train the generative weights W of each RBMs using Gibbs sampling.

3.4 Training a Deep Network

a) First train a layer of features that receive input directly from the pixels.
b) Then treat the activations of the trained features as if they were pixels and learn features of features in a second hidden layer.
c) It can be proved that each time an another layer of features are added so as to improve a variational lower bound on the log probability of the training data.

The proof is slightly complicated. But it is based on a net equivalence between an RBM and a deep directed model (described later) the generative model after learning 3 layers. To generate data:

1. Get an equilibrium sample from the top-level RBM by performing alternating Gibbs sampling for a long time.

2. Perform a top-down pass to get states for all the other layers. So the lower level bottom-up connections are not part of the generative model. They are just used for inference.

Before fine-tuning, a layer-by-layer pre-training of RBMs is performed. The outputs of a RBM are fed as inputs to the next RBM and the process repeats until all the RBMs are pre trained. This layer-by-layer unsupervised learning is critical in DBN training as practically it helps avoid local optima and alleviates the over-fitting problem that is observed when millions of parameters are used. Furthermore, the algorithm is very efficient in terms of its time complexity, which is linear to the number and size of RBMs [10]. Features at different layers contain different information about data structures with higher-level features constructed from lower-level features. Note that the number of stacked RBMs is a parameter predetermined by users and pre-training requires only unlabeled data (for good generalization).

There are other variations for pre-training: instead of using RBMs, for example, stacked de-noising auto-encoders and stacked predictive sparse coding are also proposed for unsupervised feature learning. Furthermore, recent results show that when a large number of training data is available, a fully supervised training using random initial weights instead of the pre-trained weights (i.e., without using RBMs or auto-encoders) will practically work well. For example, a discriminative model starts with a network with one single hidden layer (i.e., a shallow neural network), which is trained by back propagation method. Upon convergence, a new hidden layer is inserted into this shallow NN (between the first hidden layer and the desired output layer) and the full network is discriminatively trained again. This process is continued until a predetermined criterion is met (e.g., the number of hidden neurons).

3.5 Convolutional Neural Networks

A typical CNN is composed of many layers of hierarchy with some layers for feature representations (or feature maps) and others as a type of conventional neural networks for classification [5]. It often starts with two altering types of layers called convolutional and sub sampling layers: convolutional layers perform convolution operations with several filter maps of equal size, while sub sampling layers reduce the sizes of proceeding layers by averaging pixels within a small neighborhood (or by max-pooling [5], [7]).

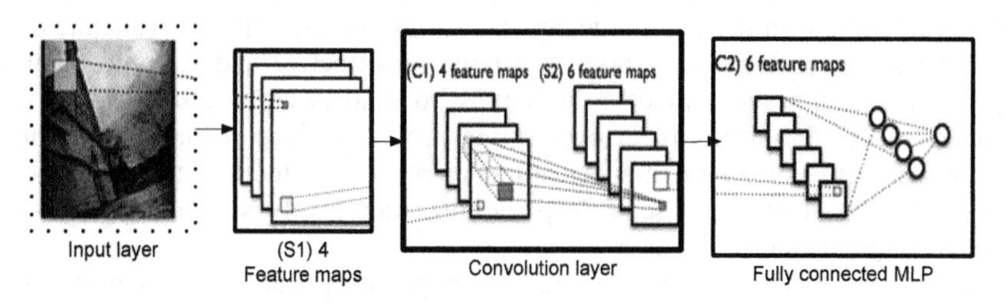

Fig. 5. Typical CNN architecture composed of layers for feature representations (or feature maps)

Fig. 5 shows a typical architecture of CNNs. The input is first convoluted with a set of filters (C layers in Fig. 5). These 2D filtered data are called feature maps. After a nonlinear transformation, a sub sampling is further performed to reduce the dimensionality (S layers in Fig. 5). The sequence of convolution/sub sampling can be repeated many times (predetermined by users).

As illustrated in Fig. 5, the lowest level of this architecture is the input layer with 2D N×N images as our inputs. With local receptive fields, upper layer neurons extract some elementary and complex visual features. Each convolutional layer (labeled C_x in Fig. 5) is composed of multiple feature maps, which are constructed by convolving inputs with different filters (weight vectors).

The sub-sampling layer reduces the spatial resolution of the feature map (thus providing some level of distortion invariance). In general, each unit in the sub-sampling layer is constructed by averaging a 2*2 area in the feature map or by max pooling over a small region.

4 Deep Learning - Based Recognition

The final goal of character recognition is to obtain the class labels of character patterns. On segmenting characters from document images, the task of recognition becomes assigning each character to a class out of a predefined set. A variety of pattern recognition methods are available, and many have been used for handwriting recognition. In this paper Deep belief network is used for recognizing handwritten Tamil characters.

The research of Deep Learning has seen a boom from the mid-2010s, and the application of Deep Learning systems to pattern recognition has yielded state-of-the-art performance. In offline handwritten character recognition they have been proved to be comparable or even superior to the standard techniques like SVM classifiers or multistage perceptron's.

5 Deep Learning for Massive Amounts of Data

While deep learning has shown impressive results in many applications, its training is not a trivial task for Big Data learning due to the fact that iterative computations inherent in most deep learning algorithms are often extremely difficult to be parallelized. Thus, with the unprecedented growth of commercial and academic data sets in recent years, there is a surge in interest in effective and scalable parallel algorithms for training deep models. In contrast to shallow architectures where few parameters are preferable to avoid over fitting problems, deep learning algorithms enjoy their success with a large number of hidden neurons, often resulting in millions of free parameters. Thus, large-scale deep learning often involves both large volumes of data and large models

In this survey, a focus on some recently developed deep learning frameworks that takes advantage of great computing power available today. Take Graphics Processors Units (GPUs) as an example: as of August 2013, NVIDIA single Precision GPUs exceeded 4.5 TeraFLOP/s with a memory bandwidth of near 300 GB/s [18]. They are particularly suited for massively parallel computing with more transistors devoted for data proceeding needs. These newly developed deep learning frameworks have shown significant advances in making large-scale deep learning practical.

A typical CUDA-capable GPU can be built schematically with four multi-processors. Each multi-processor (MP) consists of several streaming multiprocessors (SMs) to form a building block. Each SM has multiple stream processors (SPs) that share control logic and low-latency memory. Furthermore, each GPU has a global memory with very high bandwidth and high latency when accessed by the CPU (host). This architecture allows for two levels of parallelism: instruction (memory) level (i.e., MPs) and thread level (SPs). This SIMT (Single Instruction, Multiple Threads) architecture allows for thousands or tens of thousands of threads to be run concurrently, which is best suited for operations with large number of arithmetic operations and small access times to memory. Such levels of parallelism can also be effectively utilized with special attention on the data flow when developing GPU parallel computing applications. One consideration, for example, is to reduce the data transfer between RAM and the GPU's global memory [19] by transferring data with large chunks. This is achieved by uploading as large sets of unlabeled data as possible and by storing free parameters as well as intermediate computations, all in global memory. In addition, data parallelism and learning updates can be implemented by leveraging the two levels of parallelism: input examples can be assigned across MPs, while individual nodes can be treated in each thread (i.e., SPs). A schematic of the overlapping patches model. Two patches A and B in the input image are shown, with each patch connected to a different set of hidden units.

6 Deep Learning for High Velocity of Data

Emerging challenges for Big Data learning also arose from high velocity: data are generating at extremely high speed and need to be processed in a timely manner. One solution for learning from such high velocity data is online learning approaches. Online learning learns one instance at a time and the true label of each instance will soon be available, which can be used for refining the model [15] [16]. This sequential learning strategy particularly works for Big Data as current machines cannot hold the entire dataset in memory. While conventional neural networks have been explored for online learning.

7 Conclusion

In this paper, a survey of OCR work done on Tamil Language scripts was presented. Here, at first, different methodologies applied in OCR development in international scenario and then different work done for Indian and Tamil language scripts recognition has been discussed. Finally, we discussed steps needed for better Indian script OCR development using Deep Learning technology. It is believed that our survey will strongly encourage activities of automatic document processing and OCR of Indian language scripts combining with Big Data Analytics.

There are no sufficient number of studies on Indian language character recognition, although there are 12 major scripts in India. Most of the pieces of existing work are concerned about Devnagari and Bangla script characters, the two most popular languages in India. Some studies are reported on the recognition of other languages like Telugu, Oriya, Kannada, Panjabi, Gujrathi etc. Structural and topological features-based tree classifier and neural network classifiers are mainly used for the recognition of Indian scripts in offline. The present review discusses different methodologies in OCR development as well as research work done on the recognition of different Indian scripts using deep learning methodology.

References

1. Siromoney, G., Chandrasekaran, R., Chandrasekaran, M.: Computer recognition of printed Tamil character. Pattern Recognition 10, 243–247 (1978)
2. Chinnuswamy, P., Krishnamoorthy, S.G.: Recognition of hand printed Tamil characters. Pattern Recognit. 12, 141–152 (1980)
3. Suresh, R.M., Ganesan, L.: Recognition of hand printed Tamil characters using classification approach. In: ICAPRDT, Kolkata, pp. 63–84 (1999)
4. Hewavitharana, S., Fernando, H.C.: A two stage classification approach to Tamil handwriting recognition. In: Tamil Internet 2002, California, USA, pp. 118–124 (2002)
5. Bhattacharya, U., Ghosh, S.K., Parui, S.K.: A two stage recognition scheme for handwritten Tamil characters. In: Proceedings of the Ninth International Conference on Document Analysis And Recognition (ICDAR 2007), pp. 511–515. IEEE Computer Society, Washington, DC (2007)
6. Sutha, J., Ramaraj, N.: Neural network based offline Tamil handwritten character recognition system. In: Proceedings of the International Conference on Computational Intelligence and Multimedia Applications, vol. 2, pp. 446–450. IEEE Computer Society, Washington, DC (2007)
7. Shanthi, N., Duraiswamy, K.: Preprocessing algorithms for the recognition of Tamil handwritten characters. In: Third International CALIBER 2005, Kochi, pp. 77–82 (2005)
8. Casey, R.G., Lecolinet, E.: A survey of methods and strategies in character segmentation. IEEE Trans. Pattern Anal. Mach. Intell. 18(7), 690–706 (1996)
9. Plamondon, R., Srihari, S.N.: On-line and off-line handwriting recognition: a comprehensive survey. IEEE Trans. Pattern Anal. Mach. Intell. 22(1), 63–84 (2000)
10. Hinton, G., Salakhutdinov, R.: Reducing the dimensionality of data with neural networks. Science 313(5786), 504–507 (2006)
11. Bengio, Y.: Learning deep architectures for AI. Found. Trends Mach. Learn. 2(1), 1–127 (2009)
12. Nair, V., Hinton, G.: 3D object recongition with deep belief nets. In: Proc. Adv. NIPS, vol. 22, pp. 1339–1347 (2009)
13. LeCun, Y., Bottou, L., Bengio, Y., Haffner, P.: Gradient-based learning applied to document recognition. Proc. IEEE 86(11), 2278–2324 (1998)
14. Rumelhart, D., Hinton, G., Williams, R.: Learning representations by back-propagating errors. Nature 323, 533–536 (1986)
15. Cirean, D.C., Meier, U., Masci, J., Gambardella, L.M., Schmidhuber, J.: Flexible, high performance convolutional neural networks for image classification. In: Proc. 22nd Int. Conf. Artif. Intell., pp. 1237–1242 (2011)

16. Scherer, D., Müller, A., Behnke, S.: Evaluation of pooling operations in convolutional architectures for object recognition. In: Diamantaras, K., Duch, W., Iliadis, L.S. (eds.) ICANN 2010, Part III. LNCS, vol. 6354, pp. 92–101. Springer, Heidelberg (2010)
17. CUDA C Programming Guide, PG-02829-001_v5.5, NVIDIA Corporation, Santa Clara, CA, USA (July 2013)
18. Le, Q., et al.: Building high-level features using large scale unsupervised learning. In: Proc. Int. Conf. Mach. Learn. (2012)
19. Bottou, L.: Online algorithms and stochastic approximations. In: Saad, D. (ed.) On-Line Learning in Neural Networks. Cambridge Univ. Press, Cambridge (1998)
20. Blum, A., Burch, C.: On-line learning and the metrical task system problem. In: Proc. 10th Annu. Conf. Comput. Learn. Theory, pp. 45–53 (1997)
21. Tellache, M., Sid Ahmed, M.A., Abaza, B.: Thinning Algorithms for Arabic OCR. In: IEEE Pacific Rim Conference on Communications, Computers and Signal Processing 1993, May 19-21, vol. 1, pp. 248–251 (1993)
22. Kwon, J.-S., Gi', J.-W., Kang, E.-K.: An enhanced thinning algorithm using parallel processing. In: Proceedings of the 2001 International Conference Image Processing, October 7-10, vol. 3, pp. 752–755 (2001)
23. Crego, E., Munoz, G., Islam, F.: Big data and deep learning: Big deals or big delusions? Business (2013), http://www.hufngtonpost.com/george-munoz-frank-islamand-ed-crego/big-data-and-deep-learnin_b_3325352.html
24. Bengio, Y., Bengio, S.: Modeling high-dimensional discrete data with multi-layer neural networks. In: Proc. Adv. Neural Inf. Process. Syst., vol. 12, pp. 400–406 (2000)
25. Marc'Aurelio Ranzato, Y., Boureau, L., LeCun, Y.: Sparse feature learning for deep belief networks. In: Proc. Adv. Neural Inf. Process. Syst., vol. 20, pp. 1185–1192 (2007)
26. Dahl, G.E., Yu, D., Deng, L., Acero, A.: Context-dependent pretrained deep neural networks for large-vocabulary speech recognition. IEEE Trans. Audio, Speech, Lang. Process. 20(1), 30–41 (2012)
27. Hinton, G., et al.: Deep neural networks for acoustic modeling in speech recognition: The shared views of four research groups. IEEE Signal Process. Mag. 29(6), 82–97 (2012)
28. Salakhutdinov, R., Mnih, A., Hinton, G.: Restricted Boltzmann machines for collaborative _ltering. In: Proc. 24th Int. Conf. Mach. Learn., pp. 791–798 (2007)
29. Cirean, D., Meler, U., Cambardella, L., Schmidhuber, J.: Deep, big, simple neural nets for handwritten digit recognition. Neural Comput. 22(12), 3207–3220 (2010)
30. Collobert, R., Weston, J., Bottou, L., Karlen, M., Kavukcuoglu, K., Kuksa, P.: Natural language processing almost from scratch. J. Mach. Learn. Res. 12, 2493–2537 (2011)

Tackling Supply Chain Management through Business Analytics: Opportunities and Challenges

Prashant R. Nair

Department of Computer Science & Engineering, Amrita School of Engineering, Amrita Vishwa Vidyapeetham University, Amrita Nagar P.O, Coimbatore, 641112 India
prashant@amrita.edu

Abstract. Information and Communication Technology (ICT) tools and technologies are revolutionizing enterprise workflow and processes. Deployment of ICT tools for supply chain planning and execution has resulted in greater agility, robustness, collaboration, visibility and seamless integration of all stakeholders in the enterprise and also extended enterprise consisting of suppliers and customers. Business analytics is emerging as a potent tool for enterprises to improve their profitability and competitive edge. Business analytics aims at building fresh perspectives and new insights into business performance using data, statistical methods, quantitative analysis and predictive modeling. Advanced analytics is being employed for several processes in supply chain planning and execution like demand forecasting, inventory management, production & distribution planning etc. Enterprise case studies of successful deployments of business analytics for supply chain management are showcased. Some challenges faced in business analytics usage like high cost, need for data aggregation from multiple sources etc are also highlighted. An integration of business analytics with disruptive and game-changing technologies like social media, cloud computing and mobile technologies in the form of the SMAC - Social, Mobile, Analytics and Cloud stack holds tremendous promise to be the next wave in enterprise computing with wide-ranging advantages like improved supply chain planning, collaboration, execution and stake-holder engagement.

Keywords: Supply Chain Management (SCM), Business Analytics, Big Data, SMAC.

1 Introduction

Information and Communication Technology (ICT) tools and technologies are revolutionizing enterprise workflow and processes. Deployment of ICT tools for supply chain planning and execution has resulted in greater agility, robustness, collaboration, visibility and seamless integration of all stakeholders in the enterprise and also extended enterprise consisting of suppliers and customers. ICT early adoption and deployment across the supply chain has become a force multiplier and determinant of competitive advantage for many enterprises [1]. The tools that are

S.C. Satapathy et al. (eds.), *Emerging ICT for Bridging the Future – Volume 1*,
Advances in Intelligent Systems and Computing 337, DOI: 10.1007/978-3-319-13728-5_64

extensively used include ERP, Supply Chain Management (SCM) Software packages, RFID [2], decision support systems [1], software agents, transportation & inventory management systems [3].

Some of the major supply chain managed challenges are directly or indirectly linked to the availability of real-time data as also accuracy of the trends and forecasts. Major challenges include the following:

- Visibility and transparency of supply chain planning and execution
- Accurate Demand forecasting
- Handling cost overruns

2 Business Analytics

A paradigm shift in enterprise computing is the advent of business analytics. Business analytics aims at building fresh perspectives and new insights into business performance using data, statistical methods, quantitative analysis and predictive modeling. At a lower level, there is business intelligence that is standard measures for comparing past performance for future improvement based on enterprise data and statistical analysis [4]. Some common usages are retailers using business analytics to predict consumer behavior and buying patterns, [5] understanding of citizen needs for government to provide better delivery of services and banks & financial institutions detecting and preventing fraudulent transactions or categorizing their customers based on their credit history.

Business analytics are primarily classified into three types namely descriptive, predictive and prescriptive [4].

- Descriptive Analytics tries to derive insights from historical data with using reporting techniques, balanced scorecards etc
- Predictive analytics tries to derive insights using statistical and machine learning techniques.
- Prescriptive analytics tries to derive insights using optimization and simulation techniques

Business analytics is estimated to be an industry worth US $ 50 billion by 2016. Gartner estimates that the amount of corporate data will grow forty four fold in the period between 2008 and 2020 with the amount of data in the order of 35 zeta bytes by 2020 [5]. Corporate data generated and managed by enterprises amounts to billions of gigabytes of data. This includes from internal and external sources. This explosion of information is now dubbed as Big Data. The key to unlocking the value of this Big Data is to have better methods to access it as also its analysis. These analytics can point towards actionable intelligence so as to help enterprises tweak their strategy [6]. Predictive analytics also have a role to play for enterprises to build scenarios in various difficult situations. Gartner also predicts Business Analytics and intelligence to be the top concern for enterprise computing especially for sectors like healthcare, retail, transportation, finance and government.

Various sources of big data which provide fodder for analytics include [5]:

- Internet data
- Primary research
- Secondary research
- Location data
- Image data
- Supply chain data
- Device data

Business Analytics provide dashboards after mining the enterprise Big Data available through several sources. The advent of the cloud and associated technologies like virtualization has also contributed to the Big Data trend. Other technologies that have a complementary effect include social media, and next generation storage technologies.

Netflix, the premier entertainment and video distribution company has over 36 million subscribers, who watch 4 billion hours of programs every quarter. Netflix through its premium, Cinematch engine uses analytics of consumer preferences to recommend movies. Another example of analytics usage from a common man's perspective is Apple iTunes [7], which analyses user experiences to facilitate quicker, downloads of the popular musical numbers and scores.

Supply-chain operations reference-model (SCOR), a new process framework is emerging as a tool and metric for evaluating the efficiency of the supply chain. SCOR focuses on business process modeling, performance metrics and good practices. Analytics can significantly improve the following four areas of SCOR [8], i.e.

- Plan (Demand and supply planning activities)
- Source (Procurement activities)
- Make (Production and Manufacturing activities)
- Deliver (Transportation and warehousing activities)

3 Business Analytics for SCM

Traditionally, ERP systems as well as SCM packages have a limited amount of analytical capabilities primarily aimed at transactional data generated. Supply Chain data is being analyzed for inventory re-ordering and management as well as demand forecasting and prediction. This could be in the form of reports, queries, alerts and forecasts. Sometimes, the accuracy of these forecasts are however suspect, considering the complex nature of business as well as unpredictable external factors like weather patterns, price and economic volatility. This calls for advanced business analytics for SCM or supply chain analytics which promises several benefits like better customer engagement, improved productivity, improved responsiveness, decision-making aid and cost reduction as a result of better inventory visibility and accurate demand forecasts. This reinforces the fact that supply chains are increasingly becoming very complex with multiple partners, suppliers and stakeholders.

Supply Chain Analytics allows companies to deconstruct new forms of data as also analyze the data to give actionable intelligence, which was not hitherto evident. Analytics will give us the ability to extract, tweak, modify, cleanse and integrate data from multiple and multifarious data sources and points. Using analytics, the value of the data is greatly enhanced and business workflow and processes can be better aligned to enterprise goals. Supply chain analytics render supply-chains with advanced capabilities like trend analysis, drilldown views, accurate forecasting, scenario & what-if analysis, simulation and optimization. These greatly improve decision-making and interpretation of situations which is very crucial for enterprises in the complex business environment influenced by trends like liberalization, globalization, outsourcing, ever-changing customer preferences and pricing pressures. A data analytics solution will provide real-time, updated and comprehensive view of all aspects of the supply chain as well as all the information generated and captured by various sub-systems in the supply chain. It also lets them collect, collate, correlate and mine data from diverse sources and applications. Timely information through analytics greatly impacts raw material sourcing, manufacturing, goods delivery and return.

Some of the application areas in SCM, where Business Analytics increased operational efficiency include [8]:

- Inventory Planning, Reordering, Optimization and Management
- Sourcing and Procurement Planning
- Demand Planning & Forecasting
- Logistics and Distribution Planning & Management
- Sales and Operations Planning
- Production planning
- Vendor & Supplier Selection and Evaluation
- Product Failure analysis
- Plant and manufacturing productivity
- Customer Relationship Management (CRM)
- Route optimization
- Fleet sizing
- Risk Management
- Benchmarking

Business analytics has potential to impact wide-ranging improvements in SCM both at the strategic and operational levels thereby improving operational efficiency and creating customer value. However some challenges faced include the high cost of such solutions. This is all the more significant, when we consider enterprises that have made substantial investments into ERP, CRM or SCM packages and solutions. This is further complicated by lack of understanding on huge benefits accruing from analytics especially by top management.

Aggregation of data from multiple sources is another challenge [9]. For example production data from the factory shop floor is difficult to gather and collate as most of these systems use legacy and proprietary applications and systems with multiple data formats. Almost 85% of big data is unstructured and needs to be translated to an understandable format. Nevertheless, several companies like SAS, Genpact and Capgemini have started hawking SCM Analytics solutions. Expert systems based on statistical analysis, have also been developed for optimization of supply chain processes like procurement, planning, execution, warehousing and transportation [10]. This renders enterprises to be robust, adaptive, flexible and achieve faster cycle times.

For some critical sectors like health care and pharma, inventory shaping is very crucial. Advanced analytics can be applied to accurately predict demand and to monitor supply and reordering replenishment policies. This also has bearing on reducing cost overruns as excess inventory in terms of work-in-progress or finished goods incur unavoidable costs for enterprises.

4 Enterprise Implementations and Deployments

- Several companies like SAS, Genpact and Capgemini are offering Supply Chain Management analytics solutions to companies across all verticals [9].
- AmBev, Latin America's largest beverage company, is a company with a wide sales, distribution and production network with 50 production units, 11,000 retail partners, 16,000 trucks and 1 million points of sale retail points. AmBev uses SAS advanced analytics for demand planning and forecasting [11]. The solution generates weekly forecasts for setting sales goals, production levels, and distribution plan. The company estimates that company's product turnover rate has improved by 50 percent after using this solution.
- IBM uses its own Buy Analysis Tool (iBAT) as its channel collaboration solution for large partners across North America and Europe. This primarily focuses on inventory management for optimized replenishment decisions under price protection. The company estimates considerable business savings due to tool.
- Retail giant, Tesco uses advanced analytics on its supply chain data to reduce waste, optimize promotions and dynamic replenishing of stock [6]. The company estimates savings of over 100 million pounds in annual supply chain costs.
- Wal-Mart Labs, the research wing of the retail bellwether is using capacities of its new acquisition, predictive analytics firm Inkiru to improve customer engagement and experience [6]. Key focus areas include site personalization, search fraud prevention and marketing.
- In India, RedBus, the online travel aggregator is now using an analytics tool to look into booking, seat availability and inventory data across their system of hundreds of bus operators serving more than 10,000 routes.
- Cisco extensively uses business analytics with respect to sourcing and procurement [10]. Several processes are outsourced so that Cisco focuses on product innovation for its networking products. Several of Cisco contract manufacturers are closely integrated into its order fulfillment systems. As a

result half of the orders are directly sent to the end user or distributor without Cisco ever taking physical possession of the product.
- The ERP world leader, SAP has its in-house BusinessObjects analytics solution. This platform has in-built scenario analysis, risk management, alerts and monitoring features for factors like order fulfillment, payments, shipment tracking etc

5 SMAC Stack

An integration of disruptive and game-changing technologies in the form of the SMAC - Social, Mobile, Analytics and Cloud stack also holds tremendous promise to be the next wave in enterprise computing. By 2020, IDC estimates that ICT spending worldwide could touch US$5 trillion mark with fourth-fifth of this driven by the SMAC stack, which is the seamless intersection of the SMAC technologies [5]. These game-changing technologies integrated together as a stack can deliver a force-multiplier effect. SMAC usages remove the use of technologies as silos or islands and the integration delivers a force-multiplier effect [7].

Forrester Research estimates that cloud computing industry will grow from $40.7 billion in 2010 to more than $241 billion by 2020 All ERP service providers are offering cloud-based versions of their products, which inevitably bring down the Total Cost of Ownership (TCO). Social Media continues to capture the user imagination with number of users hovering around 1.5 billion. At last count, number of mobile phones topped 4.5 billion, majority of them smart phones [5]. Retail giants are spending between 20 to 25 percent of their advertising budget on Facebook and other social media channels. Every major enterprise is building a profile in major social media with Facebook pages, Twitter handles and Linked-in accounts with nearly 90 percent of premier global banks using social networking extensively. SMAC stack is a next-gen architecture which leverages and intersects these technologies to form a new product-service combo. This combo promises to bring about greater operational synergies without the baggage of legacy systems and applications. IT departments are usually grappled with diverse legacy hardware, software and networking technologies & platforms. Mobile technologies and cloud computing can easily integrate diverse hardware and storage devices. Social media can facilitate instant dialogue collaboration. Business Analytics provide dashboards after mining the enterprise Big Data available through several sources. SMAC technologies can be easily accessed by all with the recent trend of enterprises encouraging their employees to Bring Your Own Device (BYOD). This will be complemented by services like PayPal, which use mobile transactions for cashless payments using tablets and smart phones.

With the number of app downloads expected to touch 183 billion annually by 2015, the proliferation of apps for mobile phones like Android, I-phone etc is another healthy trend for SMAC stack to become more popular and widespread [6]. The importance of the human element is brought out though the use of social media. Communication, collaboration and information dissemination is instant thereby unlocking their value. Business analytics bring out actionable intelligence about our suppliers and customers. Cloud Computing and associated technologies like virtualization, Software-as-a-service (SAAS) etc powers the force-multiplying transformative combination of social media, mobile apps and analytics.

One of the earliest adopter of the SMAC stack is Netflix. Movies recommended by Netflix are hosted on the cloud through the provider, Amazon Web services. The movies can be easily accessed using tablets, smart phones and/or television. Customer choices and preferences with regard to the movie selection and recommendation are also in Facebook. This is a classic example of a SMAC stack building a collaborative eco-system, where technologies complement each other and brings value to the customer. The ultimate beneficiary is the consumer, who gets a personalized experience of watching movies.

Another popular enterprise implementation is the Oracle Social Network, which is a cloud-based enterprise social network & collaboration tool [12] with various features for real-time dialogue & updates, content creation, transactional workflow and business analytics & dashboards etc. The separation between enterprise, collaborative and social media applications disappears in this tool. It uses stream-based conversations between stakeholders and applications along the process workflow.

The ubiquitous, Google Apps is yet another example of SMAC stack integration [13]. Google Apps has various value-added services like Calendar, Document Management E-mail, Chat groups, bulletin boards, websites etc.

Among IT bellwethers, Cognizant is championing the SMAC stack, using Cognizant 2.0 platform, which is based on it. Cognizant claims that the platform has improved its efficiency of client delivery by 17% [5]. This platform has a friendly UI and combines the project and knowledge management activities. Some of the features of this platform include blogs, micro-blogs, bulletin boards & discussion forums and wikis; activity stream that aggregates notifications; Ispace, a platform for crowd sourcing and ideas and Integrated Learning Management System (LMS)

With Internet of Things (IoT) emerging in a big way, SMAC stack opens new vistas and limitless possibilities to create a future generation of social machines, devices and consumer products. Perceivable Benefits for enterprise computing and supply chain management are as follows [14]:

- Improved use of legacy systems, processes and applications for both supply chain planning and execution
- Better collaboration and dialogue between various stakeholders
- Better customer interactions
- Improved sales, inventory and distribution management

6 Conclusion

Business analytics is emerging as an effective technology for business enterprises for improving their bottom-line as well as maintain their competitive edge over their rivals. Business analytics aims at building fresh perspectives and new insights into business performance using data, statistical methods, quantitative analysis and predictive modeling. Advanced analytics have found applications in various supply chain processes like demand forecasting, inventory management, production & operations planning, transportation management etc. Enterprises like IBM, Cisco and Tesco have successfully deployed business analytics for supply chain management processes. Some challenges faced in business analytics usage like high cost, need for data aggregation from multiple sources etc are also highlighted. Business analytics coupled with disruptive and game-changing technologies like social media, cloud computing and mobile technologies in the form of the SMAC - Social, Mobile, Analytics and Cloud stack holds tremendous promise to be the next wave in enterprise computing with many benefits like improved supply chain planning, collaboration, execution and stake-holder engagement.

References

1. Nair, P.R., Balasubramaniam, O.A.: IT Enabled Supply Chain Management using Decision Support Systems. CSI Comm. 34(2), 34–40 (2010)
2. Nair, P.R.: RFID for Supply Chain Management. CSI Comm. 36(8), 14–18 (2012)
3. Nair, P.R., Raju, V., Anbuudayashankar, S.P.: Overview of Information Technology Tools for Supply Chain Management. CSI Comm. 33(9), 20–27 (2009)
4. Davenport, T.H.: Competing on Analytics. Harv. Busi. Rev. (2006)
5. Frank, M.: Don't Get SMACked: How Social, Mobile, Analytics and Cloud Technologies are reshaping the Enterprise. Cognizant Future of Work (2012)
6. Chandrasekharan, R., Udhas, P.: The SMAC Code: Embracing new technologies for future business, CII and KPMG Report Information (2013), http://www.kpmg.com/IN/en/IssuesAndInsights/ArticlesPublications/Documents/The-SMAC-code-Embracing-new-technologies-for-future-business.pdf
7. Nair, P.R.: The SMAC effect towards Adaptive Supply Chain Management. CSI Comm. 38(2), 31–33 (2014)
8. Trkman, P., et al.: The impact of business analytics on supply chain performance. Dec. Sup. Syst. (2010)
9. Sahay, B.S., Ranjan, J.: Real time business intelligence in supply chain analytics. Inform. Mgmt. & Comp. Secu. 16(1), 28–48 (2008)
10. Oliveira, M.P.V., McCormack, K., Trkman, P.: Business analytics in supply chains – the contingent effect of business process maturity. Exp. Sys. with Appln. (2012)
11. SAS Company Information, http://www.sas.com
12. Oracle Social Network, Oracle white paper information, http://www.oracle.com/technetwork/middleware/webcenter/socialnetwork/overview/wp-oracle-social-network-2011-513258.pdf
13. Google Apps for Business Information, http://www.google.com/enterprise/apps/business
14. Nair, P.R.: The SMAC effect towards Adaptive Supply Chain Management. CSI Comm. 38(2), 31–33 (2014)

A Multiple Search and Similarity Result Merging Approach for Web Search Improvisation

Vijayalakshmi Kakulapati[1], Sudarson Jena[2], and Rajeswara Rao[3]

[1] Dept of CSE, JNTU, Hyderabad
vldms@yahoo.com
[2] Dept of IT, Gitam University, Hyderabad
skjena2k5@gmail.com
[3] Dept of CSE, JNTU, Vijayanagaram
raob4u@yahoo.com

Abstract. The increase of dependency on web information demands an improvisation in web search for accurate and highly similar results. Most search engines retrieves results based on its web crawled index database it maintains, which limited the scope of search. It also performs similar search technique for both sort and long query. To deliver consistently superior results one must understand the exact intent of the query and each keyword in the query strength. This paper focuses on proposing a multiple search and similarity result merging framework for web search improvisation with identification of strength of keywords of user query. Based on the strength of query computation multiple searches are made in different search engine for obtaining multiple results. To merge the obtain result a similarity conversion based method is proposed which provides high accurate and similar results for search improvisation.

Keywords: Web Search, Search engine, Similarity measures, Query processing, Multiple Searching.

1 Introduction

Searching is the second most popular activity on the web after emails service. Search engines are prominent tools which are used for searching. Any search engine hardly capable of covering of huge resource of web for searching [1][4]. Searching is becoming complex due to variation of user query length for search. A user query can be a single word, phrases or questions. In most cases, the correct result is depends on how efficiently the user query processed and related to the information retrieved [3]. A study on information retrieval to evaluate the overlapping of results retrieved of three most popular search engines namely Google, Yahoo, Bing ,Ask, AltaVista and Excite reveals that 85% results are repeating, 80% results were found common in all search engines and 3% of results were unique. The high percentage of repeating and unique result infer that only a single search engine result may not be sufficient to provide the relevant and required results, and at the same time an efficient processing of user query to find the strength is important for accurate web search. Even if user

S.C. Satapathy et al. (eds.), *Emerging ICT for Bridging the Future – Volume 1*,
Advances in Intelligent Systems and Computing 337, DOI: 10.1007/978-3-319-13728-5_65

perform same search in different search engine its leads to more dissatisfaction as same results is repeating.

A user search queries are often based on an approximation and synopsis of information needs. Accurately matching against the terms in the search query is a woefully inadequate method for finding the correct or even correlated information [19]. In contrast, user approximation can leverage the associations between these terms in order to understand what the user really wants to find. These terms may be vague and not even be correct for the specific type of information, especially since the user may only have a rough approximation of the target information or be unfamiliar with the mechanism of search engines. So, it's very important to analyze user queries before putting to search engine for result retrieval. The proposed approach suggests using multiple search engine based on the strength of user query. The search result return by multiple search engines needs to merge effectively to obtain accurate result.

2 Related Works

Many research studies are made on search queries on web search engines. One of the oldest research works [9] described the analysis of a large AltaVista query log and describes the correlation analysis of the log entries, studying the interaction of terms within queries. Beeferman D et.al [12] describes how the queries issued and the URLs clicked from the results can be viewed as a bipartite graph and how agglomerative clustering can be applied on the bipartite graph to discover clusters of similar queries and similar URLs. Baeza Yates et.al [5] describes a way to represent queries in a vector space based on a graph derived from the query click bipartite graph and shows how the representation can be used to infer some semantic relationships.

Measuring the semantic similarity[18][20] between two texts has been studied extensively in the information retrieval and natural language processing communities such as stemming [8][13], translation models [11] and query expansion [6][14]. However, there is no research has done on the assessing the strength of a query posed by the users for information retrieval.

Border A. [15] analyzed that user's need behind their queries to a search engine. The study reveal that based on the need user queries can be classified into three groups such as Navigational queries, Informational queries and Transactional queries. The analysis states that 48% are informational queries, 30% are transactional queries and 20% are navigational queries. So, it is very important to understand what kind of need behind a user posing a query. Our proposal computes the queries keywords weight with relate to user query logs history of a user, and based on the overall keywords weight the no of multiple searches needs to make are decided.

3 Framework for Multiple Searching

The proposed framework describes multiple searching shown in Figure-1. It consists of *Query processing, Keywords Weight, Multiple Searching* and *Results Ranking and merging* as major components. Each component combined to perform to improvise the web search.

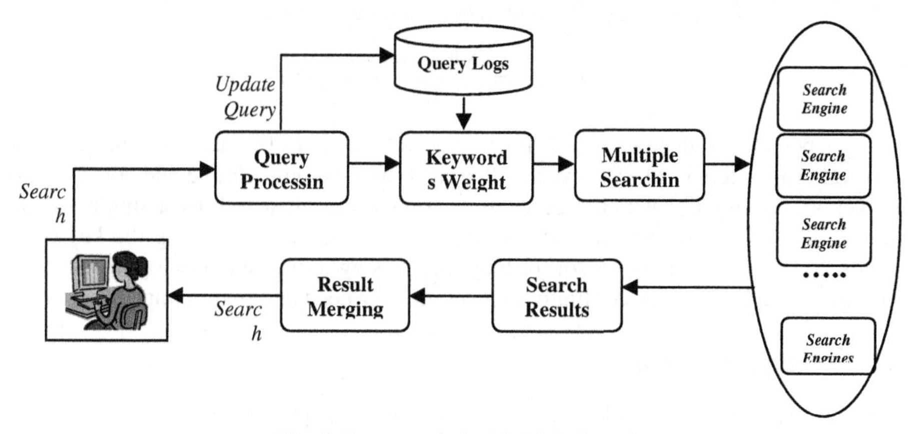

Fig. 1. Framework for Multiple Search

3.1 Query Processing

As described that user generally submit a query bases on need or approximation which generally consists of lot of determiners, preposition, adjectives and nouns. It is necessary filter out the key words from the user input query. Query processing is a text processing block which does the extraction of keywords and filtration of determiners and prepositions from user query phrase submitted for search. For example if a user submit a query as "*recent improvement in medical treatment for cancer*", the keywords extracted are as "recent", "improvement", "medical", "treatment", "cancer" and "in", "for" are filtered out by the query processor. Query processer maintains a filter library to filter the determiners and prepositions from the query. It also updates the user query to Query Log database.

3.2 Keywords Weight

The weight of keywords in a query explains the hardness of the query and intension of the user information need implicitly. Here hardness is used to make a decision for number of search need to make for efficient information retrieval. Keywords weight block of the framework computes the each keywords weight using user history query logs.

Suppose a query Q represents a vector of keywords, K as $(k_1, \dots, k_n)$ and L is a vector of query logs represent as $(l_1,.., l_n)$ for a user. If a keyword, k does not appear in L its weight will be zero. If a keyword is present in L, its weight are computed based on two factors as *Keyword frequency* (*kf*) and *Log frequency* (*lf*). The *kf* of a keyword based on the number of times it appears in the query logs and *lf* is the number of logs records related to keywords K. So, the keyword weight (*kw*) will be compute as

$$kw = \frac{\sum_{i=1}^{n} k \in l_i}{|L|} \quad (1)$$

Using the value of *kw* we will compute the hardness percentage of a query as Q_{hard},

$$Q_{hard} = \frac{\sum_{i=1}^{n}(kw_i)}{|K|} \times 100 \quad (2)$$

Q_{hard} percentage decides the number of search needs to be made for information retrieval. Silverstein et al. [9] revealed that most users submit simple and sort queries on web search queries, which receive most common result search by a single search engine. The proposed approach improvise this drawback by computing the keyword weight and query hardness for sort or long query using user past query log history and perform multiple search accordingly to meet the user need and search improvisation.

3.3 Multiple Searching

Multiple searching is a novel concept proposed in this paper based on the hardness of query. Current Meta Search engine are developed based on the multiple search engines but most the Meta search engines are focus on effective merging techniques of search engine results rather than the user information need which some time present very high irrelevant results. As described in section-1, that 85% of results are repeating for same query in different search engine, a novel approach is proposed to overcome the blind search on multiple search engines and a decision based search approach has made based on the query hardness. To take a decision we compute the query hardness using methods (1) and (2).

To support multiple searching as M, lets assume n search engines as $S = (S_i, \ldots .S_n)$ are integrated in the framework and each search engine has rank based on there popularity and coverage scope. For effectively selection of number of search engines for multiple searching we make four ranges of query hardness as shown in Table-1.

Table 1. Query Hardness Range and Multiple Searching

Hardness Range (Q_{hard})	Multiple Searching (M)
0 – 25%	*M=n/4*
>25 – 50%	*M=n/3*
>50 – 75%	*M=n/2*
>75 – 100%	*M=n/1*

Based on the M value number of searching is made and the selection of search engine is based on there rank score. If $M=1$, then highest ranked search engine are used for searching and if $M=2$, then top 2 ranked search engine are used for searching. So, based on the hardness range and multiple searching effective results are retrieved. As multiple search engines are used for searching in turn multiple results are obtained. The obtained results undergo a result merging process to present the results.

3.4 Search Results and Result Merging

Search results block collect the multiple search results returned by multiple search engines and the collected results processed for result merging by result merging block of the framework. Several result merging algorithms are proposed [10][17] but merging based on the local rank of results is simple and yields good performance. Similarity conversion based methods for converting ranks to similarities proposed by Lee.J [16]. We propose a modified similarity conversion methods for effective ranking and merging of the result to improvise the search result.

Let's assume for a query Q is submitted to n number of search engine which participating in multiple searching. Top 10 results of each search engine are collected as $R_n = (r_1, \ldots, r_{10})$. A unique set of results are collected from search engines through verifying with buffered search results. The verification eliminates the duplicate results and continues the search till top 10 non duplicates result are collected. To effectively merge the collected result we follow two algorithms. First, we calculate each result similarity rank and second, we find each result relevancy similarity to the query to find the final rank of the search result.

First Algorithm: To rank the unique results, a modified similarity conversion method derived for each result as *sim_rank (r)* based on the local result rank and search engine rank score value.

$$sim\_rank(r) = \left(1 - \frac{r_{rank-1}}{No.of\ collected\ Results}\right) + S_{score} \tag{3}$$

where, r is result, r_{rank} is the local rank of the result r and S_{score} is search engine rank score value. Search engine rank score is a constant value assigned depends on the popularity survey. The value of S_{score} is high as 1 for highly popular and low as 0 for lower popular search engine.

Second Algorithm: The ordered results might be correct in relate of rank conversion but in case of query relevancy it may differ. We calculate the relevancy based on the results content information. The search results of search engine contain rich information about the retrieved results [2], especially the title and snippets of search result can reflects the high relevancy to the corresponding document of the query. This approach utilizes the title and snippets of the results to find the relevancy similarity.

Let's assume, D_T and D_S are number of distinct query terms in title as T and snippets as S for a query q, and D_Q as distinct . QD_{len} is the total number of distinct terms in query. N_T and N_S are total number of query terms in the title and snippets. T_{len} and S_{len} are total number of terms in title and snippets. T_{Order} as query terms order in title or snippets, if terms order same as query then $QT_{Order}=1$ else will be 0. Then, to compute the relevancy similarity of title and snippets following method will be used,

$$sim\_relv(T,q) = \left(\frac{D_T + 1}{QD_{len}}\right) \times \left(QT_{Order} + \frac{N_T}{T_{len}}\right) \tag{4}$$

$$sim\_relv(S,q) = \left(\frac{D_S + 1}{QD_{len}}\right) \times \left(QT_{Order} + \frac{N_S}{S_{len}}\right) \tag{5}$$

To find the final similarity of the result and query we merged the value as follows,

$$sim\_relv(R,q) = \left(\frac{TD_{ts}}{QD_{len}}\right) \times (\, sim\_rel(T,q) + sim\_rel(S,q)) \tag{6}$$

Where, R is result and TD_{ts} is total number of distinct query terms.

$$result\_rank(R) = sim\_rank(r) + sim\_relv(R, q) \tag{7}$$

The computed value of $result\_rank(R)$ of a result will be used for final ranking. Higher the value, higher will be the rank of the result.

4 Experiment and Results

For experiment evaluation we developed the designed framework using Java Enterprise Technology, Tomcat Web Server and DAT File for storing user query logs having a three column data structure for user id, query and timestamp. Five search engines as Yahoo, Bing, Ask, AltaVista and Excite are integrated with framework for evaluation. To test the developed framework we submit multiple repetitive queries to the designed framework to build query log database of 3000 records. We assign rank score to the search engine as Yahoo=1, Bing=0.8, Ask=0.6, AltaVista=0.4 and Excite=0.2.

To find the effectiveness of the web search improvisation we find the results relevancy to the query submitted and rank of those results to measure the recall and precision in *with* and *without multiple search* integration. Recall is measured in the proportion of the relevant results that are retrieved to a query and precision is measured in the proportion of the retrieved results that are relevant as described below.

$$Recall\ Ratio = \frac{Relevant\ search\ Results}{Relevant\ results\ to\ query} \tag{8}$$

$$Precision\ Ratio = \frac{Relevant\ search\ Results}{Retrieved\ results} \tag{9}$$

To test the proposed framework we submit a query without multiple search approach selecting a particular search engine for retrieving results, here we select the search engine having highest search score and same query we repeated with multiple search approach also. The experiment repeated this test several times submitting various ambiguities query to evaluate the effectiveness and improvisation.

The experiment observe that all search engines repeats the same results on posing same query or in a little modified query, where as the proposed framework improvise this search on repetitive query or modified query as the hardness of queries increases with repeat search query which retrieve more relevant results filtering out the duplicate results in support of multiple search.

4.1 Input Query

Query is a collection of keywords pose by a user to search its needs. But it is very important to understand a input query to make a successful retrieval. To evaluate this work we tested with a sample query as "*chocolate*" as it not previous log in relevant to not found so hardness is between 0-25% which make to search in 1 search engine. The improvisation the query to "*chocolate gift*" computes the hardness of the query of between 25-50% as user previous log found in user log which makes to call 2 search engines for searching. Similarly, we repeated this with new query as "*chocolate gift online*" and "*chocolate gift online shopping*" etc., where different numbers of search engine are invoked which is depends on the hardness query computed. The obtained results in precision and recall in relevant to query is discussed below.

4.2 Results

To measure the improvisation we compute recall and precision on different queries in each run with observing the three required factors as *Retrieved results, Relevant results to Query* and *Relevant search results.* As we presents top 10 from the merged result to the user so *retrieved results factor* always remains 10 only. Other two factors are observed on each run and an analysis is shown Figure-2 and Figure-3.

Figure-2 and 3 describes the precision and recall comparison between with and without multiple search approach. The result shows that with multiple searches the precision and recall ratio is higher compare to without multiple searches. The high ratio of precision with multiple searches is due to more relevant result retrieved from the retrieved results and the high ratio of recall is due to more relevant results retrieval to the query. As recall measuring both factor are high for relevant result and relevant result to query results a higher ratio in comparison.

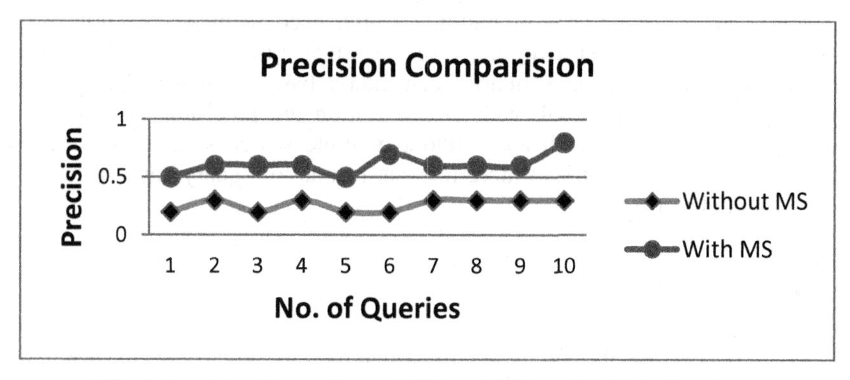

Fig. 2. Precision Comparison of With and Without Multiple Search

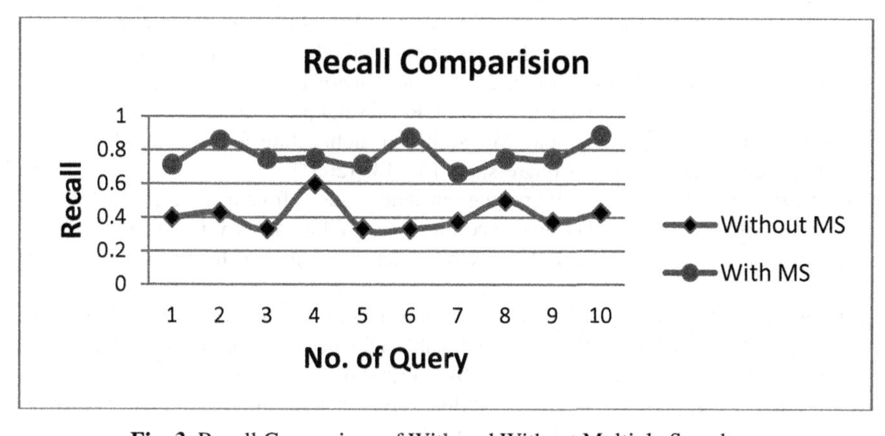

Fig. 3. Recall Comparison of With and Without Multiple Search

The low result ratio of without multiple searches is due to the same result repeating for the same query repeating multiple times or even for different or modified query. Same result retrieval minimizes the relevant results and relevant result to query which directly low down the ration of precision and recall.

So, high the precision and recall on retrieved results improvise the user needs and satisfaction. It might possible the retrieve result may not be relevant to query or the user need, but repetitively submitting same or similar query improvise it's search results with increasing the hardness of the same query and leads to multiple searches for presenting more relevant and non-duplicate results.

5 Conclusion

In this paper we propose a novel framework to improvise the web search. Most search engines retrieve same results on same or similar query which create dissatisfaction to the users and degrading web search. We improvise this drawback with proposing a novel multiple search and result merging approach. The proposed approach measures the hardness of the query in support of past users query log. Hardness of a query computed based on the keywords weight present in query. Based on the hardness a decision is made for number of multiple search need to make. Multiple searches results are processed to retrieve unique results which further processed for effective ranking and merging using similarity conversion based methods. The experiment results support our proposal with improvisation in precision and recall ratio comparison measures of with and without multiple searches. In future we like to extend our approach to retrieve more relevant results to a query and user using user behavior model construction.

Refrences

[1] Chen, H., Lin, M., Wei, Y.: Novel association measures using web search with double checking. In: Proc. of the COLING/ACL 2006 (2006)

[2] Sahami, M., Heilman, T.: A web-based kernel function for measuring the similarity of short text snippets. In: Proc. of 15th International World Wide Web Conference (2006)

[3] Jarvelin, K., Kekalainen, J.: IR Evaluation Methods for Retrieving Highly Relevant Documents. In: Proceedings of the 23rd Annual International ACM SIGIR Conference on Research and Development in Information Retrieval, pp. 41–48. ACM Press (2000)

[4] Risvik, K., Aasheim, Y., Lidal, M.: Multi-tier architecture for web search engines. In: First Latin American Web Congress, pp. 132–143 (2003)

[5] Baeza Yates, R., Tiberi, A.: Extracting semantic relations from query logs. In: Proceedings of the 13th ACM SIGKDD Conference, San Jose, California, USA, pp. 76–85 (2007)

[6] Lavrenko, V., Croft, W.B.: Relevance based language models. In: Proceedings of SIGIR 2001, pp. 120–127 (2001)

[7] Spink, A., Jansen, B.J., Blakely, C., Koshman, S.: Overlap Among Major Web Search engines. In: ITNG 2006 Third International Conference on Information Technology: New Generations 2006, April 10-12, pp. 370–374 (2006)

[8] Church, K., Hanks, P.: Word association norms, mutual information and lexicography. Computational Linguistics 16, 22–29 (1991)

[9] Silverstein, R., Helzinger, M., Marais, H., Moricz, M.: Analysis of a very large AltaVista query log. SRC Technical Note, 1998-014 (October 26, 1998)

[10] Shokouhi, M., Zobel, J.: Robust result merging using merging using sample-based score estimates. ACM Trans. Information System 27(3), 1–29 (2009)

[11] Berger, A., Lafferty, J.: Information retrieval as statistical translation. In: Proceedings of SIGIR 1999, pp. 222–229 (1999)

[12] Beeferman, D., Berger, A.: Agglomerative clustering of a search engine query log. In: Proceedings of the Sixth ACM SIGKDD International Conference, Boston, Massachusetts, United States, pp. 407–416 (2000)

[13] Krovetz, R.: Viewing morphology as an inference process. In: Proceedings of SIGIR 1993, pp. 191–202 (1993)

[14] Zhai, C., Lafferty, J.: Model-based feedback in the language modeling approach to information retrieval. In: Proceedings of CIKM 2001, pp. 403–410 (2001)

[15] Border, A.: Taxonomy of web search. ACM SIGIR Forum 36(2), 3–10 (2002)

[16] Lee, J.: Analyses of Multiple Evidence Combination. In: ACM SIGIR Conference on Research and Development in Information Retrieval, pp. 267–276 (1997)

[17] Craswell, N., Hawking, D., Thistlewaite, P.: Merging results from isolated search engines. In: Proceedings of the 10th Australasian Database Conference, pp. 189–200 (1999)

[18] Resnik, P.: Using information content to evaluate semantic similarity in taxonomy. In: Proc. of 14th International Joint Conference on Artificial Intelligence (1995)

[19] Vlez, B., Wiess, R., Sheldon, M., Gifford, D.: Fast and effective query refinement. In: Proc. of 20th Annual International ACM-SIGIR Conference on Research and Development in Information Retrieval (1997)

[20] Lin, D.: An information-theoretic definition of similarity. In: Proc. of the 15th ICML (1998)

System for the Detection and Reporting of Cardiac Event Using Embedded Systems

Maheswari Arumugam[1] and Arun Kumar Sangaiah[2]

[1] Sambhram Institute of Technology, Bangalore, India
maheswari.a@gmail.com
[2] School of Computing Science and Engineering, VIT University, Vellore, India
sarunkumar@vit.ac.in

Abstract. Premature death and disability from sudden cardiac arrest continue to be a serious public health burden. Electrocardiography (ECG) is a ubiquitous vital sign health monitoring method used in the healthcare systems. The early detection of abnormality in ECG signal for cardiac disease leads to timely diagnosis and prevents the person from death. Normally the surgeons have to study a large amount of ECG data to search for abnormal beat in ECG. If the surgeons fail to note the abnormal cycles that are very less in number, it leads to fatigue. In the proposed research work, an intelligent ECG digital system is designed to interpret these abnormal signals which reduce the tedious work of interpreting ECG. The proposed work monitors the heart beat continuously in a convenient manner and with maximum accuracy for diagnosing. The ECG signals are acquired in real time and amplified through an Instrumentation amplifier. The resulting signals are processed for the cancellation of noise through a low pass filter and notch filter. Further processing of these signals in the microcontroller detects the abnormalities of cardiac arrest. The result is communicated through GSM which reduces the burden of the doctors to a greater extent. These signals are stored in the Secure Data (SD) card to have a complete history of the signals before and after the occurrence of the cardiac event. The proposed research work combines the capabilities of real-time monitoring of ECG signals and the abnormal symptom-reporting systems.

Keywords: Real-time ECG, cardiac occurrence, GSM communication, analysis, detection, reporting.

1 Introduction

Problems in heart rhythm are called arrhythmias. An arrhythmia is an irregular heartbeat. If the path of conduction is damaged or blocked, the rhythm of heart changes. When the electrical impulses to the heart which forms heartbeats are not properly working, arrhythmias occur. In this condition the heart may beat too fast, slow or irregularly. This affects tremendously the ability of the heart to pump blood around the body. Depending on the nature of heart beat, four conditions of arrhythmia normally occur. If the heart beat is too fast, it leads to a condition called tachycardia.

S.C. Satapathy et al. (eds.), *Emerging ICT for Bridging the Future – Volume 1*,
Advances in Intelligent Systems and Computing 337, DOI: 10.1007/978-3-319-13728-5_66

If it is too slow, the occurrence is called bradycardia. The other two conditions occur due to too early changing of heart rhythm which is premature contraction and irregular heart rhythm known as fibrillation.

The main cause of death in the middle-to-old aged people, worldwide, is cardiovascular disease. A survey taken recently shows that children too are affected by this disease. The World Health Organization (WHO) estimates that nearly 16.7 million people around the globe face death every year due to cardiovascular disease. This can be controlled effectively and efficiently if diagnosis is done early. Regular monitoring of cardiac arrhythmias from continuous ECG is the normal procedure adopted by Heart Specialists to detect the arrhythmias. A statistical report recently released by the Cardiac arrest association of WHO in 2014, indicates the severity of this disease and the need for research. In the year 2013 it was recorded that around 359,400 out of hospital cardiac arrest incidences took place. Here the survival rate was just 9.5%. These diseases are generally difficult to detect by physicians without a continuous monitoring of the change in a person's vital signs. Therefore, close and continuous monitoring is needed, which can assist healthcare providers to identify whether a patient is in a healthy condition or not. New research direction in cardiac healthcare is the need of the hour to prevent such untimely deaths.

The traditional methods used for monitoring and diagnosing cardiac events are based on the detection of the presence of specific signal features by a human observer. Due to the present lifestyle conditions and the need for continuous observation of such conditions, an automated arrhythmia detection of attempt to solve this problem. Such techniques work by transforming the mostly qualitative diagnostic criteria into a more objective quantitative signal feature classification problem.

The standard procedure adopted in observing cardiac patients is the ECG recording for a long time ranging from 24 hours to 48 hours. The primary tool used to realize the severity of a myocardial infarction or heart attack is ECG. The survival of the patient, suffered from cardiac arrest, depends mainly on identifying the minuscule changes in ECG waveform in a quick manner accurately. The current research work focuses on design and development of a system dedicated for detection of abnormality and analysis of ECG data. These ECG data are then transmitted applying the advancements in wireless technology to overcome the distance gap between patient and doctor. The previous research results indicate clearly that by reducing the time between assessment and hospital admission reduces treatment time by one-third.

The current studies have shown that acquisition and analysis of ECG signal for diagnosis of abnormalities are still challenging tasks in spite of the technological advances. The improvements in the ECG system included in this research allow a heart specialist to diagnose and recognize the severity of a heart attack much earlier. It has been proved that the survival rate of cardiac patients is 37 minutes faster in the case ECG assessments done before their arrival to the hospital. This clearly indicates that it is important to reduce the delay in getting a heart infraction signal to a health care. If this delay is reduced to a large extent by using the technological advancements of today, the chances of survival rate of an individual can be increased. The focus of this research is to provide a resourceful way for individuals to have a continuous contact with healthcare professionals irrespective of their location and time. This research work is based mainly on the method used for optimized measurement and reporting the findings of the cardiac event.

2 Related Works

Traditionally Holter recording [1] has been used during normal patient activity to record ECG signals. A Holter is a small, mobile and light device that records, during a period of 24, 48 hours or 72 hours, ECG signals which are later analyzed in the hospital. Here the processing of ECG signals is done with special software at a later stage after completing the ECG recording. A diagnostic report that is generated later helps the cardiologists to do further analysis. But the HOLTER recording system indicates that it does not give accurate results as the critical cardiac abnormalities do not necessarily occur during these 72 hours. The serious drawback of the system is that if the patient suffers from a serious rhythm, the Holter only records it, but it does not react to it on time. This system does not detect the abnormalities of cardiac arrest when patients are out of the hospitals. Moreover, such a system cannot automatically transmit information at the moment when abnormal cardio activity is present.

In the year 2006, a small wireless ECG with communication through a blue tooth [2] to a Personal Digital Assistant was developed to monitor the cardiac signals. In this system an ECG sensor system of small size was created and connected wirelessly to handheld devices which graphically present the ECG signals. A small embedded ECG sensor system prototype was developed here. The major drawback of this system is that ECG analysis and interpretation are not done in real time continuously.

Many of the research works carried out previously [3] and [5] concentrates on the different methods adopted for the acquisition of ECG signals. All these works focuses on reducing the number of leads and also to gather complete information as in 12 lead ECG system.

A few research works [4] communicate the result to the needy person through wireless communications. This is a system based on a microcontroller with various sensors to continuously measure the heartbeat, body temperature of the patient. The same is displayed on the LCD continuously. In this system the information about patient's health is provided within every prescribed interval of time to the Doctor. This data is provided to doctor via Wireless GSM modem. In this work most of the heart parameters are measured and communicated in real time but analysis is not made.

In the implementation [6] of a wireless patient monitoring system the distance between the patient and the physician is not at all a restrictive parameter. The system also enhances their mobility. The system is a generalized system where the analysis of the cardiac abnormality is based on the comparison with the threshold value. In real time the threshold value changes for a child and an adult. The work becomes effective if it covers every aspect of cardiac event.

The system developed [7] processes the ECG using MATLAB to detect the abnormalities and provides the online information about the patient status such as the patient's heart beat rate, ECG and the patient history every 30 minutes. Within these 30 minutes if any abnormality in ECG occurs, the system does not have a backup plan to report the occurrence.

In the research work [8] ECG processors with on-sensors are designed for the heart monitoring in real time. The analysis of ECG signals is carried out in their developed system-on-chip (SoC) which gives a timely warning against the fatal vascular signs. The comparison of traditional ECG processing with on chip ECG processing along with deviations in the result is not highlighted in this work.

An algorithm for R peak detection and R-R interval [9] is implemented using filter and discrete wavelet transform. The information about the R peak obtained is used in the classification and analysis of arrhythmia detection. The R peak location and R-R interval provides approximate information about heart beats.

The prototype [10] developed is useful for measuring the heart rate and temperature of a person. This information is transmitted to the medical advisory for the preliminary precautions so that patient can be under control and also prevented from serious situation before reaching the hospital. This system does not include the transmission of abnormal signals which is essential for analysis by medical fraternity.

The model [12] is a general framework which is based on the functioning of a heart beat sensor. If the heart beat sensor fails the whole system of monitoring collapses.

The developed system [15] measures and transmits heart rate continuously to a central monitoring station. Here the cardiac arrhythmia monitoring system solely depends on heart rate measurement which gives only an approximate result.

The proposed research work combines all the features of acquisition of the ECG signals, processing of the signals and communicating the abnormalities to the doctor in the nearest hospital in real time with minimum time delay.

3 Proposed Methodology

The most prescribed diagnostic procedure in the medical world is the ECG (electrocardiogram). The ECG waveform helps the physicians to get the information about the electrical activity associated with the different aspects of a heart rhythm. This is useful in assessing an individual's cardiac rhythm and heart health. These cardiac electrical potentialities are routinely used to diagnose heart disease. Normally the pattern of beats analysis in ECG identifies irregular cardiac rhythms (arrhythmias). This information is then used to evaluate the effects of drugs and to monitor surgical procedures.

An ECG is a quasi-periodical signal which rhythmically repeats and is synchronized by the mechanical function of the heart. It consists of P, QRS, T and U wave of specific amplitudes and duration. The P wave gives a knowledge about the atria depolarization that spreads from the sinoatrial (SA) node throughout the atria. The QRS complex is the result of ventricle depolarization. The T wave is due to the ventricle repolarization. In an ECG, U wave may or may not be present. The time interval between the onset of atrial depolarization and the onset of ventricular depolarization is PR interval. The entire ventricle depolarization is called ST interval.

3.1 Biosignal Processing

The basic requirements for the analysis of the biomedical signals are acquisition and processing. The presence or lack of presence of P, R, T waves as well as the QT interval and PR interval are meaningful parameters in the screening and diagnosis of cardiovascular diseases. The different functions of the heart during a cardiac cycle are characterized fundamentally by 5 waves in the ECG. The amplitude of a typical ECG signal is 1 mV and a bandwidth of 0.05 Hz to 100 Hz. This signal is subjected to several sources of noise, bad contacts and movement of the electrodes, muscles contraction, and electromagnetic interference due to inductive coupling and capacitive coupling through the power line. This last source is the most dominant.

In the proposed research work, a framework as in fig.1 is designed to capture the ECG signals in real-time using an ECG acquisition system comprising of an instrumentation amplifier, LPF and a notch filter. The electrical signal derived from the electrode is in the order of 1mv peak-peak. In order to perform ECG signal processing for detecting the abnormalities, this signal needs amplification.

The acquired ECG signal is given to Instrumentation amplifier for amplification. The amplified signal is a mixture of raw ECG and amplified noise signals. Sometimes the noise can completely dominate the ECG and makes the amplified signal useless. Therefore an Instrumentation amplifier is used here. The sampled ECG signal contains some amount of line frequency noise. This noise can be removed by low pass filtering the signal. The amplified ECG signal is given to the MSP430 which is having internal on chip analog to digital converter. The output signal of the analog unit is then applied to the ADC of the Microcontroller unit (MCU), and it is sampling performed.

The Microcontroller unit includes primarily a MSP430 microcontroller which forms the hard core of the entire structure and a micro SD (Secure Digital) memory card for storing samples of the digitized ECG. The filtered digital signal can be sent as the output to the display unit by the DAC of MSP 430 or it can be transmitted to the PC using UART of the MSP430. The complete framework for ECG acquisition and diagnosis process is depicted in figure 1. In this work the ECG signals are stored in the SD cards for getting a complete history of the ECG signals at a later stage. The reporting of the detected abnormality is carried out by the communication unit to the doctor through GSM module.

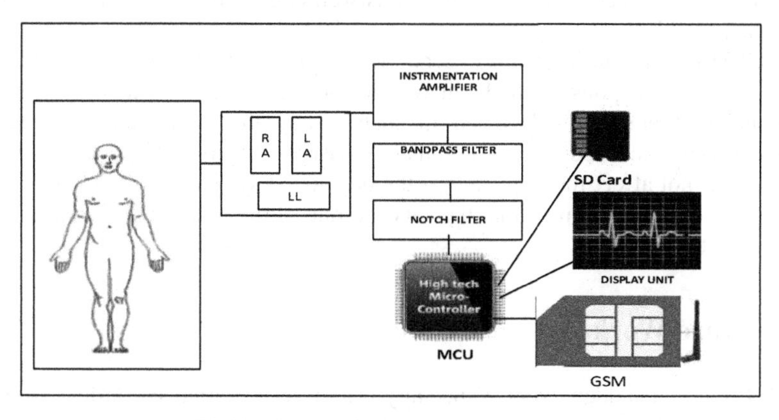

Fig. 1. ECG acquisition and diagnosis unit

Table 1. Normal Parameters of ECG

Type of wave	Amplitude in mv	Duration in seconds
P	< = 0.2	Interval PR is 0.12
R	1.6	0.07
T	0.1 – 0.5	Interval QT is 0.35
ST interval	-	0.015
U	< 0.1	Interval TU is 0.12

The analysis and detection of the abnormality in the ECG wave for various types of cardiac arrest are performed by the MCU (Microcontroller Unit). The built-in interface present in the microcontroller, MSP430 allows the serial communication with a PC or GSM. As stated in the medical documents the heart normal rhythm at rest is between 60 and 100 bpm (beats per minute). Beyond this range there is cardiac arrhythmia. The normal values for the waves of ECG and their intervals are indicated in table 1. The analysis unit checks these normal values during processing of each waveform.

3.2 Abnormality Report

The reduction in the time delay of communicating the processed result is a very important factor in increasing the survival rate of an individual. This is effectively done in the communication unit. This module gets the output from the ECG acquisition and diagnosing unit. This unit or module consists of a GPS network to locate the individual who is about to get cardiac arrest. This information is also carried over to the web server through the cloud.

The server network has a database server for recording the data from the acquisition and diagnosing unit. It also has an application server where an exe is written to analyze the signal if some failure occurs in the ECG processing unit. For this, the signals can directly be taken from SD card.

To improve further, the effectiveness of this system, the processed results can be sent to a call center network. This network is well equipped so that no time is lost in saving the individual.

The GSM module further helps to send the information to many doctors. This helps to get a quick response which increases the efficiency of this system. The communication module is designed in such a way that communication with all networks is possible. This system architecture ensures that a failure in any part of the network does not affect the objective of the developed system. A rescue team can be alerted through this system so that the survival rate of the individual is increased to a greater extent.

4 Future Work

Cardiac event is a sudden event. Especially if a driver experiences cardiac arrest, the passengers travelling in the vehicle is put to a lot of unknown troubles. Future research can focus on this issue and can develop a wheel mounted ECG that detects and records the abnormalities in ECG. This system can be made to work along with an intelligent robot which can stop the vehicle once the ECG device detects the abnormalities in ECG.

An Embedded system can also be designed and interfaced with the steering wheel system so that, if the driver shows an abnormality in his ECG an alert system can be triggered to inform about the driver's condition to the passengers and stop the vehicle by turning off the ignition with all safety precautions of the passengers.

Also an enhancement can be made in the proposed research work to trigger a demand mode pacemaker in case when heart loses its ability to generate pulses.

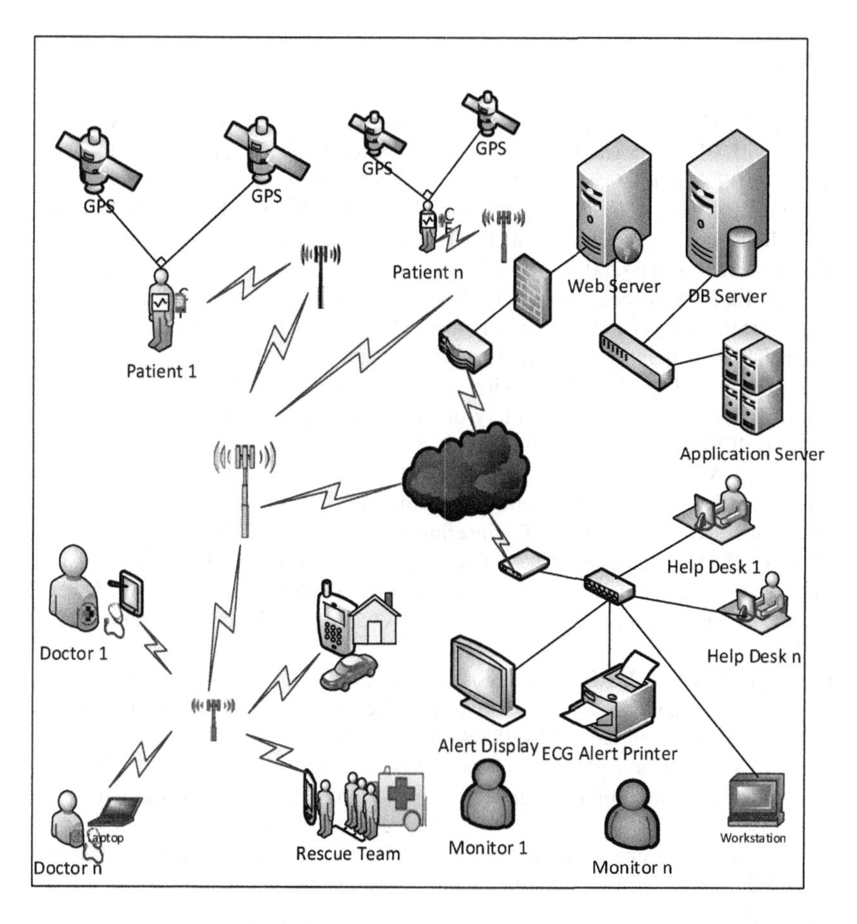

Fig. 2. Diagnosis and communication unit

All the features used in the proposed research work can be combined and put into a (System-on chip) SoC chip and made as a single device which can be described as the ultimate cardiac arrest diagnostic tool.

5 Conclusion

A unique embedded system that can be used for ECG storage along with a display unit and an arrhythmia monitor for individuals who are at the risk of heart problems is proposed. The various research findings of the previous works on monitoring of cardiac arrest are carefully studied. The proposed research work reduces to a larger extent the problems encountered in the research works on diagnosis of cardiac arrest. The proposed work enables online service for signals to be processed in real time and automatic alert of abnormality upon its detection. The processing procedures on onboard of the proposed system give immediate results even when there is outage in other parts of the framework. There is continuous monitoring of patient's cardiac status.

References

1. Lucani, D., Cataldo, G., Cruz, J., Villegas, G., Wong, S.: A portable ECG monitoring device with Bluetooth and Holter capabilities for telemedicine applications. In: Proc. 28th IEEE EMBS Ann. Int. Conf., New York City, USA (2006)
2. Romero Legameta, I., Addisson, P.S., Grubb, N., Robertson, C.E., Fox, K.A.A., Watson, J.N.: Real-time classification of ECGs on a PDA. IEEE Trans. Inform Technol. Biomed. 30, 565–568 (2003)
3. Cao, H., Li, H., Stocco, L., Leung, V.: Design and evaluation of a novel wireless three-pad ECG system for generating conventional 12-lead signals. In: Proceedings of the 5th International Conference on Body Area Networks (2010)
4. Akay, Y.M., Akay, M., Welkowitz, W., Semmlow, J.L., Kostis, J.B.: Noninvasive coustical detection of coronary artery disease: a comparative study of signal processing methods. IEEE Trans. Biomed. Eng. 40(6), 571–578 (1993)
5. Al-Omary, A., El-Medany, W., Al-Hakim, R.: Heart Disease Monitoring System Using Web and Smartphone. International Journal of Advanced Research in Electrical, Electronics and Instrumentation Engineering 3(4), 8265–8273 (2014)
6. Chen, H.-H., Chiang, C.-Y., Chen, T.-C., Liu, C.-S., Huang, Y.-J., Lu, S.S., Lin, C.-W., Chen, L.-G.: Analysis and Design of On-sensor ECG Processors for Realtime Detection of Cardiac Anomalies Including VF, VT, and PVC. J. Sign. Process. System 65, 275–285 (2011)
7. Dhir, J.S., Panag, N.K.: ECG Analysis and R Peak Detection Using Filters and Wavelet Transform. International Journal of Innovative Research in Computer and Communication Engineering 2(2) (February 2014) ISSN (Online): 2320-9801
8. Mamun, A., Alqahtani, M.: A microcontroller-based automatic heart rate counting system from fingertip. Journal of Theoretical and Applied Information Technology 62(3), 597–604 (2014)
9. Maheswari, A., Ramachandran, V.: System for Detection of Vital Signals with an Embedded System. African Journal of Information and Communication Technology 6(1), 22–30 (2011)
10. Rotariu, C., Manta, V., Ciobotariu, R.: Remote Cardiac Arrhythmia Monitoring System Using Wireless Sensor Networks. In: 11th International Conference on Development and Application Systems, Suceava, Romania, May 17-19, pp. 75–78 (2012)
11. Thakor, A., Kher, R.: Wearable ECG Recording and Monitoring System based on MSP430 Microcontroller. International Journal of Computer Science and Telecommunications 3(10), 40–44 (2012)

An Early Warning System to Prevent Human Elephant Conflict and Tracking of Elephant Using Seismic Sensors

D. Jerline Sheebha Anni[1] and Arun Kumar Sangaiah[2]

[1] VIT University, Vellore, Tamil Nadu, India
jerline_dhanaraj@yahoo.co.in
[2] School of Computing Science and Engineering, VIT University,
Vellore, Tamil Nadu, India
sarunkumar@vit.ac.in

Abstract. Human Elephant Conflict has been a major issue in the forest border areas, where the human habitat is troubled by the entry of wild elephants. This makes HEC a major real time environmental based on research problem. The aim of this paper is to reduce HEC, by identifying the nature of the elephants as proposed by many ecology professors and researchers. The conflict varies depending on the field and the habitation of human and elephant. Hence the objective is to take a survey of elephant tracking using different methodologies and to help both human and the elephant. This article completely focus on the field based on survey, caused by both human and elephant and the technical and Non-technical methodologies used for elephant tracking. This paper also has a proposed methodology using seismic sensors (Vibration) with high quality video cameras. These methodologies illustrate a crystal clear view of elephant path tracking. The outcome of the proposed methodology expects to produce an early warning system, which tries to save the life of both human and elephants.

Keywords: Human-Elephant Conflict, Ecology, Elephant tracking, Seismic sensors.

1 Introduction

The main objective of this paper is to reduce Human –Elephant Conflict. Human-elephant conflict affects two broad categories. One is problem faced by humans, another one is problem faced by elephants. The solution for these problems targets to save our life from elephants attack as well as to protect elephant from elephant attackers (elephant poaching). In general, there are two kind of loses, human death and elephant death. Thinking in a short while, the root cause of these losses is caused by both human and elephants. There is no compotator behind these events.

1.1 Problem Faced by Humans

HEC describes any situation where elephants cause problems for people. Elephants may damage people's crops, they may destroy people's houses and property, and

S.C. Satapathy et al. (eds.), *Emerging ICT for Bridging the Future – Volume 1*,
Advances in Intelligent Systems and Computing 337, DOI: 10.1007/978-3-319-13728-5_67

they may even endanger people's lives. Human settlement areas were once used by elephants and this becomes the only reason for the conflict of elephants with human [6]. The other reason is that, human settlement areas targets around the sources of permanent water and this makes the elephant to enter into human habitats to drink, especially during dry season.

1.2 Problem Faced by Elephant

Senseless mining in the hills and forest has resulted in elephants entering human habitations in search of food and water. The drying up of water resource in the summer is also to be blamed for it. Due to squeezing of forest area, wild animals are losing their habitat. The animals are facing food and water scarcity due to massive deforestation [9],[16]. Hence elephants often come out of forest in search of food and water, which are always available in human habitats.

The only solution to get rid of HEC is intelligent elephant tracking system. Tracking is the science of observing animal paths and signs. The aspiration of tracking is to gain clear knowledge about the animal (being tracked) and it also depends on the environmental surroundings. This makes elephant tracking mandatory to know the status of HEC. Because of HEC more number of deaths occurred in last decades. Human paths and their habitation are well known by everyone. But it is difficult to trace the elephant habitation. Hence elephant tracking is must to minimize HEC. Elephant tracking involves technical and non-technical methodologies which were discussed in the next section.

2 Literature Review

In India many measures have been practiced to alleviate HEC. Few practices had given short-term solutions, but not last for long-term. And one more problem with some practices is related to the cost. Methods to avoid elephant's entry are categorized into two, namely non-technical and technical.

2.1 Non-Technical Methods

Many of the Non-Technical methods were utilized and still utilizing the same. These methods are followed by the farmers for protesting their crops and to save their life from elephant attacks. The following are the list of various non-technical methods.

Method Name: Crop Guarding

Through this method people are constructing the huts on fields and on trees to get a clear view of elephant arrival, according to the perspective of the view, the people may gathered and avoid the elephant arrival [4].

Method Name: Noise and Throw

Group of people gathered making noise like shouting in a huge voice and throwing objects. These kind of activates shows that humans aggressiveness. According to this method indicates to elephant that their presence is detected. So it will not come to that particular area [10].

Method Name: Fire

Lighting fires is one of the traditional method from our ancient periods for guarding crops against elephants and other wild animals [1],[3].

Method Name: Alarm

Assigning the alarms in the crop fields, they rang the sound in a sequence time gaps at night times. This will help to avoid the elephant entry.

Method Name: Teams, Kumkies

To protect crop fields from elephant through two ways according to this methodology, one is to make a team, because of single human will not have that much courage to swipe/remove the elephant from the field. So the teamwork will pay a big role to remove the elephant from the crop field [7].

Second one is kumkies, Kumkies are one kind of elephant, and its nature is practiced by humans to fight with aggressive elephants.

Method Name: Dung Balls

Dung balls counts are useful to get an approximate estimate of the number of elephants residing in that area [2].

2.2 Technical Methods

The following are the list of various technical methodologies proposed by different researchers for elephant tracking. These are the popular methodologies used worldwide, but each had its own merits and demerits. The demerits made these methodologies to provide a short term solution for elephant tracking.

Method name: Amplified Monitoring Devices (Video/Standalone Cameras)

These monitoring devices are used even now for visualizing the movements of wild animals lively by using video and standalone cameras [15].

Method Name: Information Systems (GPS/GSM, Collars)

Information systems helps to observe the movements of elephants in a meticulous location through radio collars attached on elephants [11],[20].

Method Name: Wireless Sensors

Wireless sensors are used to collect information about the wild animals. These sensors can be directly attached to the animal or environment. Seismic sensors are widely used in the following areas in various articles namely, environmental monitoring, animal tracking, vehicle tracking, weather monitoring, medical care and for other seismic detections [17]. Seismic sensors are also used for collecting census and to monitor the animal population in remote areas [19].

3 Research Gap

A number of approaches have been used to address HEC including destroying problem elephants and translocation of elephants. Alongside these centralized interventions, there are also traditional (Non-Technical) methods that farmers use to defend their crops from elephants such as lighting fire, making loud noises and throwing various kinds of missiles. The construction of various barriers like electric fences and radio collars were sophisticated due to the expense and frequent failure.

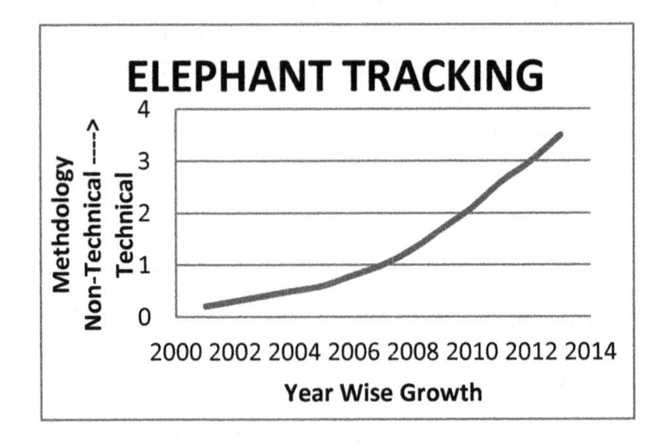

Fig. 1. Research Gap of Elephant Tracking

The graph shown in Figure 1 represents the year wise growth of elephant tracking as per various publications [3],[5],[8],[12],[13],[14],[18]. The X-axis points to the year of occurrence and the Y-axis points to the technical and non-technical methodologies which were briefly explained in the previous section. According to the graph technical methodologies were widely used in recent years. Because technical aspect always considered to be harmless. The proposed methodology targets to fill this research gap because it tries to provide a long term solution for HEC.

4 Objectives and Contribution of Proposed Work

The aim of this proposed research is to reduce HEC and the objective is to help both human and elephants without endangering their lives. This target can be achieved by the proposed methodology using wireless sensors (seismic) with high quality cameras. The outcome of the proposed methodology expects to produce an early warning system, which tries to save the life of both human and elephants.

5 Proposed Methodology

The proposed methodology for elephant tracking focuses on wireless sensors with high quality cameras. Wireless sensors in specific to be used for this proposed work is seismic sensors. Seismic sensors are one type of vibration sensors. Seismos is a Greek word meaning shaking. These sensors are sensitive to up-down motions of the earth and detects the following,

- Measures motion of the ground including earthquakes, tsunami and volcanic eruptions etc.
- Detects vibrations caused by the movement of animals or human beings and Monitors protected areas to restrict entry of unwanted persons or animals.

Seismic sensor in our project is used to detect the vibration caused by elephants. The result of the seismic sensor will be displayed as a graph (seismogram). Seismogram includes various sized waves for different animals. Among them elephant, the mega herbivore will be easily detected because of its large size and weight (a lengthy wave is displayed). If it is detected, then it alerts the forest authorities and the public using an alarm. The authority should take action for chasing the elephant and to bring them back to the forest. The public after hearing the alarm, it is advised to be in a safe place.

Here, one of the main criteria is to think about the location of Seismic sensor. It should be located at the boundaries of forest and human settlements (few kilometers inside the forest) and to place them at the boundaries of water body. Seismic sensors are buried inside the land, so it is not harmed by animals or humans.

5.1 Architectural Diagram with Work

The architectural diagram shown in Figure 2 of the proposed methodology includes 3 stages,

Stage 1 represents the problem faced by HEC.
Stage 2 represents the solution (Elephant Tracking) for reducing HEC.
Stage 3 represents the outcome of elephant tracking.

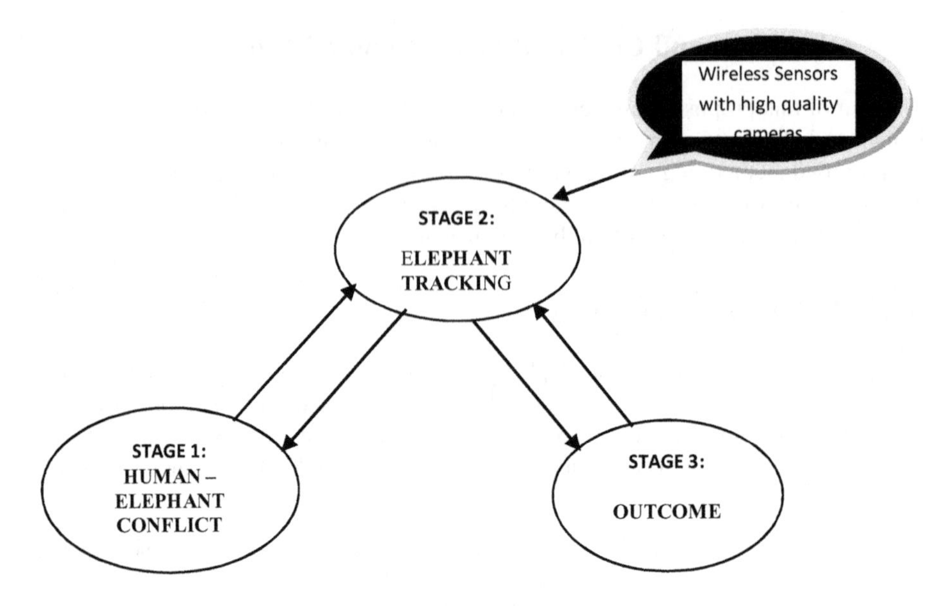

Fig. 2. Architectural diagram

Stage 1: Human-Elephant Conflict

Asian countries like India, attains a massive raise in human population. The survival of this population is only based on agricultural and industry. The growth in these two fields converts the forest areas into human habitats. This leads the wild animals to face sensitive lack of resources (food and water). In search of their resources animals are forced to enter into human settlements. Among various wild animals elephant (Elephas maximus) the mega herbivore plays a major role. India experiences major loss annually by elephants. This leads to the increase of Human-Elephant Conflict (HEC).

Stage 2: Elephant Tracking

The only solution to get rid of HEC is intellectual elephant supervision and tracking system. The target of this tracking system is to help both human and elephants. Many researchers and scientist had been published and implemented different methodologies and their real time solutions for elephant tracking.

Stage 3: Outcome of Tracking System

The outcome of the tracking system aims to help both human and elephants by,

- Providing prior information to the forest authorities.
- Providing alert or warning to the people.
- The outcome targets to decrease the HEC without endangering life of human and elephant.

5.2 Expected Outcome

The proposed research on the development of seismic sensors for elephant tracking expected to provide,

- Early warning system for HEC.
- Hold up for long term HEC administration.
- Avoids false alarm rates and reduce negative HEC.

6 Future Work

This survey was used to identify the basic idea behind Human Elephant Conflict and Elephant Tracking. As well as this research is not end with this survey. In future, there is a plan to track the elephant through the vibration sensors. Hopefully, this survey will help to redirect the path in a proper manner. The main role of this survey is to find long term solution, low cost tracking system and should be harmless to both human and elephants.

7 Conclusion

This survey of study will help to evaluate different technical and non-technical methodologies used for elephant tracking and summarizes the research gap between non-technical and technical methodologies. A research made for detecting the root cause for the conflict between human and elephants. Forced and suggested a proposed methodology to provide harmless long term solution for both human and elephants.

References

1. Archie, E.A., Chiyo, P.I.: Elephant behavior and conservation: social relationships, the effects of poaching, and genetic tools for management. Molecular Ecology 21, 765–778 (2012)
2. Arivazhagan, C., Ramakrishnan, B.: Conservation perspective of Asian Elephants (Elephasmaximus) in Tamil Nadu, Southern India. International Journal of Bio-Technology (Special Issue), 15–22 (2010)
3. Census of India, `http://www.censusindia.net`
4. Chiyo, P.I., Moss, C.J., Archie, E.A., Albert, S.: Using molecular and observational techniques to estimate the number and raiding patterns of crop-raiding elephants. Journal of Applied Ecology 48, 788–796 (2010)
5. Choudhury, A.U.: Human-Elephant Conflicts in Northeast India. Human Dimensions of Wildlife 9, 261–270 (2004)
6. Edirisinghe, R., Dias, D., Chandrasekara, R., Wijesinghe, L., Siriwardena, P., Sampath, P.K.: WI-ALERT: A Wireless Sensor Network Based Intrusion Alert Prototype For HEC. International Journal of Distributed and Parallel Systems (IJDPS) 4, 14–19 (2012)
7. Fernando, P., Janaka, H.K., Prasad, T., Pastorini, J.: Identifying Elephant Movement Patterns by Direct Observation. Gajah 33, 41–46 (2010)

8. Gureja, N., Menon, V., Sarkar, P., Kyarong, S.S.: Ganesha to Bin Laden: Human-elephant conflict inSonitpur District of Assam, pp. 45–51. Wildlife Trust of India, New Delhi (2002)
9. Hobson, K.A., Johnson, R.B., Cerling, T.: Using Isoscapes to Track Animal Migration, pp. 120–129. Springer Science (2010)
10. Manakadan, R., Swaminathan, S., Daniel, J.C., Desai, A.: A Case History of Colonization in the Asian Elephant: Koundinya Wildlife Sanctuary (Andhra Pradesh, India). Gajah 33 01, 17–25 (2010)
11. Mayilvaganan, M., Devaki, M.: Elephant Localization And Analysis of Signal Direction Receiving in Base Station Using Acoustic Sensor Network. International Journal of Innovative Research in Computer and Communication Engineering 2(2) (2014)
12. Nyhus, P.J., Tilson, R., Sumianto, P.: Crop-raiding elephants and conservation implications at Way Kambas National Park, Sumatra, Indonesia. Oryx 34, 262–274 (2000)
13. Ogra, M., Badola, R.: Compensating Human-Wildlife Conflict in Protected Area Communities: Ground-Level Perspectives from Uttarakhand, India. Human Ecology 36, 717–729 (2008)
14. Ogra, M.V.: Human-wildlife conflict and gender in protected area borderlands: A case study of costs, perceptions, and vulnerabilities from Uttarakhand (Uttaranchal), India. Geoforum 39, 1408–1422 (2008)
15. Olivier, P.I., Ferreira, S.M., Aarde, R.J.: Dung survey bias and elephant population estimates in southern Mozambique. African Journal of Ecology 47, 202–213 (2010)
16. Sanchez, G., Sanchez, F., Losilla, F., Kulakowski, P., Palomares, F.: Wireless Sensor Network Deployment for Monitoring Wildlife Passages. Sensors, 72–78 (2010)
17. Santhosh Kumar, D.: Design and Deployment of Wireless Sensor Networks. Master's thesis, Department of Computer Science and engineering, IIT Bombay, 1–33 (2009)
18. Sukumar, R.: A brief review of the status, distribution and biology of wild Asian elephants Elephasmaximus. International Zoo Yearbook 40, 1–8 (2006)
19. Wood, J.D., O'Connell-Rodwell, C.E., Klemperer, S.: Using seismic sensors to detect elephants and other large mammals: a potential census technique. J. Appl. Ecol. 42, 587–594 (2005)
20. Zeppelzauer, M.: Automated detection of elephants in wildlife Video. Journal on Image and Video Processing, 24–31 (2013)

Impact of ICT Infrastructure Capability on E-Governance Performance: Proposing an Analytical Framework

Deepak Dahiya and Saji K. Mathew

Department of Management Studies, Indian Institute of Technology Madras, Chennai, India
deepak.dahiya@smail.iitm.ac.in, saji@iitm.ac.in

Abstract. The impact of IT infrastructure capability on the performance of a firm has been addressed quite sufficiently in the past literature but how these factors when coupled together to study the impact on E-Governance performance is yet to garner the main focus of attention. Effective E-Government readiness is characterised by efficient deployment of ICT infrastructure capability and the same has been recognized worldwide in the various studies on E-Government. A two phased study was proposed to meet the objectives defined for this research. This paper describes the work completed in the first phase. In the first phase of the study, an exploratory study was conducted to identify, articulate and contextualize variables extracted from literature survey and field survey. Finally, based on the exploratory study, the author's work lays the groundwork for the development of a conceptual framework along with suggested methodology for understanding how ICT infrastructure capability affects E-Governance performance and direction for future work.

Keywords: Infrastructure, Service, E-Governance, Interoperability, Cloud and Framework.

1 Introduction

Several studies have examined how information technology investments contribute to firm performance in the business context [10][15][24]. Using resource based view [16] it has been suggested that firms should focus on creating enterprise-wide ICT capability rather than merely invest in IT. This enterprise-wide ICT capability to achieve sustainable competitive advantage is created by integrating three aspects of a firm's ICT capability: ICT infrastructure, human IT skills and IT enabled intangibles. ICT infrastructure refers to the physical IT assets consisting of the computer and communication technologies and the shareable technical platforms and databases[10][17]. Although the individual components of the IT architecture have become commodity like, the architecture that provides interoperability and provides the flexibility to respond to the changing needs of an organization is of strategic significance to the organization.

S.C. Satapathy et al. (eds.), *Emerging ICT for Bridging the Future – Volume 1*,
Advances in Intelligent Systems and Computing 337, DOI: 10.1007/978-3-319-13728-5_68

Key elements of ICT infrastructure capability in e-business have been identified in prior studies as reliability, inter-operability and flexibility [1][17-18]. The impact of IT infrastructure capability on the performance of a firm has been addressed quite sufficiently in the past literature but how these factors when coupled together to study the impact on E-Governance performance is yet to garner the main focus of attention. Effective E-Government readiness is characterised by efficient deployment of ICT infrastructure capability and the same has been recognized worldwide in the various studies on E-Government[9]. Emergence of infrastructure as a service deployment model has opened an economical service delivery model for business and government [12]. However it is not clear as yet how this new form of IT services delivery in infrastructure would fulfil the requirements of ICT infrastructure capability and in turn provide better governance.

This work seeks to study how ICT infrastructure capability dimensions impact E-Governance performance under the influence of service delivery models and a control group. Section 2 provides a detailed literature review on the factors under consideration and provides the necessary motivation and the necessity for this research, section 3 highlights the methodology by identifying dimensions and then actually proposing the integrated conceptual framework. Sections 4 and 5 summarizes the conclusion and the future direction of work. Finally, section 6 lists the references.

2 Theoretical Background

E-Governance performance has been the focal point of several previous studies. Westerback [4] analyzed 18 cases from different federal bodies in the United States including both the private and the public sector to understand the applicability of the strategic information management (SIM) best practices to measure performance in terms of improved service delivery and positive return on government technology investments. Key practices identified by the research include, using performance measures as a proxy for return on investment; taking advantage of leading edge technology; making use of proven project management techniques and identification of "current" best practices.

2.1 Frameworks for E-Governance Performance Study

Some scholars have developed frameworks to measure E-Government success. Zakarya [3] provide an integrated architecture framework (IAF) for E-Government and presents a critical analysis of barriers experienced in public sector organizations. Combining the core elements of Strategy, Technology, Organization and Environment, the STOPE model [2] has been proposed for the development and assessment of E-Government in the digital age. Esteeves et al. [5] propose a three dimensional framework for the assessment of E-Government initiatives. The components of E-Government Assessment Framework (EAM) include constructs from both a social and technical perspective. The assessment dimensions of EAM are based on the STOPE model with two additional assessment dimensions outside of the STOPE framework: operational and services.

The E-Governance Division of Ministry of Communications and Information Technology (MoCIT), Government of India [20] felt it necessary to create a rational framework for assessing e-Government projects on various dimensions and thus the E-Governance Assessment Frameworks (EAF Version 2.0) took shape [8]. The objective of the framework was to justify significant investments in E-Government projects, subjective assessment and value judgement, facilitate funding agencies to take a rational view and channelize future efforts in the right direction.

2.2 ICT Infrastructure Capability

A firm's ICT infrastructure capability has been generally understood as its integrated set of reliable ICT infrastructure services available to support an organization's existing and future initiatives [17]. Following the resource based view of the firm, IT infrastructure is one resource that could provide sustainable competitive advantage to the firm [10][15]. First, IT infrastructure as a valuable resource must be reliable to ensure IT services to be delivered under warranty conditions defined by service level agreements [17-18]. Reliability is assured by redundancy in the provisioning of IT infrastructure. Second, flexibility is a key attribute of an IT infrastructure to adapt fast to changes that might be required in future scenarios. Byrd and Turner [1] based on a quantitative study covering major fortune 1000 firms showed that there is a positive association between flexibility of IT infrastructure and competitive advantage of the firm. Third, interoperability is an essential attribute of IT infrastructure to be valuable to an organization. Proprietary technologies without openness to operate with other systems could adversely affect systems integration and ability to expand [21].

To summarize, most studies on the effect of ICT infrastructure capability have been conducted in business setting where an organization invests in IT with profit as motivation. However E-Government investment outcomes cannot be measured by profits and hence requires a separate investigation to understand how various factors lead to desired E-Governance performance. Further, E-Governance performance studies have been characterized by frameworks that are intuitive or arbitrary and do not report validation results [5]. Further, E-Governance performance and its determinants have been scarcely addressed in academic literature with adequate theoretical approach.

The research work builds on the existing literature on ICT infrastructure capability that will extend the work on improving performance in case of E-Governance.

In this study we address some of these gaps by developing a conceptual model to study E-Governance performance and empirically test how IT infrastructure capability influences E-Government services.

3 Methodology

A two phased study was proposed to meet the objectives defined for this research. In the first phase of the study, an exploratory study was conducted to identify, articulate and contextualize variables extracted from literature survey and field survey. Based on the exploratory study, the authors developed a conceptual framework for understanding how ICT infrastructure capability affects E-Governance performance.

This paper describes the work completed in the first phase. Subsequently propositions and specific hypotheses shall be formulated.

3.1 Identifying Dimensions of ICT Infrastructure Capability and E-Governance Performance

The study explores the different dimensions of ICT Infrastructure capability that are important in the choice of technology in E-Governance. In addition the service delivery model and the control group are identified that would play the role of moderating and control variables in the proposed analytical framework. The extensive literature review carried out for identifying the attributes of ICT Infrastructure capability and for studying the various E-Governance frameworks for better performance leads us to table 1 that depicts the various dimensions identified for our research.

Table 1. Identifying Dimensions of ICT Infrastructure Capability and E-Governance Performance

Key Dimension	
ICT Resources	ICT Asset
	ICT Capability
Service Delivery model	
E-Governance Performance (Impact)	
Controls	

3.2 Proposed Analytical Framework

Based on the exploratory study, the authors developed a conceptual framework for understanding how ICT infrastructure capability affects E-Governance performance. Fig. 1 depicts the proposed analytical framework for our research.

3.2.1 ICT Infrastructure Capability: Reliability, Flexibility, Scalability and Innovation

First, IT infrastructure as a valuable resource must be reliable to ensure IT services to be delivered under warranty conditions defined by service level agreements [17 – 18]. Second, flexibility is a key attribute of an IT infrastructure to adapt fast to changes that might be required in future scenarios [1]. Thirdly, the scalability factor implies that the system can withstand elastic load subject to increase in user base with addition of hardware and software and also by managing new design changes and

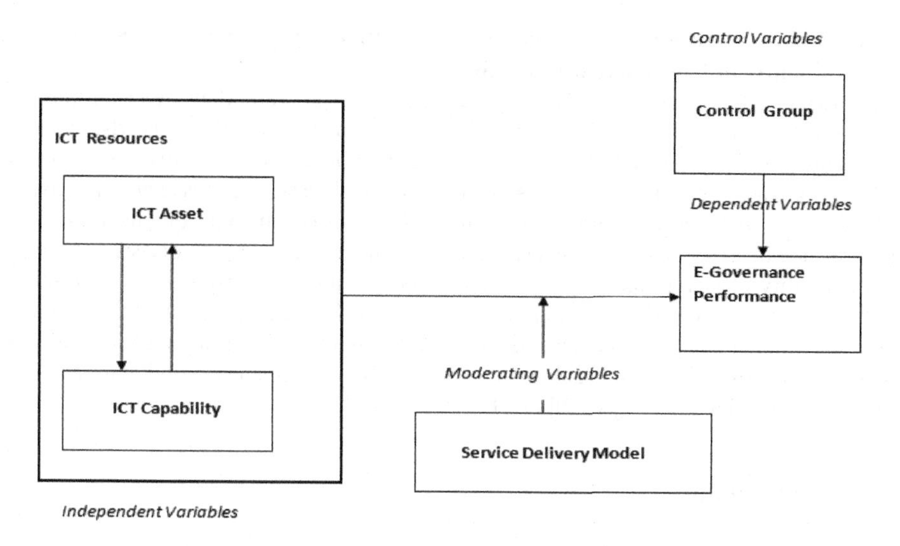

Fig. 1. Proposed Analytical Framework

allowing infrastructure portability [3][8]. Finally, innovation i.e. transformation is Citizen (external user) related, Information Related (Internal user), People related (Internal user), Process related (Internal user) and technology related which concerns with kinds of citizen services are improved [22 – 23]. Proprietary technologies without openness to operate with other systems could adversely affect systems integration and ability to expand thereby being non-interoperable [21].

3.2.2 Services Delivery Models: Cloud Deployment and SOA Deployment

The service delivery model comprises of the service oriented architecture deployment and cloud deployment.

The service oriented architecture deployment provides attributes at both horizontal and vertical interoperability levels. The cloud deployment model includes attributes like resource elasticity implying Platform as a Service (PaaS), infrastructure deployment implying Infrastructure as a Service (IaaS), application deployment implying Software as a Service (SaaS) along with transference of risk for quality of service and technology trend to support third party hardware and software [12 – 14].

3.2.3 Control Group: Stakeholders, IT Intangibles, Organization and Environment

Attributes include stakeholders implying decision makers and technical staff, politicians and bureaucrats, IT intangibles implies consolidation of past and current data, improved citizen service, increased citizen participation, better coordination between system users and suppliers in IT system evaluation and shared IT practices for system development. Organization incorporates size, structure, culture, data handling mechanism and government policy relevant common standards and policies. Finally, Environment includes influencing factors like IT literacy, sustainability and Inter-departmental communication [2][9][10-11][25].

3.2.4 E-Governance Performance: Return on Investment, Quality of Service, Quality of Governance and Value

The attributes that impact E-Governance Performance were identified as return on investment (RoI) that includes factors like quick response to citizens needs, increase population coverage, speed of service delivery, time bound service, citizen role clarity and participation, providing services at rural base with ease of access and 24x7 availability with problem resolution mechanism. The second attribute quality of service (QoS) includes secure and cheaper services with mechanism for disaster recovery while quality of governance (QoG) includes process audits, transparency and replicability using data warehousing and data mining. Lastly, value as an attribute includes activities like create new businesses, attract investments, generate employment, risk assessment, cost reduction, achieving efficiency, effectiveness and within time citizen service delivery [2][4 - 9][11][13][25].

4 Conclusion

This paper broadly highlights the work done by researchers to study the relationship between ICT infrastructure capability and its impact on E-Governance performance. In addition the dimensions of service delivery model and the control group are identified that also influence the E-Governance performance directly or indirectly in the proposed analytical framework. A two phased study was proposed to meet the objectives defined for this research. This paper describes the work completed in the first phase. In the first phase of the study, an exploratory study was conducted to identify, articulate and contextualize variables extracted from literature survey and field survey. Finally, based on the exploratory study, the author's work lays the groundwork for the development of a conceptual framework along with suggested methodology for understanding how ICT infrastructure capability affects E-Governance performance and direction for future work.

5 Future Direction of Work

With phase 1 of the exploratory study complete, the future activities of the research work include formulating hypotheses, proceeding to quantitative study phase that will include developing and administering a survey questionnaire, validation of the instrument and testing of hypotheses. Lastly, the data analysis phase will include inferences from the study and implications for policy makers.

References

1. Byrd, A.T., Turner, D.E.: Measuring the Flexibility of Information Technology Infrastructure: Exploratory Analysis of a Construct. Journal of Management Information Systems 17(1), 167–208 (2000)
2. Bakry, S.H.: Development of E-Government: A STOPE view. Wiley's International Journal of Network Management 14(5), 339–350 (2004)

3. Zakareya, E., Zahir, I.: E-Government adoption: architecture and barriers, Business Process Management Journal. Business Process Management Journal 11(5), 589–611 (2005)
4. Westerback, L.K.: Towards Best Practices for Strategic Information Technology Management. Elsevier Government Information Quarterly 17(1), 27–41 (2000)
5. Esteves, J.R.C.: A comprehensive framework for the assessment of E-Government projects. Paper published in the Elsevier International Journal on Government Information Quarterly 25, 118–132 (2008)
6. Bhatnagar, S., Rama Rao, T.P., Singh, N., Vaidya, R., Mandal, M.: Impact Assessment Study of E-Government Projects in India. Report prepared by Centre for e-Governance, IIM Ahmedabad for Department of Information Technology, Government of India, New Delhi, pp. 1–98 (2007)
7. National eGovernance Plan: Impact Assessment of e-Governance Projects, Department of Information Technology, Ministry of Communications and Information Technology, Government of India & IIM Ahmedabad, pp. 1–108 (2008)
8. Rama Rao, T.P., Venkata Rao, V., Bhatnagar, S.C., Satyanarayana, J.: E-Governance Assessment Frameworks, E-Governance (Assessment & Replication) Division, E-Governance and E-Rural Group, Department of Information Technology, Government of India, http://unpan1.un.org/intradoc/groups/public/documents/APCITY/UNPAN023017.pdf (last accessed March 24, 2013)
9. UNDP, E-Governance & E-Government, http://www.apdip.net/projects/E-Government (last accessed February 06, 2014)
10. Bharadwaj, S.A.: A Resource-Based Perspective on Information Technology Capability and Firm Performance: An Empirical Investigation. MIS Quarterly 24(1), 169–196 (2000)
11. Aral, S., Weill, P.: IT Assets, Organizational Capabilities, and Firm Performance: How Resource Allocations and Organizational Differences Explain Performance Variation. informs Organization Science 18(5), 763–780 (2007)
12. Armbrust, M., Fox, A., Griffith, R., Joseph, D.A., Katz, H.R., Konwinski, A., Lee, G., Patterson, A.D., Rabkin, A., Stoica, I., Zaharia, M.: Above the Clouds: A Berkley View of Cloud Computing. Technical Report, University of California at Berkley, http://www.eecs.berkeley.edu/Pubs/TechRpts/2009/EECS-2009-28.pdf (last accessed on May 13, 2014)
13. ReddyRaja, A., Varma, V.: Cloud and E-Governance, imaginea white paper, https://www.socialbyway.com/wp-egovcloud (last accessed on May 13, 2014)
14. DEIT, Government of India's GI Cloud (Meghraj) Strategic Direction Paper, Ministry of Communications and IT, http://deity.gov.in/content/gi-cloud-initiative-meghraj (last accessed on May 13, 2014)
15. Mata, J.F., Fuerst, W., Barney, J.: Information Technology and Sustained Competitive Advantage: A Resource-Based Analysis. MIS Quarterly 19(4), 487–505 (1995)
16. Barney, J.B.: Firm Resources and Sustained Competitive Advantage. Journal of Management 17, 99–120 (1991)
17. Weill, P., Vitale, M.: What IT Infrastructure Capabilities are Needed to Implement E-Business Models? MIS Quarterly Executive 1(1), 17–34 (2002)
18. Austin, R.D., Applegate, L.M., Deborah, S.: Corporate Information Strategy and Management: Text and Cases, 8th edn. McGraw-Hill (2008)
19. National eGovernance Plan, Impact Assessment of e-Governance Projects, Department of Information Technology, Ministry of Communications and Information Technology, Government of India & IIM Ahmedabad, pp. 1–108 (2008)

20. Ministry of Communications and Information Technology, Governmant of India web site, http://www.mit.gov.in/content/rd-e-governance (last accessed on January 11, 2013)
21. Shapiro, C., Varian, H.: Information Rules: A Strategic Guide to the Network Economy, 1st edn., 368 pages. Harvard Business Review Press (1998) ISBN-10: 087584863X
22. Satyanarayana, J.: Managing Transformation: Objectives to Outcomes, 1st edn., 312 pages. PHI Learning Private Ltd., Delhi (2012) ISBN-978-81-203-4537-9
23. Satyanarayana, J.: E-Government: The Science of the possible, 4th edn., 312 pages. PHI Learning Private Ltd., Delhi (2011) ISBN-978-81-203-2608-8
24. Dahiya, D., Mathew, S.K.: Review of Strategic Alignment, ICT Infrastructure Capability and E-Government Impact. Paper Published in the Web Proceedings of the 5th ICT Innovations Conference 2013, Ohrid, Macedonia, pp. 42–51 (2013) (ISSN 1857-7288)
25. UN, Plan of Action: E-Government for Development, http://www.itu.int/wsis/docs/background/themes/egov/action_plan_it_un.doc (last accessed on July 06, 2014)

Towards Identifying the Knowledge Codification Effects on the Factors Affecting Knowledge Transfer Effectiveness in the Context of GSD Project Outcome

Jagadeesh Gopal[1], Arun Kumar Sangaiah[2], Anirban Basu[3], and Ch. Pradeep Reddy[1]

[1] School of Information Technology and Engineering, VIT University, Vellore, India
profgjagadeesh@gmail.com
[2] School of Computing Science and Engineering, VIT University, Vellore, India
arunkumarsangaiah@gmail.com
[3] Department of Computer Science & Engineering, East Point College of Engineering, Bangalore, India
anbasu@pqrsoftware.com

Abstract. Global software development (GSD) is a knowledge intensive process of offshore/onsite teams that helps in planning and designing a coherent software system to meet the business needs. The aim of this research is to reveal the influence of GSD teams' (offshore/onsite) knowledge codification factors on knowledge transfer (KT) effectiveness with respect to the outcome of GSD projects. Knowledge codification is the process of converting the tacit knowledge to explicit knowledge. In the GSD projects, where offshore/onsite teams' are distributed and working in various geographic locations, the process of knowledge codification has influenced by several factors. Thus, our objective of this paper is to address the knowledge codification factors on knowledge transfer effectiveness in the context of GSD project outcome perceived by GSD teams. Moreover, this study explores to integrate effectiveness of knowledge transfer in GSD project outcome relationship from the service provider perspective in following dimensions: product success, successful collaboration, and personal satisfaction. This research study employs survey methods to empirically validate the research model in Indian software companies for their view of GSD projects.

Keywords: Knowledge codification, Knowledge Transfer (KT), Global Software Development (GSD), Offshore/onsite team, Project outcome.

1 Introduction

GSD is primarily an outsourcing technique in which on-site teams' are working in the client location, understanding, and auditing the client requirements. Whereas, offshore teams operating at different regions, is executing the requirement based on the inputs provided by the on-site teams. In today's world GSD teams' (offshore/on-site) knowledge codification factors have created a significant impact on knowledge transfer (KT) effectiveness with relate to the outcome of a GSD projects. Subsequently, the

S.C. Satapathy et al. (eds.), *Emerging ICT for Bridging the Future – Volume 1*,
Advances in Intelligent Systems and Computing 337, DOI: 10.1007/978-3-319-13728-5_69

literature on the knowledge management theory reveals that the role of knowledge codification factors and their significant impact on project success [1,2,3]. However, the earlier studies of knowledge management theory focused on the relationship between organizational elements and performance of knowledge transfer [4,5,6]. In addition, many studies reveal the significance of knowledge transfer success in Information System (IS) outsourcing [7,8,9,10,11,12].

Moreover, codification effects on factors affecting the knowledge transfer among GSD teams' have not been sufficiently investigated despite their importance in IS outsourcing context. In this study, our aim is to overcome the limitations of earlier studies which only investigated the effects of knowledge transfer success. Thus, this research explores a deeper insight into the codification factors on GSD teams' knowledge transfer effectiveness at individual (team and knowledge context) and organizational level (technology and organization context) with relate to the GSD project outcome.

2 Research Question

This research aims to develop a comprehensive methodology for measuring codification effects on factors affecting the knowledge transfer in the context of the GSD project outcome perceived by GSD teams. To achieve this we have specified the following research objectives:

Research Objective: To explore how GSD teams' knowledge-related factors, team-related factors, technology-related factors, and organizational-related factors may be created a significant impact on KT effectiveness and to investigate their effect on GSD project outcome. This forms the basis for following research questions:

RQ 1: What are factors affecting the knowledge transfer in the context of GSD projects as reported in literature?

RQ 2: What are codification effects on the factors affecting a knowledge transfer among GSD teams' from a GSD project outcome perspective?

To address this research questions the empirical study is planning to carry out in Indian software companies for analyze the codification effects on the factors affecting the knowledge transfer perceived by GSD teams. The rest of this paper is organized as follows: Section 3 presents the overview of theoretical foundation and Section 4 presents the proposed model used in this research. Sections 5 and 6 present the methodology and conclusion of the study respectively.

3 Overview of Theoretical Foundation

This section presents the earlier studies and develops a research framework that is used in this paper. Previous study [13] has defined knowledge transfer as: "a process of exchange of explicit or tacit knowledge between two agents, during which one agent purposefully receives and uses the knowledge provided by another''. In this research, agent can be referred as offshore and onsite team of an organizations. Recently, many researchers have investigated the effects of knowledge transfer in IS outsourcing success [7,8,10,14,15]. Moreover, these studies have addressed the knowledge transfer in IS outsourcing success in software provider and service receiver point of view in a matured relationship.

However, there is no prior research for measuring knowledge codification effects on the factors affecting a knowledge transfer in the context of GSD project outcome. This research therefore seeks to identify the significance of knowledge codification factors that affect the knowledge transfer of GSD teams' and integrating knowledge characteristics with organizational elements through various dimensions: knowledge-context, team-context, technology-context, and organization-context. Overall, this research on the above potential factors and their measurement is collectively identified through existing studies. The morphology of codification effects on successful knowledge transfer among GSD teams' for GSD project outcome initiatives are classified under four dimensions as shown in Appendix-A. Each of the dimension having two or more constructs has been elaborated in the following sections.

3.1 Knowledge Context

Earlier studies have addressed the characteristics of knowledge and significance of knowledge transfer in IS outsourcing success [7,8,9,11,12]. Moreover, earlier researchers [4,5,6] have integrated the advantage of individual and organizational knowledge for effectiveness of knowledge transfer. In addition, the significance of SECI (Socialization, Externalization, Combination, and Internalization) in knowledge management theory is greatly acknowledged in literature [16,17,18].

However, there is no prior research for measuring knowledge codification effects on the factors affecting GSD teams' knowledge transfer effectiveness in GSD project outcome relationship on the basis of SECI. Hence, this research seeks to provide an insight to integrate knowledge characteristics at the team and organization level factors for knowledge transfer effectiveness with relate to GSD project outcome from SECI process indicators.

3.2 Team Context

Drawing upon the previous literatures [14,19,20,21,22] team characteristics have measured with four indicators: Meta cognition, cohesion, mutual support, and interpersonal communication. The focus of this study is to investigate how knowledge codification process at the offshore/onsite team level that influences the knowledge transfer effectiveness in GSD project context. Here effectiveness of knowledge transfer refers to degree in which GSD teams' fulfill the expectations regarding the GSD project outcome. Whenever offshore and onsite teams interact for the purpose of common goals, team performance is important in order to achieve overall outcome of

GSD project [7]. Subsequently, number of studies [7,12,15] has investigated that significance of team performance in the context of knowledge transfer to make the project successful. Based on this context, we have related that the quality of teamwork is likely to increase the knowledge transfer effectiveness of teams in GSD projects.

3.3 Technology Context

Previous studies [6,23] have addressed the impact of Information and Communication Technologies (ICT) in knowledge management with respect to sharing, exchanging disseminating knowledge and technologies. In addition, the earlier study [4] examined the support and essential role of ICT in software development process. Drawing upon the literatures [11,24] technology factors for knowledge transfer effectiveness of GSD teams' characteristics have measured with two indicators: ICT tools and ICT infrastructure. Hence this research has given insight into the relationship between ICT and knowledge transfer effectiveness of GSD teams' for achieving successful outcome of projects goals.

3.4 Organization Context

According to the earlier studies [4,7,25,26] organizational elements are central aspects for making effective knowledge transfer. This implies focusing organizational structure characteristics for GSD teams in order to perform knowledge related task addressed in this study. Moreover, organizational structure has been addressed in terms of its influence of GSD teams' knowledge transfer in the context of GSD project outcome.

4 Research Model

The overview of theoretical perspectives is identified through literature reviews. To address the research gap, the following research model is proposed as shown in Fig.1. In this model GSD project outcome is the dependent variable, which is hypothesized as being influenced by GSD teams knowledge transfer (KT) factors. In addition, KT is measured in terms of GSD teams' knowledge, team, technology, and organization dimensions. The phenomenon of codification effects on the factors affecting KT is hypothesized as moderating the relationship between the various dimensions and GSD project outcome.

This paper specifies the codification effects on the KT and its influential factors in each dimension are measured using existing indicators. Moreover, this study presents the results from KT effectiveness of GSD teams' with relate to the outcome of GSD projects from service provider perspective into three dimensions: product success, personal satisfaction, and successful collaboration.

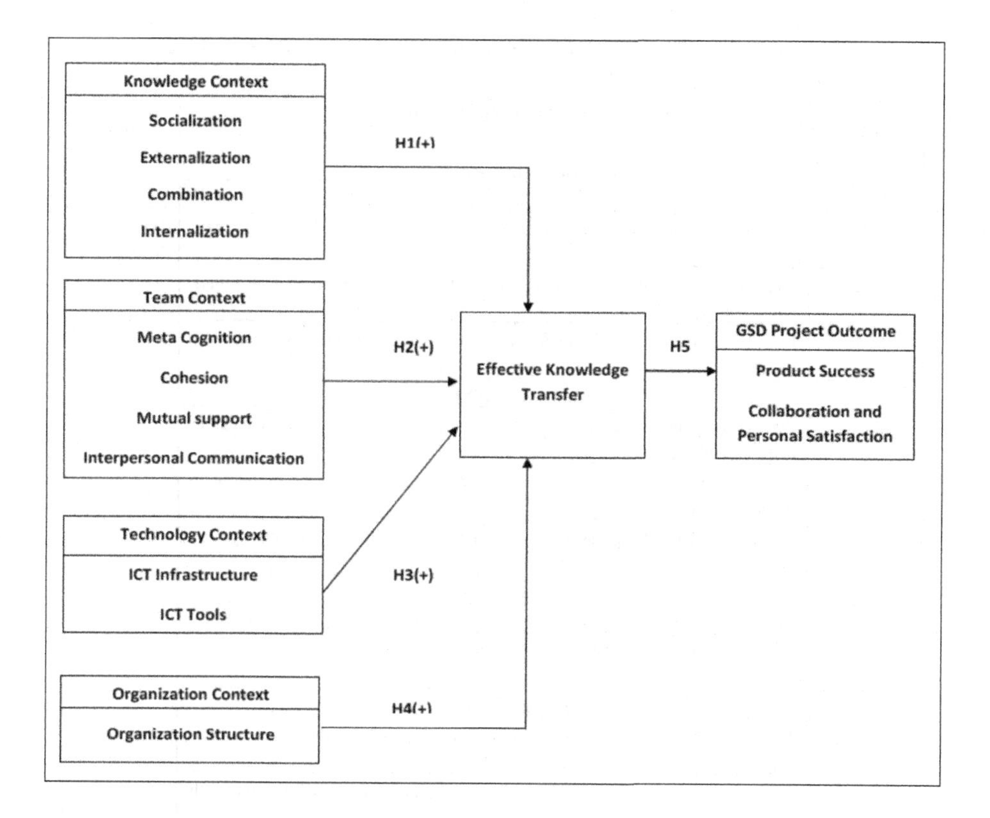

Fig. 1. The GSD Teams' Knowledge Transfer Effectiveness in the context of GSD Project Outcome

5 Research Methodology

Our research objective is integrating organizational elements and knowledge characteristics for the effectiveness of GSD teams' knowledge transfer with relate to GSD project outcome. To achieve this research objective empirical study is being planned to carry out to measure codification effects on the factors affecting knowledge transfer effectiveness on GSD project outcome. The details of operational constructs and their related literatures are summarized in Table-1. The following hypotheses are proposed to test the research model.

H1(a): Knowledge context which consists of socialization aspect will have positive impact on the effectiveness of knowledge transfer by GSD teams.

H1(b): Knowledge context which consists of externalization aspect will have positive impact on the effectiveness of knowledge transfer by GSD teams.

H1(c): Knowledge context which consists of combination aspect will have positive impact on the effectiveness of knowledge transfer by GSD teams.

H1(d): Knowledge context which consists of internalization aspect will have positive impact on the effectiveness of knowledge transfer by GSD teams.

Table 1. The Overview of Measurement Indicators from the Literature

Construct Item	Indicators of Existence	Sources of Item
	Knowledge Context	
	Socialization (SI)	
SI-1	I intend to be involved in gathering information and experiences from others within my organization.	[16,17,27,28,29,30,31]
SI-2	I intend to be involved in sharing information and experiences with others within my organization.	
SI-3	I will tell to others what I think to make sure my understanding is the same as theirs.	
	Externalization (EX)	
EX-1	I can describe professional or technical terms with conversational language to help communication in a team.	
EX-2	I will help others to clearly express what he/she has in mind by encouraging them to continue what they are saying.	
EX-3	When others cannot express themselves clearly, I usually help them clarify their points.	
	Combination (CO)	
CO-1	I intend to build up materials by gathering literature and technical information.	
CO-2	When coming across problems, I tend to use my experience to help solving problems.	
	Internalization (IN)	
IN-1	I intend to share and try to understand management vision through communications with colleagues.	
IN-2	I intend to be involved in searching and sharing new values and thoughts with colleagues.	
	Team Context	
	Meta Cognition (MC)	
MC-1	Do your team members feel they are cooperating rather than competing with each other in fulfilling the goals?	[4,18,19,20,21,22,29]
MC-2	Do your team members feel personally involved in the implementation and committed to devote themselves?	
	Cohesion (CH)	
CH-1	It was important to the members of our team to be part of this project. All members were fully integrated in our team.	
CH-2	The team members were contributing to the achievements of the teams' goals in a accordance to their specific potentials.	
CH-3	The team members were strongly attached and integrated as a team to the project	
	Mutual Support (MS)	
MS-1	The team members helped and supported each other as best they could.	
MS-2	Suggestions and contributions of team members were discussed and further developed.	

Table 1. *(continued)*

Interpersonal Communication (IC)		
IC-1	The team members were happy with the precision and usefulness of the information received from other team members.	
IC-2	Project-relevant information was shared openly by all team members.	
Technology Context		
ICT Infrastructure (ICT)		
ICT-1	We have blended hardware and /or software assets with business capabilities to generate a novel process, product, or service in our organization.	[4,6,11,21,23,24,25]
ICT-2	The selection of the appropriate communication channels and mechanisms is of vital importance for the success of the KT task.	
ICT Tools (IT)		
IT-1	We improved our ability to manage the technology /applications from time to time	
IT-2	The organization provides various tools and technologies to facilitate knowledge sharing and exchange	
Organization Context		
Organization Structure (OS)		
OS-1	Knowledge (know-how, technical skill, or problem solving methods) is well codified in my company	[4,6,18,23,24]
OS-2	Information flows easily throughout the organization regardless of employee roles or other boundaries	
OS-3	Certain tasks require the information of teams with members from different departments in order to be accomplished.	
GSD Project outcome (GSD)		
GSD-1	We are satisfied with our overall benefits from IT outsourcing	[7,32,33,34]
GSD-2	Our objective is to improve the capability of IT to support the needs of business operations	
GSD-3	Our objective was to improve quality of services	
GSD-4	How satisfied was your organization with the project performance regarding expected functionality	
GSD-5	How satisfied was your organization with the overall outcome of a project	

H1(e): Knowledge context which consists of socialization, externalization, combination and internalization aspect will have positive impact on the effectiveness of knowledge transfer by GSD teams.

H2(a): Team context which consists of Meta cognition aspect will have positive impact on the effectiveness of knowledge transfer by GSD teams.

H2(b): Team context which consists of cohesion aspect will have positive impact on the effectiveness of knowledge transfer by GSD teams.

H2(c): Team context which consists of mutual support aspect will have positive impact on the effectiveness of knowledge transfer by GSD teams.

H2(d): Team context which consists of interpersonal communication will have positive impact on the effectiveness of knowledge transfer by GSD teams.

H2(e): Team context which consists of Meta cognition, cohesion, mutual support, and interpersonal communication will have positive impact on the effectiveness of knowledge transfer by GSD teams.

H3(a): An extensive use of ICT infrastructure among GSD teams will have positive impact on the effectiveness of knowledge transfer.

H3(b): An extensive use of ICT tools among GSD teams will have positive impact on the effectiveness of knowledge transfer.

H3(e): Technology context which consists of ICT infrastructure and ICT Tools will have positive impact on the effectiveness of knowledge transfer by GSD teams.

H4(a): There exists a positive relationship between communication flow of GSD teams in an organization and the performance of knowledge transfer.

H4(b): There exists a positive relationship between organizational context and the performance of knowledge transfer by GSD teams.

H5: Effectiveness of knowledge transfer which is measured in knowledge, team, technology, and organization context perceived by GSD teams will have a positive impact on GSD project outcome.

To test the hypothesis against research model, the empirical survey among Indian Software Companies has been carried out to investigate the effectiveness of knowledge transfer in the context of GSD project outcome is being planned to address in our future work.

6 Conclusion

The previous research findings on knowledge transfer effectiveness of organizational performance are carefully studied. The main contribution of this research is focused on integrating codification effects on the factors affecting knowledge transfer perspective with prior research in knowledge transfer and organizational performance, particularly focusing effectiveness of GSD teams' knowledge transfer with relate to GSD project outcome. In this process, we have integrated a set of measures on knowledge, team, organization, and technology context for GSD project outcome initiative research. Moreover, to address relationships between codifications effects on GSD teams' KT effectiveness with relate to GSD project outcome initiatives needs further research.

References

1. García-Muiña, F.E., Pelechano-Barahona, E., Navas-López, J.E.: Knowledge codification and technological innovation success: Empirical evidence from Spanish biotech companies. Technological Forecasting and Social Change 76(1), 141–153 (2009)
2. Albino, V., Garavelli, A.C., Schiuma, G.: A metric for measuring knowledge codification in organisation learning. Technovation 21(7), 413–422 (2001)
3. Prencipe, A., Tell, F.: Inter-project learning: processes and outcomes of knowledge codification in project-based firms. Research Policy 30(9), 1373–1394 (2001)

4. Syed-Ikhsan, S.O.S., Rowland, F.: Knowledge management in a public organization: a study on the relationship between organizational elements and the performance of knowledge transfer. Journal of Knowledge Management 8(2), 95–111 (2004)
5. Kang, J., Rhee, M., Kang, K.H.: Revisiting knowledge transfer: Effects of knowledge characteristics on organizational effort for knowledge transfer. Expert Systems with Applications 37(12), 8155–8160 (2010)
6. Susanty, A., Handayani, N.U., Henrawan, M.Y.: Key Success Factors that Influence Knowledge Transfer Effectiveness: A Case Study of Garment Sentra at KabupatenSragen. Procedia Economics and Finance 4, 23–32 (2012)
7. Sangaiah, A.K., Thangavelu, A.K.: An adaptive neuro-fuzzy approach to evaluation of team level service climate in GSD projects. Neural Computing and Application 25(3-4), 573–583 (2013), doi:10.1007/s00521-013-1521-9)
8. Al-Salti, Z., Hackney, R.: Factors impacting knowledge transfer success in information systems outsourcing. Journal of Enterprise Information Management 24(5), 455–468 (2011)
9. Mohamed, A., Arshad, N.H., Abdullah, N.A.S.: Knowledge transfer success factors in IT outsourcing environment. Science 6(6), 916–925 (2009)
10. Nidhra, S., Yanamadala, M., Afzal, W., Torkar, R.: Knowledge transfer challenges and mitigation strategies in global software development—A systematic literature review and industrial validation. International Journal of Information Management 33(2), 333–355 (2013)
11. Aziati, A.H., Juhana, S., Hazana, A.: Knowledge Transfer Conceptualization and Scale Development in IT Outsourcing: The Initial Scale Validation. Procedia-Social and Behavioral Sciences 129, 11–22 (2014)
12. Blumenberg, S., Wagner, H.T., Beimborn, D.: Knowledge transfer processes in IT outsourcing relationships and their impact on shared knowledge and outsourcing performance. International Journal of Information Management 29(5), 342–352 (2009)
13. Kumar, J.A., Ganesh, L.S.: Research on knowledge transfer in organizations: a morphology. Journal of Knowledge Management 13(4), 161–174 (2009)
14. Gang, Q., Bosen, L.: Research on model of knowledge transfer in outsourced software projects. In: 2010 International Conference on E-Business and E-Government (ICEE), pp. 1894–1899. IEEE (May 2010)
15. Beulen, E., Tiwari, V., van Heck, E.: Understanding transition performance during offshore IT outsourcing. Strategic Outsourcing: An International Journal 4(3), 204–227 (2011)
16. Nonaka, I., Toyama, R.: The knowledge-creating theory revisited: knowledge creation as a synthesizing process. Knowledge Management Research & Practice 1(1), 2–10 (2003)
17. Nonaka, I., Toyama, R.: The theory of the knowledge-creating firm: subjectivity, objectivity and synthesis. Industrial and Corporate Change 14(3), 419–436 (2005)
18. Choi, B., Lee, H.: An empirical investigation of KM styles and their effect on corporate performance. Information & Management 40(5), 403–417 (2003)
19. Hoegl, M., Praveen Parboteeah, K., Gemuenden, H.G.: When teamwork really matters: task innovativeness as a moderator of the teamwork–performance relationship in software development projects. Journal of Engineering and Technology Management 20(4), 281–302 (2003)
20. Hoegl, M., Weinkauf, K., Gemuenden, H.G.: Interteam coordination, project commitment, and teamwork in multiteam R&D projects: A longitudinal study. Organization Science 15(1), 38–55 (2004)

21. Zeng, Q.Y., Wang, Z.Y.: Evaluation of measurement items involved in the assessment of personal tacit knowledge. In: 2011 IEEE 18Th International Conference on Industrial Engineering and Engineering Management (IE&EM), pp. 1868–1871. IEEE (September 2011)
22. Hoegl, M., Gemuenden, H.G.: Teamwork quality and the success of innovative projects: A theoretical concept and empirical evidence. Organization Science 12(4), 435–449 (2001)
23. Duan, Y., Nie, W., Coakes, E.: Identifying key factors affecting transnational knowledge transfer. Information & Management 47(7), 356–363 (2010)
24. Lindner, F., Wald, A.: Success factors of knowledge management in temporary organizations. International Journal of Project Management 29(7), 877–888 (2011)
25. Wong, K.Y.: Critical success factors for implementing knowledge management in small and medium enterprises. Industrial Management & Data Systems 105(3), 261–279 (2005)
26. Al-Alawi, A.I., Al-Marzooqi, N.Y., Mohammed, Y.F.: Organizational culture and knowledge sharing: critical success factors. Journal of Knowledge Management 11(2), 22–42 (2007)
27. Huang, J.C., Wang, S.F.: Knowledge conversion abilities and knowledge creation and innovation: a new perspective on team composition. In: Proceedings of the Third European Conference on Organizational Knowledge, Learning, and Capabilities, pp. 5–6 (April 2002)
28. Karim, N.S.A., Razi, M.J.M., Mohamed, N.: Measuring employee readiness for knowledge management using intention to be involved with KM SECI processes. Business Process Management Journal 18(5), 777–791 (2012)
29. George, B., Hirschheim, R., von Stetten, A.: Through the Lens of Social Capital: A Research Agenda for Studying IT Outsourcing. Strategic Outsourcing: An International Journal 7(2), 2–2 (2014)
30. Grimaldi, R., Torrisi, S.: Codified-tacit and general-specific knowledge in the division of labour among firms: a study of the software industry. Research Policy 30(9), 1425–1442 (2001)
31. Tseng, S.M.: Knowledge management system performance measure index. Expert Systems with Applications 34(1), 734–745 (2008)
32. Arun, K.S., Thangavelu, A.K.: Factors affecting the outcome of Global Software Development projects: An empirical study. In: International Conference on Computer Communication and Informatics (ICCCI), IEEE (2013), doi:10.1109/ICCCI.2013.6466113
33. Sangaiah, A.K., Thangavelu, A.K.: An exploration of FMCDM approach for evaluating the outcome/success of GSD projects. Central European Journal of Engineering 3(3), 419–435 (2013)
34. Kumar, S.A., Thangavelu, A.K.: Exploring the Influence of Partnership Quality Factors towards the Outcome of Global Software Development Projects. International Review on Computers & Software 7(5) (2012)

Formulation and Computation of Animal Feed Mix: Optimization by Combination of Mathematical Programming

Pratiksha Saxena[1] and Neha Khanna[2]

[1] Department of Applied Mathematics
Gautam Buddha University
Greater Noida, India
pratiksha@gbu.ac.in
[2] Gautam Buddha University
Greater Noida, India
neha15khanna@gmail.com

Abstract. The aim of this paper is to develop the models for ruminant ration formulation for different weight classes. Least cost ration and better shelf life are taken as main objectives of the study. Firstly the linear programming (LP) models have been developed for obtaining least cost ration. Then stochastic programming (SP) models have been developed to incorporate nutrient variability.

1 Introduction

Ration formulation is one of the basic requirements of dairy industry for better yield as performance of the animal is directly dependent on the nutritionally balanced ration. The main objective of ration formulation is to meet the nutrient requirement of the animal at different stages of the production at minimum cost.

To achieve this objective linear programming has been used for last many years. A review was presented based on programming techniques used for animal diet formulation chronologically [1]. Linear programming technique was first used by Waugh in 1951 that defined the feeding problem in linear form and optimized the ration [2]. In many real life situations the decision maker may have multiple objectives as cost minimization, nutrient maximization, better shelf life etc. [3]. Rehman and Romero discussed about limitations of linear programming for rigidity of constraint set and singularity of objective function [4]. Several decision criteria have been considered by using Goal programming [5]. GP has been used to formulate the ration using one hundred and fifty food raw materials which satisfy the daily nutritional requirements of Thais. The results obtained by GP showed a marked improvement over those of LP [6]. A model has been developed to optimize the feed rations of active and trained sport horses [7]. Nonlinear programming techniques were used for weight gain in sheep [8]. Linear and nonlinear programming techniques were compared for animal diet formulation [9]. To reduce the effect of variability in nutrient content methods referred as safety margin(SM) and right hand side

S.C. Satapathy et al. (eds.), *Emerging ICT for Bridging the Future – Volume 1*,
Advances in Intelligent Systems and Computing 337, DOI: 10.1007/978-3-319-13728-5_70

adjustment(RS) has been given ([10], [11]). Another method to consider this variation in nutrient content is mentioned by using linear and stochastic programming technique [12].

The objective of this paper is to obtain the least cost ration for dairy cattle at different stages of livestock as well as to increase the shelf life of the feed ingredients used in the ration by using linear programming. Stochastic programming has been used to consider the variability in nutrient content at 80% level of probability.

The main objectives of this paper can be classified as follows:

i. To develop the models with the objectives of minimizing cost and water content in the ration and find the LP solution for least cost ration and minimum water content.
ii. To develop the above models taking into account the variation of nutrient content with Probability level of 80% and find the solutions.

The paper is organized as follows:

In Section 2, input data for dairy cattle such as composition of feed ingredients, minimum nutrient requirements etc. has been discussed. General linear programming models for least cost ration with the satisfaction of nutrient requirement of dairy cows at different stages of livestock have been developed. Models for minimizing water content have also been developed at different stages of livestock in order to maximize shelf life of ration. Then stochastic programming models have been developed for the same objectives as discussed above. The analysis of obtained results has been presented in section 3 while conclusion has been presented in section 4.

2 Material and Methods

The objective of diet formulation is to provide a least cost ration which satisfy the animal nutrient requirement as well as provide better shelf life. The data required for the formulation of models are as follows: Cost of the feed ingredients, Water content of feed ingredients, Minimum nutrient required at different stages of weight, Chemical composition of different feed ingredients. The main objectives in this paper are

i. To achieve least cost ration with the satisfaction of minimum nutrient requirement at different weight classes
ii. To achieve the ration with minimum water content with the satisfaction of minimum nutrient requirement at different weight classes

In formulation of above models, additional constraints are added. The minimum quantity of barley grain and rice bran has been taken as 5 grams to get more balanced diets in terms of nutritive value. Lowest price is the criteria for the selection of these feed ingredients. First linear programming models have been developed for achieving these objectives. Then Stochastic Programming models have been developed with the same set of objective and constraints to deal with nutrient variability. The probability level has been taken as 80%.

2.1 Input Data for Determination of Feed Blend

Let us introduce the following notations:

z objective function,

c_j per unit cost of feed ingredient j ,

x_j quantity of jth feed ingredient in the feed mix ,

a_{ij} amount of nutrient i available in the feed ingredient j ,

b_i minimum requirement of ith nutrient ,

i index identifying feed nutrient components with i = 1,2,….m. j index identifying feed components with j =1,2,………n. Six nutrients and sixteen feed ingredients have been used in these models for optimization of feed blend. Input data for cost of feed ingredients, water content and nutritional composition of feed ingredients has been shown in table 1. Composition of feed ingredients and water content is taken from recommendations of NRC and feedipedia [13, 14]. Minimum nutrient requirement at different weight of dairy cattle to reach at 680 Kg weight has been given in table 1.

Table 1. Minimum Requirement for Different Nutrients at Different Weight of Dairy Cattle to reach at 680 kg Weight

Nutrient	200 kg	300 kg	450 kg	680 kg
ME	8.54	9.54	7.49	6.28
CP	127	123	94	155
NDF	315	315	315	300
DM	5200	7100	11300	23600
Ca	11.3	15	13	10
P	9.1	10.6	13	5

2.2 Linear Programming Model for Determination of Feed Blend for Cost and Water Content Minimization

$$\begin{aligned} &Min\ z = \sum c_j x_j \\ &s.t. \sum_{j=1}^{n} a_{ij} x_j \geq b_i \\ &x_2 \geq 5, \\ &x_{11} \geq 5, \\ &x_j \geq 0, b_i \geq 0 \end{aligned} \tag{1}$$

Linear programming models have been formulated for obtaining the optimum values of feed ingredients to achieve weight at different stages of livestock. Models have been sub-divided into four models to: (i) minimize the per kg cost of ration of dairy cow of weight 680 kg, (ii) minimize the per kg cost of ration of dairy cow of weight 200 kg to reach the weight 680 kg, (iii) minimize the per kg cost of ration of dairy cow of weight 300 kg. to reach the weight 680 kg, (iv) minimize the per kg cost of ration of dairy cow of weight 450 kg to reach the weight 680 kg.

Linear programming models have also been formulated to maximize the feed blend quality in terms of its shelf life. It can be done by minimizing the water content of the feed blend. These models have been sub-divided into four models. These models have been formulated to obtain optimum values of feed components to (i) minimize the water content of ration of dairy cow of weight 680 kg, (ii) minimize the water content of ration of dairy cow of weight 200 kg to reach the weight 680 kg, (iii) minimize the water content of ration of dairy cow of weight 300 kg to reach the weight 680 kg (iii) minimize the water content of ration of dairy cow of weight 450kg to reach the weight of 680 kg.

These eight models have been formulated to find optimal value of feed components to reach different stages of livestock. Then to deal with the variability of nutrition values of feed components stochastic programming models have been established.

2.3 Stochastic Programming Model for Determination of Feed Blend for Cost and Water Content Minimization

$$\min\ z = \sum c_j x_j$$

$$s.t.\ \sum_{j=1}^{n}\left(a_{ij} - z\left(\sqrt{\sum_{j=1}^{n}\sigma_{ij}^{2}}\right)\right)x_j \geq b_i \tag{2}$$

$$x_2 \geq 5,$$

$$x_{11} \geq 5,$$

$$x_j \geq 0, b_i \geq 0$$

Stochastic models have been developed to include nutrient variability These models are similar to LP models to some extent, The objective function is same as in LP. σ_{ij}^{2} represents variance of nutrient i in ingredient j and it is included with a certain probability level, rest of the variables are defined as above . This model has been formulated with 80% assumed probability which takes the variability of nutritional values of feed component. This assumption implies that there is 80% probability that a ration contains the desired level of nutrients. Value of z is 2.33 for this level of probability. The requested probability determines the nutrient concentration for ration formulation. Variability of nutrient is taken into account as a nonlinear term of variance in each feed component. Again eight models have been formulated for sixteen feed components to obtain optimum value of variables to reach at specific levels of livestock. Water content is again minimized to increase the shelf life of ration. Stochastic programming models have been formulated for determination of optimum values of feed ingredients to achieve weight at different stages of livestock. Models have been sub-divided into four models to (i) minimize the per kg cost of ration of dairy cow of weight 680 kg, (ii) minimize the per kg cost of ration of dairy cow of weight 200 kg to reach the weight 680 kg, (iii) minimize the per kg cost of ration of dairy cow of weight 300 kg. to reach the weight 680 kg, (iv) minimize the per kg cost of ration of dairy cow of weight 450 kg to reach the weight 680 kg.

Stochastic programming models have also been formulated to maximize the feed blend quality in terms of shelf life of ration. It can be done by minimizing the water content of the feed blend. This model is sub divided into four models. These models have been formulated to obtain optimum values of feed components to (i) minimize the water content of ration of dairy cow of weight 680 kg, (ii) minimize the water content of ration of dairy cow of weight 200 kg to reach the weight 680 kg (iii) minimize the water content of ration of dairy cow of weight 300 kg to reach the weight 680 kg (iv) minimize the water content of ration of dairy cow of weight 450kg to reach the weight 680 kg.

3 Results and Discussion

All diet formulation system modeled by linear programming is solved by TORA. Optimum values of feed ingredients for minimum cost and for minimum water content to reach different weight class of dairy cattle are obtained by these model solutions. Stochastic programming models also provides the minimum cost and minimum water content but the difference between two results are that stochastic programming results includes variability of nutrient content. That goal of adding nutrition variability to animal diet is achieved by introducing the variance term in the stochastic programming models. 16 models are presented and analyzed by the blend of linear programming and stochastic programming. Table 2 represents minimum cost and minimum water content for different level of livestock by using the blend of the two techniques linear programming and stochastic programming.

Table 2. Minimum cost and minimum water content for different weight gain range

Weight gain range (Kg.)	Minimum Cost (Rs./Kg.)		Minimum Water Content	
	Linear Programming	Stochastic Programming	Linear Programming	Stochastic Programming
0-680	213.35	255.21	2.61	2.74
200-680	104.34	173.31	1.19	1.73
300-680	107.79	178.41	1.22	1.77
450-680	118.73	175.66	1.39	1.74

Optimum values of feed ingredients for minimization of cost and water content by linear and stochastic programming is shown in table 3. From table 3, it can be concluded that although linear programming models are providing feed mix at lower values of costs but stochastic programming models are also considering variation of nutrient contents which is one of the very important aspects of animal diet.

Table 3. Optimal Solution value of feed ingredients for cost minimization by linear and stochastic programming

Feed Ingredients	Linear Programming				Stochastic Programming			
	0-680	200-680	300-680	450-680	0-680	200-680	300-680	450-680
Alfalfa hay	0	0.31	0.56	0	5.66	5.24	5.60	5.40
Barley grain	5	5	5	5	5	5	5	5
Rice bran(fibre 11-20%)	5	5	5	5	5.11	5	5	5
Wheat straw	16.19	0	0	2.68	10.69	0	0	0

Larger amount of nutrient component is included in the feed mix by using stochastic models as compared to linear models and this variability is due to the standard deviation inclusion in SP models.

Results obtained for cost minimization shows that when cost is optimized by stochastic programming technique, it is providing the better results in the sense of nutritional variability as compared to linear programming technique. It also includes variability of nutrient components at slightly higher cost. Graphical representation of comparison of results by linear and stochastic programming is shown by fig. 1. Table 3 represents the total costs for a feed blend to reach dairy cattle through each weight class at specific growth rate. It is clear from the results shown in table 4, that more number of variables are included in the diet, if modelled by stochastic programming technique. It should also be pointed out that marginal values of variables are also greater in this case. Linear programming models and stochastic models are used with the constraints such that all nutrients at least achieve the NRC requirements.

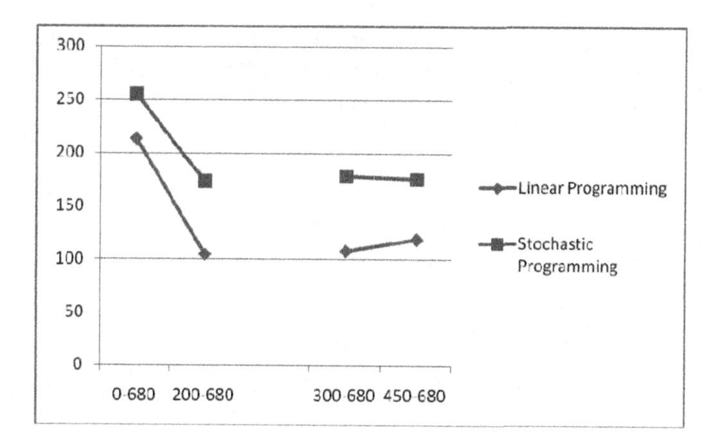

Fig. 1. Optimum cost for all weight ranges by LP and SP

Graphical view of results for minimization of water content from feed mix for better shelf quality of meal for different stages of livestock is shown by fig. 2. It is clear from the figure that both programming techniques are giving very close results for all the weight classes, however it is greater when modeled by LP. Therefore it can be concluded that better shelf life is obtained by SP models, as it is at minimum side with inclusion of nutritional variability in diet for all weight classes.

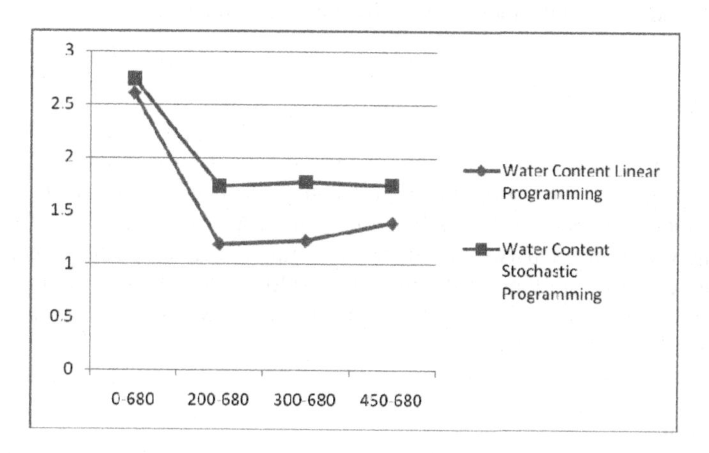

Fig. 2. Comparison of results for minimum water content for all weight ranges by LP and SP

16 models are discussed for cost minimization and better shelf life of feed mix. Fig. 3 represents optimum share of nutrient ingredients in feed mix for cost minimization by linear as well as stochastic programming. This graphical view provides results for share of optimum nutrient ingredients for different stages of livestock. It shows the optimum quantity to be included in animal diet to reach the different weight classes at minimum cost, which is different from the previous work in this area that takes the objective of optimum weight gain at minimum cost. It shows

the superiority of SP models which has greater value of feed components. It also provides importance of additional constraint to the LP and Sp models as due to these additional constraints, more variables are included in the feed mix.

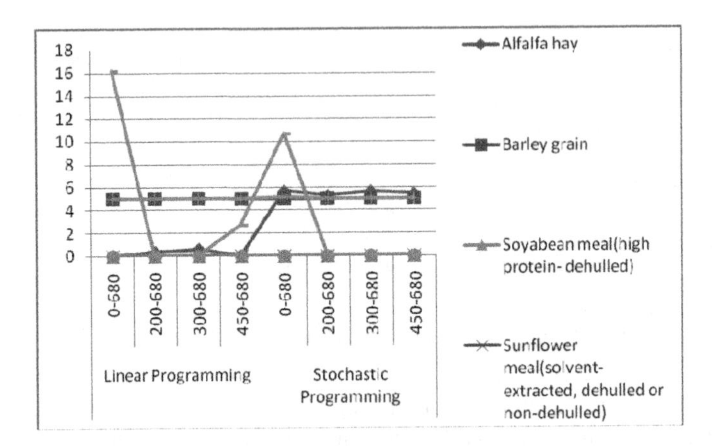

Fig. 3. Optimum values of nutrient ingredients for different stages of livestock for cost minimization (LP models and SP models)

Similarly, results are obtained for water content and value of objective functions by LP models and SP models does not vary by marginal values.

4 Conclusion

This paper proposes a combination of mathematical programming (Linear and Stochastic programming) to optimize the livestock feed mix. The paper represents a bi-criteria model and successful application of SP programming models. Introduction of nutrient variability and two additional constraints affects the cost, shelf life and ingredient content of the feed mix. It explores the blend of mathematical programming to reach different stages of livestock which is different from the previous work in this area.

References

1. Saxena, P., Chandra, M.: Animal Diet Formulation: A Review (1950-2010). CAB Reviews: Perspectives in Agriculture, Veterinary Science, Nutrition and Natural Resources 6(57), 1–9 (2011)
2. Waugh, F.V.: The minimum cost dairy feed. Journal of Farm Economics 33, 299–310 (1951)
3. Lara, P., Romero, C.: Relaxation of nutrient requirements on livestock rations through interactive multigoal programming. Agricultural Systems 45, 443–453 (1994)
4. Rehman, T., Romero, C.: Multiple- criteria decision making technique and their role in livestock formulation. Agriculture System 15, 23–49 (1984)

5. Rehman, T., Romero, C.: Goal programming with penalty functions and livestock ration formulation. Agriculture System 23, 117–132 (1987)
6. Anderson, M., Earle, M.D.: Diet planning in the third world by linear and goal programming. Journal of Operational Research Society 34(1), 9–16 (1983)
7. Prišenk, J., Pažek, K., Rozman, C., Turk, J., Janžekovič, M., Borec, A.: Application of weighted goal programming in the optimization of rations for sport horses. Journal of Animal and Feed Sciences 22, 335–341 (2013)
8. Saxena, P.: Application of nonlinear programming in the field of animal nutrition: A problem to maximize the weight gain in sheep. National Academy Science Letter 29(1-2), 59–64 (2006)
9. Saxena, P.: Comparison of Linear and Nonlinear Programming Techniques for Animal Diet. Applied Mathematics 1(1), 106–108 (2011)
10. Roush, W.B., Cravener, T.L., Zhang, F.: Computer formulation observations and caveats. Journal of Applied Poultry Res. 5, 116–125 (1996)
11. St. Pierre, N.R., Harvey, W.R.: Incorporation of uncertainty in composition of feeds into least-cost ration models. 1.Single-chance constrained programming. Journal of Dairy Science 69(12), 3051–3062 (1986)
12. Tozer, P.R.: Least cost ration formulations for Holstein dairy Heifers by using linear and stochastic programming. Journal of Dairy Science 83, 443–451 (2000)
13. National Research Council. Nutrient requirements of Dairy Cattle. 6th Rev. edn. Natl. Acad. Sci., Washington, DC (1989)
14. http://www.feedipedia.org/

Resource Grid Architecture for Multi Cloud Resource Management in Cloud Computing

Chittineni Aruna[1] and R. Siva Ram Prasad[2]

[1] Department of CSE, KKR & KSR Institute of Technology and Sciences,
Acharya Nagarjuna University,
Guntur, Andhra Pradesh, India
chittineni.aruna@gmail.com
[2] Department of CSE and Head, Department of IBS,
Acharya Nagarjuna University, Guntur, Andhra Pradesh, India

Abstract. Multi Cloud Architecture is an emerging technology in cloud computing to improve the security of data and process management in single clouds. Dynamic Resource Allocation is a proven technique for efficient Resource Management in Single Cloud Architecture to mitigate resource overflow and underflow problems. Resource Management in Single Cloud Architecture is an improving area and still it has to address many problems of resource management like resource bottlenecks. Resource Management at single cloud always recommends the additional resources adoption due to frequent resource underflow problems. In this paper we proposed a Resource Grid Architecture for Multi Cloud Environment to allocate and manage the resources dynamically in virtual manner. This architecture introduces a resource management layer with logical resource grid and uses the virtual machines to map the resources against physical systems of single cloud. Experiments are proving that our Resource Grid Architecture is an efficient architecture to manage the resources against multiple clouds and supports the green computing.

Keywords: Resource Grid Architecture, Dynamic resource management, Multi Cloud Environment, Cloud Computing, Green Computing.

1 Introduction

Cloud computing is the new age technology offers on-demand, pay-per-use, reliable and QoS (Quality of Service) resources as IaaS, PaaS and SaaS over internet. Cloud migration is increasing rapidly in cloud computing to achieve the cloud benefits from non-cloud based applications. Day to Day improving features and advancements of cloud computing are assuring the future and attracting the organizations and stake holders towards cloud adoption. This rapid growth of cloud computing needs high level security and huge resources like memory, process and storage to satisfy the SLA [1].

Cloud data security, Process tampering prevention[3] and resource management are the three important aspects of cloud computing to concern. Recent researches[4, 5 and 6] raised numerous issues on cloud data security and trustworthiness of cloud

S.C. Satapathy et al. (eds.), *Emerging ICT for Bridging the Future – Volume 1*,
Advances in Intelligent Systems and Computing 337, DOI: 10.1007/978-3-319-13728-5_71

service providers at cloud data center level. George and Pavlos[6] et al, assumes that present cloud is following the unsecured Honest but Curious model for data and process management. To mitigate the security issues and to prevent the process tampering in single cloud architecture, recently multi cloud architecture [7, 8] emerged as an alternative. This approach distributes the encrypted data and process among multiple clouds to resist from disclose of complete data by compromising of any single cloud. Federating multiple clouds not only improves the security, but also provides the additional benefits to cloud like high availability, auto elasticity, transparency and trustworthiness. Efficient resource management in cloud architecture can save the 15 to 25 percent of resources from additional resource budget. Present multi cloud architectures are still suffering from resource bottlenecks [2] and scalability issues of SLA due to decentralized resource management. This inefficient resource management needs frequent extension and wastage of process, memory and storage resources at every single cloud level and causes to resource overflow and underflow problems.

In this paper to address the above discussed problems, we proposed a Resource Grid Architecture (RGA) for Multi Cloud Environment to allocate and manage the resources dynamically in a virtual manner. RGA federates the resources of each single cloud of multi cloud environment to create the centralized virtual resource pool. This pool is called as Resource Grid (RG), is a collection of virtual machines (VM's) controlled by Virtual Resource Manager (VRM). Every virtual machine is a representation of Physical System (PM) at single cloud operation level to identify the resource allocation needs and management. This architecture introduces a new resource management layer with logical resource grid and uses the virtual machines to map the resources against physical systems of single cloud. RGA includes the load management system as an integral part to assess the resource requirements of multiple clouds to mitigate resource overflow and underflow problems. Experiments are proving that our RGA is an efficient architecture to manage the resources against multiple clouds and supports the green computing.

The rest of the paper is organized as follows. Section 2 discusses about multi cloud architecture, resource management techniques and section 3 provides the implementation of RGA architecture. Section 4 depicts the experimental setup, simulations and section 5 concludes this paper. Section 7 lists the main references of this paper to design.

2 Related Work

In this section we discuss about the need of multi cloud architecture and resource management efficiency in cloud.

Today public clouds are offering the three prominent cloud service layers SaaS, IaaS and PaaS over internet, which made the user sensitive data asset availability from intranet environment to internet environment for remote access flexibility and data globalization. This aspect raises various data and process security issues [7, 8], due to the less control on data to data owner and full control on data to service provider. Cloud service provider may be honest but curious [6] also under some circumstances. Several researches were introduced the Third Party Auditing [9] and Client Side Data Encryption [10] to mitigate the unauthorized cloud data access and process tampering in cloud. These encryption techniques are having scalability problems and complex to manage to satisfy SLA. To avoid these problems in cloud, Bohli and Jensen [7] et al proposed multi cloud architecture by federating the single clouds of cloud computing environment. This architecture becomes an alternative to single cloud to improve the security for data and application by using multiple clouds simultaneously. In order to resist the adversary attacks this architecture uses the data and process distribution among multiple clouds. The single cloud and multi cloud architectures are shown in Figure 1.

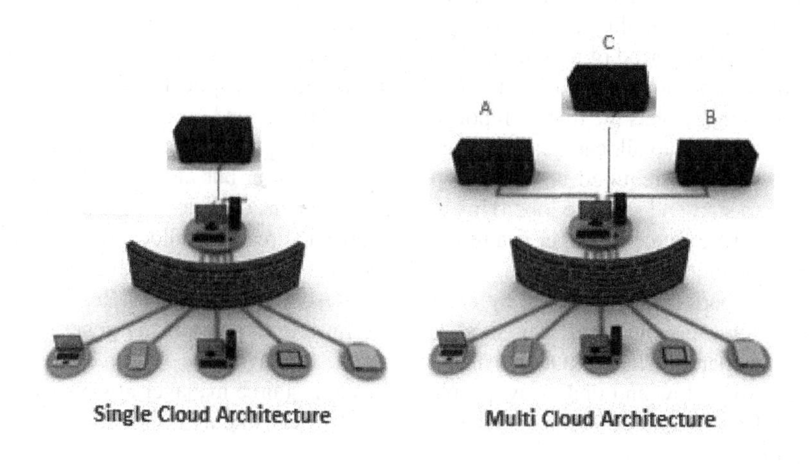

Fig. 1. Single and Multi Cloud Architectures

Resource management is an important concern in cloud computing to manage the primary resources like memory, process and storage etc. Rapid growth of cloud computing demands high amount of resources to rent cloud users and to run the cloud applications. This is a frequent requirement for every single cloud depends on the resource usage and prediction. Static allocation of resources [12] in cloud cannot be used effectively because of resource need variations for an application over time. These variations of resource usage create overflow and underflow problems in resource allocation. To avoid these problems resources should be adjusted dynamically for efficient resource management in cloud architecture. Dynamic resource management [11] reduces the resource contention, scarcity of resources, resource fragmentation, and overflow, underflow problems. In this research we create a resource pool to manage multiple cloud resources effectively through corresponding virtual machines.

3 Resource Grid Architecture

This section gives the comprehensive explanation and implementation of our Resource Grid Architecture for Multi Cloud Environment. RGA is logically cross cut into four layers for efficient resource management in multi cloud environment as shown in figure 2.The layers are i) Cloud Layer ii) Network layer iii) VM layer iv) Resource Grid Layer.

3.1 Cloud Layer

This layer is a collection of multiple remote clouds with available physical resources like storage, network, process, memory. Resource Scheduler, Utility Tracker, Local Resource Manager and physical resources are the parts of this layer. Local Resource Manager is the head of Resource Scheduler and Utility Tracker. Utility Tracker is a tool to track the resource utilization at single cloud level based on certain interval period of time. This tool enquires the allocated resources to cloud application to track the record of resource utilization and sends this information to cloud Resource Prediction Unit for hotspot and cold spot detection. Utility Tracker is a prescheduled program which runs automatically over the specified period of time to track usage and to upload this data to Local Resource Manager. In order to allocate the additional resources to any application of cloud, Local Resource Manager gets the tracker report from utility tracker and gives the instructions to resource scheduler to allocate resources and starts the new job. Resource scheduler is not only allocating the required resource to running application but also working as a job scheduler to start and stop the jobs. This is simply an extension to job scheduler program of OS. Local Resource Manager is a module to verify the resource allocation and utilization at single cloud level and is an agent of Global Resource Manager. Extracted tracking information periodically updated by each Local Resource Manager to Resource Grid Layer for allocation analysis. This module is responsible to periodically generate a Resource Occupancy Chart (ROC), which summarize the available resources, allocated resources, over flow and underflow problems. Local resource manager always works as per the Global Resource Manager instructions and works like as a virtual node in this architecture.

3.2 Network Layer

This layer is the conjunction between cloud layer and the resource grid layer and having collection of routers, firewalls, HDMI interfaces etc. Secured communication relevant all aspects are implemented in this layer along with https like secured protocols. This layer contains the high speed wired and wireless network components to support high speed commutation among clouds. This layer is the backbone for this architecture because intra cloud communication is highly demanded in this implementation for resource sharing and process sharing.

3.3 VM Layer

This layer is a list of virtual machine groups where each group is a set of virtual machines for physical system mapping. In this layer we proposed a separate virtual machine group for single cloud architecture to achieve the operational feasibility. Each VM is a virtual representation of a cloud physical machine at VM Layer level. This is the prominent way to perform mapping between logical resources usage with physical requirements. Every VM is having the OS part for virtual process management and the data part for data manipulations. Virtual machines are always available to resource grid layer monitoring and resource mapping.

3.4 Resource Grid Layer (RGL)

RGL is a collection of management modules of this architecture like Global Resource Manager (GRM), Load Balancing System (LBS), Monitoring System, Resource Allocation Tracker, Resource Prediction Unit (RPU) etc as shown in figure 2. All of these modules are interlinked together to monitor, analyze, allocate and track the resource information of multiple clouds at a centralized virtual location. This layer is called as a grid because of this is a centralized pool of logical resources, which predicts the requirement of resources based on past statistics. In this architecture, we introduce the logical federation of resources from multiple single clouds instead of physical collaboration to achieve the resource management. With the inspiration of water Grid Management System from Indian government, that collaborate all available small water plants together to adjust the water to avoid drinking water problem in all seasons. Single cloud environments are not good enough to manage the resources, due to the problems of limited resources, inefficient resource management system and frequently occurring resource overflow and under flow problems. To mitigate all these problems at single cloud level we are implemented a multi cloud resource management layer to share the resources over clouds. In this case, our RGL connects all individual clouds and adjusts one cloud resources with another cloud via network layer to support Green Computing [13].

Global Resource Manager is the vital role in this layer and to make decisions of resource allocation and to mitigate resource congestion problem in this architecture. Monitoring System is the management level monitoring module, which collects the periodic reports from local resource manager of each single cloud. This system prepares the Cloud Health Charts for each cloud of multi cloud architecture to represent the current status of resource usage and availability. Resource prediction unit is the integral part of the Resource Grid Layer to assess the future requirements of individual cloud based on their recent utilization statistics at cloud applications and physical machines level. In the same way this will calculate at all clouds level and generates a Resource Occupancy Chart (ROC) to give the clarity about every cloud level available resources, allocated resources, overflow of resources and underflow of resources. This module prepares another report is Future Resource Prediction Chart (FRPC) by estimating future requirements at each cloud level and handover this to Global Resource Manager. By considering this report GRM suggests the additional resources requirement for future to individual clouds. Load balancing System takes the ROC as input from Resource Prediction Unit to level the overflow and underflow of resources at single clouds. This is a knowledgeable system with an efficient Dynamic Resource Allocation Unit to adjust the resources and processes among multiple clouds by mapping with respective virtual machines. Load Balancing System finds the resource hot spots and cold spots at every cloud level to balance the resources to mitigate resource overflow and underflow problems. After mapping the resources to VM's, allocation information will send to allocation tracker at resource grid level. This tracker is having a dashboard to display utilization ratio at cloud level, grid level and displays the resource savings of each cloud by becoming a part at Resource Grid.

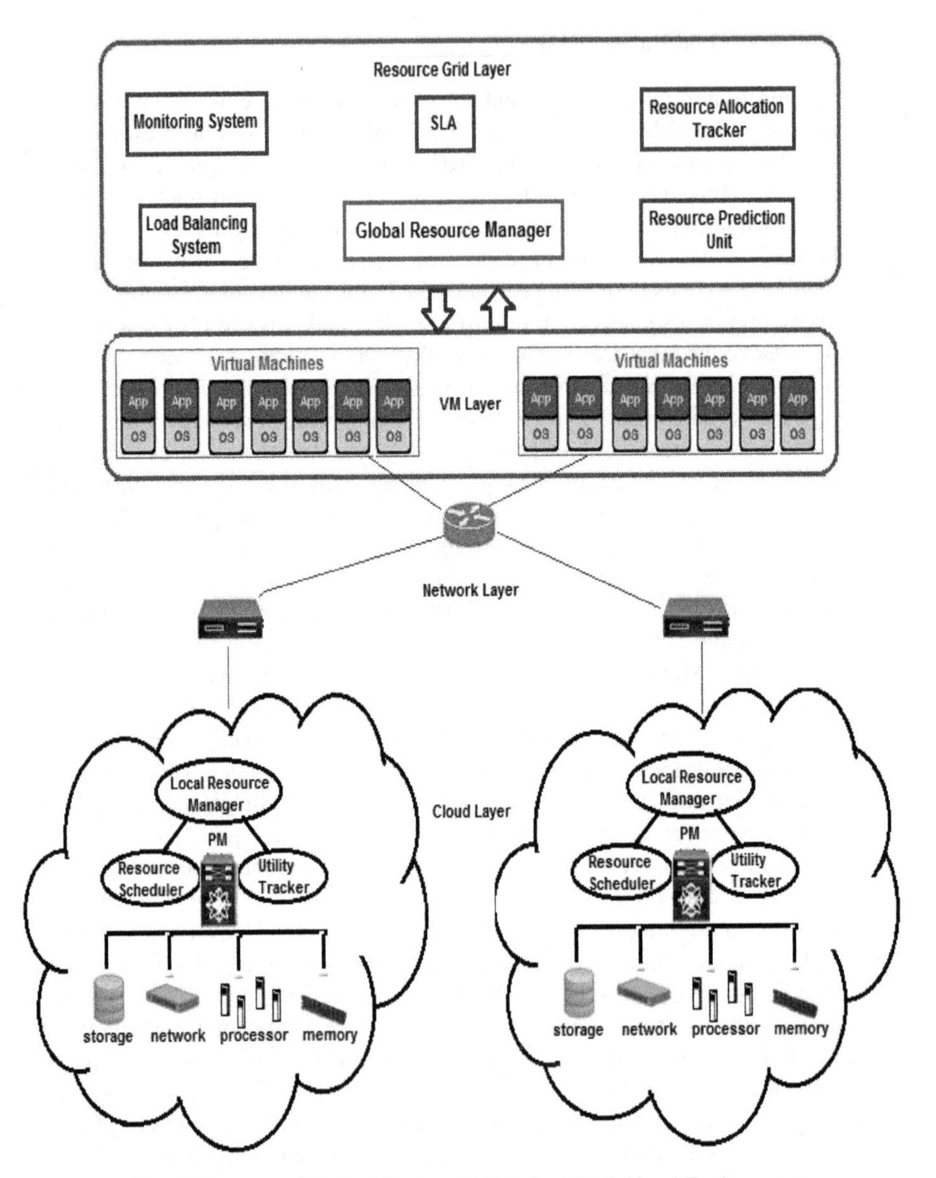

Fig. 2. Resource Grid Architecture (RGA) for Multi Cloud Environment

4 RGA Experimental Setup and Process Flow at a Glance

In this section we discuss about the experimental setup of RGA and the process flow along with the obtained result information at glance.

4.1 Experimental Setup

As the architecture is having multiple clouds as a part of this, it is important to create a large scale clouds for experiments. However it is extremely complex to setup the large scale clouds for experiments, we created 5 different single clouds with each cloud having more than 100 physical nodes with commodity hardware at every system level. These 5 clouds are complete heterogeneous and connected through a high speed wireless network system for intra communication to each other and the resource grid layer. We implemented the resource grid layer as a part of one cloud of 5 clouds and allocated an independent processor and memory pool for that module. Every commodity (Physical Machine) system is having the Linux (Ubuntu), 8 GB RAM, 1 TB hard drive and AMD8 Processor with 3.5 GHZ speed. To simulate the cloud platform we used the CloudSim Toolkit[14], which creates a Virtual Machine (VM) environment to map with the physical systems of cloud. Each VM is running an application with various workloads and having the individual process execution environments mapped with available static resources. A set of VM's of same cloud are forming a VM cluster to deviate from cloud to cloud. After a trial run of individual cloud with static resources, resources are adjusted as per resource requirement. This allocation may need to update at runtime depends on application workload leads to resource overflow or under flow. We run this experimental setup for 3 days to find hidden flaws to make this architecture more reliable.

4.2 RGA Process Flow

While running the experiments the process flow of this architecture is described in this section. As per the above discussion every cloud is deployed with physical resources to map with VM's which are running the user applications. After the specific period amount of time the report generation and updating to next higher level will start at every possible module of this architecture. This time vary from module to module subject to complexity and need. This reporting is divided to multi hop environment, where at each hop is connected to its next level and having various modules to generate different reports for process. Initially the utility tracker generate the resource availability and allocation at cloud individual physical machine level and submits to the first hop manager is Local Resource Manager. After trail run execution we set the periodical reporting time difference as 5 min. This report will generate at every cloud level automatically and sends to Local Resource Manager for every 5min. At this stage, Local Resource Manager will connect to the second hop and updates the report to monitoring system through network layer. This system will get the similar report from multiple clouds and forwards this to Resource Prediction Unit to generate ROC and FRPC reports. Resource Prediction Unit forwards the ROC to Load Balancing System and FROC to GRM. As per the information of ROC, Load Balancing System adjusts the resources of one cloud with another by considering the overflow and underflow of resources at each cloud level. At the same time, concurrently this will updates the adjusted information to Resource Allocation Tracker, which maintains a dash board to display the statistics of resource utilization. For every 15 min this dash board will update with the available resources at grid, allocated resources at grid and overflow, underflow problems at every individual cloud level as shown in below graphs.

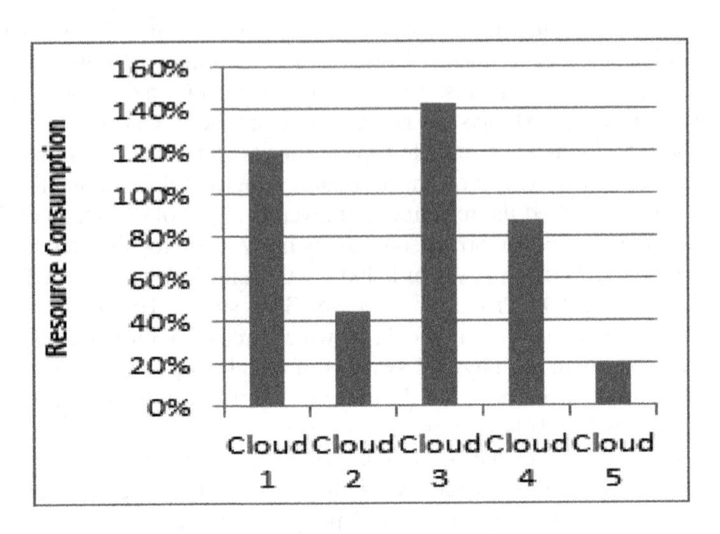

Graph 1(a). Resource Occupancy Chart by RPU

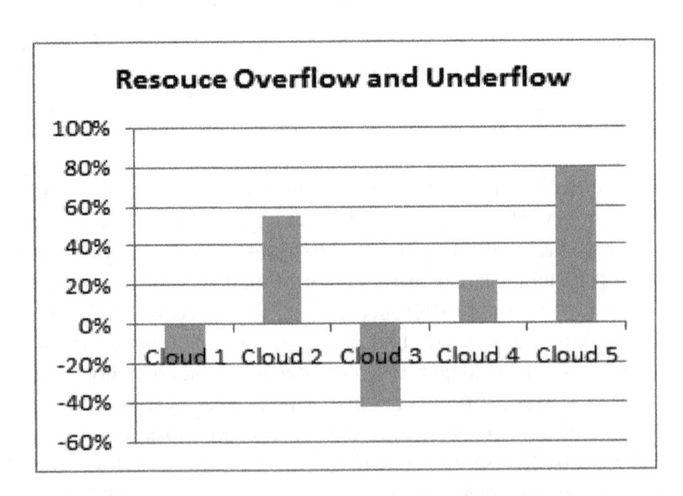

Graph 1(b). Overflow and underflow graph by RAT at multi cloud level

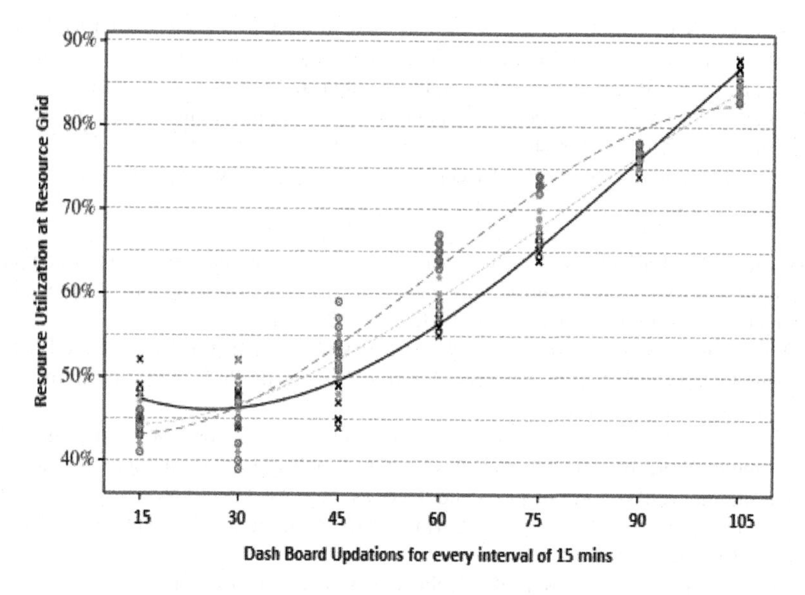

Graph 2. Dashboard resource utilization updates at Resource Grid Level

5 Conclusion

In this paper we are introducing an efficient new architecture to implement the resource management in cloud computing by creating a Resource Grid at Multi Cloud level. Logical integration of cloud resources are proven to mitigate the resource overflow and underflow problems at every individual cloud level, without adopting new resources. Virtual Machines layer with executing processes are mapped with physical resources to adjust resources dynamically as per the load balancing system. This approach is having multi hop reporting mechanism to send the various analysis and utilization statistics to next level for efficient resource management at multiple clouds level. Experimental results are showing that RGA is having the high scalability than resource management at individual cloud environments. Our future work concentrates on how to recommend the additional resources adoption to a cloud owner and updating the billing system for an individual cloud to pay for usage of other cloud resource usage.

References

1. Bobroff, N., Kochut, A., Beaty, K.: Dynamic placement of virtual machines for managing sla violations. In: Proc. of the IFIP/IEEE International Symposium on Integrated Network Management (IM 2007) (2007)
2. Chase, J.S., Anderson, D.C., Thakar, P.N., Vahdat, A.M., Doyle, R.P.: Managing energy and server resources in hosting centers. In: Proc. of the ACM Symposium on Operating System Principles (SOSP 2001) (October 2001)

3. Jensen, M., Schwenk, J., Gruschka, N., Lo Iacono, L.: On Technical Security Issues in Cloud Computing. In: Proc. IEEE Int'l Conf. Cloud Computing, CLOUD-II (2009)
4. Chow, R., Golle, P., Jakobsson, M., Shi, E., Staddon, J., Masuoka, R., Molina, J.: Controlling data in the cloud:outsourcing computation without outsourcing control. In: Proceedings of the 2009 ACM Workshop on Cloud Computing Security, pp. 85–90. ACM (2009)
5. Vijayan, J.: Vendors tap into cloud security concerns with new encryption tools, `http://www.cio.com.au/article/376252/vendorstapintocloudsecurityconcernsnewencryptiontools/`
6. Drosatos, G., Efraimidis, P.S., Athanasiadis, I.N.: A privacy-preserving cloud computing system for creating participatory noise maps. In: IEEE 36th International Conference on Computer Software and Applications, 0730-3157/12. IEEE (2012)
7. Bohli, J.-M., Jensen, M., Gruschka, N., Schwenk, J., Iacono, L.L.L.: Security Prospects through Cloud Computing by Adopting Multiple Clouds. In: Proc. IEEE Fourth Int'l Conf. Cloud Computing (CLOUD) (2011)
8. Bernstein, D., Ludvigson, E., Sankar, K., Diamond, S., Morrow, M.: Blueprint for the Intercloud—Protocols and Formats forCloud Computing Interoperability. In: Proc. Int'l Conf. Internet andWeb Applications and Services, pp. 328–336 (2009)
9. Hulawale, S.S.: Cloud Security Using Third Party Auditing and Encryption Service. M.S Thesis (July 2013)
10. Li, M., Yu, S., Zheng, Y., Ren, K.: Scalable and Secure Sharing of Personal Health Records in Cloud Computing using Attribute-based Encryption. IEEE Transactions on Parallel and Distributed Systems 56 (2012)
11. Song, Y., Wang, H., Li, Y., Feng, B., Sun, Y.: Multi-Tiered On-Demand resource scheduling for VM-Based data center. In: Proceedings of the 2009 9th IEEE/ACM International Symposium on Cluster Computing and the Grid, vol. 00, pp. 148–155 (2009)
12. Harchol-Balter, M., Downey, A.: Exploiting process lifetime distributions for load balancing. ACM Transactions on Computer Systems (TOCS) 15(3), 253–285 (1997)
13. `http://searchdatacenter.techtarget.com/definition/green-computing`
14. Calheiros, R.N., Ranjan, R., Beloglazov, A., Rose, C.A.F.D., Buyya, R.: CloudSim:a toolkit for modeling and simulation of cloud computing environments and evaluation of resource provisioning algorithms. Software: Practice and Experience 41(1), 23–50 (2011)

An Efficient Framework for Building Fuzzy Associative Classifier Using High-Dimensional Dataset

S. Naresh[1], M. Vijaya Bharathi[2], and Sireesha Rodda[3]

[1,2] GMR Institute of Technology, Rajam, Srikakulam, A.P., India
[3] GITAM University, Visakhapatnam, A.P., India
[1] PG –M.Tech, Dept. of CSE
csenaresh8@gmail.com

Abstract. Association Rule Mining (ARM) with reference to fuzzy logic is used to further data mining tasks for classification and clustering. Traditional Fuzzy ARM algorithms have failed to mine rules from high-dimensional data efficiently, since those are meant to deal with relatively much less number of attributes or dimensions. Fuzzy ARM with high-dimensional data is a challenging problem to be addressed. This paper uses a quick and economical Fuzzy ARM algorithm FAR-HD, which processes frequent item sets using a two-phased multiple-partition approach especially for large high-dimensional datasets. The proposed algorithm is an extension to the FAR-HD process in which it improves the accuracy in terms of associative soft category labels by building a framework for fuzzy associative classifier to leverage the functionality of fuzzy association rules. Fuzzy ARM represent latent and dominant patterns in the given dataset, such a classifier is anticipated to supply superb accuracy particularly in terms of fuzzy support.

Keywords: Fuzzy Associative Classifier, Fuzzy Clustering, High Dimensional Data, SURF Vectors, Fuzzy Association Rule Mining.

1 Introduction

Association Rule Mining extracts frequent patterns within the sort of any dataset, the obtained patterns are generated based on their frequencies. Using association rule mining to process the high-dimensional dataset will not work efficiently and definitely there is a loss of information while getting the results and which are also not accurate, by using fuzzy logic can able to scale back the loss of information called sharp boundary problem.

Fuzzy ARM has been extensively employed in relational/transactional [3] dataset with less to medium number of attributes or dimensions. Rules can be derived from high-dimensional datasets like the domain of images in order to train fuzzy associative classifier.

The traditional approaches are failed to mine association rules in high-dimensional dataset [17], [18] but FAR-HD [1] [2] given better performance when compare to the traditional fuzzy association rule mining algorithms. FAR-HD specially designed

S.C. Satapathy et al. (eds.), *Emerging ICT for Bridging the Future – Volume 1*,
Advances in Intelligent Systems and Computing 337, DOI: 10.1007/978-3-319-13728-5_72

algorithm for high-dimensional data with large datasets. So by using the existing efficient approach the implementation of classifiers with association rules and fuzzy logic delivers better accurate results.

2 Related Work

Fuzzy Association Rule Mining (ARM) based classification gives good results instead of taking directly a trained dataset. The new techniques of fuzzy ARM provides good performance with increase in datasets like image.

Mangalampalli and Pudi[1][2] proposed a Fuzzy ARM algorithm specifically for large high-dimensional datasets. FAR-HD [1] and FAR-Miner [2] has been compared experimentally with Fuzzy Apriori and FAR-Miner is to work efficiently on large datasets, both the approaches are faster than fuzzy Apriori. FAR-HD provides better performance with comparison of FAR-Miner. F-ARMOR [4] is another approach which also gives fast and efficient performance on large datasets but not on high dimensional datasets.

Rajendran and Madheswaran[19] proposed a Novel Fuzzy Association Rule Image Mining Algorithm for Medical Decision Support System. Novel Fuzzy Association Rule Mining (NFARM) gives the diagnosis keywords to physicians for making a better diagnosis system.

Usha and Rameshkumar[20] provides a complete survey on application of Association Rule Mining on Crime Pattern Mining, in which an analyst can analyze the criminal tactics based on their behavior like crime reports, arrest reports and other statistics by using association rules.

Wei Wang and Jiong Yang explained, with the rapid growth of computational biology and e-commerce applications, high-dimensional data [21] become very common. The emergence of various new application domains, such as bioinformatics and e-commerce, underscores the need for analyzing high dimensional data. They both presented a few techniques for analyzing high dimensional data, e.g., frequent pattern mining, clustering, and classification. Many existing works have shown through experimental results that fuzzification of association rules provide better performance.

3 Fuzzy Preprocessing

This preprocessing approach consists of two phases, the generation of numerical vectors from the input dataset is comes under the first phase and therefore the second phase is the process of conversion of numerical feature vectors in to a fuzzy clusters representation.

Fuzzy preprocessing [1], [2] is the process of converting feature vectors in the form of SURF vectors. FAR-HD uses the concept of SURF vectors and Fuzzy C-Means clustering algorithm is to search out the fuzzy association rules with efficiency.

3.1 SURF (Speeded-Up Robust Features)

Here input dataset is images, the features of the images in the form of SURF values. SURF [4] is an advanced image processing approach to find the matching points between the two similar scenes or objects of images and is part of many computer vision applications. Each SURF vector consists of 64 dimensions like size and resolution.

The applications of SURF are used in Bio-informatics like Face, Palm, Finger, Iris, Knuckle recognition. SURF is 3-times better faster than SIFT(Scale Invariant Feature Transform).First it will take each image and found interesting points in the image and by using Fast-Hessian Matrix approach it will generate the eigen vectors called matching points between two similar scenes of images. The matching is often done based on a distance between the feature vectors. The SURF values are then applied to the FCM algorithm to find the fuzzy clusters.

3.2 Fuzzy C-Means Clustering (FCM)

Fuzzy C-Means (FCM) [8] is an algorithm for fuzzy clustering [9] which allows one piece of data belong to two or more clusters, it is an extension of K-Means algorithm [13], [14]. In the 70's mathematicians introduced spatial term FCM algorithm to improve the accuracy of fuzzy clustering under noise. Initially in the year 1973 Dunn developed FCM algorithm and then improved by Bezdek in 1981, especially used in pattern recognition and the minimization of the objective function of FCM is as follows in (eq.1). It's a very important tool for processing the image in clustering objects from image. Identifying exactly the number of clusters is a critical task because each feature vector belongs to each cluster to some degree instead of whole belonging to just single cluster. It reduces the lexical ambiguity (polysemy) and semantic relation (synonymy) that happens in crisp clustering.

$$\sum_{i=1}^{N}\sum_{j=1}^{C}\mu_{ij}^{m}\,||x_i - c_j||^2 \quad , \quad 1 \leq m < \infty \tag{1}$$

where m is any real number such that $1 \leq m<\infty$, μ_{ij} is the degree of membership of x_i in the cluster j, x_i is the ith d-dimensional measured data, c_j is the center of the cluster, the fuzziness parameter m is an arbitrary real number ($m>1$). The amount of fuzziness can be controlled using an appropriate value ($\approx$ 1.1–1.5) of the fuzziness parameter m (Eq. 1).

4 Fuzzy Association Rule Mining in High Dimensional Dataset (FAR-HD)

4.1 Algorithms

FAR-HD algorithm uses two multiple phases of algorithms in a partition based approach to generate fuzzy association rules. Fuzzy versions of apriori algorithms do not perform fast against large datasets, even for crisp ones. FAR-HD with comparisons of fuzzy adaptation of association algorithms provides an efficient performance, 7 to 15 times faster when compare with most popular fuzzy apriori and 1.1 to 4 times faster with FAR-Miner [4] approach, but all versions of fuzzy algorithms are works with only fewer attributes.

4.1.1 FAR-HD First Phase

After getting the clusters with respect to fuzzy logic, the first phase of algorithm takes the individual fuzzy cluster data as input and generates frequent itemsets [3] by using support (0.0 to 1.0) and confidence (%) and then generates transaction list by picking up fuzzy support from singletons to large number of frequent itemsets.

First Phase of algorithm scans each current partition of the dataset from the individual formed clusters, and builds a tidlist for each singleton found. After all singletons in the present partition have been enumerated, the transaction id lists (tidlists) [1], [2], [4] of singletons which are not frequent are removed.

4.1.2 FAR-HD Second Phase

The second phase of algorithm is quite different from first phase which extracts the frequent itemsets from the whole fuzzy clusters by using the same support and confidence measures and then derives fuzzy association rules. For each remaining itemset, the algorithm identifies singletons and then obtains the tidlist by intersecting the tidlists of all the singletons. Thus, it alternate between outputting and deleting itemsets, and creating new itemsets, until no more itemsets are left. Then the process of algorithm will now terminate.

4.2 Support and Confidence

Support can be calculated by using fuzzy logic 0.0 to 1.0, referred as fuzzy support and the confidence in terms of percentage (%) using Eq. 2 and 3, can be referred in the section 6. During the evaluation of algorithms the t-norms [1], [2], [3] called support value and confidence value will be given by the user in the interval of [0.0, 1.0] and (0 % to 100 %) respectively for one time, the given support and confidence values will be there throughout during the itemsets generation and frequent itemsets generation and then finally for the entire fuzzy association rules generation.

$$\text{Support (A => B)} = \frac{\sum_{X \in D}(A \cap_T B)(x)}{|X|} \quad (2)$$

$$\text{Confidence (A => B)} = \frac{\sum_{X \in D}(A \cap_T B)(x)}{\sum_{X \in D}(A)(x)} \quad (3)$$

5 Fuzzy Associative Classifier Using High-Dimensional Data

The major difference between a non- fuzzy classifier and fuzzy classifier is, instead of assigning a class label to each pattern, a soft class label with degree of membership in each class is attached to each pattern. Fuzzy Associative Classifier [5], [12] uses Fuzzy Association Rules as input and generates class labels based on the measures such as

Support and Confidence the class labels are generated and then derive Associative soft class labels. Associative soft class labels can be derived by using previous rules called fuzzy association rules with respect to a fuzzy support value, such value provides an accurate result when compare to traditional classifier approaches.

5.1 Applications of FAC Framework

By extending this framework of Fuzzy Associative Classifier [15], [19] can be used in the face, iris, knuckle, finger, palm authentication with a simple fuzzy value and also in the weather prediction, temperature prediction, and disease prediction in a faster way with a fuzzy accurate support value. Fuzzy Association Classifier with Fuzzy Support [15] value gives a better accurate result by using a fast and efficient algorithm called FAR-HD with respect to this proposed algorithm called Fuzzy Associative Classifier.

5.2 Proposed Algorithm

The algorithm searches for the rules from the images using an advanced image processing concept SURF to achieve feature vectors and by applying fuzzy clustering called fuzzy c means algorithm to represent the feature vectors in the form of fuzzy clusters then by applying the efficient process of two phases of algorithms, [1] the fuzzy rules can be derived. The output rules which is having confidence in between 70-100% comes under one group of class and calculate the support of the range of rules and assign it as fuzzy support and similarly for other ranges between 40-69 % as one group and 0 to 39% as another group of class. Thus experts prefer to work with fuzzified data (i.e. linguistic variables) rather than with the exact numbers. It is in a way, a quantization of the numerical attributes into categories directly understandable by the experts. Hence closing the gap between the data analyst and the domain expert with an accurate value has been done. Thus the classifier is able to produce accurate classifications from high dimensional data (image), can be used to make well informed decisions.

```
Algorithm: Fuzzy Associative Classifier for High
Dimensional Dataset
Inputs: Image Dataset (D), supp
Outputs: Classifiers (C)
1 Initialize FI to denote Frequent Itemsets
2 FI = GetFI(D, supp)
3 For each item it in FI
      Generate association rule and add to R
4 END FOR
5 For each rule r in R
     If confidence is in range 70-100 then
        assign class label C1
        assign supp
     else if confidence is in range 40-69 then
        assign class label C2
        assign supp
     else assign class label C3
        assign supp
6 END FOR
7 RETURN C
```

```
Procedure: GetFI(Image Dataset, supp)
1 Initialize FI
2 Extract Surf Values (S)
3 For each unique item s in S
     Compute support supp1
     If(supp1>=supp)
        Add item to FI
     END IF
  END FOR
4 return FI
```

6 Results

The following are the experimental results of the framework of fuzzy associative classifier [18] which was implemented using java platform and are represented as below figures. The accurate results of fuzzy associative classifier are shown in figure: 8 by comparing with the fuzzy association rules with class labels in figure: 7. Figure: 1 represents the GUI for input images and applying SURF to calculate interest points of each image as shown in figure: 2. Figure: 3 displays the matching points between the two images as graphical lines, figure: 4 represents the matching points between the images as values and then click fuzzy clustering button it will asks for a file name with extension .num to save the clusters in it, as shown in the figure: 5. The final process is to load that clustered file for the association rules it will ask for the measures of support in the interval [0.0-1.0] and confidence as 0-100 % , it will process the frequent itemsets as shown in the section 5.1 and the results of the itemsets are shown in Figure: 6. Figure: 7 display all the rules with its respective confidence and class labels. Finally the proposed algorithm applicable here and generates simple class labels with a fuzzy support value (refer section 4.2). Thus figure: 8 represent the soft class labels with accurate fuzzy value using association rules.

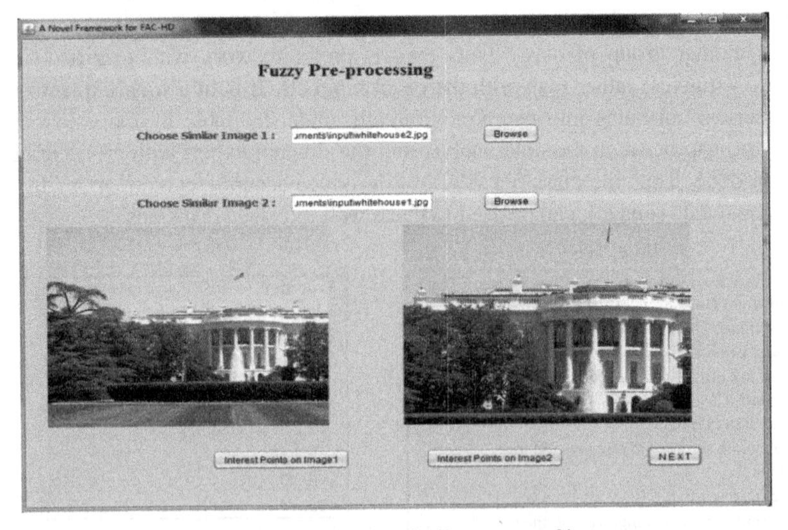

Fig. 1. GUI - Input of two similar scenes of images

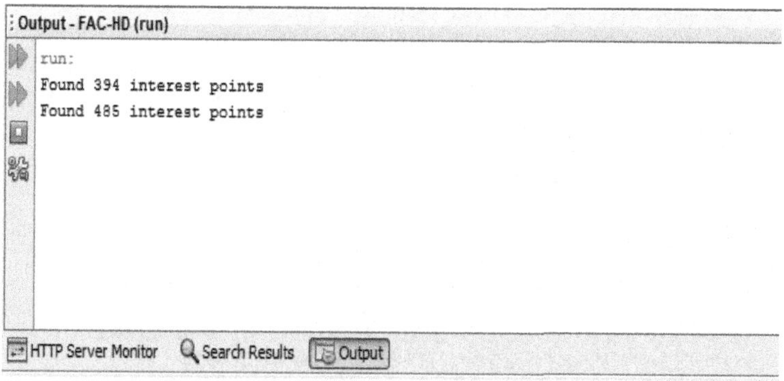

Output - FAC-HD (run)

```
run:
Found 394 interest points
Found 485 interest points
```

HTTP Server Monitor   Search Results   Output

Fig. 2. Interesting points of each image

Fig. 3. GUI-Matching points between the input images

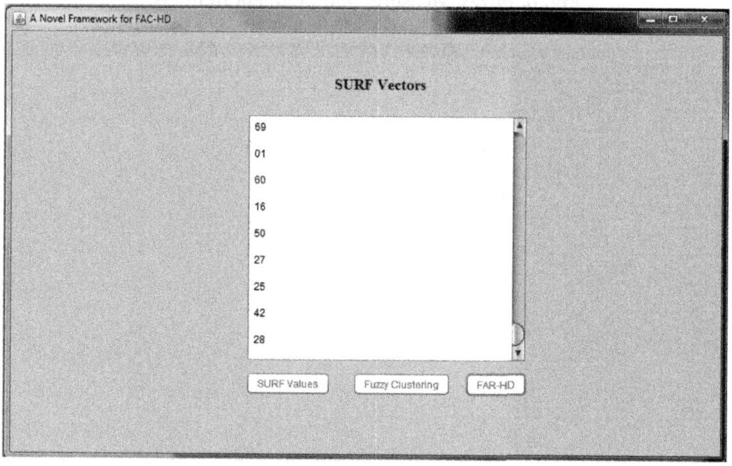

A Novel Framework for FAC-HD

SURF Vectors

69
01
60
16
50
27
25
42
28

SURF Values   Fuzzy Clustering   FAR-HD

Fig. 4. SURF matching vectors in the text box

<3,1.0> <6,0.08> <11,1.0> <20,1.0> <26,1.0> <35,1.0> <41,0.89> <48,1.0>
<3,0.4> <10,1.0> <11,0.02> <16,0.83> <17,0.17> <25,1.0> <27,0.9> <31,1.0> <41,0.11>
<5,1.0> <7,0.28> <8,0.22> <15,1.0> <19,0.83> <20,0.17> <25,1.0> <26,0.84> <27,0.13> <35,1.0> <40,1.0> <45,1.0>
<5,1.0> <6,1.0> <15,1.0> <20,1.0> <25,1.0> <29,0.47> <30,0.53> <35,1.0> <40,1.0> <42,0.09> <43,0.91> <50,1.0>
<4,1.0> <8,1.0> <13,1.0> <16,0.7> <25,1.0> <28,0.72> <29,0.28> <32,1.0> <39,1.0> <45,1.0>
<2,0.5> <6,0.04> <11,1.0> <16,0.1> <24,0.83> <26,1.0>
<6,1.0> <11,0.66> <12,0.13> <20,1.0> <26,1.0>
<5,1.0> <13,0.19> <14,0.21> <24,0.83> <30,1.0> <31,1.0> <40,1.0> <41,0.89> <49,1.0>
<5,1.0> <6,1.0> <19,0.8> <20,0.2> <25,1.0> <27,1.0> <31,1.0> <40,1.0> <45,1.0>
<5,1.0> <6,0.68> <14,0.29> <15,0.36> <18,0.71> <19,0.29> <25,1.0> <30,1.0> <35,1.0> <39,1.0> <45,1.0> <48,1.0>
<5,1.0> <7,1.0> <11,0.52> <12,0.2> <20,1.0> <25,1.0> <29,0.98> <30,0.02> <32,0.4> <33,0.6> <40,1.0> <45,1.0>
<5,1.0> <7,0.77> <16,0.2> <25,1.0> <32,1.0> <39,1.0> <41,1.0>
<2,0.5> <10,1.0> <14,1.0> <16,0.7> <23,0.67> <28,0.93> <29,0.07> <31,1.0> <42,0.38> <43,0.62> <48,1.0>
<5,1.0> <7,1.0> <14,0.85> <20,1.0> <25,1.0> <29,1.0> <35,1.0> <40,1.0> <44,1.0> <49,1.0>
<10,1.0> <14,0.88> <15,0.06> <18,1.0> <23,0.67> <29,1.0> <37,1.0> <41,0.22> <48,1.0>
<5,1.0> <6,0.74> <20,1.0> <25,1.0> <28,0.89> <31,1.0> <38,1.0>

Fig. 5. Fuzzy Clustering for SURF vectors

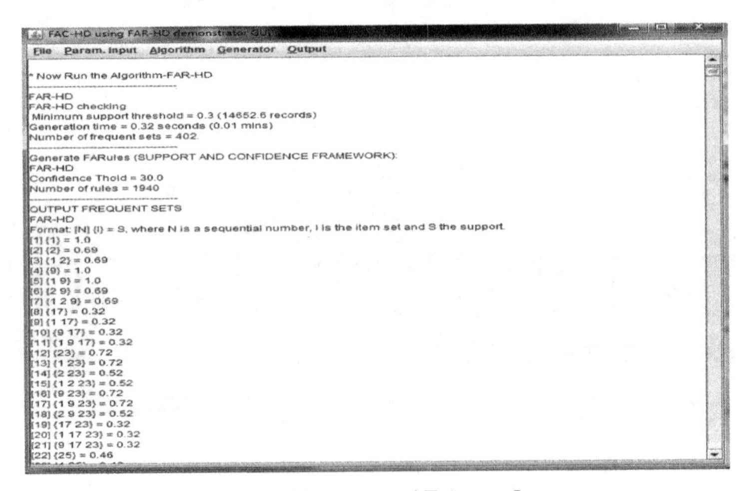

Fig. 6. Fuzzy Measures and Frequent Itemsets

```
FAC-HD using FAR-HD demonstrator GUI
File  Param. Input  Algorithm  Generator  Output
(1206) {97} -> {2 56} 70.62-------------->C1
(1207) {66 97} -> {2} 70.55-------------->C1
(1208) {23} -> {46 97} 70.53-------------->C1
(1209) {61} -> {23 56} 70.47-------------->C1
(1210) {97} -> {2 52} 70.46-------------->C1
(1211) {51 97} -> {2} 70.45-------------->C1
(1212) {1} -> {23 56} 70.45-------------->C1
(1213) {9} -> {23 56} 70.44-------------->C1
(1214) {45 51} -> {23} 70.33-------------->C1
(1215) {45} -> {23 61} 70.22-------------->C1
(1216) {23} -> {2 52} 70.14-------------->C1
(1217) {52 61} -> {2} 70.13-------------->C1
(1218) {52 56} -> {2} 70.07-------------->C1
(1219) {56} -> {23 52} 69.97-------------->C2
(1220) {56 61} -> {2} 69.95-------------->C2
(1221) {45 52} -> {2} 69.93-------------->C2
(1222) {46 61} -> {2} 69.83-------------->C2
(1223) {9 52} -> {2} 69.82-------------->C2
(1224) {52} -> {1 2} 69.82-------------->C2
(1225) {1 52} -> {2} 69.82-------------->C2
(1226) {52} -> {2} 69.82-------------->C2
(1227) {52} -> {2 9} 69.8-------------->C2
(1228) {51 61} -> {97} 69.75-------------->C2
(1229) {45 56} -> {2} 69.72-------------->C2
(1230) {61} -> {2} 69.72-------------->C2
(1231) {1 61} -> {2} 69.72-------------->C2
(1232) {61} -> {1 2} 69.72-------------->C2
(1233) {9 61} -> {2} 69.72-------------->C2
(1234) {61} -> {2 9} 69.7-------------->C2
(1235) {45} -> {23 56} 69.67-------------->C2
(1236) {9 56} -> {2} 69.63-------------->C2
(1237) {56} -> {1 2} 69.63-------------->C2
(1238) {1 56} -> {2} 69.63-------------->C2
(1239) {56} -> {2} 69.63-------------->C2
(1240) {51} -> {97} 69.62-------------->C2
(1241) {1 51} -> {97} 69.62-------------->C2
(1242) {51} -> {1 97} 69.62-------------->C2
(1243) {9 51} -> {97} 69.62-------------->C2
(1244) {56} -> {2 9} 69.61-------------->C2
```

Fig. 7. Rules with confidence and class labels

Table 1. Fuzzy Association Rules with Class Labels

Rules	Support	Class Label
R1	1.0	C1
R2	1.0	C1
R3	0.9	C1
R4	0.8	C1
R5	0.72	C1
R6	0.75	C1
R7	0.66	C2
R8	0.58	C2

6.1 Measures

Table 1 represents the sample rules having support with class labels, with respect to this table the measures of confidence and support can be calculated by using Eq. 2 and 3. Actual results are shown in fig.7; the table is to understand the calculation part of support and confidence.

6.1.1 Sample Calculation

Confidence for C1= (1.0+1.0+0.8+0.9+0.72+0.75)/ (1.0+1.0+0.8+0.9+0.72+0.75+0.6+0.58) =81.41

Confidence for C2= (0.6+0.58)/ (1.0+1.0+0.8+0.9+0.72+0.75+0.6+0.58) =18.6

Support for C1= (1.0+1.0+0.8+0.9+0.72+0.75)/8=0.65

Support for C2= (0.6+0.58)/ 8=0.15

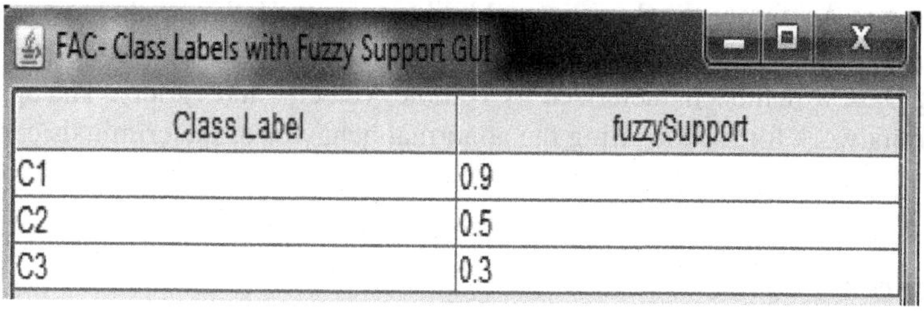

FAC- Class Labels with Fuzzy Support GUI

Class Label	fuzzySupport
C1	0.9
C2	0.5
C3	0.3

Fig. 8. Fuzzy Associative Classifier with soft labels

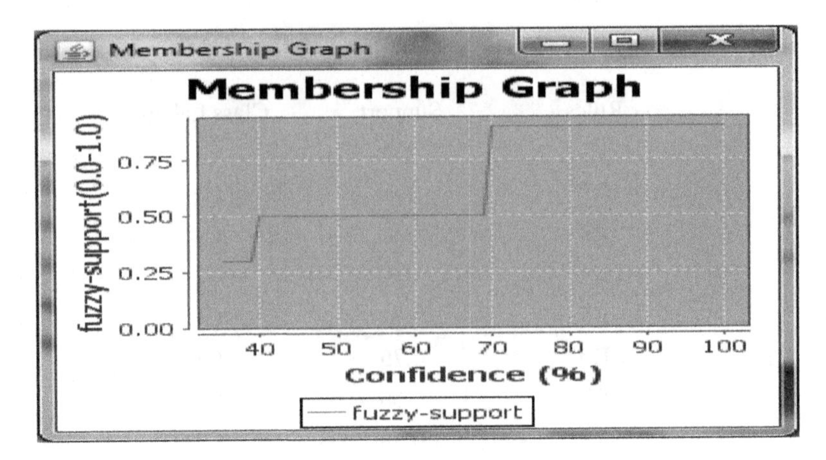

Fig. 9. Membership Graph

7 Conclusions and Future Work

In this paper association rule mining was studied for very large high-dimensional data in the image domain. The framework by name "Fuzzy Associative Classifier using images" was implemented using Java platform. The algorithm is capable of modeling fuzzy association rules extracted from image dataset. The algorithm is meant for classifying future datasets with the model that has been built. The underlying fuzzy association rules from which the classifier is trained, provide hidden patterns or relationships among attributes of the dataset. The application of this framework uses in Bioinformatics, it helps to identify the authentication with the help of categorical values .Experts prefer to work with fuzzified data (i.e. linguistic variables) rather than with the exact numbers. It is in a way, a quantization of the numerical attributes into categories directly understandable by the experts. Hence closing the gap between the data analyst and the domain expert with an accurate value has been done. Thus the classifier is able to produce accurate classifications from high dimensional data (image), can be used to make well informed decisions. The empirical results are encouraging. As future work, we would like to extend the applications of fuzzy associative classifiers to other domains. Another direction is to explore the classifier with Big Data which is characterized by volume, velocity and variety. The application of this framework for investigating the abnormal behavior of the criminals can also be explored.

References

[1] Mangalampalli, A., Pudi, V.: FAR-HD: A Fast And Efficient Algorithm For Mining Fuzzy Association Rule In Large High-Dimensional Datasets. In: 2013 IEEE Fuzzy Systems (FUZZ) (2013)

[2] Mangalampalli, A., Pudi, V.: FAR-miner: a fast and efficient algorithm for fuzzy association rule mining. IJBIDM 7(4), 288–317 (2012)
[3] Rajput, D.S., Thakur, R.S., Thakur, G.S.: Fuzzy Association Rule Mining based Frequent Pattern Extraction from Uncertain Data. 2012 IEEE Department of Computer Applications (2012) 978-1-4673-4805-8/12
[4] Mangalampalli, A., Pudi, V.: Fuzzy association rule mining algorithm for fast and efficient performance on very large datasets. In: FUZZ-IEEE, pp. 1163–1168 (2009)
[5] Bay, H., Ess, A., Tuytelaars, T., Gool, L.J.V.: Speeded-up robust features (SURF). Computer Vision and Image Understanding 110(3), 346–359 (2008)
[6] Thabtah, F.A.: A review of associative classification mining. Knowledge Eng. Review 22(1), 37–65 (2007)
[7] Agrawal, R., Imielinski, T., Swami, A.N.: Mining association rules between sets of items in large databases. In: SIGMOD Conference, pp. 207–216 (1993)
[8] Agrawal, R., Srikant, R.: Fast algorithms for mining association rules in large databases. In: VLDB, pp. 487–499 (1994)
[9] Bezdek, J.C.: Pattern Recognition with Fuzzy Objective Function Algorithms. Kluwer Academic Publishers, Norwell (1981)
[10] Fung, B.C.M., Wang, K., Ester, M.: Hierarchical document clustering using frequent itemsets. In: SDM (2003)
[11] Malik, H.H., Kender, J.R.: High quality, efficient hierarchical document clustering using closed interesting itemsets. In: ICDM, pp. 991–996 (2006)
[12] Delgado, M., Marn, N., Sánchez, D., Miranda, M.A.V.: Fuzzy association rules: General model and applications. IEEE Transactions on Fuzzy Systems 11, 214–225 (2003)
[13] Veloso, A., Meira Jr., W., Zaki, M.J.: Lazy Associative Classification. In: ICDM 2006: International Conference on Data Mining, pp. 645–654. IEEE Computer Society (2006)
[14] Fung, B.C.M., Wang, K., Ester, M.: Hierarchical document clustering using frequent itemsets. In: SDM (2003)
[15] Zhuang, L., Dai, H.: A maximal frequent itemset approach for web document clustering. In: 2004 IEEE International Conference on Computer and Information Technology, CIT(2004) 0-7695-2216-5/04
[16] Mangalampalli, A., Chaoji, V., Sanyal, S.: I-FAC: Efficient Fuzzy Associative Classifier for Object Classes in Images. In: 2010 IEEE Fuzzy Systems (FUZZ) (2010)
[17] Alcalá-Fdez, J., Alcalá, R., Herrera, F.: A Fuzzy Association Rule-Based Classification Model for High-Dimensional Problems with Genetic Rule Selection and Lateral Tuning. IEEE Transactions on Fuzzy Systems 9(5), 857–872 (2011)
[18] Guvena, E., Buczaka, A.L.: An OpenCL Framework for Fuzzy Associative Classification and Its Application to Disease Prediction. Conference by Missouri University 2013-Baltimore, MD
[19] Rajendran, P., Madheswaran, M.: Novel Fuzzy Association Rule Image Mining Algorithm for Medical Decision Support System. International Journal of Computer Applications (0975 - 8887) 1(20) (2010)
[20] Usha, D., Rameshkumar, K.: A Complete Survey on application of Frequent Pattern Mining and Association Rule Mining on Crime Pattern Mining. International Journal of Advances in Computer Science and Technology 3(4) (April 2014)
[21] Wang, W., Yang, J.: Mining High-Dimensional Data. In: Data Mining and Knowledge Discovery Handbook, pp. 793–799 (2005)

A Survey on Access Control Models in Cloud Computing

RajaniKanth Aluvalu[1] and Lakshmi Muddana[2]

[1] Department of Computer Engineering,
RK University Rajkot, India
rajanik.rkcet@gmail.com
[2] Department of Information Technology GITAM University,
Hyderabad, India
lakshmi.muddana@gitam.edu

Abstract. Cloud computing, is an emerging computing paradigm, enabling users to remotely store their data in a server and provide services on-demand. In cloud computing cloud users and cloud service providers are almost certain to be from different trust domains. Cloud computing's multitenancy and virtualization features pose unique security and access privilege challenges due to sharing of resources among potential un trusted tenants. Hence Privacy, trust and access control are the critical issues met in cloud computing. Access control is of vital importance in cloud computing environment, since it is concerned with allowing a user to access various cloud resources. A secure user enforced data access control mechanism must be provided before cloud users have the liberty to outsource sensitive data to the cloud for storage. Heterogeneity of services in cloud computing environment demands to adopt varying degrees of granularity in access control mechanism with an efficient encryption system. In this paper, we are going to analysis various Access control models for cloud computing and possible solutions for their limitations

Keywords: Risk aware, Accountability, cloud security, data sharing, Attribute based encryption, cipher text policy, fine grained access control.

1 Introduction

Cloud Computing is a subscription-based service where you can obtain networked storage space and computer resources. In cloud computing model customers plug into the cloud to access IT resources which are priced and provided on demand services. This cloud model composed of five essential characteristics, four service models and four deployment models. Users can store their data in cloud and there is a lot of personal information and potentially secure data that people store on their computers and this information is now being transferred to the cloud. Here we must ensure the security of user's data, which is in cloud. Users prefer only the cloud which can be trusted [29]. In order to increase the trust in cloud storage, the concept of accountability can be used. Accountability is likely to become a core concept in cloud that increase the trust in cloud computing. It helps to trace the user's data, protecting sensitive and confidential information, enhancing user's trust in cloud computing [29].

S.C. Satapathy et al. (eds.), *Emerging ICT for Bridging the Future – Volume 1*,
Advances in Intelligent Systems and Computing 337, DOI: 10.1007/978-3-319-13728-5_73

1.1 Essential Characteristics

a. **On-demand service-**consumers can use web services to access computing resources on-demand as needed automatically

b. **Broad network access-can access** Services from any internet connected device.

c. **Resource Pooling-**customers can share a pool of computing resources with other customers.

d. **Rapid Elasticity-**enables computing resources or user account to be rapidly and elastically provisioned

e. **Measured Service-**control and optimize services based on metering and automatically monitor the resources [29].

Characteristics			
Broad Network Access	Resource Pooling	Rapid Elasticity	On demand self service
Service models			
Infrastructure as a service (IaaS)	Platform as a service (PaaS)	Software as a service (SaaS)	Security as a service (SECaaS)
Deployment Models			
Public Cloud	Private Cloud	Hybrid Cloud	Community Cloud

Fig. 1. Cloud Computing Model

1.2 Service Models

i. **Software as a Service (SaaS):** The capacity provided to the consumer is to use the provider's applications running on a cloud infrastructure. The application is accessible from client devices through web browser.

ii. **Platform as a Service (PaaS):** The capability provided to the consumer is to deploy onto the cloud infrastructure

iii. **Infrastructure as a Service (IaaS):** The capability provided to the consumer is to provision processing, storage, network and other fundamental computing resources. where the consumer is able to deploy and run arbitrary software, which can include operating system and application.

iv. **Security as a Service (SECaaS):** security is delivered as a service without requiring on-premises hardware avoiding substantial capital outlays. These security services often include authentication, anti-virus, anti-malware/spyware, intrusion detection [29].

1.3 Deployment Models

a. **Public Clouds:** Public cloud computing services are provided off-premise to the general public and the computing resources are shared with the provider's other customers.

b. Community Clouds: Cloud infrastructure shared by several organization that have shared concern, managed by organization or third party.
c. Private Clouds: Cloud infrastructure for single organization only, may be managed by the organization or third party, on or off premise.
d. Hybrid Clouds: It uses public clouds for general computing while customer data is kept within a private cloud.

2 Survey

2.1 Cloud Federation

It is the practice of interconnecting the cloud computing environments of two or more service providers for the purpose of load balancing traffic and accommodating spikes in demand. Cloud federation requires one provider to wholesale or rent computing resources to another cloud provider [9, 12]. Those resources become a temporary or permanent extension of the buyer's cloud computing environment, depending on the specific federation agreement between providers. Cloud federation offers two substantial benefits to cloud providers [14]. First, it allows providers to earn revenue from computing resources that would otherwise be idle or underutilized. Second, cloud federation enables cloud providers to expand their geographic footprints and accommodate sudden spikes in demand without having to build new points-of-presence (POPs).Service providers strive to make all aspects of cloud federation from cloud provisioning to billing support systems (BSS) and customer support transparent to customers. When federating cloud services with a partner, cloud providers will also establish extensions of their customer-facing service-level agreements (SLAs) into their partner provider's data centers [10].

2.2 Security and Accountability in Cloud

Cloud stores huge amount of user's data, there is a critical need of providing security for the data stored on cloud. The owner of the data does not aware about where their data is stored and they do not have control of where data is placed. Here it explores the security challenges in cloud. Some of the security risks include secure data transfer, secure software interface, secure stored data, user access control, data separation [26]. To promote privacy and security concern of end users accountability mechanism is used. Here the basic concept is that user's private data are sent to the cloud in an encrypted form, and then with the encrypted data processing is carried out. Accountability become a core concept in cloud that helps to increase trust in cloud computing. The term Accountability refers to a narrow and imprecise requirement that met by reporting and auditing mechanisms. Accountability is the agreement to act as a responsible proctor of the personal information of others, to take responsibility for protection and appropriate use of that information beyond legal requirements, and to be accountable for misuse of that information.

2.3 Access Control Models

Cloud computing has rapidly become a widely adopted paradigm for delivering services over the internet. Therefore cloud service provider must provide the trust and security, as there is valuable and sensitive data in large amount stored on the clouds. Cloud computing environment is widely distributed and highly dynamic. Static policies will not be efficient for cloud access models. We require access models with dynamic policies. For protecting the confidentiality of the stored data, the data must be encrypted before uploading to the cloud by using some cryptographic algorithms [17, 18]. We will be discussing various access control models that support dynamics policies, attribute based access models using encryption scheme and its categories.

A. *Attribute Based Encryption (ABE)*

An attribute based encryption scheme introduced by Sahai and Waters in 2005 and the goal is to provide security and access control. Attribute-based encryption (ABE) is a public-key based one to many encryptions that allows users to encrypt and decrypt data based on user attributes. In which the secret key of a user and the ciphertext are dependent upon attributes (e.g. the country she lives, or the kind of subscription she has). In such a system, the decryption of a ciphertext is possible only if the set of attributes of the user key matches the attributes of the ciphertext. Decryption is only possible when the number of matching is at least a threshold value *d*. Collusion-resistance is crucial security feature of Attribute-Based Encryption .An adversary that holds multiple keys should only be able to access data if at least one individual key grants access[5].

The problem with attribute based encryption (ABE) scheme is that data owner needs to use every authorized user's public key to encrypt data. The application of this scheme is restricted in the real environment because it uses the access of monotonic attributes to control user's access in the system.

B. *Key Policy Attribute Based Encryption (KP-ABE)*

It is the modified form of classical model of ABE. Users are assigned with an access tree structure over the data attributes. Threshold gates are the nodes of the access tree. The attributes are associated by leaf nodes. To reflect the access tree Structure the secret key of the user is defined. Ciphertexts are labeled with sets of attributes and private keys are associated with monotonic access structures that control which ciphertexts a user is able to decrypt. Key Policy Attribute Based Encryption (KP-ABE) scheme is designed for one-to-many communications [17].

KP-ABE scheme consists of the following four algorithms:

Setup: Algorithm takes input K as a security parameter and returns PK as public key and a system master secret key MK.PK is used by message senders for encryption. MK is used to generate user secret keys and is known only to the authority.

Encryption: Algorithm takes a message M, the public key PK, and a set of attributes as input. It outputs the ciphertext E.

Key Generation: Algorithm takes as input an access structure T and the master secret key MK. It outputs a secret key SK that enables the user to decrypt a message encrypted under a set of attributes if and only if matches T.

Decryption: It takes as input the user's secret key SK for access structure T and the ciphertext E, which was encrypted under the attribute set. This algorithm outputs the message M if and only if the attribute set satisfies the user's access structure T.

The KP-ABE scheme can achieve fine-grained access control and more flexibility to control users than ABE scheme.

The problem with KP-ABE scheme is the encryptor cannot decide who can decrypt the encrypted data. It can only choose descriptive attributes for the data; it is unsuitable in some application because a data owner has to trust the key issuer.

C. Cipher Text Policy Attribute Based Encryption

Another modified form of ABE called CP-ABE *introduced by Sahai.* In a CP-ABE scheme, every ciphertext is associated with an access policy on attributes, and every user's private key is associated with a set of attributes. A user is able to decrypt a ciphertext only if the set of attributes associated with the user's private key satisfies the access policy associated with the ciphertext. CP-ABE works in the reverse way of KP-ABE [24]. The access structure of this scheme or algorithm, it inherits the same method which was used in KP-ABE to build. And the access structure built in the encrypted data can let the encrypted data choose which key can recover the data; it means the user's key with attributes just satisfies the access structure of the encrypted data [19]. And the concept of this scheme is similar to the traditional access control schemes. The encryptor who specifies the threshold access structure for his interested attributes while encrypting a message. Based on this access structure message is then encrypted such that only those whose attributes satisfy the access structure can decrypt it. The most existing ABE schemes are derived from the CP- ABE scheme.

CP-ABE scheme consists of following four algorithms:

Setup: This algorithm takes as input a security parameter K and returns the public key PK as well as a system master secret key MK. PK is used by message senders for encryption. MK is used to generate user secret keys and is known only to the authority.

Encrypt: This algorithm takes as input the public parameter PK, a message M, and an access structure T. It outputs the ciphertext CT.

Key-Gen: This algorithm takes as input a set of attributes associated with the user and the master secret key MK. It outputs a secret key SK that enables the user to decrypt a message encrypted under an access tree structure T if and only if matches T.

Decrypt: This algorithm takes as input the ciphertext CT and a secret key SK for an attributes set. It returns the message M if and only if satisfies the access structure associated with the ciphertext CT.

It improves the disadvantage of KP-ABE that the encrypted data cannot choose who can decrypt. It can support the access control in the real environment. In addition, the user's private key is in this scheme, a combination of a set of attributes, so an user only use this set of attributes to satisfy the access structure in the encrypted data.

Drawbacks of the most existing CP-ABE schemes are still not fulfilling the enterprise requirements of access control which require considerable flexibility and efficiency. CP- ABE has limitations in terms of specifying policies and managing user attributes. In a CP-ABE scheme, decryption keys only support user attributes that are organized logically as a single set, so the users can only use all possible combinations of attributes in a single set issued in their keys to satisfy policies. After that ciphertext-policy attribute- set based encryption (CP-ASBE or ASBE for short) is introduced by *Bobba, Waters et al* [7]. ASBE is an extended form of CP-ABE. It organizes user attributes into a recursive set based structure and allows users to impose dynamic constraints on how those attributes may be combined to satisfy a policy. The CP-ASBE consists of recursive set of attributes. The challenge in constructing a CP-ASBE scheme is in selectively allowing users to combine attributes from multiple sets within a given key. There is challenge for preventing users from combining attributes from multiple keys.

D. Attribute-Based Encryption Scheme with Non-monotonic Access Structures

Previous ABE schemes were limited to expressing only monotonic access structures and there is no satisfactory method to represent negative constraints in a key's access formula. Ostrovsky et al. proposed an attribute-based encryption with non-monotonic access structure in 2007[17].

Non-monotonic access structure can use the negative word to describe every attributes in the message, but the monotonic access structure cannot.

This scheme contains four algorithms:

Setup (d). In the basic construction, a parameter d specifies how many attributes every ciphertext has.

Encryption (M, γ,PK). To encrypt a message M ε GT under a set of d attributes γ *C* Zp, choose a random value s ε Zp and output the ciphertext E.

Key Generation (˜A,MK,PK). This algorithm outputs a key D that enables the user to decrypt an encrypted message only if the attributes of that ciphertext satisfy the access structure ˜A

Decrypt (CT;D): Input the encrypted data *CT* and private key *D*, if the access structure is satisfied it generate the original message M.

It enables Non-monotonic policy, i.e. policy with negative attributes. The problem with Attribute-based Encryption Scheme with Non- Monotonic Access Structures is that there are many negative attributes in the encrypted data, but they don't relate to the encrypted data. It means that each attribute adds a negative word to describe it, but these are useless for decrypting the encrypted data. It can cause the encrypted data overhead becoming huge. It is inefficient and complex each cipher text needs to be encrypted with d *attributes*, where *d* is a system-wise constant.

E. Hierarchical Attribute-Based Encryption

This scheme Hierarchical attribute-based encryption (HABE) is derived by Wang et al The HABE model (Fig 2) consists of a root master (RM) that corresponds to the third trusted party (TTP),multiple domain masters (DMs) in which the top-level DMs correspond to multiple enterprise users, and numerous users that correspond to all personnel in an enterprise. This scheme used the property of hierarchical generation of keys in HIBE scheme to generate keys [22].

Then, HABE scheme is defined by presenting randomized Polynomial time algorithms as follows:

Setup (K)→(params,MK0): The RM takes a sufficiently large security parameter K as input, and outputs system parameters params and root master key MK0. ***CreateDM(params,MKi, PKi+1) → (MKi+1):*** Whether the RM or the DM generates master keys for the DMs directlyunder it using params and its master key. ***CreateUser(params,MKi, PKu, PKa) → (SKi,u, SKi,u,a):***The DM first checks whether U is eligible for a, which is administered by itself. If so, it generates a user identity secret key and a user attribute secret key for U, using params and its master key; otherwise, it outputs "NULL".

Encrypt(params; f ;A; {PKa|a E A})→(CT): A user takes a file f, a DNF access control policy A, and public keys of all attributes in A, as inputs, and outputs a ciphertext CT. ***Decrypt(params,CT,SKi,u,{SKi,u,a|aECCj}→(f):***A user, whose attributes satisfy the j-th conjunctive clause CCj, takes params, the ciphertext, the user identity secret key, and the user attribute secret keys on all attributes in CCj, as as inputs, to recover the plaintext.

This scheme can satisfy the property of fine grained access control, scalability and full delegation. It can share data for users in the cloud in an enterprise environment. Furthermore, it can apply to achieve proxy re-encryption [4]. But in practice, it is unsuitable to implement. Since all attributes in one conjunctive clause in this scheme may be administered by the same domain authority, the same attribute may be administered by multiple domain authorities.

F. Multi-authority Attribute Based Encryption

V Bozovic, D Socek, R Steinwandt, and Vil-lanyi, introduce Multi-authority attribute-based encryption. In this scheme it use multiple parties to distribute attributes for users. A Multi-Authority ABE system is composed of K attribute authorities and one central authority. Each attribute authority is also assigned a value dk. The system uses the following algorithms:

Setup: A randomized algorithm which must be run by some trusted party (e.g. central authority). Takes as input the Security parameter. Outputs a public key, secret key pair for each of the attribute authorities, and also outputs a system public key and master secret key which will be used by the central authority.

Attribute Key Generation: A randomized algorithm run by an attribute authority. Takes as input the authority's secret key, the authority's value dk, a user's GID, and a set of attributes in the authority's domain AkC. (We will assume that the user's claim of these attributes has been verified before this algorithm is run). Output secret key for the user. ***Central Key Generation:*** A randomized algorithm runs by the central authority. Takes as input the master secret key and a user's GID and outputs secret key for the user. ***Encryption:*** A randomized algorithm runs by a sender. Takes as input a set of attributes for each authority, a message, and the system public key. Outputs the ciphertext. ***Decryption:*** Deterministic algorithms run by a user. Takes as input a cipher-text, which was encrypted under attribute set AC and decryption keys for an attribute set Au. Outputs a message m if |Ak C ∩Ak u| > dk for all authorities k.

It allows any polynomial number of independent authorities to monitor attributes and distribute private keys and tolerate any number of corrupted authorities. In this model, a recipient is defined not by a single string, but by a set of attributes. Complication in multi-authority scheme required that each authority's attribute set be disjoint. The below table gives the comparison of each schemes in the attribute based encryption.

Setup: Algorithm takes input K as a security parameter and returns PK as public key and a system master secret key MK.PK is used by message senders for encryption. MK is used to generate user secret keys and is known only to the authority.

Encryption: Algorithm takes a message M, the public key PK, and a set of attributes as input. It outputs the ciphertext E.

Key Generation: Algorithm takes as input an access structure T and the master secret key MK. It outputs a secret key SK that enables the user to decrypt a message encrypted under a set of attributes if and only if matches T.

Decryption: It takes as input the user's secret key SK for access structure T and the ciphertext E, which was encrypted under the attribute set. This algorithm outputs the message M if and only if the attribute set satisfies the user's access structure T.

The KP-ABE scheme can achieve fine-grained access control and more flexibility to control users than ABE scheme. The problem with KP-ABE scheme is the encrypter cannot decide who can decrypt the encrypted data. It can only choose descriptive attributes for the data; it is unsuitable in some application because a data owner has to trust the key issuer.

G. Cipher Text Policy Attribute Based Encryption

Another modified form of ABE called CP-ABE *introduced by Sahai.* In a CP-ABE scheme, every cipher text is associated with an access policy on attributes, and every user's private key is associated with a set of attributes. A user is able to decrypt a ciphertext only if the set of attributes associated with the user's private key satisfies the access policy associated with the ciphertext. CP-ABE works in the reverse way of KP-ABE [19]. The access structure of this scheme or algorithm , it inherit the same method which was used in KP-ABE to build and the access structure built in the encrypted data can let the encrypted data choose which key can recover the data, it means the user's key with attributes just satisfies the access structure of the encrypted data. And the concept of this scheme is similar to the traditional access control schemes. The encryptor who specifies the threshold access structure for his interested attributes while encrypting a message. Based on this access structure message is then encrypted such that only those whose attributes satisfy the access structure can decrypt it. The most existing ABE schemes are derived from the CP- ABE scheme.

H. Adding Attributes to RBAC

ABAC requires upto 2^n rules for n attributes, attempting to implement the same controls in RBAC could; in a worst case require 2^n roles [30].

There are three approaches of RBAC-A to handle relationship between roles and attributes.

1. *Dynamic Role:* implementation of dynamic roles might let the user's role be fully determined by the front end attribute engine, while others might use the front end only to select from among a predetermined set of authorized users.
2. .*Attribute Centric :* A role name is just one of many attributes, in contrast with conventional RBAC, the role is not a collection of permissions but the name of an attribute called role.
3. *Role Centric:* attributes are added to constrain RBAC, constraint rules can only reduce permissions available to the user, not expand them.

I. Risk Aware Access Control Model

The increasing need to share information in dynamic environments such as cloud computing has created a requirement for risk-aware access control systems.

The standard RBAC model is designed to operate in a relatively stable, closed environment and does not include any support for risk [1]. Risk-aware RBAC models that differ in the way in which risk is represented and accounted for in making access control decisions.

The core goal of risk aware access control (RAAC) systems is to provide a mechanism that can manage the trade-off between the risks of allowing unauthorized access with the cost of denying access when the inability to access resources may have profound consequences. This approach is particularly useful to allow some risky access in an emergency situation.

3 Analysis

Techniques/Parameters	ABE	KP-ABE	CP-ABE	HASBE	MA-ABE
Fine grained access control	Low	Low	Average	Flexible	Good
Efficiency	Average	Average	Average	Flexible	Scalable
Computational Overhead	High	High	Average	High	Average
Collusion Resistant	Average	Good	Good	Good	High

Fig. 2. Analysis of access models

4 Implementation of Access Models

Access control models can be implemented by using XACML, which stands for extensible Access Control Markup Language. The standard defines a declarative access control policy language implemented in XML and a processing model describing how to evaluate access requests according to the rules defined in policies [4].

As a published standard specification, one of the goals of XACML is to promote common terminology and interoperability between access control implementations by multiple vendors. XACML is primarily an Attribute Based Access Control system (ABAC), where attributes (bits of data) associated with a user or action or resource are inputs into the decision of whether a given user may access a given resource in a particular way. Role-based access control (RBAC) can also be implemented in XACML as a specialization of ABAC.

5 Conclusion

In this paper, we had analysed trust based cloud federation and effective access models required for cloud federation. We had analysed Risk aware role based access control model, Integrating RBAC with ABAC, various attribute based encryption schemes: ABE, KP-ABE, CP-ABE, HABE etc. we proposed risk aware role based access control model integrated with hierarchical attribute set based access model for trust based cloud federation. Our proposed access model is trust based, will be effective and ensures scalability.

References

[1] Bijon, K.Z., Krishnan, R., Sandhu, R.: Risk-Aware RBAC Sessions. In: Venkatakrishnan, V., Goswami, D. (eds.) ICISS 2012. LNCS, vol. 7671, pp. 59–74. Springer, Heidelberg (2012)

[2] Chen, L., Crampton, J.: Risk-Aware Role-Based Access Control. In: Meadows, C., Fernandez-Gago, C. (eds.) STM 2011. LNCS, vol. 7170, pp. 140–156. Springer, Heidelberg (2012)

[3] Kandala, S., Sandhu, R., Bhamidipati, V.: An Attribute Based Framework for Risk-Adaptive Access Control Models. In: 2011 Sixth International Conference on Availability, Reliability and Security (ARES) (2011)

[4] Brossard, D.: XACML 101 – a quick intro to Attribute-based Access Control with XACML (September 30, 2010), http://www.webframer.eu

[5] Bobba, R., Khurana, H., Prabhakarn, M.: Attribute sets a practically Motivated Enhancement to attribute based Encryption (July 27, 2009)

[6] Cheng, Y., Park, J.-H.: Relationship-Based Access Control for Online Social Networks: Beyond User-to-User Relationships. In: 2012 International Conference on Privacy, Security, Risk and Trust (PASSAT), 2012 International Conference on Social Computing (SocialCom), September 3-5 (2012)

[7] Zissis, D., Lekkas, D.: Addressing cloud computing security issues. Future Generation Computer Systems (March 2012)

[8] Sood, S.K.: A combined approach to ensure data security in cloud computing. Journal of Network and Computer Applications (November 2012)

[9] Singhal, M., Chandrasekhar, S., Tingjian, G., Sandhu, R., Krishnan, R., Gail-Joon, A., Bertino, E.: Collaboration in multicloud computing environments: Framework and security issues. Computer 46(2) (February 2013)

[10] Alhamad, M., Dillon, T., Chang, E.: SLA-Based Trust Model for Cloud Computing. In: 13th International Conference on Network-Based Information Systems (2010)

[11] Gohad, A., Rao, P.S.: 1 * N Trust Establishment within Dynamic Collaborative Clouds. In: 2012 IEEE International Conference on Cloud Computing in Emerging Markets (CCEM) (2012)

[12] Sato, H., Kanai, A., Tanimoto, S.: A Cloud Trust Model in a Security Aware Cloud. In: 2010 10th IEEE/IPSJ International Symposium on Applications and the Internet (SAINT) (July 2010)

[13] Mon, E.E., Naing, T.T.: The privacy-aware access control system using attribute-and role-based access control in private cloud. In: 2011 4th IEEE International Conference on Broadband Network and Multimedia Technology (IC-BNMT) (2011)

[14] Farcasescu, M.R.: Trust Model Engines in cloud computing. In: 14th International Symposium on Symbolic and Numeric Algorithms for Scientific Computing (2012)

[15] Muchahari, M.K., Sinha, S.K.: A New Trust Management Architecture for Cloud Computing Environment. In: 2012 International Symposium on Cloud and Services Computing (2012)

[16] Varadharajan, V., Tupakula, U.: TREASURE: Trust Enhanced Security for Cloud Environments. In: 2012 IEEE 11th International Conference on Trust, Security and Privacy in Computing and Communications (2012)

[17] Priyadarsini, K., Thirumalai Selvan, C.: 'A Survey on Encryption Schemes for Data Sharing in Cloud Computing. IJCSITS 2(5) (October 2012) ISSN: 2249-9555

[18] Antony, N., Melvin, A.A.R.: A Survey on Encryption Schemes in the Clouds for Access Control. International Journal of Computer Science and Management Research 1(5) (December 2012)

[19] Betten Court, J.: Ciphertext-policy attribute based encryption. In: Proceedings of IEEE Symposium on Security and Privacy, pp. 321–334 (2007)

[20] Zhu, Y., Huy, H., Ahny, G.-J., Huangy, D., Wang, S.: Towards Temporal Access Control in Cloud Computing. In: INFOCOM 2012 (2012)

[21] Yu, S., Wang, C., Ren, K., Lou, W.: Achieving Secure, Scalable, and Fine-grained Data Access Control in Cloud Computing. In: INFOCOM 2010 (2010)
[22] Wang, G., Liu, Q., Wub, J., Guo, M.: Hierarchical attribute-based encryption and scalable user revocation for sharing data in cloud servers (July 1, 2011)
[23] Begum, R., Kumar, R.N., Kishore, V.: Data Confidentiality Scalability and Accountability (DCSA) in Cloud Computing 2(11) (November 2012)
[24] Hota, C., Sankar, S.: Capability-based Cryptographic Data Access Control in Cloud Computing. Int. J. Advanced Networking and Applications 03(03), 1152–1161 (2011)
[25] Kandukuri, R., Paturi, V.R., Rakshit, A.: Cloud security issues. In: Proceedings of the 2009 IEEE International Conference on Services Computing, pp. 517–520 (September 2009)
[26] Buyya, R., Yeo, C.S., Venugopal, S., Broberg, J., Brandic, I.: Cloud computing and emerging IT platforms: vision, hype, and reality for delivering computing as the 5th utility. Future Generation Computer Systems 25(6), 599–616 (2008)
[27] Salesforce.com, Inc., Force.com platform, http://www.salesforce.com/tw/ (retrieved December 2009)
[28] SAP AG, SAP services: maximize your success, http://www.sap.com/services/index.epx (retrieved January 2010)
[29] BOOK: Cloud computing Bible: Author: Sosinky, Barrie Edition. John Wiley & Sons Publication (2011)
[30] Richardkun, D., et al.: Adding attributes to role based access control. IEEE-Computer 43(6)

Author Index